AF530692

Biology and Biotechnology of Anoxygenic Phototrophic Bacteria

Biology and Biotechnology of Anoxygenic Phototrophic Bacteria

Dr. K. Rajyalaxmi
Dr. S. Girisham
Dr. S.M. Reddy

Department of Microbiology
Kakatiya University
Warangal – 506009, Telangana

Published by

SCIENTIFIC PUBLISHERS (INDIA)
5 A, New Pali Road
P.O. Box 91
Jodhpur - 342 001 INDIA
E-mail: info@scientificpub.com
Website: http://www.scientificpubonline.com

ISBN: 978-93-88043-73-1 (Hardback)
ISBN: 978-93-88043-94-6 (E-book)

Printed in India

PREFACE

Anoxygenic phototrophic bacteria, a small group of bacteria, colonize a variety of substrates deficient in oxygen. They are distinct groups with a variety of physiological and ecological characters. Though they are limited in number but differ widely in morphology only features and metabolism. They can be cultured by enrichment. They are not only unique in their photosynthetic metabolism but also form a tool in understanding the true mechanism and evolution process. They are equally diverse in central carbon metabolism and sulfur metabolism. Despite their being extremophilic nature proved to be biotechnologically potential and play a crucial role in the environment. The characterization and classification of these organisms proved to be challenging and deserved more intensive studies. The details of this important group of bacteria are discussed in this book. The morphology and classification of this unique group of bacteria formed in chapter 1. While isolation and culturing of these bacteria are discussed in chapter 2. The taxonomy of different anoxygenic phototrophic bacteria was discussed in the 3rd chapter. Process of photosynthesis, photometabolism of central carbon metabolism and sulfur metabolism are discussed in the 4th, 5th and 6th chapters. Subsequent chapters 7, 8 and 9 cover the special applications of these specialized organisms in the service of mankind. The significant role played by these bacteria in bioremediation in oxygen-deficient ecological niches is discussed in chapter 10. Finally, biotechnological applications of these specialized organisms are analyzed in chapter 11.

The contribution made by our research students paved the way for writing this book. A task of this magnitude needs the support of several people. The book of specialized nature needs to borrow literature and it was not possible to quote all the references and only limited and more essential are quoted. We have liberally borrowed the materials from different authors for which we gratefully acknowledge. Author thanks the encouragement received from the UGC in the form of financial assistance and University authorities for providing facilities and encouragement. Thanks are also due to the Head Department of Microbiology for providing necessary facilities. Finally, we thank Mr. Tanay Sharma of Scientific Publishers, Jodhpur for timely publication in a nice get-up.

Authors

ABBREVIATIONS

(Re)/(Si)-CS, (Re)/(Si) – Citratesynthase

2DEAS – 2-dimensional evolutionary absorption spectra

2DES – 2-dimensional electronic spectroscopy

2D – Two-dimensional

3D – 3-dimensional

3HOP – 3-hydroxypropionate

A – Acceptor

AAPB – Aerobic anoxygenic phototrophic bacteria

ABC – aerobic bacteriochlorophyll-containing bacteria

ACL – ATPcitratelyase

Af. – Afifiella

Alc – Allochromatium

Amp – Ampicillin

ANAPB – Anaerobic anoxygenic phototrophic bacteria

Anc – Ancalochloris

APAT – Adenylylsulfatephospateadenylyl transferase

APB – Anoxygenic phototrophic bacteria

APC – Allophycocyanin

APS – Adenosine-5u-phosphosulfate

Atb – Acidithiobacillus

ATCC – American type culture collection

ATP – Adenosine Triphosphate.

B800 – Bulk Bchl with a maximal absorbance at 800 nm

B820 – Bulk Bchl with a maximal absorbance at 820 nm

B850 – Bulk Bchl with a maximal absorbance at 850 nm

BA/B – Monomeric Bchl on branch A or B

Bchl – Bacteriochlorphyll

B – Factors indication of mobility of an atom

Blc – Blastochloris

Bpheo – Bacteriopheophythin

C – Cytochrome

CAR – Carotenoid

Cba – Chlorobaculum

CCL – Citryl-CoAlyase

CCS – Citryl-CoAsynthetase

CD – Circular dichroism

Cfl – Chloroflexus

Chl – Chlorobium

Chp – Chloroherpeton

Chr – Chromatium

Cln – Chloronema

CMC – Critical micelle concentration

D – Donor

DDH-DNA – DNA Hybridization

DE52 – Diethylaminoethyl cellulose

DLS – Diamond light source

DMNG – 3 Decyl 3-maltose neopentyl glycol

DMSO – Dimethyl sulfoxide

DMSP – dimethylsulfoniopropionate

DNA – Deoxyribonucleic acid

DS – Down stream

DSMZ – Deutsche Sammlung von Mikroorganismen und Zellkulturen GmbH

Dsv – Desulfovibrio

DTT – Dithiothreitol

E. coli – Escherichia coli

Ect – Ectothiorhodospira

ED – Entner–Doudoroff

EDTA – Ethylenediaminetetraacetic acid

EM – Electron microscopy

EMP – Emden–Meyerhof–Parnas

EPS – extracellular polymeric substances

ESA – Excited state absorption

ET – Energy transfer

Ets – Ectothiorhodosinus

FAPB – Filamentous anoxygenic phototrophic bacteria

FBP – Fructose1,6-bisphosphat

Fd – Ferredoxin

Fe-S – Iron-sulfur

FMO – Fenna–Matthews–Olson

FWHM – Full width half maximum

G+C – Guanine and Cytosine

GAP – d-glyceraldehyde-3-phosphate

Glc – Glucose

GPCRs – G-protein coupled receptors

GSB – Green sulfur bacteria

GSB – Ground state bleach

GSNB – Green non-sulfur bacteria

H – Heavy subunit of RC

HA/B – Bacteriopheophytin on branch A or B of RC

Hba – Heliobacillus

Hbt – Heliobacterium

Hch – Halochromatium

HEPES – N-2-hydroxyethylpiperazine-N-2-ethane sulfonic acid

HL – High light

Hlr. – Halorhodospira

Hph – Heliophilum

HPLC – High performance liquid chromatography

Hrs – Heliorestis

Htr – Heliothrix

I – Spectral overlap of D emission and A absorption

IAA – Isoamyl alcohol

IC – Internal conversion

IC21 – Internal conversion from S2 to S1

ICM – Intra cytoplasmic membrane

IMAC – Immobilised metal affinity chromatography

Isc – Isochromatium

JB – JellyBody REFMAC5

K – Strength of interactions between D and A

Kan – Kanamycin

KDPG – 2-keto-3-deoxy-6-phosphogluconate

KO – Knock-out

L – Light subunit of RC

LB – Luria Broth

LC – Liquid chromatography

LC-MS-MS – Liquid chromatography-tandem Mass spectrometry

LDAO – Lauryldimethylamine oxide

LGT – lateral gene transfer

LH – light harvesting

LH1:RC – Light harvesting 1 complex : reaction centre or Core complex

LH-II – Light harvesting complex II

LH-I – Light harvesting complex 1

LL – Low light

LLG – Log-likelihood gain

Lpb – Lamprobacter

Lpc – Lamprocystis

M – Medium-subunit of RC

MAG – Monoacylglycerol

Mch – Marichromatium

MCS – Multiple cloning site

MES – 2-ethanesulfonic acid

MK – Menaquinone

MPT – molybdopterin

MR – Molecular replacement

MWCO – Molecular weight cut off

N – Length of the conjugated carbon chain in Car molecules

NADPH – Nicotinamide adenine dinucleotide phosphate

NGS – Next generation sequencing

NIR – Near infra red

NM – Nonyl-maltoside

NMR – Nuclear magnetic resonance

NOPA – Non collinear optical parametric amplifier

OAA – oxaloacetate

OCP – Orange carotenoid protein

OD – Optical density

OPP – Pathway,oxidativepentosephosphate pathway

OR – Optical ratio

Osc – Oscillochloris

OTCA/RTCAcycle – Oxidative/reductivetricarboxylicacidcycle

PA/B – Special pair of Bchl molecules on brance A or B of RC

PAPS – 3'-phosphoadenosine-5'-phosphosulfate

PBS – Phycobilisomes

PC – Phycocyanin

PCR – polymerase chain reaction

Pcs – Paracoccus

PE – Phycoerythrin

PEG – Polyethylene glycol

PG – Phosphoglyceratemutase

Phs – Phaeospirillum

Pi – Isoelectric point

Pld – Pelodictyon

PNSB – Purple non-sulphur bacteria

PS^- – Photosynthetic minus phenotype

PS^+ – Photosynthetic positive phenotype

PSB – purple sulfur bacteria

PSI – Photosystem I

PSII – Photosystem II

PSR – polysulfide reductase

PSU – Photosynthetic unit

PSU – Photosynthetic unit

Ptc – Prosthecochloris

Q/QH2 – Quinone/Quinol exchange

QA/B – Quinone binding site on branch A or B of the RC

R – Distance between the centre of D to the centre of A molecules

R5P – Ribose-5-phosphate

Rba – Rhodobacter

Rbc – Rhabdochromatium

Rbi – Rhodobium

Rbl – Rhodoblastus

RC – Reaction center

Rca – Rhodobaca

Rcs – Rhodocista

Rcy – Rhodocyclus

Rdv – Rhodovulum

Rfactor – Reliability factor

Rfx. – Rhodoferax

RFZ – *Rotational Z- factor*

RG – Rhodopin glucoside

Rhv – Rhodovibrio

RLP – RuBisCO-like protein

Rmi – Rhodomicrobium

RNA – Ribonucleic acid

Rof – Roseiflexus

ROS – Reactive oxygen species

Ros – Roseospira

Rpi – Rhodopila

Rpl. – Rhodoplanes

RPM – Rate per minute

Rps – Rhodopseudomonas

RQ – Rhodoquinone

rRNA – ribosomal RNA

Rsa – Rhodospira

Rsp – Rhodospirillum

Rss – Roseospirillum

RT – Room temperature

Rts – Rhodothalassium

RuBisCO – ribulose 1,5-bisphosphate carboxylase/oxygenase.

Rvi. – Rubrivivax

SCX – Strong cation exchange

SDG – Sucrose density gradients

SDS – PAGE Sodium dodecyl sulphate polyacrylamide gel electrophoresis

SE – Stimulated emission

SEC – Size exclusion chromatography

SGP – sulfur globule protein

SMS – Single molecule spectroscopy

SOE – Splice overlap extension

SQR – sulfide:quinone reductase

Succs- C – Succinate media without casamino acids

Tac – *Thioalkalicoccus*

Tba – *Thiobaca*

Tca – *Thiocapsa*

TCA – Tricarboxylic acid

Tca – *Thiocapsa*

Tch – *Thermochromatium*

Tco – *Thiococcus*

Tcs – *Thiocystis*

Tdc – *Thiodictyon*

TFA – Trifluoroacetic acid

Tfc – *Thioflavicoccus*

TFZ – Translational Z-factor

Thb – *Thiobacillus*

Thc – *Thiohalocapsa*

Tlp – *Thiolamprovum*

Tm – Melting temperature

Tms – *Thiomicrospira*

Tpd – *Thiopedia*

Trc – *Thiorhodococcus*

Trs – *Thiorhodospira*

Trv – *Thiorhodovibrio*

Tsp – *Thiospirillum*

UB – Ubiquinone

UHR – TOF Ultra-high resolution time of flight

US – Up stream

U – Ubiquinone

Wol – Wolinella

WT – Wild type

α-KG – α-ketoglutarate

β-DDM – β- Dodecyl maltoside

β-DM – β- Decylmaltoside

τd – Life time of Donor molecule

yd – Flouresence quantum yield

CONTENTS

1.	**Introduction**	**1-12**
	1.1. History	1
	1.2. Ecological diversity	3
	1.3. Biodiversity	5
	1.4. Biotechnological applications	6
2.	**Isolation and Cultivation**	**13-43**
	2.1. Direct Isolations	13
	2.2. Isolation Procedures of Purple Bacteria	14
	2.2.1. Enrichment and Isolation of Purple Sulfur Bacteria	14
	2.2.2. Enrichment and Isolation of *Ectothiorhodospiraceae* members	17
	2.2.3. Enrichment and Isolation of Purple Non-Sulfur Bacteria	19
	2.3. Isolation Procedures of Green Bacteria	36
	2.3.1. Enrichment and Isolationof Green Sulfur Bacteria	36
	2.3.2. Enrichment and Isolation of Green Non-Sulfur Bacteria	39
	2.4. Isolation Procedures of *Heliobacteriaceae* Members	41
3.	**Taxonomy**	**44-304**
	3.1 Purple Non-Sulfur Bacteria	62
	3.2 Purple Sulfur Bacteria	166
	3.3 Green Sulfur Bacteria	250
	3.4 Green Non-Sulfur Bacteria (Multicellular Filamentous Bacteria)	274
	3.5 Firmicutes	291
4	**Photosynthesis**	**305-344**
	4.1 Anaerobic Photosynthetic Bacteria	307
	4.2 Aerobic Anoxygenic Phototrophic Bacteria	310
	4.3 Photosynthetic Apparatus- Light-Harvesting Pigments	310
	4.4 Properties of Light	320

4.5 Electron Transport in Purple Bacteria 323
4.6 The Structure of the Light-Harvesting Complexes 325
4.7 The function of Light-Harvesting Complexes 328
4.8 The Reaction Centre (RC) 331
4.9 Photosynthetic Unit (PSU) 333
4.10 Electron Transport 338
4.11 Photophosphorylation in Halophilic Phototrophic Bacteria 340
4.12 Photochemical Reaction Centre 342
4.13 Photosynthetic Electron Transport Chain 342
4.14 The pattern of Electron Flow 342

5. Carbon Metabolism 345-387

5.1 Acetate Assimilation 348
5.2 The flow of Reducing Equivalent in Photoautotrophs and Photoheterotrophs 350
5.3 Anaplerotic CO_2 Assimilation 352
5.4 Carbohydrate Metabolism 353
5.5 Aldolases Employed in the EMP and ED Pathway 356
5.6 The Tricarboxylic Acid Cycles (Forward, Reverse and Branched) 358
5.7 Enzymes Involved in Citrate Metabolisms 358
5.8 Citrate Metabolism and The TCA Cycle 362
5.9 Evolutionary Perspectives 363
5.10 Carbon metabolism in Photoorganotrophs 364
5.11 Autotrophic CO_2 fixation (Dark phase of photosynthesis) 365
5.12 Photometabolism in Anoxygenic Phototrophic Bacteria 369
5.13 Hydrogen and Inorganic Sulfur Compounds 372
5.14 Acetate Photometabolism 373
5.15 Pyruvate Photometabolism 375
5.16 Formate Photometabolism 376
5.17 Succinate Photometabolism 376
5.18 Butyrate Photometabolism 379
5.19 Anaerobic Tricarboxylic Acid and Reductive CO_2 Fixation 379
5.20 Acetone and Alcohol Photometabolism 383
5.21 Glycine Photometabolism 384
5.22 Methane Photometabolism 384
5.23 Aromatic Compounds Metabolism 384
5.24 Photochemical Nitrogen Fixation 385
5.25 The Electron Acceptor in Purple Bacteria 386

6 Sulfur Metabolism **388-421**

6.1 Introduction 388

6.2 Transformations of Sulfur Compounds 390

6.3 Oxidative Sulfur Metabolism 397

6.4 Sulfate Assimilation 414

6.5 Uptake of Sulfate 418

6.6 Conclusion 420

7 Ecology of Anoxygenic Phototrophic Bacteria **422-434**

7.1 The Role of Hydrogen Sulfide 423

7.2 The Relations to Oxygen 423

7.3 The Role of Light 424

7.4 Habitats of Purple Sulfur Bacteria 424

7.5 Habitats of Purple Non-Sulfur Bacteria 429

7.6 Habitats of Green Sulfur Bacteria 432

7.7 Habitats of *Heliobacteriaceae* 434

8 Hydrogen - Non-Conventional energy **435-456**

8.1 WastewaterTreatment and Hydrogen Production 437

8.2 Light-Driven Hydrogen Production 443

i. Direct Biophotolysis 443

ii. Indirect Biophotolysis 444

8.3 Photo-Fermentation 445

8.4 Dark Fermentation 453

8.5 Hybrid Reactor System 454

8.6 Water–Gas Shift Reaction of Photoheterotrophic Bacteria 454

9 Single Cell Protein **457-477**

9.1 Microorganisms 459

9.2 Raw Materials 467

9.3 Mixed Cultures 468

9.4 Steps in SCP production 469

9.5 Fermentation Processing for SCP Production 470

9.6 Nutritional Status of SCP 471

9.7 Toxicological Status of SCP 473

9.8 Nucleic Acid Status 474

9.9 Improvement of Quality of SCP 474

9.10 Downstream Process 475

9.11 Utilization of SCP 475

9.12 Prospects 477

10 Bioremediation of Wastewater 478-516

10.1 Dissolved Oxygen Concentration as an Indicator of Water Quality 479

10.2 Site Surveys 479

10.3 The Strength of Fermentation Effluents 480

10.4 Waste Disposal and Treatment 481

10.4.1 Disposal 482

10.4.2 Treatment 484

11 Biotechnological Applications 517-545

11.1 Biosynthesis of Carotenoids and Other Terpenoids 520

11.2 Production of Functional Membrane Proteins 521

11.3 Ubiquinones (Coenzymes Q) 521

11.4 Vitamins 524

11.5 Hormones 525

11.6 Enzymes 527

11.7 Antibiotics/Antiviral Substances 528

11.8 Ammonia (NH_3) 529

11.9 Nitrogen Fixation 529

11.10 Bacteriocins 531

11.11 Denitrification 532

11.12 Hopanoids 532

11.13 Biofertilizers 534

11.14 Biodegradable Polyesters 538

11.15 Biopesticides 541

11.16 Enhancement of Calorific Value of Biogas 543

11.17 Enhancement of Indigo Extraction 544

11.18 Microbial Enhanced Pulping of Secondary Fibers 544

11.19 Miscellaneous Applications 544

References 546-640

Chapter 1

INTRODUCTION

Anoxygenic Phototrophic Bacteria (APB) are a unique assemblage of predominantly aquatic bacteria that are able to grow under anaerobic conditions by photosynthesis that lacks photosystem-II and does not produce O_2. They are physiologically and phylogenetically diverse groups and contain a variety of bacteriochlorophylls (*a,b,c,d,e,g*) and carotenoids of spirilloxanthin, rhodopinal, sphaeroidene (yellowish-brown in the absence of O_2 and reddish-brown in the presence of O_2) and okenone series. However, all species contain cytochrome, quinones and non-heme iron protein of the ferredoxin type. They are mostly Gram (–) negative bacteria that are spherical, spiral, rods, or vibriod shaped and unicellular. However, uniseriately multicellular filamentous forms also occur. In the presence of sulfide and light, cells of purple bacteria form highly refractile globules of elemental sulfur. They are widely distributed in terrestrial and diverse aquatic ecosystems including extremes of light, anoxic conditions, sulfide concentration, pH, temperature and salinity.

1.1. HISTORY

Since the discovery of Anoxygenic Phototrophic Bacteria (APB) in the 19th century, a large number of species have been described from a wide variety of planktonic and benthic environments. These are widely distributed and abundant in all kinds of stagnant water bodies (Imhoff, 2006c). Ecological niches of these bacteria are anoxic parts of waters and sediments that receive light of sufficient quality and quantity to allow phototrophic development. The purple bacteria form colored blooms that are perceptible to the naked eye, while purple sulfur bacteria (PSB) don't develop color and not visible to the naked eye. These are seen in the sediments or water samples containing H_2S (Imhoff, 2001a). The most important properties that determine the development and distribution of phototrophic bacteria in nature are the availability of light, the concentration of sulfide and oxygen which forms counter-current gradients in most of the natural habitats (Imhoff, 2006c).

Winogradsky (1988) gave a comprehensive description of different morphological types of purple bacteria that he observed in different samples from nature and raw cultures which formed the basis for the identification of purple sulfur bacteria in visible enrichment and natural habitats by later workers. Some of the historical events that led to the understanding of these bacteria are briefly discussed. Lankester (1876) described red-colored crust formed by pigmented bacteria as *Bacterium rubescene* and attempted to characterize pigment as bacterioporphyrin. Winogradsky (1888) described the purple sulfur bacteria, *Chromatium* by employing the Winogradsky column, which was designed as a large glass cylinder packed with mud mixed with organic matter, calcium carbonate, calcium sulfate and overlaid with lake, pond, or ditch water.

Molisch (1907) described *Rhodonostoc capsulatum* subsequently named it *Rhodopseudomonas capsulatus*, which exhibits heterotrophic growth and does not produce oxygen. He also isolated two pigments and named them bacteriopurpurin for red pigment and bacteriochlorin for green pigment. Shibata (1931) proposed a theory on carbon or nitrogen assimilation by purple sulfur bacteria. Hans Molisch (1856-1936) described anoxygenic phototrophic non-sulfur (APNS) bacteria, *Rhodonostoc capsulatum* and *Rhodopseudomonas capsulatus*. He demonstrated that oxygen is not produced during their photosynthesis (Light reaction) and heterotrophic growth of these bacteria. He also separated two pigments bacteriopurpurin (red pigment) and bacteriochlorin (green pigment) from these bacteria. Van Neil (1931) proposed that purple sulfur bacteria can use H_2S as a hydrogen donor, during the reduction of CO_2in the process of photosynthesis. He published papers on photosynthesis in purple non-sulfur bacteria (PNSB) like *Rhodopseudomonas capsulatus* and *Rhodobacter sphaeroides*. Gest and Kamen (1949) reported that *Rsp. rubrum* can use H_2 and CO_2 for photoautotrophic growth. Light-dependent H_2 production and N_2 reduction are catalyzed by an enzyme complex, nitrogenase. Vernon (1953) recorded large quantities of cytochrome *c* in *Rhodopseudomonas rubrum* growing photo autotrophically under anaerobic conditions. Cohen-Bazire *et al.* (1964) observed unique structures, chromosomes in electron micrographs in green sulfur bacteria (GSB). Evans *et al.* (1966) proposed a reductive TCA cycle in *Chlorobium.* Sirevag (1995) provided a missing link in the reductive TCA cycle. Reed and Clayton (1968) have isolated photosynthetic reaction center complex from *Rhodopseudomonas palustris*. Lascelles and Wertlleb (1971) developed a mutant that can produce a large amount of bacteriochlorophyll and carotenoids under aerobic conditions. Pierson and Castenholz (1974) reported anoxygenic phototrophic gliding filamentous bacterium *Chloroflexus aurantiacus*. Sato *et al.* (1989) observed that certain aerobic heterotrophic bacteria with bacteriochlorophyll '*a*' are unable to grow anaerobically with light as an energy source. Thomas Beatty (2002) described aerobic anoxygenic phototrophic bacteria. Gest and Favinger (1983) isolated for the first time *Heliobacterium chlorum* which contains bacteriochlorophyll '*g*'. It contains only photosystem-I and its moiety consists of a single homo-dimeric polypeptide. Youvan *et al.* (1984) published nucleotide sequences of the reaction center and LH1 structural genes of *Rhodobacter capsulatus*. Deisenhofer *et al.* (1988) have been awarded the Nobel Prize in chemistry for the determination of the structure of the photosynthetic reaction center. Widdel *et al.* (1993) discovered autotrophic growth of purple bacteria on CO_2 with Fe^{+2} as an electron donor. This gave support for the hypothesis of early life on earth which explains the deposition of banded iron formation in an assumed anoxic biosphere in Archean times. Mc Dermatt *et al.* (1994) observed phototactic behavior of *Rhodospirillum centenum.* Kobayashi *et al.* (2000) identified the primary electron acceptor of green sulfur bacteria as a chlorophyll derivative. Lang and Beatty (2000) reported the structural gene sequence and organization of *Rhodobacter capsulatus* gene transfer agents which are similar to that of double-stranded DNA phages but the expression of these genes is regulated by a cellular two-component signal transduction system. Lang *et al.* (2002) feel that GTA structural gene clusters are similar to that found in *Alpha-proteobacteria* suggesting a long evolutionary history of GTA-like elements and this group of bacteria.

The APB cell cultures form distinct colorations. The colors of the cell suspensions are purple-violet to purple-red, orange-brown, yellowish-brown, brown and green due to the presence of different photosynthetic pigments which are located either in the cytoplasmic membrane or the internal membrane systems. In purple bacteria, these intracellular membranes consist of small finger-like intrusions, vesicles, tubules, or lamellae parallel to or at an angle to the cytoplasmic membrane. They carry the photosynthetic apparatus, the reaction centers (RC), and light-harvesting (LH) pigment-protein complexes surrounding the reaction center in the plane of the membrane (Drews and Imhoff, 1991).

The photosynthetic metabolism of these organisms differs from the rest of the photosynthetic organisms in that, water cannot serve as an electron donor but utilize an external source of electron donors such as reduced sulfur compounds, molecular hydrogen, or organic compounds. The carbon assimilation is either through the reductive pentose phosphate cycle or the reductive citric acid cycle. Some members are reported to fix atmospheric nitrogen. The photosynthetic pigments and the components of the photosynthetic apparatus are located within a more or less extended system of internal membranes, originating from and being continuous with the cytoplasmic membrane.

1.2. ECOLOGICAL DIVERSITY

APB is widely distributed in nature and is found in aquatic and a few terrestrial environments where sufficient quality and quantity of light is available under both anaerobic and aerobic environments. From these environments, they are isolated by enrichment techniques. Lakes (Jansen and Harfoot, 1991), waste lagoons, intertidal zones (Glaeser and Overmann 1999), sediments of ditches, moist soils of ponds and rivers (Sucharita *et al.,* 2010) are some of the habitats where these organisms flourish. These are found in diverse aquatic habitats such as freshwater (Overmann *et al.,* 1998), hypersaline (Milford *et al.,* 2000), alkaline (Bryantseva *et al.,* 2000), hot springs (Madigan, 2003) and acid water (Hirashi *et al.,* 2000). They are also reported from the arctic (Roeselers *et al.,* 2007) and Antarctic (Karr *et al.,* 2003) regions.

The light intensity, salinity, temperature, pH and concentration of sulfide and oxygen are the important environmental factors which influence their occurrence in different habitats and type of APB growing (Drews and Imhoff, 1991). These bacteria are reported to play an important role in the carbon, nitrogen, hydrogen and sulfur cycles (Sasikala and Ramana, 1999). The abundance and diversity of purple bacteria are influenced by sulfide concentration, light quality and quantity, pH and temperature (Imhoff, 2006c).

The Purple Sulfur Bacteria (PSB) are widely distributed in all kinds of stagnant water bodies depending on the availability of light, the light intensity and the concentration of sulfide and oxygen. Unlike PNSB, blooms of PSB often develop in sulfide aquatic ecological niches exposed to light. However, mixed communities of PSB and GSB can be observed, where PNSB regularly accompany the mass development of phototrophic sulfur bacteria. The sulfide produced by the sulfate-reducing bacteria in sediments of productive stratified lakes diffuses upwards from the sediment and triggers the growth of the PSB. These bacteria can easily be identified by our naked eye by their characteristic colored blooms (Overmann *et al.,* 1999). The PSB is also most common in microbial mat forms in marine, hypersaline environments and in the effluents of thermal spring (Castenholz and Pierson, 1995). Species of *Thiocapsa* and *Allochromatium* were observed in different layers of marine microbial mats because of their metabolic diversity (Van Gemerden and Mas, 1995). Members of *Chromatiaceae* specifically inhabit different types of freshwater, marine and hypersaline habitats (Imhoff, 2006c). Some other PSB such as *Allochromatium vinosum, Alc. renukae* were recorded in brackish water (Anil Kumar *et al.,* 2008a), while *Marichromatium bheemlicum* in marine aquaculture pond (Anil Kumar *et al.,* 2007b), *Halochromatium roseum* in saltern(Anil Kumar *et al.,* 2007b) and *Marichromatium fluminis* in river waters were recorded (Sucharita *et al.,* 2010).

Several species of PSB such as *Ectothiorhodospira, Halochromatium, Halorhodospira, Marichromatium* and *Thiohalocapsa* are halophilic (Imhoff *et al.,* 1979), while *Ectothiorhodospira saline* is extremophilic and grows under hypersaline conditions (Venkata Ramana *et al.,* 2010). PSB is also reported from hot, cold, acidic and alkaline environments apart from fresh water and marine coastal environments (Madigan, 2003). Though most of the purple bacteria are mesophilic, some of the members of *Chromatiaceae* are thermophilic and psychrophilic (Petri and Imhoff, 2001).

The PSB such as *Thermochromatium tepidum* occurs in extreme environments and carries out photosynthesis at 57°C (Madigan, 1986) and down to 0°C, pH from 3 to 11 and at salinities up to saturated solutions of NaCl (32% w/v). Members of *Chromatiaceae* and *Ectothiorhodospiraceae* were identified from samples of salt lakes of Chile based on metagenomic studies (Thiel *et al.,* 2010).

Purple Non-Sulphur Bacteria (PNSB) occurs in all ecological niches where anoxic conditions prevail (freshwater, polluted water and marine waters), microbial mats and sediments that receive sufficient quality and quantity of light.

Members of PNSB widely occur in all kinds of stagnant water bodies including wastewater ponds, slaughterhouse waste, lakes, coastal lagoons and other aquatic habitats. They are also reported in sediments, moist soils and paddy fields. The development of PNSB occurs in those aquatic habitats where a significant amount of soluble organic matter and low O_2 concentration is available, but rarely forms colored blooms like those of PSB because of their sulfide sensitivity (Madigan and Jung, 2009). A large number of PNSB are reported in mud and water of eutrophic ponds, ditches and lakes and stratified environments (Biebl and Drews, 1969). They are found in pelagic water environments along with PSB (Biebl, 1973).

The PNSB also occurs in marine, hypersaline environments, sediments of intertidal zones, salt marshes and polluted harbor basins, hot springs but not in the open sea. The marine species of PNSB have a very high tolerance to sulfide when compared to freshwater isolates. They have a very low tolerance to sulfide and they even use sulfide and thiosulfate as photosynthetic electron donors. This is an adaptation to the environment, which has a characteristically high activity of sulfate reduction. Most of the PNSB is typically marine bacteria (Imhoff, 1988). On the other hand, *Rhodopseudomonas palustris* , *Rps. vannielli,* and *Rhodobacter maris* which are recorded from marine habitats, are not typical marine bacteria. Halophilic species such as *Rhodovibrio salinarum* and *Rhodothalasium salexigens* are well adapted to hypersaline environments (Rodriguez-Valera *et al.,* 1985). *Rhodovibrio sodomensis* was reported from Dead Sea sediments (Mack *et al.,* 1993). Some PNSB possesses tolerance to a wide temperature range and *Rhodobacter aestuarii* (estuarine environment) and *Rhodobacter johrii* is an endospore producing cryptic species (semiarid tropical soils) are reported from a wide range of environments (Kulichevskaya *et al.,* 2006; Arunasri *et al.,* 2008; Venkata Ramana *et al.,* 2009; Girija *et al.,* 2010).

The purple non-sulfur bacteria from colored microbial mats in swine wastewater ditches (Okubo *et al.,* 2006), the edge of lakes, water under the ice (Karr *et al.,* 2003), conventional sewage plants (Siefert *et al.,* 1978) and industrial effluents (Vasavi, 2001; Srinivas *et al.,* 2001; Ramchander *et al.,* 2007). The extremophilic PNSB like *Rhodovibrio, Rhodobium, Rhodothalassium* , *Rhodovulum* and *Rhodospira* are reported to grow at 20% NaCl concentration (Imhoff, 2001a) and *Rhodoferax antarcticus,* a typical psychrophilic APB, grows at 0°C (Madigan *et al.,* 2000). Some PNSB is psychrotolerant and reported by a large number of workers (Lakshmi *et al.,* 2009; Srinivas *et al.,* 2012; Shalem Raj *et al.,* 2013; Venkata Ramana *et al.*, 2013 and Lakshmi *et al.,* 2013).

1.3. BIODIVERSITY

The task of studying the diversity of living organisms can be made easier and more effective if the various organisms are arranged in an orderly manner by classification which is a part of the taxonomy. Taxonomy is almost synonymous with systematic or biosystematics is traditionally divided into three parts, (i)

Classification, the orderly arrangement of organisms into taxonomic groups based on similarity; (ii) **Nomenclature**: the labeling of the units defined in (i); (iii) **Identification**: The process of determining whether an organism belongs to one of the units defined in (i) and (ii). The methods used in the bacterial taxonomy are being continuously improved since their invention, the widespread use of the electron microscope, introduction of computers, the analysis of DNA of an organism and DNA–DNA hybridization (Stackebrandt *et al.,* 2002) brought a qualitative change in characterization and the classification of purple bacteria. Therefore, an attempt is made to give a comprehensive account of the taxonomy of anoxygenic phototrophic bacteria based on the morphological, physiological and molecular basis.

1.4. BIOTECHNOLOGICAL APPLICATIONS

Anoxygenic Phototrophic Bacteria with diverse metabolic pathways grow by various modes of growth (photoautotrophy, chemoautotrophy, heterotrophy and fermentation) coupled with high growth rate and amendable to easy manipulation are potential organisms for a variety of biotechnological applications. Since light is the source of energy, these bacteria proved to be highly economical and beneficial in different fields of biotechnology (Becker, 1994). As APB has high protein content with a good amount of essential amino acids, vitamins, biological co-factors and a lower amount of nucleic acids. Many workers have recommended anoxygenic purple non-sulfur bacteria as a source of vitamin B_{12} (Goldrick *et al.,* 2005), Ubiquinone Q_{10} (Sasaki *et al.,* 2002), SCP for pisciculture and poultry industry (Salma *et al.,* 2007), metal ion uptake (Sanaa *et al.*, 2006), hopanoids (Nagumo *et al.,* 1991) and therapeutic S-adenosyl- L-cystine (Yamada *et al.,* 1986).

Some of these bacterial strains may prove to be an economical source of carbohydrate and methane production (Mitsui *et al.*, 1985; Vatsala and Balaji, 1987). Production of 5-Amino levulinic acid, a potent herbicide, produced by anoxygenic photosynthetic bacteria proved to be highly useful (Sasaki *et al*,. 2002). Carotenoids produced by *Rba. sphaeroides* are used as natural dyes in foods (Quian *et al.*, 1991).

APB are also being exploited in production of hypocholesterolemic substances (Salma *et al.*, 2007; Ramchander *et al.,* 2011), hydrolytic enzymes (Scavetta *et al.*, 2000; Srinivas *et al.,* 2002, 2004, 2005, 2007 and Ramchander *et al.,* 2008, 2010), IAA (Rajyalaxmi and Girisham, 2015), FAA (Rajyalaxmi *et al*,. 2015), biodegradable plastics (Poly-β-hydroxyAlkanoates (PHA) and Poly β-hydroxy butyrate (PHB)) (Sangkharak and Prasertsan, 2008; Waligorska *et al.* , 2009; Ramachander *et al.,* 2010a, 2010b, 2012) and hydrogen production (Zabut *et al.,* 2005; Ramchander *et al.*, 2010, 2011, 2012a, b, c and 2013a, b; Srinivas *et al.,* 2014a, b; Vasavi *et al.,* 2014 and Rajyalaxmi *et al.*, 2018), biodegradation of organic and inorganic substances in wastewater (Vasavi *et al.,* 2008; Ramchander *et al.,* 2007, 2014 and 2011a,, Rajyalaxmi *et al.,* 2015; Rajyalaxmi *et al.*, 2017), Cytokinins (Serdyuk *et al.,* 1993), Rhodestrin (Sunayana *et al.,* 2005),

Rhodethrin (Ranjith *et al.,* 2007a), Phenols and esters (Ranjith *et al.,* 2010a), Rubrivivaxin cytotoxic and cyclooxygenase-i inhibitory metabolites (Ranjith *et al.,* 2011) and ammonia leaching (Sasikala and Ramana, 1990; Ramchander, 2009).

Natural waters are reported to be contaminated with various heavy metals arising from a variety of industrial effluents (Grousset *et al.,* 1999). Unlike organic contaminants which can be degraded to harmless chemical species, heavy metals cannot be degraded. Members of *Rhodospirillaceae* show a high resistance towards toxic heavy metal oxides and oxyanions (Moore and Kaplan, 1992). Therefore, many of these bacteria are being exploited for remediation of heavy metal pollution (Watanabe *et al.,* 2003; Rajyalaxmi *et al.,* 2018).

Carbohydrates which represent the most diverse class of biopolymers are receiving growing interest to study their diversity and biomedical applications. It is widely known that the diversity of carbohydrates is much more in micro-organisms than in mammals. Both Gram (+) positive and Gram (-) negative bacteria produce extracellular polysaccharides which form either as a discrete capsule covalently attached to the cell envelope or as slime weakly bound to the cell surface. These various glycoconjugates and polysaccharides on the surface of the cell often contain the antigenic determinants that initiate an immunogenic response in a host. In addition, these surface carbohydrates provide recognition elements for pathogens such as bacteriophages. Photosynthetic bacterial strains proved to be an economical source of carbohydrate source for biological methane production.

Increasing population and limitations on cultivable land and water resources have forced mankind to search for an alternative source of food and feed. Microorganisms are a potential source of food and feed either directly or indirectly in pisciculture and poultry (Azad *et al.,* 2002). As anoxygenic phototrophic bacteria meet the criteria of nutritional quality (Litchfield, 1980), these organisms are being used as good alternative food and feed (Azad *et al.,* 2002; Ponsano *et al.,* 2002).

Many APB and PNSB proved to be potential source of single-cell protein. It was observed that cellular yield based on substrate ,*Rhodocyclus gelatinosus* has grown aerobically in the dark on starch was comparable with that of several yeasts (Noparatnaraporn *et al.,* 1983). The different essential amino acids of anoxygenic phototrophic bacteria are comparable to those of soybean and egg proteins (Ponsano *et al.*, 2003) and can be used as a superior source of single-cell protein (Azad *et al.*, 2001). Biomass of *Rps. palustris* is used as a diet for aquaculture (Kim and Lee, 2000). Ponsano *et al.* (2003) reported the nutritional richness of biomass of *Rcy. gelatinosus* has grown on poultry slaughterhouses. Several PNSB such as *Rps. palustris* (72-74%) (Kim and Lee, 2000) and strain DK6 of phototrophic PNSB (65.2%) (Kantachote *et al.*, 2005) are reported to be a rich source of protein.

Ammonia production by microorganisms at ambient temperature and pressure is expected to become an alternative process for commercial and eco-friendly production of ammonia. APB is reported to leach out ammonia during their growth on amino acids. Ammonia leaching by resting cells of *Rhodobacter sphaeroides* O.U.001 (Sasikala and Ramana, 1990), mixed culture of photosynthetic bacteria (Hiroo *et al.,* 2004) and some of the PNSB (Ramchander, 2009) were reported.

The colour of phototrophic bacteria mostly depends on the nature and quantitative composition of photosynthetic pigments. The pigments can be recognized by absorption spectra of intact cells. The absorption maxima at < 450 nm and at 650-1000 nm are due to chlorophylls, while absorption maxima in the range of 400-500 nm are due to carotenoids. The carotenoid synthesis by APB was discussed elegantly by Jonathan *et al.* (2007). Carotenoids produced by *Rba. sphaeroides* are being used as dyes in the food industry (Qian *et al.,* 1991). Nubel *et al.* (2001) have reported a variety of carotenoids from phototrophic bacteria growing in hypersaline mats.

Auxins are the major growth regulators which regulate the growth of plants, particularly in cell enlargement and cell differentiation. Indole acetic acid, the main auxin which is synthesized in leaves and shoot tips of higher plants through the shikimic acid pathway of tryptophan, is also synthesized by these bacteria. Although several PNSB is reported to produce IAA, it is still underexplored. IAA, which is a common product of tryptophan metabolism, is reported to be produced by APB (Rajyalaxmi and Girisham, 2015).

Photobio transformation of IAA by some of the PNSB was reported to play a major role in the production of phytohormones (Sasikala and Ramana, 1995). A phytohormone, Rhodestrin, was isolated as a metabolite of anthranilate Photobio transformation by *Rba. sphaeroides* O.U5, induced profuse and early rooting in tissue cultured plants (Sunayana *et al.,* 2005). Ranjith *et al.* (2007a) reported the production of novel indole terpenoid ether, Rhodethrin, by *Rba. sphaeroides* have phytohormonal activity. Mujahid *et al.* (2010) reported the production of Indole-3-acetic acid (*Rvi. benzoatiliticus* JA2), Indole-3-aldehyde (*Rvi. benzoatiliticus* and *Rba. sphaeroides* DSM 158), Indole-3-ethanol (*Rsp. rubrum* ATCC 1170) and Anthranilic acid (*Rba. sphaeroides* DSM 158) by different APB. Ranjith *et al.* (2011) for the first time reported the production of rubrivivaxin, a new phenol terpenoid ester having a cytotoxic and cyclooxygenase-I inhibitory effect, by *Rvi. benzoatiliticus* JA2.

Poly-β-hydroxybutyrate (PHB) is a unique intracellular polymeric material that gets accumulated under unbalanced growth conditions in a wide variety of bacteria (Reddy *et al.,* 2003; Lenz and Marchesault, 2005). PHB is a reserve material to dissipate excess energy. It is a biodegradable, thermoplastic polymer and has industrial applications such as biodegradable carriers for long-term dosage of drugs. Moreover, it can be used to produce surgical pins, sutures, staples, swabs and wound dressing (Yigit *et al.,* 1999).

Brandl *et al.* (2001) reported that *Rba. sphaeroides* produced PHA as the major component (97%) and a small amount of PHV (3%) under anaerobic light conditions. Boeree *et al.* (1993) have developed a biodegradable composite material composing of PHB with hydroxyapatite from APB. Yamane (1992) excellently reviewed the biodegradable polyester metabolism of APB. Noparatnaraporn *et al.* (2001) reported the production of PHA by *Rba. sphaeroides* and *Rba. sphaeroides* IL 106 from acetic acid. The PHA production by photosynthetic bacteria at 37- 40ºC in an aerobic dark condition was reported to be advantageous (Lorrungruang *et al.*, 2006).

Naturally occurring surface-active molecules derived from microorganisms are known as biosurfactants (Gautam and Tyagi, 2006). The major classes of biosurfactants are glycolipids, phospholipids, neutral lipids, fatty acids, lipopolysaccharides, lipoprotein-lipopeptides, flavolipids, rhamnolipids, complete cell surface itself and those not fully characterized (Desai and Banat, 1997; Bodour *et al.,* 2003). These are amphiphilic. The hydrophobic and hydrophilic moieties of the biosurfactants confer them the ability to accumulate between two fluid phases, thereby reducing surface and interfacial tension (Dehghan-Noudeh *et al.*, 2010). The identification and characterization of biosurfactants produced by various microorganisms including APB are discussed excellently by Reis *et al.* (2013). These form an interesting group of materials for application in areas like agriculture, mining, food processing, chemistry, cosmetics, pharmaceutics, oil industries, waste utilization and environmental pollution control (Rodrigues *et al.*, 2006; Thavasi *et al.*, 2011). The use of various biosurfactants for the degradation of pesticides in soil and water has also gained importance. Of various biosurfactants, rhamnolipids produced by APB are widely used (Zhang and Miller, 1992; Providenti *et al.*, 1995; Banat *et al.*, 2000).

Interest in large-scale production of enzymes by APB has increased in food processing, agricultural and pharmaceutical industries. Proteases produced by different APB were investigated by several workers (Srinivas *et al.*, 2007 and Ramchander *et al.,* 2010). Out of two types of proteases produced by *Chloroflexus aurantiacus*, type I protease is stable at 70°C and pH between 4.0 and 11.0 (Watanabe *et al.,* 1993). A novel protein modifying enzyme of *Rhodocyclus gelatinosus* is used in food preparations (Isaki and Kamio, 1992). Serine protease from PNSB is useful in different areas of public interest (Hyun *et al.,* 1989). Amylases of *Cfl. aurantiacus* (Ratnakhanokchai *et al.,* 1992) and other APB are useful in the degradation of glycogen and pullulan (Buranakarl*et al.,* 1988).

Production of lipases (Srinivas *et al.,* 2005), L- asparaginase (Ramchander *et al.,* 2008) by *Rba. capsulatus, Rps. palustris, Rps. acidophila* and *Rcy. gelatinosus* isolated from leather industry effluents can be exploited. Phytase which is used as a feed additive, antioxidant, strong chelators and also essential for treating metabolic disorders (Konietzny and Greiner, 2004) is produced by a number of APB (Sasirekha *et al.,* 2012; Kumar *et al.,* 2013; Suhas Kumar and Arun Kumar, 2014).

The availability of phosphorus to plant is limited for various reasons and the application of phosphorus in the form of chemical fertilizers is responsible for the deterioration of the soil quality (Tewari *et al.,* 2004). Therefore the present scenario is shifting from the chemical ion of fertilizers to natural fertilizers (Biofertilizers) for sustainable agriculture. The application of phosphate solubilizing bacteria proved to be more beneficial in the sustenance of soil fertility (Narula *et al.,* 2000; Whitelaw, 2000). The positive effect of phosphate solubilization by anoxygenic phototrophic bacteria on food and fodder crops has been reported by Mittal *et al.* (2008) and Gulati *et al.* (2009).

Anaerobic treatment of domestic and industrial wastewater releases a large amount of phosphorus and nitrogen and responsible for the eutrophication of rivers, lakes and seas all over the world (Trepanier *et al.,* 2002). APB is reported to be efficient in the recycling of phosphorus in an ecofriendly manner (Nagadomi *et al.,* 2000). Efficient removal of phosphorus from wastewater by *Rhodocyclus* spp. (Julie *et al.,* 2002), *Rhodopseudomonas palustris, Rps. acidophila, Rhodobacter capsulatus, Rhodospirillum rutila*, *Rsp. rubrum* (Ramachander *et al.,* 2008) and *Allochromatium* sp. GSKRLMBKU-01 (Rajyalaxmi *et al.,* 2015) has been reported.

The diminishing natural resources, ever-growing human population and the increased demand for energy, necessitated the need for newer, safer, cleaner and renewable energy. Hydrogen is likely to fulfill all the requirements of safer energy. Hence, the search for a safe and sustainable source for its production has been intensified. Hydrogen by its wide desired attributes (Ahmet Lokurlu *et al.,* 2003) is used as feedstock in the chemical and metallurgical industries (Venkataramani, 2002). The application of hydrogen as a fuel is limited to about 5% due to a lack of advanced production, storage and transportation and utilization technology.

Biological systems provide a wide range of sources to generate hydrogens such as direct bio photolysis, indirect bio photolysis, Photo fermentation and dark fermentation (Hallen beck and Benemann, 2002 and Nath and Das, 2004). Though hydrogen can be produced by several biological processes, the photobiological production of hydrogen offers potential, clean and renewable energy (Weaver *et al.,* 1980). Prospects of biological hydrogen production by APB are being investigated all over the world (Sasikala *et al.,* 1991).

Microbial hydrogen production by fermentation under dark conditions (Oh *et al.,* 2008; Lo *et al.,* 2008), photoheterotrophic conditions (Willison *et al.,* 1984) photo-fermentation under dark conditions (Ghirardi *et al.,* 2005, Li and Fang, 2008) and mixed cultures of hetero- and/or phototrophs (Wang and Wan,2009) has been reported from different parts of the world. Hydrogen production using different types of reactors such as batch (Venkatamohan *et al.,* 2007; O-Thong *et al.,* 2007), semi-continuous (Gavala *et al.,* 2006) and continuous modes of operation (Lee *et al.,* 2006, Zhang *et al.,* 2007, Younesi *et al.,* 2008) are in operation. Hydrogen production using wastewater such as dairy wastewater (Venkataramana and Vatsala 1990) and distillery effluents (Vatsala *et al.,* 2008) by APB proved to be more beneficial.

Among APB, purple non-sulfur bacteria are reported to have high potentials for hydrogen production (Ramchander *et al.*, 2010; Srinivas *et al.*, 2014a,b and Vasavi *et al.*, 2014). Utilization of wastewater for photobiological hydrogen generation by photosynthetic purple non-sulfur bacteria (Vasavi *et al.,* 2008; Ramchander *et al.,* 2010) proved to be more promising. Immobilization of APB for hydrogen production has an added advantage (Ramchander *et al.,* 2013a, b) for its sustainability. Purple sulfur bacteria *Allochromatium* sp. GSKRLMBKU-01 produced a sustainable amount of hydrogen under phototrophic conditions was also reported (Rajyalaxmi *et al.*, 2018).

Microorganisms present in wastewater besides interacting with other microflora and fauna, act on different constituents of wastewater and bring a significant qualitative change in water quality (Robertson and Edberg, 1997). Treatment of wastewater and industrial effluents by anaerobic organisms especially APB offers several advantages over anaerobic treatment. Further, APB plays a crucial role in biogeochemical cycles such as anaerobic catabolism of aromatic compounds (Sasikala and Ramana, 1998; Carmona *et al.,* 2008). By virtue of these properties, APB is being exploited for treating different types of wastewater (Panwichian *et al.,* 2010; Madukasi *et al.,* 2010).

APB is efficient in the degradation of organic wastes of agricultural and food industries (Sasikala *et al.,* 1995; Ogbonna *et al.,* 2000) and oil containing sewage water (Takeno *et al.,* 2005). The denitrification and deammonification capabilities of APB are of special interest because of the deleterious effect of the accumulation of ammonia and nitrates (Bashan and Bashan, 2004). *Rhodovulum sulfidophilum* is being used for the treatment and utilization of sardine processing wastewater (Azad *et al.,* 2003).

APB are ideal organisms for the treatment of waste water under their ability to grow at high BOD, needs no dilution of wastewater, require small treatment space, no sludge disposal problem, easy maintenance, tolerance to cold, the feasibility of nitrogen, grease removal and production of biomass rich in protein and vitamins (Perrotta and Harwood, 1994; vanderWoude*et al.,* 1994).

Microbial reduction of toxic heavy metals to less soluble forms represents a useful detoxification process. Photosynthetic purple bacteria have been shown to be resistant to different heavy metals. Smiejan *et al.* (2003), Watanabe *et al.* (2003), Gad El-Rab *et al.* (2006) and Buccolieri *et al.* (2006) have successfully used different APB for treating industrial wastewater rich in different heavy metals. Hydrogenases isolated from *Thiocapsa roseopersiciana* could reduce the metal ions and oxidize the metals (Zadvorny *et al.,* 2006). Many bacteria including *Rba. sphaeroides* (Nepple *et al.,* 2000), *Rba. capsulatus* KU002 (Ramchander *et al.,* 2011), *Rps. palustris* KU003 (Ramchander *et al.,* 2013) and *Rhodobacter* sp. GSKRLMBKU-03 (Rajyalaxmi *et al.,* 2017) are reported to reduce chromium.

APB comprises members of Purple Sulfur Bacteria (PSB), Purple Non-Sulfur Bacteria (PNSB), Green-Sulfur Bacteria (GSB), Green Non-Sulfur Bacteria (GNSB), and *Heliobacteria* (Yurkov and Beatty, 1998). Based on 16S rRNA gene sequence analysis, they are classified into five distantly related phyla; *Proteobacteria* (Purple bacteria), *Chlorobi* (Green-sulfur bacteria), *Chloroflexi* (Green Non-Sulfur bacteria) and *Firmicutes* (Gram +ve positive *Heliobacteria* .) Bryant *et al.* (2007) described monotypic phototrophic bacteria *Chloroacidobacterium thermophilum* which is included in the phylum *Acidobacteria* .

■■■■

Chapter 2

ISOLATION AND CULTIVATION

As far back as the primary isolation of purple sulfur bacteria by Winogradsky (1888), it has turned out to be standard to separate these microorganisms by improvement culture. Winogradsky sections are set up in tall glass chambers with plant buildups, calcium sulfate, anaerobic mud and water of characteristic natural surroundings and incubated in diminish light for a few days which results in the development of a few layers of various hues demonstrate the development of indicated organisms. The phototrophic bacteria from various layers are streaked and permitted to develop required microbes. Therefore, a few changes to this segment method is made. The arrangement and created just as the focal points and confinements of such enduring crude societies are very much examined by several workers. Most purple sulfur bacteria that flourish in Winogradsky segments additionally can be developed and isolated by direct techniques in synthetic media. Phototrophic sulfur bacteria might be specifically enriched from the most normal environment in suitable media under anoxic conditions as in light.

The primary reasonable enhancement and isolation methodology for purple- and green sulfur bacteria was developed by van Niel (1931). Further, endeavors to develop in unadulterated culture the huge celled purple sulfur bacteria such as *Chromatium okenii* and *Thiospirillum jenense*, prompted the formulation of a synthetic medium that contains nutrients vitamin B_{12} (Pfennig, 1965; Pfennig and Lippert, 1966). A propelled formula of this medium is given below (Medium 1). With a slight change, this culture medium permits the isolation and cultivation of most purple-and green sulfur microorganisms that occur in freshwater, estuarine and marine natural habitats.

2.1. DIRECT ISOLATION

As an option in contrast to enrichment followed by pure culture isolation, pure cultures might be acquired from normal samples without earlier improvement by legitimately inoculating agar media. This technique must be utilized at whatever point the examination of the common diversity has endeavored and data on the natural abundance and distribution of the species in a sample is desire. Media with low selectivity are required for this methodology. At the point when the incubation time is adjusted to the development of gradually developing strains or of those not well adjusted to the given conditions. Separation in or on agar enables these cells to develop into small colonies that can be used for further moves.

Profound agar dilution series (see below) ought to be prepared legitimately from the sample without earlier enrichment in liquid culture. Indeed, even for this situation, be that as it may, the incubation conditions for the agar cultures ought to resemble the conditions utilized for liquid enrichment cultures of the ideal phototrophic bacteria, *i.e.,* utilization of low temperatures and low light forces as demonstrated above is recommended. Strategies for direct isolation of the phototrophic bacteria from a characteristic sample use agar dilution series or inoculation of agar plates to isolate the cells preceding incubation. For inoculation, a sample of water, mud, slop, or even soil might be used as a homogeneous suspension in medium or filtered-sterilized water from the habitats. Samples containing under 10 cells/mL should be concentrated by centrifugation (agar dilution series) or filtration (agar plates). All methods for direct isolation are suitable for the determination of living cell counts when known amounts of the sample are used in appropriate dilutions.

2.2. ISOLATION PROCEDURES OF PURPLE BACTERIA

2.2.1. Enrichment and Isolation of Purple Sulfur Bacteria

The composition and preparation of two different culture media are described below for the enrichment and isolation of purple sulfur bacteria. The Medium 1 is suitable for the development of almost all purple bacteria in laboratory culture, including those species that are most difficult to grow such as *Thiopedia rosea, Thiospirillum jenense ,Chromatium okenii* and *Thiodictyon elegans*. With minor modifications, this culture medium composition was published by Pfennig (1965), Pfennig and Lippert (1966), Truper (1970), Eichler and Pfennig (1988) and Pfennig and Trüper (1992). In addition, Medium 2 is a very similar culture medium, easier to prepare and allows the cultivation of many common purple sulfur bacteria (Biebl and Pfennig, 1978).

2.2.1.1. *Preparation of Medium 1* (Eichler and Pfennig, 1988)

The medium is prepared in a 5.0-liter bottle which contains four openings at the top. Two openings are connected to tubing, they are 1. A short, gas inlet tube with a sterile cotton filter and 2. An outlet tube for the medium, which reaches the bottom of the central part of the vessel at one end and at the other end it has a silicon rubber tube with a pinchcock and a bell for aseptic dispensing of the medium into bottles. The other two openings have gas-tight screw caps. One of these openings is for the addition of sterile solutions and the other can serve as a gas outlet.

i. **Preparation of Solution 1 (Eichler and Pfennig, 1988) (g/L):** KH_2PO_4, 1.7; NH_4Cl, 1.7; KCl, 1.7; $MgSO_4 \cdot 7H_2O$, 2.5; $CaCl_2 \cdot 2H_2O$,

1.25 g; Distilled water, 4,900 mL. Solution 1 is autoclaved for 45 min at 121°C in the 5.0 liter bottle with a Teflon-coated magnetic stir bar. It is cooled to room temperature under N_2 atmosphere with a positive pressure of 0.05–0.1 atm (a manometer for low pressures is required). The cold medium is saturated with CO_2 by magnetic stirring for 30 min under a CO_2 atmosphere of 0.05–0.1 atm. The sterile solutions 2 through 5 are added through one of the screw-cap openings against a stream of either N_2 gas or a mixture of 95% N_2 and 5% CO_2, while the medium is magnetically stirred. For enrichment and isolation of pure cultures from marine or estuarine habitats, 100 g of NaCl is added to solution 1 and the contents of $MgSO_4.7H_2O$ are increased to 15.0 g.

ii. **Preparation of Solution 2 (Vitamin B_{12} Solution):** A sterile-filtered stock solution containing 2.0 mg of vitamin B_{12} in 100 mL of distilled water is prepared and 5.0 mL is added to solution 1 .

iii. **Preparation of Solution 3 (Trace Element Solution SL-12):** Distilled water, 1.0 liter; Ethylene diamine tetraacetate-Na_2, 3.0 g; $FeSO_4 \cdot 7H_2O$, 1.1 g; H_3BO_3, 300 mg; $CoCl_2 \cdot 6H_2O$, 190 mg; $MnCl_2 \cdot 4H_2O$, 50.0; $ZnCl_2$, 42.0 mg; $NiCl_2 \cdot 6H_2O$, 24.0 mg; $Na_2MoO_4 \cdot 2H_2O$; 18.0 mg; $CuCl_2 \cdot 2H_2O$, 2mg. The salts are dissolved in the order given, the pH is adjusted to 2–3 with HCl, the solution is sterilized and 5.0 mL are added to solution 1.

iv. **Preparation of Solution 4 (Sodium bicarbonate ($NaHCO_3$) Solution):** A solution of 7.5% $NaHCO_3$ is prepared, saturated with CO_2, autoclaved under a CO_2 atmosphere in a tightly closed bottle and 100 mL is added to solution 1.

v. **Preparation of Solution 5 (Sodium Sulfide Solution):** A 10% $Na_2S.9H_2O$ solution is prepared in a screw-cap bottle and after replacement of the air by N_2, the bottle is tightly closed and autoclaved. 20 mL of the sterile solution was added to the solution 1.

In the wake of combining and cautiously mixing solution 1 through 5, the pH of the medium is balanced by stirring under an atmosphere of CO_2 (0.5 bar pressure) for around 40 min to pH 7.2. The medium is then quickly administered aseptically under the pressure of N_2 (0.05–0.1 atm) into sterile 100 mL bottles with metal screw caps containing autoclavable rubber stoppers. A tiny pea-sized air bubble is left in each bottle to meet conceivable pressure changes. The firmly fixed screw-cap bottles can be store for several weeks to a month in dark. During the initial 24 h, some trace metals chiefly iron and sulfides of the medium precipitate as fine dark particles. No other sediments ought to emerge again alternatively in the clear medium.

vi. Preparation of Supplement Solutions

a. **Sulfide Solution for Feeding of Batch Cultures:** The amount of $Na_2S.9H_2O$ was primarily added to Medium 1 (higher initial concentrations may be inhibitory for certain species) will only produce very constrained growth. After the sulfide and sulfur are oxidized, the bacteria stop their development and might be damaged by further illumination. To keep the cultures growing and to acquire high cell yields, it is important to feed the cultures many times with sterile, partially neutralized sulfide solution. Two different sulfide feeding solutions are prepared as follows:

Dissolve 3.6g of $Na_2S \cdot 9H_2O$ in 100 mL of distilled water (a 0.15 M solution) in a 250 mL screw-cap bottle. After substitution of the air by N_2, the bottle is tightly closed and autoclaved. To prepare the neutralized feeding solution, a deliberately measured amount of this solution is added to a sterile Erlenmeyer flask containing a magnetic stir bar. The solution is brought to achieved pH 7.3 by drop-wise addition of sterile 1M H_2SO_4 on a magnetic stirrer. If too much acid is added, the sulfide solution becomes turbid attributable to the precipitation of elemental sulfur. The nearly neutralized solution is immediately utilized for nourishing 100 mL bottle cultures. Depending on the population density, 1.0–2.0 mL is used for *Chromaticeae*. Before the addition, an equivalent amount of culture medium is aseptically removed from the bottle culture.

Preferably the method of Siefert and Pfennig (1984) offers a ready-to-be-used, neutralized sulfide feeding solution. This solution is prepared in a 500 mL bottle with a small outlet at the bottom, associated by a rubber tube with a pinchcock to a test tube firmly closed by a screw cap (Siefert and Pfennig, 1984). In a volume of 250 mL of distilled water, 7.0 g of $Na_2S.9H_2O$ and 2.65g of sodium carbonate (Na_2CO_3) are dissolved and autoclaved under a CO_2 atmosphere. The sterile solution is saturated with CO_2 under pressure (0.8 bar) until the pH is approximately 7.3. The solution is stored under CO_2 pressure and ready for use. For nourishment of cultures, a small volume is pressure-released from the bottle into the test tube and transferred from this tube into the cultures by using a sterile pipette.

b. **Preparation of Thiosulfate Solution for Isolation of *Chromatiaceae*:** Phototrophic cultures of purple sulfur bacteria can be supplemented with 0.1% of thiosulfate from a stock solution (10 g of $Na_2S_2O_3.5H_2O$ dissolved in 100 mL of distilled water). This solution is prepared in a 200 mL screw-cap bottle and autoclaved. 1.0 mL of this solution is added aseptically to 100 mL of the culture medium.

c. **Preparation of Acetate Solution for Isolation of *Chromatiaceae*:** The growth of purple sulfur bacteria can be enhanced by the addition of acetate as a readily assimilated carbon source. The ammonium and magnesium salts of acetate are utilized to avoid strong pH changes during growth. 2.5 g of Ammonium acetate 2.5 g of magnesium acetate are dissolved in 100 mL of distilled water and the solution is autoclaved. The standard application of this solution about 1.0 mL is added aseptically to 100 mL of culture medium. The addition of acetate solution to media used for deep agar dilution series of

Chromatiaceae is highly recommended. To liquid cultures, acetate should be added only if these are free of challenging contaminants such as purple non-sulfur bacteria.

2.2.2. Medium for the Enrichment and Isolation of *Ectothiorhodospiraceae* Members

i. Basal Medium for the Isolation of *Ectothiorhodospira* (per liter medium)

Contains KH_2PO_4, 0.8 g; Trace element solution SLA, 1.0 mL; 1M sodium carbonate (Na_2CO_3) buffer with pH 9.0, 200 mL and NaCl, 30.0 g. This solution is sterilized in an autoclave. Solutions of 20.0% NH_4Cl (4 mL/L), 1.0% $CaCl_2.2H_2O$ (5 mL/L), 2.0% $MgCl_2.6H_2O$ (5 mL/L) and 5.0% $Na_2S.9H_2O$ (5–10 mL/L) are sterilized separately and added after cooling to room temperature. 1.0 g/L sodium thiosulfate ($Na_2S_2O_3$) is used as an electron donor, and 1.0-2.0 g/L sodium acetate may be added to the basal medium before sterilization. The final salinity is adjusted by altering the concentrations of sodium chloride (NaCl), keeping under consideration the constant level of carbonate buffer, which is equivalent to ~2.0% total salts. The growth occurs in medium with 5% total salts. Part of the sodium chloride (10% of the total final salt content by weight) may be replaced by sodium sulfate (Na_2SO_4). To adjust the salinity to a total of 20% salts, either 180 g NaCl or 20 g Na_2SO_4 and 160 g NaCl are added to a final volume of the 1.0-liter medium. The medium is transferred into tightly closed screw-capped bottles that are filled and stored until use. Earlier to inoculation, it is important to preincubate the bottles, after removal of an appropriate small volume, at the incubation temperature to achieve growth development.

OR

Basal Medium for the Enrichment and Isolation of *Ectothiorhodospiraceae* (g/L)

Contains, KH_2PO_4, 0.5 g; $CaCl_2.2H_2O$, 0.05 g; NH_4Cl, 0.80 g; $MgCl_2.6\ H_2O$, 0.10 g; NaCl, 180.00 g; Na_2SO_4, 20.00 g; Na_2,CO_3, 6.00 g; Trace element solution (SLA) 1.00 mL; Yeast extract, 0.50 g; Na-succinate, 1.00 g; Distilled water, 800.00 mL. Adjust pH to 8.5. Bubble with nitrogen gas and fill in tubes with a rubber septum under a stream of nitrogen gas. Autoclave at 121°C for 15 min. After sterilization add 1.0 mL/L filter-sterilized vitamin solution (B), 200 mL of filter-sterilized solution Bicarbonate solution (C) and 10 mL/L sterilized $Na_2S.9H_2O$ solution (D). pH 8.5.

Vitamin Solution (B): Biotin, 10.0 mg; Nicotinamide, 35.0 mg; Thiamine-HCl x 2 H_2O 30.0 mg; p-Aminobenzoic acid, 20.0 mg; Pyridoxal hydrochloride, 10.0 mg; Ca-pantothenate, 10.0 mg; Vitamin B_{12}, 5.0 mg; Distilled water 100 mL.

Bicarbonate Solution (C): $NaHCO_3$ 14.0 g; Distilled water, 200 mL.

$Na_2S.9H_2O$ Solution (D): $Na_2S.9H_2O$ 10.0 g, Distilled water, 100 mL.

The Trace Element Solution SLA: Contains $FeCl_2.4H_2O$, 1.8 g; $CoCl_2.6H_2O$, 250 mg; $NiCl_2.6H_2O$, 10 mg; $CuCl_2.5H_2O$, 10 mg; $MnCl_2.4H_2O$, 70 mg; $ZnCl_2$, 100 mg; H_3BO_3, 500 mg; $Na_2MoO_4.2H_2O$, 30 mg; and $Na_2SeO_3.5H_2O$, 10 mg. These components are dissolved in 1.0 liter of double distilled water. The pH of the solution is adjusted with HCl to 2.0–3.0.

ii. Basal Medium for the Isolation of *Ectothiorhodospira magna* (g/L)

Contains NH_4Cl, 0.50 g; $MgCl_2.6\ H_2O$, 0.20 g; Na-acetate, 1.0 g; $Na_2S_2O_3$, 0.50 g; $Na_2S.9H_2O$, 0.50 g; $NaHCO_3$, 5.0 g; Na_2CO_3, 5.0 g; Yeast extract, 0.10 g; Vitamin B_{12},10.0 µg; Trace elements solution, 1.0 mL; Distilled water 1.0 L and the pH was adjusted to pH 9.0-9.5. Prepare the medium without the $NaHCO_3$, Na_2S x 9 H_2O, Na_2CO_3 and vitamin B_{12}, under a nitrogen atmosphere. Autoclave and add the vitamin B_{12} from a filter-sterilized stock solution and then $NaHCO_3$ (autoclaved in sealed, half-full vessels), $Na_2S.9H_2O$, Na_2CO_3 from sterile stock solutions. The final pH of the medium should be 9.0-9.5.

The Trace Elements Solution (Pfennig and Lippert, 1966) (g/L): EDTA 5.00 g; $FeSO_4.7\ H_2O$ 2.2 g; $ZnSO_4.7H_2O$. 0.10 g; $MnCl_2.4H_2O$ 0.03 g; H_3BO_3, 0.03 g; $CoCl_2.6\ H_2O$ 0.20 g; $CuCl_2.2H_2O$, 0.03 g; $NiCl_2.6H_2O$, 0.03 g; $Na_2MoO_4.2H_2O$, 0.03 g; Distilled water 1000 and pH was adjusted to pH 3.0-4.0.

iii. Medium for the Isolation of *Ectothiorhodospira variabilis* (g/L)

Contains NaCl, 50 g; Na_2CO_3, 5.0 g; $NaHCO_3$ 15.0g; KCl, 0.1 g; K_2HPO_4, 0.50 g; $MgCl_2.6\ H_2O$, 0.2 g; NH_4Cl, 0.50 g; $Na_2S.9H_2O$, 0.5 g; $Na_2S_2O_3.5H_2O$, 0.5 g; Yeast extract 0.1 g; Na acetate, 0.5 g; Trace element solution, 1.00 ml; Vitamin B_{12}, 20.0 µg; Distilled water 1.0 L. Adjust pH to 9.0-9.5. Prepare and boil the medium in the absence of $Na_2S.9H_2O$, $NaHCO_3$, Vitamin B_{12}, $CaCl_2.2H_2O$ and $MgCl_2.6H_2O$. Cool under a stream of nitrogen and add the remaining components, except vitamin B_{12} and $Na_2S_2O_3.5H_2O$. Transfer into the tubes or bottle fitted with rubber stoppers and autoclave for 15 min at 15 psi pressure-121°C. Cool to room temperature and then add the vitamin B_{12} (filter sterilized) and $Na_2S_2O_3.5H_2O$ from sterile stock solutions.

Trace Element Solution: HCl (25%; 7.7 M) 10.0 mL; $FeCl_2.4H_2O$ 1.5 g; $ZnCl_2$, 70.0 mg $MnCl_2.4H_2O$, 100 mg; H_3BO_3, 6.0 mg; $CoCl_2.6\ H_2O$, 190 mg; $CuCl_2.2H_2O$, 2.0 mg; $NiCl_2.6H_2O$, 24.0 mg; $Na_2MoO_4.2H_2O$, 36.0 mg; Distilled water, 990.0 mL. First dissolve $FeCl_2$ in the HCl, then dilute in water, add and dissolve the other salts. Finally make up the volume to 1.0 L.

iv. Medium for the isolation of *Ectothiorhodospira abdelmalekii* (g/L)

KH_2PO_4, 0.8 g; NH_4Cl, 0.8 g; $CaCl_2.2H_2O$, 0.05 g; $MgCl_2.6H_2O$ 0.1 g; NaCl, 120.0 g; Na_2SO_4, 15.0 g; $NaHCO_3$, 10.0 g; Na_2CO_3, 4 - 5.0 g; Vitamin solution VA, 1.0 mL; Trace element solution SLA (see above SLA solution (d)), 1.0 mL, $Na_2S.9H_2O$, 0.5 g; Na-acetate, 1.0 g; Distilled water, 1.0 L. Adjust pH to 8.5, filter sterilize and add an appropriate amount of a sterile Na_2S solution. The medium was prepared without the addition of Na acetate and $Na_2S.9H_2O$ and may be added to the sterile medium from sterile stock solutions. Prepare the medium under an atmosphere of N_2. Transfer into screw-capped or crimp top bottles fitted with a rubber septum and sterilize at 121°C for 15 minutes in screw-caped bottles. After sterilization add 1.0 mL/L filter-sterilized vitamin solution. Incubate at 30°C at a light intensity of 1500 lux under anaerobic conditions.

Vitamin solution (VA): Biotin 10.0 mg; Nicotinic acid amide 35.0 mg; Thiamine-HCl.$2H_2O$, 30.0 mg; p-Aminobenzoic acid, 20.0 mg; Pyridoxal hydrochloride, 10.0 mg; Ca-pantothenate, 10.0 mg; Vitamin B_{12}, 5.0 mg; Distilled water 100.0 mL.

v. Medium for the Isolation of *Thiorhodospira* (g/L distilled water)

Contains KH_2PO_4, 0.5 g; NH_4Cl, 0.5 g; NaCl, 0.5 g; $MgCl_2.6H_2O$, 0.2g; $CaCl_2.2H_2O$, 0.05 g; $NaHCO_3$, 2.5 g; Na_2CO_3, 2.5 g; Sodium acetate, 0.5 g; Sodium malate, 0.5 g; yeast extract, 0.1 g; $Na_2S.9H_2O$, 0.7 g and trace element solution SL8, 1.0 mL. The pH is adjusted to 9.0–9.5 (Bryantseva *et al.,* 1999b).

2.2.3. Enrichment and Isolation of Purple Non-Sulfur Bacteria

i. Culture Medium and Carbon Sources

Aside from the enrichment in Molisch columns, a synthetic medium must be prepared for all enrichment and purification steps. An enormous number of media, which differed in the composition of the minerals as well as in the organic substrates, was proposed (van Niel, 1944; Drews, 1965; Kaiser, 1966; Haskins and Kihara, 1967; Pratt and Gorham, 1970 and Swoager and Lindstrom, 1971). The medium has given here (Pfennig, 1978a) was intended to keep the vital supplement concentration low. It has been effectively utilized for about 15 years. Yeast extract is added only in a small amount as a source of vitamins. Prior media for *Rhodospirillaceae* did not contain vitamin B_{12}, but it was recently observed to be necessary for various strains (Pfennig, 1978a; Siefert *et al.,* 1978) and should routinely be added.

ii. Medium1 for Isolation of *Rhodospirillaceae* (Biebl and Pfennig, 1981) (g/L)

The medium contains per liter of distilled water, KH_2PO_4, 0.5; $MgSO_4.7H_2O$, 0.2; NaCl, 0.4; NH_4Cl, 0.4; $CaCl_2.2H_2O$, 0.05; Organic carbon substrate (succinic acid), 1.0; Yeast extract, 0.2; Fe-citrate solution (0.1 g/100 mL), 5.0 mL; Trace element solution SL-7, 1.0 mL; Cyanocobalamin (vitamin B_{12}) solution (1.0 mg/100 mL), 1.0 mL.

Preparation of Trace Element Solution SL-7 (g/L): Contains $ZnCl_2$, 0.07; $MnCl_2.4H_2O$, 0.1; H_3BO_3, 0.06; $CoCl_2.6H_2O$, 0.2; $NiCl_2.6H_2O$, 0.02; $CuCl_2.2H_2O$, 0.02; $NaMO_4.2H_2O$, 0.04 and HCl (25% v/v), 1.0 mL. Add components to 1.0 L distilled/ deionized water and mix thoroughly and sparge with 100% N_2. Autoclave for 15 min at 15 psi pressure-121°C.

Preparation of Ferric Citrate Solution: Add 0.1 g of ferric citrate to 100 mL distilled/deionized water and mix thoroughly. Autoclave for 15 min at 15 psi pressure-121°C.

Preparation of Vitamin B_{12} Solution: Add 1.0 mg vitamin B_{12} to 100 mL distilled/ deionized water and mix thoroughly. Filter sterilize.

Preparation of Medium: Add components except for ferric citrate solution, vitamin B_{12} solution and trace elements solution SL-7 to distilled/deionized water and bring volume to 993.0 mL. Succinic acid or glutaric acid may be used for the dicarboxylic acid substrate and mix thoroughly. Adjust pH to 6.8 ± 0.2. Autoclaved for 15 min at 15 psi pressure-121°C. Aseptically add 5.0 mL sterile ferric citrate solution, 1.0 mL sterile vitamin B_{12} solution and 1.0 mL sterile trace elements solution. Mix thoroughly and aseptically distribute into sterile screw-capped tubes to maintain anaerobic conditions. The temperature was maintained at 27 ± 2°C.

Several media have been used for the enrichment and isolation of *Rhodospirillaceae* species (Biebl and Pfennig, 1981; Imhoff and Truper, 1992). Among these, a mineral medium has been used for isolation of the majority of purple non-sulfur bacteria (Imhoff 2005).

iii. Medium 2 for Isolation of *Rhodospirillaceae* Members (g/L)

Contains Fatty acid substrate, 1.0; KH_2PO_4, 0.5; NaCl, 0.4; NH_4Cl, 0.4; $MgSO_4.7H_2O$, 0.2; Yeast extract, 0.2; $CaCl_2.2H_2O$, 0.05; $NaHCO_3$ solution, 40.0 mL; Ferric citrate solution, 5.0 mL; Trace element solution SL-7, 1.0 mL; Vitamin B_{12} solution, 1.0 mL. pH was adjusted to 7.3 ± 0.2 and temperature 27 ± 0.2°C.

Preparation of Trace Elements Solution SL-7 (g/L): See Medium ii for isolation of *Rhodospirillaceae.*

Preparation of Ferric Citrate Solution: See Medium ii for isolation of *Rhodospirillaceae*

Preparation of Vitamin B_{12} Solution: See Medium ii for isolation of *Rhodospirillaceae.*

Preparation of $NaHCO_3$ Solution: Add 5.0g $NaHCO_3$ to 100 mL distilled/deionized water and mix thoroughly.

Preparation of Medium: Add components except for $NaHCO_3$ solution to distilled/deionized water and bring the final volume to 1.0 L. Acetate, propionate, or butyrate salts may be used for the fatty acid substrate. Mix thoroughly and adjust the pH to 7.3 ± 0.2. Distribute the media into 50 mL volume flasks and autoclave for 15 min at 15 psi pressure-121°C. Cool to room temperature and immediately add 2.0 mL of sterile $NaHCO_3$ solution to each flask containing 50 mL of the medium before inoculation.

iv. Medium3 for Isolation of *Rhodospirillaceae* Members (g/L)

This media was especially used for enrichment and isolation of *Rhodopseudomonas acidophila* and *Rhodomicrobium vannielii*. Contains fatty acid or dicarboxylic acid substrate, 1.0; KH_2PO_4, 0.5; NaCl, 0.4; NH_4Cl, 0.4; $MgSO_4.7H_2O$, 0.2; Yeast extract, 0.2; $CaCl_2.2H_2O$, 0.05; Ferric citrate solution, 5.0 mL; Trace element solution SL-7, 1.0 mL; Vitamin B_{12} solution, 1.0 mL. pH was adjusted to 5.2-5.5 and temperature 27 ± 0.2°C.

Preparation of Trace Elements Solution SL-7 (g/L): See Medium 1 for isolation of *Rhodospirillaceae.*

Preparation of Ferric Citrate Solution: See Medium 1 for isolation of *Rhodospirillaceae.*

Preparation of Vitamin B_{12} Solution: See Medium 1 for isolation of *Rhodospirillaceae.*

Preparation of Medium: Add components to distilled/ deionized water and adjust the final volume to 1.0 L. Acetate, propionate, or butyrate salts may be used for the fatty acid substrate. Succinic acid or glutaric acid may be used for the dicarboxylic acid substrate. Lactate or ethanol may be used as an alternate substrate. Mix thoroughly and adjust the pH to 5.2-5.5. Dispense the media into tubes or flasks. Autoclave for 15 min at 15 psi pressure-121°C.

v. Modified Medium for Isolation of *Rhodospirillaceae* Members (g/L)

This media is used for the isolation and enrichment of *Rhodobacter capsulatus, Rba. sphaeroides, Rhodovulum sulfidophilum (Rba. sulfidophilus), Rhodocyclus tenuis, Rhodopseudomonas blastica, Rps. marina, Rps. palustris, Rps. rosea, Rps. viridis, Rhodospirillum rubrum, Phaeospirillum molischianum (Rsp. molischianum), Phs. fulvum, (Rsp. fulvum), Pararhodospirillum photometricum (Rsp. photometricum), Rhodothalassium salexigens (Rsp. salexigens)* and *Rubrivivax gelatinosus*.

Contains ammonium acetate, 1.5; KH_2PO_4, 0.5; $MgSO_4.7H_2O$, 0.4; NaCl, 0.4; NH_4Cl; 0.4; Yeast extract, 0.3; Disodium succinate, 0.25; $CaCl_2.2H_2O$; 0.05; Ethanol, 0.5 mL; Ferric citrate solution, 5.0 mL; Trace elements solution SL-6, 1.0 mL; Vitamin B_{12} solution, 0.4 mL and Neutralized sulfide solution, variable. pH was adjusted to 6.8 ± 0.2 and temperature was maintained at 27 ± 2°C.

Preparation of Ferric Citrate Solution: Add 10.0 mg ferric citrate to 10.0 mL distilled/deionized water. Mix thoroughly and sparge with 100% N_2 for 3 min. Autoclave for 15 min at 15 psi pressure-121°C. Store under nitrogen.

Trace Elements Solution SL-6 (g/L): Contains $MnCl_2.4H_2O$, 0.03; H_3BO_3, 0.3; $CoCl_2.6H_2O$, 0.2; $ZnSO_4.7H_2O$, 0.1; $Na_2MoO_4.2H_2O$, 0.03; $NiCl_2.6H_2O$, 0.02; $CuCl_2.2H_2O$, 0.01. Add components to distilled/deionized water and make up the volume to 1.0 L. Mix thoroughly and adjust the pH to 3.4.

Vitamin B_{12} Solution: Add 10.0 mg vitamin B_{12} to 100 mL distilled/deionized water. Mix thoroughly and sparge with 100% N_2 for 3 min. Autoclave for 15 min at 15 psi pressure-121°C and store under N_2 gas.

Neutralized Sulfide Solution: Add 1.5g of $Na_2S.9H_2O$ to 100 mL distilled/ deionized water containing 250 mL screw-capped bottle fitted with a butyl rubber septum. Add a magnetic stir bar and mix thoroughly. Sparge with 100% N_2 for 3 min and autoclave for 15 min at 15 psi pressure-121°C. Cool to room temperature. Adjust the pH to 7.3 with sterile 2M H_2SO_4. Without opening the bottle, H_2SO_4 was added to the bottle using a sterile syringe. Stir the solution continuously to avoid precipitation of elemental sulfur. The final solution should be clear and yellow.

Preparation of Medium: Add components, except neutralized sulfide solution to distilled/deionized water. Make up the volume to 1.0 L and mix thoroughly. Gently heat the media and bring to boiling for 3-4 min under a stream of 100% N_2. Distribute 45.0 mL of a prepared medium into 50.0 mL screw-capped tubes that are flushed with 100% N_2. Autoclave for 15 min at 15 psi pressure-121°C. Cool to room temperature. Before inoculation,0.25-0.5 mL of neutralized sulfide solution was added aseptically and anaerobically to each tube.

vi. Modified Medium I for Isolation of *Rhodospirillaceae* Members (g/L)

These media is mainly used for the enrichment and isolation of *Rhodomicrobium vannielii, Rhodopseudomonas acidophila* and *Rhodobacter capsulatus*. Contains di-sodium succinate, 1.0; KH_2PO_4, 0.5; $MgSO_4.7H_2O$, 0.4; NaCl, 0.4; NH_4Cl, 0.4; Yeast extract, 0.2; $CaCl_2.2H_2O$. 0.05; Ferric citrate solution (0.1%), 5.0 mL; Trace element solution SL-6, 1.0 mL. Adjust pH to 5.7 ± 0.2 and temperature 27 ± 0.2°C.

Preparation of Trace Elements Solution SL-6 (g/L): See modified medium for isolation of *Rhodospirillaceae* members.

Preparation of Medium: Add components to distilled/deionized water and bring the volume to 1.0L. Mix thoroughly and adjust pH to 5.7. Distribute 40.0 mL media into tubes or bottles. Sparge with 100% N_2 for 1-2 min. Autoclave for 15 min at 15 psi pressure- 121ºC.

vii. Modified Medium II for Isolation of *Rhodospirillaceae* Members (g/L)

This medium is mainly used for the isolation and maintenance of *Rhodobacter sphaeroides, Rhodocyclus tenuis, Rhodopseudomonas blastica, Rps. palustris, Rps. viridis, Rhodospirillum fulvum, Rsp. molischianum, Rsp. photometricum, Rsp. rubrum* and *Rubrivivax*. Contains Yeast extract, 1.0; Disodium succinate, 1.0; KH_2PO_4, 0.5; $MgSO_4.7H_2O$, 0.4; NaCl, 0.4; NH_4Cl, 0.4; $CaCl_2.2H_2O$, 0.05; Ferric citrate (0.1%) solution, 5.0 mL; Trace element solution SL-6, 1.0 mL; Ethanol, 0.5 mL. Adjust the pH to 6.8 ± 0.2 and temperature 27 ± 0.2°C.

Preparation of Trace Elements Solution SL-6 (g/L): See modified medium for isolation of *Rhodospirillaceae* members.

Preparation of Medium: Add components to distilled/deionized water and adjust the volume to 1.0 L. Mix thoroughly and adjust pH to 6.8. Distribute 40.0 mL media into tubes or bottles and sparge with 100% N_2 for 1-2 min. Autoclave for 15 min at 15 psi pressure- 121°C.

viii. Medium for Enrichment and Isolation of *Rhodospirillum* species (g/L)

AT Medium contains (g/L): Contains KH_2PO_4, 1.0; $MgCl_2.6H_2O$, 0.5; NaCl, 1.0; NH_4Cl, 1.0; $CaCl_2.2H_2O$, 0.1; $NaHCO_3$, 3.0; Na_2SO_4, 0.7; Organic carbon compound (Sodium malate, sodium succinate, sodium acetate and sodium pyruvate), (10 mM) and for oxygen sensitive strains sodium ascorbate, 0.5g or thioglycolate, 0.25g were added separately. Sulfate free trace element solution SLA, 1.0 mL; Vitamin solution (1.0 mg/100 mL), 1.0 mL. The final pH of the media was adjusted to 6.9.

The Composition of Trace Elements Solution SLA Contains (mg/L): Contains $FeCl_2.4H_2O$, 1.8g; $ZnCl_2$, 100; $MnCl_2.4H_2O$, 70; H_3BO_3, 30; $CoCl_2.6H_2O$, 250; $NiCl_2.6H_2O$, 10; $CuCl_2.2H_2O$, 10; $NaMO_4.2H_2O$, 30 and $Na_2SeO_3.5H_2O$, 10. The pH of the solution is adjusted to 2.0-3.0 with HCl.

Vitamin Solution VA (mg/L): Prepared in double distilled water, Biotin, 0.01%; Niacinamide, 0.035%; Thiamine dichloride, 0.03%; Para-aminobenzoic acid, 0.02%; Pyridoxal hydrochloride, 0.01; Calcium pantothenate, 0.01%; Vitamin B_{12}, 0.005%.

Phototrophic cell cultures can be maintained by standard techniques in liquid nitrogen at -80°C (Imhoff 2005).

ix. Medium for the Isolation of *Rhodospirillum fulvum* (g/L)

Contains KH_2PO_4, 1.5; $MgSO_4.7H_2O$, 0.4; NaCl, 0.4; NH_4Cl, 0.4; $CaCl_2.2H_2O$, 0.05; Yeast extract, 0.2; Disodium succinate, 1.0; Ferric citrate solution, 5.0 mL; Trace element solution, 1.0 mL; Ethanol, 0.5 mL. pH was adjusted to 6.8 ± 0.2 and temperature 27 ± 0.2°C.

Preparation of Ferric Citrate Solution: Add 0.01 g of ferric citrate to 10.0 mL distilled/deionized water and mix thoroughly. Filter and sterilize.

Preparation of Trace Element Solution (g/L): Contains H_3BO_3, 0.3; $CoCl_2.6H_2O$, 0.2; $ZnSO_4.7H_2O$, 0.1; $MnCl_2.4H_2O$, 0.03; $NaMO_4.2H_2O$, 0.03; $NiCl_2.6H_2O$, 0.02; $CuCl_2.2H_2O$, 0.01. Add components to distilled/ deionized water and make up the volume to 1.0 L. Mix thoroughly and sterilize.

Preparation of Medium: Add components except for ferric citrate solution, trace elements solution and ethanol to distilled/deionized water and make up the final volume to 1.0 L. Mix thoroughly and autoclave for 15 min at 15 psi pressure-121°C. Aseptically add 5.0 mL sterile ferric citrate solution, 1.0 mL sterile trace elements solution and 0.5 mL filter-sterilized ethanol. Mix thoroughly and dispense into sterile tubes or flasks under aseptic conditions.

x. Medium for the Isolation of *Rhodospirillum salinarum* (g/L)

Contains NaCl, 100; KCl, 5.0; $MgCl_2.6H_2O$, 5.0; $MgSO_4.7H_2O$, 5.0; NH_4Cl, 5.0; Peptone solution, 30.0 mL; Yeast extract solution, 30.0 mL; Ferric citrate solution, 10.0 mL; Trace element solution, 5.0 mL.

Preparation of Peptone Solution: Add 15.0 g peptone to 100 mL distilled/deionized water and bring volume to 100 mL. Mix thoroughly, filter and sterilize.

Preparation of Yeast Extract Solution: Add 15.0 g yeast extract to 100 mL distilled/deionized water. Mix thoroughly, filter and sterilize.

Preparation of Ferric Citrate Solution: Add 0.1 g ferric citrate to 10.0 mL distilled/deionized water. Mix thoroughly, filter and sterilize.

Preparation of Trace Element Solution (g/L): Contains H_3BO_3, 0.3; $CoCl_2.6H_2O$, 0.2; $ZnSO_4.7H_2O$, 0.1; $MnCl_2.4H_2O$, 0.03; $NaMO_4.2H_2O$, 0.03; $NiCl_2.6H_2O$, 0.02; $CuCl_2.2H_2O$, 0.01. Add components to distilled/deionized water and make up the final volume to 1.0 L. Mix thoroughly and sterilize.

Preparation of Medium: Add components except for ferric citrate solution, trace elements solution and peptone solution to distilled/deionized water and bring the final volume to 1.0 L. Mix thoroughly and autoclave for 15 min at 15 psi pressure-121°C. Aseptically add 10.0 mL sterile ferric citrate solution, 5.0 mL sterile trace elements solution and 30.0 mL sterile peptone solution and 30.0 mL sterile yeast extract solution. Mix thoroughly and distribute into sterile tubes or flasks under aseptic conditions.

xi. Medium for Isolation of *Pararhodospirillum* species

Species of *Pararhodospirillum* can be isolated through standard techniques for anaerobes in agar dilution series and on agar plates, anaerobically, as these are oxygen-sensitive (Biebl and Pfennig, 1981; Imhoff and Truper, 1992). These conditions can be accomplished by adding 0.5g/L of sodium ascorbate or 0.2g/L thioglycolates to the growth medium in filled screw-capped bottles. Cell cultures can be maintained by standard techniques in liquid nitrogen at -80°C (Imhoff, 2005).

xii. Medium for Isolation of *Phaeovibrio sulfidiphilus* (g/L)

Contains KH_2PO_4, 0.5; $MgSO_4.7H_2O$, 0.5; NaCl, 0.4; NH_4Cl, 0.68; $CaCl_2.2H_2O$, 0.25; Sodium pyruvate, 0.5; Sodium succinate, 0.5; Sodium acetate, 0.5; Yeast extract, 0.5; Ferric citrate solution (0.1%), 5.0 mL; $NaHCO_3$ (10%), 15.0 mL; $Na_2S.9H_2O$ (1M), 1.0 mL; Vitamin B_{12} solution (2mg/100 mL), 1.0 mL; Trace element solution SL-7, 1.0 mL and Distilled water, 1.0 L. Medium was prepared under nitrogen gas (Lakshmi *et al.,* 2011a).

Preparation of Trace element solution SL-7 (mg/L): Contains HCl (25%, v/v), 1.0 mL; $ZnCl_2$,0.07; $MnCl_2.4\ H_2O$, 100.0 mg; H_3BO_3, 60.0 mg; $CoCl_2.6H_2O$, 200.0 mg; $CuCl_2.2H_2O$, 20.0 mg $NiCl_2.6H_2O$, 20.0 mg; Na_2MoO_4. $2H_2O$, 0.04; Distilled water, 1.0L.

Preparation of Medium: Add components to distilled/deionized water and make up the volume to 1.0 L. Mix thoroughly and autoclave for 15 min at 15 psi pressure-121°C. $NaHCO_3$ solution (autoclaved in sealed, half-full vessels under CO_2), $Na_2S.9H_2O$ solution (autoclaved) and vitamin B_{12} solution (filter sterilized) is added to the autoclaved medium and the final pH was adjusted to 7.0 ± 0.4.

xiii. Medium for Isolation of *Rhodovibrio* species (g/L)

Species of *Rhodovibrio* require high salt concentrations and complex nutrients for growth (Imhoff, 2015). Thus complex media with salt concentration of approximately 10% and anaerobic incubation in the light constitute selective conditions for the enrichment of *Rhodovibrio* species named as DSIC or SAL (Mack *et al.,* 1993) contains, Yeast extract, 1.0; Sodium acetate, 1.0; NaCl, 125; K_2SO_4, 2.5; KH_2PO_4, 0.6; $MgCl_2.6H_2O$, 10.0; NaCl, 0.4; NH_4Cl, 0.5; $CaCl_2.2H_2O$, 0.2; $NaHCO_3$, 1.0; $Na_2S_2O_3.5H_2O$, 0.1; MOPS buffer 2.1; Trace element solution SLA, 1.0 mL; Cyanocobalamin (vitamin B_{12}), 20 µg and distilled water, 1.0 L.

Standard techniques for isolation of anaerobic bacteria in agar dilution series and on agar plates can be applied for *Rhodovibrio* species (Imhoff, 1988; Imhoff and Truper, 1992). The medium of Nissen and Dundas (1984) is suitable for the growth of *Rhodovibrio salinarum* and can also be used for *Rhodovibrio sodomensis*. The medium contains (g/L): NaCl, 80; $MgCl_2 \cdot 6H_2O$, 0.4; $CaCl_2 \cdot 2H2O$, 0.1; $(NH_4)_2SO_4$, 0.5; KH_2PO_4, 2.5; sodium acetate, 1.0; sodium glutamate 1.0; Distilled water, 1.0 L and Trace element solution SLA, 1.0 mL. The initial pH is adjusted to 7.0.

xiv. Medium for Isolation of *Rhodospira* species (g/L)

Standard techniques for the isolation of anaerobic bacteria in agar dilution series can be applied for the isolation of *Rhodospira* species. Members of *Rhodospira* were isolated by an enrichment medium Biebl and Pfennig (1981) medium with the following modifications. Contain $MgSO_4.7H_2O$, 1.0; $CaCl_2.2H_2O$, 0.15; NaCl, 20.0; NH_4Cl, 0.5; KH_2PO_4, 0.35; supplemented with $Na_2S_2O_3.5H_2O$ (0.5mM), $NaHCO_3$ (0.1% w/v) and sodium pyruvate (0.3% w/v). Vitamin solution, 1.0 mL (Pfennigs and Truper, 1981); Trace element solution (SL-12B), 1.0 mL, pH of the media was adjusted to 7.0. *Rhodospira truperi* can be grown on Pfennigs medium used for isolation of purple sulfur bacteria with the addition of 2.0% NaCl and vitamins or yeast extract.

Preparation of Trace Element Solution SL-12B: Distilled water, 1.0 liter; Ethylene diamine tetraacetate-Na_2, 3.0 g; $FeSO_4 \cdot 7H_2O$, 1.1 g; H_3BO_3, 300 mg; $CoCl_2 \cdot 6H_2O$, 190 mg; $MnCl_2 \cdot 4H_2O$, 50.0; $ZnCl_2$, 42.0 mg; $NiCl_2 \cdot 6H_2O$, 24.0 mg; $Na_2MoO_4 \cdot 2H_2O$; 18.0 mg; $CuCl_2 \cdot 2H_2O$, 2mg. The salts are dissolved in the order given, the pH was adjusted to 6.0 and the solution is sterilized by autoclaving for 15 min at 15 psi pressure-121°C.

xv. Medium for the Isolation of *Roseospira* species (g/L)

Contains KH_2PO_4, 0.3; NH_4Cl, 0.5; $CaCl_2.2H_2O$, 0.05; $MgCl_2.6H_2O$, 1.0; $MgSO_4.7H_2O$, 0.5; NaCl, 20.0; Na-acetate, 5mM and/or di-Na-succinate, 5 mM; yeast extract, 0.5; Trace elements solution (Wolfe's Mineral Solution), 1.0 mL and Distilled deionized water, 1.0 L. Add components to distilled water and autoclave at 15 psi pressure at 121°C for 15 minutes. Cool under nitrogen gas. Then add 1.0 mL of Solution A. Add 10.0 mL of Solution B and Add 10.0 mL of Solution C. Adjust pH to 6.8-7.0. Distrubute as desired under nitrogen gas.

Preparation of Wolfe's Mineral Solution (g/L): Contains Nitrilotriacetic acid, 1.5; $MgSO_4.7H_2O$, 3.0; $MnSO_4.H_2O$, 0.5; NaCl, 1.0; $FeSO_4.7H_2O$, 0.1; $CoCl_2$. $6H_2O$, 0.1; $CaCl_2$, 0.1; $ZnSO_4.7H_2O$, 0.1; $CuSO_4.5H_2O$, 0.01; $AlK(SO_4)_2.12H_2O$, 0.01; H_3BO_3, 0.01; $Na_2MoO_4.2H_2O$, 0.01 and Distilled water, 1.0 L. Add nitrilotriacetic acid to approximately 500 mL of distilled water and adjust pH to 6.5 with KOH to dissolve the components. Bring volume to 1.0 L with remaining water and add remaining components one at a time.

Preparation of Solution A (Wolfe's Vitamin Solution) (mg/L): Contains Biotin, 2.0; Folic acid, 2.0; Pyridoxine hydrochloride, 10.0; Thiamine HCl, 5.0; Riboflavin, 5.0; Nicotinic acid; 5.0; Calcium D-(+)-pantothenate, 5.0; Vitamin B_{12}, 0.1; para-aminobenzoic acid, 5.0; Thioctic acid, 5.0; Distilled water, 1.0 L. Filter-sterilize.

Preparation of Solution B: Dissolve 1.0 g of Sodium ascorbate in 20.0 mL distilled/deionized water. Filter-sterilize.

Preparation of Solution C: Dissolve 3.0 g of $NaHCO_3$ in 20.0 mL of distilled/deionized water. Filter-sterilize.

Preparation of Medium: Media were autoclaved and cooled under N_2/CO_2 (90/10, v/v). Vitamin solution, sodium ascorbate and $NaHCO_3$ were then aseptically added to the medium. The final pH was adjusted to 6.8–7.0 and the medium was dispensed into sterile 50 mL screw-capped bottles. Organic substrates (5 mM Na-acetate and/or 5 mM di-Na-succinate) were added as substrates before use. In addition, $Na_2S.9H_2O$ (0.02% w/v) was also added to the medium before utilization. Pure cultures were grown in 50 mL screw-capped bottles and stored at +4°C in the dark for preservation.

xvi. Medium for Isolation of *Roseospirillum* species (g/L)

Pure cultures were obtained by repeated deep-agar dilution series (Pfennig 1978a). Purity was checked microscopically and by growth tests in basal medium supplemented with casamino acids and yeast extract. The basal medium CR used for isolation and the initial cultivation of the organism contained per liter of distilled water contains KH_2PO_4, 0.25; NH_4Cl, 0.34; KCl, 0.34; NaCl, 20.0; $MgSO_4.7H_2O$, 2.8; $CaCl_2$. $2H_2O$, 0.25; $NaHCO_3$, 1.5; Na_2S. $9H_2O$, 0.60; Vitamin mixture solution (Pfennig 1978a), 1.0 mL and Trace element solution (SL-12) (Overmann *et al.* 1992), 1.0 mL. The medium pH was adjusted to 7.2-7.3. Autoclave for 15 min at 15 psi pressure-121°C. Then the medium was aseptically distributed into culture vessels as described by Pfennig and Truper (1989). After initial growth tests, substrate tests were performed in a modified version of the medium for purple non-sulfur bacteria (RS medium). This medium contained per liter of distilled water contains KH_2PO_4, 1.0; NH_4Cl, 0.50; NaCl, 20.0; $MgSO_4.7H_2O$, 2.80; $CaCl_2.2H_2O$, 0.02; yeast extract, 0.20; $Na_2S.\ 9H_2O$, 0.60 and Trace element solution (SL12), 1.0 mL. The pH was also set to 7.2–7.3. Stock cultures were grown photosynthetically in 50 mL screw-capped tubes or bottles at 100 mmol quanta m^{-2} s^{-1} light intensity of a tungsten lamp bulb. Repeated addition of neutralized sulfide solution (Siefert and Pfennig 1984) was used to obtain higher cell yields. Stock cultures were stored at 4°C in the dark.

Preparation of Trace Element Solution SL-12: Distilled water, 1.0 liter; $FeSO_4{\cdot}7H_2O$, 1.1 g; H_3BO_3, 300 mg; $CoCl_2{\cdot}6H_2O$, 190 mg; $MnCl_2{\cdot}4H_2O$, 50.0; $ZnCl_2$, 42.0 mg; $NiCl_2{\cdot}6H_2O$, 24.0 mg; $Na_2MoO_4{\cdot}2H_2O$; 18.0 mg; $CuCl_2{\cdot}2H_2O$, 2mg. The salts are dissolved in the order given, the pH is adjusted to 6.0 and filter-sterilize.

xvii. Medium for Isolation of *Rhodopila globiformis* (g/L)

Contains Mannitol, 1.5; Sodium gluconate, 0.56; KH_2PO_4, 0.4; NaCl, 0.4; NH_4Cl, 0.4; $MgCl_2.6H_2O$, 0.4; $Na_2S_2O_3.5H_2O$, 0.2; $CaCl_2.2H_2O$, 0.05; Ferric citrate, 0.005; Vitamins solution, 1.0 mL; Trace elements solution (SL-6), 1.0 mL. pH of the media was adjusted to 4.9 ± 0.2 and temperature was maintained at 27 ± 2°C.

Preparation of Vitamins Solution (g/500 mL): Contains Nicotinamide, 0.175; Thiamine HCl, 0.15; p-Aminobenzoic acid, 0.1; Biotin, 0.05; Pyridoxine HCl, 0.05; Calcium pantothenate, 0.05; Cyanocobalamin, 0.025. Add components to distilled/deionized water and bring the final volume to 500 mL. Mix thoroughly and filter-sterilize.

Preparation of Trace Elements Solution SL-6 (g/L): Contains H_3BO_3, 0.3; $CoCl_2.6H_2O$, 0.2; $ZnSO_2.7H_2O$, 0.1; $MnCl_2.4H_2O$, 0.03; $NiCl_2.6H_2O$, 0.02; $CuCl_2.2H_2O$, 0.01. Add these components to the distilled/deionized water and adjust the final volume to 1.0 L. Mix thoroughly and adjust the pH to 3.4. Autoclave for 15 min at 15 psi pressure-121°C.

Preparation of Medium: Add components to distilled/deionized water and bring the final volume to 1.0 L. Mix thoroughly and adjust the pH to 4.9. Distribute the media to screw-capped tubes and flasks and autoclave for 15 min at 15 psi pressure-121°C. Vitamin and trace elements solution was added aseptically after sterilization.

xviii. Medium for Isolation of *Rhodopseudomonas* species, *Rps. viridis* and *Rps. acidophila* (g/L)

The basal medium ATCC 650 used for isolation and the initial cultivation of the organism contained per liter of distilled water, Sodium succinate, 1.5; KH_2PO_4 , 1.0 g; NH_4Cl, 0.5g; $MgSO_4.7H_2O$, 0.4g; NaCl, 0.4; $CaCl_2.2H_2O$, 0.05; Distilled water, 1.0 L and Trace elements solution (mg/100 mL), 10.0 mL. pH was maintained at 5.6-6.0 and temperature at 27 ± 2°C.

Preparation of Trace Elements Solution (g/100 mL): Contain ferric citrate, 0.3; $MnSO_4.H_2O$, 0.002; H_3BO_3, 0.001; $CuSO_4.5H_2O$, 0.001; $(NH_4)_6 Mo_7O_2.4H_2O$, 0.002; $ZnSO_4$, 0.001; EDTA, 0.05; $CaCl_2.2H_2O$, 0.02; Distilled water, 100 mL. Add components to distilled/deionized water and bring the final volume to 100 mL. Mix thoroughly and filter-sterilize.

Preparation of Medium: Add components, except trace elements solution to distilled/deionized water and adjust the final volume to 990 mL. Mix thoroughly and autoclave for 15 min at 15 psi pressure-121°C. Cool to room temperature and add 10.0 mL of trace elements solution under aseptic conditions. Mix thoroughly and transfer to sterile tubes or flasks.

xix. Medium for Isolation of *Rhodopseudomonas rutila* and *Rhodopseudomonas palustris* (g/L)

Contain Sodium glutamate, 2.0; Sodium L-malate, 2.0; Yeast extract, 2.0; KH_2PO_4, 1.0; $NaHCO_3$, 0.5; $MgSO_4.7H_2O$, 0.2; $CaCl_2.2H_2O$, 0.1; $MnSO_4.H_2O$, 0.002; Biotin, 0.001; Nicotinic acid, 0.001; Thiamine-HCl, 0.001; $CoCl_2.6H_2O$, 0.0005; $FeSO_4.7H_2O$, 0.0005. Add components to distilled/deionized water and adjust the final volume to 1.0 L. pH was adjusted to 6.8 ± 2. Mix thoroughly and transferred into the tubes or flasks and autoclaved for 15 min at 15 psi pressure-121°C. Incubated under light and the temperature was maintained at 27 ± 2°C.

xx. Medium for Isolation of *Rhodopseudomonas sulfoviridis* (g/L)

Contain Ammonium acetate, 1.5; Sodium malate, 1.0; KH_2PO_4, 0.5; $Na_2S_2O_3$, 0.5; $MgSO_4.7H_2O$, 0.4; NaCl, 0.4; NH_4Cl; 0.4; Yeast extract, 0.3; Disodium succinate, 0.25; $CaCl_2.2H_2O$; 0.05; Ferric citrate solution, 5.0 mL; Trace elements solution SL-6, 1.0 mL; Ethanol, 0.5 mL; Vitamin B_{12} solution, 0.4 mL; Neutralized sulfide solution, variable. pH was adjusted to 6.8 ± 0.2 and temperature was maintained at 27 ± 2°C.

Ferric Citrate Solution: Add 10.0 mg ferric citrate to 10.0 mL distilled/ deionized water. Mix thoroughly and sparge with 100% N_2 for 3 min. Autoclave for 15 min at 15 psi pressure-121°C and store under nitrogen.

Trace Elements Solution SL-6 (g/L): Contains $MnCl_2.4H_2O$, 0.5; H_3BO_3, 0.3; $CoCl_2.6H_2O$, 0.2; $ZnSO_4.7H_2O$, 0.1; $Na_2MoO_4.2H_2O$, 0.03; $NiCl_2.6H_2O$, 0.02; $CuCl_2.2H_2O$, 0.01. Add components to distilled/deionized water and make up the volume to 1.0 L. Mix thoroughly, filter-sterilize and store under nitrogen.

Vitamin B_{12} Solution: Add 10.0 mg vitamin B_{12} to 100 mL distilled/deionized water. Mix thoroughly and sparge with 100% N_2 for 3 min. Autoclave for 15 min at 15 psi pressure-121°C and store under N_2 .

Neutralized Sulfide Solution: Add 1.5g of $Na_2S.9H_2O$ to 100 mL distilled/ deionized water containing 250 mL screw-capped bottle fitted with a butyl rubber septum. Add a magnetic stir bar and mix thoroughly. Sparge with 100%

N_2 for 3 min and autoclave for 15 min at 15 psi pressure-121°C. Cool to room temperature. Adjust the pH to 7.3 with sterile 2M H_2SO_4. Without opening the bottle H_2SO_4, was added to the bottle using a sterile syringe. Stir the solution continuously to avoid precipitation of elemental sulfur. The final solution should be clear and yellow.

Preparation of Medium: Add components, except neutralized sulfide solution to distilled/deionized water. Make up the volume to 1.0 L and mix thoroughly. Gently heat the media and bring to boiling for 3-4 min under a stream of 100% N_2. Distribute 45 mL of the prepared medium into 50 mL screw-capped tubes that have been flushed with 100% N_2. Autoclave for 15 min at 15 psi pressure-121°C. Cool to room temperature. Ferric citrate solution, trace elements solution, vitamin B_{12} are added to the medium under aseptic conditions. Before inoculation aseptically and anaerobically adds 0.5 mL of neutralized sulfide solution to each tube.

xxi. Medium for Isolation of *Rhodomicrobium vannielii* and *Rhodoblastus acidophilus* (g/L)

Contains Di-sodium succinate, 1.0; KH_2PO_4, 0.5; $MgSO_4.7H_2O$, 0.4; NaCl, 0.4; NH_4Cl, 0.4; Yeast extract, 0.2; $CaCl_2.2H_2O$, 0.05; Ferric citrate solution, 5.0 mL; Trace elements solution, 1.0 mL. Adjust the pH to 5.7 ± 0.2 and temperature was maintained at 27 ± 2°C.

Preparation of Ferric Citrate Solution (g/100 mL): Add 0.1g of ferric citrate to distilled/ deionized water and make up the final volume to 100 mL. Mix thoroughly and autoclave for 15 min at 15 psi pressure-121°C.

Preparation of Trace Elements Solution (g/L): Contains H_3BO_3, 0.3; $CoCl_2.6H_2O$, 0.2; $ZnSO_2.7H_2O$, 0.1; $Na_2MoO_4.2H_2O$, 0.03; $MnCl_2.4H_2O$, 0.03; $NiCl_2.6H_2O$, 0.02; $CuCl_2.2H_2O$, 0.01. Add these components to distilled/ deionized water and make up the final volume to 1.0 L. Mix thoroughly and autoclave for 15 min at 15 psi pressure-121°C.

Preparation of Medium: Add components to distilled /deionized water and bring volume to 1.0 L. Mix thoroughly. Distribute 40 mL medium into 50 mL crew capped bottles/tubes and flush each bottle with nitrogen gas for 1-2 min. Immediately close the bottles with rubber septa and screw caps. Autoclave for 15 min at 15 psi pressure-121°C. Sterile syringes are used to inoculate and remove the samples. Incubate under light using a tungsten lamp. The ferric citrate solution and trace elements solution was added to the medium under aseptic conditions.

xxii. Medium for Isolation of *Rhodobium* species (g/L)

Contains NaCl, 20.0; $MgCl_2.6H_2O$, 1.0; NH_4Cl, 0.6; KH_2PO_4, 0.5; $CaCl_2.2H_2O$, 0.15; Sulfide solution, 10.0 mL; Yeast extract solution, 10.0 mL; Sodium pyruvate solution, 10.0 mL; Ferric citrate solution, 5.0 mL; Trace elements solution SL-7, 1.0 mL. pH was adjusted to 7.2 ± 0.2 and the temperature was maintained at 27 ± 2°C.

Preparation of Ferric Citrate Solution: Dissolve 0.1 g ferric citrate in 100 mL distilled/deionized water and mix thoroughly. Autoclave for 15 min at 15 psi pressure-121°C.

Preparation of Trace Elements Solution SL-7 (mg/L): Contains $CoCl_2.6H_2O$, 200.0; $MnCl_2.4H_2O$, 100.0; $ZnCl_2$, 70.0; H_3BO_3, 60.0; $Na_2MoO_4.2H_2O$, 40.0; $CuCl_2.2H_2O$, 20.0; $NiCl_2.6H_2O$, 20.0; HCl (25%), 1.0 mL. Add components to distilled water/deionized water and make up the final volume to 1.0 L. Mix thoroughly and autoclave for 15 min at 15 psi pressure-121°C.

Preparation of Vitamin Solution: Dissolve 2.0 mg vitamin B_{12} in distilled water/deionized water and make up the volume to 1.0 L. Mix thoroughly, filter-sterilize.

Preparation of Sodium Pyruvate Solution: Add 3.0 g of sodium pyruvate to distilled/deionized water and make up the volume to 10.0 mL. Mix thoroughly and autoclave for 15 min at 15 psi pressure-121°C.

Preparation of Yeast Extract Solution: Dissolve 0.4g of yeast extract in 10.0 mL of distilled/deionized water and mix thoroughly. Autoclave for 15 min at 15 psi pressure-121°C.

Preparation of Sulfide Solution: Dissolve 0.25g of $Na_2.9H_2O$ in 10.0 mL distilled/deionized water and mix thoroughly. Autoclave under 100% N_2 for 15 min at 15 psi pressure 121°C. Cool to room temperature.

Preparation of Medium: Add components, except ferric citrate, trace elements, sulphide, vitamin, sodium pyruvate and yeast extract solutions to distilled/ deionized water and make up the volume to 963.0 mL. Mix thoroughly, adjust the pH to 7.2. Gently heat and bring to boiling. Autoclave for 15 min at 15 psi pressure-121°C and cool to room temperature. Aseptically add ferric citrate, trace elements, vitamin, sulphide, sodium pyruvate and yeast extract solutions. Mix thoroughly and distribute into culture vessels at aseptic conditions.

xxiii. Modified Medium IV for Isolation of *Rhodobacter sulfidophilus* (Hiraishi and Ueda 1994) (g/L)

Contain NaCl, 25.0; Yeast extract, 1.0; Disodium succinate, 1.0; KH_2PO_4, 0.5; $MgSO_4.7H_2O$, 0.4; NH_4Cl, 0.4; $CaCl_2.2H_2O$, 0.05; Ferric citrate (0.1%) solution, 5.0 mL; Trace element solution SL-6, 1.0 mL; Ethanol, 0.5 mL. Adjust pH to 6.8 ± 0.2 and temperature 27 ± 0.2°C.

Preparation of Trace Elements Solution SL-6 (g/L): Contains H_3BO_3, 0.3; $CoCl_2.6H_2O$, 0.2; $ZnSO_4.7H_2O$, 0.1; $MnCl_2.4H_2O$, 0.03; $Na_2MoO_4.2H_2O$, 0.03; $NiCl_2.6H_2O$, 0.02; $CuCl_2.2H_2O$, 0.01. Add components to distilled/deionized water and make up the volume to 1.0 L. Mix thoroughly and adjust pH to 3.4.

Preparation of Medium: Add components to distilled/deionized water and make up the final volume to 1.0 L. Mix thoroughly and adjust pH to 6.8. Distribute 40 mL media into tubes or bottles. Sparge with 100% N_2 for 1-2 min. Autoclave for 15 min at 15 psi pressure-121°C. Cool to room temperature and trace elements solution was added aseptically to the medium.

xxiv. Medium for Isolation of *Rhodobacter adriaticus* (Hiraishi and Ueda, 1994) (g/L)

Contains NaCl, 25.0; $NaHCO_3$, 3.0; K_2HPO_4, 1.0; NH_4Cl, 1.0; $MgCl_2.6H_2O$, 0.5; Sodium ascorbate, 0.5g; $CaCl_2.2H_2O$, 0.1; Trace elements solution SLA, 1.0 mL; Vitamin solution, 1.0 mL; pH was adjusted to 7.0 ± 0.2 and temperature was maintained at 27 ± 2°C.

Preparation of Trace Elements Solution SLA (g/L): Contains $CuCl_2.2H_2O$, 10.0; $FeCl_2.4H_2O$, 10.0; $FeCl_2.4H_2O$, 1.8; H_3BO_3, 0.5; $CoCl_2.6H_2O$, 0.25; $ZnCl_2$, 0.1; $MnCl_2.4H_2O$, 70.0 mg; $Na_2MoO_4.2H_2O$, 30.0 mg; Na_2SeO_3. $5H_2O$, 10.0; $NiCl_2.6H_2O$, 10.0 mg. Add components to distilled/deionized water and make up the volume to 1.0L. Mix thoroughly and adjust the pH to 2.0-3.0. Autoclave for 15 min at 15 psi pressure-121°C.

Preparation of Vitamin Solution (mg/L): Nicotinamide, 35.0; thiamine HCl, 30.0; p-Aminobenzoic acid, 20.0; Pyridoxal-HCl, 10.0; Calcium DL-pantothenate, 10.0; Biotin, 10.0; Vitamin B_{12}, 5.0. Add components to distilled/ deionized water and make up the final volume to 1.0 L. Mix thoroughly and filter-sterilize.

Preparation of Medium: Add components except for vitamin solution to distilled /deionized water and make up the volume to 1.0Litre. Mix thoroughly. Autoclave for 15 min at 15 psi pressure-121°C. Aseptically add 1.0 mL of sterile vitamin solution and mix thoroughly. Aseptically, distribute the medium into sterile tubes or flasks.

xxv. Medium for Isolation of *Rhodobacter changlensis* (g/L)

Contains Yeast extract, 0.4; Sodium pyruvate, 3.0; NH_4Cl, 0.6; $MgCl_2.6H_2O$, 0.5; K_2HPO_4, 0.5; NaCl, 0.4; $CaCl_2.2H_2O$, 0.05; Trace elements solution SL7, 1.0 mL; Vitamin solution, 1.0 mL. pH was adjusted to pH 7.2 ± 0.2 and temperature was maintained at 28 ± 2°C.

Preparation of Trace Elements Solution SL-7 (mg/L): Contains $CoCl_2.6H_2O$, 200.0; $MnCl_2.4H_2O$, 100.0; $ZnCl_2$, 70.0; H_3BO_3, 60.0; $Na_2MoO_4.2H_2O$, 40.0; $CuCl_2.2H_2O$, 20.0; $NiCl_2.6H_2O$, 20.0; HCl (25%), 1.0 mL.Add components to distilled/deionized water and make up the final volume to 1.0 L. Autoclave for 15 min at 15 psi pressure-121°C.

Preparation of Vitamin Solution: Add 20.0 mg of vitamin B_{12} to distilled/ deionized water and make up the final volume to 100 mL. Mix thoroughly and filter sterilize.

Preparation of Medium: Add components, except vitamin solution to distilled/ deionized water and make up the final volume to 1.0 L. Mix thoroughly and adjust the pH to 7.2. Gently heat and bring to boiling for 15 min at 15 psi pressure-121°C. Cool to room temperature and add vitamin solution under aseptic conditions. Mix thoroughly, transfer to tubes and vessels in aseptic conditions. All the tubes were incubated anaerobically in the presence of light (2400 lx) at temperature 28±2°C.

xxvi. Medium for Isolation of *Rhodobacter capsulatus* and *Rhodobacter sphaeroides* (g/L)

Contains Yeast extract, 1.0; Disodium succinate, 1.0; KH_2PO_4, 0.5; $MgSO_4 . 7H_2O$, 0.4; NaCl, 0.4; NH_4Cl, 0.4; $CaCl_2.2H_2O$, 0.05; Ferric citrate solution, 5.0 mL; Trace elements solution, 1.0 mL and Ethanol, 0.5 mL. Medium pH was adjusted to 5.8 ± 0.2 and the temperature maintained at 27 ± 2°C.

Preparation of Ferric Citrate Solution (g/mL): Add 0.1g ferric citrate to distilled/deionized water and make up the volume to 100 mL and mix thoroughly. Autoclave for 15 min at 15 psi pressure-121°C.

Preparation of Trace Elements Solution (g/L): Contains H_3BO_3, 0.3; $CoCl_2.6H_2O$, 0.2; $ZnSO_4.7H_2O$, 0.1; $Na_2MoO_4.2H_2O$, 0.03; $MnCl_2.4H_2O$, 0.03; $NiCl_2.6H_2O$, 0.02; $CuCl_2.2H_2O$, 0.01. Add these components to distilled/deionized water and make up the volume to 1.0 L. Mix thoroughly and autoclave for 15 min at 15 psi pressure-121°C.

Preparation of Medium: Add components to distilled/deionized water and adjust the final volume to 1.0 L and mix thoroughly. Distribute 40.0 mL medium into 50.0 mL screw-capped bottles and flush each bottle with N_2 gas for 1 to 2 min. Close immediately with rubber septa and screw caps and autoclave for 15 min at 15 psi pressure-121°C. Sterile syringes are used to inoculate and remove the samples. The ferric citrate and trace elements solution was added to the medium aseptically. Incubate bottles under light using a tungsten lamp.

xxvii. Medium for Isolation of *Rhodobacter veldkampii* (Hansen and Imhoff 1985)

Contains Solution 1, 1500 mL; Solution 3, 1000 mL, Solution 4, 120 mL and Solution 5, 120 mL. Adjust the medium pH to 4.0 ± 0.1 and temperature maintained at 27 ± 2°C.

Preparation of Solution 1: Add 2.0 g $CaCl_2.2H_2O$ was added to distilled/deionized water and make up the final volume to 2.5 L. Mix thoroughly and autoclave under 100% N_2 for 15 min at 15 psi pressure-121°C.

Preparation of Solution 2 (g/100 mL): Contains Sodium ascorbate, 2.4; KH_2PO_4, 1.0; KCl, 1.0; NH_4Cl, 0.8; $MgCl_2.6H_2O$, 0.8; Heavy metal solution, 50 mL; Vitamin solution, 15 mL and Vitamin B_{12} solution, 3.0 mL. Add components to distilled/deionized water and make up the final volume to 100 mL. Mix thoroughly and autoclave for 15 min at 15 psi pressure-121°C.

i. **Preparation of Heavy Metal Solution (g/L):** EDTA, 1.50; $FeSO_4.7H_2O$, 0.2; $ZnSO_4.7H_2O$, 0.1; $MnCl_2.7H_2O$, 0.02; Modified Hoagland Trace Element Solution, 6.0 mL. Add these components to distilled/deionized water and make up the volume to 1.0 L and mix thoroughly. Autoclave for 15 min at 15 psi pressure-121°C

Preparation of Modified Hoagland Trace Elements Solution (g/3.6L): Contains H_3BO_3, 11.0; $MnCl_2.4H_2O$, 7.0; $ZnCl_2$, 1.0; $CuCl_2$, 1.0; $NiCl_2.6H_2O$,

1.0; $CoCl_2.6H_2O$, 1.0; $AlCl_3$, 1.0; KI, 1.0; KBr, 0.5; LiCl, 0.5; $SnCl_2.2H_2O$, 0.5; $BaCl_2$, 0.5; $Na_2MoO_4.2H_2O$, 0.5; $Na_2SeO_3.5H_2O$, 0.5; $NaVO_3.H_2O$, 0.1. Add components sequentially to distilled/ deionized water and bring final volume to 3.6L. Mix thoroughly after adding each component until dissolved. Adjust pH to just below 7.0. Adjust the final pH to 3.4. The flaky yellow precipitate which is formed after mixing and transform into a very fine white precipitate after standing for one or a few days. Mix thoroughly before use.

ii. **Preparation of Vitamin Solution (mg/100 mL):** Contains Pyridoxamine 2HCl, 5.0; Nicotinic acid, 2.0; Thiamine, 1.0; Pantothenic acid, 0.5; Biotin, 0.2; Para-aminobenzoic acid, 0.1. Add these components to distilled/deionized water and make up the final volume to 100 mL. Mix thoroughly and sparge with 100% CO_2 for 30 minutes.

iii. **Preparation of Vitamin B_{12} Solution:** Add 2.0 mg Vitamin B_{12} to distilled/deionized water and make up the volume to 3.0 mL and mix thoroughly. Filter-sterilize.

Preparation of Solution 3 (g/L): Contains $NaHCO_3$, 4.5; solution 2, 100 mL. Add $NaHCO_3$ to distilled/deionized water and adjust the volume to 900 mL and mix thoroughly. Sparge with gaseous CO_2 for at least 30 min. Add 100 mL solution 2. Immediately filter and sterilize using CO_2 pressure to push the liquid through (no suction).

Preparation of Solution 4: Add $Na_2S.9H_2O$ to distilled/ deionized water in a flask with a magnetic stirrer and make up the final volume to 200 mL. Mix thoroughly and autoclave under 100% N_2 for 15 min at 15 psi pressure-121°C. Cool to room temperature. All the solutions were added to the medium under aseptic conditions.

xxviii. Medium for Isolation of *Rhodovulum sulfidophilum* (Hiraishi and Ueda 1994) (g/L)

Contains NaCl, 25.0; yeast extract, 1.0; Disodium succinate, 1.0; KH_2PO_4, 0.5; $MgSO_4.7H_2O$, 0.4; NH_4Cl, 0.4; $CaCl_2.2H_2O$, 0.05; Ferric citrate (0.1%) solution, 5.0 mL; Trace element solution, 1.0 mL and Ethanol, 0.5. Adjust pH to 5.8 ± 0.2 and temperature 27 ± 0.2°C.

Preparation of Ferric Citrate Solution: Add 0.1g ferric citrate to 100 mL distilled/deionized water and mix thoroughly. Autoclave for 15 min at 15 psi pressure-121°C.

Preparation of Trace Elements Solution SL-6 (g/L): Contains H_3BO_3, 0.3; $CoCl_2.6H_2O$, 0.2; $ZnSO_4.7H_2O$, 0.1; $MnCl_2.4H_2O$, 0.03; $Na_2MoO_4.2H_2O$, 0.03; $NiCl_2.6H_2O$, 0.02; $CuCl_2.2H_2O$, 0.01. Add components to distilled/deionized water and make up the final volume to 1.0 L. Mix thoroughly and autoclave for 15 min at 15 psi pressure-121°C.

Preparation of Medium: Add components to distilled/deionized water and adjust the final volume to 1.0 L. Mix thoroughly and distribute 40.0 mL medium in 50.0 mL screw capped bottles. Flush each bottle for 1 to 2 min with nitrogen gas and then close immediately with rubber septa and screw caps. Autoclave for 15 min at 15 psi pressure-121°C. Sterile syringes are used to inoculate and remove the samples. The ferric and trace elements solution were added to the medium under aseptic conditions. Incubate bottles in light using a tungsten lamp.

xxix. Medium for Isolation of *Rhodovulum iodosum* and *Rhodovulum robiginosum* (Straub *et al.*, 1999) (g/L)

Contains KH_2PO_4, 0.4; NaC1, 26.4; NH_4C1, 0.25; $MgSO_4.7H_2O$, 6.8; $MgC1.6H_2O$, 5.7; $CaCl.2H_2O$, 1.5; KC1, 0.66; KBr, 0.09; Distilled water, 1000 mL. After autoclaving and cooling under an atmosphere of N_2/CO_2 (90:10, v/v) the following components were added (per liter) from sterile stock solutions described in detail (Widdel and Bak, 1992): Trace element mixture containing sodium- EDTA, 1.0 mL; 1 M $NaHCO_3$ solution (84g/L autoclaved under CO_2 atmosphere), 30 mL; selenite/tungstate solution, 1.0 mL; vitamin mixture (para-aminobenzoic acid, biotin, nicotinate, pantothenate and pyridoxine), 1.0 mL; thiamine solution, 1.0 mL and vitamin B_{12} solution, 1.0 mL (all vitamins were filter sterilized). The pH was adjusted to 6.8. The medium was aseptically transferred to 20 mL volume tubes or bottles (100, 250 and 500 mL volume) Sparge with N_2-CO_2 (90:10 v/v) gases and sealed with black rubber stoppers.

Preparation of FeS Solution: The suspension of FeS was prepared by aseptic mixing of equal volumes of 0.2 M $FeSO_4$ and 0.2 M Na_2S solutions aseptically under an atmosphere of N_2. The black precipitate was washed aseptically by repeated sedimentation, removal of the supernatant and addition of sterile medium. Finally, the suspension was adjusted to 0.2 mol of FeS per liter. Aliquots were taken with sterile pipettes while the suspension was magnetically stirred under N_2 .

Preparation of Trace Elements Solution (g/L): Sodium-EDTA, 5.20; H_3BO_3 , 0.3; $CoCl_2.6H_2O$, 0.19; $ZnSO_4.7H_2O$, 0.144; $MnCl_2.4H_2O$, 0.1; $NaMO_4.2H_2O$, 0.036; $NiCl_2.6H_2O$, 0.024; $CuCl_2.2H_2O$, 0.002. Add components to distilled/ deionized water and make up the volume to 1.0 L. Adjust pH to 6.0. Mix thoroughly and sterilize.

Selenite/Tungstate Solution (mg/L): Contain $Na_2SeO_3.5H_2O$, 6.0; $Na_2WO_4.2H_2O$, 8.0; NaOH, 20.0. Add components to distilled/deionized water and make up the final volume to 1.0 L. Mix thoroughly and filter-sterile.

Preparation of Vitamin Solution (mg/L): Contains Para-amino benzoic acid, 4.0; Biotin, 1.0; Nicotinate, 10.0; Pantothenate, 5.0; Pyridoxin, 15.0. Add components to 100 mL distilled/deionized water and mix thoroughly. Filter-sterilize.

Preparation of Vitamin B_{12} Solution: Add 5.0 mg vitamin B_{12} to 100 mL distilled/deionized water and mix thoroughly. Filter-sterilize.

Vitamin B_1 Solution: Add 10.0 mg thiamine to 100 mL distilled/deionized water and mix thoroughly. Filter-sterilize.

Preparation of Medium: To the medium $Na_2S_2O_3$, 0.5 mM was added as a source of reduced sulfur from a filter-sterilized stock solution (1 M). From a 1.0

M $FeSO_4$ anoxic stock solution (autoclaved and maintained under N_2) was added to the culture medium by the addition of $FeSO_4$ to the medium (final concentration 10 mM), a white precipitate (a mixture of ferrous carbonate and ferrous phosphate) formed. In the initial stages of growth, the precipitate turns orange/brown as the ferrous ions are oxidized to ferrihydrite. An anoxic suspension of FeS was prepared according to previous studies (Ehrenreich and Widdel, 1994). 1.0 M sodium sterile sulfide solution was added to the media. Culture tubes and bottles were incubated horizontally on a Plexiglass plate and illuminated with two tungsten lamps (25 W) placed above and below the cultures at a distance of 30 cm. Enrichment cultures were incubated at 20ºC and gently shaken every other day to allow even distribution of bacteria and ferrous iron minerals.

xxx. Medium for Isolation of *Rhodobaca barguzinensis* (Boldareva *et al.*, 2008)(g/L)

Contains KH_2PO_4, 0.5; NaCl, 20.0; $NaNO_3$, 0.4; NH_4Cl 0.4; Na_2SO_4, 0.5; $MgCl_2$. $6H_2O$, 0.2; Yeast extract, 1.0; $Na_2S_2O_3$. $5H_2O$. 1.0; KCl, 0.5; Na_2CO_3, 10.0; $NaHCO_3$, 5.0; Trace element solution SL-7, 1.0 mL; Vitamin B_{12} (20 mg/L), 0.5 mL and Distilled water, 1.0 L. Adjust pH to 9.0-9.5 .

Preparation of Trace Element Solution SL-7 (g/L): Contains HCl (25%, v/v), 1.0 mL; $ZnCl_2$, 0.07; $MnCl_2$. 4 H_2O, 0.1; H_3BO_3, 0.06; $CoCl_2$. $6H_2O$, 0.2; $CuCl_2$. 2 H_2O, 0.02; $NiCl_2.6H_2O$, 0.02; Na_2MoO_4. $2H_2O$, 0.04; Distilled water 1.0 L. Add the components to distilled/deionized water and make up the final volume to 1.0 L. Mix thoroughly and autoclave for 15 min at 15 min at 15 psi pressure-121°C.

Preparation of Medium: Add the components except $Na_2S_2O_3.5H_2O$, $NaHCO_3$, Vitamin B_{12}, $MgCl_2.6H_2O$ to the distilled/deionized water. Cool under a stream of nitrogen and add the remaining components, except vitamin B_{12} and $Na_2S_2O_3.5H_2O$. Distribute into tubes or bottles fitted with rubber stoppers and autoclave. Add the vitamin B_{12} (filter sterilized) and $Na_2S_2O_3.5$ H_2O from sterile stock solutions.

xxxi. Modified Medium III for Isolation of *Rhodocyclus purpureus* (Pfennig, 1978b) (g/L)

Contains Yeast extract, 1.0; Ammonium acetate, 0.5; KH_2PO_4, 0.5; $MgSO_4$. $7H_2O$, 0.4; NaCl, 0.4; NH_4Cl,, 0.4; $CaCl_2.2H_2O$, 0.05; Vitamin B_{12}, 0.02; Ferric citrate (0.1%) solution, 5.0 mL; Trace element solution SL-6, 1.0 mL and Ethanol, 0.5 mL. Adjust pH to 6.8 ± 0.2 and temperature to 27 ± 0.2°C.

Preparation of Trace Elements Solution SL-6 (g/L): Contains $MnCl_2.4H_2O$, 0.5; H_3BO_3, 0.3; $CoCl_2.6H_2O$, 0.2; $ZnSO_4.7H_2O$, 0.1; $Na_2MoO_4.2H_2O$, 0.03; $NiCl_2.6H_2O$, 0.02; $CuCl_2.2H_2O$, 0.01. Dissolve these components in distilled/ deionized water and make up the final volume to 1.0 L. Mix thoroughly and filter-sterilize.

Preparation of Vitamin B_{12} Solution (mg/100 mL): Add 10.0 mg of vitamin B_{12} to distilled/deionized water and make up the final volume to 100 mL. Mix thoroughly and sparge with 100% N_2 gas for 3 min.

Preparation of Ferric Citrate Solution: Add 10.0 mg ferric citrate in 10.0 mL distilled/deionized water. Mix thoroughly and sparge with 100% N_2 gas for 3 min.

Preparation of Medium: Add components to 1.0 L distilled/deionized water and mix thoroughly. Gently heat and bring to boiling. Boil for 3-4 min under the steam of 100% N_2. Distribute 45 mL of the prepared medium into 50 mL screw-capped tubes that are flushed with 100% N_2. Autoclave for 15 min at 15 psi pressure-121°C. Cool to room temperature. Sterile syringes are used to inoculate and remove samples. The trace elements solution, vitamin solution and ferric citrate solution were added to the medium under aseptic conditions. Incubate tubes under light using a tungsten lamp.

2.3. ISOLATION PROCEDURES OF GREEN BACTERIA

The members of green sulfur bacteria are isolated by enrichment techniques studied by Pfennig and Truper (1981), Biebl and Pfennig (1978) and Bose (1963). The primary enrichment and isolation methodology for specific species of purple and green sulfur bacteria were developed by van Niel (1931). The culture conditions for green sulfur bacteria of the *Chlorobium* were highly improved by Larsen (1952) and Bose (1963). Further endeavors to develop the huge purple sulfur bacteria, *Chromatium okenii* and *Thiospirillum jenense,* in pure culture, prompted the elaboration of another synthetic medium that contains vitamin B_{12} (Pfennig, 1965; Pfennig and Lippert, 1966; van Niel, 1971). With slight changes, this culture medium (Medium 1) provides rather non-selective development conditions and permits the cultivation of most purple and green sulfur bacteria that develop in freshwater, estuarine and marine habitats. Therefore, this medium will be described first in the present article.

2.3.1. Enrichment and Isolation of Green Sulfur Bacteria

i. Enrichment and Isolation of *Chlorobium* species

Chlorobium species may be enriched and isolated from anoxic water, mud samples of freshwater and brackish water and marine and saline habitats. The source and species composition of the inoculums is of the primary significance of the results of enrichment experiments. In liquid enrichment cultures, strains/species may finally be obtained that were not directly detected in the original sample. A *Chlorobium* species which are the dominant type in given natural habitat outgrow by a species that occur only incidentally in the habitat. When certain colour or cell types were detected microscopically in a given sample are to be isolated by agar shake dilution. The cultured media used for the isolation of *Chlorobium* species are studied by Larsen (1952), Sirevag and Ormerod (1977) and Pfennig and Truper (1981).

ii. Defined Medium for Isolation of Green Sulfur Bacteria (g/L)

The medium has the following composition include KH_2PO_4, 0.30; NH_4Cl, 0.34; KCl, 0.34; $CaCl_2$. $2H_2O$, 0.15; NaCl, 20.0; $MgSO_4.7H_2O$, 0.5 (for seawater 3.0) and trace element solution, 1.0 mL. After the medium has been autoclaved and cooled under an atmosphere of 90% N_2 plus 10% CO_2, the following components are added aseptically from sterile stock solution, 10% (w/v) solution of $NaHCO_3$ (saturated with CO_2 and autoclaved under a CO_2 atmosphere), 10% (w/v) solution of $Na_2S.9H_2O$ (autoclaved under an N_2 atmosphere) and Vitamin B_{12} solution 1.0 mL (2.0 mg in 100 mL distilled water). Anaerobic conditions were created by continuous flushing with the gas mixture of 90% N_2 plus 10% CO_2. The pH of the medium is adjusted to 6.7 with a sterile 2M H_2SO_4 or 2M Na_2CO_3 solution while the medium is magnetically stirred. The medium is then dispensed aseptically into sterile 100 mL bottles with metal screw tops containing autoclavable rubber seals. A small air bubble is left in each bottle to meet possible pressure changes.

The Trace Element Solution (mg/L): 25% (w/v) HCl; $FeSO_4.7H_2O$, 2.0; $CoCl_2.6H_2O$, 190; $MnCl_2.4H_2O$, 100; $ZnCl_2$, 70; $NaMO_4.2H_2O$, 36; $NiCl_2.6H_2O$, 24; H_3BO_3, 6.0; $CuCl_2.2H_2O$, 2.0. Initially the $FeSO_4$ is dissolved in 10 mL of HCl, followed by distilled water and other components. Autoclave for 15 min at 15 psi pressure-121°C.

Pure cultures of all the phototrophic green bacteria can be maintained in the defined mineral medium. 100 mL screw-capped bottles are preferentially used as culture vessels. The freshly grown cultures should have used up the transiently formed elemental sulfur. At this stage, the stock cultures keep well if they become too dim light at room temperature. After consumption of sulfide and sulfur, the cultures are transferred to a fresh culture medium.

Long-term preservation of GSB is successfully carried out by storage in liquid nitrogen. For this purpose, heavy cell suspensions of liquid cultures are supplemented with DMSO as protective agents to a final concentration of 5%. Such suspensions are filled into 2.0 mL plastic amplus, sealed and freeze stored. Some strains were successfully preserved by lyophilization.

iii. Medium for Enrichment and Isolation of *Prosthecochloris* species

Prosthecochloris species occurs in the hydrogen sulfide-containing mud of shallow saline water bodies. It can be isolated by direct inoculation of a mud sample with dilution into 0.8% agar medium of Pfennig (1965) or medium used for *Chlorobium* containing the following components NaCl, 2%; $Na_2S.9H_2O$, 700-1500 mg/L; NH_4-acetate, 400 mg/L; (replace NH_4Cl, 330 mg/L) and Vitamin B_{12}, 20 µg/L. Incubated in screw-capped glass bottles in a luminist at 25-35°C and a light intensity of 500-2000 lx.

Prosthecochloris can be maintained in the liquid cultures if transferred 2-3 months. Grown cultures are stored at 4°C and very low light intensity.

iv. Medium for Enrichment and Isolation of *Pelodictyon* species

The liquid culture medium and agar shake dilution method used for isolation of *Chlorobium* species is also used for isolation of species *Pelodictyon*. Not more than 1.5 mM sulfide and 3 mM acetate are added to the medium. The light intensity should be limited to 50-200 lux from a tungsten lamp. The cultures are incubated at about 20°C (Pfennig and Cohen-Bazire, 1967). Before the second dilution, the suspension of green and brown colonies isolated from the first dilution series was observed microscopically checked for the presence of gas vacuoles. Only cell suspensions of gas vacuolated cells are cultivated further. The three-dimensional cell aggregates characteristic of few species may develop only in liquid culture media and dependent on cultural conditions. The methods used for the maintenances of this genus are the same as described for the genus *Chlorobium* .

v. Medium for Enrichment and Isolation of *Ancalochloris* species

Isolation and enrichment of these cultures can be obtained by using the same culture medium as described for *Chlorobium* with low H_2S content (300 mg $Na_2S.9H_2O$/L) (pH 6.8-7.0). Cultures are incubated at low light intensity (100-200 lux) and a temperature between 18 and 20°C. Vitamin B_{12} (20 µg/L) should be added. Pure culture isolation should be tried by dilution in liquid media. Hence *Ancalochloris* does not grow in agar shake dilution media. *Ancalochloris* can be maintained in liquid culture if transferred every 2-3 months. Grown cultures are stored at 4°C and very low light intensity.

vi. Medium for Enrichment and Isolation of *Chloroherpeton* species

Chloroherpeton species occur in littoral marine muds containing large quantities of decaying vegetation when brought into the laboratory and incubated in a glass container in dim light at 20-25°C develops green and red or purple patches on the container walls. This bacterium forms a significant component of the microbial population of such green areas that appears first and can be identified by its morphological characteristics. Isolation can be achieved by repeated shake cultures in medium that contains: NaCl, 170 mM; $MgCl_2$, 15 mM; KCl, 4 mM; $CaCl_2$, 1 mM; NH_4Cl, 5 mM; KH_2PO_4, 0.1 mM; $NaHCO_3$, 30 mM; $Na_2S.9H_2O$, 2.5 mM; Trace element solution (Widdel and Pfennif, 1981); Vitamin B_{12}, 0.05 µg/100 mL) and 0.8% washed agar, pH 6.9. The inoculated tubes are sealed with Vaseline: paraffin oil mixture (1:1) and incubated 50-100 cm from a 40W incandescent lamp. Light green fluffy colonies should appear within a week and are picked when they are 1-2 mm in diameter.

The cultures may be maintained illuminated in liquid culture as described above by additions of CO_2-neutralized $Na_2S.9H_2O$ (pH 6.5-6.8) to a final concentration of 2 mM. Cultures should be maintained in a fresh liquid medium once a month. Long-term preservation in liquid nitrogen is carried out with freshly grown cultures after the addition of 5% DMSO.

2.3.2. Enrichment and Isolation of Green Non-Sulfur Bacteria

i. Medium for Enrichment and Isolation of *Chloroflexus* species

Chloroflexus species are developed from freshwater. Mesophilic variety of *Cfl. aurantiacus* in lake bottom water and sediments columns are isolated by adding mineral salts, trace elements, Na_2SO_4, 0.6 g/L; $Na_2S.9H_2O$, 0.1 g/L; yeast extract, 0.025 g/L; and casein hydrolysate, 0.025 g/L. Low light intensity (<3000 lx) and a temperature of 25-30°C (Castenholz and Pierson, 1981) were maintained.

Maintenance of Cultures: Cultures may be maintained under anaerobic or semi-anaerobic conditions in liquid or agar-solidified medium. A variety of media would suffice but the author and several others have used the medium given below with the chelatornitrilotri acetic acid, 0.1 g/L and the buffer glycylglycine, 0.8 g/L (Castenholz and Pierson, 1981). Yeast extract, 2-3 g/L or yeast extract, 1.0 g/L and casamino acids, 2.0 g/L are used as main substrates. Screw capped tubes or flasks nearly filled with hot sterile medium provide sufficiently anaerobic conditions for massive pigment synthesis and photo heterotrophic growth. Incandescent lamps provide good growth at low light intensities (1-3000 lx). Pigment containing trichomes of the common forms of *Chloroflexus* will develop on agar plates in incubators at 45-55°C with or without an overlay procedure. Plate cultures richer in bacteriochlorophylls require incubation in anaerobic conditions. These cultures must be subcultured every 7-14 days. Agar shake or subcultures may also be used.

Chloroflexus may be stored in ampul at -196°C in liquid nitrogen for a few years, but the results have been variable. Dense (not aged) cultures should be used and suspended in the medium.

ii. Medium for Enrichment and Isolation of *Chloroflexus* species(g/L)

Contains Nitrilotriacetic acid, 0.1; $CaSO_4.2H_2O$, 0.06; $MgSO_4.7H_2O$, 0.10; NaCl, 0.008; KNO_3, 0.10; $NaNO_3$, 0.70; Na_2HPO_4, 0.11; Glycylglycine, 0.8; NH_4Cl, 0.2; Yeast extract, 1.0-2.0; Vitamin free casamino acids, 2.0; Micronutrient solution 0.5 mL; $FeCl_3$ solution (0.29 g/L), 1.0 mL and distilled water 1.0 L.

Micronutrient Solution Preparation (g/L): Contains Conc. H_2SO_4, 0.5 mL; $MnSO_4.H_2O$, 2.28; $ZnSO_4.7H_2O$, 0.50; H_3BO_3, 0.50; $CuSO_4.5H_2O$, 0.025; $Na_2MoO_4.2H_2O$, 0.025, $CoCl_2.6H_2O$, 0.045 and distilled water, 1.0 L. Autoclave for 15 min at 15 psi pressure-121°C.

The medium is prepared as a 20 fold concentrated stock which is stored unautoclaved in darkness at 4°C. The micronutrients and $FeCl_3$ solution are included in the 20 fold stock. The pH is adjusted to 8.2 with 2M NaOH after medium dilution. Agar (15g/L) if used is added after pH adjustment. Values are in final concentrations.

iii. Medium for Enrichment and Isolation of *Heliothrix* species

Heliothrix species grow in a relatively few hot springs in Oregon, where it forms a visible orange mat about a few millimeters in vertical thickness which is overlayed by a greenish layer composed of several cyanobacteria. The pH is about 8.5. In hot springs the population constitutes a spectacular enrichment of *Htr. oreogonensis* may comprise >95%. The masses of filaments commonly appear as a swollen upper layer which can be easily collected by a syringe with a 17-18 gauge needle. Because the filaments also aggregate rapidly, the collected material can be washed and allowed to aggregate repeatedly, thus the material can be purified by washing out unicellular cyanobacteria and other bacteria (Pierson *et al.,* 1984a). The filaments are large enough to be seen as individual at low power with a compound microscope or at high power with a dissecting microscope thus it helps in allowing to pick off individual filaments. Filaments of *Htr. oreogonensis* are manually removed from an agar plate on which filaments have been gliding and transferred to the edges of a colony growing in the light at 45°C on IMC medium (see below).

Maintenance of these bacteria has been achieved by monthly transfer on slants or plates of Agar-solidified IMC medium (pH 8.1). They are incubated at 45°C and under moderate light intensity 3000 lx cool white fluorescent lamps in an atmosphere of 95% air and 5% CO_2. It is not known whether the co-cultures survive lyophilization or storage at the temperature of liquid nitrogen.

Preparation of Medium IM: Contains 250 mL of solution A; Agar (for plates), 15 g; Distilled water, 650 mL. Autoclaved for 15 min at 15 psi pressure-121°C and then cooled to 50°C. This medium was added to 100 mL of solution B (Pierson *et al.,* 1984a).

i. **Preparation of Solution A(per litre):** Contains $CaCl.2H_2O$, 320 mg; $MgSO4.7H_2O$, 400 mg; KCI, 500 mg; Na Cl, 1 g; $(NH_4)_2SO_4$, 500 mg; KH_2PO_4, 300 mg; $FeCl_3$, 0.292 mg; micronutrient solution SL7 (Pfennig and Truper 1981), 10 mL; vitamin B_{12}, 0.005 mg. Adjust the pH to 7.6 with 1 to 2N NaOH. The precipitate which formed was removed by filtration with Whatman no. 1 filter paper.

ii. **Preparation of Solution B:** Sterile 0.05M $NaHCO_3$ (in water) and was prepared by autoclaving, followed by bubbling vigorously with sterile CO_2 for 1 hour.

Preparation of Medium IMC: This medium was prepared by adding components to 1.0 liter of IM medium before autoclaving which contains glucose, 0.25 g; vitamin-free Casamino Acids, 0.25 g and vitamin solution, 0.5 mL (Pierson *et al.,* 1984a).

Preparation of Vitamin Solution: It is composed of Nicotinic acid, 2.0 mg/mL; Thiamine hydrochloride, 1.0 mg/mL; Para-aminobenzoic acid, 0.2 mg/mL and Biotin, 0.02 mg/mL.

iv. Medium for Enrichment and Isolation of *Oscillochloris* species

Oscillochloris species was isolated from also bacterial mats on the surface of fresh-water mud containing hydrogen sulfide by serial dilutions in a test tube with the modified agar medium used for isolation of green sulfur bacteria which contains additional anhydrous Na_2SO_4, 330 mg/L; Yeast extract, 25 mg/L; Tryptone, 25 mg/L; Vitamin B_{12}, 20 μg/L; Trace elements; Agar, 0.2-0.4% and $Na_2S.9H_2O$, 300 mg/L, pH was adjusted to 7.5. The inoculum was incubated at a light intensity of 2000 lx and at a temperature of 20-35°C. Isolated colonies were isolated from the highest dilutions and purified by further dilution series. Therefore, associated phototrophic bacteria mainly purple non-sulfur bacteria were removed after 3-4 dilution series. Monocultures still contained facultative anaerobic fermenting bacteria producing hydrogen from lactate and sulfur from sulfate-reducing bacteria. Good growth of cultures free of other phototrophic bacteria was obtained in the medium described above. The concentration of cell material increased 10 to 20 fold. *Osc. chrysea* did not grow in semi-solid agar media. Therefore, it was not possible to obtained purified cultures. Cultures of *Oscillochloris* may be stored in a refrigerator at 4°C in liquid medium for 1-2 months at a low light intensity.

2.4. ISOLATION PROCEDURES OF *HELIOBACTERIACEAE* MEMBERS

Enrichments for *Heliobacteria* can be developed in a 0.25% (w/v) yeast extract liquid medium or in various defined or even complex media. Pure cultures (except for the alkaliphiles *Hrs. daurensis* and *Hrs. baculata*) typically grow well in medium PMS or PYE. These media can also be used for enrichment and isolation of *Heliobacteria* but have the disadvantage that they also support good growth of a variety of non-phototrophic fermentative bacteria as well. For most species of *Heliobacteria* that have been isolated so far, warm (~40°C) enrichment temperatures and high light intensities seem to favor their development and help to select against purple non-sulfur bacteria. The latter can be major competitors in heliobacterial enrichments in moist soil/mud or sediments are used as inoculum. Because most of the *Heliobacteria* also fix N_2 (Kimble and Madigan, 1992a; Kimble and Madigan, 1992b; Stevenson *et al.,* 1997), deleting the ammonia from medium PMS (or its variations) makes it more selective for *Heliobacteria*, especially if the pasteurization step is also employed. Medium PYE is a rich medium that supports luxuriant growth of almost pure cultures of *Heliobacteria*, but it is not particularly successful in primary enrichments.

i. Pyruvate Mineral Salts (PMS) Medium (per Liter)

Contains EDTA, 10 mg; $MgSO_4.7H_2O$, 200 mg; $CaCl_2.2H_2O$, 75 mg; NH_4Cl 1.0 g; K_2HPO_4 0.9 g;KH_2PO_4 0.6 g; Sodium pyruvate, 2.2 g; Distilled water, 1.0 mL; Trace element solution, 1.0 mL; Biotin, 15 μg; Vitamin B_{12}, 20 μg. Adjust the pH to 6.8 with NaOH or HCl and sterilize by autoclaving for 15 min at 15 psi pressure-121°C. After cooling, briefly add either methionine, cysteine, or thiosulfate to a final concentration of 1 mM and transfer to an anoxic chamber until used. If thiosulfate is used, it can be added to the medium before autoclaving.

ii. Modified PMS Medium

LMS medium is a pyruvate-free PMS medium containing sodium lactate at a final concentration of 20 mM.

AMS medium is pyruvate-free PMS medium containing 20 mM sodium acetate plus 0.1% (w/v) $NaHCO_3$.

PMS-N medium is PMS medium minus NH_4Cl. The gas in the headspace of the enrichment should be N_2:CO_2 (95:5). This medium is suitable for enrichments using nitrogen fixation as a selective tool.

iii. Mixed Acid Mineral Salts (MAMS) Medium (per Litre)

This semi-defined enrichment medium avoids the use of pyruvate and contains three less easily fermented organic acids. EDTA, 10.0 mg; NaCl, 0.4 g; $MgSO_4.7H_2O$, 200 mg; $CaCl_2.2H_2O$, 75 mg; NH_4Cl, 0.8 g; K_2HPO_4, 0.45 g; KH_2PO_4, 0.3 g; $NaHCO_3$, 1.0 g; Trace elements solution (see below) 1.0 mL ; Yeast extract 0.2 g; Vitamin B_{12}, 20 μg; Sodium acetate, 1.0 g; Sodium malate, 2.0 g; Sodium lactate, 1.0 g. Adjust the pH to 6.8 and sterilize by autoclaving for 15 min at 15 psi pressure-121°C. Add $NaHCO_3$ (after cooling) from a filter-sterilized stock solution and either methionine, cysteine, or thiosulfate to a final concentration of 1 mM and transfer to an anoxic chamber until used. If thiosulfate is used, it can be added to the medium before autoclaving.

Modified MAMS (MAMS-N) Medium, MAMS medium minus NH_4Cl (MAMS-N) is suitable for *Heliobacteria* and other anoxygenic phototroph enrichments using N_2-fixation as a selective tool (Gest *et al.,* 1985). The headspace of enrichment should be filled with N_2 : CO_2 (95:5).

iv. Pyruvate-Yeast Extract (PYE) Medium (per Liter)

This medium is suitable for the growth of pure cultures of all species of *Heliobacteria* except those of *Heliorestis*. Contains K_2HPO_4, 1.0 g; $MgSO_4$. $7H_2O$, 200 mg; $CaCl_2.2H_2O$, 20.0 mg; $Na_2S_2O_3.5H_2O$, 100 mg; Sodium pyruvate 2.2 g; Yeast extract, 4.0 g and Double distilled water 1.0 L. Adjust the pH to 7.0. Autoclave for 15 min at 15 psi pressure-121°C and store in sealed tubes or bottles in an anoxic chamber until used. Alternatively, 1–2 mM sulfide was added (final pH 6.8) and the media was dispensed to sterile screw-cap tubes or bottles.

v. Medium Hr for Isolation of *Heliorestis daurensis* and *Heliorestis baculata* (g/L)

For enrichment and culture of alkaliphilic *Heliobacteria*, a modification (designated medium Hr) of the medium described by Bryantseva *et al.* (1999) is used. 0.6 liter distilled water contains EDTA, 5.0 mg; Trace elements solution, 1.0 mL; KH_2PO_4, 0.5 g; NH_4Cl, 0.5 g; $MgSO_4.7H_2O$, 0.2 g; $CaCl_2.7H_2O$,0.75g; Na_2CO_3, 2.5 g; $NaHCO_3$, 2.5 g; Sodium acetate, 1.0 g; Yeast extract, 0.1 g; $Na_2S{\cdot}9H_2O$, 0.5 g. The carbonate and bicarbonate are autoclaved, dry in a 0.5–1.0 L bottle and dissolved in 300 ml of sterile distilled water after autoclaving, this solution is then added to the main nutrient solution with gentle stirring. Sulfide is prepared as a separate solution (100 mL) and also added to them in solution after autoclaving. The main salt solution is prepared in a 2–10 L dispensing jar containing a spigot for dispensing media into tubes or bottles. After autoclaving for 15 min at 15 psi pressure at 121°C, all solutions are mixed. pH is adjusted with sterile HCl or NaOH to 9.0–9.5. The final assembled medium is immediately transferred to screw-cap tubes or bottles and capped tightly. A slight grayish-black precipitate forms in the medium.

Trace Element Solution (g/L): 1.0 liter of distilled water contains EDTA (sodium salt), 5.2 g; $FeCl_2.4H_2O$, 1.5 g; $ZnCl_2$, 0.7 g; $MnCl_2.4H_2O$, 0.1 g; H_3BO_3 , 0.06 g; $CoCl_2.6H_2O$, 0.19 g; $CuCl_2.2H_2O$,0.017 g; $NiCl_2.6H_2O$, 0.025 g; Na_2MoO_4. $2H_2O$, 0.18 g; $VSO_4.2H_2O$, 0.03g and $NaWO_4.2H_2O$, 0.02 g. Add compounds in the above order; ethylene diamine tetra-acetic acid (EDTA) should be totally dissolved before adding the other components and store at 4°C.

■■■■

Chapter 3

TAXONOMY

Anoxygenic phototrophic bacteria (APB) are a unique assemblage of predominantly aquatic bacteria that can grow under anaerobic conditions by photosynthesis that lack photosystem-II and do not produce oxygen. They are physiologically and phylogenetically diverse group and contain a variety of bacteriochlorophylls such as Bchl *a, b, c, d, e* and *g* and carotenoids such as spirilloxanthin, sphaeroidene, sphaeroidenone, chloroxanthin, lycopene, lycopenal, rhodopin, rhodopinal, chlorobactene, isorenieratene, β-carotene, γ-carotene, dimethyl sphaeroidene and demethyl sphaeroidenenone as major components of spirilloxanthin, okenone, rhodopinal, chlorobactene series and unusual carotenoids. However, all species contain cytochrome, quinones and non-heme iron protein of the ferredoxin type. The photosynthetic metabolism of these organisms differs from the rest of the photosynthetic organisms in that water cannot serve as an electron donor but utilize an external source of electron donors such as reduced sulfur compounds, molecular hydrogen, or organic compounds. The carbon assimilation is either through the reductive pentose phosphate cycle or reductive citric acid cycle and some members are reported to fix atmospheric nitrogen. They are Gram (–) negative bacteria that are spherical, spiral, rods or vibriod shaped and unicellular or uniseriately multicellular filamentous forms also occur. In most cases, multiplication is by binary fission, while some species have a polar type of cell growth and multiply by budding. In the presence of sulfide and light, cells of purple bacteria form highly refractile globules of elemental sulfur. They are widely distributed in terrestrial and diverse aquatic ecosystems including extremes of light, anoxic conditions, sulfide concentration, pH, temperature and salinity.

The anoxygenic phototrophic bacterial cell cultures form distinct colorations. The colors of the cell suspensions range from purple-violet to purple-red, orange-brown, yellowish-brown, brown and green due to the availability of various photosynthetic pigments which are located either in the cytoplasmic membrane or the internal membrane systems.

In APB the photosynthetic pigments and the photosynthetic apparatus are located within a more or less extended system of internal membranes that is originated from the cytoplasmic membrane and maintain continuity with the cytoplasmic membrane. The intracellular membrane consists of small finger-like intrusions, vesicles, tubules, or lamella parallel to (or) at an angle to the cytoplasmic membranes.

They contain photosynthetic apparatus such as the reaction centers (RC) and Light-harvesting (LH) pigment-protein complexes surrounding the reaction center in the plane of the membrane (Drews and Imhoff, 1991). In green bacteria, the photosynthetic pigments are located in the cytoplasmic membrane and in the chromosomes that underlie and are attached to the cytoplasmic membrane, while *Heliobacteria* contain no membrane structures.

Anoxygenic phototrophic bacteria comprise members of purple sulfur bacteria (PSB), Purple non-sulfur bacteria (PNSB), Green sulfur bacteria (GSB) (Overmann, 2001), Green non-sulfur bacteria (GNSB), *Heliobacteria* (Yurkov and Beatty, 1998) and aerobic anoxygenic phototrophic bacteria (Imhoff, 1995; Yurkov and Beatty, 1998). Based on 16S rRNA gene sequence analysis, they are classified into five distantly related phyla (**Fig. 3.1** ;) *Proteobacteria* (PSB and PNSB), *Chlorobi* (Green-sulfur bacteria), *Chloroflexi* (Filamentous Green non-sulfur bacteria) and *Firmicutes* (Gram (+) positive *Heliobacteria*). The single species phototrophic members such as *Chloroacidobacterium thermophilum* were described as a new genus and included in the phylum *Acidobacteria* (Bryant *et al.,* 2007).

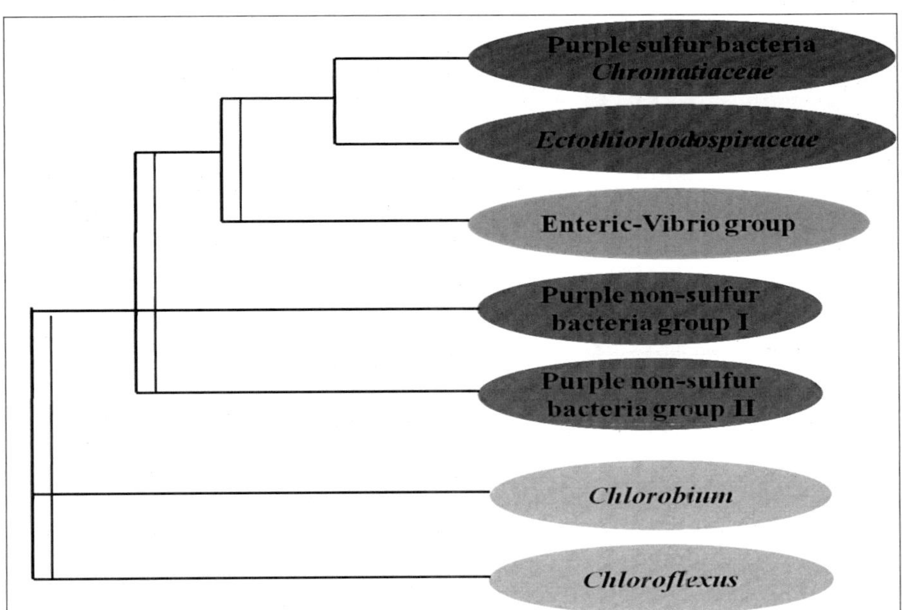

Fig. 3.1. Dendrogram showing relationship among APB and some Non-Phototrophic Relative

Proteobacteria are again classified into 3 classes; *Alphaproteobacteria* , *Betaproteobacteria* and *Gammaproteobacteria* (**Fig. 3.2**). Phylogenetic analyses of purple bacteria based on comparative 16S rRNA sequencing have shown that

PNSB is included either in *Alphaproteobacteria* or *Betaproteobacteria*, while PSB is included in *Gammaproteobacteria* (Imhoff *et al.*, 2005).

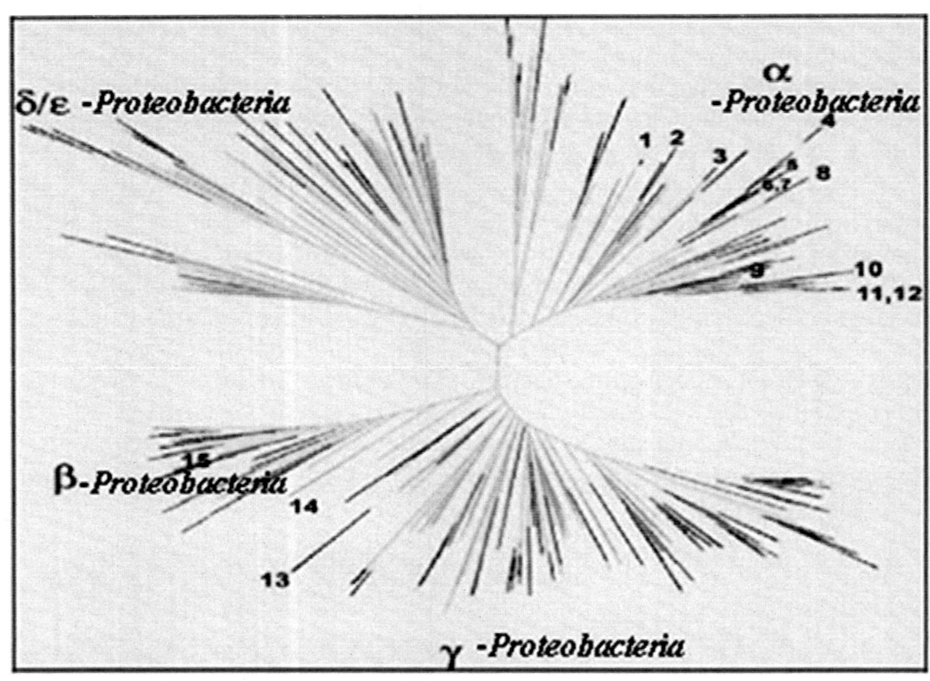

Fig. 3.2. 16S rRNA Tree of the *Proteobacteria*, Based on Sequences from the Ribosomal Database [http://rdp.cme.msu.edu].

For visualization, 16S rRNA gene data are filtered so that only lineages at 97% or less 16S rRNA gene distance are shown. Bold lineages are those proteobacterial species with one or more genome-sequenced members. Bold, numbered lineages indicate the positions of sequenced purple/phototrophic *Proteobacteria* in the tree. Other lineages have 16S rRNA data available from isolates or environmental samples. Proteobacterial classes are shown in Greek letters. Numbered purple proteobacteria are 1. *Rhodospirillum rubrum*, 2. *Rsp. centenum*, 3. *Erythrobacter* NAP1, 4. *Roseobacter denitrificans* (multiple strains), 5. *Roseovarius* sp. (multiple strains), 6. *Loktanella*, 7. *Jannaschia* sp. CCS1, 8. *Rhodobacter sphaeroides* (incl. *Rba. capsulatus*), 9. *Fulvimarina*, 10. *Methylobacterium*, 11. *Bradyrhizobium* (multiple species), 12. *Rhodopseudomonas palustris* (multiple strains), 13. *Thermochromatium tepidum* (proprietary sequence), 14. *Halorhodospira* sp. SL1, 15. *Rubrivivas gelatinosus* (photosynthetic gene cluster only, no genome) (Swingley *et al.*, 2009).

Classification of different APB is given in **Table 3.1** and the key to the identification of different groups of APB as provided in Bergey's Manual of Systematic Bacteriology (Second Edition) is given below.

Table 3.1. Classification of Different Anoxygenic Phototrophic Bacteria

S.No.	Groups	Families	Genus
I)	**Purple Bacteria**		
a)	**Purple Non-Sulfur Bacteria**	*Rhodospirillaceae*	*Rhodospirillum, Phaeospirillum, Phaeovibrio, Rhodocista, Rhodospira, Rhododvibrio, Roseospira*
		Acetobacteraceae	*Rhodopila, Rhodovastum*
		Bradyrhizobiaceae	*Rhodopseudomonas, Rhodoblastus*
		Hyphomicrobiaceae,	*Rhodoplanes, Blastochloris, Rhodomicrobium*
		Rhodobiaceae	*Afifella, Rhodobium, Roseospirillum*
		Rhodobacteraceae	*Rhodobacter, Rhodobaca, Rhodovulum, Rubribacterium*
		Rhodothalassiaceae	*Rhodothalassium*
		Comamonadaceae,	*Rhodoferax, Rubrivivax*
		Rhodocyclaceae	*Rhodocyclus*
b)	**Purple Sulfur Bacteria**	*Chromatiaceae*	*Allochromatium* *Chromatium* *Halochromatium* *Isochromatium* *Imhoffiella* *Lamprobacter* *Lamprocystis* *Marichromatium* *Phaeochromatium* *Phaeobacterium* *Rhabdochromatium* *Thermochromatium* *Thioalkalicoccus*

S.No.	Groups	Families	Genus
			Thiocapsa
			Thiocystis
			Thiococcus
			Thiodictyon
			Thioflavicoccus
			Thiohalocapsa
			Thiolamprovum
			Thiopedia
			Thiorhodococcus
			Thiorhodovibrio
			Thiospirillum
			Thiobaca
			Thiophaeococcus
		Ectothiorhodospiraceae	*Ectothiorhodospira*
			Ectothiorhodosinus
			Halorhodospira
			Thiorhodospira
II)	**Green Bacteria**		
a)	Green Sulfur Bacteria	*Chlorobiaceae*	*Chlorobium,*
			Prosthecochloris,
			Chlorobaculum,
			Pelodictyon
			Ancalochloris
			Chloroherpeton ,
b)	Multicellular Filamentous	*Chloroflexaceae*	*Chloroflexus*
	Green Non-Sulfur Bacteria	*Oscillochloridaceae*	*Oscillochloris, Chloronema*
		Roseiflexaceae	*Roseiflexus, Heliothrix*
III	**Genera Incertae Sedis** ***Firmicutes***	*Heliobacteriaceae*	*Heliobacterium, Heliobacillus, Heliophilum, Heliorestis, Candidatus Heliomonas lunata*

Key to the Groups and Families of the Anoxygenic Phototrophic Bacteria

I. Cells can grow by a photoautotrophic and/or photoorganotrophic metabolism under anaerobic conditions. .... **Purple Bacteria**

A. Cells preferably grow by a photoorganotrophic metabolism with simple organic substrates. Most species also able to grow photolitho-autotrophically with molecular hydrogen as an electron donor. Some species capable of using sulfide or thiosulfate as electron donors, if elemental sulfur is formed as an oxidation product, sulfur globules appear only outside the cells, never inside. Elemental sulfur rarely oxidized further to sulfate. Most species require one or more growth factors. ....**Purple Non- Sulfur Bacteria**

1. Unicellular bacteria, cells are spherical/ ovoid/ short or long rods/ spirals of various widths. Non-motile or motile by means of polar, sub-polar flagella, or peritrichous flagella, they may have lateral flagella, divide by binary fission or show polar growth and budding. Containing bacteriochlorophyll *a* or *b* and various types of carotenoids. They are mostly able to grow photoheterotro-phically under anaerobic conditions in light. ...... ***Rhodospirillaceae***

2. Cells are spherical/ovoid/rods. Motile utilizing single polar flagella, divide by binary fission. Containing bacteriochlorophyll, *a* or *b* and various types of carotenoids. They are mostly able to grow photoheterotrophically under anaerobic conditions in light. Facultative phototrophs grow anaerobically in light and aerobically in dark. .... ***Acetobacteraceae***

3. Cells are rod-shaped, motile utilizing flagella, show polar growth, budding and asymmetric cell division. Containing bacteriochloro-phyll, *a* or *b* and various types of carotenoids. Photoheterotrophic growth occurs with several organic compounds as carbon and electron donor. Photoautotrophic growth is possible under anae-robic conditions with molecular hydrogen, thiosulfate or sulfide as the electron donor. Some species grow chemotropically under microaerobic to aerobic conditions. .... ***Bradyrhizobiaceae***

4. Cells are rod/ovoid-shaped, motile by means of polar/sub-polar flagella, show polar growth, budding and asymmetric cell division. Intracytoplasmic membranes present as lamellae underlying and parallel to the cytoplasmic membrane containing bacteriochloro-phyll *a* or *b* and various carotenoids of spirilloxanthin series. Mostly grow photoheterotrophically under anaerobic conditions and microaerophilic growth in the dark is also possible. ....---- ***Hyphomicrobiaceae***

5. Cells are ovoid to rod-shaped/vibriod to spiral-shaped, motile utilizing single/ bipolar flagella. Multiply by budding and asymmetric cell division. The intracytoplasmic membrane of the lamellar type and contain bacteriochlorophyll *a* and carotenoids of spirilloxanthin and lycopenal series. Facultative photoheterotrophs that grow anaerobically in light or aerobically in dark...-.....................--- ***Rhodobiaceae***

6. Cells are rod/ovoid/spiral/vibrioid shaped, motile utilizing polar flagella, multiply by binary fission. The intracytoplasmic photosynthetic membrane system is vesicular/ lamellar type lying parallel to a cytoplasmic membrane containing bacteriochlorophyll *a* and carotenoids of sphaeroidene and spirilloxanthin series. Growth occurs preferably photoheterophically under anoxic conditions in the light, chemotrophic growth is also possible under aerobic conditions in the dark. .... ***Rhodobacteraceae***

7. Cells are vibrioid to spiral-shaped, motile utilizing flagella, multiply by binary fission. Intracytoplasmic membranes are lamellar stacks lying parallel to the cytoplasmic membrane containing bacteriochlorophyll *a* and carotenoids of spirilloxanthin series are present. Facultative or obligatory anaerobic growth occurs... ***Rhodothalassiaceae***

8. Cells are rods/vibrioid, motile utilizing polar/ bipolar/peritrichous flagella. Multiply by binary fission. Intracytoplasmic membranes may be absent. Bacteriochlorophyll and carotenoids of sphaeroidene or spirilloxanthin series are present. Under anaerobic conditions in the presence of light, the growth is possible photoheterotrophically with organic substrates, photoautotrophically with H_2 and CO_2. Chemoheterotrophic growth is possible under aerobic conditions in dark. .... ***Comamonadaceae***

9. Cells are slender/curved/straight rods, non-motile or motile by means of polar flagella, multiply by binary fission. Internal photosynthetic membranes are present in the form of small, single finger-like intrusions containing bacteriochlorophyll *a* and carotenoids of spirilloxanthin or sphaeroidene series. Photoautotrophic growth possible with molecular hydrogen, photoheterotrophic growth possible with different organic substrates under anaerobic conditions in the light. Chemotrophic growth occurs under aerobic conditions in the dark.................... ***Rhodocyclaceae***

B. Cells contain bacteriochlorophyll *a* or *b* and various carotenoids of groups. Photosynthetic pigments are located in the cytoplasmic membrane and in intracytoplasmic membrane systems of different fine structures, which are continuous with the cytoplasmic membrane. They are able to grow phototrophically under anoxic conditions in the light... **Purple Sulfur Bacteria**

1. Cells are able to grow with sulfide and elemental sulfur as the sole photosynthetic electron donor. In the presence of both sulfide and

light, globules of elemental sulfur are formed inside the cells and further oxidized to sulfate. .... ***Chromatiaceae***

2. Cells able to grow by photoautotrophic and /or photoorganotrophic metabolism. Sulfide and elemental sulfur are used for photoautotrophic growth. In the presence of both sulfide and light, globules of elemental sulfur are formed outside the cells and may be further oxidized to sulfate. Growth was dependent on saline and alkaline growth conditions. ....***Ectothiorhodospiraceae***

II. Cells contain low concentrations of bacteriochlorophylls *a* and in most cases, *c, d,* or *e* as the major bacteriochlorophyll components, as well as various carotenoids. The light-harvesting photosynthetic pigments are located in the chlorosomes which underlie and are attached to the cytoplasmic membrane. **Green Bacteria**

A. Cells can grow with sulfide and elemental sulfur as the sole photosynthetic electron donor. In the presence of both sulfide and light, globules of elemental sulfur are formed outside the cells, never inside. All species are strictly anaerobic and obligately phototrophic. Simple organic substrates may be photoassimilates in the presence of both sulfide and light; therefore cells are potentially mixotrophic. .... ***Chlorobiaceae* (Green Sulfur Bacteria)**

B. Cells are arranged in multicellular filaments that may be capable of gliding motility. Species with bacteriochlorophyll and various carotenoids. ....---- ***Chloroflexaceae* (Green Non- Sulfur Bacteria)**

1. Cells are arranged in multicellular flexible filaments with gliding motility. The trichrome formation is seen, which appears green or yellowish-green. Cells may form gas vesicles, contain bacteriochlorophyll *c* and various carotenoid pigments. Chlorosomes are the light-harvesting photosynthetic apparatus. Growth occurs as photolithoautotrophic or photolithoheterotrophic under anoxic conditions. Some strains grow at microaerobic conditions.. ***Oscillochloridaceae***

2. Cells are unbranched straight filamentous shaped, appear reddish-brown to reddish-orange colour. Cells show gliding motility. Chlorosomes are absent. Contain the only bacteriochlorophyll *a* and derivatives of γ-carotene and keto-myxocoxanthin. Facultative phototrophs. ----***Roseiflexaceae***

III. A. Gram (–) negative, long rod-shaped bacteria with gliding motility and strictly anaerobic, obligately photoorganotrophic metabolism. Cells lack intracytoplasmic membranes or chlorosomes but contain bacteriochlorophyll *g* and carotenoids. The Colour of the cell suspension was brownish-green. .... ----***Heliobacteriaceae* (General Incertae Sedis)**

Anoxygenic phototrophic purple bacteria were classified under the phylum *Proteobacteria* which is proposed for all phototrophic bacteria and chemotropic relatives (Stackebrandt *et al.,* 1988).The recognition of the close genetic relationship between phototrophic bacteria and chemotropic bacteria is mainly based on 16S rRNA oligonucleotide catalogs and 16S rRNA gene sequences, which led Woese to call them "The purple bacteria and their relatives". Among all the five phyla of photosynthetic prokaryotes (*Proteobacteria, Acidobacteria, Chlorobi, Chloroflexi* and *Firmicutes*), the purple *Proteobacteria* are the most metabolically diverse.

Photosynthesis in purple bacteria occurs under anoxic conditions. Major photosynthetic pigments are bacteriochlorophylls *a/b* and various types of carotenoids. Based on genetic relationships and 16S rRNA, these bacteria are separated into *Alpha-proteobacteria* and *Beta-proteobacteria* and called the purple bacteria and their relatives (Imhoff and Truper, 1989). Later bacteria within these groups were rearranged accounting to phylogeny, chemotaxonomic characteristics and eco-physiological properties. These are most metabolically diverse and utilize hydrogen and a number of small organic molecules as photosynthetic electron donors and use as an electron donor and a diverse number of metabolic pathways for energy generation, carbon assimilation and nitrogen, sulfur and phosphorus metabolism (Imhoff *et al.*, 2005; Madigan and Jung, 2009). Growth with reduced iron as electron donor has been demonstrated in some phototrophic purple bacteria by Widdel *et al.* (1993) and Ehrenreich and Widdel (1994). Purple sulfur bacteria are strong photoautotrophs and capable of limited photoheterotrophy, but they are poorly equipped for dark metabolism and growth. By contrast, purple non-sulfur bacteria, nature's preeminent photoheterotrophs capable of photo autotrophy and possess diverse capacities for dark metabolism and growth. Several members of these bacteria inhabit extreme environments including extreme temperature, pH and salinity. These bacteria are important phototrophs because of the following reasons.

1. Consume a toxic substance H_2S and contribute organic matter to the anoxic environment by their autotrophic capacity.
2. Consume organic compounds, primarily non-fermentable organic compounds in their role as photoheterophs and
3. Offer scientists for the study of diverse photosynthetic metabolisms

Purple sulfur bacteria differ from purple non-sulfur bacteria on both metabolic and phylogenetic grounds. In certain habitats particularly favorable for their development, purple bacteria have also been shown to be significant primary producers (Czeczuga, 1968; Takahashi and Ichimura, 1968; Overmann *et al.*, 1994, 1996, 1999).

Recently APB classification is revised (Garitty *et al.,* 2005) and they have been spread out into different groups of bacteria as precised in **Table 3.2** and **Fig. 3.3** .

Table 3.2: Revised Classification of Anoxygenic Phototrophic Bacteria under the Classes *Alphaproteobacteria* (PNSB), *Betaproteobacteria* (PNSB) and *Gammaproteobacteria* (PSB).

S.No.	Taxa (Three letters Abbreviation), Earlier Heterotypic Synonym and Other Synonym	Name of Genus and Species
I.	***Alphaproteobacteria:***	
i.	***Rhodospirillales***	
a.	***Rhodospirillacea***	
	Rhodospirillum (Rsp.)	*Rhodospirillum rubrum*
	Pararhodospirillum	*P. photometricum* *P. sulfurexigens* *P. oryzae*
	Phaeospirillum (Phs.)	*Pheospirillum fulvum* *(Rsp. fulvum)* *Phs. molischianum* *(Rsp. molischianum)* *Phs. chandramohanii* *Phs. tilakii* *Phs. oryzae*
	Phaeovibrio (Phv.)	*Phv. sulfidiphilus*
	Rhodocista (Rcs.)	*Rcs. centenaria* *Rcs. pekingensis*
	Rhodospira (Rsa.)	*Rsa. trueperi*
	Rhodovibrio (Rhv.)	*Rhv. salinarum* *(Rsp. salinarum)* *Rhv. sodomensis* *(Rsp. sodomenses)*
	Roseospira (Ros.)	*Ros. mediosalina* *(Rsp. mediosalina)* *Ros. marina* *Ros. navarrensis* *Ros. goensis* *Ros. visakhapatnamensis* *Ros. thiosulfatophila*
b.	***Acetobacteraceae***	
	Rhodopila (Rpi.)	*Rpi. globiformis* *(Rps. globiformis)*
	Rhodovastum	*Rhodovastum atsumiense*

S.No.	Taxa (Three letters Abbreviation), Earlier Heterotypic Synonym and Other Synonym	Name of Genus and Species
ii.	***Rhizobiales***	
a.	***Bradyrhizobiaceae***	
	Rhodopseudomonas (Rps.)	*Rps. palustris*
		Rps. julia
		Rps. rhenobacensis
		Rps. faecalis
		Rps. harwoodiae
		Rps. parapalustris
		Rps. pseudopalustris
		Rps. pentothenatexigens
		Rps. thermotolerans
		Rps. telluris
	Rhodoblastus (Rbl.)	*Rbl. acidophilus (Rps. acidophila)*
		Rbl. sphagnicola
b.	***Hyphomicrobiaceae***	
	Rhodoplanes (Rpl.)	*Rpl. roseus*
		Rpl. elegans
		Rpl. pokkaliisoli
		Rpl. serenus
		Rpl. piscinae
		Rpl. oryzae
		Rpl. azumiensis
		Rpl. tepidicaeni
		Rpl. tepidamans
	Blastochloris (Blc.)	*Blc. viridis (Rps. viridis)*
		Blc. sulfoviridis (Rps. sulfoviridis)
		Blc. gulmargensis
	Rhodomicrobium (Rmi.)	*Rmi. vannielii*
		Rmi. udaipurense
c.	***Rhodobiaceae***	
	Afifella (Af.)	*Afi. marina (Rbi. marinum) (Rps. marina)*
		Afi. pfennigii (Rhodobium Pfennigii)

S.No.	Taxa (Three letters Abbreviation), Earlier Heterotypic Synonym and Other Synonym	Name of Genus and Species
	Rhodobium (Rbi.)	*Rbi.Orientis, Rbi. gokarnense*
	Roseospirillum (Rss.)	*Rss. parvum*
iii.	***Rhodobacterales***	
a.	***Rhodobacteraceae***	
	Rhodobacter (Rba.)	*Rba. capsulatus (Rps capsulata)*
		Rba. maris
		Rba. megalophilus
		Rba. aestuarii
		Rba. azollae
		Rba. lacus
		Rba. sediminis
		Rba. viridis
		Rba. changlensis
		Rba. flagellates
		Rba. sediminicola
		Rba. ruber
		Rba. tardus
		Rba. thermarum
	Luteovulum	*Luteovulum sphaeroids (Rba. sphaeroides)*
		Luteovulum johrii (Rba. johrii)
		Luteovulum ovatum (Rba. ovatus)
		Luteovulum alkalitolerans (Rba. alkalitolerance)
		Luteovulum azotoformans (Rba. azotoformans)
	Phaeovulum	*Phaeovulum vinykumarii (Rba. vinaykumarii)*
		Phaeovulum veldkampii (Rba. veldkampii)
	Fuscovulum	*Fuscovulum blasticum (Rba. blasticus)*
	Rhodovulum (Rdv.)	*Rdv. sulfidophilum (Rps. sulfidophila)*
		Rdv. adriaticum (Rps. adriatica)
		Rdv. euryhalinum (Rba. euryhalinus)

S.No.	Taxa (Three letters Abbreviation), Earlier Heterotypic Synonym and Other Synonym	Name of Genus and Species
		Rdv. strictum
		Rdv. iodosum
		Rdv. robiginosum
		Rdv. marinum
		Rdv. imhoffii
		Rdv. visakhapatnamense
		Rdv. kholense
		Rdv lacipunicei
		Rdv. steppense
		Rdv. phaeolacus
		Rdv. salis
		Rdv. viride
		Rdv. tesquicola
		Rdv. aestuarii
		Rdv. algae
		Rdv. bhavnagarence
		Rdv. mangrove
	Rhodobacca (Rca.)	*Rca. bogoriensis*
		Rca. barguzinensis
	Rhodobaculum	*Rhodobaculum claviforme*
	Rubribacterium	*Rubribacterium polymorphum*
iv.	***Rhodothalassiales***	
a.	***Rhodothalassiaceae***	
	Rhodothalassium (Rts.)	*Rts. salexigens*
II.	***Betaproteobacteria:***	
i.	***Burkholderiales***	
a.	***Comamonadaceae***	
	Rhodoferax (Rfx.)	*Rfx. fermentans*
		Rfx. antarcticus
		Rfx. saidenbachensis
	Rubrivivax (Rvi.)	*Rvi. gelatinosus*
		(Rcy. gelatinosus)
		(Rps. gelatinosa)
		Rvi. benzoatilyticus
ii.	***Rhodocyclales***	
a.	***Rhodocyclaceae***	

S.No.	Taxa (Three letters Abbreviation), Earlier Heterotypic Synonym and Other Synonym	Name of Genus and Species
	Rhodocyclus (Rcy.)	*Rcy. tenuis (Rsp. tenue)*
		Rcy. purpureus
III.	***Gammaproteobacteria***	
i.	***Chromatiales***	
a.	***Chromatiaceae***	
	Allochromatium (Alc.)	*Alc. vinosum (Chr. vinosum)*
		Alc. minutissimum (Chr. minutissimum)
		Alc. warmingii (Chr. warmingii)
		Alc. renukae
		Alc. phaeobacterium
		Alc. humboldtianum
	Chromatium (Chr.)	*Chr. okenii*
		Chr. weissei
	Halochromatium (Hch.)	*Hch. salexigens (Chr. salexigens)*
		Hch. glycolicum
		Hch. roseum
	Isochromatium (Isc.)	*Isc. buderi (Chr. buderi)*
	Imhoffiella	*Imhoffiella bheemlicus*
	Lamprobacter (Lpb.)	*Lpb. modestohalophilus*
	Lamprocystis (Lps.)	*Lps. roseopersicina*
		Lps. purpurea (Amoebobacter purpureus)
	Marichromatium (Mcr.)	*Mcr. gracile (Chr. gracile)*
		Mcr. purpuratum (Chr. purpuratum)
		Mcr. indicum
		Mcr. bheemlicum
		Mcr. littoris
		Mcr. chrysaorae
	Phaeochromatium (Phc.)	*Phc. fluminis*
	Phaeobacterium (Phb.)	*Phaeobacterium nitratireducens*
	Rhabdochromatium (Rbc.)	*Rbc. marinum*

S.No.	Taxa (Three letters Abbreviation), Earlier Heterotypic Synonym and Other Synonym	Name of Genus and Species
	Thermochromatium (Tch.)	*Tch. tepidum*
	Thioalkalicoccus (Tac.)	*Tac. limnaeus*
	Thiocapsa (Tca.)	*Tca. roseopersicina*
		Tca. litoralis
		Tca. pendens
		Tca. rosea
		Tca. marina
	Thiocystis (Tcs.)	*Tcs. minor (Chr. minus)*
		Tcs. violascens (Chr. violascens)
		Tcs. violacea
		Tcs. gelatinosa
		Tcs. cadagnonensis
		Tcs. chemoclinalis
	Thicoccus (Tco.)	*Tco.pfennigii (Tca. pfennigii)*
	Thiodictyon (Tdc.)	*Tdc. Elegans*
		Tdc. bacillosum
	Thioflavicoccus (Tfc.)	*Tfc. mobilis*
	Thiohalocapsa (Thc.)	*Thiohalocapsa halophile (Thiocapsa halophila)*
		Thc. marina
	Thiolamprovum (Tlp.)	*Tlp. pedioforme*
	Thiopedia (Tpd.)	*Tpd. rosea*
	Thiorhodococcus (Trc.)	*Trc. minor*
		Trc. mannitoliphagus
		Trc. kakainadensis
		Trc. modestalkaliphilus
		Trc. fuscus
		Trc. alkaliphilus
	Thiorhodovibrio (Trv.)	*Trv. winogradskyi*
	Thiospirillum (Tsp.)	*Tsp. jenense*
	Thiobaca (Tba.)	*Tba. trueperi*
	Thiophaeococcus (Tpc.)	*Thiophaeococcus mangrove*
		Tpc. fuscus

S.No.	Taxa (Three letters Abbreviation), Earlier Heterotypic Synonym and Other Synonym	Name of Genus and Species
b.	***Ectothiorhodospiraceae***	
	Ectothiorhodospira (Ect.)	*Ect. mobilis*
		(Ect. marismortu)
		Ect. vacuolata
		(Ect. shaposhnikovii)
		Ect. haloalkaliphila
		Ect. marina
		Ect. variabilis
		Ect. saline
		Ect. magna
	Ectothiorhodosinus (Ets.)	*Ets. mongolicus*
	Halorhodospira (Hlr.)	*Hlr. halophila (Ect. halophila)*
		Hlr. halochloris (Ect. halochloris)
		Hlr. abdelmalekii
		(Ect. abdelmalekii)
		Hlr. neutriphila
	Thiorhodospira (Trs.)	*Trs. sibirica*
IV.	**Green Sulfur Bacteria**	
i.	***Chlorobiales***	
a.	***Chlorobiaceae***	
	Chlorobium (Chl.)	*Chl. limicola*
		Chl. phaeobacteroides
		Chl. phaeovibroides
		Chl. clathratiforme
		Chl. luteolum
		Chl. ferroxidans
		Chl. chlorochromatii
	Prosthecochloris (Ptc.)	*Ptc. aestuarii*
		Ptc. vibrioformis
		Ptc. indica
	Chlorobaculum (Cba.)	*Cba. macestae*
		Cba. limnaeum
		Cba. parvum
		Cba. tepidum
		Cba. thiosulfaophilum

S.No.	Taxa (Three letters Abbreviation), Earlier Heterotypic Synonym and Other Synonym	Name of Genus and Species
		Cba. chlorovibrioides
	Pelodictyon (Pld.)	*Pld. phaeum*
	Anacalochloris (Anc.)	*Ancalochloris perfilievii*
	Chloroherpeton (Chp.)	*Chp. thalassium*
V.	**Green Non- Sulfur Bacteria**	
i.	***Chloroflexales***	
a.	***Chloroflexaceae***	
	Chloroflexus (Cfl.)	*Cfl. aggregans*
		Cfl. aurantiacus
		Cfl. islandicus
b.	***Oscillochloridaceae***	
	Oscillochloris (Osc.)	*Osc. chrysea*
		Osc. trichoides
	Chloronema (Cln.)	*Cln. giganteum*
		Cln. spiroideum
c.	***Roseiflexaceae***	
	Roseiflexus (Rof.)	*Rof. castenholzii*
	Heliothrix (Htr.)	*Htr. oreogonensis*
VI.	**Genera Incertae Sedis**	
	Firmicutes	
i.	***Heliobacteriaceae***	
	Heliobacterium (Hbt.)	*Hbt. chlorum*
		Hbt. gestii
		Hbt. modesticaldum
		Hbt. sulfidophilum
		Hbt. undosum
	Heliobacillus (Hba.)	*Hba. mobilis*
	Heliophilum (Hph.)	*Hph. fasciatum*
	Heliorestis (Hrs.)	*Hrs. daurensis*
		Hrs. baculata
		Hrs. convoluta
		Hrs. acidaminivorans
	Candidatus Heliomonas	*Candidatus Heliomonas lunata*

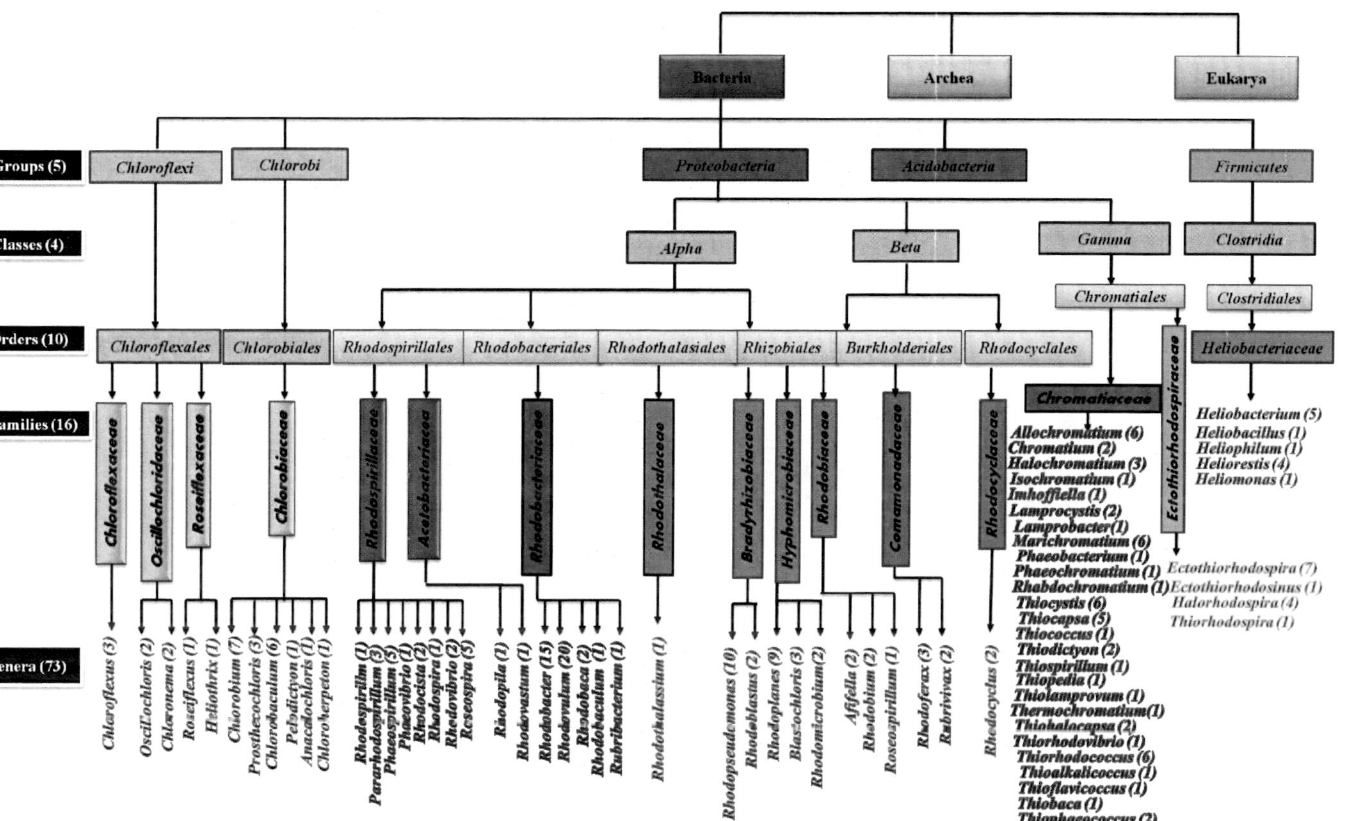

Fig. 3.3. Phylogenetic Distribution of Anoxygenic Phototrophic Bacteria (No. of Species for each Genus is Given in Bracket)

3.1 PURPLE NON-SULFUR BACTERIA

CLASS I. *ALPHAPROTEOBACTERIA* (Garrity *et al.,* 2005) Garrity *et al.* (2015)

The class *Alphaproteobacteria* was circumscribed for this volume based on phylogenetic analysis of 16S rRNA sequences. The class contains the orders *Rhodospirillales, Rhizobiales, Rhodobacteriales* and *Rhodothalassiales* .

Order 1. *Rhodospirillales* (Pfennig and Truper, 1971) Garrity *et al.* (2015)

The order *Rhodospirillales* is morphologically, metabolically and ecologically different and outlined based on phylogenetic analysis of 16S rRNA sequences. It includes chemolithotrophs, chemoorganotrophs and facultative photoheterotrophs. Few of them are also able to grow photoautotrophically. Further, some other species can grow methylotrophically. *Rhodospirillaceae* and *Acetobacteraceae* are included in this order. The key to the families of *Rhodospirillales* is given below.

1. Unicellular bacteria, cells are spherical/ ovoid/ short or long rods/ spirals of various widths. Non-motile or motile utilizing polar, sub-polar flagella or peritrichous flagella. They may have lateral flagella. Divide by binary fission or show polar growth and budding. Contain bacteriochlorophyll *a* or *b* and various types of carotenoids. They were mostly able to grow photoheterotrophically under anaerobic conditions in light. .... **---- *Rhodospirillaceae***
2. Cells are spherical/ovoid/rods. Motile utilizing single polar flagella, divide by binary fission. Containing bacteriochlorophyll *a* or *b* and various types of carotenoids. They were mostly able to grow photoheterotrophically under anaerobic conditions in light. Facultative phototrophs grow anaerobically in light and aerobically in dark... ***Acetobacteriaceae***

Family1. *Rhodospirillaceae* (Pfennig and Truper, 1971), Garrity *et al.* (2015)

The family *Rhodospirillaceae* is described based on phylogenetic analysis of 16S rRNA sequences. It contains the genera *Rhodospirillum* (Type genus), *Azospirillum, Inquilinus, Levispirillum, Magnetospirillum, Pararhodospirillum, Phaeospirillim, Rhodocista, Rhodospira, Rhodovibrio, Roseospira, Roseos-pirillum, Skermanella, Thalassospira . Tistrella* , *Inquilinus, Thalassospira* and *Tistrella* were proposed and they are not described here (Coenye *et al.*, 2002). Among them, purple non-sulfur bacteria were included in the following genera *Rhodospirillum, Pararhodospirillum, Phaeospirillum, Phaeovibrio, Rhodocista, Rhodospira, Rhodovibrio* and *Roseospira.* Cells are spherical or ovoid or short or long rods or spirals of various widths. Motile utilizing polar- or sub-polar or peritrichous flagella, may have lateral flagella or non-motile. These are divided by binary fission or show polar growth and budding. Gram (–) negative, internal photosynthetic membranes are continuous with the cytoplasmic membrane and consist of finger-like intrusions, vesicles, or lamella. Colour of cell suspension is green, yellowish-brown, brown, brownish-red, red, purple, purple-red, or purple-violet. Photosynthetic pigments are bacteriochlorophyll *a* or *b* (esterified with phytol or geranylgeranyl) and various types of carotenoids which are located in the cytoplasmic membrane and in the internal membrane systems. In most species, the formation of pigments and the internal membrane system are repressed under aerobic conditions but become derepressed at oxygen tension below a certain level.

The preferred mode of growth for most genera is photoheterotrophic under anoxic conditions in the presence of light with different organic substrates or few grow as photoautrophically with either molecular hydrogen or in few species they utilize sulfide, thiosulfate, or elemental sulfur as electron donor and CO_2 as sole carbon source. Growth factors are required in most cases. They also grow chemotrophically in the dark under microaerobic to aerobic conditions. Some grow as chemoautotrophs. *Azospirillum* ,*Magnetospirillum* and *Skermanella* are chemoorganotrophic. Some species may metabolize anaerobically in the dark with sugars and either nitrate, DMSO, or trimethylamine-N-oxide as electron sink or poorly with metabolic intermediates as electron acceptors. Ammonia, dinitrogen and many organic compounds are generally used as nitrogen sources. Nitrate is also utilized by some strains. The majority of the species can assimilate sulfate as the sole sulfur source, only a few species depend on the presence of reduced sulfur compounds as a sulfur source. Sulfide is inhibitory for several species even at low concentrations. Whereas, some species exhibit a remarkable tolerance towards this compound. One or more vitamins are essentially required as growth factors. Biotin, thiamine, niacin and para-aminobenzoic acid (PABA) are most commonly utilized by these bacteria. These compounds are rarely needed by species of the *Chromatiaceae* and *Ectothiorhodospiraceae*, which may require vitamin B_{12} as the sole growth factor. The growth of several species is enhanced by a small amount of yeast extract, while some other species have a complex nutrient requirement. Polysaccharides, poly-β-hydroxybutyric acid (PHB) and polyphosphate are formed as storage material.

Members of this group are widespread and found not only in freshwater, marine and hypersaline environments but also in moist soils and paddy fields. They exist preferably in aquatic habitats with significant amounts of soluble organic matter and low oxygen tension, but they rarely form coloured blooms, which are characteristically formed by representatives of other families of anoxygenic phototrophic bacteria. The mol% G+C of the DNA is 52.0-73.2 (Bd and T_m .) *Rhodospirillum* is the type genus. The family *Rhodospirillaceae* includes 9generaof purple non-sulfur bacteria. The detailed description and key to the identification of these genera is given below.

Key to the Genera of *Rhodospirillaceae*

1. Cells are vibrioid or spiral, 0.7-1.5 μm wide, motile utilizing polar flagella. Internal photosynthetic membranes are present as vesicles or lamella but not as small finger intrusions of the cytoplasmic membrane. .... ***Rhodospirillum***

2. Cells are spiral-shaped, motile, multiply by binary fission. Internal photosynthetic membranes are present as lamellar stacks forming a sharp angle to the cytoplasmic membrane. Photosynthetic pigments are bacteriochlorophyll and carotenoids of lycopene and rhodopin. .... ***Pararhodopsirillum***

3. Cells are spherical, 0.5-1.0 µm wide, motile utilizing polar flagella. Cell division by binary fission, the internal photosynthetic membrane is vesicles containing bacteriochlorophyll *a* and carotenoids of lycopene and rhodopin. .... ***Phaeospirillum***

4. Cells are vibrioid shaped, 1.2-2.5 µm wide, cell division by binary fission. Chimeric internal membranes of lamellar stacks and vesicles present in a single cell, containing bacteriochlorophyll *a* and carotenoids of rhodopinal series. .... ***Phaevibrio***

5. Cells are ovoid to spiral-shaped, 1.0-2.0 µm wide, showing polar growth and cell division by binary fission. The internal photosynthetic membrane is present as lamellae containing bacteriochlorophyll *a* and carotenoids of the spirilloxanthin series. .... ***Rhodocista***

6. Cells are vibrioid to spirilloid-shaped, 0.6-0.8 µm wide, multiply by binary fission. The internal photosynthetic membrane system is vesicular type contains bacteriochlorophyll *b* and carotenoids of spirilloxanthin series. .... ***Rhodospira***

7. Cells are vibrioid to spiral-shaped, 0.6-0.9 µm wide. Motile, the inte-rnal photosynthetic system is a vesicular type containing bacteriochl-orophyll *a* and carotenoids of the spirilloxanthin series.. ***Rhodovibrio***

8. Cells are vibrioid to spiral-shaped, motile, multiply by binary fission. Intracytoplasmic photosynthetic membranes are present as vesicles, contain bacteriochlorophyll *a* and various carotenoids. ....***Roseospira***

Molecular Characterization

Phylogenetically, the family *Rhodospirillaceae* is a member of *Rhodospirillales* , subclass *Alphaproteobacteria* of phylum *Proteobacteria* (Stackbrandt *et al.* , 1988). According to 16S rRNA gene sequence compositions of the type species, the next similar family is *Acetobacteriaceae* (Gillis and De Ley, 1980). The genera *Rhodospirillum* , *Phaeospirillum* , *Phaeovibrio* , *Rhodocista* , *Rhodospira* , *Rhodovibrio* , *Roseospira* and *Roseospirillum* are included in purple non-sulfur bacteria.

Phylogenetic Structure of the Family and its Genera

Extensive 16S rRNA gene sequence analyses of type species and strains constitute the phylogenetic structure within the family *Rhodospirillaceae* .

1. *Rhodospirillum* (*Rsp.*)(Molisch, 1907) Imhoff (2015)

Cells are vibriod to spiral-shaped, 0.7-1.5 μm wide and motile utilizing bipolar flagella. They divide by binary fission and are gram (-) negative. Internal photosynthetic membranes are present as vesicles or lamella lying parallel to the cytoplasmic membrane or forming a sharp angle to it. Photosynthetic pigments are bacteriochlorophyll *a* (esterified with phytol or geranylgeraniol is the major component of *Rsp. rubrum*) and carotenoids of the spirilloxanthin series with spirilloxanthin itself lacking in some species and contain ubiquinone (Q-8/Q-10) and rhodoquinones(RQ-8/RQ-10) with 8 or 10 isoprene units are present. Major cellular fatty acids are $C_{18:1}$, $C_{16:1}$ and $C_{16:0}$ with $C_{18:1}$ as the dominant component (51-55% of total fatty acids). Three different types of intracytoplasmic membranes are present in various species. *Rsp. rubrum* and *Rsp. photometricum* has a large type of cytochrome C_2, while the other species have a small type. The major quinones are the same in *Rsp. fulvum* and *Rsp. molischianum* (ubiquinone (Q-9) and menaquinone (MK-9) with a side chain of nine isoprenoid units) but different in all other species.

Growth occurs preferably photoheterotrophically under anaerobic conditions in the presence of light. All species have various growth factors requirements. Some species are very sensitive to oxygen (except *Rsp. rubrum* and *Rsp. salexigens*), while others grow(Madigan *et al.*, 1984). The nitrogenase of *Rsp. rubrum* is subject to post-translational inactivation by ADP-ribosylation under energy-limiting conditions and equally well aerobically in the dark and under phototrophic conditions and anaerobically in the light. Molecular hydrogen and sulfide may be used as photosynthetic electron donors during photoautotrophic growth. Fermentation and oxidant-dependent growth may occur. Sulfate is reduced through 3'-phosphate adenosine 5'-phosphosulfate in *Rsp. rubrum* and *Rsp. fulvum*. Polysaccharides, poly-β-hydroxybutyric acid and phosphates may be present as storage products. *Rhodospirillum* species can grow with ammonia or N_2 as the sole nitrogen source (Ludden and Roberts, 1995). Ammonia assimilation is mediated by the glutamine synthetase/glutamate synthase reactions, which are NADPH-dependent in *Rsp. rubrum* (Brown and Herbert, 1977). However, a nitrate reductase source is not possible (Ketchum and Sevilla, 1973; Klemme, 1979). Purines can be used as a nitrogen source under anoxic conditions in the presence of light and under oxic conditions in the dark (Aretz *et al.,* 1978).

The optimum growth temperature and pH are 25-35°C and pH 6.8-7.2 respectively. The heterogeneity of this genus may call for further separation into various genera. However, this seems inappropriate until more strains of each species are studied in detail and more data become available to justify the formation of new genera. The mol% G+C of the DNA is 60.5-65.8 (Bd) and 62.1-63.5 (T_m). Validly published species of the genus are *Rhodospirillum rubrum, Rsp. photometricum ,Rsp. molischianum ,Rsp. fulvum ,Rsp. salexigens , Rsp. salinarum ,Rsp. sodomens ,Rsp. mediosalinum ,Rsp. tenue, Rsp. cenetenum , Rsp. sulfurexigens* and *Rsp. oryzae*. The type species is *Rsp. rubrum* **(Fig. 3.4)**.

Table 3.3. Differentiating Characteristics of Species of the *Rhodospirillum* and *Pararhodospirillum*

Characteristics	***Rhodospirillum rubrum***	***Pararhodospirillum photometricum***	***Pararhodo-spirillum sulfurexigens***	***Pararhodo-spirillum oryzae***
Habitat	Freshwater	Freshwater	Freshwater	Soil
Cell shape	Vibrio to spiral	Spiral	Spiral	Spiral
Cell width (µm) (Length and width of a complete spiral turn (µm))	0.8-1.0 (7-10 x 1.5-2.5)	1.1-1.5 (4-7 x 2.5-4.0)	1.0-1.3 (8-9 x 2.5-3.0)	1.0-3.0 x 4.0-16.0
Internal membrane system	Vesicles	Lamellae stacks	Lamellae stacks	Lamellae stacks
Flagellation	Bipolar	Polar tufts	Bipolar tufts	Amphitrichous
Motility	+	+	+	+
Colour	Pink-deep red	Brown-orange to brown-red/dark brown	Pink	Brown
Bacterioclorophyll	*a*	*a*	*a*	*a*
Major carotenoids	Spirilloxanthin, Rhodovibrin	Rhodopin and lycopene	Rhodovibrin	Lycopene, Rhodopin, Rhodopin - glucoside
Quinones (Ratio)	Q-10, RQ-10 (9:1)	Q-8, RQ-8 (8:2)	Q-8, RQ-8 (9:1)	Q-8, RQ-8 (9:1)
Cytochrome *c* size	Large	Large	Large	Large
Salt tolerance (%)	0-0.5	0-1.5	0-1.5	Not required
Optimum Tempe-rature (range) (˚C)	30-35 (25-35)	25-30 (25-35)	30 (25-35)	30 (25-40)
pH optimum (range)	6.8-7.0 (6.0-8.5)	7.0 (6.5-7.5)	7.0 (ND)	7.0 (6.0-8.0)
Growth factors	Biotin	Nicotinamide	Thiamine	Thiamine, niacin and para- amino benzoic acid
Photolithoautotrophic growth	-	-	-	+
Dark aerobic growth	-	-	-	+
Nitrogen fixation	+	-	-	-
PHB accumulation	+	NA	NA	NA
Oxidation of sulfide	+	-	-	+

Characteristics	*Rhodospirillum rubrum*	*Pararhodospirillum photometricum*	*Pararhodo-spirillum sulfurexigens*	*Pararhodo-spirillum oryzae*
Whole cell absorption maxima (nm)	375-377, 510-517, 546-550, 590-595, 807-808, 881-885	NA	374, 473, 500, 530, 590, 791 and 848	374, 491, 527, 791 and 851
DNA G+C content (mol%)	63.8–66.8	64.74	64.7	60.2

Symbols: +, Substrate utilized or present; (-), substrate not utilized or absent; (+), weak growth; ±, strain-dependent reaction; ND, not determined; NA, no data available; Q-8, ubiquinone; Q–10, ubiquinone 10; RQ–8, rhodoquinones 8; RQ-10, rhodoquinone 10. Data for *Rhodospirillum rubrum* is taken from Molisch (1907) and for species of *Pararhodospirillum* is taken from Lashmi *et al.* (2014).

Table 3.4. Carbon Sources /Electron Donors and Nitrogen Sources Utilized by Species of *Rhodospirillum* and *Pararhodospirillum*

Carbon source/ Electron donors and Nitrogen sources used for growth	*Rhodospirillum rubrum*	*Pararhodo-spirillum photometricum*	*P. sulfure-xigens*	*P. oryzae*
Acetate	+	+	+	+
Arginine	+	-	-	ND
Aspartate	+	±	-	ND
Benzoate	-	-	-	ND
Butyrate	+	+	+	+
Caproate	+	-	+	-
Caprylate	ND	-	-	-
Crotonate	+	-	-	-
Citrate	-	-	-	-
Ethanol	+	+	-	+
Formate	-	-	±	+
Fructose	±	+	(+)	-
Fumarate	+	+	+	+
Glucose	-	+	(+)	(+)
Glutamate	+	±	(+)	(+)
Glutarate	-	±	±	+
Glycerol	-	+	-	-
Glycolate	ND	+	-	-
Lactate	+	+	(+)	+
Malate	+	+	+	+

Carbon source/ Electron donors and Nitrogen sources used for growth	***Rhodospirillum rubrum***	***Pararhodo-spirillum photometricum***	***P. sulfure-xigens***	***P. oryzae***
Mannitol	-	+	-	+
Methanol	±	+	-	(+)
Pelargonate	ND	±	±	±
Propionate	+	±	-	-
Pyruvate	+	+	+	+
Succinate	+	+	+	+
Tartarate	±	-	-	+
Valerate	+	+	+	+
Electron donors				
Hydrogen	+	+	+	+
Sulfide	+	+	+	±
Sulfur	-	±	-	±
Thiosulfate	-	+	+	±
Nitrogen Sources				
Alanine	±	+	ND	+
Asparagine	±	+	ND	+
Ammonia	±	+	ND	+
Nitrogen	±	+	ND	+
Nitrate	±	-	ND	+
Adenine	±	ND	ND	+
Guanine	±	ND	ND	ND
Xanthin	±	ND	ND	ND
Uric acid	±	-	ND	-

Symbols: +, Substrate utilized or present; (-), substrate not utilized or absent; (+), weak growth; ±, strain-dependent reaction; ND, not determined; NA, no data available. Data for *Rhodospirillum rubrum* is taken from Molisch (1907), Data for species of *Pararhodospirillum* is taken from Lashmi *et al.* (2014)

Out of 12 species, 11 of these have been reclassified into other genera (Imhoff *et al.,* 1998). At present only one species is included in the genus *Rhodospirillum* i.e., *Rsp. rubrum*. The major differentiating characteristics between *Rhodospirillum* species and other species are précised in **Tables 3.3** and **3.4**.

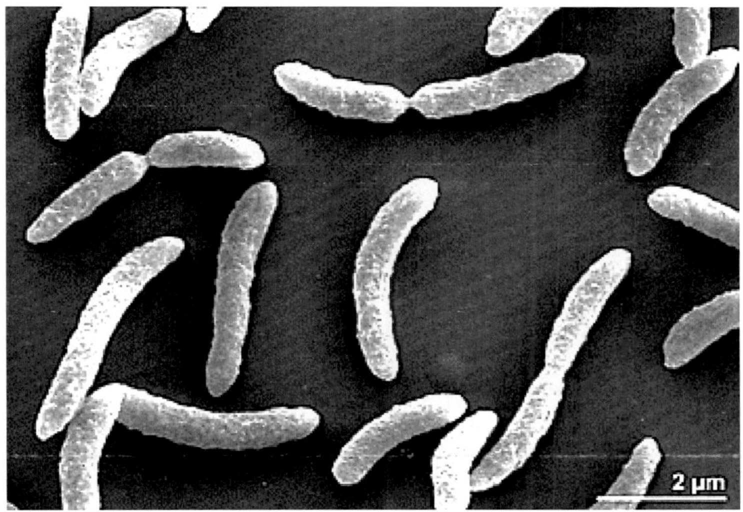

Fig. 3.4. Electron micrograph of *Rhodospirillum rubrum*(Bar=2 µm)

Differentiation of the *Rhodospirillum* from other Genera

Several chemotaxonomic properties distinguished *Rhodospirillum* species from other spiral-shaped purple non-sulfur bacteria (PNSB). They differ from *Phaeospirillum* species, their closest relatives in major quinone components and cytochrome *c* structure. Large type cytochrome c_2 are present in *Rsp. rubrum* and *Pararhodospirillum photometricum*, whereas small-type cytochrome c_2 were found in *Phaeospirillum* species (Ambler *et al.,* 1979). The phylogenetic relationship of these bacteria based on 16S rDNA sequences is shown in **Fig. 3.5** (Imhoff *et al.,* 2005). The differentiating characteristics of *Rhodospirillum* with other genera of *Rhodospirillaceae* members are presented in **Table 3.5** .

Table 3.5. Differentiating Characteristics of Genera of *Rhodospirillaceae*

Characteristics	*Rhodospirillum*	*Pararhodospirillum*	*Phaeospirillum*	*Phaeovibrio*	*Rhodocista*	*Rhodospira*	*Rhodovibrio*	*Roseospira*
Cell diameter (µm)	0.8-1.0	1.0-3.0	0.5-1.0	0.3-0.5	1.0-2.0	0.6-0.8	0.6-0.9	0.4-1.0
Internal membrane system	Vesicles	Lamellae	Lamellae	Chimeric*	Lamellae	Vesicles	Vesicles	Vesicles
Motility	+	+	+	+	+	+	+	+
Cell Shape	Vibrioid-Spirilloid	Spiral	Vibrioid - Spirilloid	Vibrioid	Vibrioid - Spirilloid	Vibrioid - Spirilloid	Vibrioid - Spirilloid	Vibrioid - Spirilloid
Colour	Red	Orange brown-Dark brown, Brown red, Pink	Brown	Brown	Pink	Beige	Red, Pink	Pink

Characteristics	***Rhodospirillum***	***Pararhodospirillum***	***Phaeospirillum***	***Phaeovibrio***	***Rhodocista***	***Rhodospira***	***Rhodovibrio***	***Roseospira***
Bacteriochlorophyll	*a*	*a*	*a*	*a*	*a*	*b*	*a*	*a*
Growth factors	Biotin	Nicotinamide, Thiamine	Para-Amino benzoic acid, Amino acids	Para-Aminobenzoic acid, Biotin	Biotin, B_{12}	Biotin, Thiamine, Pantothenate	Cobalamine	Thiamine, Para-Aminobenzoic acid, Nicotinamide
Aerobic growth	+	-	Microaerophilic	+	-	+	-	Microaerophilic
Oxidation of sulfide	+	-	-	-	ND	-	-	+
Salt requirement)%(	None	0-1.5	None	None	None	2.0 (0.5-5.0)	8.0-12.0 (3-24)	4-7 (0.5-15)
Optimal temperature (Range)	30-35	25-30	25-30	25-30	40-45	25-30 (20-35)	35-42	30-35
Optimal pH (Range)	6.8-7.0	7.0	7.0-7.3	7.0-8.0	6.8	7.3-7.5 (7.0-7.8)	7.0 -8.0 (6.0-9.0)	6.8-7.0 (6.7-9.0)
Habitat	Fresh water	Fresh water, Soil	Fresh water	Brackish water	Fresh water, Warm springs	Salt marsh	Saltern lakes	Salty Springs
Mol% G+C of the DNA	60.5-65.8 Bd	60.0-65.8	60.0-66.0	67.8-68.8	69.9-70.0 Bd	65.7 Td	65.0-69.0 Bd	66.6 Bd
Cytochrome *c* size	Large	Large	Small	Small	ND	Small	ND	ND
Major quinones	Q-10, RQ-10	Q-8, RQ-8	Q-9, MK-9	ND	Q-9	Q-7, MK-7	Q-10, MK-10	ND
Major fatty acids								ND
$C_{14:0}$	2.10	1.00	0.60	2.60	ND	7.50	1.00	ND
$C_{16:0}$	14.0	25.2	15.1	27.0	ND	27.9	7.40	ND
$C_{16:1}$	27.1	22.2	25.8	1.0 -3.6	ND	1.20	0.30	ND
$C_{18:0}$	1.80	0.40	1.20	1.50	ND	1.20	23.0	ND
$C_{18:1}$	54.8	51.0	54.5	50.0	ND	60.7	35.2	ND

Symbols: +, Positive in strains/present; -, Negative/absent; Q, ubiquinone; MK, menaquinone; RQ, rhodoquinone*, Both Lamellae and Vesicles; Bd, Bouyant density; Td, Thermal denaturation; ND, Not determined. Data for *Rhodospirillum, Phaeospirillum, Rhodovibrio* and *Roseospira* is taken from Imhoff *et al.* (1998). Data for *Phaeovibrio* and *Pararhodospirillum* is taken from Lakshmi *et al.* (2011) and (2014). Data for *Rhodocista* (Kawasaki *et al.,* 1994) and *Rhodospira* (Pfennig *et al.,* 1998), is adopted.

Taxonomic Comments

Traditionally all spiral-shaped phototrophic purple non-sulfur bacteria are assigned to the genus *Rhodospirillum* (Pfennig and Truper, 1974). Identification of a large amount of chemotaxonomic and phylogenetic diversity in the PNSB and their occurrence in various groups of the *Proteobacteria* primarily led to the taxonomic reclassification of those species belonging to the *Betaproteobacteria* . *Rsp. tenue* was assigned to *Rhodocyclus tenuis* (Imhoff *et al.,* 1984) and this reclassification was later supported by 16S rDNA sequence analyses (Hiraishi *et al.,* 1991a). After this reclassification, all species of *Rhodospirillum* were *Alphaproteobacteria.* However, the group remained very heterogeneous in phenotypic properties and genetic relationships.

The description of the new species destined to the genus *Rhodospirillum* based on their spiral shape continued till now. However, most of them were quite different from *Rsp. rubrum* (type species) of this genus. Four halophilic species were classified along the several freshwater species of *Rhodospirillum.* In addition, *Rsp. centenum* is added to this genus (Favinger *et al.,* 1989), based on significant differences from *Rsp. rubrum* has been established in the description of species. Recently, the extreme genetic distance among the spiral-shaped PNSB has been recognized in various proposals. *Rsp. centenum* was transferred to a new genus *Rhodocista centenaria* (Kawasaki *et al.,* 1992). Another new spiral-shaped species, *Rhodospira trueperi* was transferred to a new genus based on essential phenotypic and genotypic differences from *Rsp. rubrum* and other known purple non-sulfur bacteria (Pfennig *et al.,* 1997). *Rsp. photometricum, Rsp. sulfurexigens* and *Rsp. oryzae* are reclassified into a new genus *Pararhodospirillum* (Lakshmni *et al.,* 2014). *Rsp. fulvum* and *Rsp. molischianum* is reclassified into a new genus *Phaeospirillum .Rsp. salexigens* was reclassified as *Rhodothalassium salexigens* which belongs to a new separate order *Rhodothalassiales* containing the family *Rhodothalassiaceae* with genus *Rhodothalassium .Rsp. salinarum* and *Rsp. sodomense* is reclassified into a new genus *Rhodovibrio salinarum* and *Rhodovibrio sodomensis* (Imhoff *et al.,* 1998).

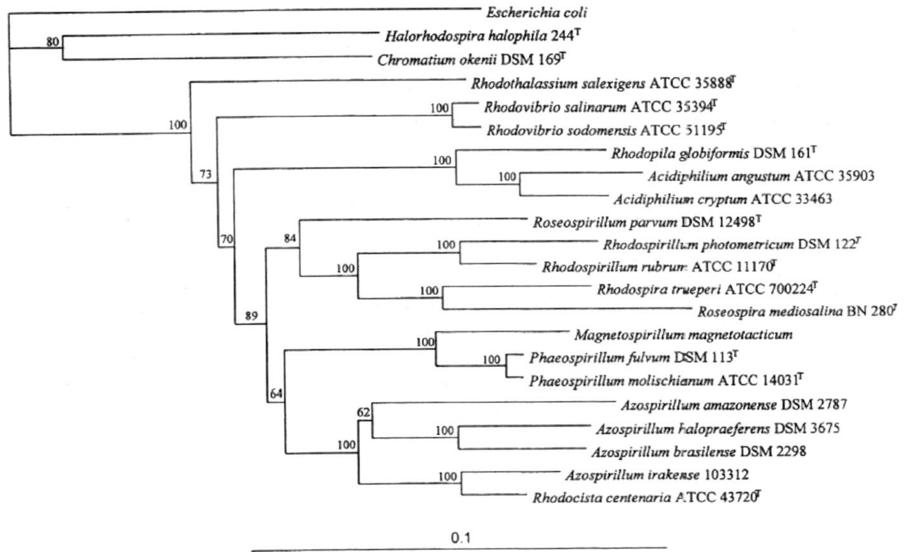

Fig. 3.5. Phylogenetic tree of *Rhodospirillum* species and related *Alphaproteobacteria* of the order *Rhodospirillales* based on 16S rDNA sequences. The sequence of *E. coli* was used as an outgroup to root the tree. Bar=10% substitution of nucleotides.

The expected heterogeneity of the genus *Rhodospirillum* became visible with the 16S rDNA sequence information of most of the known species (Imhoff *et al.,* 1998) and these data indicate the truth that the spiral-shaped *Alpha-proteobacteria* are phylogenetically quite distantly related to each other and do not justify classification in one and same genus. Therefore, the spiral-shaped phototrophic *Alpha-proteobacteria* are probably reclassified based on different phenotypic properties and 16S rDNA sequence similarities (Imhoff *et al.,* 1998) (**Fig. 3.5** .)

Major quinone components and fatty acid composition, salt requirement and phylogenetic relationship based on 16S rDNA sequences were considered as primary importance in describing and differentiating these genera. Many phylogenetic lines of salt-dependent species were identified and the salt-dependent property was considered as genus-specific (Imhoff *et al.,* 1998). Only *Rsp. rubrum* was maintained as species of the genus *Rhodospirillum* and other species were transferred to the new genera.

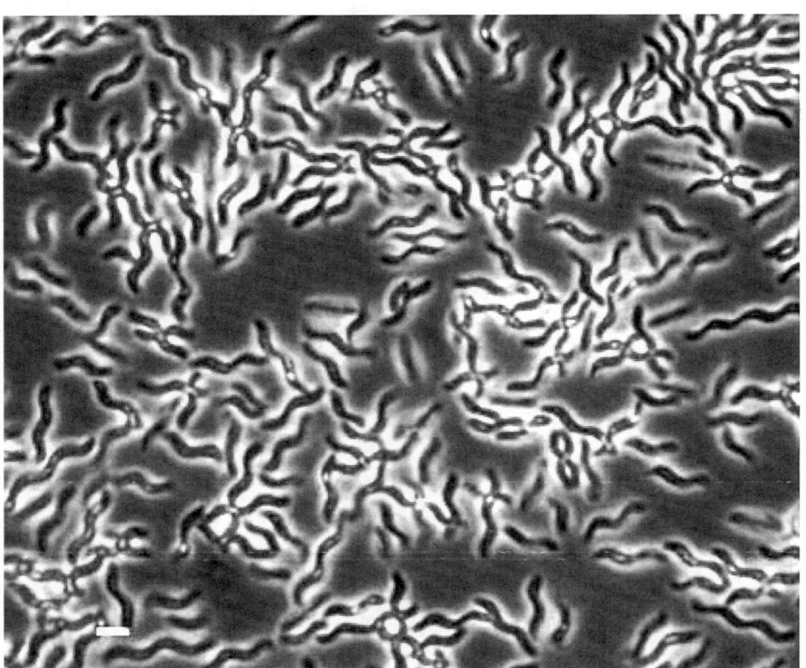

Fig. 3.6. Phase-contrast micrograph of cells of *Pararhodospirillum sulfurexigens*

2. *Pararhodospirillum* (Resembling *Rhodospirillum*) Lakshmi *et al.* (2014)

Cells are spiral-shaped, motile, mesophilic and freshwater phototrophic bacteria, which multiply by binary fission. Cells are Gram (-) negative. Internal photosynthetic membranes are present as lamellar stacks forming a sharp angle to the cytoplasmic membrane. Photosynthetic pigments are bacteriochlorophyll *a* and carotenoids lycopene and rhodopin are present. Metabolism is strictly anaerobic and obligately phototrophic. They prefer neutral pH and growth factors are required for growth. Ubiquinones (Q-8) and rhodoquinones (RQ-8) with 8 isoprene units are present. Major cellular fatty acids are $C_{16:0}$ and $C_{18:1}\omega7c$/ $C_{18:1}\omega6c$. Three hydroxy acids of C_{14}, C_{15} and C_{16} are present. The mol% G+C content of the genomic DNA is 60–65.8 (HPLC). Delineation of the genus is primarily determined by the phylogenetic information from 16S rRNA gene sequences, chemotaxonomic and phenotypic traits. *Pararhodospirillum photometricum* (Molisch, 1907) is the type species. Validly published species are *P. photometricum, P. sulfurexigens* (**Fig. 3.6**) and *P. oryzae*. The major differentiating characteristics and carbon sources/electron donors used by species of the genus *Pararhodospirillum* are précised in **Tables 3.3 and Table 3.4.**

Differentiation of the *Pararhodospirillum* from other Genera

A number of chemotaxonomic properties distinguished *Pararhodospirillum* species from other spiral-shaped purple non-sulfur bacteria (PNSB). They differ from *Phaeospirillum* and *Rhodospirillum* species, in their major quinone components and cytochrome *c* structure. They are also different in their growth factors requirement. Large type cytochrome c_2 are present in *Rsp. rubrum* , whereas small-type cytochrome c_2 were found in *Phaeospirillum* species (Ambler *et al.,* 1979). The differentiating characteristics of *Pararhodospirillum* with other genera of *Rhodospirillaceae* members are presented in **Table 3.5** .

Taxonomic Comments

The phylogenetic relationship of these bacteria based on 16S rRNA sequences is shown in **Fig. 3.6A** (Lakshmi *et al.,* 2014). The 16S rRNA sequence analysis of *Pararhodospirillum sulfurexigens* strain JA143^T ,*Rsp. rubrum* ATCC11170^T and *P. photometricum* strain DSM122^T exhibit sequence similarity of 95.72% and 95.58% respectively, justifying the description of the former as the type strain of the novel genus (Kumar *et al.*, 2008). In the same analysis, *Rhodovibrio sodomensis* strain DSI^T and *Rhodovibrio salinarum* strain ATCC35394^T were grouped into a separate clade. Based on the 16S rRNA gene sequence analysis, phylogenetic relationships of *Pararhodospirillum oryzae* strain JA318^T with *Pararhodospirillum sulfurexigens* strain DSM122^T and *Rsp. rubrum* ATCC 11170^T indicated that the *P. oryzae* clustered with type strains of *Rhodospirillum* (Lakshmi *et al.*, 2014). The highest sequence similarity of *Pararhodospirillum oryzae* strain JA318T was found with the type strain of *Pararhodospirillum sulfurexigens* (99.9%).

DNA-DNA Hybridization experiments carried out between strains JA318^T of *Pararhodospirillum oryzae*, strain JA143^T of *Pararhodospirillum sulfurexigens* and strains JA318^T of *Pararhodospirillum oryzae* and DSM122^T of *Pararhodospirillum photometricum* resulted in reassociation values of 52 ± 2% and 45.1 ± 1% (n=5) respectively. These hybridization values encouraged the classification of strain JA318^T as a different species (Lakshmi *et al.*, 2014).

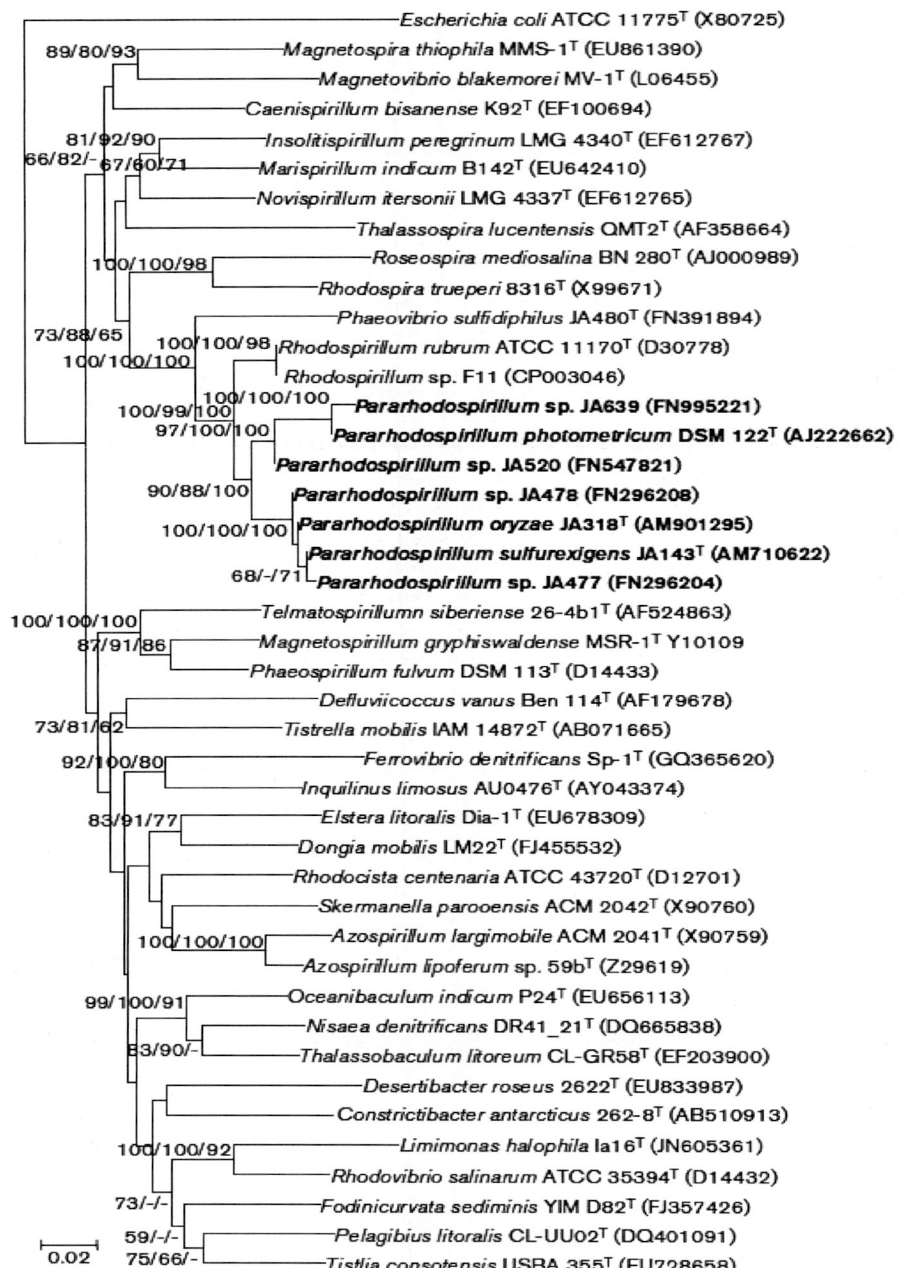

Fig. 3.6A. Phylogenetic tree of *Pararhodospirillum* species and related members of family *Rhodospirillaceae* based on 16S rRNA sequences. The sequence of *E.coli* was used as an outgroup to the root of the tree. Bar=0.02 nucleotide substitutions per position.

3. *Phaeospirillum* (*Phs.*) (Brown spiral) (Imhoff *et al.*, 1998) Imhoff (2015)

Cells are vibrioid to spiral-shaped, 0.5-1-0 μm wide, motile by means of polar flagella, multiply by binary fission and are Gram (-) negative. Intracytoplasmic photosynthetic membranes are present as lamellar stacks, contain photosynthetic pigments such as bacteriochlorophyll *a* and carotenoids of spiriiloxanthin series lacking spirilloxanthin. Contain ubiquinones and menaquinones Q-9 and MK-9 are major components. Freshwater bacteria without a requirement for NaCl or sea salt for growth. Growth occurs preferably photoheterotrophically under anoxic conditions in the presence of light but is also possible in the dark at very low oxygen tensions. Growth factors are required. They have G+C content between 60 and 66 mol%. The known species of the genus are *Phs. fulvum* is type species (van Niel, 1944) **(Fig. 3.7)**. Five species are validly published and their differential characteristics and utilization of carbon sources by the different species of *Phaeospirillum* are précised in **Tables 3.6** and **3.7** .

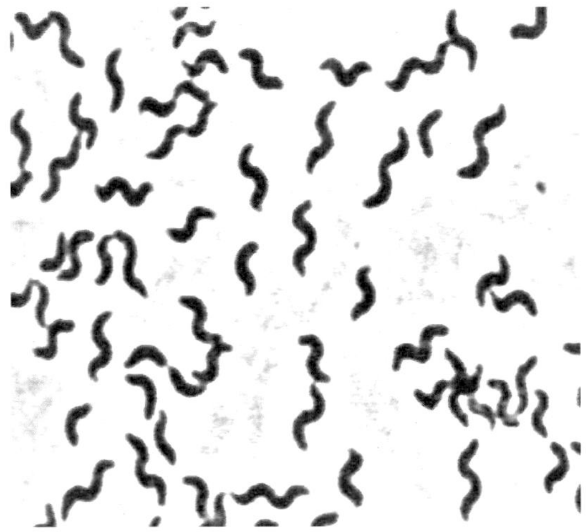

Fig. 3.7. Phase-contrast micrograph of cells of *Phs. fulvum*

Differentiation of the *Phaeospirillum* from other Genera

Besides 16S rDNA sequence similarities (Kawasaki *et al.*, 1993b; Imhoff *et al.* , 1998), the size and sequences of the small size cytochrome c_2 (Ambler *et al.* , 1979), quinone and fatty acid composition and other properties of *Phaeospirillum* with other genera of *Rhodospirillaceae* members are précised in **Table 3.5** .

Taxonomic Comments

Phaeospirillum species previously have been included in the genus *Rhodospirillum.* Due to significant differences in chemotaxonomic properties and 16S rDNA sequences, they have been reclassified into the new genus *Phaeospirillum* (Imhoff *et al.*, 1998). The name *Phaeospirillum* was first proposed by Kluyver and van Niel (1936) for brown-colored, spiral-shaped phototrophic bacteria, but was not included in the Approved List of Names and lost its standing in nomenclature. The name was renewed by Imhoff *et al.* (1998). Due to high similarity in 16S rDNA sequences (greater than 99% similarity), both *Phaeospirillum* species could be treated as a single species. This was not proposed due to well-recognized phenotypic differences between the two species. The phylogenetic relationships of these bacteria based on 16S rDNA and 16S rRNA sequences are shown in **Fig. 3.5** (Imhoff *et al.*, 2005) and **3.7A** (Shalem Raj *et al.*, 2012).

Table 3.6. Differentiating Characteristics of Species of the *Phaeospirillum*

Characteristics	*Phs. tilakii*	*Phs. chandra-mohanii*	*Phs. fulvum*	*Phs. molischi-anum*	*Phs. oryzae*
Habitat	Aquatic sediments	Freshwater	Freshwater	Freshwater	Paddy soil
Cell shape	Spiral shaped	Spiral shaped	Spiral shaped	Spiral shaped	Spiral shaped
Size (μm)	0.5–0.8 x 2.0–6.0	0.8–1.0 x 4.0–8.0	0.5–0.7 x 3.5	0.7–1.0 x 4.0–6.0	0.8–1.2 x 2.0–6.0
Width of one complete turn of a spiral (μm)	Forms long chains of 10–30 μm	6–8. forms chains of 30 μm	1.0–1.6	1.5–2.5	–
Internal membrane system	Lamellae	Lamellae	Lamellae	Lamellae	Lamellae stacks
Spheroplast formation	–	–	–	+	+
Colour of cell suspensions	Brown	Browm–orange	Deep brown	Brown–orange to brown-red/dark brown	Brown
Bacteriochlorophyll	*a*	*a*	*a*	*a*	*a*
Carotenoids and its composition (%)*	Ly (7-10), Rp (66-69), HLYG (22), DHLYDG (2-3)	Ly (10), RP, HLY (77), HLYG (11), DHLYDG (2)	Ly (11), Rp (84), RPG (5), HLYG	Ly (17), Rp (77), RPG (1) HLYG (5), DHLYDG (1)	Ly (12), Rp (81), RPG (7)
Major quinone	Q-9, MK-9	Q-9, MK-9	Q-9, MK-9	Q-9, MK-9	Q-9, MK-9
Major quinone ratio (Q–9:MK–9	7 : 3	9 : 1	8 : 2	8 : 2	ND
Salt requirement *	NR	NR	NR	NR	NR
DNA G+C content (mol%)	62.7±1.1	60.54	64.3-65.3	60.5-64.8	63.3
16S rRNA gene sequence similarity with novel strains)%(	99.6–100†	96.5-96.7	96.7-96.9	96.5-96.7	100
Temperature range (optimum) °C	25-35 (30)	25-35 (30)	ND (25-30)	ND (30)	25-35 (30)
pH range (optimum)	5.0-9.0 (6.0-6.5)	6.5-8.0 (7.0)	6.0-8.5 (7.3)	6.0-8.5 (7.3)	6.0-8.0 (7.0)
Growth factors	NR	NR	pABA	Amino acids	NR
Whole cell absorption maxima (nm)	377, 464, 491, 527, 590, 803 and 854	383, 407, 461, 527, 593, 803, 860	NA	375, 465, 488-491, 520-528, 590-595, 803-807, 850-855	374, 491, 527 and 851

Characteristics	*Phs. tilakii*	*Phs. chandra-mohanii*	*Phs. fulvum*	*Phs. molischi-anum*	*Phs. oryzae*
Cytochrome *c* size	ND	ND	Small	Small	ND
Aerobic growth in dark	–	–	Micro-oxic	Micro-oxic	–

All strains were motile. Organic substrate utilization was tested during photoheterotrophic growth. Pyruvate was utilized by all strains. Arginine, citrate, tartrate and fructose were not utilized by any of the taxa. Major quinones (>70%) were Q–9 and MK–9 in all taxa. +, Substrate utilized or present; –, substrate not utilized or absent; NA, information not available; ND, not determined/not detected; NR, Not required; pABA, para-amino benzoicacid. *, Optimal salt concentration (range of salt concentration tolerated); †, Type strain is considered as 100%; §t, Trace amounts (< 0.5%); *, DHLYDG- Dihydroxylycopene diglucoside; HLYG-hydroxylycopene glucoside; Ly- lycopene; Rp- rhodopin. Data for *Phs. tilakii* (Raj *et al.*, 2012), *Phs. chandramohanii* (Anil Kumar *et al.*, 2009), *Phs. fulvum* and *Phv. molischianum* (Imhoff *et al.*, 1998), and *Phs. oryzae* (Lakshmi *et al.*, 2011) is adopted.

Table 3.7. Carbon Sources /Electron Donors and Nitrogen Sources Utilized by Species of *Phaeospirillum*

Carbon sources/ Electron donors and Nitrogen sources used for growth	*Phs. tilakii*	*Phs. chandra-mohanii*	*Phs. fulvum*	*Phs. molischianum*	*Phs. oryzae*
Acetate	–	(+)	+	+	+
Arginine	–	ND	–	–	–
Aspartate	–	–	+/–	+/–	–
Benzoate	–	–	+	–	–
Butyrate	+	(+)	+	+	–
Caproate	+	(+)	+	+	–
Caprylate	+	+	+	+	–
Citrate	–	ND	–	–	–
Crotonate	+	+	–	+	–
Ethanol	–	–	+	+	–
Formate	–	–	–	ND	–
Fructose	–	ND	–	–	–
Fumarate	–	+	+	+	+
Glucose	–	–	+/–	–	–
Glutamate	–	–	(+)	+/–	–
Glycerol	+ / –	–	–	–	–
Lactate	+	–	–	+/–	–
Malate	(+)/–	+	+	+	–
Mannitol	–	(+)	–	–	–
Methanol	–	–	+/–	–	–
Pelargonate	–	–	+	+	–
Propionate	+	–	+	+	–
Pyruvate	+	+	+	+	+
Succinate	–	–	+	+	+
Tartarate	–	–	ND	–	–
Valerate	–	(+)	+	+	–

Electron donors					
Hydrogen	ND	ND	ND	ND	ND
Sulfide	ND	+	–	–	ND
Sulfur	ND	+	–	–	+
Sulfate	ND	+	+	–	+
Thiosulfate	+	ND	–	–	+
Nitrogen sources utilized	Ammonium chloride	Ammonium chloride, Dinitrogen	Ammonia, Nitrogen	Ammonia, Nitrogen, Yeast extract and vitamin free casamino acids	Dinitrogen, ammonium chloride, glutamate, urea
Reductants used for growth	ND	ND	Ascorbate, thioglycolate	Ascorbate, thioglycolate	ND

Symbols: +, Substrate utilized or present; (-), substrate not utilized or absent; (+), weak growth; ±, strain-dependent reaction; ND, not determined; NA, no data available. Data for *Phs. tilakii* (Raj *et al.*, 2012), *Phs. chandramohanii* (Anil Kumar *et al.*, 2009), *Phs. fulvum* and *Phv. molischianum* (Imhoff *et al.*, 1998), is adopted.

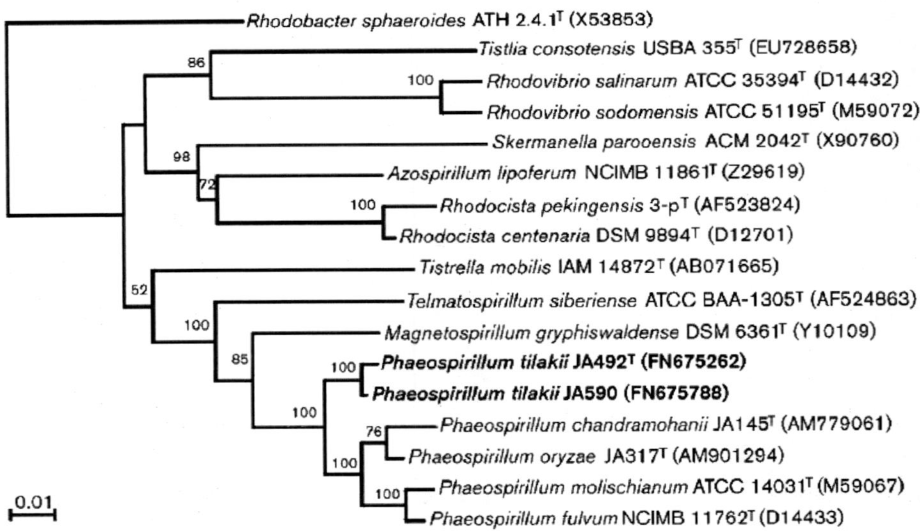

Fig. 3.7A. Phylogenetic tree of *Phaeospirillum* species and related members of the *Rhodospirillaceae* based on 16S rRNA sequences. The sequence of *Rhodobacter sphaeroides* ATH 2.4.1^T was used as an outgroup to root the tree. Bar=1% nucleotide substitution per 100 nucleotides.

The phylogenetic positioning of the *Phaeospirillum* type species is distributed in three sister clades. *Phs. fulvum* and *Phs. molischianum* forms one clade with 99.1% 16S rRNA gene sequence similarity. Another clade encompasses *Phs. chandramohanii* and *Phs. oryzae* which shares 98.2% similarity. The third clade contains *Phs. tilakii* type strain sharing approximately 97% similarity with *Phs. chandramohanii* or *Phs. oryzae* clade and around 96.7%, while *Phs.*

fulvum or *Phs. molischianum* and *Phs. chandramohanii* or *Phs. oryzae* shares approximately 97%. The 16S rRNA gene sequence similarities between *Phs. tilakii* JA 492^T and the other *Phaeospirillum* strains ranged between 96.5% and 97.4%. Besides the neighbors, *Phs .fulvum* DSM 113^T and *Phs. molischianum* ATCC 14031^T share the same branch with a 99.1% 16S rRNA gene sequence similarity.

4. *Phaeovibrio* (*Phv.*) (Brown vibrio) Lakshmi *et al.* (2011)

Cells are vibrioid, motile by polar flagella and multiply by binary fission. Chimeric internal membranes of lamellar stacks and vesicles are present in a single cell (**Fig. 3.8**). Photosynthetic pigments are bacteriochlorophyll *a* and carotenoids of rhodopinal series are present. They are mesophilic and occur in freshwater. Metabolism is strictly anaerobic and obligately phototrophic. A few organic substrates are photo assimilated by these bacteria. Growth factors are required. The major fatty acids are $C_{18:1}\omega 7c$ and $C_{16:0}$. The mol% of the G+C content of genomic DNA is 67.8–68.8 (HPLC). The description of the genus is determined primarily by phylogenetic information from 16S rRNA gene sequences. *Phaeovibrio sulfidiphilus* (Lakshmi *et al.,* 2014) is the type species. Only one species is validly published. The distinguishing characteristics are précised in **Table 3.8.**

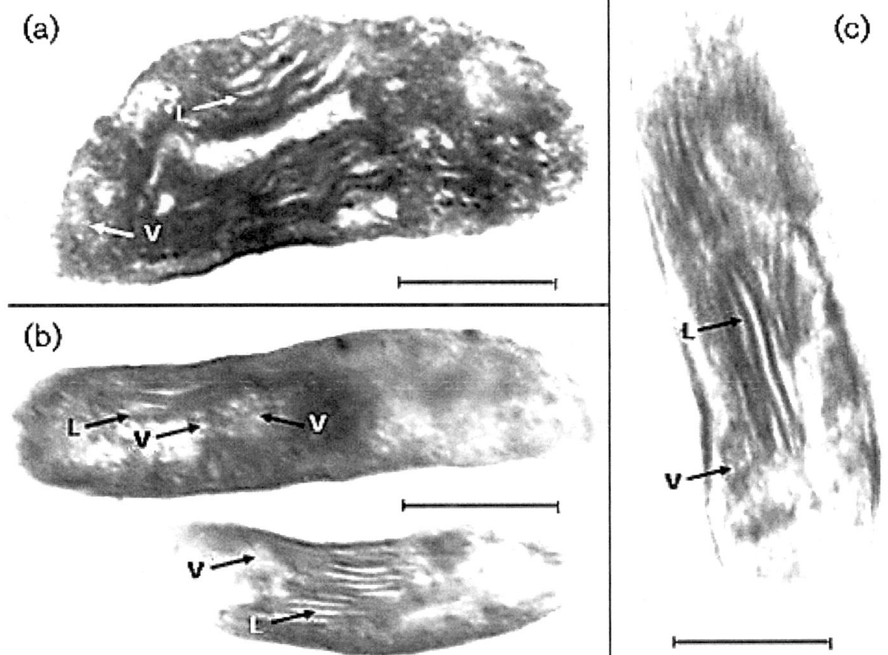

Fig. 3.8. Electron micrographs of ultrathin section of cells of *Phaeovibrio* (Lakshmi *et al.,* 2011), showing chimeric ICM structures. V, Vesicular; L, lamellar. Bars, 333 (a), 300 (b) and 330 (c) nm.

Differentiation of the *Phaeovibrio* from other Genera

Besides 16S rDNA sequence similarities (Lakshmi *et al.,* 2011), morphological characteristics, quinone and fatty acid composition and other properties of *Phaeovibrio* with other genera of *Rhodospirillaceae* members are précised in **Table 3.5**. Similarly, the differential characteristics and utilization of carbon sources by the species of *Phaeospirillum* and *Rhodocista* , *Rhodospira* and *Rhodovibrio* are précised in **Table 3.8** .

Table 3.8. Differentiating Characteristics of Species of *Phaeovibrio* ,*Rhodocista* , *Rhodospira* and *Rhodovibro*

Charact-eristic	***Phv. sulfidophilus***	***Rhodocista centenaria***	***Rcs. pekin-gensis***	***Rhodospira trueperi***	***Rhodovibrio salinarum***	***Rhodovibrio sodomensis***
Habitat	Brackish water	Freshwater, Alkaline hot springs/ Paper factory effluents	Wastewater	Saltmarsh sediments/ laminated microbial mats	Saltern lakes	Salt lakes/ Dead seawater
Cell shape	Vibrioid	Vibroid–Spiral	Vibroid–Spiral	Vibroid–Spiral	Rods-Spiral	Vibroid–Spiral
Cell size (μm)*	0.3-0.5 x 1.2–2.5	0.5-1.0 x 1.0–2.0	0.6-0.8 x 0.8-1.5	0.6-0.8 x 1.5-3.0	0.3–0.5 x 1.0-3.0	0.6-0.7 x 1.6-2.5
Colour	Light brown	Pink	Pink	Beige to Peach	Pink-Red	Pink
Bacterio-chlorophyll	*a*	*a*	*a*	*b*	*a*	*a*
Major carotenoids	Rhodopin	Spirillo-xanthin	Spirillo-xanthin	Tetrahydros pirillo-xanthin	Spirillo-xanthin	Spirillo-xanthin
Major quinones	ND	Q-9	Q-9	Q-7, MK-7	Q-10, MK-10	ND
Temperature range (opti-mum) °C	20-40 (25-30)	39-47 (40-45)	25-44 (31-42)	20-35 (25-30)	20-45 (42)	25-47 (35-40)
pH range (optimum)	7.0-8.0 (7.0)	5.7-7.0 (6.8)	6.5-8.0 (7.0)	7-7.8 (7.3-7.5)	ND (7.5-8.0)	ND (7.0)
NaCl tolera-nce (Optim-um) (%, w/v)	0–0.5	NR	NR	2.0 (0.5–5.0)	12-18 (4.0)	6.0-20 (12.0)
Internal membrane system	Chimeric §	Lamella	Lamella	Vesicles	Vesicles	Vesicles
PHB formation	ND	+	–	ND	ND	+
DNA G+C content (mol%)	67.8–68.8	69.9	68.8	65.7	67.4	66.2-66.6
Growth factor(s)	Biotin, Para-Aminoben-zoic acid	Biotin, vitamin B_{12}	Thiamine, vitamin B_{12}	Biotin, Thiamine, Pantoth-enate	Cobalamine	Cobalamine

Charact-eristic	*Phv. sulfidophilus*	*Rhodocista centenaria*	*Rcs. pekin-gensis*	*Rhodospira trueperi*	*Rhodovibrio salinarum*	*Rhodovibrio sodomensis*
Photolithoautotrophic growth	–	–	ND	–	–	–
Aerobic growth	ND	+	+	–	+	+
Oxidation of sulfide	+	ND	–	+	–	ND
Nitrogen fixation	–	+	–	ND	ND	–
Absortion maxima of living cells (nm)	377-380, 488-491, 524-527, 593, 794 and 863	475-550, 587, 800 and 875	378, 516, 548, 594, 798 and 885	397, 458, 490, 600, 689, 801, 889 and 986	380, 595, 800 and 870	377, 421, 516, 551, 591 and 875-880
Carbon source/electron donor used for growth						
Acetate	+	+	+	+	+	+
Aspartate	ND	+	+	ND	ND	ND
Benzoate	–	ND	–	ND	ND	–
Butyrate	–	+	ND	+	ND	–
Caproate	–	+	ND	ND	ND	ND
Caprylate	–	+	ND	ND	ND	ND
Ethanol	–	ND	–	–	ND	ND
Fructose	–	–	ND	ND	+	–
Fumarate	–	–	ND	+	+	ND
Glucose	–	–	–	ND	+	–
Glutamate	–	+	+	ND	+	ND
Glycerol	–	ND	+	ND	+	–
Lactate	–	+	+	+	+	+
Malate	–	–	–	+	+	+
Mannitol	–	ND	–	ND	+	ND
Methanol	–	ND	ND	ND	ND	ND
Pyruvate	+	+	+	+	+	+
Propionate	–	ND	ND	+	+	–
Succinate	+	–	+	+	+	+
Valerate	–	+	ND	+	ND	ND
Sulfur sources	Sulfide	ND	Sulfate	Sulfide	ND	ND
Nitrogen sources utilized	Ammonium salts and Glutamine	Ammonia, Nitrogen, Glutamate, Aspartate and Alanine	Ammonium salts and Glutamate	ND	ND	Ammonia, Yeast extract

All strains are motile; Organic substrate utilization was tested during photoheterotrophic growth; Pyruvate and acetate were utilized by all strains; arginine and citrate were not utilized by any of the strains; Symbols: +, Substrate utilized/Present; –, Substrate not utilized /absent; (+), Weak growth; NR, Not required; ND, Not determine/No data available; For spiral cells, dimensions are length and width of one complete turn of the spiral; §, Lamellar Stacks and vesicles present in a single cell.Data for *Phaeovibrio* (Lakshmi *et al.,* 2011), *Rhodocista* (Kawasaki *et al.,* 1994), *Rhodospira* (Pfennig *et al.,* 1998) and *Rhodovibrio* (Imhoff *et al.,* 1998) is adopted.

Taxonomic Comments

Phaeovibrio sulfidophilus branches separately from other *Rhodospirillaceae* genera sharing the highest similarities of the 16S rRNA gene sequences sharing approximately 91-92% similarity. The phylogenetic relationships of these bacteria based on 16S rRNA sequences are shown in **Fig. 3.8A**. The highest sequence similarities are found with *Rhodospirillum* species such as *Rsp. sulfurexigens* (92.2%), *Rsp. rubrum* (91.5%) and *Rsp. photometricum* (91.4%). Whereas similarities to other species of the genera include *Rhodocista, Phaeospirillum*, *Rhodovibrio, Rhodospira* and *Roseospira* was only 80–91% (Lakshmi *et al.,* 2011).

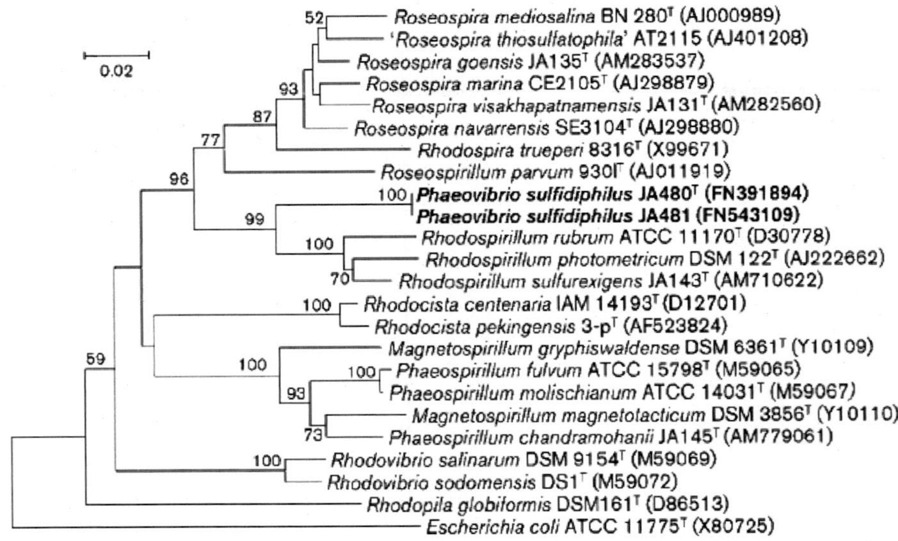

Fig. 3.8A. Phylogenetic tree of *Phaeovibrio* species and related *Alphaproteobacteria* of the order *Rhodospirillales* based on 16S rRNA sequences. The sequence of *E.coli* was used as an outgroup to the root of the tree. Bar=2 substitution per 100 nucleotide positions.

5. *Rhodocista* (*Rcs.*) (Red basket) (Kawasaki *et al.*, 1994) Imhoff (2015)

Cells are vibrioid to the spiral shape, 3.0 μm long and 1.0 to 2.0 μm wide. Motile by means of long polar flagellum and are Gram (-) negative. The internal photosynthetic membrane is lamellae lying parallel to the cytoplasmic membrane. Growth is possible photoheterotrophically under anoxic conditions in the presence of light and also shows chemotrophic growth under micro-oxic conditions in the dark. Photosynthetic pigments are bacteriochlorophyll *a* and carotenoids of the spirilloxanthin series. The cells are converted to thick-walled cysts which are resistant to desiccation and heat(**Fig. 3.9**) (Zhang *et al.,* 2003), when aerobically grown on agar media. The major quinone is Q-9. The G+C content of the DNA is 69-70 mol%. *Rhodocista centenaria* is the type species

(Kawasaki *et al.,* 1994). *Rhodocista pekingensis, Rhodocista centenaria* is the validly published species. The distinguished characteristics and utilization of carbon sources or electron donors by the two species are précised in **Table 3.8.**

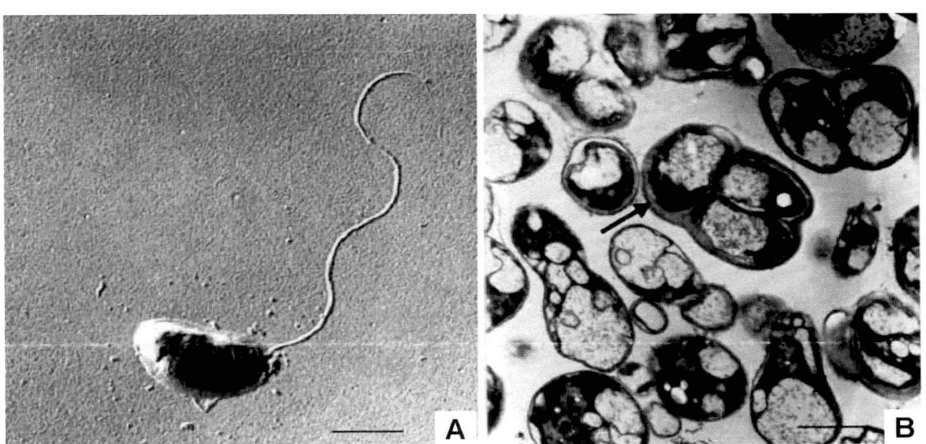

Fig. 3.9. Electron micrographs showing the morphology and cysts of *Rhodocista pekingensis*. A. Top, single polar flagellum; B, cysts formed during cultivation (Bars= 1 μm).

Differentiation of the *Rhodocista* from other Genera

The differential characteristics of *Rhodocista* with other genera of *Rhodospirillaceae* members are précised in **Table 3.5** .

Taxonomic Comments

The 16S rDNA gene sequence analysis of the genus *Rhodocista* indicated that it forms a distinct phylogenetic branch of *Rhodospirillaceae*. In fact, the phylogenetic positioning of spiral-shaped purple non-sulfur bacteria on the basis of 16S rDNA gene sequences indicates the heterogeneity of *Rhodospirillum* which is clarified by Kawasaki *et al.* (1992). The *Rsp. centenum* was transferred to *Rhodocista centenaria* (Kawasaki *et al.*, 1992) based on recognition of great genetic distance between *Rhodospirillum rubrum* and *Rsp. centenum.* The phylogenetic relationships of these bacteria based on 16S rDNA sequences are shown in **Fig. 3.5** and **3.9A** (Zhang *et al.*, 2003). The phylogenetic positioning of the *Rhodocista* species indicates a close cluster with *Azospirillum irakense* and *Azospirillum amazonense* but quite far away from the other *Azospirillum* species (Zhang *et al.*, 2003). This very close relationship was also exhibited by the second species *Rhodocista pekingensis* . *Rhodocista* species and *Azospiriilum irakense* and *Azospirillum amazonense* share about 96-97% sequence similarity. However, the *Azospiriilum irakense* is not phototrophic.

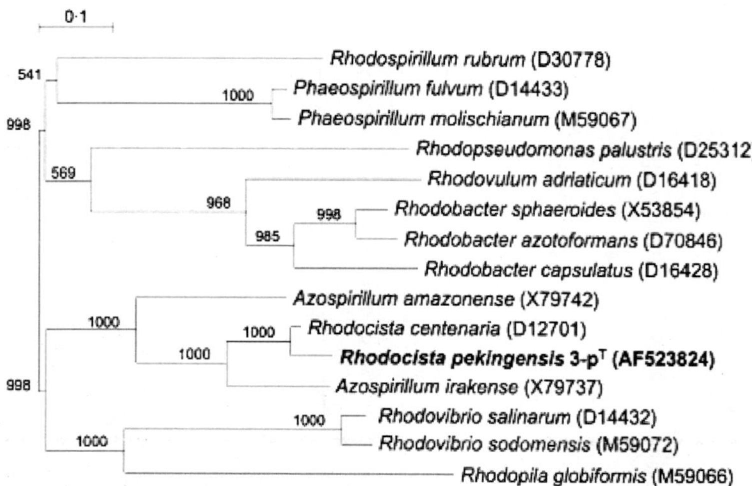

Fig. 3.9A. Phylogenetic tree of *Rhodocista* species and related members of the family *Rhodospirillaceae* based on 16S rDNA sequences. The sequence of *E. coli* was used as an outgroup to root the tree. Bar=0.1% sequence divergence.

6. *Rhodospira* (*Rsa.*) (The rose spiral) (Pfennig *et al.,* 1998) Imhoff (2015)

Cells are vibrioid to spirilloid shaped **(Fig. 3.10)** multiply by binary fission. Motile by means of bipolar tufts of flagella, Gram (-) negative. The intracytoplasmic membrane system is of vesicular type, contains bacteriochlorophyll *b* and carotenoids of spirilloxanthin series as photosynthetic pigments. In vivo absorption maxima of bacteriochlorophyll with longer wavelength was observed at 986 nm. Growth occurs preferably as photoorganotrophically under anoxic conditions in the light but is also possible under microoxic conditions in the dark. Growth factors are required. Sulfate is required for growth which is further oxidized to elemental sulfur. Optimal growth occurs at a temperature of 25–30°C and pH 7.3–7.5 at 2% NaCl. Growth ranges are 0.5–5 % NaCl, pH 7.0–7.8 and 20–35°C. Microbial mats in a salt marsh are suitable habitats for their growth. According to the 16S rDNA sequence of the type strain of its type species. *Rhodospira trueperi* (Pfennig *et al.,* 1998) is the type species. Single species is validly published and its characters are précised in **Table 3.8.**

Differentiation of the *Rhodospira* from other Genera

The members of *Rhodopsira* species are characterized by salt requirement specifically of marine bacteria, the requirement for sulfide as a reduced sulfur source, uncommon absorption spectra due to the presence of bacteriochlorophyll *b* and as well as ubiquinones (Q-7) and menaquinone (MK-7) with seven isoprene units and carbon sources utilized (**Table 3.8**). The differential characteristics of *Rhodospira* from other genera of *Rhodospirillaceae* members are précised in **Table 3.5** .

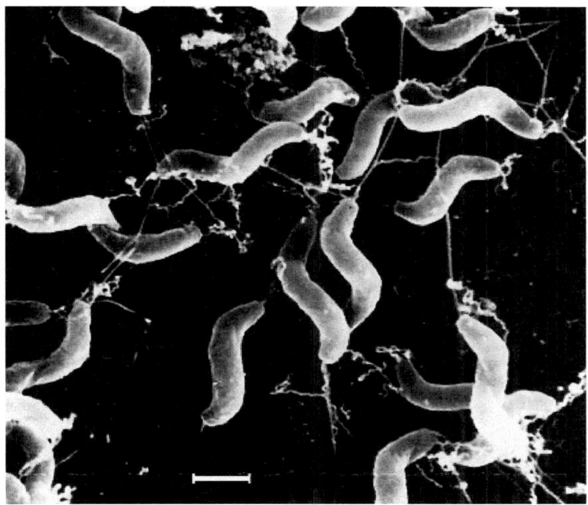

Fig. 3.10. Scanning electron micrograph showing the cell shape of *Rhodospira* (Bar=1 μm).

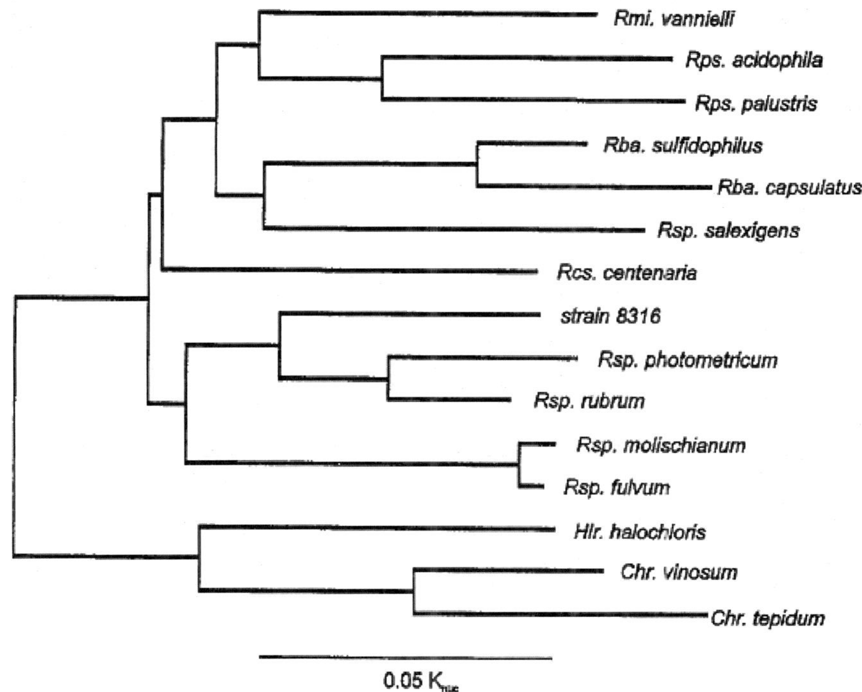

Fig. 3.10A. Phylogenetic tree of *Rhodospira* species and related members of the family *Rhodospirillaceae* based on 16S rRNA sequences. The relatedness of *Rhodospira trueperi* strain 8316 to other phototrophic purple bacteria is shown (Pfennig *et al.,* 1997).

Taxonomic Comments

The phylogenetic relationship of these bacteria based on 16S rDNA and 16S rRNA sequences are illustrated in **Fig. 3.5** and **3.10A** (Pfennig *et al.,* 1997). Phylogenetic positioning of the *Rhodospira* signifies its close relationship with the genus *Roseospira* with 16S rRNA gene sequence similarity of about 93-94%. *Rhodospira trueperi* strain 8316 also showed the closest sequence similarity with *Rsp. rubrum* (92.8%) and *Rsp. photometricum* (90.6%) which is widely accepted for differentiating between genera. Based on fatty acid and lipid composition, unique pigment composition, reduced sulfur compounds, the requirement of salt for optimal growth strongly suggests placing this bacterium in a new genus and a new species for which the name *Rhodospira trueperi* is proposed (Pfennig *et al.,* 1997).

7. *Rhodovibrio* (*Rhv.*) (Rose *vibrio*) (Imhoff *et al.*, 1998) Imhoff (2015)

Cells are vibrioid- to spiral-shaped, 0.6-0.9 μmwide, motile by means of polar flagella (**Fig. 3.11 A & B**), multiply by binary fission and are Gram (-) negative. Intracytoplasmic photosynthetic membranes are present as vesicles contain photosynthetic pigments, bacteriochlorophyll *a* and carotenoids of the spirilloxanthin series. Contain ubiquinones (Q-10) and menaquinones(MK-10) with 10 isoprenes units as major components. Major cellular fatty acids are $C_{18:1}$ and $C_{18:0}$. Halophilic bacteria that require NaCl or sea salt for growth. Have salt optima above seawater salinity and tolerate up to 20% total salts. Growth occurs preferably photoheterotrophically under anoxic conditions in the presence of light, chemotrophic growth is also possible under microoxic to oxic conditions in the dark. Complex nutrients are required. Poly-β-hydroxybutyrate produced as storage polymer (**Fig. 3.11C**). They have a G+C content between 65 and 69 mol%. *Rhodovibrio salinarum* (Imhoff *et al.,* 1998) (type species) and *Rhodovibrio sodomensis* are validly published species and their characteristics are presented in **Table 3.8.**

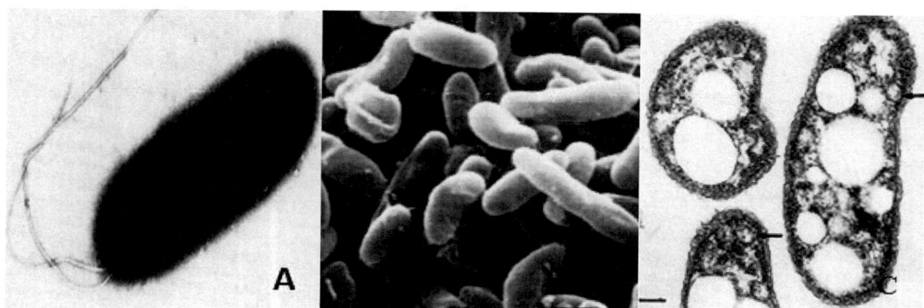

Fig. 3.11. A.Scanning electron micrograph of cell of *Rhodovibrio salinarum* with flagella (16000x). B. Scanning electron micrograph of *Rhodovibrio sodomensis* (Bar = 1 μm), C. Transmission electron micrograph, arrows showing large droplets of PHB granules and occasional chromatophores in *Rhodovibrio sodomensis* (Bar = 0.5 μm).

Differentiation of the *Rhodovibrio* from other Genera

The important differentiating properties for *Rhodovibrio* and other phototrophic spiral-shaped *Rhodospirillaceae* members are given in **Table 3.5** .

Taxonomic Comments

The genus *Rhodovibrio* contains two species *Rhodovibro salinarum*, a type species and *Rhv. sodomensis* (Imhoff *et al.*, 1998). Based on their shape and phototrophic capacity, *Rhodovibrio* species were originally consigned to the genus *Rhodospirillum* (Nissen and Dundas, 1984; Mack *et al.*, 1993). Based onthe identification of high phenotypic and chemotaxonomic diversity of spiral-shaped phototrophic *Alphaproteobacteria* and the great genetic distance between most of these bacteria, a rearrangement of the species of this group became essential. Many distinct differences among *Rhodovibrio* species and *Rhodospirillum rubrum,* led to their reclassification in the new genus *Rhodovibrio* (Imhoff *et al.*, 1998). The genus name *Rhodovibrio* has been originally proposed by Molisch (1907) for bacteria *Rhodovibrio parvus* later recognized as belonging to *Rhodopseudomonas palustris* as this name was not included in the Approved List of Names, which does not have standing nomenclature and could be reviewed (Imhoff *et al.*, 1998). The phylogenetic relationships of *Rhodovibrio* from other genera based on 16 rDNA sequences are illustrated in **Fig. 3.5** (Imhoff *et al.,* 2005).

The 16S rDNA sequence analysis of *Rhodospirillum rubrum* ATCC11170^{T} , *Pararhodospirillum photometricum* strains DSM122^{T} and E11, *Rhodovibrio sodomensis* strain ATCC51195^{T} and *Rhodovibrio salinarum* strain ATCC35394^{T} resulted in the separation of the last two species into a separate phylogenetic clade and the proposal of new genus *Rhodovibrio* (Imhoff *et al.*, 1998) is justified.

8. *Roseospira* (*Ros.*) (Rosy spiral) (Imhoff *et al.,* 1998) Imhoff and Gorlenko (2015)

Cells are vibrioid to spiral-shaped, 0.4-1.0 μm wide, motile by means of polar flagella, multiply by binary fission and are Gram (-) negative **(Fig. 3.12)** . Intracytoplasmic photosynthetic membranes are present as vesicles, contain bacteriochlorophyll *a* and carotenoids of spirilloxanthin series as photosynthetic pigments. Halophilic bacteria that require NaCl or sea salt for growth. Have salt optima above seawater salinity and tolerate up to 15% total salts. Growth occurs preferably photoheterotrophically under anoxic conditions in the presence of light and chemotrophic growth is also possible under micro-oxic conditions in the dark. Growth factors are required. Thane a G+C content between 65 and 68 mol%. *Roseospira mediosalina* (Imhoff *et al.,* 1998) is the type species. Warm sulfur springs are the selective habitats for their isolation. Six species are validly published. All six described species of the genus can be distinguished by the following characteristics (**Table 3.9** .)

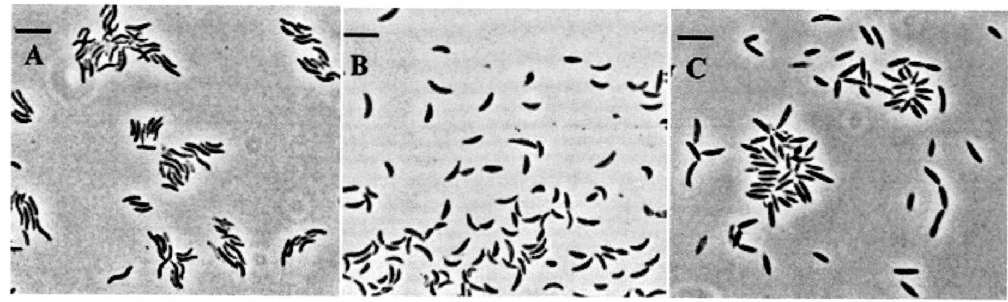

Fig. 3.12. Phase-contrast photomicrographs of *Roseospira marina* (A), *Ros. navarrensis* (B) and *Ros. thiosulfatophila* (C).

Table 3.9. Differentiating Characteristics of species of *Roseospira* and *Roseospirillum*

Characteristic	***Roseospira marina***	***Roseospira navarrensis***	***Roseospira thiosulfatophila***	***Roseospira mediosalina***	***Roseospira visakhapatnamensis***	***Roseospira goensis***	***Roseospirillum parvum***
Cell morphology	Spiral-Vibrioid	Vibrioid-Rod, pairs	Spiral-Vibrioid	Vibrioid-Spiral	Vibrioid	Vibrioid-crescent shaped	Vibrioid -Spiral
Cell size (μm)	0.4–8.0 x 1.5–6.0	0.6–0.9 x 3.5–6.5	0.5–0.8 x 2.5–6.5	0.8–1.0 x 2.2–6.0	0.5–0.9 x 2.0–6.0	0.8–1.0 x 3.0–8.0	0.4-0.6 x 1.8-2.6
Motility	+	+	+	+	–	+	+
Flagellation	Bipolar tufts of flagella	Bipolar tufts of flagella	A polar tuft of flagella	A polar tuft of flagella	Non-motile	Pair of monopolar flagella	Bipolar flagella
Colour	Red	Brown–Red	Red	Brown–Red	Red–Brown	Red–Brown	Pink-Pinkish red
Bacteriochlorophyll	*a*	*a*	*a*	*A*	*a*	*a*	*a*
Major carotenoid(s)	Rhodovibrine, Rhodopine	Rhodopine, Spirilloxanthin	Rhodovibrine, Spirilloxanthin	Rhodopine, Lycopene	Rhodovibrine	Rhodovibrine	Spirilloxanthin, Lycopenal
Absortion maxima of living cells (nm)	375, 500, 530, 589, 800 and 863	377, 501, 530, 590, 801 and 863	378, 501, 530, 590, 798, 883 and 909	379, 474, 504, 534, 591, 803 and 860	371, 473, 503, 540, 590, 803, 847 and 900	377, 473, 503, 540, 590, 803 and 857	380, 492, 515, 549, 595, 806 and 911
Shoulder peak at 909 nm	–	–	+	–	+	–	+
Optimal NaCl concentration (Range) (%, w/v)	2.0–4.0 (0.5-10)	3.0–4.0 (1.0-10)	0.5 (0.2-5.0)	4.0–7.0 (0.5-15)	2.0 (1.0-5.0)	1.0–3.0 (1.0-5.0)	1.0-2.0 (1.0-6.0)

Characteristic	*Roseospira marina*	*Roseospira navarrensis*	*Roseospira thiosulfatophila*	*Roseospira mediosalina*	*Roseospira visakhapatnamensis*	*Roseospira goensis*	*Roseospirillum parvum*
Optimal temperature (°C)	30–35	30–35	30–35	30–35	30	30	30
Optimum pH for growth (Range)	6.7–6.8 (5.3–8.4)	6.8–7.0 (6.0–8.5)	6.8–7.0 (5.6–8.6)	7.0	7.0	7.5	7.9
Internal membrane system	Vesicles	Vesicles	Vesicles	Vesicles	Vesicles	Vesicles	Lamellae
DNA G+C content (mol%)	68.8–69.4	66.8	71.9–72.3	66.6	67.0	71.0	71.2
Growth factor (s)	N,T, PABA	YE	YE	N,T, PABA	N,T, PABA	N,T, PABA	Cobalamine/ Complex
Assimilatory SO_4^{2-}	+	–	–	+	–	–	+
Dark aerobic growth	+	μ	μ	μ	+	+	μ
Nitrogen fixation	+	+	+	+	-	-	ND
Photolithoautotrophic growth$	–	H_2S	H_2S, S_2O_3	H_2S	–	–	H_2S, S_2O_3
Carbon source/electron donor used for growth							
Acetate	+	+	+	+	–	(+)	+
Aspartate	+	+	+	+	+	–	ND
Benzoate	–	+	–	–	–	–	ND
Butyrate	+	+	+	+	–	+	+
Casamino acids, Y.E	+	+	+	ND	–	–	+
Citrate	–	+	–	–	–	–	ND
Crotonate	+	+	+	ND	–	–	ND
Cysteine	(+)	ND	ND	ND	+	–	ND
Formate	(+)	–	–	–	–	(+)	ND
Fructose	+	–	–	–	–	–	+
Fumarate	+	+	-	+	–	(+)	+
Glucose	-	-	+	-	–	(+)	+
Glutamate	+	+	+	+	–	-	+
Glycerol	+	+	+	+	–	+	ND
Lactate	+	+	–	+	–	(+),+,–	+
Malate/ Succinate	+	+	–	+	+	–	+

Characteristic	*Roseospira marina*	*Roseospira navarrensis*	*Roseospira thiosulfatophila*	*Roseospira mediosalina*	*Roseospira visakhapatnamensis*	*Roseospira goensis*	*Roseospirillum parvum*
Mannitol	+	+	-	-	–	+	ND
2–Oxoglutarate	+	+	+	ND	-	+	+
Propionate	+	+	–	+	–	–	+
Sulfide	–	+	+	+	-	–	+
Thiosulfate	–	–	+	–	–	–	+
Valerate	+	+	+	–	–	+	+
Habitat	Anoxic sediments, water from coastal and environments	Microbial mats in coastal and marine environments	Anoxic sediments exposed to light in inland saline springs	Marine	Marine water	Marine salterns	Marine sediments

Symbols: +, Substrate utilized or present; –, substrate not utilized or absent; (+), weak growth; m, growth under micro–oxic conditions; ND, not determined. Pyruvate was utilized by all the strains. Caproate, caprylate,glycolate, tartrate, methionine, ethanol and propanol were not utilized by any of the strains. ND, No data available; NR, not reported. PABA, p–Aminobenzoic acid; N, Niacin; T, thiamine; YE, Yeast Extract.*Organic substrate utilization was tested during photoheterotrophic growth. $ Photolithotrophic growth by H_2S, Hydrogen sulfide; S_2O_3, Thiosulfate. Data for *Ros. marina*, *Ros. navarrensis*, *Ros. thiosulfatophila* is taken from Guyoneaud *et al.* (2002); Data for *Ros. mediosalina* is taken from Imhoff *et al.* (1998); and *Ros. visakhapatnamensis* and *Ros. goensis* is taken from Kalyan Chakravarthy *et al.* (2007). Data for *Roseospirillum parvum* is adopted from Glaeser and Overmann (2001).

Differentiation of the *Roseospira* from other Genera

The major differentiating properties for *Roseospira* from other genera of *Rhodospirillaceae* members are presented in **Table 3.5** .

Taxonomic Comments

Based on their spiral shape and phototrophic capacity, these species were originally consigned to the genus *Rhodospirillum*. However, DNA homology to *Rsp. rubrum* was insignificant (Kompantseva and Gorlenko, 1984). Based on the identification of high phenotypic and chemotaxonomic diversity of spiral-shaped phototrophic *Alphaproteobacteria* and the great genetic distance among these bacteria (Kawasaki *et al.*, 1993b) led to the rearrangement of the species of this group became essential (Imhoff *et al.*, 1998). The great differences between *Roseospira* and *Rhodospirillum* led to their reclassification (Imhoff *et al.*, 1998). The phylogenetic relationships of these bacteria from other genera based on 16S rDNA and 16S rRNA sequences are illustrated in **Figs. 3.5** and **3.12A.**

Comparative 16S rDNA sequence analysis of *Roseospira marina* CE2105, *Ros. navarrensis* SE3104, *Ros. thiosulfatophila* AT2115, *Ros. mediosalina* ,*Ros. visakhapatnamensis* JA131 and *Ros. goensis* JA135 showed that these strains formed a subgroup together with *Rhodospira trueperi* within the *Rhodospirill-*

aceae family of the *Alphaproteobacteria*, which is well separated from *Rhodospirillum* genus and their closest relatives. *Roseospira marina* CE2105, *Ros. navarrensis* SE3104 and *Ros. thiosulfatophila* AT2115 have a similar salt requirement that is phylogenetically correlated with 16S rDNA similarity ranging from 97.6% to 96.5%, whereas *Roseospira mediosalina* requires a higher optimal salt concentration and similarity ranging from 95.6% to 94.7% in comparison with the former three species (Guyoneaud *et al.*, 2002). Comparison of 16S rDNA sequence of *Roseospira visakhapatnamensis* JA131 and *Ros. goensis* JA135 showed that they are closer to *Ros. navarrensis* (95.9%), *Ros. marina* (95.5%), *Ros. mediosalina* (94.2%) and *Ros. thiosulfatophila* (96.1%). Sequence similarity between *Ros. visakhapatnamensis* JA131 and *Ros. goensis* JA135T is 96.6% (Chakravarthy *et al.*, 2007).

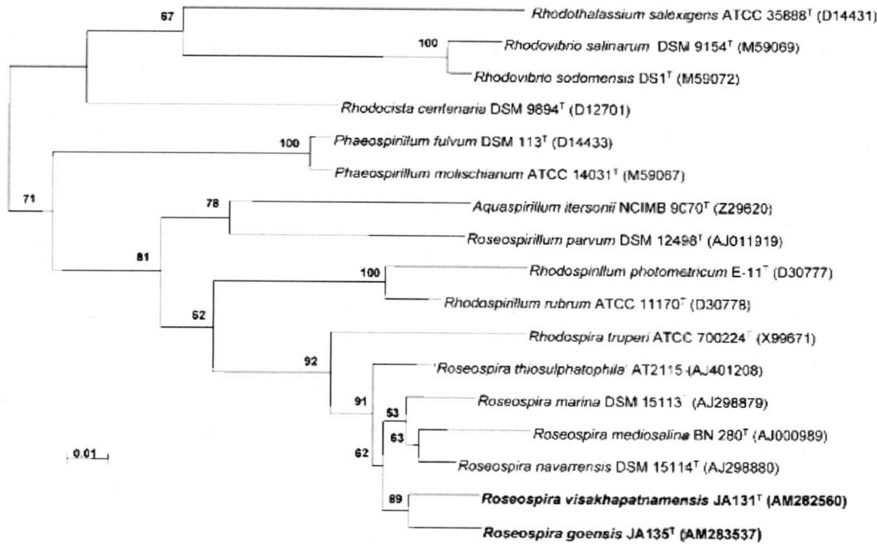

Fig. 3.12A. Phylogenetic tree of *Roseospira* species and related members of the family *Rhodospirillaceae* based on 16S rRNA sequences. Bar=1 nucleotide substitution per 100 nucleotides (Chakravarthy *et al.,* 2007).

DNA-DNA hybridization (DDH) studies of *Roseospira marina* CE2105, *Ros. navarrensis* SE3104 and *Ros. thiosulfatophila* AT2115 showed low homologies between them and supported the separation into three species (Guyoneaud *et al.*, 2002). No data are available for other species or genera.

Family 2. *Acetobacteraceae* (Gillis and De Ley, 1980) Sievers and Swings (2015)

The members of the family *Acetobacteraceae* are taxonomically included in the order *Rhodospirillales* of the class *Alphaproteobacteria* and include 32 genera are validly published. They are Gram (-) negative, acetic acid bacteria, has undergone many taxonomic changes in respect to their genus status. Several species of acetic acid bacteria were newly described. Relationships among the acetic acid bacteria were studied genotypically by DNA-rRNA hybridization. DNA-DNA hybridization and ribosomal RNA gene sequences (5S rRNA, 16SrRNA and 23S rRNA). *Gluconacetobacter ,Acidomonas, Asasia* as well as *Gluconobacter* species. They possess Q-10 ubiquinone as the major quinone component. The Q-10 equipped species contain besides minor amounts of ubiquinone of the Q-9 type (Gillis and GDe Ley, 1980). The ubiquinone of *Acetobacter pasteurianus ,Acetobacter pomorum* and *Acetobacter aceti* is Q-9 along with minor components of ubiquinone of Q-8. *Acetobacter* is the type genus. Among 32 genera only 2 genera are considered as phototrophic purple non-sulfur bacteria and are discussed in this chapter. The detailed description and key to the identification of these two genera are given below.

Key to the Genera of *Acetobacteraceae*

1. Cells are spherical to ovoid-shaped (cocci), motile utilizing polar flagella. Cell division by binary fission, internal photosynthetic membranes as vesicles containing bacteriochlorophyll *a* and grow mostly photoheterotrophically under anaerobic conditions in light. They grow at low pH (pH 4.8-5.0). .... ***Rhodopila***
2. Cells are ovoid to rod-shaped, motile by means of polar flagella. Cell division by binary fission, internal photosynthetic membranes as vesicles; containing bacteriochlorophyll *a* and carotenoids of spirilloxanthin series. Facultative phototrophs grow anaerobically in light and aerobically in dark. .... ***Rhodovastum***

1. *Rhodopila* (Red-spherical bacterium) (Imhoff *et al.,* 1984) Madigan and Imhoff (2015)

Formerly this species (*Rhodopila globiformis*) was included in the genus *Rhodopseudomonas.* Cells are spherical to ovoid **(Fig. 3.13)** (Pfennig, 1974), 1.6-1.8 μm in diameter under optimal growth conditions and motile utilizing polar flagella. Divide by binary fission and the cultures are purple-red. They have vesicular intracytoplasmic membranes and the photosynthetic pigments are bacteriochlorophyll *a* and carotenoids. Ubiquinone (Q-9/10), rhodoquinone (RQ-9/10) and menaquinone (MK-9/10) with 9 and 10 isoprene units are present. Cells grow preferably photoheterotrophically under anaerobic conditions in the presence of light. Not able to use most of the carbon sources utilized by other purple non-sulfur bacteria. Cells are sensitive to oxygen but grow under microaerobic conditions in the dark. They grow only at low pH and the optimum pH is 4.8-5.0. Biotin and para-aminobenzoic acid are required as growth factors. Ammonia is the best nitrogen source, dinitrogen, urea and a number of amino acids can also serve as the sole nitrogen source. Nitrate is not utilized. Ammonia is assimilated via glutamine synthetase or glutamate synthase reaction (NADH-dependent). Growth occurs with sulfate as a sole sulfur source at low concentrations. Sulfate is reduced via adenosine-5'-phosphosulfate. Cysteine and thiosulfate are good sulfur sources, but sulfite and sulfide are growth inhibitory even at low concentrations. The only isolate was obtained from warm acidic sulfur springs in Yellowstone National Park (Wyoming, USA). The genus is monotypic with the publication of *Rhodopila globiformis* (Pfennig, 1974). Characteristics of the species of *Rhodopila* are precised in **Table 3.10** .

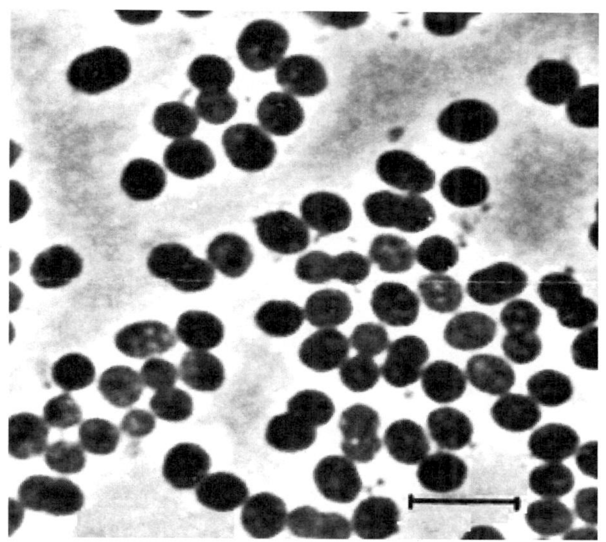

Fig. 3.13. Phase-contrastmicrograph of *Rhodopila globiformis* (Bar= 5μm).

Table 3.10. Differentiating Characteristics of Species of *Rhodopila* and *Rhodovastum*

Characteristics	***Rhodopila globiformis***	***Rhodovastum atsumiense***
Habitat	Sulfur spring/Acidic freshwater	Paddy soil
Cell shape	Cocci	Ovals, rods
Cell diameter (μm)	1.6–1.8	2.0–3.0
Motility	+	+
Near IR peak for BChl *a*	865	844
Absortion maxima of living cells (nm)	378, 504, 813, 862-865 and 890	375, 465, 492, 529, 592, 804 and 844
Colour of cell suspensions	Purple–Red	Red–Brown
Carotenoids	ND	Spirilloxanthin
Quinones	Q–9/10, RQ-9/10, MK–9/10	Q–10, RQ–10
Internal membrane system	Vesicular	Vesicular
Cytochrome *c* size	Small	ND
DNA G+C content (mol%)	66.3 (Bd)	67.8 (HPCL)
Anaerobic growth in light	+	+

Characteristics	*Rhodopila globiformis*	*Rhodovastum atsumiense*
Aerobic growth in dark	–	+
Temperature range (°C) (Optimum)	25-35 (30-35)	20-40 (30-35)
pH range (Optimum)	4.2-6.5 (4.2-6.5)	5.0-8.5 (6.0-6.5)
Vitamins required	Biotin, PABA	Biotin, PABA
Carbon sources/Electron donors used for growth		
Acetate	–	+
Aspartate	–	+
Asparagine	–	+
Benzoate	–	-
Butyrate	–	-
Caproate	–	+
Casamino acids	–	+
Ethanol	+	+
Glycerol	–	–
Glutamate	–	+
Glutamine	–	+
Glucose	+	+
Lactose	–	-
Lactate	–	+
Malate	+	+
Mannose	–	+
Pyruvate	+	+
Succinate	+	+
Sucrose	–	-
Sorbitol	–	+
Tartarate	+	–
Peptone (0.05%)	+	+
Valerate	–	+

Symbols: +, Present /utilized; –, absent/not utilized; ND, not determine; PABA,Para-aminobenzoic acid; Q-9/10, ubiquinone with 9/10 isoprene units; RQ-10, rhodoquinone with 10 isoprene units; MK-9/10, menaquinone with 9/10 isoprene units; Bd- Buoyant density. 1.The substrates utilized by all species are fructose, glucose, mannitol, ethanol, pyruvate, gluconate, fumarate, malate and yeast extract. 2. The substrates utilized by none of the species are formate, propionate, butyrate, caprylate, citrate, benzoate, rhamnose, lactose, sucrose, ducitol, methanol and leucine. Data for *Rhodpila* and *Rhodovastum* is adopted from Imhoff *et al.* (1984) and Okamura *et al.* (2018).

Differentiation of the *Rhodopila* from other Genera

The genus *Rhodopila* is characterized by a single species, *Rhodopila globiformis* which differs from all other APB of *Alphaproteobacteria* in its outstanding physiological and chemotaxonomic properties and phylogenetic position based on 16S rDNA sequence analysis. The intracytoplasmic membrane structure is similar to that of *Rhodobacter* species, but a number of significant differences exist between these two genera. *Rhodopila* reduces adenosine 5'- phosphosulfate instead of 3'-phosphate adenosine 5'-phosphosulfate. It has different quinone, carotenoid and polar lipid composition and has a small cytochrome C_2. It stands apart from *Rhodobacter* and *Rhodopseudomonas* species based on DNA-rRNA hybridization studies and 16s rDNA analyses. It has in common with *Rhodomicrobium vannielii* in the reduction of sulfate.

First and foremost, acidophilic (Pfennig, 1974) and unique carotenoid composition (Schmidt, 1978) characterize *Rpi. globiformis*. However, cytochrome c_2 structure (Dickerson, 1980b), sulfate assimilation via adenosine–5'-phosphosulfate (Imhoff, 1982), quinone and polar lipid composition (Imhoff and Bias Imhoff, 1995), presence of highly potential iron-sulfur protein (HiPIP) and 16S rDNA sequence analysis of *Rpi. globiformis* shows a clear relationship to both phototrophic and chemotrophic species. A particularly close relationship exists to the acidophilic chemotrophic bacteria including *Acidiphilum* species **(Fig 3.5).** This suggests that the acidophilic lifestyle of *Rpi. globiformis* is a fundamental physiological property rooted in its phylogenetic ties to other acidophilic bacteria. The close relationship to *Acidiphilum* species is of particular interest because these bacteria contain an extraordinary bacteriochlorophyll with Zn instead of Mg as the chelated metal ion (Wakao *et al.*, 1996). Due to the presence of a small type of cytochrome C_2, substantial amounts of a high-potential iron-sulfur protein, and the presence of unusual arninolipids as major components of the polar lipids. Therefore, *Rhodopila* is significantly different from all other genera and should be recognized as a separate genus .

Taxonomic Comments

Several distinct characteristics of *Rpi. globiformis* warrant its treatment as a separate genus (Imhoff *et al.*, 1984). This species was originally described as *Rhodopseudomonas globiformis* (Pfennig, 1974) and later reclassified as *Rpi. globiformis*, due to their unique physiological, chemotaxonomic and phylogenetic characteristic (Imhoff *et al.*, 1984).

2. *Rhodovastum* (The vast rose) Okamura *et al.* (2018)

Cells are ovoid to rods shaped (**Fig. 3.14**) (Okamura *et al.,* 2009), multiply by binary fission. Motile utilizing a single polar flagellum and are Gram (-) negative. Intracytoplasmic membranes are of the vesicular type with photosynthetic pigments as bacteriochlorophyll *a* and carotenoids of the normal spirilloxanthin series. Phototrophic cell cultures are red to brown in colour. Mesophilic, slightly acidophilic and non-halophilic. Facultative phototrophs are able to grow anaerobically in the presence of light and aerobically in the dark. Photoorganotrophic growth also occurs with simple organic compounds. Sulfide may not be utilized. The major cellular fatty acids are $C_{18:1}\omega7c$. 3-OH $C_{18:0}$.

Ubiquinone (Q-10) and rhodoquinone (RQ-10) with 10 isoprene units are the major quinones. The phylogenetic position is in the family *Acetobacteriaceae.* The genus is monotypic with single validly published species *Rhodovastum atsumiense* (Okamura *et al.,* 2009) is the type species. The characteristics of this species are given in **Table 3.10.**

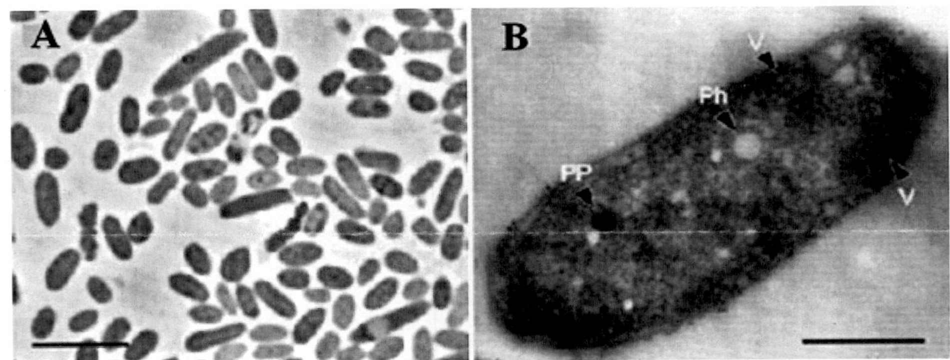

Fig. 3.14. Cell morphology and ultrastructure of *Rhodovastum atsumiense* (A) Phase-contrast micrograph of intact cells (Bar = 10 μm); (B) electron micrograph of ultrathin section of cell (Bar = 1.0 μm). Arrows with letters V, Ph and PP indicate vesicular intracytoplasmic membranes, poly-β-hydrobutyric acid like body and polyphosphate like granule respectively.

Differentiation of the *Rhodopila* from other Genera

The phenotypic and genotypic characteristics of *Rhodovastum* with other genera are précised in **Table 3.10**. The only species of this genus *Rhodovastum astumiense* is characterized with the largest cell diameter among other purple non-sulfur bacteria so far described. This species also can be differentiated from *Acs. rubrifaciens* and *Rpi. globiformis* in a number of phenotypic characters such as colour of the cell suspension, bacteriochlorophyll a absorption maxima at the near-infrared region, optimum pH, carbon sources and quinone composition. Based on the above properties this species was reclassified into a separate genus.

Taxonomic Comments

The 16S rRNA gene sequence analysis showed the sequence similarity between *Rhodovastum atsumiense* strain G2-11^{T}and *Acs. rubrifaciens* and *Rpi. globiformis* appears low (<94%) enough to classify these species into a novel genus (Okamura *et al.,* 2009). The phylogenetic relationships of these bacteria from other genera of *Acetobacteraceae* based on 16S rRNA sequences are illustrated in **Fig. 3.15** (Okamura *et al.,* 2009).

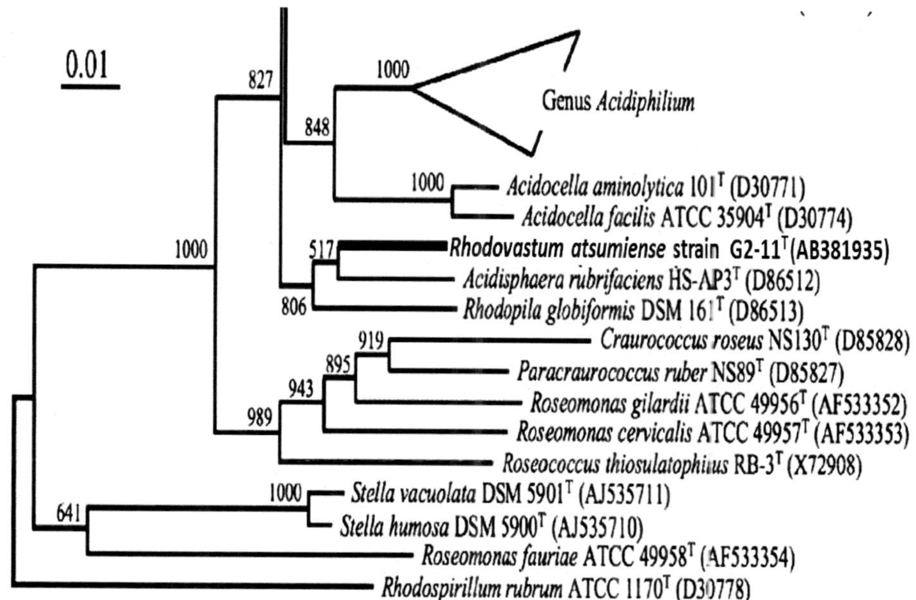

Fig. 3.15. The phylogenetic tree showing the relationship of *Rhodovastum* species to other members of the family *Acetobacteraceae* based on 16S rRNA gene sequence analysis.

Order 2. *Rhizobiales* Kuykendall (2015)

The order *Rhizobiales* is a phenotypically heterogeneous assemblage of Gram (-) negative bacteria and is based solely on 16S rRNA gene sequence analysis. *Rhizobium* is the type genus. A total of 10 families are included in *Rhizobiales* (*Bartonellaceae, Beijerimckiaceae, Bradyrhizobiaceae, Brucellaceae, Hyphomicrobiaceae, Phyllobacteriaceae, Methylobacteriaeae, Methylocystaceae*, *Rhizobiaceae* and *Rhodobiaceae* .) *Bradyrhizobiaceae, Hyphomicrobiaceae* and *Rhodobiaceae* contain purple non-sulfur bacteria. Hence, a detailed description of only these 3 families is described below.

Family 1. *Bradyrhizobiaceae* (Garrity *et al.*, 2005) Garrity *et al.* (2015)

The family *Bradyrhizobiaceae* was determined on the basis of phylogenetic analysis of 16S rRNA sequence. This family is diverse phenotypically, metabolically and ecologically. The family contains genera *Bradyrhizobium* (type genus), *Afipia, Agromonas, Blastobacter, Bosea, Nitrobacter, Oligotropha, Rhodoblastus* and *Rhodopseudomonas*. This family consists of organisms that fix N_2, photosynthetic organisms, organisms capable of aerobic and/or anaerobic respiration and human pathogens. *Bradyrhizobium* is the type genus. *Rhodoblastus* and *Rhodopseudomonas* are phototrophs. Hence, these two genera are described in this chapter.

Table 3.11. Differentiating Characteristics of Genera of *Rhizobiales*

Characteristics	*Rhodopseudomonas*	*Rhodoblastus*	*Rhodoplanes*	*Blastochloris*	*Rhodomicrobium*	*Afifeilla*	*Rhodobium*	*Roseospirillum*
Habitat	Fresh water	Slightly acidic fresh water ponds	Fresh water	Fresh water	Fresh water	Sea water	Saltern lakes	Salt marsh
Cell diameter (μm)	0.6-0.9	1.0-3.0	0.8-1.0	0.6-0.9	1.2-1.5	0.7-0.9	0.7-0.9	0.4-0.6
Internal membrane system	Lamellae	Lamellae	Lamellae	Lamellae	Lamellae	Lamellae	Lamellae	Lamellae
Type of budding	Tube	Sessile	Sessile	Tube	Tube	ND	Sessile	ND
Motility	+	+	+	+	+	+	+	+
Cell Shape	Rods	Rods	Rods	Rod-Ovoid	Ovoid-elongated ovoid	Rods	Ovoid-Rod	Vibrioid, spirilloid
Rosette formation	+	-/+	-	+	Complex aggregates	Budding	+/-	-
Colour	Red, Brown-Red	Purple-Red to Orange-Brown	Pink	Green to Olive green	Salmon pink to Orange red	Pink, Red	Pink-Red	Pink, Red
Carotenoids	Spirillo-xanthin	Spirillo-xanthin, Rhodopin, Rhodopinal	Spirillo-xanthin sries	1,2-dihydroneurosporene, 1,2-dihydrolycopene	Spirillo-xanthin series	Spirillo-xanthin	Spirillo-xanthin series	Spirillo-xanthin, Lycopenal
Bacteriochlorophyll	*a*	*a*	*a*	*b*	*a*	*a*	*a*	*a*
Growth factors	pABA, Biotin	NR	Niacin	pABA	NR	Complex	Biotin, pABA	Complex
Aerobic growth	+	+	+	+	+	+	- /+	+
Sulfate assimilation	+ (PAPS)	+ (APS)	ND	-	+ (APS)	ND	ND	+
Denitrification	+/-	-	+	-	ND	-	+	ND
Photoautotrophic growth with	H_2, Thiosulfate, Sulfide	H_2	Thiosulfate	Thiosulfate or Sulfide	H_2 , Sulfide	-	Thiosulfate	
Salt requirement (%)	Not required	upto 1%	Not required	Not required	Not required	upto 5%	upto 5%	1-2%
Optimal temperature (Range)	30-37	25-30	30	25-30	30	25-30 (20-35)	30-35	30
Optimal pH (Range)	6.9	5.5-6.0	7.0-7.5	7.0-8.0	6.0	6.9-7.1	7.0-7.5	7.9

Characteristics	*Rhodopseudomonas*	*Rhodoblastus*	*Rhodoplanes*	*Blastochloris*	*Rhodomicrobium*	*Afifeilla*	*Rhodobium*	*Roseospirillum*
Mol% G+C of the DNA	64.8-66.3 Bd	62.2-66.8 Bd	66.8-69.7 (HPLC)	66.3-71.4 Bd	61.8-63.8 Bd	61.5-64.1	65.2-65.7	71.2
Cytochrome *c* size	Large	Small	ND	Small	Small	ND	ND	ND
Major quinones	Q-10	Q-10, MK-10, and RQ-10	Q-10, RQ-10	Q8-Q10 and MK-7-MK9	Q-10, RQ-10	Q-10, MK-10	Q-10, MK-10	ND

Symbols: +, Positive in strains/present; -, Negative/absent; +/-, variable in different strains; Q-7/8/9/10, ubiquinone with 8/9/10 isoprene units; MK-7/8/9/10, menaquinone with 7/8/9/10 isoprene units; RQ-10, rhodoquinone with 10 isoprene units; Bd, Bouyant density; Td, Thermal denaturation; NR, Not required; ND, Not determined. (APS), via adenosine-5′-phosphosulfate; (PAPS), via 3′-phosphoadenosine-5′-phosphosulfate, pABA, Para-amino benzoic acid.

1. *Rhodopseudomonas* (*Rps.*) (Rod shaped) (Czurda and Maresch 1937) Imhoff (2015)

Cells are rod-shaped (**Fig. 3.16**), 0.6-2.5 μm wide and 0.6-5.0 μm long, motile utilizing flagella or non-motile. They show polar growth, budding and asymmetric cell division. The internal photosynthetic membrane is present as lamella underlying and parallel to the cytoplasmic membrane. Bacteriochlorophyll *a* or *b* and carotenoids of spirilloxanthin series are present as Photosynthetic pigments. Photoheterotrophic growth occurs with a number of organic compounds as carbon /and electron donors. Photoautotrophic growth is possible under anaerobic conditions with molecular hydrogen, thiosulfate or sulfide as the electron donor. Some species grow chemotrophically under microaerobic to aerobic conditions. Straight-chain monounsaturated $C_{18:1}$ is the major component of cellular fatty acids, contains ubiquinones (Q-10) with 10 isoprene units as major quinone. Growth factor requirements and sensitivity to oxygen vary with the species. *Rps. palustris* differs from other species by containing a large type of cytochrome c_2. The cytochrome structures of the other species in this genus are not known. Studies of lipopolysaccharides have revealed 2,3-diamino-2,3-deoxyglucose unique structural components of *Rps. palustris.* Data on the lipopolysaccharide structures of other species are not available.

The quinones present in various species of this genus are also different and they differ in the length of the isoprenoid chain. The pathway of sulfate assimilation in *Rps. palustris* reduce 3'-phosphate adenosine 5'-phosphosulfate. Sulfide can serve as the sole sulfur source; sulfide is oxidized to elemental sulfur and thiosulfate to sulfate. The optimum growth temperature is 25-35°C. *Rps. palustris* is the type species (Molisch, 1907).The major characteristics and utilization of carbon sources by species of *Rhodopseudomonas* are presented in **Tables 3.12** and **3.13** .

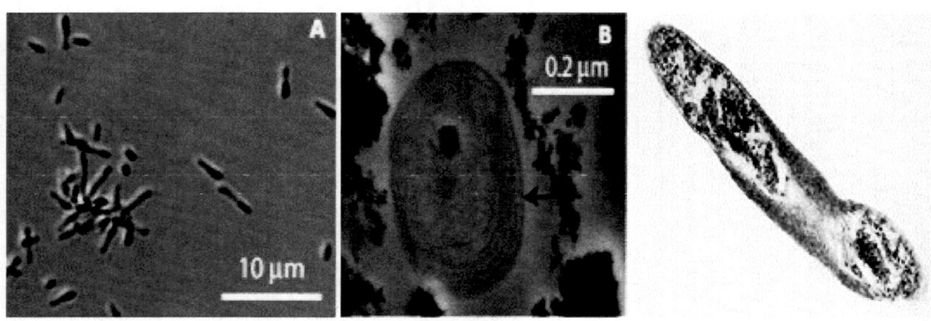

Fig. 3.16. (A) Phase-contrast micrographs showing the cell morphology of *Rps. palustris* (Bar 10 µm). (B) and (C) Lamellar intracytoplasmic membranes (arrow) of a cell from a 1 week-old culture grown with Fe (II) as the electron donor. Fe (III) precipitates formed outside the cell.

Table 3.12. Differentiating Characteristics of Species of the *Rhodopseudomonas*

Characteristics	***Rps. palustris***	***Rps. rhenobacensis***	***Rps. faecalis***	***Rps. harwoodiae***	***Rps. parapalustris***	***Rps. pseudopalustris***	***Rps. pentothenatexigens***	***Rps. thermotolerans***	***Rps. telluris***
Isolation source	Soil/ Sludge/ pond water	Eutrophic pond	Chicken faeces	Salt-pan sediments	Soil	Soil/ pond water	Paddy soil samples	Soil	Paddy soil
Cell shape	Rods with rounded ends	Rods	Vibrio	Rods	Rods	Rods	Rods	Rods	Rods
Cell size (width X length in µm)	0.4-1.0 X 1.5-3.0	0.4-0.6 X 1.5-2.0	0.6-0.8 X1.0-2.0	0.8-1 X 2.0-3.5	0.8-1.2 X 2.0-4.0	0.5-1.0 X 2.0-6.0	0.5-1.0 X 2.0-6.0	0.8-1.0 X 2.0-5.0	0.8-1.5 X 2.0-4.5
Type of Budding	Tube	Sessil	Sessil	Tube	Sessile	Sessile	Tube	Sessile	Tube
Color of cultures	Red-brown	Red	Red	Red	Red-brown	Red-brown	Orange	Pink	Peach brown
Optimal pH (range)	6.5-8.5 (6-9)	5.5 (5.0-7.0)	7 (6.5-7.0)	6-7.5 (5-9)	6.5-7 (6- 9)	6-8 (7)	7.0-8.0 (6.0-8.0)	6.5-7.5 (6.0-8.0)	6.3-7.0 (5.0-9.0)
Temperature optimum (range; °C)	30 (25-40)	25-35 (30)	30 (25-35)	30 (25-40)	30 (15-35)	30 (20-35)	30 (25-50)	30-35 (25-60)	30-35 (25-40)
Absorption maxima of living cells (nm)	803, 862, 863	805, 878	804-805, 870-874	805, 863	803, 862,-863	800, 861-863	806, 863	806, 863	377, 496-498, 529, 590, 804 and 860
Aerobic dark growth	+	+	-	+	+	+	-	+	+
Photoautotrophic growth with	H_2, Thiosulfate, Sulfide	-	H_2, Sodium bi-carbonate	Thiosulfate	-	Thiosulfate	Thiosulfate	Thiosulfate	-

Characteristics	*Rps. palustris*	*Rps. rhenobacensis*	*Rps. faecalis*	*Rps. harwoodiae*	*Rps. parapalustris*	*Rps. pseudopalustris*	*Rps. pentothenatexigens*	*Rps. thermotolerans*	*Rps. telluris*
Growth factors	PABA	PABA	Thiamine	PABA	Thiamine, PABA	PABA	Thiamine, PABA, Pantothenate	PABA	PABA
Major Polar lipids	DPG, PG, PE, PC, OL, AL1-3,GL1-2, L1-7	DPG, PG, PE, PC, OL, AL1-3, GL1-2, L1-3 & 7	DPG, PG, PE, PC, OL, AL1-3, GL1, L1-3	DPG, PG, PE, PC, AL1 & 3, GL1-3, L1-3 & 7,8	DPG, PG, PE, PC, AL1 & 3, GL2, L1-3	DPG, PG, PE, PC, OL, AL1-3, GL1-3, L1-3 & 7	DPG, PG, PE, PC, AL1; L1–4,6,8,9	DPG, PG, PE, PC, Li-4,6	DPG, PE, PC, PG

Symbols: +, Present /utilized; –, absent/not utilized; ND, not determine; PABA, Para-Aminobenzoic acid; DPG, Diphosphatidylglycerol, PG, phosphatidylglycerol, PE, phosphatidylethanolamine, PC, phosphatidylcholine; an unidentified aminolipid (AL2) and an aminohopanoid (AH). AL1–9, unknown aminolipids; L1–5, unknown lipids. Numbering of polar lipids is adopted from Kampfer *et al.* (2006). Data for *Rhodopseudomonas palustris* (van Neil, 1944), *Rps. rhenobacensis* (Hougardy *et al.*, 2000), *Rps. faecalis* (Zhang *et al.*, 2002), *Rps. harwoodiae, Rps. parapalustris, Rps. pseudopalustris* (Venkata Ramana *et al.*, 2012), *Rps. pentothenatexigens, Rps. thermotolerans* (Kumar *et al.*, 2013), *Rps. telluris* (Hiraishi and Okamura (2017) is adopted.

Table 3.13. Carbon Sources and Electron Donors used by Species of *Rhodopseudomonas*

Organic substrates	*Rps. pentothen-atexigens*	*Rps. thermotolerans*	*Rps. harwoodiae*	*Rps. parapalustris*	*Rps. pseudopalustris*	*Rps. palustris*	*Rps. faecalis*	*Rps. rhenobacensis*	*Rps. telluris*
Acetate	-	+	-	+	-	-	+	-	+
ΨBenzoate	-	-	+	-	+	+	-	-	-
Butyrate	+	+	+	-	+	+	+	+	(+)
Caproate	-	-	+	-	+	(+)	(+)	-	(+)
Caprylate	-	-	-	-	+	-	-	-	-
Ethanol	-	-	-	-	+/-	+	-	+	(+)
Formate	-	-	-	-	+	-	-	+	ND
Fumarate	-	-	(+)	+	+	+	(+)	+	+
D- Glucose	-	-	-/+	+	-/+	)+(	-	-	-
Glutamate	-	-	-	-	+	-	-	-	(+)
Glycerol	-	-	(+)/-	+	+	+	-	(+)	+
Glycolate	-	-	-	-	+	-	-	-	ND
Malate	-	+	-	+	+	+	+	+	+
Malonate	-	-	(+)/-	-	+	+	-	-	+
Propanol	-	-	-	-	+	-	-	-	ND
Propionate	-	-	(+)	-	+	+	(+)	-	(+)
Sorbitol	-	-	-	+	+	-	-	-	ND
Succinate	+	(+)	(+)	+	+	+	+	+	+
Tartarate	-	-	-	-	-	-	-	+	-

Valerate	+	+	+	+	+	-	-	-	+
Oxogl-utarate	-	+	+	+	+	-	-	-	ND
Molecular features									
$DDH(%) relatedness									
1	100								
2	35.4+5.2	100							
3	43.6±2.2	24.6±4.5	100						
4	43.3±5.3	44.1±6.7	33.1±.2	100					
5	35.8±4	28.2±5.5	45.0±9.2	48.0±2.2	100				
6	32.9±8	42.5±6.2	47.5±4.4	45.3±2.8	32.8±3.8	100			
7	41.9±6	25.4±3.3	28.4±2.9	32.1±2.2	41.0±8.3	40.7±2	100		
8	43.8±3.3	33.7±6.0	27±4.6	24.1±3.8	32.5+14.5	38.6±4	3.5±4.2	100	
9	ND	ND	ND	ND	ND	ND	ND	ND	ND
DNA G+C content (mol%) HPLC	64.8	66.2	62.4	63.8	65.8	64.9	64	65.4	66.3

Symbols: +, Present /utilized; –, absent/not utilized; (+), weak growth; (+)/-, weak growth or absent; ND, not determine; DDH, DNA-DNA hybridization. Data for *Rhodopseudomonas palustris* (van Neil, 1944), *Rps. rhenobacensis* (Hougardy *et al.,* 2000), *Rps. faecalis* (Zhang *et al.,* 2002), *Rps. harwoodiae, Rps. parapalustris, Rps. pseudopalustris* (Venkata Ramana *et al.,* 2012), *Rps. pentothenatexigens, Rps. thermotolerans* (Kumar *et al.,* 2013), *Rps. telluris* (Hiraishi and Okamura (2017) is adopted.

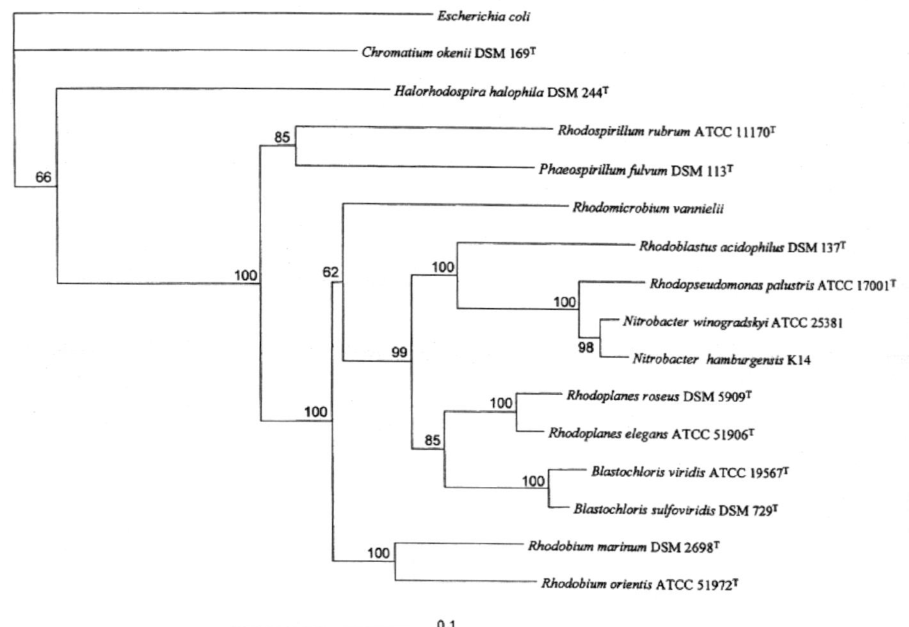

Fig. 3.17. Phylogenetic tree of *Rhodopseudomonas* species and related *Alphaproteobaceria* of the order *Rhizobiales* based on 16S rDNA sequences. The sequence of *E.coli* was used as an outgroup.

Differentiation of the *Rhodopseudomonas* from other Genera

Differential characteristics of the genus *Rhodopseudomonas* from other genera of *Rhizobiales* are précised in **Table 3.11**. Differentiation of the genera and species of *Rhodopseudomonas* , *Rhodoblastus* , *Rhodoplanes* , *Rhodobium, Rhodomicrobium* and *Blastochloris* based on 16S rDNA sequence analysis is illustrated in **Fig. 3.17** (Imhoff *et al.,* 2005).

Taxonomic Comments

After removal of purple non-sulfur bacteria that contained vesicular internal photosynthetic membranes and *Betaproteobacteria* from the genus *Rhodopseudomonas,* the remaining genera had lamellar internal membrane structure and a budding model of growth and reproduction (Imhoff *et al.*, 1984). However, after the removal of bacteria such as *Rhodopila, Rhodobacter, Rhodovulum* and *Rubrivivax,* the genus *Rhodopseudomonas* still represented a heterogeneous assemblage of species (Imhoff *et al.*, 1984). Based on 16S rDNA sequence data (Kawasaki *et al.*, 1993b), isolation and description of new species and addition of data, *Rhodopseudomonas rosea* was transferred to the genus *Rhodoplanes* as *Rpl. roseus* (Hiraishi and Ueda, 1994b), *Rhodopseudomonas marina* to *Rhodobium* as *Rhodobium marinum,* and *Rps viridis* and *Rps. sulfoviridis* to the genus *Blastochloris* as *Blc. viridis* and *Blc. sulfoviridis* (Hiraishi, 1997). *Rhodopseudomonas acidophila* was transferred to *Rhodoblastus acidophilus* (Imhoff, 2001).

Rhodopseudomonas blastica has many chemotaxonomic properties in common with bacteria in the informal group *Alphaproteobacteria* ,*Rhodobacter, Rhodovulum* and *Paracoccus denitrificans.* Its 16S rDNA sequence is most similar to and clusters with the *Rhodobacter* species . Therefore, it was transferred to *Rhodobacter blasticus* (Kawasaki *et al.*, 1993b). Hiraishi *et al.* (1992b) considered *Rhodopseudomonas rutila* identical to *Rps. palustris* and the species name is a later subjective synonym of *Rps. palustris,* although it has been described as a separate species (Akiba *et al.*, 1983). In the original species description, *Rps. palustris* was not included as a reference organism property of the new isolates were reinvestigated later (Hiraishi *et al.*, 1992b). These authors established identity between the type strains of *Rps. palustris* and *Rps. rutila.* In particular, regarding the utilization of carbon sources or electron donors such as acetate, propionate, benzoate, fructose sulfide and thiosulfate, Hiraishi *et al.* (1992b) disagreed with Akiba *et al.* (1983) and found conformity with the properties of six *Rps. palustris* strains. The former authors also found that *Rhodopseudomonas rutile* requires Para-aminobenzoic acid as a growth factor. Both bacteria were identical in quinone, lipid and fatty acid composition and 16S rDNA sequence. DNA-DNA similarity between *Rps. rutila* and *Rps. palustris* strains were between 80-90% (Imhoff *et al.,* 1984).

In addition to *Rps. palustris,* nine species of *Rhodopseudomonas* are known. *Rps. rhenobacensis* is a bacterium notably reducing nitrate (Hougardy *et al.* , 2000). *Rps. faecalis* (Zhang *et al.,* 2002), *Rps. harwoodiae* ,*Rps. parapalustris* , *Rps. pseudopalustris* (Venkata Ramana *et al.,* 2012), *Rps. pantothenatexigens* and *Rps. thermotolerans* (Kumar *et al.,* 2013), *Rps. telluris* (Hiraishi and Okamura, 2017) and *Rhodopseudomonas julia* (Kornpantseva, 1989a) are validly published species. *Rhodopseudomonas cryptolactis* was also isolated from a hot spring and is characterized by a lack of autotrophic growth and a quite restricted substrate spectrum (Stadrwalddemchick *et al.*, 1990). This bacterium has not been validly published.

2. *Rhodoblastus* (*Rbl.*) (The budding rose) (Imhoff, 2001) Imhoff (2015)

Cells are rod-shaped to elongate-ovoid, slightly curved, motile utilizing flagella, show polar growth, budding and asymmetric cell division **(Fig. 3.18A, B&C)**. Gram (-) negative. Internal photosynthetic membranes appear as lamellae underlying and parallel to the cytoplasmic membrane(**Fig. 3.18 D**). Straight-chain monounsaturated $C_{18:1}$ and $C_{16:1}$ and saturated $C_{16:0}$ are present as major cellular fatty acids. Ubiquinones (Q-10), rhodoquinones (RQ-10) and menaquinones (MK-10) with 10 isoprene units are present. Grow preferably photoheterotrophically under anoxic conditions in the presence of light. Photoautotrophic growth may also possible under anoxic conditions with hydrogen as the electron donor. Chemotrophic growth possible under microoxic to oxic conditions, but with some substrates, growth occurs anaerobically by fermentation. Growth factors are not required. Mesophilic freshwater bacteria prefer acidic pH . Slightly acidic freshwater ponds are the substrates. The G+C content of the DNA is 62.2-66.8 mol% (buoyant density). *Rhodoblastus acidophilus* (Pfennig 1969)is the type species. *Rbl. acidophilus* and *Rbl. sphagnicola* are validly published species. The distinguishing characteristic and carbon utilization by these species are presented in **Tables 3.14** and **3.15**.

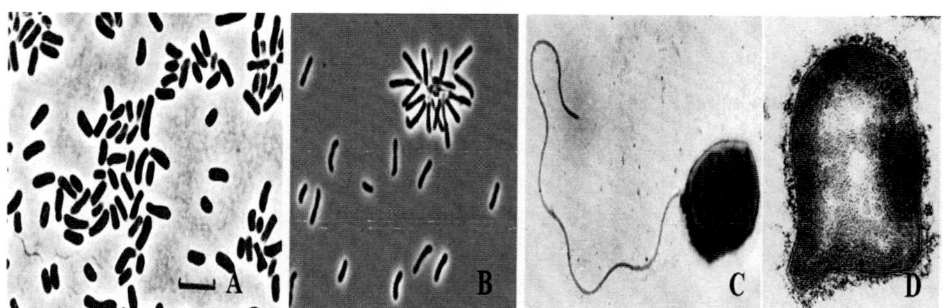

Fig. 3.18. (A) Phase-contrast micrograph of *Rbl. acidophilus*, polar type of cell growth is observed, tufts of detached flagella can be seen (Pfennig, 1969). (B) Phase-contrast micrograph of *Rbl. sphagnicola.* (C) Negatively stained young cell of *Rbl. sphagnicola* showing a single polar flagellum, (D) Electron micrograph of an ultrathin section showing lamellar intracytoplasmic membranes (Kulichevskaya *et al.,* 2006). Bars: 5 μm (A), 10 μm (B) and 0.5 μm (C, D).

Table 3.14. Differentiating Characteristics of Species of the Genus *Rhodoblastus*

Characteristics	*Rbl. acidophilus*	*Rbl. sphagnicola*
Cell diameter (µm)	1.0-1.3 x 2.0-5.0	0.8-1.0 x 2.0-6.0
Cell shape	Rods	Rods/Slightly curved
Rosette formation	(+)	+
Colour of cultures	Red to orange-red	Purplish- red
Bacteriochlorophyll	*a*	*a*
Carotenoids	Glucosides of Rhodopin and Rhodopinal	Rhodopin, Rhodopinal, Lycopene, Lycopenal
Absorption maxima of living cells (nm)	375, 460, 490, 525, 590, 805, 855, and 890 nm	377, 463, 492, 527, 592, 806 and 867 nm
Salt requirement (%)	None	Upto 1%
Optimum pH	5.5-6.0	5.2-5.5
Optimum temperature (◦C)	25-30	25-30
Sulfate assimilation	+ (APS)	ND
Citrate photo assimilation	W	-
Glycerol photo assimilation	-	+
Phototrophic growth on butyrate	Poor	Good
Aerobic dark growth	+	-
Denitrification	-	-
Fermentation of fructose	-	-
Photoautotrophic growth with	H_2	H_2
Growth factors	Not required	Not required
DNA G+C content (mol%)	62.2-66.8 (Bd)	62.6
Cytochrome C_2 size	Small	-
Major quinones	Q-10. MK-10, RQ-10	Q-10, Q-9

Symbols: +, positive; -, negative; W, weakly positive; Bd, Buoyant density; MK-10, Menaquinone with 10 isoprene units; Q-9/10, ubiquinone with 9/10 isoprene units; RQ-10, Rhodoquinone with 10 isoprene units. Data for *Rbl. acidophilus* and *Rbl. sphagnicola* is adopted from Imhoff (2001) and Kulichevskaya *et al.* (2006).

Differentiation of *Rhodoblastus* from other Genera

Differential characteristics of *Rhodoblastus* and other genera of *Rhizobiales* such as *Rhodopseudomonas* , *Rhodoplanes* , *Blastochloris* , *Rhodomicrobium* , *Afifeilla* and *Rhodobium*are shown in **Tables 3.11**. Their phylogenetic relationships are shown in **Fig. 3.17** (Imhoff *et al.,* 2017).

Table 3.15. Utilization of Carbon Sources/Electron Donors by species of *Rhodoblastus*

Carbon source	*Rbl. acidophilus*	*Rbl. sphagnicola*
Acetate	+	+
Aspartate	-	ND
Benzoate	-	-

Carbon source	*Rbl. acidophilus*	*Rbl. sphagnicola*
Butyrate	+/-	+
Caproate	+/-	+
Caprylate	-	ND
Citrate	+/-	-
Ethanol	+	+
Formate	+/-	+
D-Fructose	-	-
Fumarate	+	ND
D-Glucose	+/-	-
Glutamate	–	–
Glycerol	+/–	+
Glycolate	+/–	ND
Lactate	+	+
Malate	+	+
Malonate	+/–	+
Mannitol	–	ND
Methanol	+/–	+
Propanol	ND	ND
Propionate	+	+
Pyruvate	+	+
Sorbitol	+	
Succinate	+	+
Tartarate	+/–	–
Valerate	+	ND
Thiosulfate	–	–
Sulfide	–	–

Symbols: +, positive/ growth present in most strains; – negative/ growth absent in most strains; +/– variable in different strains; ND, not determined. Data for *Rbl. acidophilus* and *Rbl. sphagnicola* is adopted from Imhoff (2001) and Kulichevskaya *et al.* (2006).

Taxonomic Comments

Rhodopseudomonas acidophila has characteristics that are distinct to *Rhodopseudomonas palustris* and other species of *Rhodopseudomonas* . *Rps. acidophila* assimilate sulfate via adenosine 5'-phosphosulfate (Imhoff, 1982), presence of small-sized cytochrome c_2, menaquinones and rhodoquinones, different fatty acid composition and glucosamine in the lipid A (not 2,3-diamino-2,3-dideoxy'hexose) (Tegtmeyer *et al.*, 1985). DNA-DNA hybridization studies between *Rps. palustris* and *Rps. acidophila* showed only less than 15% similarity. Based on the above essential properties which are in contrast to *Rhodopseudomonas palustris* and to *Blastochloris* species (formerly assigned to *Rhodopseudomonas* ,) *Rps. acidophila* has been transferred to another genus *Rhodoblastus* as *Rhodoblastus acidophila* (Imhoff, 2005).

Family 2. *Hyphomicrobiaceae* (Babudieri, 1950) Garrity *et al.* (2015)

The family *Hyphomicrobiaceae* is described on the basis of phylogenetic analysis of 16S rRNA sequences. The family is morphologically, metabolically and ecologically different. It contains18 genera: Are *Hyphomicrobium* (type genus), *Ancalomicrobium, Ancylobacter, Angulomicrobium, Aquabacter, Azorhizobium* , *Blastochloris, Devosia, Dichotomicrobium, Filomicrobium, Gemmiger, Labrys, Methylorhabdus, Pedomicrobium, Prosthecomicrobium, Rhodomicrobium, Rhodoplanes, Seliberia, Starkeya* and *Xanthobacter*. Cells are Gram (-) negative, many members form hyphae or prosthesis, a number reproduce by budding and include organisms that are photosynthetic, facultatively methylotrophic, facultatively chemolithoautotrophic and chemoheterotrophic. The mol% G+C of the DNA varies between 49-72%. *Rhodoplanes* , *Blastochloris* and *Rhodomicrobium* are the phototrophic bacteria, hence they are described below. Based on 16S rRNA sequence comparisons, a number of taxonomic rearrangements have been made in the past.

1. Reclassification of organisms formerly classified within the *Hyphomicrobiaceae* into other families.
2. Placement of organisms earlier attributed to other groups into new members of the family.
    i. The reclassification of *Rhodopseudomonas rosea* to the type species of the new genus *Rhodoplanes roseus* (Hiraishi and Ueda 1994).
    ii. The reclassification of *Rhodopseudomonas cryptolactis* (Stadwald-Demchick *et al.,* 1990), a name that was effectively published but not validated, as *Rhodoplanes cryptolactis* (Okamura *et al.,* 2007).
    iii. The transfer of *Rhodopseudomonas viridis* (Drews and Giesbrecht 1966) and *Rhodopseudomonas sulfoviridis* as genus *Blastochloris* as *Blastochloris viridis* and *Blastochloris sulfoviridis* respectively (Hiraishi, 1997).

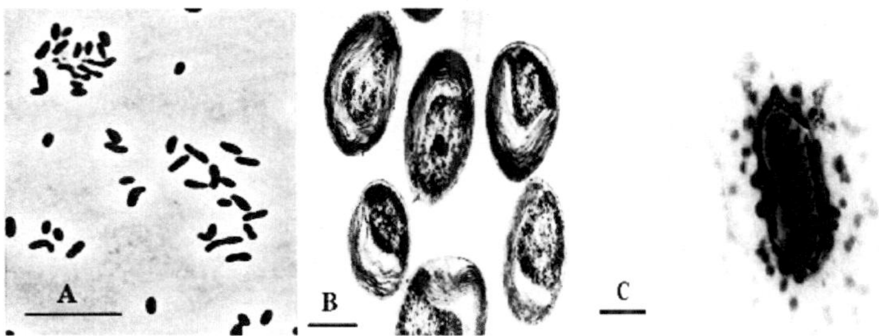

Fig. 3.19. A. Phase-contrast micrographs of *Rpl. rosea* (Bar=8 μm), B. Thin-section of electron micrograph showing lamellar type internal photosynthetic membrane system (Bar=0.5 μm) and C. Electron micrograph of a negatively stained *Rpl. roseus* with polar flagellum (Bar= 1 μm).

1. *Rhodoplanes* (*Rpl.*) (A red wanderer) (Hiraishi and Ueda, 1994) Hiraishi and Ueda (2015)

Cells are gram-negative, rod-shaped, multiply by budding and asymmetric cell division **(Fig. 3.19)** (Hiraishi and Ueda, 1994). Motile by means of polar, sub-polar, or randomly distributed flagella. Photosynthetic cultures are pink, while aerobic chemotrophic cultures are colorless. Photographically grown cells contain intracytoplasmic membranes of the lamellar type together with bacteriochlorophyll *a* and carotenoids of the normal spirilloxanthin series. $C_{18:1}$ and $C_{18:0}$ are predominant straight chain monounsaturated components of the cellular fatty acids. Ubiquinone (Q-10) and rhodoquinone (RQ-10) are present as major quinones. Mesophilic, neutrophilic and non-halophilic. Facultative phototrophs growing anaerobically in the light or aerobically in the dark. Anaerobic growth in the dark by nitrate respiration is possible, whereas aerobic growth in the dark by denitrification is possible. Some representatives are thermotolerant. Growth in the presence of NaCl at a concentration of 1.0% or more is inhibited.

Growth is preferentially photoheterotrophic under anoxic conditions in the light, with simple organic substrates as carbon and energy sources. Photoautotrophic growth occurs with thiosulfate as the electron donor in the presence of 0.01% yeast extract. Sulfide may not be utilized. Chemotrophic growth is possible under oxic conditions in the dark at full atmospheric oxygenic tension. Hydrolysis of gelatin and Tween 80 is negative. The G+C content of DNA ranges from 66.8 to 69.7 mol%. The phylogenetic position is the alpha-2 subclass of *Proteobacteria.* Grow in freshwater and wastewater environments. *Rhodoplanes roseus* is the type species. *Rhodoplanes roseus, Rpl. elegans ,Rpl. piscinae ,Rpl. pokkaliisoli , Rpl. serenus, Rpl. oryzae, Rpl. tepidamans, Rpl. tepidicaeni* and *Rhodoplanes azumiensis* are validly published species. The characteristics of nine species are precised in **Table 3.**

Table 3.16. Differential Characteristics of Species of the *Rhodoplanes*

Characteristic	*Rpl. roseus*	*Rpl. elegans*	*Rpl. serenus*	*Rpl. pokkaliisoli*	*Rpl. piscinae*	*Rpl. oryzae*	*Rpl. tepidamans*	*Rpl. tepidicaeni*	*Rpl. azumiensis*
Isolation source	Fresh water lake sediments	Activated sludge	Pond water	Paddy Soil	Fish pond	Paddy soil	Hot spring	Hot spring	Hot spring
Cell length (µm)	1.8–2.5	2.0–3.5	2.0–3.0	3.0–5.0	2.0–3.0	2.0–3.0	2.0–4.0	1.5–3.5	2.0–4.0
Type of budding	Sessile	Tube	Tube	Sessile	Tube	Sessile	Tube	Tube	Tube
Colour of cell culture	Pink	Pink	Pinkish red	Yellowish–brown	Pinkish red	Peach brown	Pink	Pink	Brownish red
Motility	+	+	+	+	+	+	+	+	+
Rosette formation	–	+	+	–	+	+	+	+	+

Characteristic	*Rpl. roseus*	*Rpl. elegans*	*Rpl. serenus*	*Rpl. pokkaliisoli*	*Rpl. piscinae*	*Rpl. oryzae*	*Rpl. tepidamans*	*Rpl. tepidicaeni*	*Rpl. azumiensis*
Intracytoplasmic photosynthetic membrane	Lamellae	Lamellae	Lamellae	Lamellae	Lamellae	Lamellae	Lamellae	Lamellae	Lamellae
Bacteriochlorophyll	*a*	*a*	*a*	*a*	*a*	*a*	*a*	*a*	*a*
Carotenoids	Spirilloxanthin series	Spirilloxanthin series	Spirilloxanthin series	Rhodopin, methylrhodovibrin and lycopene	Rhodopin, spirilloxanthin, anhydrorhodovibrin and rhodovibrin	Rhodopin, spirilloxanthin, anhydrorhodovibrin and rhodovibrin	Rhodopin lycopene spirilloxanthin, and anhydrorhodovibrin	Rhodopin, spirilloxanthin, anhydrorhodovibrin, rhodovibrin and lycopene	Rhodopin, spirilloxanthin, anhydrorhodovibrin, rhodovibrin and lycopene
Major quinones	Q–10, RQ–10	Q–10, RQ–10	Q–10, RQ–10	Q–10, RQ–10	Q–10, RQ–10	Q–10, RQ–10	Q–10, RQ–10	Q–10, RQ–10	Q–10, RQ–10
Absorption maxima of living (nm)	373–375, 468, 493, 527–530, 591–593, 799–801 & 850–854	373–375, 466, 491, 527–530, 591–593, 799–801 and 854–856	378, 482, 512, 550, 590, 800 and 850	380, 464, 491, 527, 596, 800 and 866	515–517, 550, 590, 802–803 and 821–824 shoulder peak at 875–878	377, 434, 533, 803 and 860	800, 821–824, and 875–877	377, 495–498, 533, 594, 800, 859–860, 892–895	376, 498, 529, 594, 800, 859 and 892
NaCl Requirement (%)	Not required	Not required	Upto 1.0%	0–2.5	0–0.5	Not required	< 1.0	0–1.5	0–1.0
pH range (optimum) for growth	6.5–8.0 (7.0–7.5)	6.0–8.5 (7.0)	5.5–9.0 (7.5–8.0)	7.0 (6.0–7.5)	5.5–9.0 (6.5–8.0)	7.0–7.5 (7.0)	6.4–8.5 (6.8–7.5)	5.5–9.0 (6.8–7.2)	5.5–8.5 (6.8–7.2)
Temperature range (optimum) for growth (°C)	30	30–35	20–43 (35)	35 (30)	25–40 (30)	30–35 (30)	25–46 (40)	25–45 (40)	25–46 (40)
Vitamin requirement (s)	Niacin	Thiamine, pABA	Not required	Niacine, pABA, Pantothenate	Niacine	Niacine, pABA	Niacin, pABA and Vit B_{12}	Niacine, Vit B_{12}	Niacin
Sulfate assimilation	+	+	+	+	+	+	–	–	+
Denitrification	+	+	+	+	+	ND	+	+	–
Electron donor/ Carbon source									
Acetate	+	+	+	+	+	+	+	(+)	+
Benzoate	–	–	–	–	–	–	–	–	–
Butyrate	+	+	+	+	+	–	+	–	–
Caproate	–	+	+	–	+	–	–	–	–

Charact-eristic	*Rpl. roseus*	*Rpl. elegans*	*Rpl. sere-nus*	*Rpl. pokkal-iisoli*	*Rpl. piscinae*	*Rpl. oryzae*	*Rpl. tepi-damans*	*Rpl. tepidi-caeni*	*Rpl. azumi-ensis*
Caprylate	ND	ND	–	–	–	–	–	–	–
Citrate	+	+	+	–	+	+	+	+	+
Ethanol	–	–	+	–	+	–	–	–	–
Formate	–	–	–	–	–	–	–	ND	ND
Fumarate	+	+	+	+	+	+	+	(+)	+
Glycolate	–	–	–	–	–	ND	–	ND	–
Gluconate	ND	ND	–	–	–	ND	–	(+)	–
Glycerol	–	–	+	+	+/–	–	–	–	–
Propionate	+	+	+	–	+	–	+	–	–
Lactate	+	+	+	+	+	–	+	(+)	+
Malonate	–	–	+	ND	ND	ND	+	(+)	+
Methanol	–	–	–	–	–	–	–	–	–
Malate	+	+	+	+	+	+	+	+	+
Nicotinate	ND	ND	–	ND	ND	ND	ND	ND	ND
Pelargonate	ND	ND	–	ND	ND	ND	–	–	–
Palmitate	ND	ND	–	ND	ND	ND	ND	ND	
Propanol	–	–	–	–	(+)	–	–	ND	–
Pyruvate	+	+	+	+	–	+	+	+	+
Sucrose	ND	ND	–	ND	–	–	ND	–	ND
Succinate	+	+	+	+	+	+		(+)	+
Tartrate	+	+	+	–	(+)	–	+	(+)	–
Trehalose	ND	ND	–	ND	ND	ND	ND	ND	ND
Valerate	+	+	+	–	+	+	–	–	–
Aspara-ginase	–	–	+	ND	ND	ND	ND	ND	–
Asparatate	–	–	+	–	–	–	–	–	–
Glutamate	–	–	+	+	+	+	–	–	–
Glutamine	ND	ND	+	ND	ND	+	+/–	+/–	+/–
Leucine	–	–	–	–	–	ND	ND	ND	ND
Methionine	ND	ND	–	+	ND	ND	–	–	–
Peptone	+	+	+	–	(+)	+	+	(+)	–
Casamino acids	+	+	+	–	+	+	–	(+)	–
Thiosulfate	+	+	+	–	+	+	+	+	–
DNA G+C content (mol%) (HPLC)	66.8	69.6-69.7	69.5	67.2	71.3	68.7	69.9	67.7	70.4

Symbols: +, Substrate utilized or present; -, Substrate not utilized or absent; (+), weak growth; +/- and (+)/-, variable in different strains. Data for *Rpl. roseus, Rpl. elegans* (Hiraishi and Ueda, 1994b), *Rpl. serenus* (Okamura *et al.,* 2009), *Rpl. pokkaliisoli* (Lakshmi *et al.,* 2009), *Rpl. piscinae* (Kalyan Chakravarthy *et al.,* 2012), *Rpl. oryzae* (Srinivas *et al.,* 2014), *Rpl. tepidamans* (Hiraishi and Okamura 2017), and *Rpl. tepidicaeni, Rpl. azumiensis* (Hiraishi, 2017) is adopted.

Differentiation of the *Rhodoplanes* from other Genera

The differentiating characteristics of *Rhodoplanes* from other genera of *Rhizobiales* are presented in **Table 3.11**. The genus *Rhodoplanes* is phenotypically similar to other genera of the budding phototrophic bacteria, especially the genus *Rhodopseudomonas*, but is separated from the genus by its phylogenetic position and outstanding physiological and chemotaxonomic characteristics. Characteristic features of the *Rhodoplanes* are denitrification, inability to grow on sugars, long straight-chain fatty acids (>C_s) and sulfide, as well as the production of both Q–10 and RQ-10. In addition, *Rhodoplanes* species are differentiated from the phenotypically similar *Rhodopseudomonas* species by cell size, vitamin requirement and carbon nutrition. It is quite easy to distinguish *Rhodoplanes* from its nearest phylogenetic neighbors the *Blastochloris* species because the latter produce bacteriochlorophyll *b* and have either Q-9 and MK-9 or Q-8 (Q-10) and MK-7 (MK-8) as the major quinones. The phylogenetic relationship of these bacteria based on the 16S rDNA sequence is shown in **Fig. 3.17.**

Taxonomic Comments

Rhodoplanes roseus was first described as *Rhodopseudomonas rosea* (Janssen and Harffot, 1991). Later, based on 16S rRNA sequence analysis and distinct chemotaxonomic information, *Rps. rosea* was transferred to the new genus *Rhodoplanes* and designated as the type species of this genus as *Rpl. roseus* (Hiraishi and Ueda, 1994). Similarly, *Rhodoplanes elegans* was described as the second species of this genus. However, another isolate strain IL245 classified as *Rhodopseudomonas* sp. (Kawasaki *et al.*, 1993b), is closely related to the two *Rhodoplanes* species with 97% sequence similarity of the 16S rDNA and may represent another species of *Rhodoplanes* .*Rhodoplanes* form a major cluster in the *Alphaproteobacteria* together with the genus *Blastochloris* and the non-phototrophic genus *Ancylobacter* based on phylogenetic properties.

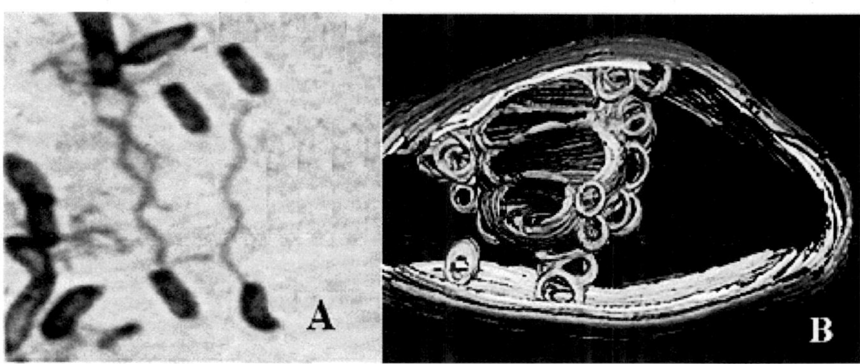

Fig. 3.20. A. Phase contrast of negatively stained *Blc. viridis* showing partially anchored subpolar flagella, B. Electron micrograph of the photosynthetic system in *Blastochloris virdis* .

2. *Blastochloris* (*Blc.*) (Green bud shoot) (Hiraishi, 1997) Imhoff (2015)

Cells are rod-shaped to ovoid-shaped, motile by means of sub-polar flagella (**Fig. 3.20A**) (Drews and Giesbrecht, 1966) and exhibit polar growth, budding and asymmetric cell division. Rosette-like cell aggregates sometimes occur. Gram (-)

negative. Photographically growing cells contain intracytoplasmic membranes that are present as lamellae underlying and parallel to the cytoplasmic membrane **(Fig. 3.20B)**. Photosynthetic pigments are BChl *b* and carotenoids. The colour of photosynthetic cultures is green to olive-green. Both ubiquinones (Q-8 to Q-10) and menaquinones (MK-7 to MK-9) are present. Photoheterotrophic growth occurs under anaerobic conditions in the presence of light. Microaerophilic growth in the dark is also possible. Anaerobic photoorganotrophic growth occurs preferentially with a number of simple organic compounds as carbon and electron donors. Anaerobic photoautotrophic growth may occur with thiosulfate or sulfide as electron donors. Chemotrophic growth is also possible under microoxic conditions in the dark. The G+C content of the genomic DNA ranges from 66-71 mol%. 16s rDNA sequence information places the genus in the alpha-2 subgroup of the *Proteobacteria* and the genus *Rhodoplanes* is the closest relative among the photosynthetic genera. Mesophilic freshwater bacteria prefer neutral pH for their growth. *Blastochloris viridis* is the type species.

Blastochloris gulmargensis (Venkata Ramana *et al.,* 2011), *Blastochloris sulfoviridis* (Keppen and Gorlenko 1975) and *Blastochloris viridis* (Drews and Giesbrecht 1966) are the validly published species and their characteristic features are précised in **Table 3.17.**

Table 3.17. Differential Characteristics of Species of *Blastochloris*

Characteristics	*Blc. gulmargensis*	*Blc. viridis*	*Blc. sulfoviridis*
Habitat	Cold sulfur springs	Freshwater	Freshwater
Cell size (μm)	1.0–1.5 x 3–5	0.6–0.9 x 1.2–2.0	0.5–1.0 x 1.0–1.8
Cell shape	Rods	Rods–Ovoid	Rods
Colour of cell suspension	Yellow–Green	Green to Olive green	Olive green
Motility	+	+	+
Type of budding	Sessile	Tube	Sessile
Bacteriochlorophyll	*b*	*b*	*b*
Internal membrane system	Lamellae	Lamellae	Lamellae
Optimal pH for growth	6.5–7.5	6.5–7.0	7.0
Optimal temperature for growth (range)	25 (15–40)	30 (25–35)	28–30 (25–40)
Growth factors	pABA, Biotin	pABA, Biotin,	pABA, Biotin, Pyridoxal phosphate
Absorption maxima of living cell (nm)	404, 455, 485, 605, 812 and 1022	400, 420, 451, 483, 604, 835 and 1020	ND
Sulfate assimilation	+	+ (PAPS)	–
Aerobic dark growth	+	(+)	(+)
Photolithoautotrophy	–	–	Thiosulfate, sulfide
Major Quinones	Q–8, MK–8	Q–9, MK–9	Q–10, MK–8

Characteristics	*Blc. gulmargensis*	*Blc. viridis*	*Blc. sulfoviridis*
Carotenoid composition (mol%)			
Neurosporene	10	9	4
1,2–Dihydroneuro-sporene	57	71	70
Lycopene	2	3	–
1,2–Dihydrolycopene	5	12	1
1,2–Dihydro–3,4–didehydrolycopene	9	5	2
7,8,11,12–Tetrahydrolycopene	17	–	23
Carbon/ Electron donor			
Acetate	–	+	–
Sulfide	–	–	+
Thiosulfate	–	–	+
Fumarate	+	+	(–)
α–ketoglutarate	+	+	–
Malate	+	(+)	+
Fructose	–	–	+
Glucose	–	(+)	+
Ethanol	–	(+)	(+)
Glutamate	–	(+)	(+)
Maltose	(+)	–	+
Sucrose	–	(+)	+
Glycerol	–	–	+
Propanol	–	–	+
G+C content of genomic DNA (mol%) (HPLC)	63.8	61.8	65.5

Carotenoids and fatty acids are expressed as percentages of the total. All strains utilize pyruvate and succinate but not benzoate, citrate, tartrate and formate; +, Positive in most strains; –, Negative in most strains; +/– variable in different strains; (+), Weak growth; pABA- para aminobenzoic acid; (PAPS), via 3′-phosphoadenosine-5′-phosphosulfate; Q–8/9/10, Ubiquinone with 8/9/10 isoprene units; MK–8/9, Menaquione with 8/9 isoprene units. Data for species of *Blastochloris* is adopted from Venkata Ramana *et al.* (2011) and Hiraishi (1997).

Differentiation of the *Blastochloris* from other Genera

Based on 16S rDNA sequence analysis the genus is placed in *Alphaproteobacteria* . *Rhodoplanes* species are the closest relatives among phototrophic bacteria. *Blastochloris* species could be well separated from other PNSB based on morphological and cultural characteristics. Their green to olive-green coloration, long and wavelength absorption maximum above 1000 nm, is due to bacteriochlorophyll *b,* clearly differentiate *Blastochloris* species from other phototrophic *Alphaproteobacteria.* The cell shape, mode of growth and cell division, formation of rosette-like aggregates, lamellar type internal photosynthetic membranes, as well as a number of chemotaxonomic characteristics and the phylogenetic properties of 16S rDNA, differentiate *Blastochloris* species from other species **(Tables 3. 11** and **Fig 3. 21)**.

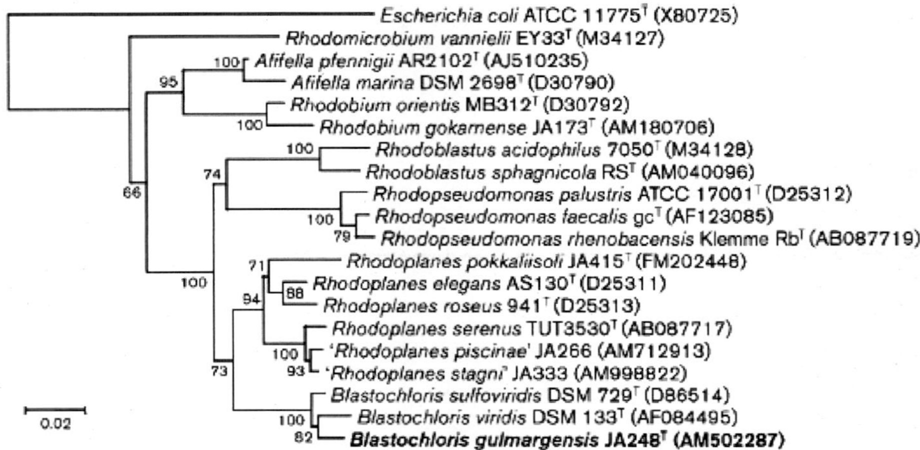

Fig. 3.21. Phylogenetic tree of *Blastochloris* species and related *Alphaproteobacteria* of the order *Rhizobiales* based on 16S rRNA sequences. The sequence of *E.coli* was used as an outgroup (Venkata Ramana *et al.,* 2011).

Taxonomic Comments

Blastochloris viridis and *Blc. sulfoviridis* of this genus have been previously designated to the genus *Rhodopseudomonas,* due to significant similarities in cell morphology and cell division to *Rhodopseudomonas palustris.* Significant differences in 16S rDNA sequences with respect to other purple non-sulfur bacteria are responsible to reconsider the taxonomic position of these bacteria and restudy the phenotypic differences from the type species of *Rhodopseudomonas,* which are essential to their reclassification into the new genus *Blastochloris* as *Blastochloris viridis* and *Blastochloris sulfoviridis* (Kawasaki *et al.*, 1993b; Hiraishi, 1997). Another species *Blc. gulmargensis* was added to the genus *Blastochloris* which shows 98.5% and 98.4% similarity with *Blastochloris sulfoviridis* and *Blc. viridis* respectively of the class *Alphaproteobacteria* (Venkata Ramana *et al.,* 2011).

3. *Rhodomicrobium* (*Rmi.*) (Red microbe) (Duchow and Douglas, 1949) Imhoff (2015)

Cells are ovoid to elongate-ovoid shaped, 1.0-1.2 μm in diameter, motile utilizing polar flagella **(Fig. 3.22)** (Duchow and Douglas, 1949) and a characteristic vegetative growth cycle occurs. This cycle includes the formation of peritrichously flagellated swarmer cells and non-motile "mother cells" that form filaments from one to several times the length of the mother cell. Daughter cells originate as spherical buds at the end of the filaments may undergo differentiation in various ways. Dry-resistant polyhedral exospores may be formed. Cells have lamellar type intracytoplasmic membrane contain bacteriochlorophyll *a* and carotenoids of spirilloxanthin series as photosynthetic pigments. The predominant cellular fatty acid is $C_{18:1}$, which consists of more than 80% of the membrane-bound fatty acids. Ubiquinone (Q-10) and rhodoquinone (RQ-10) with 10 isoprene units are present and the lipopolysaccharides are determined by glucosamine containing phosphate-free lipid A with amide bound $C_{16:0\ 3OH}$. Small type cytochrome c_2 is present. All strains utilize pyruvate and succinate but not tartrate, citrate, format, or benzoate.

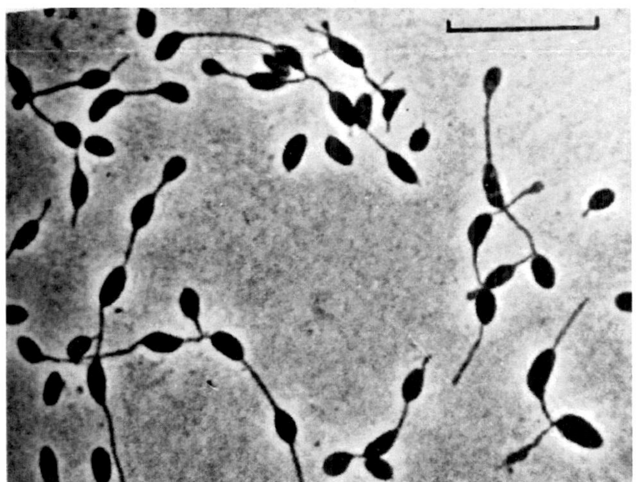

Fig. 3.22. Phase-contrast micrograph of cells of *Rhodomicrobium vannielii* with filaments and buds of various sizes (Bar = 10µm).

They grow preferably photoheterotrophically under anaerobic conditions in the presence of light. Photoautotrophic growth may be possible with hydrogen and sulfide as electron donors. Various organic substrates as carbon and electron sources, molecular hydrogen, ferrous iron and sulfide (at low concentration) may be used as photosynthetic electron donors. Growth factors are not required. Cells may also grow under microaerobic to aerobic conditions in the dark. The optimum growth temperature and pH are 25-35°C and 6.0-7.0 respectively. Mesophilic freshwater bacteria with the presence of acidic pH between 5.2 and 6.5 are suitable habitats. Dinitrogen and ammonia are the best nitrogen sources, ammonia assimilated via glutamine synthetase/glutamate synthase (NADH-dependent). Sulfate can be used as the sole sulfur source and is assimilated via adenosine-5'phosphosulfate. The mol% G+C of the DNA is 61.8-63.8 (Bd). *Rhodomicrobium vannielii* (type species) and *Rmi. udaipurense* are two validly published species and their characteristics feature are presented in **Table 3.18.**

Table 3.18. Differential Characteristics of Species of *Rhodomicrobium*

Characteristics	***Rmi. vannielii***	***Rmi. udaipurense***
Habitat	Freshwater	Freshwater
Cell size (width 6 length, µm)	1.0–1.2 X 2.0–2.8	1.2–14 X 2.8–3.5
Cell shape	Ovoid - lemon shaped	Ovoid to Swollen rods
Type of budding	Tube	Tube
Motility	+	+
Rosette formation	Complex aggregates	Complex aggregates
Colour of cell suspension	Salmon Pink-Orange brown-Red	Reddish-brown
Bacteriochlorophyll	*a*	*A*
Carotenoids	Rhodopin, β-carotene	Spirilloxanthin series
Internal membrane system	Lamellae	Lamellae
Absorption maxima of living cells	378, 461, 488–490, 522–525, 800–807, and 869–872	377, 512, 595 and 860
Optimal pH (range)	6.0 (5.2-7.5)	6.5-7.5 (5.5-8.0)
Optimum temperature (range)	30 (25-35)	30 (10-40)
Salt requirement	Not required	Not required
Sulfate assimilation	+ (APS)	+
Photoautotrophic growth with	H_2, sulfide	H_2, sulfide, thiosulfate
Aerobic dark growth	+	–
Growth factors	Not required	Complex
Major polar lipids	PE, DPG, PG, PC, PL, L1-2, AL1-9	PE, DPG, PC, PL, AL1–6,9
Major hopanoids	BHD1, BHD2, DPL, UH1–4	BHD1, BHD2, DPL, DPE, UH1-2,4,5
Carbon sources/Electron donors		
Aspartate	–	(+)
Caproate	+	–
Ethanol	+	–
Glutamate	–	+
Glycerol	+	–
Malate	(+)	+
Propanol	+	(+)
Butanol	+	(+)
Propionate	+	+
Succinate	(+)	+
Tartarate	–	+
Valerate	+	(+)

Characteristics	*Rmi. vannielii*	*Rmi. udaipurense*
Utilization of sulfur source		
Thioglycolate	–	+
Cysteine	–	+
DNA G+C content (mol%)	64.2	62.4

Symbols: +, Substrate utilized/present; 2, substrate not utilized/absent; (+), weak growth. (APS), via adenosine–5′–phosphosulfate; AL1–9, Unidentified aminolipids; PE, phosphatidylet-hanolamine; DPG, diphosphatidylglycerol; PC, phosphatidylcholine; PG, phosphatidyl phosphatidylglycerol; PL, unidentified phospholipid; L1,2, unidentified lipids. BHD1,2, Bacteriohopane derivatives; DPL, diplopterol; DPE, diploptene; UH1–5, unidentified hopanes. Data for *Rhodomicrobium udaipurense* and *Rmi. vannielii* is adopted from Venkata Ramana *et al.* (2013) and Duchow and Douglas (1949).

Differentiation of the *Rhodomicrobium* from other Genera

The polar growth mode, the budding type of multiplication and the lamellar type of the internal photosynthetic membranes are similar to those of other budding purple non-sulfur bacteria. Prostheca formation and the characteristic growth cycle are the most distinct characteristic properties. According to rRNA-DNA hybridization studies, *Rhodomicrobium* is clearly distinguished from other PNSB (Gillis *et al.*, 1982). Other essential characteristics are the compositions of lipid A, polar lipids and fatty acids. The important characteristic of *Rhodomicrobium* from other genera of *Rhizobiales* are presented in **Table 3.11**.

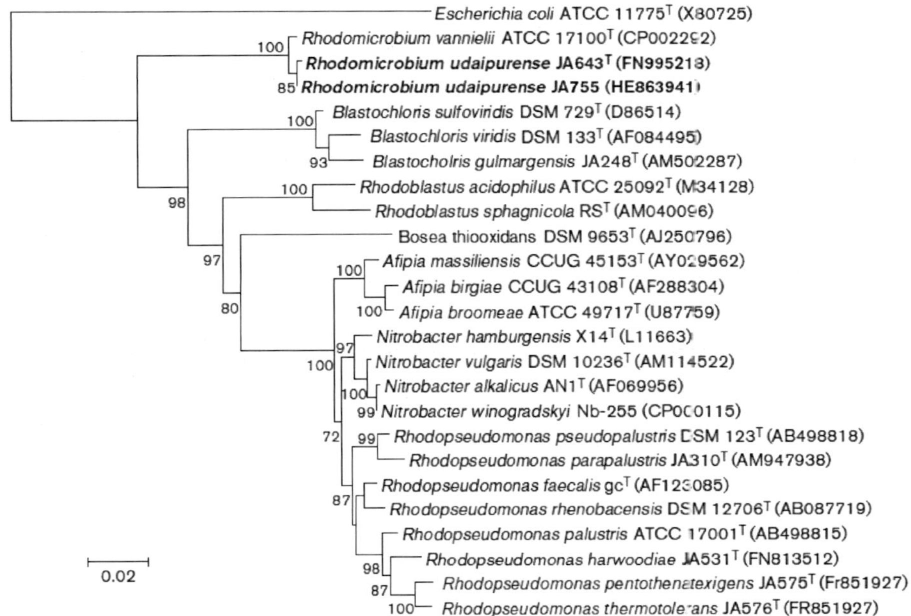

Fig. 3.23. Phylogenetic tree of *Rhodomicrobium* species and related *Alphaproteobacteria* of the order *Rhizobiales* based on 16S rRNA sequences. The sequence of *E. coli* was used as an outgroup (Venkata Ramana *et al.*, 2013).

Taxonomic Comments

Based on 16S rDNA sequence analysis, *Hyphomicrobium vulgare, Blastochloris viridis,* and *Rhodopseudomonas palustris* are the phylogenetic relatives of *Rhodomicrobium* (Kawasaki *et al.*, 1993b). Based on DNA-DNA hybridization data, cell morphology, polar lipid composition, fatty acid profile, metabolite fingerprinting, hopanoid composition and organic substrate utilization, *Rhodomicrobium* was reclassified into a separate genus. The phylogenetic relationship of these genera based on 16S rRNA sequence is shown in **Fig. 3.17**and **3.23** (Venkata Ramana *et al.,* 2013).

Family 3. *Rhodobiaceae* (Garrity *et al.,* 2006) Garrity *et al.* (2015)

The family *Rhodobiaceae* was described on the basis of phylogenetic analysis of 16S rRNA sequences (Garrity *et al.,* 2005). The family includes a diverse group of organisms, despite representing a monophyletic clade that comprises organisms that share mainly characteristics commonly encountered in many other *Alphaproteobacteria*, namely fatty acids, polar lipids and respiratory quinones. Four of the species of this family possess bacteriochlorophyll *a* are facultative photoheterotrophs and may even capable of anaerobic photo lithotrophic growth. Other species are strictly chemoorganotrophic. The morphology is also diverse ranging from rod-shaped cells to cocci that divide by budding or asymmetric cell division. Some species produce prosthecae or form irregular appendages and hyphal-like budding structures. Photosynthetic membranes present as lamellar stacks. Do not grow under aerobic conditions in the dark and under anaerobic conditions in the light. The species of the genera have ubiquinone U-9, U-10, U-11and /or menaquinone (MK-10) as the major lipo quinones. Unsaturated straight-chain fatty acids are the predominant acyl groups of the family. $C_{18:1}\omega7c$, $C_{18:1}$ $\omega9c$, $C_{19:0}$ cyclo $\omega8c$ and $C_{18:1}$ $\omega7c$ 11-methyl. $C_{16:0}$ is the major fatty acid. Hydroxy fatty acids are also present in members of this family. Require NaCl for growth. *Rhodobium* is the type genus (Hiraishi *et al.,* 1995; Garrity *et al.,* 2005). *Afifella, Anderseniella, Lutibaculum, Parvibaculum, Rhodoligotrophos, Tepidamorphus* and *Bauldia* are the other genera of the family.

The genus *Rhodobium* was initially described to include a species previously named *Rhodopseudomonas marina* (Imhoff 1983), as *Rhodobium marinum* and the newly described species named *Rhodobium orientis* (Hiraishi *et al.,* 1995). Later, Caumette *et al.* (2007) described a species that was named *Rhodobium pfennigii*. In the same year the species *Rbi. gokarnense* was described by Srinivas *et al.* (2007). These organisms have a marine origin, possess bacteriochlorophyll *a* and share morphological, physiological and biochemical characteristics but who belong to two clades. The taxonomy of the genus *Rhodobium* was later questioned because 16S rRNA gene sequence analysis showed that the validly named species of the genus *Rhodobium* did not form a monophyletic clade (Urdiain *et al.,* 2008). These authors, therefore, proposed the genus *Afifella,* not only based on phylogenetic analysis but also based on the phenotypic and chemotaxonomic characteristics for the species formerly named as *Rbi. marinum* and *Rbi. pfennigii* and changed to *Afifella marina* and *Afifella pfennigii* .

1. *Rhodobium* (*Rbi.*) (Red life) (Hiraishi *et al.*, 1995) Imhoff and Hiraishi, 2015

Cells are ovoid to rod-shaped, 0.7 to 0.9 μm wide, and 1.0 to 3.2 μm long. Motile by means of polar or randomly distributed flagella. Cells multiply by budding and asymmetric cell division **(Fig. 3.24 A&B)**. Rosette formation is rare. Gram (-) negative. Phototrophically grown cells contain intracytoplasmic membranes of the lamellar type together with bacteriochlorophyll *a* and carotenoids belonging to the spirilloxanthin series. Ubiquinone 10 and menaquinone 10 are the major quinones. Facultative photoheterotrophs that grow anaerobically in light or aerobically in dark. Photosynthetic cultures are pink to red, while aerobic chemotrophic cultures are colorless. The absorption maxima of living cells of *Rbi. orientis* and *Rbi. gokarnense* respectively are 377, 468, 500, 591, 802 and 870 and 370, 402, 488, 530, 590, 803 and 870 nm. Mesophilic and halophilic. Optimal growth occurs at 25 to 35°C, at pH 6.9 to 7.5, and in the presence of 1 to 5% NaCl. Photoorganotrophy with simple organic compounds as carbon sources is the preferred mode of growth. Straight chain saturated and monounsaturated $C_{18:1}$ and $C_{18:0}$ are the major components of cellular fatty acids with $C_{18:1}$ ω7c as predominant. Phosphatidylglycerol (PG), diphosphatidyl glycerol (DPG), posphatidylcholine (PC), phosphatidylethanolamine (PE), an unknown amino lipid (AL) and an unknown glycolipid (GL), phosphatidyl monomethyl ethanolamine (PME) are the major polar lipids. Spermidine (SPD) and spermine (SPM) are predominant polyamines. The G+C contents of the DNAs range from 61.5 to 65.7 mol%. The phylogenetic position is in the alpha-2 subclass of the *Proteobacteria* . *Rhodobium orientis* (Hiraishi *et al.,* 1995) is the type species. *Rhodobium gokarnense* and *Rhodobium orientis* are validly published species. Distinguishing characteristics of these species are precised in **Table 3.19.**

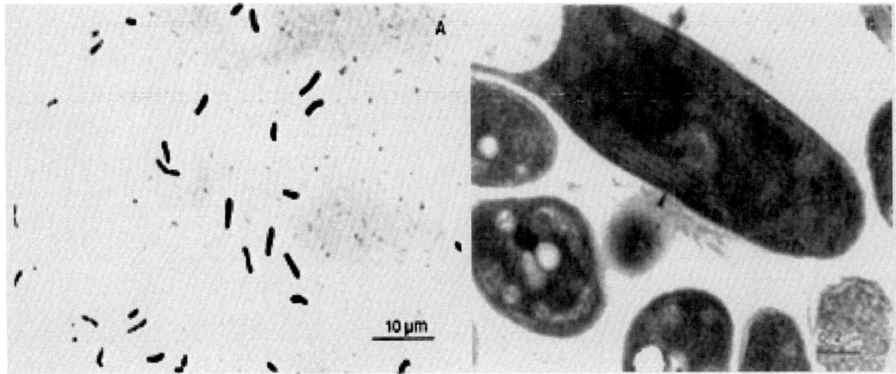

Fig. 3.24. A. Phase-contrast photomicrograph of intact cells, B. Transmission electron micrograph of an ultrathin section of cells showing the general morphology and ultrastructure of *Rhodobium orientis*.

Table 3.19. Differential Characteristics of Species of *Afifella, Rhodobium* and *Roseospirillum* (Monotypic)

Characteristics	***Rbi. gokarnense***	***Rbi. orientis***	***Afifella marina***	***Afifella pfennigii***	***Rss. parvum***
Habitat	Saltern	Seawater	Seawater	Brackish ponds	Saltmarsh
Cell size (μm)	0.5–0.6 x 1.0–2.0	0.7–0.9 x 1.5–3.2	0.7–0.9 x 1.0–2.5	0.7–0.9 x 1–2.5	0.4–0.6 x 1.8–2.6
Morphology	Rods	Rods	Rods	Rods	Vibrioid, spirilloid
Flagella	Polar	Polar	Polar	Polar	Bipolar
Tpe of budding	Sessile	Sessile	Sessile	Sessile	ND
Rosette formation	–	+/–	–	+/–	+
Motility	–	+	+	+	+
BChl *a* absorption peak (nm)	872	870	883	909	911
Colour of the cell suspension	Pink to Pinkish-red	Pink, Red	Pink, Red	Pink	Pink, Red
Bacteriochlorophyll	*a*	*a*	*a*	*a*	*a*
Carotenoids	Spirillo-xanthin	Spirillo-xanthin	Spirillo-xanthin	Spirillo-xanthin	Spirillo-xanthin, Lycopenal
Cytoplasmic membrane	Lamellae	Lamellae	Lamellae	Lamellae	Lamellae
Whole cell absorption (nm)	370, 402, 488, 530, 590, 803 and 872	377, 468, 500, 530, 591, 802, 870	375, 483, 516, 533, 590, 803, 883	395, 483, 512, 548, 594, 909	380, 492, 515, 549, 595, 806 and 911
Whole cell FTR transmission maxima (cm–1)	3420, 2969, 2926, 2857, 1728, 1651, 1541, 1404, 1231, 1059, 575	3445, 2969, 2926, 2857, 1655, 1588, 1460, 1707, 621	ND	ND	ND
NaCl range (optimum)	0.5–10 (2.0–6.0)	2.0–8.0 (4.0–5.0)	1–5 (1–3)	0–5 (1–2)	1.0–6.0 (1.0–2.0)
Optimum Temperature (◦C)	30	30–35	25-30	30–35	30
pH range (opimum)	5.0–9.0 (6.5–8.0)	6.0–8.5 (7.0–7.5)	ND (6.9–7.1)	6.3–8.5 (7.2–7.5)	ND (7.9)
Aerobic dark growth	(+)	+	(+)	+	+
Denitrification	–	+	–	+	+
Fermentative growth on fructose	–	–	+	+	+
Photolithoautotrophc growth	–	+	+	+	–

Characteristics	*Rbi. gokarnense*	*Rbi. orientis*	*Afifella marina*	*Afifella pfennigii*	*Rss. parvum*
Vitamins required	Yeast extract	Biotin, PABA	Complex	Thiamine, Yeast extract	Complex
Major quinones	Q–10, MK–10	Q–10, MK–10	Q–10, MK–10	Q–10, MK–10	ND
DNA G+C content (mol%)	72.4	65.2–65.7	61.5–64.1	67.5	71.2
Utilization of carbon/ electron donor					
Acetate	+	+	+	+	+
Aspartate	–	–	–	+	ND
Butyrate	+	+	+	+	+
Caproate	–	+	+	ND	ND
Citrate	+	–	–	–	–
Caprylate	–	–	+/–	ND	ND
Ethanol	–	–	+/–	–	–
Fumarate	+	+	+	+	+
D–fructose	–	+	+	+	+
Formate	–	(+)	(+)	–	–
Glycerol	–	(+)	(+)	–	ND
Glutamate	–	–	+	+	+
Lactate	–	+	–	+	+
Mannitol	+	+	+	–	ND
Malate	+	+	+	+	+
Proponal	–	+	+/–	–	ND
Propionate	–	+	+	+	+
Pyruvate	+	+	+	+	+
Succinate	+	+	+	+	+
Valerate	–	+	+	+	+
Casamino acids	+	+	+	+	+
2-oxoglutarate	ND	ND	ND	+	ND
Electron donors					
Sulfide	–	+/–	+	+/–	+
Thiosulfate	–	+/– +	+	+	

Symbols: +, Substrate utilized by many strains; –, substrate not utilized by many strains ; (+), weak growth /+ – and (+)/–. variable in different strains; ND, not determine/no data available. Data for species of *Afifella, Rhodobium* and *Roseospirillum* is adopted from Urdiain *et al.* (2009), Caumette *et al.* (2007), Hiraishi *et al.* (1995), Sinivas *et al.* (2007) and Glaeser and Overmann (1999).

The genus *Rhodobium* consists of species of the PNSB that are truly marine or slightly halophilic bacteria, multiply by budding or asymmetric cell division and contain internal photosynthetic membrane systems of the lamellar type. In this respect, *Rhodobium* species are similar to other bacteria in the group of phototrophic bacteria belonging *alpha-2-proteobacteria* such as *Rhodoplanes, Blastochloris* and *Rhodopseudomonas*. The genus *Rhodobium* is differentiated from the latter genera by its phylogenetic position, natural habitats and chemotaxonomic characteristics. Major differentiating properties between *Rhodobium* species with other proteobacteria such as *Afifella* and *Roseospirillum* are shown in **Tables 3.11** and **3.19**. The phylogenetic relationships of these bacteria based on 16S rDNA and 16S rRNA sequence are shown in **Fig. 3.17** and **3.21** .

Taxonomic Comments

At present, the genus *Rhodobium* consists of two species, *Rbi. orientis* and *Rbi. marinum.* After tentative assignments to *Rhodopseudomonas palustris* (Imhoff and Truper, 1976), primarily based on morphological and physiological characteristics, the latter species was described as *Rhodopseudomonas marina* (Imhoff, 1983), Later, phylogenetic analysis based on 16S rDNA sequences revealed that this species was far distant from *Rhodopseudomonas palustris,* the type species of the genus *Rhodopseudomonas,* but closely related to new isolates from marine environments (Hiraishi *et al.*, 1995d). These new isolates were assigned to the new genus *Rhodobium* and described as the type species or this genus, *Rhodobium orientis* and *Rhodopseudomonas marina* were transferred to this genus as *Rhodobium marinum* (Hiraishi *et al.*, 1995d), Mangels *et al.* (1986) isolated an N_2-fixing marine strain of PNSB and designated it *"Rhodopseudomonas marina* var, *agilis"*. Upon reinvestigation, no differences in physiological properties and 16S rDNA sequence between the type strain of *Rhodobium marinum* and *"Rhodopseudomonas marina* var. *agilis"* were found (Hiraishi *et al.*, 1995d), confirming the phenotypic and genotypic coherency of the two organisms at the species level. Therefore, "var. *agilis"* isolates should be considered as strains of *Rhodobium marinum.*

2. *Afifella* (*Af.*) Urdiain *et al.* (2008)

Afifella cells are rod-shaped, motile, divide by budding and asymmetrical cell division (**Fig. 3.25 A&C**). Phototrophically grown cells contain intracytoplasmic membranes of the lamellar type (**Fig. 3.25 B**) (Urdiain *et al.,* 2008) containing bacteriochlorophyll *a* and carotenoids of spirilloxanthin series. The absorption maxima of living cells for *A. marina* and *A. pfennigii* respectively are 375, 483, 516, 533, 590, 803 and 883 and 395, 483, 512, 548, 594 and 909 nm. The major fatty acid is $C_{18:1}\omega7c$. Aerobically grown species under dark conditions exhibit a quinone system with ubiquinone containing 10 isoprene units (U-10) as predominant quinone and a significant amount of menaquinone containing 10 isoprene units (MK-10). Sym-homospermidine (SPD) is the predominant polyamine. The major polar lipids are diphosphatidyl glycerol (DPG), phosphatidylglycerol (PG), phosphatidylcholine (PC), phosphatidylethanolamine (PE), an unknown amino lipid (AL) and an unknown glycolipid (GL). Phosphatidyl monomethyl ethanolamine (PME) is also present. Mesophilic marine bacteria that require NaCl for growth and grow in seawater and environments with higher salinities. Chemoorganotrophic growth in the dark can occur aerobically or microscopically and in some cases fermentatively. Photoheterotrophic growth occurs anaerobically in the light and photolithotrophic growth occurs with thiosulfate. The DNA G+C content varies between 62.2 and 67.5 mol%. The species of this genus have been isolated from saline environments. *Afifella marina* is the type species. Two species are validly published: *Afifella marina* (Urdiain *et al.*, 2008) and *Afifella pfennigii* (Caumette *et al.*, 2007). The distinguish in characteristic properties and carbon utilization by two species are presented in **Table 3.19.**

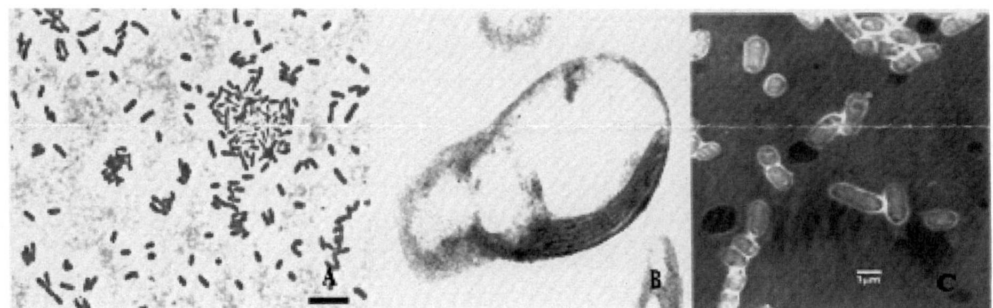

Fig. 3.25. A. Phase-contrast photomicrograph of cells of *Afifella pfennigii* (Bars= 5μm), B. Transmission electron micrograph of an ultrathin section of cells (Bars= 0.1μm) (Caumette *et al.,* 2007), C. Transmission electron micrograph of negatively stained *Afifella marina* showing inserted subpolar flagella and the asymmetric division (Bar= 0.1μm).

Differentiation of the *Afifella* from other Genera

The important characteristics of *Afifella* from other genera of *Rhizobiales* are presented in **Table 3.11**. Differences in chemotaxonomic traits such as glycolipid and polyamine compounds containing *sym*-homospermindine and 16S rRNA phylogenetic analysis, the *Rhodobium marinum* and *Rbi. pfennigii* transferred into a new genus *Afifella* as *Afifella marinum* and *Afifella pfennigii.* According to DNA-DNA hybridization studies, *Afifella* is clearly distinguished from other PNSB (Urdiain *et al.,* 2008).

Taxonomic Comments

Based on 16S rRNA sequence analysis, *Afifella* is the phylogenetic relative of *Rhodobium* and *Anderseniella* (Brettar *et al.,* 2007). Based on DNA-DNA hybridization data, fatty acid profile, unique glycolipids, polyamine pattern and organic substrate utilization, *Afifella* was reclassified into a separate genus. Based on 16rRNA sequence analysis it is clear that *Rbi. orientis* and *Rbi. gokarnense* which exhibited similarly of 94.0%, <94.4% and <94.0% with species of *Rbi. marinum, Rbi. pfennigii* and *Rhodopseudomonas julia* respectively. Hence two species such as *Rbi. marinum* and *Rbi. pfennigii* are reclassified into a new genus *Afifella* as *Afifella marina* and *Afifella pfennigii*. The phylogenetic relationship of these genera based on 16S rRNA sequence is shown in **Fig. 3.21** (Venkata Ramana *et al.,* 2011) and **3.26** (Urdiain *et al.,* 2008).

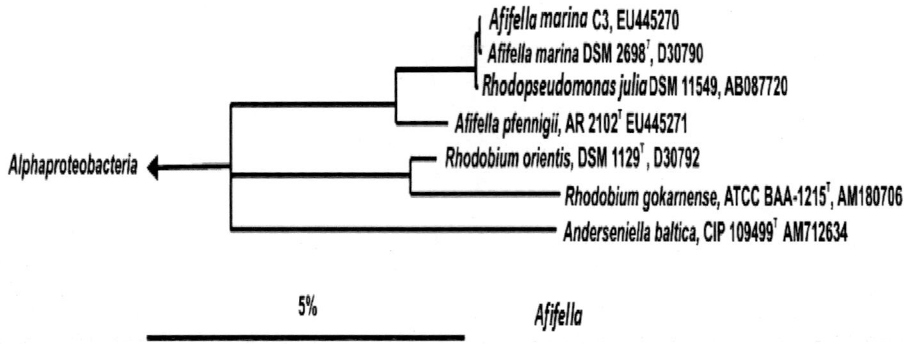

Fig. 3.26. Phylogenetic tree of *Afifella* species and *Rhodobium* species with other *Alphaproteobaceria* of the order *Rhizobiales* based on 16S rRNA sequences (Urdiani *et al.,* 2008).

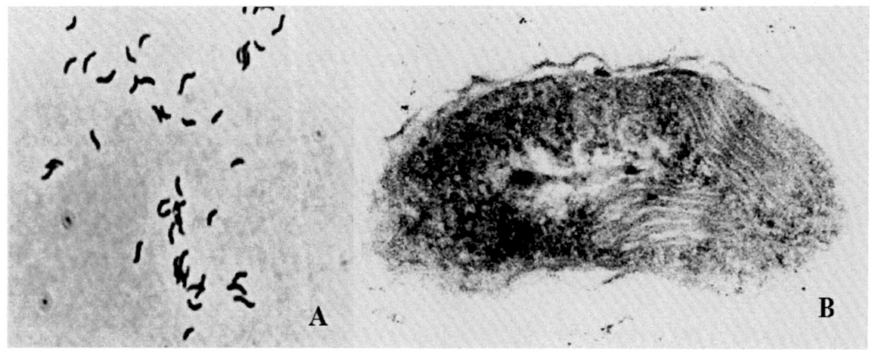

Fig. 3.27. A. Phase contrast micrograph of *Roseospirillum parvum,* B. Ultra-thin longitudinal section showing intracellular membrane system with lamellar stacks (Bar=0.3 µm).

3. *Roseospirillum* (*Rss.*) (The Rosy spirillum) (Glaeser and Overmann, 2001) Imhoff (2015)

Cells are vibrioid to spiral-shaped (**Fig. 3.27A**) (Glaeser and Overmann, 1999) and motile by means of bipolar monotrichous flagellation. Cells have an intracytoplasmic membrane of the lamellar type (**Fig. 3.27B**) and contain bacteriochlorophyll *a* and carotenoids of spirilloxanthin and lycopenal as photosynthetic pigments. The colour of the cell suspension is pink to pinkish-red. L(+)-alanine, and L(+)-glutamate are assimilated in the presence of organic substrates. Sulfide, thiosulfate, or elemental sulfur served as a reduced sulfur source for photosynthetic growth. Sulfide or thiosulfate served as an electron donor for anoxygenic phototrophic growth. Unable to oxidize elemental sulfur to sulfate. Optimum growth at pH 7.9, temperature 30ºC and salinity of 1-2% NaCl. Cells grow photolithoautotrophically and photoheterotrophically under anoxic conditions in the presence of light. The growth is also possible at microaerobic conditions in the dark (Chemotrophic growth). The G+C contents of the DNAs range are 71.2 mol%. The phylogenetic position is in the *Alphaproteobacteria* and is most closely related to *Rhodobium*. The genus is monotypic with single species *Roseospirillum parvum* (Glaeser and Overmann, 1999) and their characteristics are précised in **Table 3.19**.

Differentiation of *Roseospirillum* from other Genera

The major differentiating characteristics between *Roseospirillum parvum* from other *Rhodospirillaceae* members are tabulated in **Table 3.11**. The phylogenetic relationship of these bacteria from other phototrophic bacteria based on 16S rDNA sequences are illustrated in **Fig. 3.28** (Glaeser and Overmann, 1999).

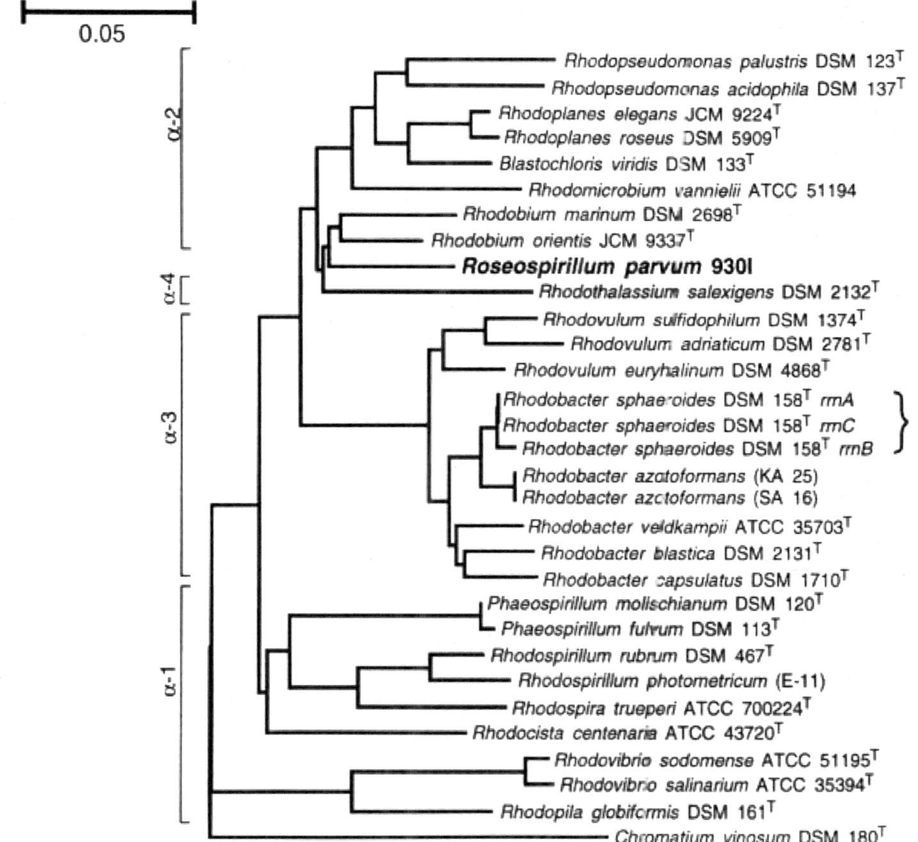

Fig. 3.28. Phylogenetic tree of *Roseospirillum* species and other *Alphaproteobacteria* based on 16S rRNA sequences (Glaeser and Overmann, 1999).

Taxonomic Comments

Based on their phylogenetic position, lipid composition and pigment composition these bacteria have been reclassified (Glaesier and Overmann, 1999). Based on 16S rRNA sequence analysis of *Roseospirillum parvum* strain 930I showed that these strain formed a subgroup together with *Rhodobium orientis* within the *Rhodobiaceae* family of the *Alphaproteobacteria*, which is well separated from *Rhodospirillum* genus and their closest relatives (**Fig. 3.28** .) *Roseospirillum parvum* strain 930I shows 93.5% similarity to the sequence of *Rhodobium orientis* JCM 9337 (Glaesier and Overmann, 1999). Similarly, *Rhodothalassium salexigens* is also showed 90.5% similarity with *Roseospirillum parvum* . Incidentally, a closer phylogenetic relationship of *Rhodothalassium salexigens* to *Roseospirillum parvum* , *Blastochloris viridis* , *Rhodomicrobium vannielii* and *Rhodobium* species than to *Rhodovibrio sodomense* and *Rhodovibrio salinarium* was observed. These findings are in contrast to a previous analysis which has been limited to the spiral-shaped phototrophic purple non-sulfur bacteria (Imhoff *et al*,. 1998).

Order 3. *Rhodobacteriales* (Garrity *et al.*, 2005) Garrity *et al.* (2015)

The order *Rhodobacteriales* was described on the basis of phylogenetic analysis of 16S rRNA sequences. The order contains the family *Rhodobacteraceae* .

Family 1. *Rhodobacteraceae* Garrity *et al.* (2015)

The family *Rhodobacteraceae* was described on the basis of phylogenetic analysis of 16S rRNA sequences. The family comprises 31 genera such as *Rhodobacter* (type genus), *Ahrensia, Albidovukum, Antarctobacter, Amaricoccus, Gemmobacter, Hirschiaphomonas, Jannaschia, Ketogulonicigenium, Leisingera, Maricaulis, Methylarcula, Octadecabacter, Pannonibacter, Paracoccus, Pseudorhodobacter, Rhodobaca, Rhodovulum, Roseibium, Roseinatronobacter, Roseivivax, Roseobacter, Roseovarius, Rubrimonas, Ruegeria, Sagittula, Staleya, Stappia* and *Sulfitobacter. Rhodobacter, Rhodobacca, Rhodovulum* and *Rubribacterium* are phototrophs, hence only 4 genera are described briefly in this chapter. *Rhodobacteraceae* are fundamentally aquatic bacteria that frequently thrive in marine environments. They comprise mainly aerobic photo- and chemoheterotrophs but also purple non-sulfur bacteria which perform photosynthesis in anaerobic environments. They are deeply involved in sulfur and carbon biogeochemical cycling and symbiosis with aquatic micro-and macroorganisms.

Family is phenotypically, metabolically and ecologically diverse. It includes photoheterotrophs which can also grow photoautotrophically or chemotrophically under suitable environmental conditions. They grow as chemoorganotrophs, which are strictly aerobic or facultatively anaerobic respiratory metabolism, facultatively fermentative organism and facultative methylotrophs. Some are denitrifying. They may require NaCl for growth. Cells are Gram (-) negative, multiply by binary fission or by budding, following monopolar growth. Motile, they exhibit flagella, positive for oxidase. Bacteriochlorophyll *a* and carotenoids of the sphaeroidene class are present as photosynthetic pigments. Carbon reserve material formed polyhydroxyalkanoates (PHA) and commonly polyhydroxy-butyrate (PHB). The major or only respiratory quinone is ubiquinone (U-10) and their cellular fatty acids are usually contain $C_{18:1}$ ω7c. Polar lipids may include phosphatidylglycerol (PG), diphosphatidyl glycerol (DPG), phosphatidylethano-lamine (PE) and several amino-phospho- and glycoprotein combinations. The mol% G+C content of DNA is above 50%, the most common range 47-76 mol%. *Rhodobacter* is the type genus.

1. *Rhodobacter* (*Rba.*) (Imhoff *et al.*, 1984) Imhoff (2015)

The genus *Rhodobacter* was established to accommodate species of purple non-sulfur bacteria with vesicular internal membranes (Except in *Rba. balsticus*) and oval to rod-shaped cells divide by binary fission. Cell diameter ranges from 0.5-1.2 μm, motile or nonmotile, motile forms have polar flagella **(Fig. 3.29)**. They may produce capsules, slime and form chains of cells. Phototrophically grown cultures are yellow-brown to yellow-green, but turn pink or red when exposed to oxygen. It contains bacteriochlorophyll *a* (esterified with phytol) and carotenoids of spheroidene series with sphaeroidene as the major components, except for *Rba. viridis* (Shalem Raj *et al.*, 2013) which contains nuerosporene as a major carotenoid. Ubiquinone (Q-10) is the major quinone, major fatty acids are $C_{18:0}$ and $C_{16:0}$ saturated and monosaturated fatty acids with $C_{18:1}$ as the predominant component. Sulfate is used as sulfur sources, photoautotrophic growth is possible with sulfide as an electron donor, while some species grow with thiosulfate and molecular hydrogen as electron donors and bicarbonate/CO_2 as carbon source (fixed through Calvin cycle). Growth occurs photohetero-trophically under anaerobic conditions in the presence of a variety of organic compounds as carbon and electron source. Most species grow chemohetero-trophically under aerobic conditions and also by denitrification, fermentation and oxidant-dependent metabolism. Polysaccharides poly-β-hydroxybutyric acid and polyphosphates may be present as storage products. They are mesophilic freshwater bacteria with optimal growth at neutral pH and in the absence of NaCl. Vitamins are required for the growth (Maybe vitamin B_{12}, biotin, niacin, thiamin, and/or para-aminobenzoic acid). Ammonium salts are the best nitrogen sources and able to fix molecular nitrogen. The optimal temperature and pH for growth are 25-35°C and pH 6.5-7.5 respectively. The mol% G+C of the DNA is 64.4 –73.2. *Rba. capsulatus* (Molisch 1907) is the type species.

At present, the species of *Rhodobacter* are separated into two clades sugge-sting a polyphyletic nature for the genus. In one hand, *Rba. capsulatus* joins *Rba. aestuarii* ,*Rba. maris* ,*Rba. vinaykumarii* and *Rba. viridis* in a group that is more closely related to *Haemotobacter* spp. then to the rest of *Rhodobacter* spp. On the other hand, the species *Rba. azotoformans* ,*Rba. johrii, Rba. megalophillus, Rba. ovatus* and *Rba. sphaeroides* form another clade that is closer to the *Gemmobacter* species. *Rba. blasticus* is marginally related to this group and *Rba.*

veldkampii occupies an isolated position among the main lineages. Recently, *Rhodobacter massiliensis* is reclassified as *Haematobacter massiliensis* .

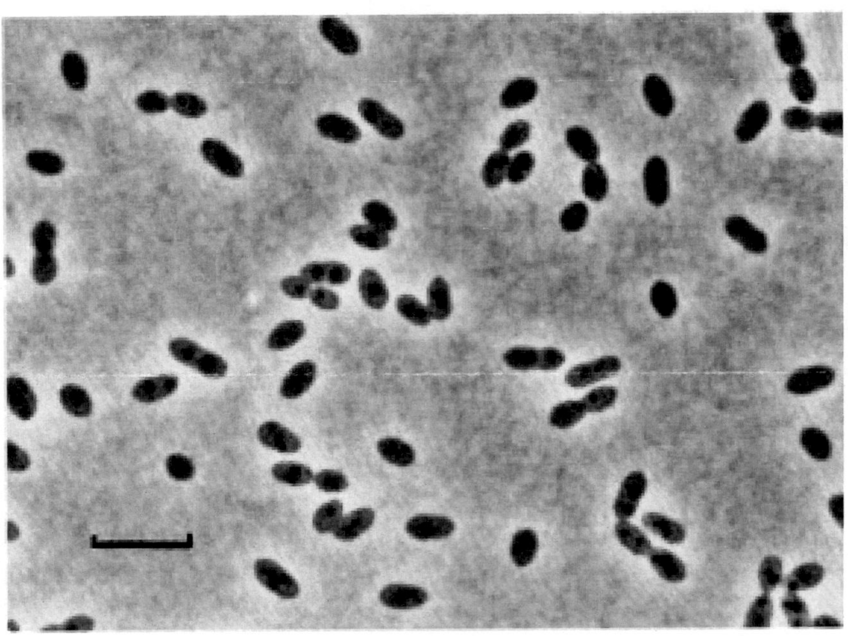

Fig. 3.29. Phase-contrast micrograph of *Rhodobacter sphaeroides* (Bar= 5 μm).

The marine representatives of the genus *Rhodobacter* is *Rba. adriaticus* (Imhoff *et al.*, 1984), *Rba. euryhalinus* and *Rba. sulfidophilus* have been transferred to the genus *Rhodovulum* based on their habitat, salt requirement and separate clustering from their freshwater counterparts and formation of a distinct clade based on 16S rRNA gene analysis. Ten more species *Rhodovulum iodosum* , *Rdv. imhoffii* , *Rdv. steppense, Rdv. lacipunicei, Rdv. kholense, Rdv. marinum* , *Rdv. robiginosum* , *Rdv. strictum* , *Rdv. visakhapatnamense* and *Rdv. phaeolacus* have been added to the genus *Rhodovulum* .

Although *Rhodobacter* spp. and *Rhodovulum* spp. differ in NaCl requirement but they share some common chemotaxonomic characteristics such as large type cytochrome C_2, sole quinone component Q-10, lipopolysaccharides with gulucos amine as sole amino sugar in their lipid A moieties, phosphate and amide-linked 3-OH-14:0 and/or 3-oxo-14:0 and ester-linked 3-OH-10:0. These characters help to distinguish from other genera of purple non-sulfur bacteria.

A total of 15 species are validly published such as *Rba. capsulatus, Rba. sphaeroides, Rba. veldkampii, Rba. aestuarii, Rba. maris, Rba. blasticus, Rba. azotoformans, Rba. vinaykumarii, Rba. johrii, Rba. megalophilus, Rba. ovatus, Rba. azollae, Rba. lacus, Rba. sediminis* and *Rba. viridis*. Differentiating characteristics of different species of the genus *Rhodobacter* is precised in **Tables 3.20** and **3.21** .

Table 3.20. Differentiating Characteristics of Species of *Rhodobacter, Luteovulum, Fuscovulum* and *Phaeovulum*

Characteristics	*Rba. capsulatus*	*Rba. megalophilus*	*Rba. changlensis*	*Rba. azollae*	*Rba. lacus*	*Rba. sediminis*	*Rba. viridis*	*Rba. flagellatus*	*Rba. maris*	*Rba. sediicola*	*Rba. ruber*	*Rba. thermarum*
Habitat	Stagnant waters, Sewage plants, Pond, Lakes	The soil of the Indian Himalayas	Snow sample of Indian Himalayas	Fresh water pond	Freshwater lake	Sediments of lagoon	Aquatic sediment	Hot spring sample	Sediments of seashore	Fresh water pond	Fresh water crocodile pond	Hot spring
Cell width (μm)	0.5–1.2	1.2-1.5	0.8–1.0	1.0-1.5	0.8-1.0	0.5-0.7	0.5-1.2	0.8-1.0	0.6-1.0	0.6-0.9	0.5-0.7	0.7-1.1
Cell length (μm)	2.0–2.5	1.5-2.0	2.0–4.0	1.0-3.0	2.0-2.2	2.1-4.0	2.0-2.5	1.4-2.2	1.0-1.5	1.0-2.0	1.2-1.8	1.8-3.4
Cell shape	O–R, C	O	O–R, C	O–R	O–R	R	O–R	O–R	O-R, C	R-O	R	O-R
Motility	+	–	–	+	+	+	+	-	+	+	-	–
Endospores	ND	–	–	–	–	–	–	+	-		ND	–
Colour of cell suspension	YB	YB	YB	YB	YB	RB	G	P	YB	YB	DR	B
Absorption maxima of living cell (nm)	376–378, 450–455, 478–480, 508–513, 590–592, 802–805, and 860–863	374, 407, 446, 476, 509, 590, 800 and 854	380, 407, 450, 470, 512, 590, 806 & 851	377, 476, 509, 595, 800 & 857	377, 493, 512, 590, 803 and 860	378, 482, 512, 590, 800 and 860	377, 456, 482, 590, 803 and 860	340, 406–416, 476–486, 586, 772 & 966	377, 455, 476, 488, 509, 590, 803 & 860	385, 449, 477, 510, 589, 810 and 851	765	375, 450, 480, 681, 759 and 846.
Carotenoid composition (mol%)												
Neurosporene	–	–	–	ND	ND	–	98	ND	ND	ND	ND	ND
DMSE	–	–	–	ND	ND	ND	–	ND	ND	ND	ND	ND
SE	83	73	67	+	ND	+	–	+	+	+	+	+
SO	11	18	21	ND	+	+	–	ND	+	+	+	+
OH–SE	4.0	2.0	12	ND	ND	+	–	ND	ND	ND	ND	ND
OH–SO	2.0	–	–	ND	ND	–	–	ND	ND	ND	ND	ND
Me–SE	–	–	–	ND	ND	ND	–	ND	ND	ND	ND	ND
Internal membrane system	Vesicles	Vesicles	Vesicles	Vesicles	Vesicles	Vesicles	Vesicles	Vesicle	Vesicles	Vesicle	Vesicle	Lamellae
Slime production	±	-	+	–	–	+	–	–		–	–	–
NaCl required (Optimum%)	– ††	–	–	Tolerate upto 2	upto 1	Tolerate upto 3.5	NR	0-1	NR (tolerate upto 3)	NR (upto 5)	Tolerate Upto 5	Upto 3
Temperature optimum (range) (°C)	30–35 (36)	30-35 (5-40)	28-30 (5–32)	30	30 (25-35)	27-35 (12-45)	30	45 (28-50)	25-30 (25-35)	30 (25-35)	25 (20-40)	37-45 (20-55)

Characteristics	*Rba. capsulatus*	*Rba. megalophilus*	*Rba. changlensis*	*Rba. azollae*	*Rba. lacus*	*Rba. sediminis*	*Rba. viridis*	*Rba. flagellatus*	*Rba. maris*	*Rba. sediinicola*	*Rba. ruber*	*Rba. thermarum*
pH range (optimum)	6.5–7.5 (7.0)	7.0 (6.8-8.0)	6.5–8.0 (6.5-7.0)	6.0-8.0 (7.0)	7.0-8.0 (7.0)	6.0-9.0 (6.0-8.0)	6.0-9.0 (6.5-8.0)	7.0-8.0 (7.0)	5.0-8.0 (6.5-7.0)	6.0-8.0(7.0)	6.5-9.0 (8.0)	6.0-8.5 7.0-7.5)
Sulfate assimilation	+	+	+	ND	+	+	–	–		-	+	–
Denitrification	–	–	–	ND	–	–	–	–		-	+	–
Vitamins required	T (B,N)	T	B, N, T	B, T	PAB, B12, T,B2, B	T	T	ND	T	N, T	B12	NR
DNA G+C content (mol%) §§	68.1–69.6	66.7	69.4	64.6	65.5	70.6	66.4	67.7	62.85	68.4	66.2	66.0

Table continued

Characteristics	*Rba. tardus*	*Luteovulum sphaeroides*	*Luteovulum azotoformans*	*Luteovulum johrii*	*Luteovulum ovatum*	*Luteovulum alkalitolerans*	*Phaeovulum veldkampii*	*Phaeovulum vinaykumarii*	*Fuscovulum blasticus*
Habitat	fresh-water lotus pond	Stagnant waters, Sewage plants, Pond, Lakes	Sewage and activated sludge	Rhizosphere soil of Jowar	Polluted ponds	Alkaline brown pond	Stagnant waters, Sewage plants, Pond, Lakes	Tidal waters	Stagnant waters, Sewage plants, Pond, Lakes
Cell width (µm)	0.6-0.8	2.0–2.5	0.6–1.0	0.8–0.9	0.9–1.2	0.6-1.0	0.6–0.8	0.8–1.2	0.6–0.8
Cell length (µm)	1.4-2.2	2.5–3.5	0.9–1.5	1.5–1.9	1.0–2.0	1.0-1.7	1.0–1.3	1.5–3.0	1.0–2.5
Cell shape	O-R	S–O	O–R	O–R	O	R-O	O–R	R	O–R
Motility	–	+	+	+	–	+	–	–	–
Endospores	–	–	–	+	–	–	ND	–	ND
Colour of cell suspension	W	GB	YB	YB	YB	YB	YB	YB	OB
Absorption maxima of living cell (nm)	367 and 755 nm	372–375, 446–450, 474–481, 507–513, 588–590, 800–805, 850–852 and 870–880.	376, 449, 476, 510, 589, 800, and 850.	374, 446, 476, 509, 590, 800 and 854.	374, 446, 476, 509, 587, 803 and 851.	377, 476, 509, 595, 800 and 857	373, 448, 477, 510, 589, 803, and 855.	377, 452, 479, 512, 590, 803 and 851.	378, 418, 476, 506, 590, 795, and 862
Carotenoid composition (mol%)									

Chara-cteristics	*Rba. tardus*	*Luteo-vulum sphaeroides*	*Luteov-ulum azotoformans*	*Luteov-ulum johrii*	*Luteov-ulum ovatum*	*Luteov-ulum alkalito-lerans*	*Phaeovulum veldkampii*	*Phaeo-vulum vinayku-marii*	*Fuscovulum blasticus*
Neuros-porene	ND	–	–	2	–	ND	ND	–	1
DMSE	ND	–	–	–	–	ND	ND	–	–
SE	ND	46	85	82	83	+	ND	20	47
SO	ND	47	–	11	12	+	ND	77	47
OH–SE	ND	–	3	2	5	ND	ND	3	5.0
OH–SO	ND	2	1	–	–	ND	ND	–	–
Me–SE	ND	–	10	–	–	ND	ND	–	–
Internal membr-ane system	Vesicles	Vesicles	Vesicles	Vesicles	Vesicles	Vesicles	Vesicles	Vesicles	Lamellae
Slime produ-ction	–	+	+	+	–	–	–	+	–
NaCl required (Optimum %)	0-0.5 (0%)	– #	–**	¶	–	NR, tolerateuu pto 3	–	1 to 4	–
Tempe-rature optimum (range) (°C)	25 (20-35)	30–34	30–35	30 (25-35)	30 (25–35)	30 (25-35)	30–35	30 (20–30)	30-35
pH range (opti-mum)	6.0-7.5 (7.0)	6.0–8.5 (7.0)	6.0–8.0 (7.0-7.5)	6.0–8.0 (7.0)	6.0–8.0 (6.5-7.0)	6.0-1.0 (7.0)	ND (7.5)	6.0–8.0 (6.0–7.5)	ND (6.5–7.5)
Sulfate assimil-ation	–	+	+	+	+	–	–	+	–
Denitri-fication	+	±	+	–	–	–	–	–	–
Vitamins required	NR	B, N, T	B, N, T	B	B, T	B, T	B, PA-BA, T	B	B, N, T, B12
DNA G+C con-tent (mol%)§§	66.0	70.8–73.2	69.5–70.2	64–65.7	70.06	65.1	64.4–67.5	68.8	65.3

Symbols: Motile species exhibit polar flagella; +, Present/detected; –, absent/not detected/No data available; ±, variable in different strains; ND, not determine; t, trace amount; C, Chains; O, ovoid; R, rod–shaped; S, spherical; W, White; GB, Greenish brown; OB, orange–brown; YB, yellowish brown; DMSE, Demethylspheroidene; SE, spheroidene; SO, spheroidenone; OH–SE, hydroxyspheroidene; OH–SO, hydroxyspheroidenone; Me–SE, methoxyneurosporene; B, Biotin; B_{12}, vitamin B_{12}; N, Niacin; PABA, Para–aminobenzoic acid; T, Thiamine; (B, N), a few strains require biotin and/or niacin; Polar, polar carotenoid; Polar carotenoid with nine conjugated double bonds; ¶, NaCl not required, but optimal growth occurs at 3% NaCl; #, Optimal growth in the absence of NaCl but growth occurs at 3% NaCl; **, Optimal growth in the absence of NaCl but growth occurs at 5% NaCl; §§, Ranges represent values for more than one strain, including the type strain. PG, phosphatidylglycerol; PE, phosphatidylethanolamine, PL1-2, unidentified phospholipids; an unidentified amino lipid (AL); L1–4, unidentified lipids. Data for Rba. capsulatus (Imhoff, 2005); Rba. megalophilus (Aruna Sri et al., 2008); Rba. changlensis (Anil Kumar et al., 2007); Rba. viridis (Shalem Raj et al., 2013); Rba. sediminis (Subhash and Lee, 2016); Rba. maris (Venkata Ramana et al., 2008); Rba. azollae and Rba. lacus (Suresh et al., 2017), Rba. sediminicola (Suresh et al., 2020); Rba. tardus (Sheu et al., 2020); Rba. flagellatus (Xian et al., 2020); Rba. thermarum (Khan et al., 2019), Rba. ruber (Chen et al.,2021), Luteovulum sphaeroides (Imhoff, 2005); Luteovulum. azotoformans (Hiraishi et al., 1996); Luteovulum johrii (Girija et al., 2010); Luteovulum ovatum (Srinivas et al., 2008); Luteovulum alkalitolerance (Suresh et al., 2018); Phaeovulum vinaykumarii (Srinivas et al., 2007b); Phaeovulum veldkampii (Hansen and Imhoff, 1985; Imhoff, 2005); Fuscovulum blasticum (Kawasaki et al., 1993, Suresh et al., 2019).was adopted.

Table 3.21. Carbon Sources and Electron Donors used by Species of the Genus *Rhodobacter*

Carbon source/ e⁻ donor	***Rba. capsulatus***	***Rba. megalophilus***	***Rba. changlensis***	***Rba. azollae***	***Rba. lacus***	***Rba. sediminis***	***Rba. viridis***	***Rba. flagellatus***	***Rba. maris***	***Rba. sediinicola***	***Rba. ruber***	***Rba. tardus***	***Rba. thermarum***
Acetate	+	+	–	+	–	+	–	+	(+)	+	+	+	+
Aspartate	±	+	–	+	+	+	–	–	–	–	+	–	–
Benzoate	–	–	–	–	–	–	–	+	–	–	–	+	+
Butyrate	+	+	–	–	–	ND	+	–	(+)	ND	ND	–	ND
Caproate	+	+	–	–	–	ND	+	–	(+)	ND	ND	ND	–
Caprylate	+	ND	–	ND	ND	ND	+	–	–	ND	ND	ND	–
Citrate	±	–	–	–	–	–	–	–	–	+	ND	–	–
Ethanol	–	+	–	–	–	ND	–	–	–	+	ND	ND	+
Formate	+	–	–	ND	ND	ND	–	–	–	+	ND	–	+
Fumarate	+	+	–	+	+	+	–	–	(+)	+	ND	–	ND
Fructose	+	+	–	+	+	+	+	–	–	–	+	+	ND
Lactate	+	+	–	–	+	ND	+	–	(+)	ND	–	–	–
Glutamate	+	+	+	+	+	+	–	–	–	+	–	–	ND
Gluconate	–	–	–	+	+	ND	–	+	–	ND	+	+	–
Glucose	+	+	+	+	+	+	+	–	–	–	+	+	+
Glycerol	–	+	–	+	+	+	–	ND	(+)	+	–	–	–
Malate	+	+	–	+	+	ND	–	ND	+	+	–	–	–
Mannitol	±	+	+	+	+	+	+	+	–	+	+	+	+
Propanol	+	–	–	–	–	ND	–	ND	–	–	–	–	ND
Propionate	+	+	–	–	+	ND	+	ND	(+)	+	–	–	ND
Pyruvate	+	+	+	+	+	+	+	–	+	+	–	+	+
Sorbitol	±	+	+	ND	ND	+	+	–	–	+	+	–	+
Succinate	+	+	+	+	+	+	–	ND	+	+	+	+	ND
Tartarate	–	+	–	ND	–	ND	–	–	–	+	–	–	+
Valerate	+	+	–	ND	–	ND	–	–	(+)	ND	–	ND	–
Hydrogen	+	–	–	–	–	+	+	ND	–	+	ND	ND	–
Sulfide	+	–	–	–	–	+	–	ND	–	+	–	–	–
Thiosulfate	–	+	–	–	–	+	+	ND	–	+	+	+	+
Sulfur	–	ND	–	ND	ND	+	ND	+	–	+	+	+	+

Table continued

Carbon source/e⁻ donor	***Luteovulum sphaeroides***	***Luteovulum azotoformans***	***Luteovulum johrii***	***Luteovulum ovatum***	***Luteovulum alkalitolerance***	***Phaeovulum veldkampii***	***Phaeovulum vinaykumarii***	***Fuscovulum blasticus***
Acetate	+	+	+	(+)	+	+	+	+
Aspartate	–	(+)	+	–	+	+	–	ND
Benzoate	–	–	–	–	–	–	–	–
Butyrate	+	+	+	+	–	+	(+)	+
Caproate	+	–	+	+	–	+	–	ND
Caprylate	+	–	–	–	–	+	–	ND

Carbon source/e⁻ donor	Luteov-ulum sphae	teov-lum zoto-formans	*Luteo-vulum johrii*	*Luteo-vulum ovatum*	*Luteovulum alkalitolerance*	*Phaeo-vulum veldkampii*	*Phaeo-vulum vinay-kumarii*	*Fusco-vulum blasticus*
		–	–	–	+	–	–	+
		–	–	–	–	–	–	–
Citrate		+	–	–	–	–	–	–
Ethanol		+	+	–	+	+	–	+
Format		+	+	–	+	–	–	+
Fum		+	+	–	+	+	–	+
F		+	+	–	–	+	+	+
	–	–	+	–	–	(+)	–	ND
	+	+	+	–	+	+	+	+
	+	+	+	+	+	–	–	+
	+	+	+	–	+	+	–	+
Mannitol	+	+	+	+	+	–	+	+
Propanol	(+)	–	–	–	–	–	+	ND
Propionate	+	+	–	(+)	–	+	–	+
Pyruvate	+	+	+	+	+	+	+	+
Sorbitol	+	+	+	+	–	–	-	+
Succinate	+	+	+	+	+	+	+	+
Tartarate	+	–	–	–	+	–	–	–
Valerate	+	–	+	+	–	+	–	ND
Hydrogen	+	–	–	–	–	–	–	+
Sulfide	+	+	–	–	+	+	–	–
Thiosu-lfate	–	–	–	–	–	+	–	–
Sulfur	–	–	–	–	+	+	–	–

For all taxa, organic substrate utilization was tested during photo–organoheterotrophic growth. All taxa utilized pyruvate, but not benzoate or arginine. +, substrate utilized; _, substrate not utilized or absent; ±, variable in different strains; (+), weak growth; ND, Not determine. Data for *Rba. capsulatus* (Imhoff, 2005); *Rba. megalophilus* (Aruna Sri *et al.,* 2008); *Rba. changlensis* (Anil Kumar *et al.,* 2007); *Rba. viridis* (Shalem Raj *et al.,* 2013); *Rba. sediminis* (Subhash and Lee, 2016); *Rba. maris* (Venkata Ramana *et al.,* 2008); *Rba. azollae* and *Rba. lacus* (Suresh *et al.,* 2017), *Rba. sediminicola* (Suresh *et al.,* 2020); *Rba. tardus* (Sheu *et al.,* 2020); *Rba. flagellatus* (Xian *et al.,* 2020); *Rba. thermarum* (Khan *et al.,* 2019), *Rba. ruber* (Chen *et al.,*2021), *Luteovulum sphaeroides* (Imhoff, 2005); *Luteovulum. azotoformans* (Hiraishi *et al.,* 1996); *Luteovulum johrii* (Girija *et al.,* 2010); *Luteovulum ovatum* (Srinivas *et al.,* 2008); *Luteovulum alkalitolerance* (Suresh *et al.,* 2018); *Phaeovulum vinaykumarii* (Srinivas *et al.,* 2007b); *Phaeovulum veldkampii* (Hansen and Imhoff, 1985; Imhoff, 2005); *Fuscovulum blasticum* (Kawasaki *et al.,* 1993, Suresh *et al.,* 2019).was adopted.

Differentiation of the *Rhodobacter* from other Genera

The major characteristic properties of *Rhodobacter* from other genera of *Rhodobacterales* are precised in **Table 3.22**. Species of *Rhodobacter* are differentiated from each other based on morphology, slime production, sulfate assimilation,denitrification, vitamin requirement, utilization of carbon source and fatty acid composition, 16S rDNA gene sequence and DNA-DNA hybridization. In enrichment cultures set up for members of this genus grow faster and out-

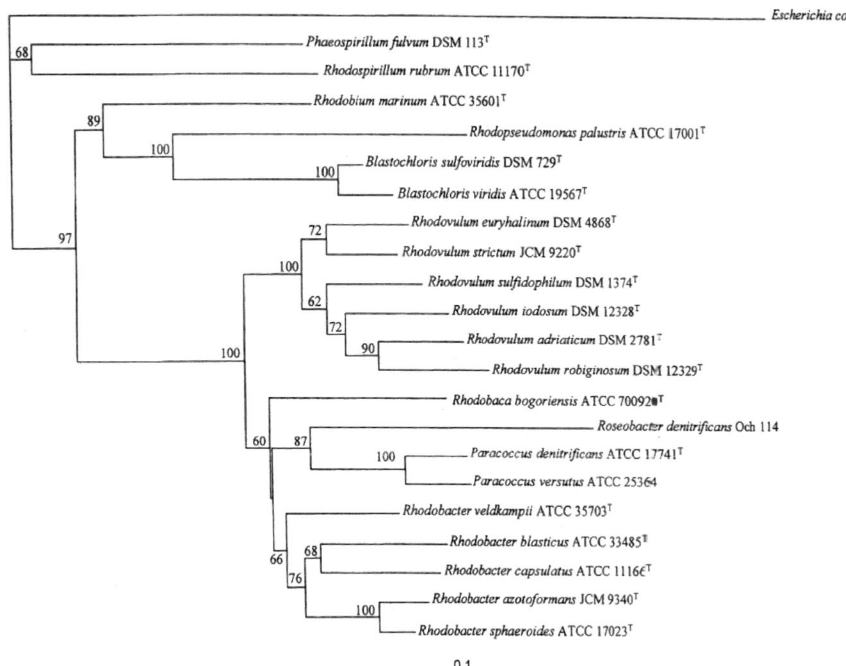

Fig. 3.30. Phylogenetic tree of *Rhodobacter* species and related *Alphaproteobacteria* of the order *Rhodobacterales* based on 16S rDNA sequences. The sequence of *E. coli* was used as an outgroup to root the tree.

Rhodobacter species have several characteristic chemotaxonomic properties that allow their distinction from other genera. All investigated species have a large-type cytochrome *c* (Ambler *et al.*, 1979) and as a major quinone component Q-10 (Imhoff, 1984a). Those species that are able to assimilate sulfate use the pathway via 3' -phosphoadenosine-5' -phosphosulfate (PAPS; Imhoff, 1982). The lipopolysaccharides of *Rba. capsulatus, Rba. sphaeroides, Rba. blasticus* and *Rba. veldkampii* contain glucosamine as sole amino sugar in their lipid A moieties, have phosphate, amide-linked 3-OH-14:0 and/or 3-oxo-14:0 and ester linked 3-OH-10:0 (Weckesser *et al.*, 1995). Differentiation of the genera and species of *Rhodobacter, Rhodovulum* and *Rhodobaca*is possible based on 16S rDNA sequences. *Rhodobacter* species are well characterized and distinct based on 16S rDNA sequence comparison (**Fig. 3.30**) (Imhoff *et al.,* 1984) and by DNA-DNA hybridization (Hiraishi *et al.*, 1996). All species of *Rhodobacter* and *Rhodovulum* were included in a study of fatty acid and polar lipid composition can be differentiated on the basis (Imhoff and Imhoff, 1995).

Taxonomic Comments

Species of the *Rhodobacter* such as *Rba. capsulatus* ,*Rba. sphaeroides* ,*Rba. blasticus* were formerly included in the genus *Rhodopseudomonas*. Based on morphological and chemo taxonomical distinct characteristics and early phylogenetic analyses led to their separation from *Rhodopseudomonas* and reclassification into the new genus *Rhodobacter* (Imhoff *et al.*, 1984; Kawasaki *et al.*, 1993a). *Rba. blasticus* is the only species of this genus that form buds during cell division and contains lamellar-type internal photosynthetic membrane (Eckersley and Dow, 1980). Based on characteristic properties of the genus *Rhodopseudomonas* at that time, this bacterium was described as a *Rhodopseudomonas* species and due to insufficient supporting data, they are retained within this genus in a reclassification of the purple non-sulfur bacteria (Imhoff *et al.*, 1984). With the availability of 16S rDNA sequences and supporting chemotaxonomic properties such as lipopolysaccharide analyses (Weckesser *et al.*, 1995) the high similarity and phylogenetic affiliation of this species of *Rhodobacter* became possible and led to its reclassification as *Rba. blasticus* (Kawasaki *et al.*, 1993a). These authors correctly proposed *Rba. blasticus* as the new name but did not give a formal description of the new combination. Unfortunately, this name came incorrectly invalidation list no. 51 of the IJSB as *Rba. blastica* and therefore a rewriting of this name was proposed by Euzeby (1997). As the genus name *Rhodobacter* is masculine, the correct name is *Rba. blasticus* .

Strains of *Rba. sphaeroides* that were confirmed as of this species by DNA-DNA hybridization studies (de Bont *et al.,* 1981) are able to denitrify (Satoh *et al.*, 1976; Pellerin and Gest, 1983). The denitrifying *Rba. azotoformans* is clearly distinct from *Rba. sphaeroides* by DNA-DNA hybridization (Hiraishi *et al.* , 1996). Based on 16S rRNA gene sequence analysis, *Rba. viridis* strain JA737^T hasthe highest sequence similarity with *Rba. capsulatus* ATCC 11166^T (98.8%), *Rba. maris* JA276^T (97%), *Rba. aestuarii* JA296^T (96.7%) and other members of the genus *Rhodobacter* (96.0%). Further, *Rba. viridis* showed 22–55% DNA–DNA relatedness with the above type strains. (98.8%) (Shalem Raj *et al.,* 2013). Recently, *Rba. azollae* and *Rba. lacus* is added to genus *Rhodobacter* based on 16S rRNA sequence and DNA-DNA hybridization studies (Suresh *et al.*, 2017). The 16S rRNA gene sequence similarity between *Rba. lacus* JA826^T and *Rba. azollae* strain JA912^{T}was 96.9%. DNA-DNA hybridization studies showed that these two species did not belong to any of the nearest species of the *Rhodobacter* similarity among themselves and type strains of nearest members <44%) (Suresh *et al.,* 2017).

Recently, the taxonomies were studied by Suresh *et al.* (2019) to resolve the taxonomic conflicts of the genus *Rhodobacter* using twelve type strains. The phylogenomic analysis showed that *Rhodobacter* spp. can be grouped into four monophyletic clusters with intermixing chemotrophs. Genomic indices (ANI and dDDH) confirmed that all the current species are well defined, except *Rhodobacter megalophilus* (Suresh *et al.*, 2019). They observed that average amino acid identity values between the monophyletic clusters of *Rhodobacter* members, as well as with the chemotrophic genera, are less than 80%, whereas the percentage of conserved protein values were below 70%, which has been observed among several genera related to *Rhodobacter*. The pan-genome analysis has revealed that there are only 1239 core genes shared between the 12 species of the genus *Rhodobacter*. The polyphasic taxonomic analysis supports the phylogenomic and genomic studies in distinguishing the four *Rhodobacter* clusters. Each cluster is comprised of one to seven species according to the current *Rhodobacter* taxonomy. Therefore, to address this taxonomic discrepancy Suresh *et al.* (2019) has proposed to reclassify the members of the genus *Rhodobacter* into three new genera, *Luteovulum* gen. nov., *Phaeovulum* gen. nov. and *Fuscovulum* gen. nov., and provide an emended description of the genus *Rhodobacter* sensu stricto (Suresh *et al.,* 2019, 2020). Also, they proposed reclassification of *Rhodobacter megalophilus* as a sub-species of *Rhodobacter sphaeroides* (Suresh *et al.,* 2019).

Rba. sphaeroides (van Niel, 1944), *Rhodobacter johrii* (Girija *et al.* 2010), *Rhodobacter ovatus* (Srinivas *et al.* 2008), *Rhodobacter azotoformans* (Hiraishi *et al.,* 1997) and *Rhodobacter alkalitolerans* (Gandham *et al.,* 2019) are reclassified into the new genera *Luteovulum* gen. nov., as *Luteovulum sphaeroides* , *Luteovulum johrii, Luteovulum ovatus* , *Luteovulum azotoformans* and *Luteovulum alkalitolerans* (Suresh *et al.,* 2019, 2020). At the same time *Rhodobacter blasticus* (Eckersley and Dow, 1980, Kawasaki *et al.,* 1993) is reclassified into the new genera *Fuscovulum* gen. nov., as *Fuscovulum blasticum* (Suresh *et al.,* 2019) is identical to that of *Rhodopseudomonas blastica* (Eckersley and Dow, 1980) and *Rhodobacter blasticus* (Imhoff *et al.,* 1984; Imhoff, 2005). On the other hand, *Rhodobacter vinaykumarii* (Srinivas *et al.,* 2007) *Rhodobacter veldkampii* (Hansen and Imhoff 1985) is reclassified into the new genera *Phaeovulum* gen. nov., as *Phaeovulum vinaykumarii* and *Phaeovulum veldkampii* (Suresh *et al.,* 2019).

2. *Luteovulum* (Small yellow egg) Suresh *et al.* (2020)

Cells are oval-to-rod shaped cells, mostly motile with a single polar flagellum and multiply by binary fission. Gram-stain-negative, Catalase and oxidase-positive. They have a vesicular internal membrane system. They contain bacteriochlorophyll *a* and carotenoids of the spheroidene series. They are primarily phototrophic and facultative aerobes and mesophilic. Phototrophic growth occurs on a number of organic substrates. The growth factors such as biotin, niacin and thiamine, alone or in combination are required for growth. NaCl requirement is not required for growth but tolerates up to 2–3% NaCl. C18:1 & 7c/ C18:1 & 6C, C18:0, C16:0, C10:0 3OH, C16:1 & 7C/C16:1 & 6c, C18:1 & 7c11 methyl are the major fatty acids. Phosphatidylethanolamine, phosphatidylglycerol, diphosphatidyl glycerol, phosphatidylcholine and an unidentified glycolipid are the major polar lipids. Hopanoids are not produced. Q10 is the major quinone. Delineation of the genus is based on 16S rRNA and rpoB gene-based phylogeny, phylogenomics, genome comparison and chemo-taxonomic differences. Members of this genera can be isolated from freshwater ponds, paddy soils, wastewater treatment plants, alkaline ponds and lake sediments. The type species is *Luteovulum sphaeroides*. The characteristic properties of different species of this genera are precised in **Table 3.20** and **3.21** .

3. ***Fuscovulum*** **(Small tawny egg), Suresh** ***et al.*** **(2020)**

Cells are rod-to-oval shaped, mostly non-motile and multiply by sessile budding. Catalase and oxidase positive. Gram (-) negative, cells have lamellar intracytoplasmic membrane architecture contain bacteriochlorophyll *a* and carotenoids of the spheroidene series. They are primarily phototrophic and growth occurs under aerobic, anaerobic and mesophilic conditions. Phototrophic growth occurs on a number of organic substrates. Nicotinic acid and thiamine are utilized as growth factors. NaCl is not required for growth. Q-10 is the major quinone. The G+C mol% of the type strain of the type species is 66.5%. The type species is *Fuscovulum blasticum*. Members can be isolated from freshwater habitats. Delineation of the genus is based on 16S rRNA and rpoB gene-based phylogeny, phylogenomics, genome comparison and chemotaxonomic differences. The characteristic properties of different species of this genera are precised in **Table 3.20** and **3.21** .

4. ***Phaeovulum*** **(Small brown egg), Suresh** ***et al.*** **(2020)**

Cells are rod to oval-shaped cells, mostly non-motile and multiply by binary fission. Catalase and oxidase positive. Gram (-) negative. Cells contain vesicular intracytoplasmic membrane architecture contain bacteriochlorophyll *a* and carotenoids of the spheroidene series. They are primarily phototrophic and growth occurs under anaerobic and mesophilic conditions. Phototrophic growth occurs on a number of organic substrates. Biotin, para-aminobenzoate and thiamine are utilized as growth factors either singly or in combination. Some strains do not contain phosphatidylcholine. Some require NaCl for growth and can tolerate up to 4%. C18:1&7c/C18:1 & 6C, C18:0, C16:0, C10:0 3OH, C16:1 & 7C/C16:1 & 6c, C18:1 & 7c11methyl are the major fatty acids. Q-10 is the major quinone. GCC mol% is 65–68.25. Members can be isolated from marine habitats including tidal waters, freshwater. Delineation of the genus is based on 16S rRNA and rpoB gene-based phylogeny, phylogenomics, genome comparison and chemotaxonomic differences. The type species is *Phaeovulum veldkampii* . The characteristic properties of different species of this genera are precised in **Table 3.20** and **3.21** .

5. ***Rhodovulum (Rdv.)*** **(Hiraishi and Ueda, 1994) Imhoff (2015)**

Cells are oval to rod-shaped, motile, or non-motile, when motile they exhibit polar growth and multiply by binary fission. Cell size is 0.5-0.9 x 0.9-3.8 μm. Phototrophically grown cells contain intracytoplasmic membranes which are vesicular type. Photosynthetic pigments are bacteriochlorophyll *a* and carotenoids of the sphaeroidene series (sphaeroidene, dimethyl sphaeroidene, hydroxysphaeroidene present in different proportions) are present. The phototrophic cultures are yellow-green to yellow-brown. The major ubiquinone is

Q-10. The major fatty acids are $C_{18:1}$ (predominant) $C_{18:0}$ and $C_{16:0}$. The polar lipid composition includes phosphatidylethanolamine (PE), phosphatidylglycerol (PG), several sulfolipids (sulfoquinovosil diglyceride, SQD) and other lipids but excludes PL. Polysaccharides, poly-β-hydroxybutyric acid, and polyphosphates may be present as storage products. They are mesophilic marine bacteria that require NaCl for growth. Growth occurs at a temperature between 25°C and 35°C and neutrophilic growing best at salinities 1.0-7.5%. Some species are alkaliphilic which grow at pH 8.5-9.0 and unable to grow at pH 7.0. Growth occurs photoheterotrophically under anoxic conditions in the presence of light. Photoautotrophic growth may possible with sulfide/thiosulfate/hydrogen and ferrous ions under anaerobic conditions in the presence of light. Chemoorganotrophic growth occurs with oxygen and organic carbon in the dark. Sulfur may be oxidized to sulfate. The utilization of carbon sources for growth is organic substrates, lactate, Kreb's cycle intermediates and short-chain fatty acids. Some amino acids and sugars as well as formate may be used by some species. Vitamin requirement is a general rule for all species, which are variable for the given combination they need. The mol% of DNA G+C content is 62-69%. Marine and hypersaline environments rich in organic matter are selective habitats for their growth. The characteristic of species of these genera is precised in **Table 3.23.** The morphology, nature photosynthetic membranes of *Rhodovulum marinum* is shown in **Fig. 3.31** (Srinivas *et al.,* 2006).

The genus *Rhodovulum* was created to separate the marine species of *Rhodobacter* from their freshwater counterparts. The genus *Rhodovulum* currently comprises twenty (20) validly published species: *Rhodovulum sulfidophilum* (*Rhodopseudomonas sulfidophila* ,) *Rdv. adriaticum* (*Rhodopseudomonas adriatica* ,) *Rdv. euryhalinum*(*Rhodobacter euryhalinus* ,) *Rdv. strictum* ,*Rdv. iodosum* ,*Rdv. robiginosum* ,*Rdv. steppense, Rdv. marinum, Rdv. lacipunicei, Rdv. kholense, Rdv. imhoffii, Rdv. viskhapatnamense* , *Rdv. tesquicola, Rdv. viride, Rdv. salis, Rdv. phaeolacus, Rdv. aestuarii, Rdv. algae, Rdv. bhavnagarense* and *Rdv. mangrovi* .

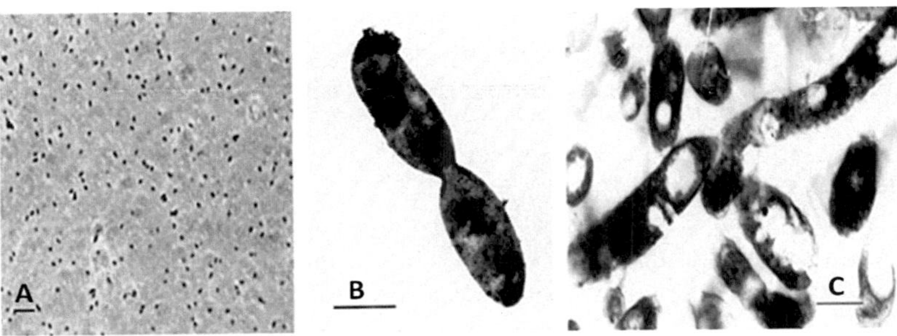

Fig. 3.31. A. Phase-contrast micrograph of cells of *Rhodovulum marinum* (Bar = 5 μm), B. Electron micrograph of negatively stained cells of *Rdv. marinum* showing binary fission division (Bar = 1 μm), C. Electron micrograph of ultrathin section of *Rdv. marinum* showing the vesicular nature of the photosynthetic membranes extending throughout the cell (Bar = 0.5 μm).

Differentiation of the *Rhodovulum* from other Genera

Differentiation of the genera and species of *Rhodobacter* , *Rhodobaca* and *Rhodovulum* are possible based on 16S rDNA and 16S rRNA sequences (**Fig. 3.30 and 3.32**) and by DNA-DNA hybridization. Differentiating phenotypic characteristics of *Rhodovulum* from other genera are given in **Table 3.22.**

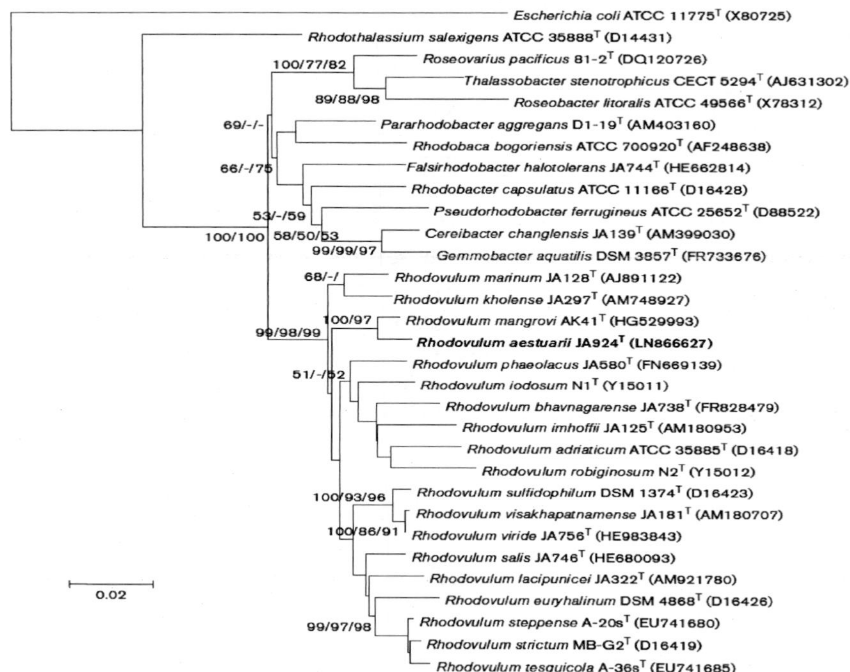

Fig. 3.32. Phylogenetic tree of *Rhodovulum* species and related *Alphaproteobacteria* of the order *Rhodobacterales* based on 16S rDNA sequences. The sequence of *E. coli* was used as an outgroup to root the tree.

Taxonomic Comments

Species of *Rhodovulum* such as *Rdv. sulfidophilum* , *Rdv. adriaticum* have formerly included in the genus *Rhodopseudomonas*. The recognition of morphologically and chemotaxonomically distinct characteristics and early phylogenetic analyses led to their separation from *Rhodopseudomonas* and classify in the genus *Rhodobacter* (Imhoff *et al.*, 1984). Based on the availability of 16S rDNA sequences it became apparent that the marine and halophilic species that depend on NaCl form a group phylogenetically separate from the freshwater species. Both groups were taxonomically differentiated and the marine species are transferred to the new genus *Rhodovulum* (Hiraishi and Ueda, 1994a). The newly described *Rba. euryhalinus* was also transferred to *Rdv. euryhalinum* (Kompantseva, 1985). Based on 16S rRNA gene sequence analyses twenty *Rhodovulum* species are validly published. Recently, *Rhodovulum aestuarii* strain

JA924[T] was included in the genus based on 16S rRNA gene sequence analyses which showed the highest similarity with *Rdv. mangrove* strain AK41T (98.19%) and other members of *Rhodovulum* (< 95.71%) (**Fig. 3.32**) (Divyasree *et al.*, 2016). Similarly, *Rdv. algae* was included in the genus which is closely related to the *Rdv. viride* JA756[T] (99.0%), *Rdv. sulfidophilum* Hansen W4T (98.9%), *Rdv. visakhapatnamense* JA181T (98.8%), *Rdv. kholense* JA297T (97.5%) and *Rdv. salis* JA746T (97.0%). However, DNA-DNA hybridization studies of this species showed only 20–45% relatedness with its phylogenetic neighbors (Ramaprasad *et al.*, 2016).

Table 3.23. Differential Characteristics of Species of *Rhodovulum*

Characteristics	***Rhodovulum sulfidophilum***	***Rdvadriaticum***	***Rdv euryhalinum***	***Rdv strictum***	***Rdv. iodosum***	***Rdv.robiginosum***	***Rdv. steppense***	***Rdv. marinum***	***Rdv. lacipunicei***	***Rdv. kholense***
Habitat	Marine sediments and Coastal marine waters	Marine sediments, Coastal marine waters,	Marine sediments, Coastal marine waters,	Tidal and sea water pool	Mud sediments	Coastal sediment	Saline soda lakes	Tidal sea-water	Saline pond water	Mud sample
Cell width (μm)	0.6-1.0 x 0.9-2.0	0.5-0.8 x 1.3-1.8	0.7-1.0 x 1.5-3.0	0.6-1.0 x 1.0-2.5	0.5-0.8 x 2.4-3.8	0.5-0.8 x 1.6-3.2	0.3-8.0 x 1.0-2.5	0.6-8.0 x 1.0-2.0	0.7-1.0	0.5-1.0 x 1.5-1.8
Cell shape *	O–R, C	R,C	O–R	YB, O–R	O–R	O–R	O–R	O–R,C	O–R	O–R
Motility	+	–	+	+	–	–	+	–	–	+
Colour of cell suspension †	YG-YB	B	YB-DB	OB	YB	YB	YB	YB	YB	Y-B
Bacteriochlorophyll	*A*	*A*	*a*	*a*	*a*	*a*	*a*	*a*	*a*	*a*
Carotenoids	Spheroidene series	Spheroidene series	Spheroidene series	Spheroidene series	Spheroidene series	Spheroidene series	Spheroidene series	Spheroidene series	Spheroidene series	Spheroidene series
Spheroidene	+	6	ND	+	+	+	ND	+	+	ND
Demethylspheroidene	–	74	ND	–	–	–	ND	–	–	ND
Hydroxyspheroidene	+	9	ND	–	+	+	ND	–	+	ND
Spheroidenone	–	4	ND	+	+	+	ND	+	+	ND
Hydroxyspheroidenone	–	–	ND	–	–	–	ND	–	–	ND
Demethylspheroidenone	–	–	ND	–	–	–	ND	–	–	ND
Neurosporene	–	7	ND	–	–	–	ND	–	–	ND
Dihydroxyneurosporene	–	–	ND	–	–	–	ND	–	–	ND
Major quinone	Q-10	ND	ND	Q-10	ND	ND	Q-10	ND	–	ND

Charact-eristics	*Rhodovulum sulfidophilum*	*Rdvadriaticum*	*Rdv euryhalinum*	*Rdv strictum*	*Rdv. iodosum*	*Rdv.robiginosum*	*Rdv. steppense*	*Rdv. marinum*	*Rdv. lacipunicei*	*Rdv. kholense*
Absorption maxima of living cells	374–378, 451–455, 480–489, 508–512, 588–592, 800–805	374–378, 447–450, 475–480, 508–512, 588–590,	378, 450–460, 479–486, 509–517, 592, 805, and	377,	ND	ND	461, 483, 515, 592, and 803	378, 402, 488, 520, 590, 802, 884	377, 449, 479, 512,	377, 468, 520,
	and 850–855	802–805 and 869	855	460, 481, 512, 590, 803 and 860			861		590, 806 and 860	590, 803 and 857
NaCl required (optimum %)	0-10 (1-6)	1.0-10 (2.5-7.5)	0.5-12.0	0.25-3.0 (0.8)	2.5-5.0	1.0-7.0 (2.5-5.0)	0.3-10 (1-5)	0.05-8.0 (1.0-3.0)	0-5.0	0.05-6.0 (1.0-4.0)
pH range (optimum)	5.0-9.0 (6.5-8.0)	6.0-8.5 (6.5-7.0)	6.0-8.5 (7.0-8.0)	7.5-9.0 (8.0-8.5)	6.3-6.8 (6.5)	6.3-6.8 (6.5)	7.5-10.0 (8.5)	5.5-7.5 (6-8)	6.5-9.0 (7-8.5)	6.0-8.0 (6-7)
Temperature range (optimum) (°C)	ND (30-35)	ND (25-30)	25-35	30-35 (ND)	ND (20-25)	ND (25-28)	(25-35)	25-35 (30)	25-35 (30)	20-35 (30)
Chemo-organotrophy	+	–	ND	+	+	+	+	+	+	+
Vitamins required ‡‡	B, N, PABA, T	B, T	B, N, PABA, T	B, PABA, T	B,N	B, B12, N	B, N, T	T	B,T	B, N,T
DNA G+C content (mol%) §§	66.3-66.6	64.9-66.7	62.1-68.6	67.3-67.7	66	69	66.1	62	68.6	63
Polar lipids	ND	PG, PE, SL1-2, L1-2, PL	ND	ND	ND	ND	ND	ND	ND	ND
Carbon/ electron donors										
Acetate	+	+	+	+	+	+	+	+	+	+
Aspartate	±	ND	+	–	–	–	(+)	–	+	–
Benzoate	–	–	–	–	–	–	(+)	–	–	–
Butyrate	+	–	+	+	+	–	+	–	+	+
Caproate	+	+	ND	+	–	–	+	–	+	+
Citrate	–	–	-	(+)	–	–	(+)	–	–	–
Cysteine	+	ND	ND	ND	+	+	+	–	+	–
Ethanol	±	+	+	–	–	0	+	+	–	–
Ferrous iron	+	–	–	ND	+	+	+	–	–	–

Characteristics	*Rhodovulum sulfidophilum*	*Rdvadriaticum*	*Rdv euryhalinum*	*Rdv strictum*	*Rdv. iodosum*	*Rdv.robiginosum*	*Rdv. steppense*	*Rdv. marinum*	*Rdv. lacipunicei*	*Rdv. kholense*
Formate	+	+	+	+	–	–	+	–	–	–
Fructose	±	–	+	+	–	–	(+)	+	+	+
Fumarate	+	+	+	+	+	–	+	+	+	+
Glutamate	+	ND	+	–	+	+	+	–	+	+
Glucose	+	+	+	+	–	–	(+)	+	+	+
Glycerol	+	+	+	–	–	+	+	+	–	–
Lactate	+	+	+	+	+	+	+	+	+	+
Malate	+	+	+	+	+	+	+	–	+	+
Mannitol	±	–	+	-	+	+	–	+	–	–
Propionate	+	+	+	+	+	+	+	–	+	–
Pyruvate	+	+	+	+	+	+	+	+	+	+
Succinate	+	+	+	+	+	+	+	+	+	+
Valerate	+	+	+	+	+	–	+	–	+	+
Hydrogen	+	–	ND	ND	+	+	–	–	–	–
Sulfide	+	+	+	+	+	+	+	–	+	+
Thiosulfate	+	+	+	+	+	+	+	–	–	+
Sulfur	+	+	+	ND	+	+	+	–	–	–

Continued

Characteristics	*Rdv. visakhapatnamense*	*Rdv. tesquicola*	*Rdv. viride*	*Rdv. salis*	*Rdv. phaeolacus*	*Rdv. aestuarii*	*Rdv. algae*	*Rdv. bhavnagarense*	*Rdv. mangrovi*
Habitat	Tidal sea water	Brackish steppe soda lakes	Pond water	Salt pan	Sediments of Brown Pond	Brackish water	Algae mat	Pink Pond	Sediment sample
Cell width (µm)	0.9-1.2	0.4-0.8 x 1.0-2.5	0.8 x 2.0-3.0	1.0 x 2.0-3.0	0.5-0.8 x 1.0-1.8	2.0-3.0	0.6-0.9 x 1.0-1.4	0.5–0.9 x1–1.8	0.5-1.0
Cell shape *	O–R	O-R	O-R	O-R	O-R	O-R	O-R	O-R	O-R
Motility	–	+	–	–	–	–	+	–	–
Colour of cell suspension †	YB	Y-B	G	YB	OB	YB	RB	RB	B
Bacteriochlorophyll	*a*	*a*	*a*	*a*	*a*	*a*	*a*	*a*	*a*
Carotenoids	Spheroidene series	Spheroidene series	Neurosporene	Spheroidene	Spheroidene	Spheroidene	Spheroidene	Spheroidene	Spheroidene
Spheroidene	ND	ND	4	41	56	55	84	16	49
Demethylspheroidene	ND	ND	–	15	12	7	9	16	43
Hydroxyspheroidene	ND	ND	5	7	8	2	–	17	1
Spheroidenone	ND	ND	–	5	17	15	–	48	4
Hydroxyspheroidenone	ND	ND	–	10	5	–	–	–	–

Characteristics	*Rdv. visakhapatnamense*	*Rdv. tesquicola*	*Rdv. viride*	*Rdv. salis*	*Rdv. phaeolacus*	*Rdv. aestuarii*	*Rdv. algae*	*Rdv. bhavnagarense*	*Rdv. mangrovi*
Demethylspheroidenone	ND	ND	–	4	–	6	–	–	–
Neurosporene	ND	ND	59	5	2	6	1	3	3
Dihydroxyneurosporene	ND	ND	15	15	–	4	–	–	–
Major quinone	ND	Q-10	Q-10	Q-10	Q-10	Q-10	Q-10	Q-10	Q-10
Absorption maxima of living cells	377, 467, 520, 590, 802 and 857	460, 483, 514, 591, 801 and 862	380, 461, 508, 593, 803 and 857	380, 455, 512, 594, 807 and 854	380, 479, 590, 803 and 863	377, 479, 593, 755, 800 and 854	375, 479, 590, 804 and 854	380, 479, 590, 803 and 863	379, 462, 592, 803 and 855
NaCl required (optimum %)	0-10	0.3-10.0 (1.0-3.0)	9.0 (1.0)	1.0-8.0 (2.0)	1.0-3.0 (1.0-1.5)	0-8.0 (2.0-4.0)	1.0-7.0 (2.0)	10.0 (1.0-3.0)	12.0 (1.0-9.0)
pH range (optimum)	4.0-9.0 (6-8)	7.5-10.0 (8.5-9.0)	6.0-8.5 (6.5-7.5)	6.0-8.5 (6.5-7.5)	6.5-8.0 (7.0-7.5)	6.0-10 (7.5)	7.5-9.0 (7.5)	7.0-9.5 (7.0-8.0)	5.0-9.0 (7.0-8.0)
Temperature range (optimum) (°C)	20-35 (30)	(25-35)	25-35 (25-30)	25-35 (30)	28-35 (30)	25-35 (30)	20-35 (30)	25-35 (28)	25-38 (28)
Chemo-organotrophy	+	+	+	+	ND	+	ND	ND	ND
Vitamins required ‡‡	B, N, T, PABA	B, PABA, T, N	Not required	T, B, PABA	Complex	Complex	B	T, P	Not required
DNA G+C content (mol%) §§	61.2	65.4	63.3	62.4	62.3	64+0.08	64.5	63.4	68.9
Polar lipids	ND	PE, PG, L1,-7	SL1, 2, PE, PG, PL1,2 L1,2	SL1, PE, PG, PL1,2, AL1,2, L1,2	ND	PG, PE, SL1, AL1, L1, 2	PG, PE, SL1-2, LP1, L1-5	PG, PE, SL1-2, L1	DPG, PE, AL1, L1-4
Carbon/electron donors									
Acetate	+	+	–	+	+	+	ND	ND	+
Aspartate	–	(+)	–	–	–	–	–	+	–
Benzoate	–	–	+	–	–	–	ND	–	–
Butyrate	+	+	+	–	–	–	ND	–	+
Caproate	+	–	+	ND	–	–	ND	–	–
Citrate	–	(+)	+	–	–	–	+	–	–
Cysteine	–	+	–	–	–	ND	ND	+	–
Ethanol	–	+	+	–	–	–	+	–	–
Ferrous iron	–	–	ND	–	–	–	ND	ND	–

Charac-teristics	*Rdv. visa-khapatna mense*	*Rdv. tesqu-icola*	*Rdv. viride*	*Rdv. salis*	*Rdv. phae-olacus*	*Rdv. aestuarii*	*Rdv. algae*	*Rdv. bha-vnagar-ense*	*Rdv. man-grovi*
Formate	–	+	ND	+	–	+	–	–	ND
Fructose	+	+	+	ND	–	+	–	ND	ND
Fumarate	+	+	–	–	+	–	–	–	+
Glutamate	+	+	+	+	+	–	–	+	–
Glucose	+	(+)	+	+	-	+	+	–	+
Glycerol	+	+	–	–	–	–	+	–	+
Lactate	+	+	+	–	–	–	+	+	–
Malate	+	+	+	+	+	–	+	+	–
Mannitol	–	–	+	+	–	–	+	–	–
Propionate	+	+	+	–	–	–	+	–	+
Pyruvate	+	+	+	+	ND	+	+	+	+
Succinate	+	+	ND	+	ND	+	ND	+	+
Valerate	+	+	+	–	–	–	–	–	+
Hydrogen	ND	–	–	–	ND	–	ND	ND	ND
Sulfide	–	+	–	–	+	+	–	+	–
Thiosulfate	+	+	–	–	+	+	–	+	+
Sulfur	–	+	–	–	ND	–	ND	+	–

Symbols: +, Present/substrate utilized; -, substrate not utilized or absent; ±, variable in different strains; (+), weak growth; YB, yellowish -Brown; Y, Yellow, Y-B, Yellow to Brown; B, Brown; OB, orange-Brown; B, Biotin; N, Niacin; t, thiamine, PABA, para -amino Benzoic acid; B_{12}, vitamin B_{12}; ND, Not determine/No data available; Polar lipids: DPG, Diphosphatidylglycerol; PG, phosphatidylglycerol; PE, phosphatidylethanolamine; SL, Sulfolipids; AL, Amino lipids, AL, Amino lipids, PL, Phospho lipids; L, Polar lipids. Data for *Rhodovulum sulfidophilum* (Hansen and Veldkamp 1973; Hiraishi and Ueda 1994); *Rdv. adriaticum* (Hiraishi and Ueda 1994); *Rdv. euryhalinum* (Kompantseva, 1989; Hiraishi and Ueda 199); *Rdv. strictum* (Hiraishi and Ueda 1995); *Rdv. iodosum* (Straub *et al.,* 1999); *Rdv. robiginosum* (Straub *et al.,* 1999); *Rdv. steppense* (Kompantseva et al., 2010); *Rdv. marinum* (Srinivas *et al*., 2006); *Rdv. lacipunice* (Chakravarthy *et al*., 2009); *Rdv. kholense* (Anil Kumar *et al.,* 2008); *Rdv. imhoffii* (Srinivas *et al*., 2007); *Rdv. visakhapatnamense* (Srinivas *et al.* 2007); *Rdv. tesquicola* (Kompantseva *et al*., 2012); *Rdv. viride* (Srinivas *et al*., 2014); *Rdv. salis* (Srinivas *et al*., 2014); *Rdv. phaeolacus* (Lakshmi *et al*., 2011); *Rdv. aestuarii* (Divyasree *et al*., 2016); *Rdv. algae* (Ramaprasad *et al*., 2016); *Rdv. bhavnagarense* (Srinivas *et al*., 2012,); *Rdv. mangrovi* (Nupur *et al*., 2014) was adopted.

6. *Rhodobaca* (*Rca.*) (A Rose berry) (Milford *et al.*, 2000) Imhoff (2015)

Cells are ovoid or rod-shaped, 0.8–1.5 μm in diameter, motile by means of polar flagella and divide by binary fission (**Fig. 3.33**). Gram-negative, belong to the *Alphaproteobacteria.* Phototrophically grown cells form vesicular-type internal photosynthetic membranes. The colour of phototrophic cultures is yellow to yellow-brown, while aerobic cultures are pink to red. Photosynthetic pigments are bacteriochlorophyll *a* and carotenoids of the spheroidene series. Photoheterotrophic growth occurs under anoxic conditions in the presence of light with a variety of organic compounds as carbon and electron source. Chemotrophic growth occurs by aerobic respiration in dark. Growth factors are required. The mesophilic bacterium grows at optimal elevated temperature 30ºC and best growth occurs at alkaline pH and in the presence of low concentration of mineral salts. Alkaline soda lakes of low to moderate salinity are selective habitats for its growth. The mol% G + C of the DNA is 58.8. *Rhodobaca* *bogoriensis* (Milford *et al.,* 2000) (type species) and *Rhodobaca barguzinensis* are the two validly published species and their characters are précised in **Table 3.24** .

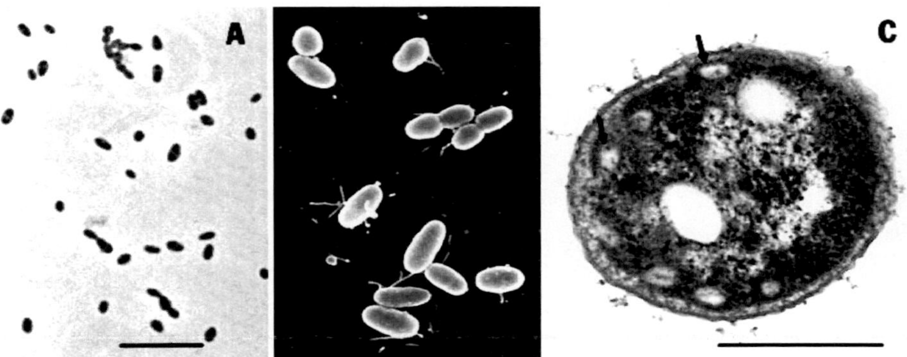

Fig. 3.33. Morphology of strain of *Rhodobaca bogoriensis*. A. Phase photomicrograph (Bar = 10 µm), B. Scanning electron micrograph (Bar = 2 µm), C. Cross-section transmission electron micrograph showing vesicular intracytoplasmic membranes (*arrows*) (in C) (Bar = 0.4 µm).

In several phenotypic properties, *Rhodobaca bogoriensis* resembles *Rhodobacter* and *Rhodovulum* species. Characteristic properties that reflect adaptation to the natural environment, African lakes (Lake Bogoria, Kenya) are with high pH (optimal at pH 9.0) and a temperature optimal at 39ºC (Milford *et al.*, 2000). No growth occurs below 30ºC. Good growth is possible from 0-6.0% NaCl with an optimum NaCl 1.0-2.0%. *Rhodobaca bogoriensis* probably able to fix dinitrogen and does not grow photoautotrophically either with hydrogen or with sulfide as electron donors. Sulfide is oxidized to extracellular elemental sulfur (Milford *et al.*, 2000). Spectral properties of whole cells indicate the presence of one light-harvesting pigment complex (LHI, core antenna complex) unlike *Rhodobacter* and *Rhodovulum* species, but similar bacteria such as *Rhodospirillum rubrum, Rhodobium mannum. Rhodocista centenaria, Rhodovibrio sodomense, Rhodospira trueperi. Roseospirillum parvum.* The sequence of the 16S rDNA clearly distinguishes *Rhodobaca* from *Rhodobacteria* and *Rhodovulum* species but also shows the association of this bacterium with these genera in the *Alphaproteobacteria* **(Fig. 3.30** and **3.32)** .

Differentiation of the *Rhodobaca* from other Genera

Characteristic properties of *Rhodobaca* are the ovoid to rod-shaped cell morphology, the presence of vesicular internal membranes and the content of carotenoids of the spaheroidene series. All these properties are shared with *Rhodovulum* and *Rhodobacter* species. Differentiation of the genera and species of *Rhodobacter* ,*Rhodobaca* and *Rhodovulum* is possible based on 16S rDNA and 16S rRNA sequences (**Fig. 3.30 and 3.32**) and by DNA-DNA hybridization.

Differentiating phenotypic characteristics of *Rhodobaca* from other genera are given in **Table 3.22.**

Taxonomic Comments

Members of the genus *Rhodobaca* including *Rca. bogoriensis* and *Rca. barguzinensis* form a phylogenetic position in a cluster with sphaeroidene containing species of the genera *Rhodobacter* and *Rhodovulum*. Further, they are rather distinct from these species showing similarity in the range of 92.2-94.5%. DNA-DNA hybridization studied showed 52% similarity between *Rca. bogoriensis* LBB1 and *Rca. barguzinensis* alga-05 (Boldareva *et al.,*2008). These two species show differences in many phenotypic characteristics such as fatty acid composition, photosynthetic pigments, organic compounds, dwelling in hypersaline soda lakes, ability to grow under low temperatures and genetic analysis. These characteristics allow this species to classify under *Rhodobaca* and separated from *Rhodobacter* and *Rhodovulum.*

7. *Rhodobaculum* (Small pink rod) Bryantseva *et al.* (2015)

Cells are polymorphic (from short rods to long spindle-shaped ones) and motile. The division is by binary constriction (**Fig. 3.34**) (Bryantseva *et al.,* 2015). The cells are gram (-) negative. They form vesicular-type photosynthetic membranes. Anaerobically and aerobically grown cell suspensions are orange-brown and pink in colour respectively. Bacteriochlorophyll *a* and carotenoids of both the spheroidene and spirilloxanthin series are the photosynthetic pigments. Both anaerobic photoheterotrophic growth and aerobic chemoheterotrophic growth in the dark are possible on several organic substrates. Growth factors are required. The organism is alkaliphilic and mesophilic. Alkaline soda lakes with moderate mineralization are selective habitats for their growth. The DNA G+C content is 67.6–69.8 mol%. A monotypic genus with single validly published species *Rhodobaculum claviforme* is the type species. The characteristic of this single species is précised in **Table 3.24.**

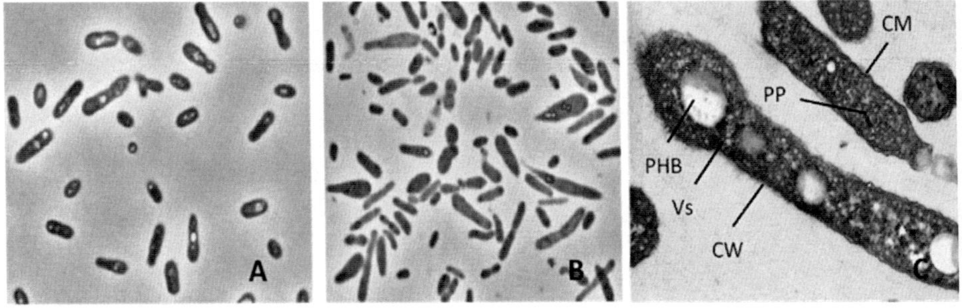

Fig. 3.34. A and B. Phase contrast micrograph of *Rhodobaculum claviforme* grown anaerobically and aerobically (Bar = 5 μm) and C. Ultra-thin structure (Bar = 1 μm). Designations: PP, polyphosphates; PHB, poly-β-hydroxybutyrate; CW, cell wall; Vs, vesicles; CM, cytoplasmic membrane.

Differentiation of *Rhodobaculum* from other Genera

Differentiation of the genera and species of *Rhodobaculum, Rhodobaca* and *Rubribacterium* is possible based on 16S rDNA and 16S rRNA sequences (**Fig. 3.35**) and by DNA-DNA hybridization. Differentiating phenotypic characteristics of *Rhodobaculum* from other genera are given in **Table 3.24.**

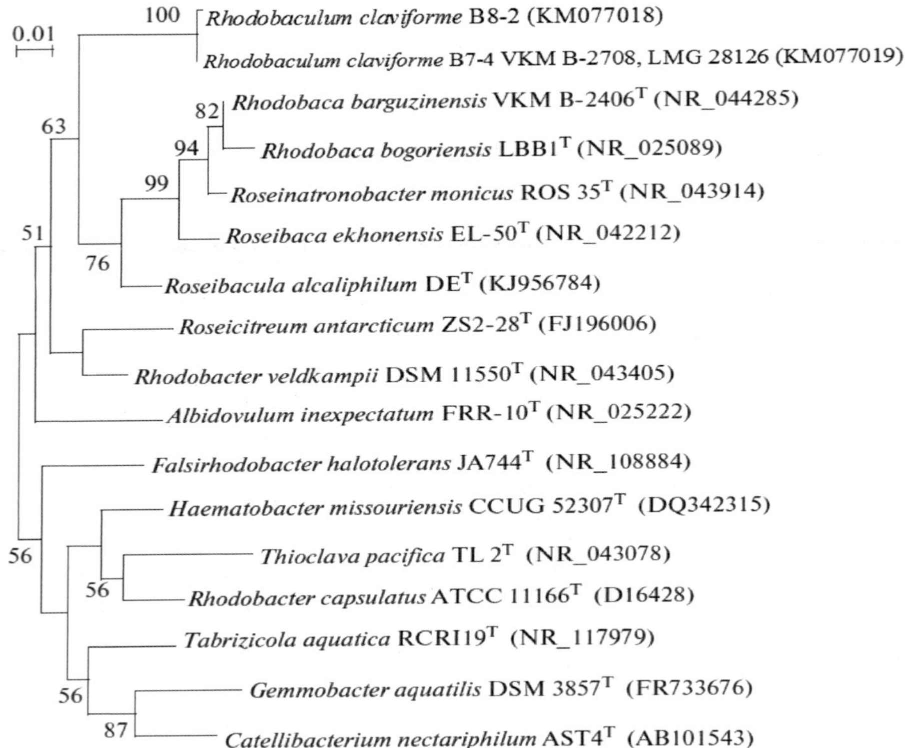

Fig. 3.35. Phylogenetic tree of *Rhodobaculum* species and related *Alphaproteobacteria* of the order *Rhodobacterales* based on 16S rRNA sequences.

Taxonomic Comments

Phylogenetic analysis of 16S rRNA gene sequencing revealed that only species of *Rhodobaculum* such as *Rhodobaculum claviforme* strain B7-4 form a separate branch in the cluster of the genera *Roseibacula, Roseibaca, Rhodobaca* and *Roseinatronobacter*(**Fig. 3.35**) (Brayntseva *et al.,* 2015). Based on genetic properties this species is closely related to alkaliphilic aerobic anoxygenic phototrophic bacteria *Roseibacula alcaliphilum* (95.2%), alkaliphillic *Rhodobaca barguzinensis* VKM B-2406^{T} (94.2%) and *Rca. bogoriensis* LBB1^{T} (93.9%). A similar level of homology was observed for the species of *Rba. veldkampii* DSM 11550^{T} (94.8%) and ABC bacteria *Roseinatranobacter monicus* ROS 35^{T} (93.5%) and *Roseicitreum antarcticum* ZS2-28^{T} (93.9%).

8. *Rubribacterium* (A Red rod) Boldareva *et al.* (2009)

The cells are oval or polymorphic, 0.7 × 1.0–1.2 μm. Under certain conditions, polymorphic cells may increase in size to 3.0 μm. Cell division is binary fission, often irregular by constriction. The cells may be motile due to polar flagella. Bacteria are gram (-) negative. Vesicular type of photosynthetic membranes distributed throughout the cell volume(**Fig. 3.36**) (Boldareva *et al.,* 2009). Bacteriochlorophyll *a* and carotenoids of both spirilloxanthin and spheroidene groups (spheroidenone, hydroxyspheroidene, dimethylspheroidene, spirilloxanthin, diketomono-dimethylspirilloxanthin, diketospirilloxanthin, anhydrorhodovibrine and lycopene) are the photosynthetic pigments. LH1 and LH2 light-harvesting systems are present. The organisms are anaerobic photoheterotrophs and facultative aerobes. Aerobic growth occurs by respiration. Ammonium is utilized as a nitrogen source for biosynthesis. A wide range of organic substrates is utilized, including organic acids, sugars and polypeptides. They are mesophilic, optimal growth occurs at temperature 20-35°C, pH 8.5-9.5 and NaCl 5-40%. An isomer of the monounsaturated $C_{18:1}\omega7$ acid was predominant (74.28% of the total fatty acids). The DNA G+C base content of the type species is 69.9 mol%. *Rubribacterium polymorphum* is the type species and its characters are précised in **Table 3.24.**

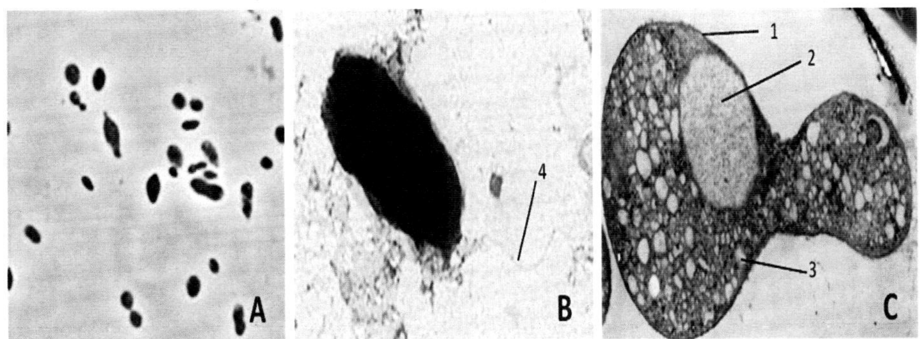

Fig. 3.36. Cell morphology and ultrastructure of *Rubribacterium* grown anaerobically in the light: A. Light microscopy (Bar = 10 μm), B. Total cells (Bar = 1 μm), C. Ultrathin section (Bar = 0.5 μm); 1. The outer membrane, 2. Storage compounds, 3. Vesicular type intracytoplasmic membrane, 4. Flagellum.

Differentiation of *Rubribacterium* from other Genera

Differentiation of the genera and species of *Rubribacterium, Rhodovulum, Rhodobaca* and *Rhodothalassium* is possible based on 16S rDNA gene sequences (**Fig. 3.37**) (Boldareva *et al.,* 2009) and by DNA-DNA hybridization. Differentiating phenotypic characteristics of *Rubribacterium* from other genera are given in **Table 3.24** .

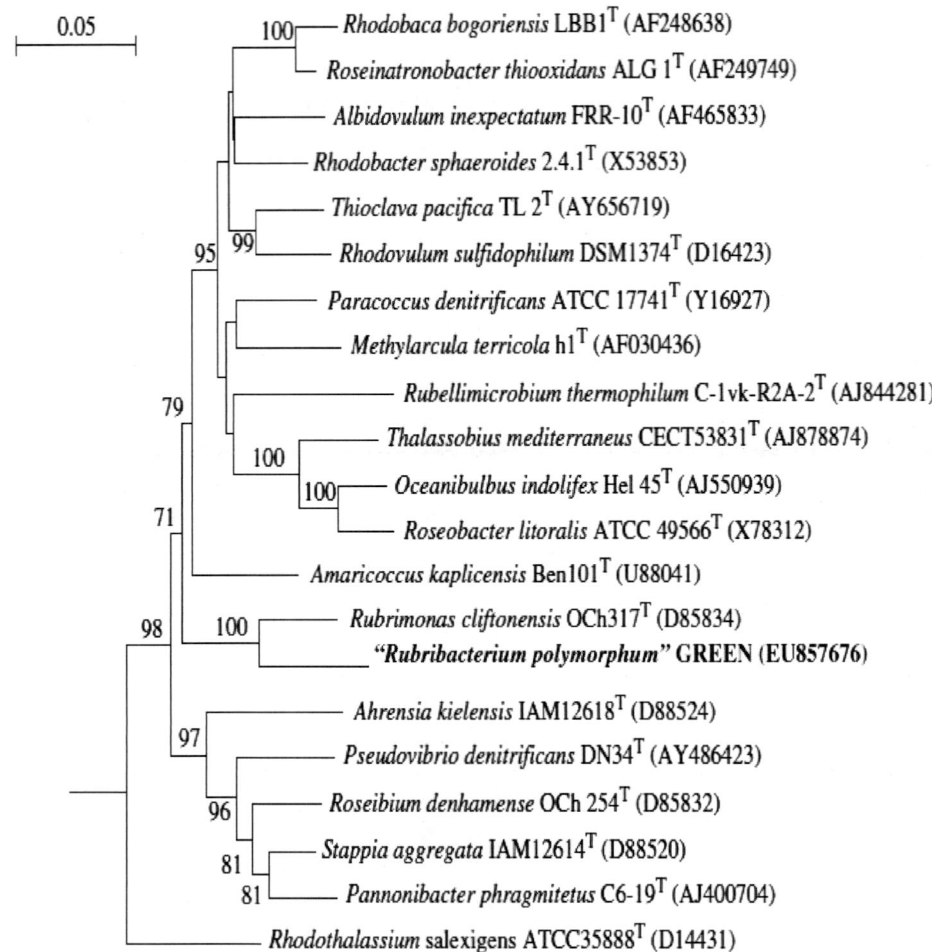

Fig. 3.37. Phylogenetic tree of *Rubribacterium* species and related *Alphaproteobacteria* of the order *Rhodobacterales* based on 16S rDNA sequences.

Taxonomic Comments

Phylogenetic analysis of 16S rDNA gene sequencing revealed that only species of *Rubribacterium* such as *Rubribacterium polymorphum* strain GREEN form a separate branch in the cluster of the genera *Rhodobaca* , *Rhodovulum* , *Rhodobacter* ,*Rhodothalassium* and other ABC bacteria (**Fig. 3.37**) (Boldareva *et al.,* 2009). The phylogenetic analysis placed this species within the family *Rhodobacteraceae* of the order *Rhodobacterales*, class *Alphaproteobacteria.* On the phylogenetic tree, *Rubribacterium polymorphum* strain GREEN showed the highest similarity with aerobic, bacteriochlorophyll *a*-containing bacteria *Rubrimonas cliftoensis.* However, the similarity level was only 93.3%. These two organisms formed a cluster with the highest level of bootstrap support (100%).

The similarity with other members of *Rhodobacteraceae* was significantly lower, not exceeding 90%. Thus, the phylogenetic analysis suggested that the position of the *Rubribacterium polymorphum* strain GREEN was within the family *Rhodobacteraceae.* Hence it is not assigned to any of the known species of this family. Based on phenotypic and genotypic characteristics it is classified into a new genus *Rubribacterium* as *Rubribacterium polymorphum* (Boldareva *et al.,* 2009).

Order 4. *Rhodothalassiales* Venkata Ramana *et al.* (2013)

The order *Rhodothalassiales* was described on the basis of phylogenetic analysis of 16S rRNA sequences. The order contains the family *Rhodothalassiaceae.*

Family 1. *Rhodothalassiaceae* Venkata Ramana *et al.* (2013)

Cells are vibrioid to spiral-shaped, Gram (-) negative, motile by means of flagella. Major fatty acids are $C_{18:1}\omega 7c$, $C_{18:1}\omega 17c$ 11-methyl and $C_{16:0}$. Q-10 and MK-10 are the major quinones. Their natural habitats are hypersaline environments, such as solar salterns. At present the family comprises the single genus *Rhodothalassium,* which is a type genus. The DNA G + C content ranges from 60.0 to 62.8 mol%.

1. *Rhodothalassium* (*Rts.*) (Rose belonging to the sea) (Imhoff *et al.,* 1998) Venkata Ramana *et al.* (2013)

Cells are vibrioid- to spiral-shaped, 0-5-1.0 μm wide and are Gram (-) negative (**Fig. 3.38A**). Motile by means of polar flagella (**Fig. 3.38C**), multiply by binary fission. Intracytoplasmic photosynthetic membranes are present as lamellar stacks lying parallel to the cytoplasmic membrane (**Fig. 3.38B**). Photosynthetic pigments are bacteriochlorophyll *a* and carotenoids of the spirilloxanthin series. The major quinone components are ubiquinones (Q-10) and menaquinones (MK-10) which contain 10 isoprene units. Major cellular fatty acids are $C_{18:1}\omega 7c$, $C_{18:1}\omega 7c11$-methyl and $C_{16:0}$ with minor amounts of $C_{14:0}$, $C_{18:0}$, $C_{18:1}\omega 5c$ and $C_{16:1}\omega 7c/C_{16:1}$ $\omega 6c$ and few strains contain $C_{19:0}cyclo\omega 8c$. Major polar lipids include diphosphatidyl glycerol (DPG), phosphatidylglycerol (PG), ornithine lipid (OL), unidentified phospholipid (PL) and unidentified amino lipid (AL). Minor amounts of unidentified lipids (L1-2) are also present. Halophilic bacteria that require NaCl or sea salt for growth. Have salt optima above seawater salinity and tolerate up to 20% or more total sales. Growth occurs preferably photoheterophically under anoxic conditions in the presence of light. Chemotrophic growth is also possible under microoxic to oxic conditions in the dark. Aerobic growth is absent in some strains. Amino acids such as niacin and thiamine may be required as growth factors. Anoxic zones of hypersaline environments such as salterns and salt lakes that are exposed to light are selective habitats for their growth. Only a single species is validly published, *Rhodothalassium salexigens* (type species) and its characteristics are précised in **Table 3.24.**

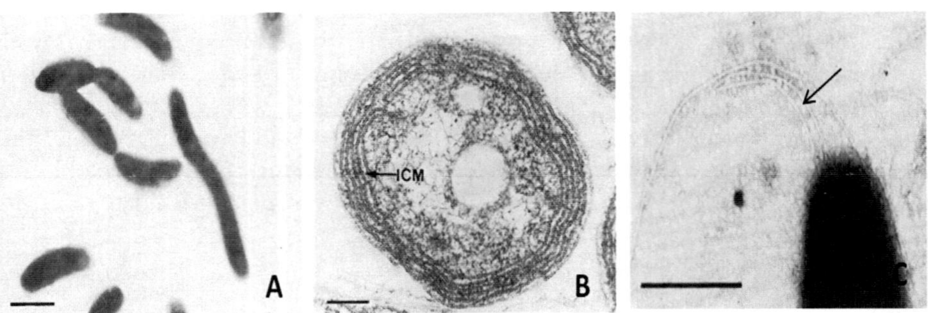

Fig. 3.38. A. Short curved or spiral shaped negatively stained cells of *Rhodothalassium salexigens*(*Rhodospirillum salexigens*), B. Ultrathin cross-section of intracytoplasmic membrane (Bar=0.1 μm), C. Cells showing three flagella in the polar region (Bar=0.5 μm) (Drews, 1981).

Differentiation of the *Rhodothalassium* from other Genera

Based on the natural environment, salt requirement, cell morphology and internal membrane structure, *Rhodothalassium* is clearly distinguished from other spiral-shaped phototrophic *Alphaproteobacteria*. In addition, it has characteristic fatty acid and quinone composition. It is clearly separated from other phototrophic purple bacteria by its 16S rRNA sequence. Major differentiating properties between *Rhodothalassium salexigens* and other phototrophic *Alphaproteobacteria* are precised in **Table 3.24**. The phylogenetic relationship of these bacteria based on 16S rRNA sequence is illustrated in **Fig. 3.39** (Venkata Ramana *et al.,* 2013).

Taxonomic Comments

Because of its spiral-shaped and phototrophic capacity, *Rhodothalasisium salexigens* was originally assigned to the genus *Rhodospirillum* comprised all spiral-shaped purple non-sulfur bacteria at that time (Drews, 1981). According to the latest edition of Bergey's Manual (Imhoff, 2005), it is included as a genus *Incertae sedis*. Phylogenetically this genus was not associated with the *Rhodospirillaes ,Rhizobiales ,Rhodobacterales* or other *Alphaproteobacteria* as it forms a distinct separate branch lineage (**Fig. 3.39**). High diversity among these bacteria is revealed by their chemical composition not only in regard to fatty acids and quinones but also in respect to the structure of internal membrane systems to growth factor requirement and salt dependency (adaptation to hypersaline habitats and high salt requirement). Because of outstanding phenotypic differences and the great genetic distance (< 90% similarity) from the type species of this genus *Rsp. rubrum* and to other purple non-sulfur bacteria it were classified into a new genus *Rhodothalassium* (Imhoff *et al.,* 1998) into the new family *Rhodothalassiaceae* and new order *Rhodothalassiales* within the *Alphaproteobacteria* (Venkata Ramana *et al.,* 2013).

Table 3.24. Differentiating Characteristics of *Rhodobacca, Rhodobaculum, Rubribacterium* and *Rhodothalasium*

Characteristics	*Rhodobaca bogoriensis*	*Rhodobaca barguzinensis*	*Rhodoba-culum claviforme*	*Rubriba-cterium polymor-phum*	*Rhodo-thallasium salexigens*
Cell size (μm)	0.8-1 X 0.8-1.5	0.8-1 0.8-1.2	1-2.5 2.5-7.0	0.7 x 1.0–3.0	0.6-0.7
Cell shape	Cocci to short rods	Ovoid/Rods	Short rods / Spindle-shaped	Oval or Polymorphic	Ovoid-spirilloid
Cell division	Binary fission	Binary fission	Binary fission	Binary fission	Binary fission
Color of cell suspension	Yellow-Yellowish brown	Yellowish-brown, Pink	Orange-brown, Pink	Yellowish-brown	Red
Bacteriochlorophyll	*a*	*a*	*a*	*a*	*a*
Carotenoids	Spheroidene, demethyl-spheroidene	Spirilloxanthin, demethyl-spheroidene	Sphaeroidene, spirilloxanthin groups	Sphaeroidene, Spirilloxanthi n	Sphaero-idene, Spirillox-anthin
Absorption maxima of living cells (nm)	475, 507, 590 and 870	492, 525, 593, 803 and 870	376, 588, 750, 803, 851 and 886	477, 507, 549, 795, 837 and 875	
Intracellular membrane system	Vesicles	Vesicles	Vesicles	Vesicle	Lamella
NaCl requirement (%)					
Range (Optimum)	1-3 (not required)	5-80.0 (20.0)	0.5-4.0 (2.0)	5–40 (10)	5-20 (6-8)
pH range (Optimum)	7.5-10 (9.0)	7.5-10 (8.0-9.0)	7.5-9.7 (8-8.8)	7.4-10 (8.5–9.5)	(6.6-7.4)
Optimum Temperature (°C)	39	20-35	25-30	30	40
Growth factors	Biotin, Thiamine	Biotin, Thiamine	Yeast extract	Biotin, Thiamine	Glutamic acid
Sulfide oxidation	S^0	S^0	No oxidation	No oxidation	No oxidation
Autotrophic growth	-	-	-	–	-
Anaerobic growth (light)	+	+	+	+	+
Aeobic growth (dark)	+	+	+	+	+
Light harvesting complex	LHI	LHI	LHI and LHII	LHI and LHII	LHI and LHII
Mol% G+C of DNA	58.8	59.8	67.6-69.8	69.9	64
Carbon sources utilized	Succinate, Malate, Pyruvate, Fumarate, Acetate, Man-nitol, Butyrate, Valerate, Hepta-noate, Xylose, Glucose, Fru-ctose and Sucrose	Simple carbon sources	Acetate, Pyruvate, Peptone, Soyton, Propionate, Malate, Lactate, Fumarate and Fructose	Simple carbon sources	Acetate, Citrate, Glucose, Glycerol, Pyruvate

Characteristics	*Rhodobaca bogoriensis*	*Rhodobaca barguzinensis*	*Rhodoba-culum claviforme*	*Rubriba-cterium polymor-phum*	*Rhodo-thallasium salexigens*
Nitrogen sources utilized	Ammonia, Aspartate, Glutamate, Asparagine, Glutamine and Casamino acids	Ammonia, Aspartate, Glutamate, Asparagine, Glutamine and Casamino acids	ND	ND	Glutamate
Habitat	Soda Lakes	Soda Lakes	Soda Lakes	Soda Lakes	Saltern lake

Symbols: +, Present or substrate utilized; –, Absent or substrate not utilized; ND, not determined. Data for *Rhodobaca bogoriensis* and *Rhodobaca barguzinensis* is adopted from Milford *et al.* (2000) and Boldareva *et al.* (2008). Data for *Rhodobaculum claviforme, Rubribacterium polymorphum* and *Rhodothallasium salexigens* is adopted from Bryantseva *et al.* (2015), Boldareva *et al.* (2009) and Venkata Ramana *et al.* (2013).

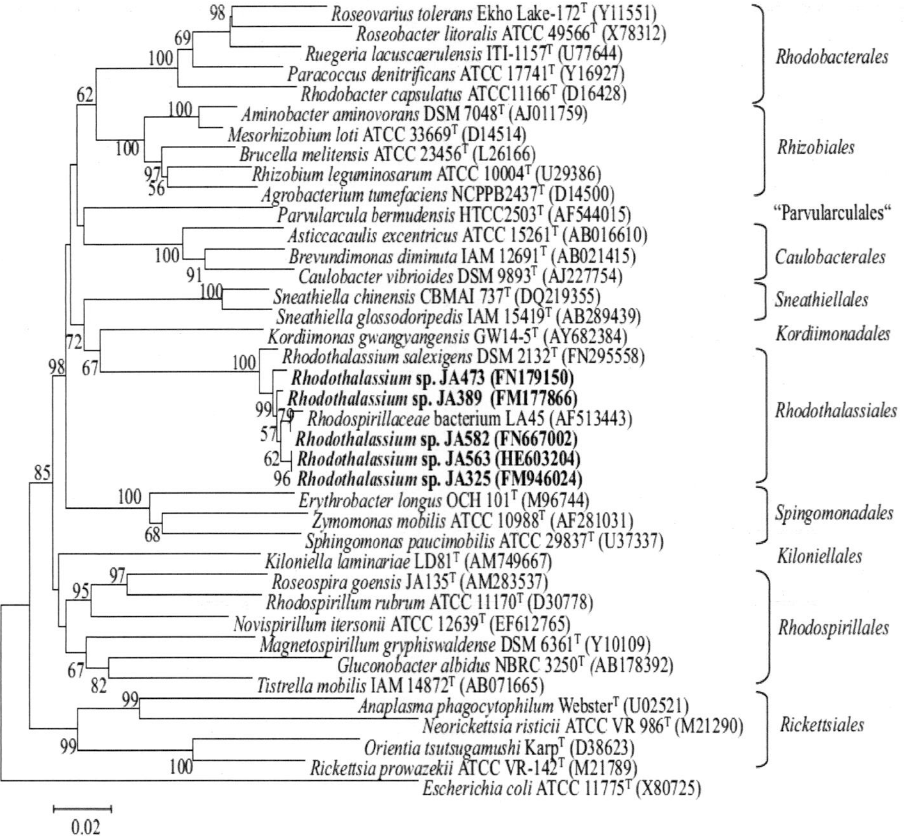

Fig. 3.39. Phylogenetic tree of *Rhodothalassium* species and related *Alphaproteobacteria* of the order *Rhodospirillaes, Rhizobiales* and *Rhodobacteriales* based on 16S rRNA sequences. The sequence of *E. coli* was used as an outgroup to root the tree.

CLASS II. *BETAPROTEOBACTERIA* (Garrity *et al.*, 2005) Garrity *et al.* (2015)

The class *Betaproteobacteria* is described on the basis of phylogenetic analysis of 16S rRNA sequences. The class contains the orders *Burkholderiales* , *Hydrogenophilales, Methylophilales, Neisseriales, Nitrosomonadales, Procabacteriales* and *Rhodocyclales* .*Burkholderiales* is the type order. UQ-8 is always found as one of the dominant respiratory quinones in all orders. In some orders, menaquinones and rhodoquinones are also found as major components. $C_{16:0}$ is typically found among the dominant fatty acids along with $C_{16:1}$ and $C_{18:1}$. Metabolically and morphologically they are distinct. G+C contents range from about 35-72 mol%. Purple non-sulfur bacteria are included in *Rhodocyclales* and *Burkholderiales*. Hence, only two orders are described.

Order 1. *Burkholderiales* Garrity *et al.* (2015)

The order *Burkholderiales* is described on the basis of phylogenetic analysis of 16S rRNA sequence. The order contains the families *Burkholderiaceae, Oxalobacteriaceae, Alcaligenaceae* and *Comamonadaceae*. Thus the order is phenotypically, metabolically and ecologically diverse and includes strictly aerobic and facultatively anaerobic chemoorganotrophs, obligate and facultative chemolithotrophs, capable of nitrogen-fixing and plant, animal and human pathogens. Anoxygenic phototrophic purple non-sulfur bacteria are included in the family *Comamonadiaceae*. Hence, only one family is described.

Family 1. *Comamonadaceae* (Willems *et al.*, 1991) Willems and Gillis (2015)

The family *Comamonadaceae* is a large and diverse bacterial family belonging to the order *Burkholderiales*. It is regularly enlarged by adding new genera and species. Currently, it comprises over 100 species in the following 29 genera such as *Acidovorax* , *Albidiferax* , *Alicycliphilus* , *Brachymonas* , *Caenimonas* , *Comamonas* ,*Curvibacter* ,*Delftia* ,*Diaphorobacter* ,*Extensimonas* ,*Giesbergeria* , *Hydrogenophaga* , *Hylemonella* , *Lampropedia* , *Limnohabitans* , *Macromonas* , *Malikia* , *Ottowia* , *Polaromonas* , *Pseudoacidovorax* , *Pseudorhodoferax* , *Ramlibacter* ,*Rhodoferax* ,*Simplicispira* ,*Tepidicella* , *Variovorax* , *Verminephrobacter* , *Xenophilus* and *Xylophilus* .

Cells are straight or slightly curved rods or spirilla. Gram (-) negative. Mostly motile by means of either single polar flagellum or bipolar tufts of one to five flagella or degenerate peritrichous flagella. No endospores or microcysts are formed. Coccoid bodies are rarely formed in some species. Chemoorganotrophic or facultatively chemolithotrophic with H_2 or CO oxidation. Oxidative metabolism using oxygen as a terminal electron acceptor. Some species can also use nitrates. The optimal growth temperature ranges from 24 to 35°C, except for *Xylophilus* spp. and (*Aquaspirillum*) *Psychrophilum*, which grow best at 4 and 18°C, respectively. Yellow insoluble pigments are produced by some species. Nitrogen fixation has been described for some species. Generally, these organisms use only very few carbohydrates, but they use a wide variety of organic acids, including amino acids. The major fatty acids are palmitoleic acid

($C_{16:1}$), palmitic acid ($C_{16:0}$) and cis-vaccenic acid ($C_{18:1}\omega7c$). The G+C contents of the DNA range from 57 to 70 mol%. These genera form a phylogenetic cluster with a relative level of 16S rRNA gene sequence. The similarity of 93–97%, harbor a remarkable phenotypic diversity that includes aerobic organography, anaerobic denitrifies and Fe^{3+}-reducing bacteria, hydrogen oxidizers, photoautotrophic, photoheterotrophic bacteria and fermentative bacteria. Most bacteria harbor in water and soil habitats. However, some *Comamonadaceae* are also pathogens and others have been described from earthworms, rumen fluid and human clinical samples. Many of them are mesophiles, a few have been found in Antarctic habitats and some in hot springs. Among 29 genera purple non-sulfur bacteria were included in *Rhodoferax*. Hence, only *Rhodoferax* is described.

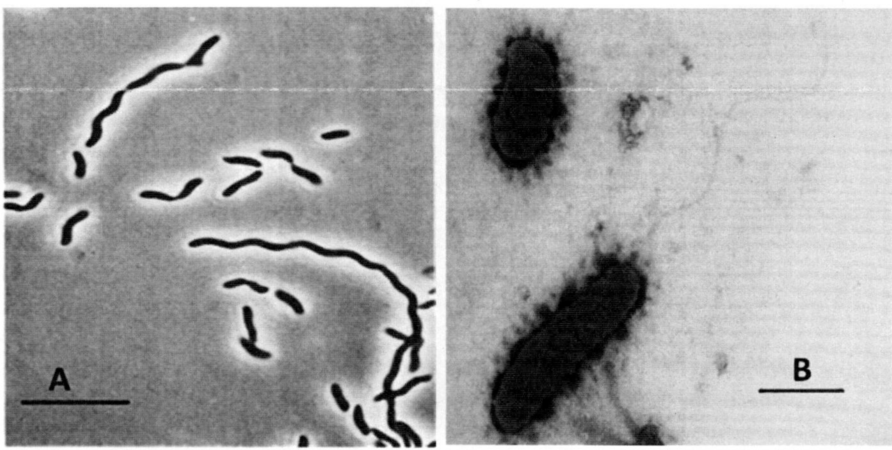

Fig. 3.40. A.Phase contrast micrograph of *Rhodoferax fermentans* showing the rod shape helical filaments (Bar=10 μm), B. Cell showing single polar flagellum (Bar=1 μm).

1. *Rhodoferax (Rfx.)* (Red and fertile) (Hiraishi *et al.,* 1991) Hiraishi and Imhoff (2015)

Cells are vibroid or curved rods, 0.2-1.0 x 1.0-5.0 μm in diameter and motile by means of polar flagella **(Fig. 3.40 A & B)** (Hiraishi *et al.,* 1991). Multiply by binary fission and Gram (-) negative. Endospore and capsule formation is not seen. Colour of photosynthetic cultures is peach-brown. Bacteriochlorophyll *a* and carotenoids with spheroidene, spirilloxanthin and OH-spheroidene as major components are present as photosynthetic pigments. Intracytoplasmic membranes may be absent. Growth is occurring probably by photosynthesis, aerobic respiration and fermentation. Photoorganotrophy with various organic compounds as carbon sources is the preferred mode of growth. They grow well on simple organic compounds as carbon sources and electron donors and in complex media substituted with peptone, yeast extract, or casamino acids. No reduced sulfur compounds serve as an energy source for photoautotrophic growth. Thiamine and biotin are required as growth factors. Ammonium salts and glutamate are used as nitrogen sources but not nitrate. Gelatin is weakly hydrolyzed. Major fatty acids are $C_{16:1}$ and $C_{16:0}$, the major components of cellular fatty acids.$C_{8:0}$3-OH is present. Ubiquinone (Q-8) and rhodoquinone (RQ-8) with 8 isoprene units are occurs as major quinone. Membrane-bound fumarate reductase activity with $FMNH_2$ as electron donor is found. The DNA G + C content is 59.5 to 61.0 mol% (HPLC). Ditchwater and activated sludge are suitable substrates for their growth. *Rhodoferax fermentans* is the type species (Hiraishi *et al.,* 1991). Four species are validly published and their characteristics properties are presented in **Table 3.25.**

Table 3.25. Characteristic Properties of Species of *Rhodoferax*

Characteristics	*Rfx. saidenbachensis*	*Rfx. fermentans*	*Rfx. ferrireducens*	*Rfx. antarcticus*
Habitat	Sediments of water reservoir	Activated sludge	Coastal aquifer sediments	Antarctic microbail mats
Cell morphology	Short Rods	Curved Rods	Rods	Curved Rods to Spiral
Flagella	Single	+	Single	Single
Cell size (μm)	0.2-0.4 x 1.0-1.8	0.6-0.9	1 x 3-5	0.6-0.9 x 1.1-3.0
Pigments	Brown	Peach- brown	Brown	Peach- brown
Carotenoids	Spaheroidene	Sphaeroidene, Spirilloxanthin, OH-spheroidene	Spaheroidene, Spirilloxanthin	Spaheroidene series
Oxygen demand	Aerobic, no photosynthesis	Aerobic respiration and fermentation, photosynthesis	Facultatively anaerobic, no photosynthesis	Phototrophic, chemo-organotrophic
Growth Temperature (°C) (optimum)	4-30	25-30	4-30 (25)	0-25
Growth at 4°C	+	NA	+	+
Growth at 37°C	-	-	-	-
pH range (optimum)	6.0-9.0	5.0-9.0	6.7-7.1 (7.0)	NA
DNA G+C (mol%)	60.3-61	59.8-60.3	59.5	61.5
Catalase	-	NA	-†	NA
Growth factors	ND	Thiamine, Biotin	ND	Biotin
Utilization of carbon sources	Lactate, glycerol, succinate, succinamic acid, b-hydroxybutyric acid, L-fucose, pyruvate, c-hydroxybutyric acid, L-alaninamide,	Acetate, valerate, pyruvate, lactate, succinate, malate, fumarate, citrate, mannitol, benzoate,	Acetate, lactate, malate, propionate, pyruvate, benzoate and succinate	Acetate, pyruvate, lactate, succinate, malate, fumarate, glucose, fructose, sucrose, citrate, asparatate

Characteristics	*Rfx. saidenbachensis*	*Rfx. fermentans*	*Rfx. ferrireducens*	*Rfx. antarcticus*
	a-hydroxybutyric acid, L-proline, cis-aconitic acid, L-asparagine, D-fructose, 2-aminoethanol and D-galactose	aspartate, glutamate, glucose, fructose, xylose, ribose, butyrate, proponol, butanol, ethanol, α-ketoglutarate		

Symbols: +, Positive/substrate utilized; -, negative/substrate not utilized; ND, data not determine; NA, data not available; †, Potentially; genes are present/absent. Data for *Rhodoferax saidenbachensis*, *Rfx. fermentans*, *Rfx. ferrireducens* and *Rfx. antarcticus* is adopted from Kaden *et al.* (2014), Hiraishi *et al.* (1991), (Ramana and Sasikala (2009) and Madigan *et al.* (2000).

Differentiation of the *Rhodoferax* from other Genera

The differentiating characteristics of *Rhodoferax* and other genera of *Betaproteobacteria* are given in **Table 3.26**. The genus *Rhodoferax* is separated from other genera of phototrophic *Proteobacteria* by its phylogenetic position and its distinct physiological and chemotaxonomic characteristics. The characteristic properties of *Rhodoferax* include fructose fermentation Q-8 and RQ-8 production, and relatively low mol% G + C contents of the DNA. The phylogenetic relationships of the phototrophic *Betaproteobacteria* are shown in **Fig. 3.41** .

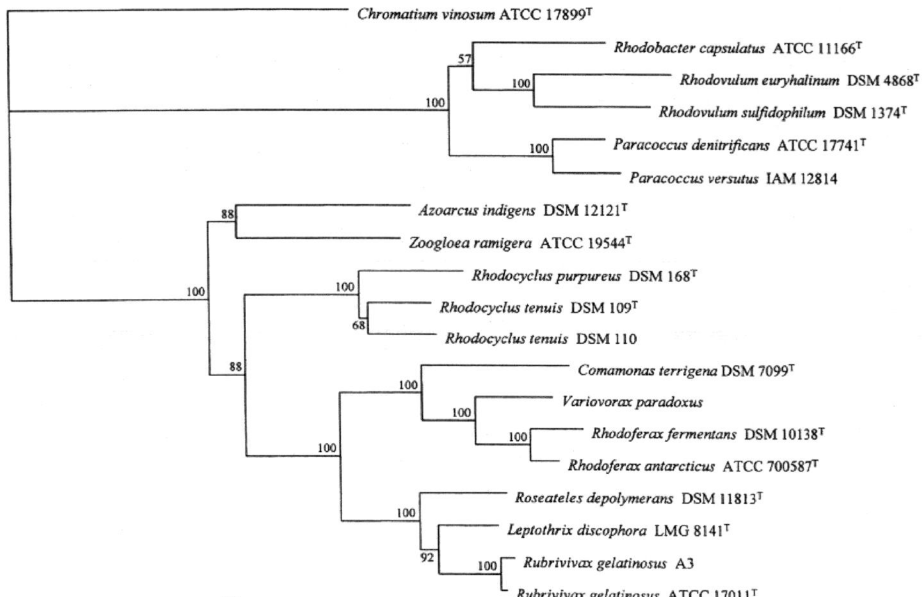

Fig. 3.41. Phylogenetic tree of *Rhodoferax* species and related *Betaproteobacteria* based on 16S rDNA sequences. The sequence of *Chromatium vinosm* was used as an outgroup to root the tree

Taxonomic Comments

The species of *Rhodoferax* such as *Rfx. fermentans* was first recognized as the *Rhodocyclus gelatinosus* (RGL group) (Hiraishi and Hoshino, 1984), due to their phenotypic properties which are similar to *Rubrivivax gelalinosus* (at that time they are known as *Rhodocyclus gelatinosus*), which resemble *Rubrivivax gelatinosus* subgroup II in their appearance (Weckesser *et al.*, 1969). More detailed physiological, chemotaxonomic and genetic studies revealed major differences between the RGL group and *Rubriviuax gelatinous* and these observations led to the proposal for the reclassification of the RGL group into the new genus *Rhodoferax* (Hiraishi *et al.*, 1991) as *Rfx. fermentans.* Later, a phylogenetic study based on 16S rDNA sequences showed that the genera *Rhodoferax, Rubrivivax* and *Rhodocyclus* are phylogenetically distinct groups within the *Betaproteobacteria* (Hiraishi, 1994). Till now four species are validly published.

Table 3.26. Differentiating Characteristics of the Genera *Rhodoferax*, *Rhodocyclus* and *Rubrivivax*

Characteristics	*Rhodoferax*	*Rubrivivax*	*Rhodocyclus*
Habitat	Ditches/activated sludge	Paddy soils	Freshwater/ sewage ditches/ swine waste lagoons
Cell diameter (µm)	0.2-1.0 x 1.0-5.0	0.4–0.7 × 1.0–3.0	0.3–1.0
Cell Shape	Curved rods/Vibroids	Curved rods	Ovoid-Rods
Intracytoplasmic membrane	-	Tube/small finger-like	Tube/small finger-like
Motility	+	+	+/-
Colour	Peach brown	Pale-peach/dirty yellowish-brown	Purple-violet to violet
Carotenoids	Spheroidene, spirilloxanthin and hydroxyspheroidene	Sphaeroidene series	Rhodopin, Rhodopinal
Bacteriochlorophyll	*a*	*a*	*a*
Major quinones	Q-8, RQ-8	Q-8, MK-8	Q-8, MK-8
Major fatty acids	$C_{16:1}$ and $C_{16:0}$	$C_{16:1}$ (Predominant) $C_{16:0}$ and $C_{18:1}$	$C_{16:1}$ (Predominant), $C_{16:0}$, and $C_{18:1}$
Storage granules	-	PHA	+
Growth factors	Thiamine, Biotin	Biotin, Thiamine	Vit B_{12}, Biotin and p-aminobenzoic acid
Aerobic growth	+	+	+
Sulfate assimilation	ND	ND	+
Denitrification	+	+	+
Photoautotrophic growth with	-	H_2, CO_2	H_2
Salt requirement (%)	Not required	Not required	Not required

Characteristics	*Rhodoferax*	*Rubrivivax*	*Rhodocyclus*
Optimal temperature (Range)	25-30 (4-30)	ND	28-30
pH Range	5.0-9.0	ND	6.5-7.5
Mol% G+C of the DNA	59.5-61.0	70.0-72.5	64.1-65.1

Symbols: +, Present; –, Absent ; +/– variable in different strains; ND, not determine/no data available. Data for *Rhodoferax*, *Rubrivivax* and *Rhodocyclus* is adopted from Hiraishi *et al.* (1991), Willems *et al.* (1991) and Pfennig (1978).

2. *Rubrivivax* (*Rvi.*) (Red living organism) Willems *et al.* (1991)

Cells are straight to slightly curved rods, 0.4 to 0.7 by 1 to 3 µm, irregularly curved cells up to 15 µm long may occur in old cultures. Sheaths are not formed. Cells are motile by means of a polar flagellum and Gram (-) negative. Bacteriochlorophyll *a* and carotenoids, sphaeroidene, hydroxyspheroidene and spirilloxanthin are present as photosynthetic pigments. Small finger-like invaginations of the cytoplasmic membrane represent photosynthetic membranes. Photoheterotrophic growth occurs under anaerobic conditions in the light, with various organic substrates as electron donors or photoautotrophic growth in the presence of H_2, CO_2 and growth factors is possible. In the dark, chemoautotrophic growth with CO as a sole carbon and energy source and aerobic chemoheterotrophic growth with various organic substrates are also possible. Nitrogen may be fixed under photoautotrophic growth conditions. Colonies are pink to deep-red because of photosynthetic pigments. Heterotrophically grown colonies are initially faintly pink. Produce poly-hydroxyalkanoates granules **(Fig. 3.42)**. The G+C values of the DNA range from 70.0 to 72.5 mol%. In DNA-rRNA hybridizations, Tm (e) values of 77 to 81.5°C are obtained versus rRNA of *Rubrivivax gelatinosus* (Molisch 1907). Two species are validly published *Rvi. benzoatilyticus* and *Rvi. gelatinosus* (type species) and their characteristics are precised in **Table 3.27.**

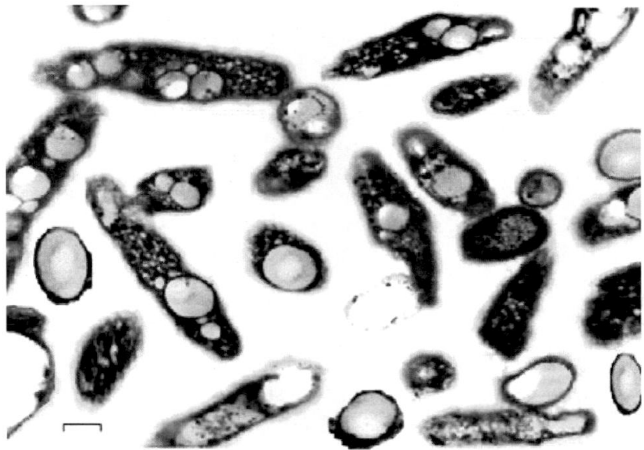

Fig. 3.42. Electron micrograph of ultrathin section of *Rubrivivax benzoatilyticus* showing poly-hydroxyalkanoate granules.

Table 3.27. Differential Characteristics of the Species of *Rubrivivax* and *Rhodocyclus*

Characteristics	***Rvi. benzoatilyticus***	***Rvi. gelatinosus***	***Rcy. purpureus***	***Rcy. tenuis***
Habitat	Paddy soil	Mud	Swine waste lagoons	Freshwater ponds/ Sewage ditches
Cell diameter (μm)	0.7-1.0 x 2.0-6.0	0.4-0.7 x 1.0-3.0	0.6-0.7 x 2.0-3.0	0.3-0.5 x 1.5-6.0
Cell shape	Rod	Rod	Half circle to circle	Curved rods
Motility	+	+	-	+
Slime production	-	-	-	+
Color	Orange-brown	Pale-peach/dirty yellowish-brown	Purple-violet to violet	Brownish-red or purple-violet
Major carotenoids	Spirilloxanthin, Sphaeroidene	Sphaeroidene series	Rhodopin, rhodopinal	Rhodopin, rhodopinal, lycopene[a]
Absorption maxima of living cells (nm)	378, 488, 520, 590, 802 and 884	ND	379, 408, 510, 535, 597, 813 and 866	378–380, 465, 492–495, 528, 592–594, 799–801, and 868–871 (Spirilloxanthin series), 377–378, 469, 495–500, 529–533, 590–592, 798–801 and 856–858 (rhodopinal series)
Growth factors	Not required	Biotin, Thiamine[b]	B_{12}, PABA, Biotin	None[c]
Intracytoplasmic membrane	–	Tubes	Tubes	Tubes
NaCl requirement (Optimum)	Not required	Not required	Not required	Not required
Temperature (°C) (Optimum)	ND	ND	30	30
pH	ND	6.6	6.5-7.5 (7.2)	6.6-7.4
Gelatin liquefaction	ND	+	–	–
Fructose fermentation	+	+	–	–
Starch hydrolysis	ND	ND	ND	ND
Tween 80 lysis	ND	ND	ND	ND
Carbon sources				
Acetate	+	+	+	+
Aspartate	ND	ND	–	–
Benzoate	+	–	+	–
Citrate	–	+	–	–
Ethanol	ND	+	–	+/–
Fructose	+	+	–	–
Fumarate	+	+	+	+
Glutamate	+	+	–	–

Characteristics	*Rvi. benzoatilyticus*	*Rvi. gelatinosus*	*Rcy. purpureus*	*Rcy. tenuis*
Glucose	+	+	–	–
Glycerol	+	–	–	–
Lactate	-	+	–	+
Malate	+	+	+	+
Mannitol	–	–	–	–
Methanol	–	+	–	–
Propionate	+	+	–	–
Pyruvate	+	+	+	+
Succinate	+	+	–	+
Sorbitol	–	-	–	ND
Tartarate	+	+	–	–
Valerate	–	–	–	+
Electron donors				
Hydrogen	+	+	+	+
Sulfide	–	–	–	+
Sulfur	–	–	–	–
Thiosulfate	–	–	–	+
N_2-fixation	–	+	–	+
Fumarate reductase activity				
With reduced methylviologen	ND	High	High	High
With $FMNH_2$	ND	Low	Low	Low
Major quinones	ND	Q-8, MK-8	Q-8, MK-8	Q-8, MK-8
Mol% G+C of DNA				
by HPLC	74.9	71.2-72.1	65.1	64.1-64.8
by Bd	–	70.5-72.4	65.3	64.8
by T_m	–	80.0-81.4	67.7	64.4-67.2

Symbols +, Positive/ utilized; -, Negative/ not utilized; Data not determined; Data not available. ND, Q-8, ubiquinone with 8 isoprene units; MK-8, menaquinone with 8 isoprene units. [a]Some strains may contain carotenoids of the spirilloxanthin series and lack rhodopinal. [b]Some strains may also require pantothenate. [c]Some strains may require vitamin B_{12}. Data for *Rhodovivax benzoatilyticus* and *Rubrivivax gelatinosus* is adopted from Ramana *et al.* (2006) and Willems *et al.* (991). Data for *Rhodocyclus purpureus* and *tenuis* is taken from Pfennig (1978) and Imhoff *et al.* (1984).

Differentiation of the *Rubrivivax* from other Genera

Rubrivivax is clearly distinguished from other phototrophic purple bacteria by its separate phylogenetic line within the *Betaproteobacteria.* The high mol% G + C content of >70 is quite characteristic. Properties to differentiate *Rubrivivax* species from other phototrophic *Betaproteobacteria* are given in **Table 3.26** and the phylogenetic relationship of the phototrophic *Betaproteobacteria* is shown in **Fig. 3.41** and **3.43.**

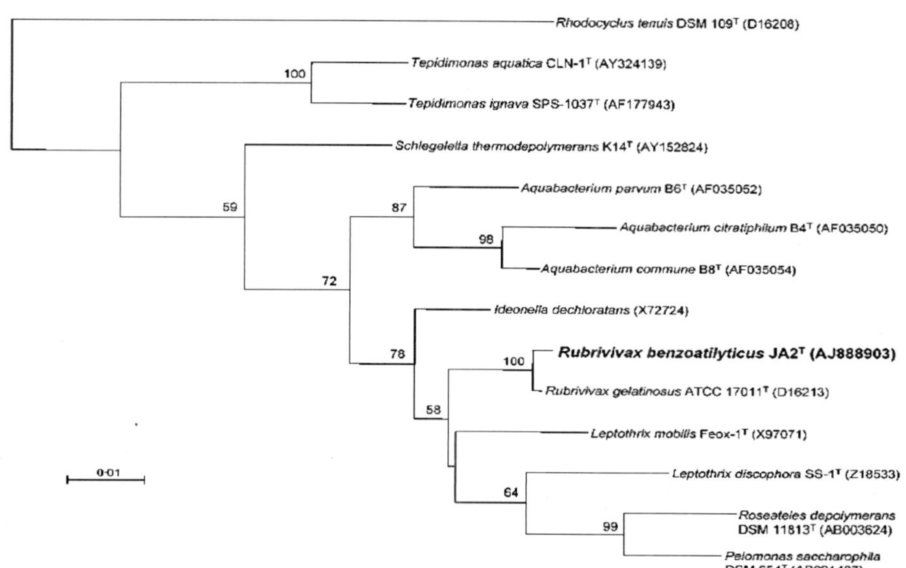

Fig. 3.43. Phylogenetic tree of *Rubrivivax* species and related *Betaproteobacteria* based on 16S rRNA sequences. The sequence of *Rhodocyclus tenuis* was used as an outgroup to root the tree.

Taxonomic Comments

Currently, two species are known. *Rubrivivax gelatinosus* and *Rubrivivax benzoatilyticus* . *Rvi. gelatinosus* has been reported to occur in two distinct morphological forms (Biebl and Drews, 1969). Form I cells are clearly curved, 0.4- 0.7 µm in diameter and produce little slime during active growth. Form II cells are more or less straight rods. However, they are sometimes also bent and produce more slime during active growth which causes sedimentation of the cells in a gelatinous layer. For instance form I cells utilize a greater variety of carbon sources and have a much shorter doubling time than form II cells (Weckesser *et al.,* 1969). Another strain of this species described by Klemme (1968), is intermediate in carbon substrate utilization with respect to the two groups established by Weckesser *et al.* (1969). However, the lipopolysaccharides of *Rvi gelatinosus* show two different serotypes, which do not cross-react with each other (Weckesser *et al.*, 1975). Detailed studies are required to clarify the taxonomic status of the strains assigned to this genus. Further information is required on the specificity of gelatin liquefaction by *Rubrivivax gelatinosus.* Siefert *et al.* (1978) have identified about half of the strains with this property as belonging to *Rhodobacter capsulatus,* while other strains, with the morphology typical of *Rubrivivax gelatinosus,* do not liquefy gelatin. Based on different phenotypic and genotypic properties, *Rhodocyclus gelatinosus* is reclassified into a new genus *Rubrivivax* as *Rubrivivax gelatinosus*. Phylogenetically they are closely related to *Rhodocyclus*(**Fig. 3.43**) (Ramana *et al.,* 2006).

Order 2. *Rhodocyclales* Garrity *et al.* (2015)

The order *Rhodocyclales* was described on the basis of phylogenetic analysis of 16S rRNA sequence. The order contains the family *Rhodocyclaceae* .

Family 1. *Rhodocyclaceae* Garrity *et al.* (2015)

The family is described on the basis of phylogenetic analysis of 16S rRNA sequence contains the genera *Rhodocyclus* (type genus) (Pfennig 1978), *Azoarcus, Azonexus, Azospira, Azovibrio, Dechloromonas, Ferribacterium, Propionibacter, Propionivibrio, Quadricoccus, Sterolibacterium, Thauera* and *Zoogloea.* Family is phenotypically, metabolically and ecologically diverse. These are photoheterotrophs, aerobes, anaerobes and facultative anaerobes utilizing a number of electron acceptors, fermentative organisms and nitrogen-fixing organisms. *Rhodocyclus* is the type genus.

1. *Rhodocyclus* (*Rcy.*) (Red circle) (Pfennig, 1978) Imhoff (2015)

Cells are slender, curved, or straight thin rods, 0.3-1.0 μm in diameter, non-motile or motile by means of polar flagella and multiply by binary fission **(Fig. 3.44 A & B)** (Pfennig, 1969). Internal photosynthetic membranes are present in the form of small, single finger-like intrusions of the cytoplasmic membrane. Bacteriochlorophyll *a* and carotenoids, rhodopsin, rhodopinal and lycopene are present as photosynthetic pigments. Phototrophic culture appears purple-violet. Ubiquinones (Q-8) and menaquinones (MK-8) with eight isoprene units are the major quinones. Straight chain $C_{16:1}$ and $C_{16:0}$ acids are the major components of cellular fatty acids. $C_{10:0}$ 3OH is present. Photoautotrophic growth with molecular hydrogen may be possible if growth factors are supplied. Preferably grow photoheterotrophically under anaerobic conditions in the presence of light with different organic substrates as the carbon and electron source. Chemotrophic growth also occurs under microaerobic to aerobic conditions in the dark. Reduced sulfur compounds are not used as photosynthetic electron donors. Assimilatory sulfate reduction is possible. Vitamins are required for growth. The optimum growth temperature and pH are 30ºC and 6.5-7.5 respectively. Mesophilic and neutrophilic freshwater bacteria. Freshwater ponds, sewage ditches, swine waste lagoons are the selective habitats for their growth. The mol% G+C of the DNA is 64.1-65.1. *Rhodocyclus purpureus* is the type species. Three species are validly published, *Rhodocyclus gelatinosus* , *Rcy. purpureus* and *Rcy. tenuis*, but recently *Rcy. gelatinosus* was transferred to *Rubrivivax gelatinosus.* Hence, only two species are present in this genus and their characteristics are presented in **Table 3.27.**

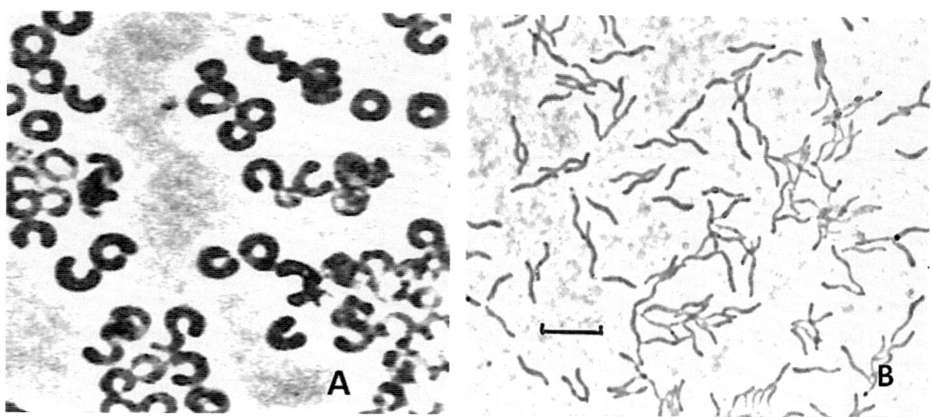

Fig. 3.44. A. Phase contrast micrograph of cells of *Rhodocyclus purpureus* (Bar=5 μm), B. *Rhodocyclus tenuis* (Bar=5 μm)

Differentiation of the *Rhodocyclus* from other Genera

Rhodocyclus species are differentiated from other phototrophic purple bacteria by their separate phylogenetic position within the class *Betaproteobacteria* (**Fig. 3.41**) . Generally, physiological properties are not significantly different from those of other purple non-sulfur bacteria. However, the utilization of benzoic acid by *Rcy. purpureus* is a unique feature among other purple non-sulfur bacteria. The requirement for vitamin B_{12} is also not common among the purple non-sulfur bacteria, which is more common among the green- and purple sulfur bacteria. Besides *Rhodocyclus purpureus,* only a single strain of *Rhodocyclus tenuis* and *Rhodopseudomonas palustris* have been reported to require vitamin B_{12} (Siefert and Koppenhagen, 1982). Properties that differentiate *Rhodocyclus* species from other phototrophic *Betaproteobacteria* are given in **Table 3.25** .

Taxonomic Comments

Currently, this genus comprises two species *Rhodocyclus purpureus and Rhodocyclus tenuis* (Formerly *Rhodospirillum tenue*) .Only a single strain of *Rcy. purpureus* is known till now. Differentiation has been observed in strains of *Rcy. tenuis* based on carotenoid composition (Schmidt, 1978), colour of the cell suspensions and absorption spectra (Biebl, 1973). Some strains of *Rcy. tenuis* have carotenoids of the spirilloxanthin series and others have carotenoids of the rhodopinal series. The type strain DSM 109 is the one that transforms the rhodopin to spirilloxanthin. Whereas another group of strains including strain DSM 110 does not form anhydro rhodovibrin and spirilloxanthin but accumulates major amounts of rhodopinal, rhodopinol and lycopenal (Schmidt, 1978). Both of these strains DSM 109 and 110 probably show a close genetic relationship (**Fig. 3.41** .)

3.2. PURPLE SULFUR BACTERIA

CLASS III. *GAMMAPROTEOBACTERIA* (Woese *et al.*, 1985, Garrity *et al.*, 2005) Garrity *et al.* (2015)

Purple sulfur bacteria (PSB) belongs to the class *Gammaproteobacteria* (Woese *et al.*, 1985). An increasing amount of chemotaxonomic and phylogenetic data led to the reclassification of the group *Gammaproteobacteria* that have 16S rRNA gene sequences related to those of the members of the order *Pseudomonadales* .*Chromatiaceae* and *Ectothiorhodospiraceae* are two families of *Chromatiales* which are also called Phototrophic Purple Sulfur Bacteria (PPSB). Cells are spherical, ovoid, spiral, rod or vibrioid shape, multiplication by binary fission. The members of PSB are distributed into 72 species representing 30 genera. In recently described species of *Ectothiorhodospira* and *Thiorhodospira sibirica,* sulfur globules remain, attached to the cells in the cell periphery or in the periplasmic space (Bryantseva *et al.*, 1999). The differences between representatives of the two families are clear in the 16S rDNA nucleotide sequences (Imhoff, 2006 c, d).

The class contains 14 orders such as *Acidithiobacillales, Aeromonadales, Alteromonadales, Cardiobacteriales, Chromatiales, Enterobacteriales, Legionellales, Methylococcales, Oceanospirillales, Pasteurellales, Pseudomonadales, Thiotrichales, Vibrionales*, and *Xanthomonadales* (Garrity *et al.*, 2015). However, anoxygenic purple sulfur bacteria are included in the order Chromatiales only and they are described.

Order I. *Chromatiales* (Imhoff, 2005) Imhoff (2015)

The *Chromatiales* are unicellular and Gram (-) negative, cells are spherical, vibrioid, spiral, or rod-shaped. Multiply by binary fission, motile by flagella or non-motile and some contain gas vesicles. These bacteria are also known as phototrophic purple sulfur bacteria. They are able to perform photosynthesis under anoxic conditions without oxygen production and also contain non-phototrophic, purely chemotrophic relatives. All *Chromatiales* grow under anoxic conditions in the presence of light. They are capable of photolithoautotrophic growth using sulfide or elemental sulfur as an electron donor. Colour of the cell suspension of these bacteria is purple-violet to purple-red, red, orange-brown, yellowish-brown, brownish-red, brown and green. These phototrophic bacteria are characterized by the presence of bacteriochlorophyll *a* or *b* and various carotenoids of the spirilloxanthin, okenone, or rhodopinal groups. Photosynthetic pigments are located in the cytoplasmic membrane and in the internal membrane system of different fine structures originating from the cytoplasmic membrane.

These organisms can grow as photolithoautotrophic or photoorganoheterotrophic metabolism under anoxic conditions. The photosynthetic metabolism differs from that of the cyanobacteria, algae and green plants in that water cannot serve as an electron donor. Photosynthetic CO_2 assimilation depends on the utilization of external electron donors such as reduced sulfur compounds, reduced iron, molecular hydrogen, or organic compounds. Elemental sulfur globules accumulate inside the cell in the presence of sulfide and light is an important property of *Chromatiaceae*, while in *Ectothiorhodospiraceae*, sulfur globules are deposited outside the cells. CO_2 is photoassimilate through the reductive pentose phosphate cycle. Fixation of N_2 has been demonstrated in many species. All species contain cytochromes, quinones, and non-heme iron proteins of the ferredoxin type. Moreover, several species of *Chromatiales* can grow chemotrophically under microoxic to oxic conditions in the dark. However, some species such as *Arhodomonas aquaeolei* (*Ectothiorhodospiraceae*), may be purely chemotrophic and unable to grow phototrophically. Two families are included in *Chromatiales* such as *Chromatiaceae* and *Ectothiorhodospiraceae* . *Chromatium* is the type genus of the family *Chromatiaceae* .

The traditional classification of *Chromatiaceae* is mainly based on morphology such as cell form and size, motility, presence of gas vesicles and formation of cell aggregate. Imhoff (1984b) has proposed the classification of *Chromatiaceae* based on 16S rDNA sequence similarities. Subsequently, *Thiorhodovibrio winogradskyi* (Overmann *et al.*, 1992b), *Rhabdochromatium marinum* (Dilling *et al.*, 1995), *Halochromatium glycolicum* (Caumette *et al.*, 1997), *Thiorhodococcus minor* (Guyoneaud *et al.*, 1997), *Thiocapsa litoralis* (Puchkova *et al.*, 2000), *Thioalkalicoccus limnaeus* (Bryantseva *et al.*, 2000), *Thioflavicoccus mobilis* (Imhoff and Pfennig, 2001) and *Thiobaca trueperi* (Rees *et al.*, 2002) have been added. With the determination of 16S rRNA gene sequence for most of the species of *Chromatium* and supported by selected phenotypic properties, members have been reclassified by Imhoff *et al.* (1998b). Details of nomenclatural changes and classification of different members of *Chromatiaceae* are précised below.

Pfennig and Truper (1971c) included *Ectothiorhodospira* as an exceptional genus in the *Chromatiaceae*, because of its ability to perform a dissimilatory sulfur metabolism. The species of *Ectothiorhodospira* showed a distinct difference from all other genera of the *Chromatiaceae*, while representatives of the purple non-sulfur bacteria were able to oxidize sulfide to extracellular elemental sulfur. Pfennig (1977) proposed to remove the genus *Ectothiorhodospira* from the *Chromatiaceae* and placed it in the purple non-sulfur bacteria. Thus phototrophic purple bacteria are grouped into three groups as *Chromatiaceae* , *Ectothiorhodospiraceae* and *Rhodospirillaceae* based on polar lipid composition, fatty acid and quinone composition (Imhoff *et al.*, 1982). A formal proposal for reclassification of genus *Ectothiorhodospira* into a family *Ectothiorhodospiraceae* .

At present, the family *Chromatiaceae* comprises 26 genera viz. *Allochromatium, Chromatium, Halochromatium, Isochromatium, Imhoffiella, Lamprocystis, Lamprobacter, Marichromatium, Phaeochromatium, Phaeobacterium, Rhabdochromatium, Thiocystis, Thiocapsa, Thiococcus, Thiodictyon, Thiospirillum, Thiopedia, Thiolamprovum, Thermochromatium, Thiohalocapsa,*

Thiorhodovibrio, Thiorhodococcus, Thioalkalicoccus, Thioflavicoccus, Thiobaca, and *Thiophaeococcus,* while the family *Ectothiorhodospiraceae* comprises four genera such as *Ectothiorhodospira, Ectothiorhodosinus, Halorhodospira* and *Thiorhodospira.*

Molecular Characterization

The available 16S rDNA sequence information for *Ectothiorhodospiraceae* (Imhoff and Suiling, 1996) and *Chromatiaceae* species (Caumette *et al.,*1997; Guyoneaud *et al.,* 1998; Imhoff *et al.,* 1998b) supports the classification of these bacteria into two different families (Imhoff, 1984b). At the same time, the existence of different genera within the family *Ectothiorhodospiraceae* (Imhoffand Suiling, 1996) and the inconsistency of the previous taxonomic classification and phylogeny of the family *Chromatiaceae* (Pfennig andTruper, 1989; Imhoff *et al.,* 1998b). In addition, the phylogenetic relatedness of both families and their distance to other bacteria of the *Gammaproteobacteria* supports their administration within one order, for which the name *Chromatiales* is proposed.

The order *Chromatiales* includes the phototrophic purple sulfur bacteria of the *Ectothiorhodospiraceae* and *Chromatiaceae* (Imhoff, 1989; Pfennig and Truper, 1989) and their purely chemotrophic relatives. The chemotrophic bacteria known so far are specifically related to *Ectothiorhodospiraceae.* The 16S rDNA sequence of *Arhodomonas aquaeolei* (Imhoff and Suiling, 1996) is similar to those of *Halorhodospira* species, but distinct at the genus level. In addition, sequences of *Nitrococcus mobilis* (Teske *et al.,* 1994) and *Nitrosococcus oceani* (Head *et al.,* 1993) place these bacteria in phylogenetic proximity to the *Ectothiorhodospiraceae,* though at a more distant level than the genus *Arhodomonas.*

Ectothiorhodospira is placed in the family *Chromatiaceaea* which forms sulfur globules outside the cells (Pfennig and Truper, 1974). With the establishment of the new family, *Ectothiorhodospiraceae* and an emended description of the family *Chromatiaceae* exclusively includes phototrophic sulfur bacteria capable of depositing globules of elemental sulfur (S^0) inside the cells. This description agrees with Molisch's (1907) definition of the *Thiorhodaceae* and is in line with their phylogenetic separation from other phototrophic purple sulfur bacteria (Imhoff and Suiling, 1996; Imhoff *et al.,* 1998b). In the past, the taxonomy of the phototrophic purple bacteria, including the family *Chromatiaceae,* was classified based on simple phenotypic characteristics (Pfennig and Truper, 1989). Until recently, little molecular and sequence-based information, such as rRNA oligonucleotide data for selected species (Fowler *et al.,* 1984) and a few 16S rDNA sequences (DeWeerd *et al.,* 1990; Wahlund *et al.,* 1991), was available for *Chromatiaceae* species. During the past few years, complete 16S rDNA sequences for most of the *Chromatiaceae* species is available (Caumette *et al.,* 1997; Guyoneaud *et al.,* 1998; Imhoff *et al.,* 1998b). While writing the eighth edition of the determinative manual, 16S rDNA sequences of few type strains of purple sulfur bacteria have not determined or not yet available. **Figs. 3.46** and **3.47** showed that separate major phylogenetic branches of the family *Chromatiaceae* contain 1. They are truly marine and halophilic species, 2. Species that are motile by polar flagella, do not contain gas vesicles and are primarily freshwater species and 3. Species with ovoid to spherical cells are primarily freshwater species, some also occur in brackish and marine habitats. The majority of them are non-motile (except *Lamprocystis roseopersicina*) and most of them contain gas vesicles. Because of the phylogenetic relationships, they are not in agreement with traditional classification. A reclassification based on 16S rDNA sequence similarity, supported by selected phenotypic properties was proposed (Guyoneaud *et al.,* 1998; Imhoff *et al.,* 1998b). A positive correlation between 16S rDNA sequence similarity values and phenotypic properties of the species is taken as a sign of the importance of these properties in achieving a phylogenetically oriented taxonomy that includes both genetic and phenotypic information (Imhoff *et al.,* 1998b).

The DNA base ratio within the family *Chromatiaceae*, expressed as the mol% G+C content, has been found to span a large scale. Thus, in the previous edition of the *Manual* the genus *Chromatium,* Pfennig (1989b) recorded extensive diversity of the G+C values within the genus *Chromatium.* A phylogenetic system would be expected to group bacteria that have much less variation in their mol% G+C content together. Indeed, after rearrangement of the species of *Chromatiaceae* according to their genetic relatedness on the basis of 16S rDNA sequence analysis (Guyoneaud *et al.,* 1998; Imhoff *et al.,* 1998b), a quite narrow range of mol% G+C content is found within many newly established genera. The ranges are 63–66% in the cluster of species of *Thiocapsa* , *Lamprocystis* and *Thiolamprovum,* while in *Marichromatium* species 69-70%, in *Halochromatium* species 64-66% and in the true *Chromatium* species 48-50%. This correlation appears to be more significant. High intrageneric and intraspecies variation of the mol% G+C content necessitates experimental examination. For example, the high intraspecies variation observed in *Allochromatium vinosum* (61.3–66.3%), *Allochromatium warmingii* (55.1–60.2%) and *Thiocystis violacea* (62.8–67.9%) suggests misclassification of some strains. However, there is a need for detailed studies to clarify these problems.

1. *Chromatiaceae* (Bavenndamm, 1924) Imhoff (2015)

Cells are spherical, ovoid, vibrioid, spiral, or rods shaped, multiply by binary fission, may contain gas vesicles, and may be motile by flagella or non-motile. Motile forms have either monotrichous or multitrichous polar flagella. Gram (-) negative and contain internal photosynthetic membranes, which are continuous with the cytoplasmic membrane and of vesicular type in most species **(Fig 3.45A**). Though some species have tubular internal membranes (**Fig 3. 45B**) others have lamellae.

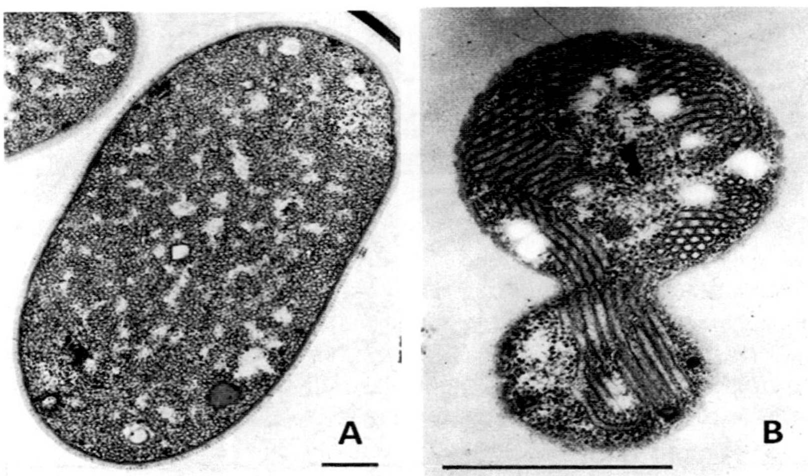

Fig 3.45. A. Ultrathin section of a cell of *Thiocystis gelatinosa* showing internal photosynthetic membranes (Bar = 1 μm). B. Cells of *Thiococcus pfennigii* showing tubular internal photosynthetic membranes in longitudinal and cross-section (Bar = 0.5 1 μm).

Photosynthetic pigments include bacteriochlorophyll *a* or *b* and carotenoids of the spirilloxanthin, okenone, or rhodopinal series are present in many species, but other carotenoids such as tetrahydrospirilloxanthin, also occur in some species. In general, the colour of cultures of different bacteria will depend on the type of carotenoid groups. Carotenoids of the spirilloxanthin group containing photosynthetic cells are orange-brown to brownish-red or pink in colour, while those with okenone are purple-red and those with carotenoids of the rhodopinal group are purple-violet. All the species are capable of photolithoautotrophic growth in the presence of light with sulfide or elemental sulfur. Elemental sulfur globules accumulate as an intermediate oxidation product inside the cells. Many species can use H_2 as an electron donor under reducing culture conditions and some use reduced iron ions. All species are potentially mixotrophic and photoassimilate a number of simple organic compounds. Acetate and pyruvate are the most widely used simple organic compounds. Some species are able to photoassimilate organic substances in the absence of sulfide or sulfur or grow photoorganoheterotrophically. Many species are strictly anaerobic and obligately phototrophic, while others are chemolithoautotrophic or chemoorganoheterotrophic growth occurs under microoxic to oxic conditions in the dark.

The pathway of autotrophic CO_2 assimilation was shown to be the reductive pentose phosphate cycle in all species tested so far. The fixation of dinitrogen has been demonstrated in several species. Storage materials are polysaccharides, poly-β-hydroxybutyrate, elemental sulfur, and polyphosphate granules. Vitamin B_{12} is required by several species. In nature, members of the *Chromatiaceae* occur in the illuminated, anoxic, and sulfide-containing parts of a large variety of aquatic environments such as sediments from moist and muddy soils, ditches, ponds, lakes, rivers, sulfur springs, estuaries, marine habitats, and even salt and soda lakes. The mol% G + C of the DNA is 45.5–70.4. *Chromatium* is the type genus.

Three major genetically distinct groups of the family *Chromatiaceae* are (i) a group of truly halophilic and salt-requiring species. It includes genera, like *Thiorhodovibrio, Marichromatium, Isochromatium, Rhabdochromatium, Thiohalocapsa, Halochromatium, Thiorhodococcus, Thiococcus, Thioflavicoccus*and *Thioalkalicoccus*, (ii) while a group that is dominated by spherical or rod-shaped nonmotile species. Many of which produce gas vesicles and primarily freshwater inhabitants (includes genera *Thiocapsa, Lamprocystis*, and *Thiolamprovum*) and (iii) the third group of motile species without gas vesicles that are primarily freshwater and include *Chromatium, Allochromatium, Thermochromatium*, and *Thiocystis*. In both of the freshwater clusters, strains or species may behave indifferently to low salt concentrations, or require only small amounts of salts. Major differentiating characteristics of genera and species of *Chromatiaceae* are precised in **Tables 3.28–3.40**. Their genetic relationship is shown in **Figs. 3.46** and **3.47** .

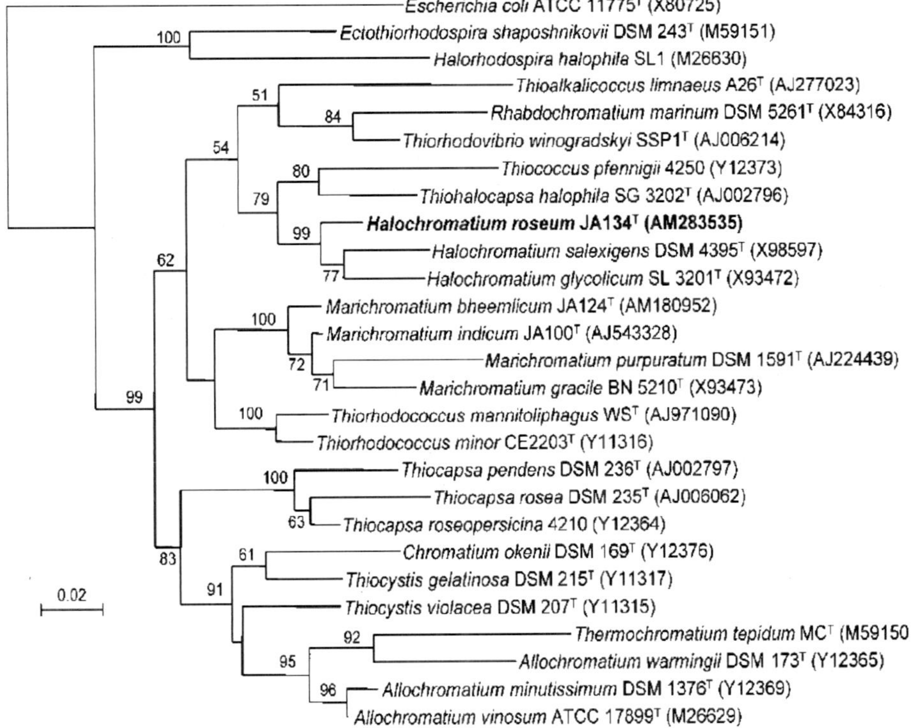

Fig. 3.46. Phylogenetic tree of family *Chromatiaceae* based on 16S rDNA. The sequence of *E. coli* was used as an outgroup to root the tree.

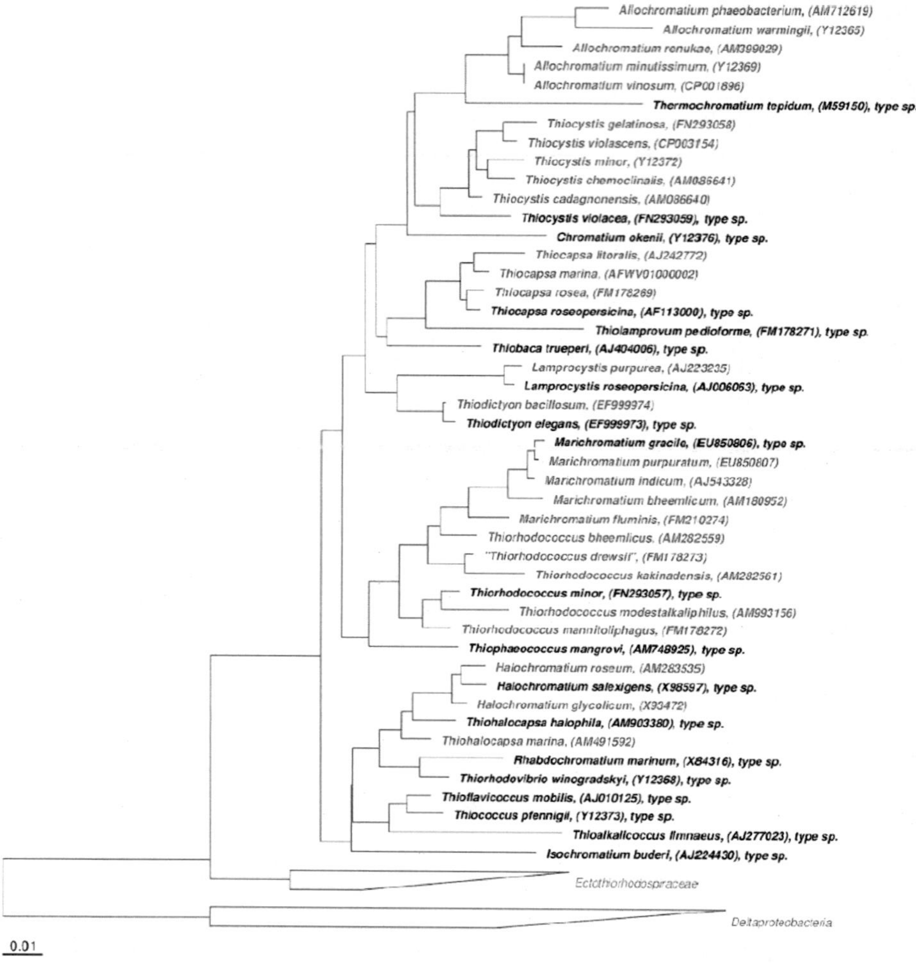

Fig. 3.47. Phylogenetic tree of family *Chromatiaceae* species of based on 16S rRNA. Representative sequences from closely related taxa were used as outgroups. The scale bar indicates estimated sequence divergence (Imhoff, 2014).

Table 3.28. Differential Characteristics of the Phototrophic Members of *Chromatiaceae* and *Ectothirhodospiraceae*

Characteristics	***Chromataiceae***	***Ectothirhodospiraceae***
Sulfur deposition	Inside cells	Outside cells
Characteristics of polar lipids	Several glycolipids	glycolipids absent
Major fatty acid:	Q8/MK8	Q7 or Q8/MK7 or MK8
$C_{16:0}$	20-35%	11-25%
$C_{16:1}$	25-37%	<10%
$C_{18:1}$	38-45%	50-75%

Characteristics	*Chromataiceae*	*Ectothirhodospiraceae*
Properties of lipopolysaccharides:		
Major aminosugars	glucosamine	diamino-dideoxy-glucose
Phosphate	present	absent
D-Mannose	+	-
D-Galacturonic acid	-	+
D-Glucuronic acid	-	+
Amide-bound fatty acid	$C_{14:0}$ 3OH	$C_{12:0}$ 3OH
Signature 16s rDNA sequence[a]		:
Position 217	T	G/A[b]
Position 234-235	CA	TG/CG[b]
Position 821-822	TC	AG
Position 876-877	GA	CT
Position 985	A	T/G[d]

[a], Numbering based on *E.coli* 16SrDNA sequence; [b] Bases above the line are found in the genera *Ectothirhodospira* and *Thiorhodospira*; bases below the line are found in the genera *Arhodomonas* and *Halorhodospia*

Family 1. *Chromatium*(*Chr.*) (Coloured one) (Party, 1852) Imhoff (2015)

Cells are straight or slightly curved rods, more than 3 µm wide, single or form in pairs, multiply by binary fission. Motile utilizing a polar tuft of multitrichous flagella that can be seen in the light microscope (**Fig. 3.48**), Gram (-) negative. Vesicular types of internal photosynthetic membranes are present containing bacteriochlorophyll *a* and carotenoids of okenone are present as photosynthetic pigments. Strictly anaerobic, obligately phototrophic and require sulfide-reduced media. Photolithoautotrophic growth occurs under anoxic conditions in the light with sulfide and elemental sulfur as electron donors. Thiosulfate and H_2 are not utilized. During oxidation of sulfide, elemental sulfur is transiently stored inside the cells in the form of highly refractile globules and sulfate is the final oxidation product. A few simple organic substrates are photoassimilates in the presence of sulfide or sulfur and bicarbonate. Assimilatory sulfate reduction is not observed. Ammonia is used as the nitrogen source, N_2 may be reduced but nitrate is not reduced. Polysaccharides, poly-β-hydroxybutyrate, elemental sulfur and polyphosphates are the storage materials. Vitamin B_{12} is required. Mesophilic freshwater bacteria with optimum growth temperatures range from 20 to 35 and without salt requirement. Water and sediment surface layers of stagnant freshwater habitats such as ditches, ponds and lakes, containing hydrogen sulfide and exposed to light are the selective habitats for their growth. The mol% G + C of the DNA is 48.0–50.0. *Chromatium okenii* (Perty, 1952) is the type species. Two species are validly published (*Chromatium okenii* and *Chromatium weissei* which can be distinguished on the basis of characters precised in **Table 3.29.**

Table 3.29. Differential Characteristics of *Chromatium* and *Allochromatium* species

Characteristics	*Chromatium okenii*	*Chr. weissei*	*Allochromatium vinosum*	*Alc. minutissimum*	*Alc. warmingii*	*Alc. renukae*	*Alc. phaeobacterium*	*Alc. humboldtianum*
Cell shape	Rod	Rod	Rod	Rod	Rod	Ovoid to rod	Rod	Ovoid to rod
Cell size (μm)	4.5-6.0 x 8.0-16.0	3.5-4.5 x 7.0-1.4	2.0 x 2.5-6.0	1.0-1.2 x 2.0	3.5-4.0 x 5.0-11.0	2.0–2.5 x 3.0–5.0	1.0-1.5 x 2.0-4.0	2.0 x 3.0
Motile by flagella	+	+	+	+	+	NR	NR	+
Color of cell suspensions	Purple red	Purple red	Brownish red	Brownish red	Purple violet	Purple - violet	Brown	Brown-red
Carotenoid group	Okenone	Okenone	Spirilloxanthin	Spirilloxanthin	Rhodopin	Lycopene	Rhodopinal	Spirilloxanthin
Mol% G + C of DNA	48	48	64.3	63.7	55.1-60.2	63.3	59.8	63.9
B12 requirement	+	+	–	–	+	–	–	–
Sulfate assimilation	+	-	+	+	–	–	+	+
Chemotrophic growth	+	-	+	+	–	–	–	+
pH optimum	7.0	7.0	7.0-7.3	7.0–7.3	7.0	7.2–8.0 (6.8–8.5)	7.5 (7.0–8.0)	6.9–8.0 (6.5–8.5)
Temperature optimum °C	20-35	20-35	25-35	25-35	25-30	28-32	30	25-35
Salt requirement	Not required	Not required	Not required[c]	Not required	Not required	Not required	Not required	3.0
Carbon sources/Electron donors								
Hydrogen	–	–	+	+	–	–	–	–
Sulfide	+	+	+	+	+	(+)	–	+
Thiosulfate	–	–	+	+	–	–	(+)	+
Sulfur	+	+	+	+	+	–	–	+
Sulfite	NR	NR	+	+	NR	–	–	–
Formate	–	–	+	+	–	+	–	NR
Acetate	+	+	+	+	+	+	+	–
Pyruvate	+	+	+	+	+	+	+	NR
Propionate	–	–	+	+	NR	–	–	+
Butyrate	–	–	±	±	–	(+)	–	NR
Lactate	–	–	NR	NR	–	–	+	+
Fumarate	–	–	+	+	–	+	+	+
Succinate	–	–	+	+	–	+	–	–
Malate	–	–	+	+	–	+	+	+

Characteristics	*Chromatium okenii*	*Chr. weissei*	*Allochromatium vinosum*	*Alc. minutissimum*	*Alc. warmingii*	*Alc. renukae*	*Alc. phaeobacterium*	*Alc. humboldtianum*
Fructose	–	–	–	–	–	–	+	+
Glucose	–	–	–	–	–	+	–	+
Ethanol	–	–	–	–	–	–	–	–
Propanol	–	–	–	–	–	–	–	–
Glycerol	–	–	–	–	–	–	–	–
Molecular features								
DDH relatedness (%)*	NA	NA	59 ± 5.44	64 ± 2.38	NA	61 ± 3.59	NA	100
16S rRNA similarity (%)*	NA	NA	98.2	98.2	96.8	97.3	98.1	100

[a]Symbols: +, positive in most strains; -, negative in most strains; ±, variable; ND, data not determined; NA, data not available; NR, not reported; [c]Marine strains may tolerate low concentrations of NaCl. Data of *Chromatium* is adopted from Perty (1852) and different species of *Allochromatium* is adopted from Imhoff *et al.* (1998), Srinivas *et al.* (2009), Anil Kumar *et al.* (2008) and Serrano *et al.* (2015)

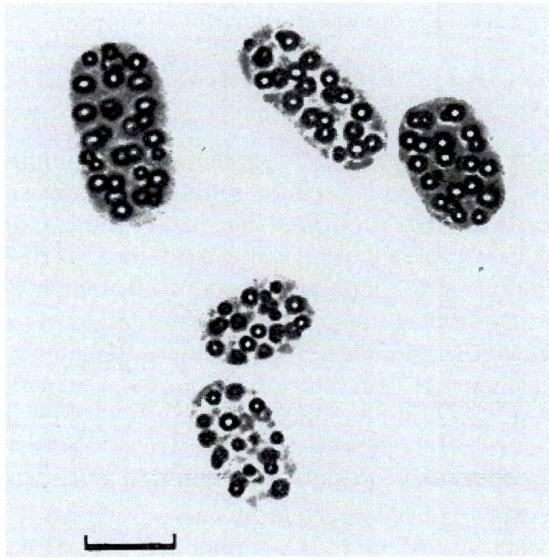

Fig. 3.48. *Chromatium okenii* DSM 169 cultured photoautotrophically with sulfide showing intracellular globules of elemental sulfur and the spiral tufts of polar flagella (Bar= 5 μm).

Differentiation of *Chromatium* from other Genera

According to 16S rDNA sequence analysis of *Chromatium* belongs to the branch of the family *Chromatiaceae* that characteristically contains freshwater species with flagellar motility. Species of this genus have very low mol% G+C values (48–50), large-sized cells (more than 3 lm wide), require vitamin B_{12} and lack the capability for chemotrophic growth. Characteristics for differentiation of *Chromatium* species from other phototrophic members of the *Chromatiaceae* are shown in **Table 3.29** and **3.40**. The genetic relationship among *Chromatium* species and other members of the *Chromatiaceae* based on 16S rDNA sequence comparison is shown in **Fig. 3.46.**

Taxonomic Comments

According to previous studies, all the rod-shaped, motile purple sulfur bacteria without gas vesicles that store elemental sulfur inside their cells are classified as *Chromatium* species (Pfennig and Truper, 1989) and 12 species are validly published in this genus. These species are not only quite distinct phenotypically but also different on the genetic basis such as the DNA base ratio within these bacteria (48.0–70.4 mol% G+C). Based on 16S rDNA sequence analysis of recognized species of *Chromatiaceae* supported for a new classification, in which the type species of the genus *Chromatium okenii*, appeared distant and separated from most of the other species previously assigned to this genus. The only other species showing sufficient phenotypic similarity to *Chr. okenii* to allow additionally to this genus is *Chr. weissei* (Imhoff *et al.,* 1998b). However, 16S rDNA sequence data for this species are not available.

2. *Allochromatium (Alc.)* (The other *Chromatium*) (Imhoff *et al.,* 1998) Imhoff (2015)

Cells are straight to slightly curved rods, single or in pairs, multiply by binary fission, motile by polar flagella, Gram (-) negative, internal photosynthetic membranes of vesicular type in which the photosynthetic pigments bacterio-chlorophyll *a* and carotenoids are located. Photolithoautotrophic growth occurs under anoxic conditions in the light with sulfide and elemental sulfur as electron donors. During oxidation of sulfide, S^0 is transiently stored inside the cells in the form of highly refractile globules (**Fig. 3.49**). Sulfate is the final oxidation product. Molecular hydrogen and thiosulfate may be used as electron donors. In the presence of sulfide and bicarbonate, organic substrates can be photo assimilated, potentially mixotrophic. Some species may grow chemoauto-trophically or chemoorganotrophically under microoxic to oxic conditions in the dark. Storage materials are polysaccharides, poly-β-hydroxybutyrate, elemental sulfur and polyphosphates. Vitamin B_{12} is required. Mesophilic bacteria grow at 25–35°C and pH 6.5–7.6. No distinct requirement for salt, except for strains that originate from the marine and brackish water environment and may tolerate or require a low concentration of salt. Grow in ditches, ponds and lakes with stagnant freshwater containing hydrogen sulfide and exposed to the light; also in sewage lagoons, estuaries, and salt marshes. The mol% G + C of the DNA is 55.1–66.3. *Allochromatium vinosum* (Ehrenberg) Imhoff *et al.* (1998b) is the type species

The genus *Allochromatium* contains 6 validly published species are *Alc. vinosum ,Alc. minutissimum, Alc. warmingii, Alc. renukae, Alc. phaeobacterium* and *Alc. humboldtianum*. These are physiologically versatile that do not require growth factors, grow chemoautotrophically and have rather high tolerance towards sulfide. They use thiosulfate, hydrogen and organic compounds as photo-synthetic electron donors. *Alc. warmingii*, on the other hand, is a specialized phototrophic species that requires vitamin B_{12} as a growth factor. Tolerates only low concentration of sulfide and is a strictly anaerobic. Differential characteristics of *Allochromatium* species are given in **Table 3.29**.

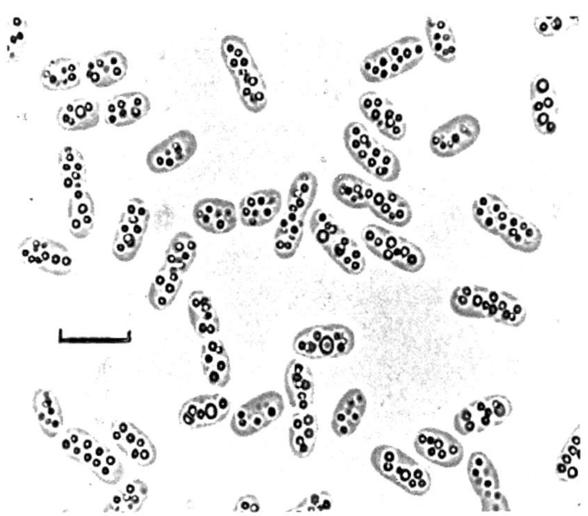

Fig. 3.49. *Allochromatium vinosum* ATCC 17899 cultured photoautotrophically with sulfide. The cells contain sulfur globules. Bright-field micrograph (Bar= 5μm).

Differentiation of *Allochromatium* from other Genera

According to 16S rDNA sequence analysis, the genus *Allochromatium* belongs to a branch of the family *Chromatiaceae* containing motile freshwater species and characterized by rod-shaped bacteria. Characteristics for differentiation of *Allochromatium* species from other phototrophic members of the *Chromatiaceae* are precised in **Table 3.29** and **3.40**. The phylogenetic relationships of *Allochromatium* species to other members of *Chromatiaceae*, based on 16S rDNA sequence comparison are illustrated in **Fig. 3.46** and **3.50.**

Taxonomic Comments

However, salt requirements are essential properties for the taxonomic characterization of *Chromatiaceae* species. The compatibility between phylogenetic position and salt response requires attention in few more species. Species such as *Allochromatium vinosum* have been reported to occur in freshwater, brackish water, and marine habitats. Large differences in the mol% G+C content of the DNA of strains assigned to this species (61.3–66.3) may indicate genetic heterogeneity. So far, clear proof of identity of various isolates from marine habitats tentatively assigned to *Allochromatium vinosum* is lacking, because of improper physiological and genetic DNA–DNA hybridization and

16S rDNA sequences characterization. Hence, more detailed studies are needed in order to reach conclusion whether differences in base composition are related to the marine or freshwater nature of these bacteria. Based on phylogenetic analysis it showed the closest similarity with species of *Marichromatium, Thiorhodococcus* and *Ectothiorhodospira* (Serrano *et al.,* 2015). Recently added *Alc. humboldtianum* strain AX1YPE form a separate branch within the genus *Allochromatium.* Based on 16S rRNA gene sequence similarity *Alc. humboldtianum* showed 98.2% with *Allochromatium vinosum* DSM 180T and *Alc. minutissimum* DSM 1376T, 98.1% with *Alc. phaeobacterium* JA144T, 97.3% with *Alc. renukae* DSM 18713T and 96.8% with *Alc. warmingii* DSM 173T. DNA–DNA hybridization values to the type strains of its closest relatives, *Alc. vinosum* and *Alc. minutissimum* is 59 and 64% respectively (Serrano *et al.,* 2015).

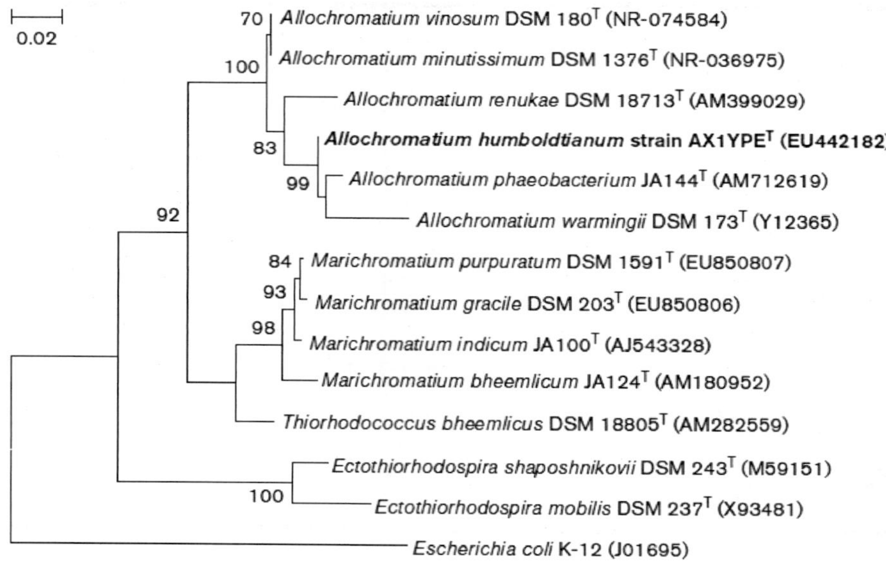

Fig 3.50. Phylogenetic tree of species of *Allochromatium* and other members of *Chromatiaceae* based on 16S rRNA. The sequence of *E.coli* was used as an outgroup to root the tree (Serrano *et al.,* 2015).

3. *Halochromatium* (*Hch.*) (Salt *Chromatium*) (Imhoff *et al.*, 1998b), Imhoff and Caumette (2015)

Cells are straight to slightly curved rods, single or in pairs, multiply by binary fission (**Fig. 3.51A**) (Caumette *et al.,* 1988). Motile by polar flagella and Gram (-) negative. Vesicular-type internal photosynthetic membranes are present with bacteriochlorophyll *a* and carotenoids of spirilloxanthin series as photosynthetic pigments (**Fig. 3.51 B**) (Caumette *et al.,* 1988). Photolithoautotrophic growth occurs under anoxic conditions in the present light with sulfide and elemental sulfur as an electron donor. During oxidation of sulfide, elemental sulfur is formed as an intermediate and stored in the form of highly refractile globules inside the cells. Sulfate is the final oxidation product. Hydrogen may be used as an electron donor. In the presence of sulfide and bicarbonate, organic substrates can be photoassimilated. Chemolithotrophic and chemoorganotrophic growth occur under microoxic conditions in the dark. Mesophilic forms grow at 20–35°C and pH 7.2–7.6. Elevated salt concentrations are required for growth. Grow in reduced sediments in saline, microbial mats in hypersaline environments and coastal lagoons with elevated salt concentrations, exposed to the light. The mol% G + C of the DNA is 64.6–66.3. *Halochromatium salexigens* (Caumette *et al* ,. 1988) is the type species. Three species such as *Hch. salexigens* (Caumette *et al.,* 1988), *Hch. glycolicum* (Caumette *et al.,* 1997) and *Hch. roseum* (Anil Kumar *et al.,* 2007) of this genus is validly published and their characteristics are précised in **Table 3.30.** Recently *Hch. roseum* is transferred into a *Lamprobacter roseus* (Gorlenko *et al.,* 2014). Hence, at present only two species are present in this genus.

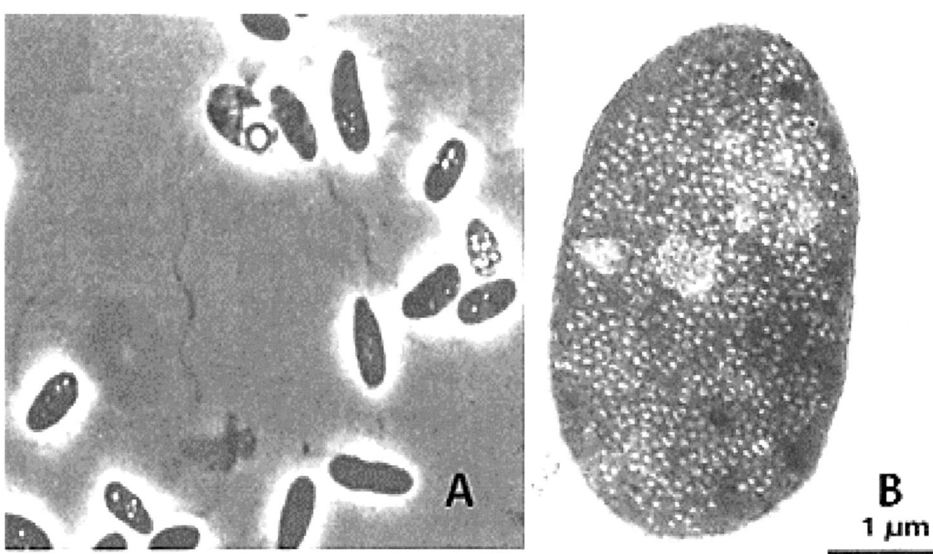

Fig. 3.51. A. Phase-contrast photomicrographs of cells of *Halochromatium salexigens* (Bar= 10 µm) and B. Ultrathin section of *Hch. salexigens* showing vesicular type intracellular membrane (Bar=5 µm).

Differentiation of *Halochromatium* from other Genera

Species of *Halochromatium* are characterized by their requirement of elevated salt concentrations and motile rod-shaped cells. The differentiating properties of *Halochromatium* from other members of *Chromatiaceae* are given in **Tables 3.30** and **3.40**. The phylogenetic relationships of *Halochromatium* and other members of *Chromatiaceae*, based on 16S rDNA sequence are illustrated in **Fig. 3.46.**

Table.3.30. Differential Characteristics of *Halochromatium, Isochromatium, Marichromatium* and *Rhabdochromatium* Species

Charact-eristics	***Halochromatium salexigens***	***Hch. glycolicum***	***Isochromatium buderi***	***Imhoffiella purpurea***	***Imhoffiella bheemlicus***
Cell shape	Rod	Rod	Rod	Single cocci	Mostly diplococci
Cell size (μm)	2.0-2.5 x 4.0-7.5	0.8-1.0 x 2.0-4.0	3.5-4.5 x4.5-9.0	1.7-2.0	4.0-6.0
Motility	+	+	+	+	+
Color of cell suspension	Pink, Rose red	Pink. Pink-red	Pink, Purple violet	Purple	Purple-violet
Carotenoid group	Spirilloxanthin	Spirilloxanthin	Rhodopin	Rhodopinal	Rhodopinal
Absorption maxima of living cells (nm)	371, 410, 515, 550, 590, 806, 836, 880 and 959	375, 483-84, 512, 547, 593, 795 and 859	ND	400, 500, 536, 594, 800 and 864	371, 460, 491, 530, 590, 803 and 857
Mol% G+C of DNA	64.6	66.1-66.5	62.2-62.8	63.1	65.5
B12 requirement	-	-	+	-	+
Sulfate assimilation	-	+	-	-	-
Chemolithotrophic growth	-	-	-	+	+
pH Optimum (Range)	7.4-7.6 (7.0-8.0)	7.2-7.4 (6.2-9.0)	7.2-7.4	7.5-8.0 (7.0-8.5)	7.0-7.5 (6.5-8.0)
Temperature Optimum (Range) (∘C)	20-30	25-35	25-35	30 (20-37)	25-30 (20-30)
NaCl range (%)	4.0-20.0	2.0-20.0	1.0-5.0	0-4.0	0.5-6.0
NaCl optimum)%(	8.0-11.0	5.0-6.0	2.0-3.0	1.0-2.0	1.0-3.0
Polar lipids	ND	ND	ND	DPG, PE, PG, L1–L6	DPG, PE, PG, APL1, L1, L4, L7
Carbon sources/Electron donors					
Hydrogen	+	+	-	-	-
Sulfide	+	+	+	(+)	+
Thiosulfate	+	+	-	(+)	+
Sulfur	+	+	+	+	-
Sulfite	+	+	-	+	-
Formate	-	(+)	-	-	-
Acetate	+	(+)	+	-	+
Pyruvate	+	(+)	+	-	+
Propionate	-	-	-	-	-
Butyrate	-	-	-	-	(+)
Lactate	-	-	-	-	+

Charact-eristics	*Halochromatium salexigens*	*Hch. glycolicum*	*Isochromatium buderi*	*Imhoffiella purpurea*	*Imhoffiella bheemlicus*
Fumarate	-	+	-	ND	ND
Succinate	-	+	-	(+)	-
Malate	-	-	-	-	-
Fructose	-	ND	-	-	-
Glucose	-	-	-	(+)	+
Ethanol	-	-	-	-	-
Propanol	-	ND	-	-	-
Glycerol	-	+	-	(+)	+
Glycolate	-	+	ND	-	+
Crotonate	-	-	-	-	-
Valerate	-	ND	-	(+)	+
Casamino acids	-	(+)	-	(+)	+

Symbols: +, positive in most strains/ substrate utilized; -, negative in most strains/substrate not utilized; ±, variable; (+), weak growth; ND, not determined. Data for *Hch. salexigens* and *Hch. glycolicum*is adopted from Caumette *et al.* (1989) and Caumette *et al.* (1997) respectively. Data for *Isochromatium buderi* is adopted from Truper and Jannasch (1968). Data for *Imhoffiella* species is adopted from Nupur *et al.* (2017).

Taxonomic Comments

Only two species of *Halochromatium* were described as species of *Chromatium* because they have rod-shaped, motile cells without gas vesicles, which were features of previously defined *Chromatium* (Pfennig and Truper, 1989). 16S rDNA sequence analyses revealed phylogenetic distant relation from true *Chromatium* species and other members of *Chromatiaceae*. Hence, they were transferred to the new genus *Halochromatium* (Imhoff *et al.,* 1998b). Species of *Halochromatium* are the most halophilic species of *Chromatiaceae* and belong to truly marine and halophilic *Chromatiaceae* species. The lastly added species of *Halochromatium* such as *Hch. roseum* branched separately but clustered with the type strains of species of the genus *Halochromatium* and was distinct from other genera of purple sulfur bacteria. The highest sequence similarities of *Hch. roseum* strain JA134T was found with the type strains of *Hch. glycolicum* (95.6%) and *Hch. salexigens* (95.4%) (**Fig. 3.46**). Apart from the 16S rRNA gene sequence dissimilarity, it showed clear phenotypic differences from other *Halochromatium* species (Anil Kumar *et al.,* 2007). *Hch. roseum* due to its weak halophilic character compared to the other two species it is transferred to the genus *Lamprobacter* as *Lamprobacter roseus* (Gorlenko *et al.,* 2014).

4. *Isochromatium* (*Isc.*) (The similar *Chromatium*) (Imhoff *et al.,* 1998b) Imhoff (2015)

Cells are straight to slightly curved rods, motile by a polar tuft of flagella and multiply by binary fission. Gram (-) negative, contain vesicular type internal photosynthetic membranes containing bacteriochlorophyll *a* and carotenoids of rhodopinal series as photosynthetic pigments. Obligately phototrophic and strictly anaerobic. Photolithoautotrophic growth occurs under anoxic conditions in the presence of light with sulfide and elemental sulfur as electron donors. During oxidation of sulfide, elemental sulfur is stored inside the cells in the form of highly refractile globules **(Fig. 3.52).** The final oxidation product is sulfate. In the presence of sulfide and bicarbonate, simple organic substrates are photo assimilated. Mesophilic with optimum growth at temperature 25–35°C and pH 6.5–7.6. Grow at salt concentrations typical for marine bacteria. Estuarine salt flats and salt marshes are selective habitats. The mol% G + C of the DNA is 62.2–62.8. *Isochromatium buderi* (Truper and Jannasch, 1968) is the type species. The monotypic genus as only one species is validly published **(Table 3.30).**

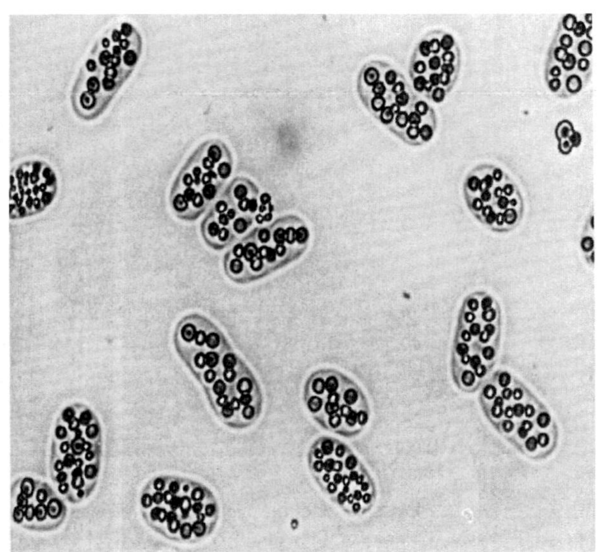

Fig. 3.52. Phase-contrast photomicrographs of cells of *Isochromatium buderi* showing gas vacuoles (Bar= 5 μm).

Differentiation of *Isochromatium* from other Genera

Characteristics of *Isochromatium* from other members of *Chromatiaceae* are presented in **Tables 3.30** and **3.40.** The genetic relationship among *Isochromatium buderi* and other *Chromatiaceae* species, based on 16S rDNA sequence is illustrated in **Fig. 3.46.**

Taxonomic Comments

Isochromatium buderi was previously described as *Chromatium buderi* but a new genus *Isochromatium* was created based on important genetic and physiological differences of species of *Chromatium* and other members of the *Chromatiaceae* (Imhoff *et al.,* 1998b). *Isochromatium buderi* is a true marine and halophilic bacterium.

5. *Imhoffiella* (Genus in honor of Johannes F. Imhoff) Nurpur *et al.* (2017)

Cells are spherical, single or diplococci and motile utilizing a single flagellum (**Fig. 3.53**) (Nupur *et al.,* 2017). Multiply by binary fission and Gram (-) negative. Vesicular-type internal photosynthetic membranes contain bacterio-chlorophyll *a* and carotenoids of the rhodopinal series. Obligately anaerobic and phototrophic. During growth on reduced sulfur sources as electron donors, elemental sulfur is an intermediately deposited number of small globules within the cell. In the presence of sulfide and bicarbonate, organic substrates can be photo assimilated. NaCl and growth factors are not required for growth. The DNA G+C content is approximately 63–65.5 mol%. Tidal water samples are the selective habitats for their growth. The type species is *Imhoffiella purpurea*. Two species are validly published and their characteristics are presented in **Table. 3.30.**

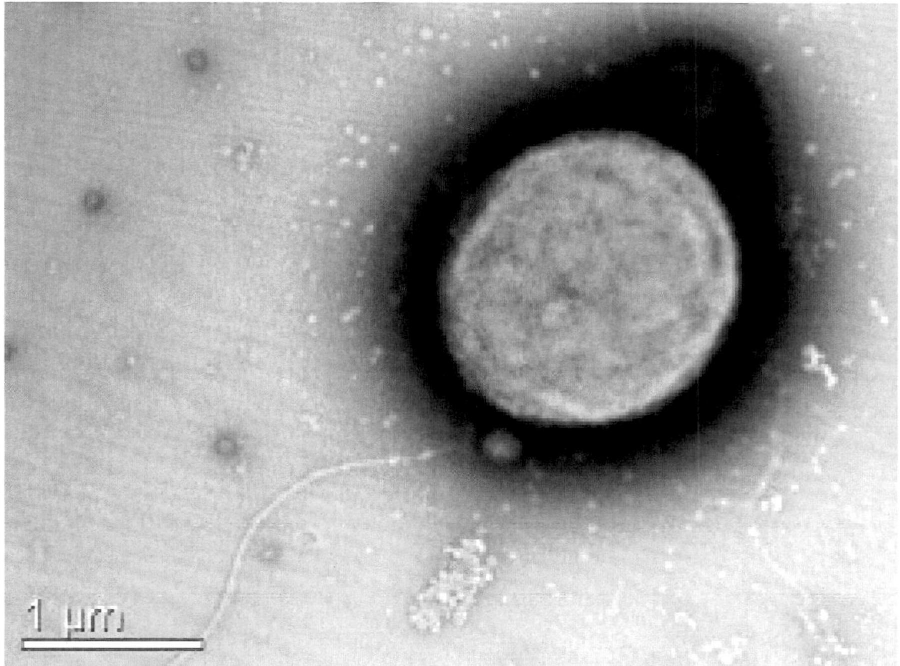

Fig. 3.53. Electron micrograph of negatively stained cells of *Imhoffiella purpurea* (Bar=1.0 µm)

Differentiation of *Imhoffiella* from other Genera

Characteristics of *Imhoffiella* from other members of *Chromatiaceae* are presented in **Table 3.30** and **3.40.** The genetic relationship among *Imhoffiella purpurea* and *Imhoffiella bheemlicus* and other *Chromatiaceae* species, based on 16S rRNA sequence is illustrated in **Fig. 3.54** (Nupur *et al.,* 2017).

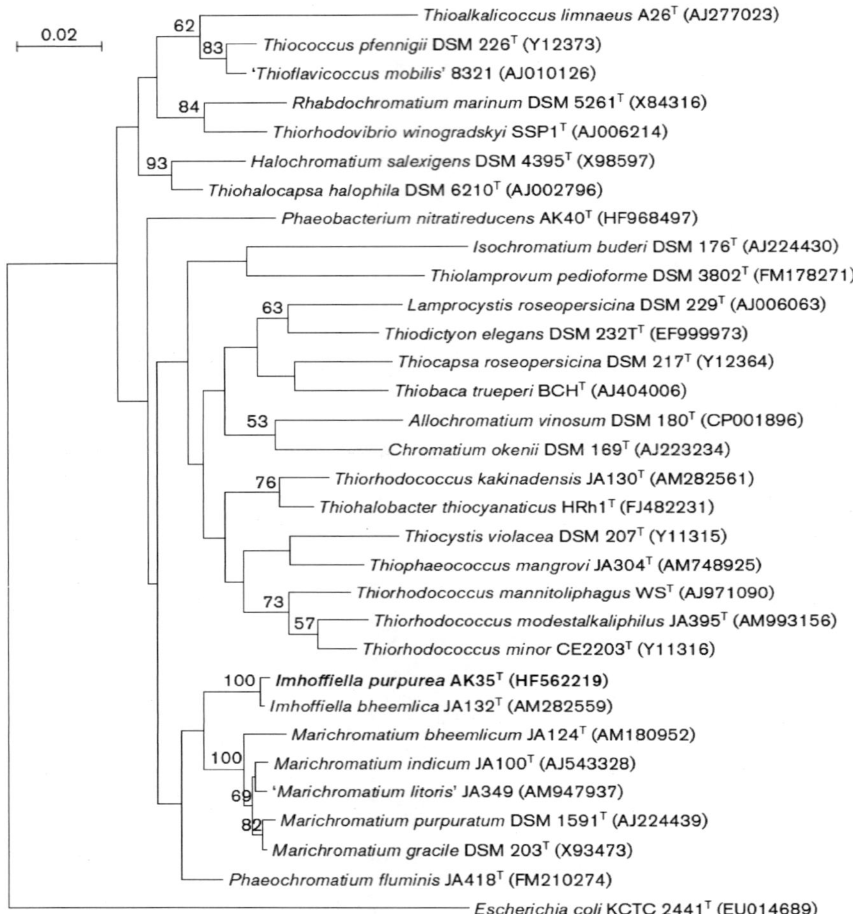

Fig. 3.54. Phylogenetic tree of *Imhoffiella* species and other genera of *Chromatiaceae* family based on 16S rRNA. The sequence of *E.coli* was used as an outgroup to root the tree.

Taxonomic Comments

Imhoffiella bheemlicus was previously described as *Thiorhodococcus bheemlicus,* but a new genus *Imhoffiella* was created based on important genetic and physiological differences such as fatty acid composition and polar lipids of this genus and other members of the *Chromatiaceae* (Nupur *et al.,* 2017). Based on the phylogenetic analysis a comparison was observed between *Imhoffiella purpurea* strain AK35^{T} ,*Thiorhodococcus bheemlicus* MTCC 8120^{T} and all other genera of family *Chromatiaceae*(**Fig. 3.54**). These differences supported the creation of a new genus *Imhoffiella*. Similarly, *Thiorhodococcus bheemlicus* was closely related to *Imhoffiella purpurea,* which leads to the reclassification of *Thiorhorodoccus bheemlicus* into *Imhoffiella bheemlicus* (Nupur *et al.,* 2017).

6. *Lamprobacter* (*Lpb.*) (Brilliant rod) (Gorlenko *et al.*, 1988) Gorlenko and Imhoff (2015)

Cells are rod-shaped or ovoid, 2.0–2.5 x 4–5 μm, multiply by binary fission and do not form aggregates. Motile by flagella during certain stages of development and motile cells normally devoid of gas vesicles (**Fig. 3.55 B**). Gas vesicles are formed in the cell periphery or in the entire cytoplasm. Gram (-) negative, Vesicular type internal photosynthetic membranes contain bacteriochlorophyll *a* and carotenoids of okenone series as photosynthetic pigments (**Fig. 3.55 C**.) Photolithoautotrophic growth occurs under anoxic conditions in the presence of light with reduced sulfur compounds as electron donors. During oxidation of sulfide or thiosulfate, elemental sulfur is transiently stored inside the cells in the form of highly refractile globules (**Fig. 3.55 A**) (Gorlenko *et al.,* 1979; Gorlenko *et al.,* 2014). In the presence of sulfide and bicarbonate, simple organic substrates are photo assimilated. Chemotrophic growth occurs under microoxic conditions in the dark. The mol% G + C of the DNA is 60–64. *Lamprobacter modestohalophilus* (Gorlenko *et al.*, 1988) is the type species.

The genus *Lamprobacter* is characterized and distinguished from other members of the family *Chromatiaceae* by rod-shaped motile cells that develop gas vesicles under certain growth conditions. Characteristics used for differentiation of *Lamprobacter* from *Lamprocystis* of the *Chromatiaceae* are précised in **Table. 3.31.**

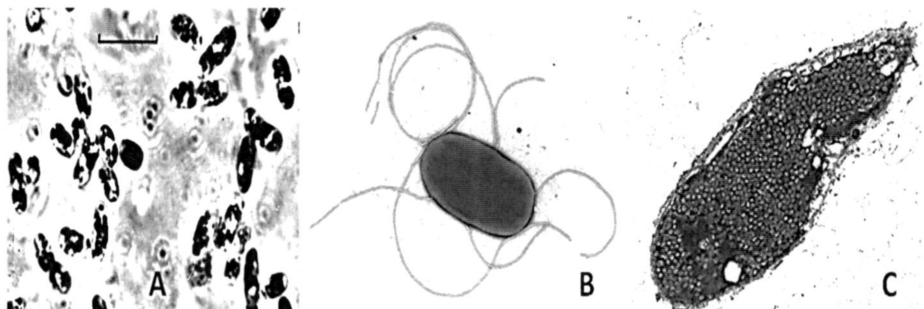

Fig. 3.55. A. Phase contrast micrograph of *Lamprobacter modestohalophilus* (Bar= 10 μm), B. Cell showing flagella (Bar=1.0 μm), C. Ultrathin section of the cells showing vesicular type internal photosynthetic membrane showing gas vesicles (Bar=0.5 μm).

Table 3.31. Differential Characteristics of *Lamprocystis* and *Lamprobacter*

Characteristics	*Lpb. modestoh-alophilus*	*Lpb. roseus*	*Lps. roseop-ersicina*	*Lps. purpurea*
Cell shape	Ovoid to rod	Rods	Sphere	Sphere
Motility	+	-	+	-
Cell diameter (μm)	2.0-2.5 x 4.0-5.0	2.0-3.0 x 3.0-5.0	2.0-3.5	1.9-2.3 x 2.0-3.2

Characteristics	*Lpb. modestoh-alophilus*	*Lpb. roseus*	*Lps. roseop-ersicina*	*Lps. purpurea*
Gas vesicles	+	+	+	+
Clumps of cells	-	-	-	+
Irregular aggregates	-	+	-	-
Irregular clumps	-	-	-	-
Irregular net-like aggregates	-	-	-	-
Platelets	-	-	-	-
Tetrads	-	-	-	-
Color of cell suspension	Purple-red	Purple–pink	Pink violet	Purple-red
Carotenoid group	Okenone	Okenone	Rhodopin	Okenone
Absorption maxima of living cells (nm)	395,485, 512, 540, 580, 798-807, 830 and 880	371, 410, 515, 550, 590, 806, 836, 880 and 959	ND	ND
Mol% G+C of DNA	60-64	64.0	63.8	63.5-65.6
B12 requirement	+	+	ND	ND
Sulfate assimilation	-	-	ND	-
Chemolithotrophic growth	+	-	-	+
pH optimum (range)	7.4-7.6	7.5 (7.0-8.0)	7.0-7.3	7.0-7.3
Temperature optimum (◦C)	23-27	27	20-30	23-25
NaCl optimum (range) (%)	1.0-4.0 (upto 9.0)	1.5-2.5 (1.0-3.0)	Not required	Not required
Carbon/electron donors				
Hydrogen	+	-	ND	-
Sulfide	+	±	+	+
Thiosulfate	+	-	ND	+
Sulfur	+	-	+	+
Formate	-	-	ND	+
Acetate	+	-	+	+
Pyruvate	+	+	+	+
Propionate	-	ND	ND	+
Butyrate	ND	ND	ND	-
Lactate	+	-	ND	+
Fumarate	-	+	ND	-
Succinate	-	+	ND	-

Characteristics	*Lpb. modestoh-alophilus*	*Lpb. roseus*	*Lps. roseop-ersicina*	*Lps. purpurea*
Malate	-	+	ND	-
Fructose	-	ND	ND	-
Glucose	ND	ND	ND	+
Ethanol	+	-	ND	-
Propanol	+	-	ND	-
Glycerol	+	-	ND	-
Glycolate	ND	-	ND	-
Crotonate	ND	ND	ND	-
Valerate	ND	ND	ND	-
Casamino acids	ND	+	ND	-

Symbols: +, positive in most strains/substrate utilized; -, negative in most strains/ substrate not utilized; ±, variable; ND, not determined or no data available. Data for *Lpb. Modestohalophilus* and *Lpb. roseum* is taken from Gorlenko and Imhoff (2005), Gorlenko *et al.* (2014) and Anil Kumar *et al.* (2007) respectively. Data for *Lps. roseopersicina* and *Lps. purpurea* is taken from Schroeter (1886) and Imhoff (2001)

Differentiation of *Lamprobacter* from other Genera

Lamprobacter is differentiated from other members of *Chromatiaceae* by rod-shaped, motile cells that develop gas vesicles under certain growth conditions. Differentiation of *Lamprobacter* from other members of *Chromatiaceae* is precised in **Tables 3.31** and **3.40**. However, detailed studies are needed to differentiate *Lamprobacter* species from other genera based on 16S rRNA sequence (**Fig. 3.56**) (Gorlenko *et al.,* 2014).

Taxonomic Comments

Lamprobacter roseus formerly described as *Halochromatium roseum.* The similarity between *Lpb. modestohalophilus* and the species *Hch. salexigens* and *Hch. glycolicum* (98.0–98.2% similarity between the 16S rRNA gene sequences), but due to their pronounced morphological andphysiological differences prevent combining them within one genus. *Hch. roseum*, which is similar to *Lpb. modestohalophilus* strains in morphology and physiology such as the presence of gas vesicles, okenone as the major carotenoid and moderate halophilic are exceptional in this respect (Kumar *et al.,* 2007). *Hch. roseum* JA134^{T} exhibits the highest similarity (99.0–99.1%) with *Lpb. modestohalophilus* strains Sivash and ShNLb02, which promote its classification within the genus *Lamprobacter* rather than *Halochromatium*. Homology between the16S rRNA gene sequences of *Hch. roseum* JA134T and those of *Hch. salexigens* and *Hch. glycolicum* was lower i.e., 98.3 and 98.4%, respectively (**Fig.3.56**). Therefore, it is proposed to accept the previous name *Lamprobacter modestohalophilus* for rod-shaped purple sulfur bacteria containing gas vesicles and forming a separate phylogenetic lineage within the halophilic purple sulfur bacteria of the family *Chromatiaceae* according to the results of 16S rRNA sequencing. Based on these genotypic and phenotypic characters, it is proposed to preserve the genus *Halochromatium* with extremely halophilic species such as *Hch. salexigens* and *Hch. glycolicum*, whereas the *Hch. roseum* is transferred to the genus *Lamprobacter* as *Lamprobacter roseus* .*Lpb. modestohalophilus* is retained as type species of the genus *Lamprobacter* (Gorlenko *et al.*, 2014).

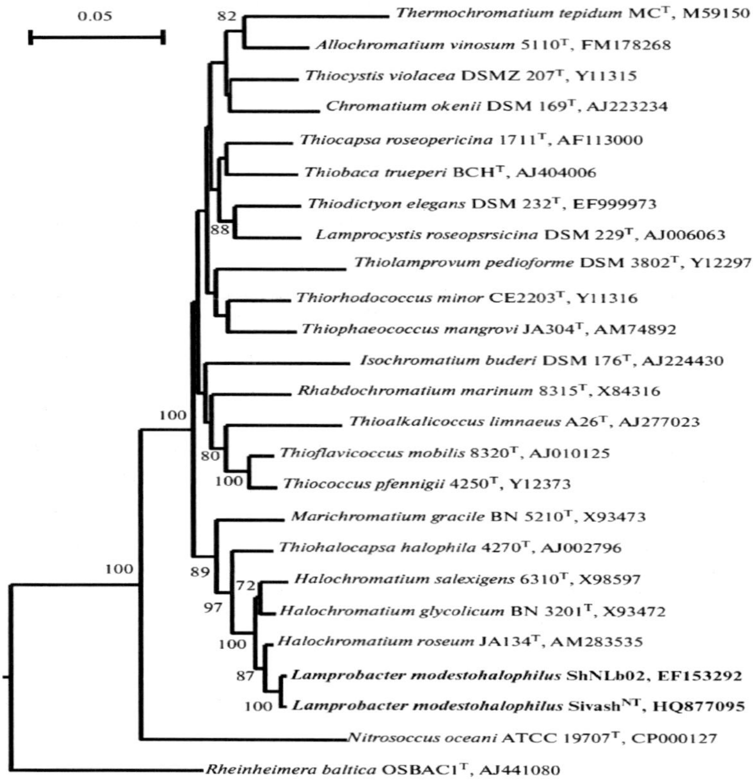

Fig 3.56. Phylogenetic tree of *Lamprobacter* species and other genera of family *Chromatiacea* based on 16S rRNA. The sequence of *E. coli* was used as an outgroup to root the tree.

7. *Lamprocystis* (*Lps.*) (Brilliant bag) (Schroeter, 1886) Imhoff (2015)

Cells are spherical to ovoid, 1.9–3.8 μm in diameter, diplococcus shaped before cell division. They may form irregular aggregates, multiply by binary fission and non-motile or motile by means of a single flagellum). Gas vesicles are formed in the central part and are Gram (-) negative (**Fig. 3.57**) (Schroeter, 1886). At high sulfide concentration (4–6 mM) and light intensity (1000–2000 lux), cells may grow as long, branching and form cell aggregates embedded in slime. Under favorable growth conditions, cell aggregates may break up into smaller clusters and spheroidal microcolonies, become motile by flagella, finally, individual motile cells are liberated. Vesicular type internal photosynthetic membranes containing the photosynthetic pigments bacteriochlorophyll *a* and carotenoids of rhodopinal and okenone. Photolithoautotrophic growth occurs under anoxic conditions in the presence of light with reduced sulfur compounds such as sulfide and elemental sulfur as electron donors. During oxidation of sulfide, elemental sulfur is transiently stored in the peripheral part of the cells that are free of gas vesicles and the final oxidation product is sulfate. In the presence of sulfide and bicarbonate, simple organic substrates are photo assimilated. Obligately phototrophic or facultatively chemotrophic under microoxic to oxic conditions in the dark. The mol% G + C of the DNA is 63.4–64.1. *Lamprocystis roseopersicina* (Kutzing, 1849; Schroeter, 1886) is the type species.

Characteristics used for differentiation of the genus *Lamprocystis* from *Lamprobacter* species of the family *Chromatiaceae* are presented in **Table 3.31.** The phylogenetic relationship of *Lamprocystis* species to other members of the *Chromatiaceae*, based on 16S rDNA sequence comparison is illustrated in **Fig. 3.47**.

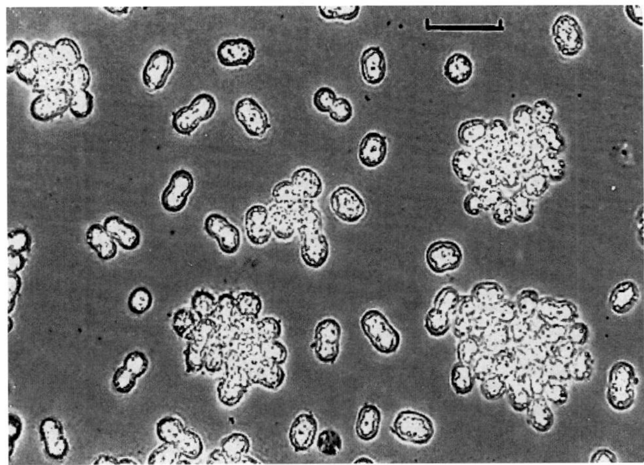

Fig. 3.57. Phase-contrast micrograph of *Lamprocystis roseopersicina* DSM 229 cultured photoautotrophically with sulfide. The irregular whitish areas within the cells are the gas vacuoles. The small spherical bodies in the cells are the sulfur globules (Bar= 10 μm).

Differentiation of *Lamprocystis* from other Genera

Characteristics used for the separation of *Lamprocystis* from other phototrophic members of *Chromatiaceae* are precised in **Tables 3.31** and **3.40.** The phylogenetic relationship of *Lamprocystis* species to other members of the *Chromatiaceae*, based on 16S rDNA sequence comparison is illustrated in **Fig. 3.47**.

Taxonomic Comments

Few strains of *Lamprocystis roseopersicina* studied in pure culture and from observations in natural samples, it is clear that cells showing the characteristics of *Lamprocystis* are quite visible but that cell of different diameters occur. Therefore, in-depth studies using a variety of pure cultures of these bacteria are necessary, to show that clusters of different strains with different cell diameters in different species.

According to 16SrRNA oligonucleotide pattern analysis of *Lamprocystis roseopersicina* (Fowler *et al.,* 1984) reveals that it is distantly related to *Thiocapsa* and *Thiodictyon* species. Sequence analyses of 16S rDNA place this species into the group of spherical to ovoid, many cells of which contain gas vesicles, non-motile freshwater species of *Chromatiaceae* (**Fig. 3.47**) and is closely related to species originally described as *Amoebobacter purpureus* (Eichler and Pfennig, 1988). In agreement to the International Code of Nomenclature of Bacteria (Lapage *et al.,* 1992) this species is transferred to *Lamprocystis* as *Lamprocystis purpurea* (Imhoff, 2001b).

The genus *Amoebobacter* with spherical, non-motile species contains gas vesicles. Two species including the type species are transferred to the genus *Thiocapsa* and a third species is transferred to the new genus *Thiolamprovum* (Guyoneaud *et al.,* 1998). As the type species is removed from the genus *Amoebobacter*, the name of other species of this genus have been changed according to rule 37a and it was not legal to retain the name *Amoebobacter purpureus* (Guyoneaud *et al.,* 1998). Hence, the transfer of this species to a new genus *Pfennigia* has been proposed (Tindall, 1999). Therefore, the high degree of similarity of 16S rDNA sequences of this bacterium and *Lamprocystis roseopersicina* determine the close relationship of these two bacteria prevent their separation into different genera (**Fig. 3.47**). Because of the importance of the genus and species name of *Lamprocystis roseopersicina* and *Pfennigia purpurea* (Tindall 1999) (*Amoebobacter purpureus*) (Eichler and Pfennig, 1988) it was transferred to *Lamprocystis purpurea* (Imhoff, 2001b).

Lamprocystis purpurea (*Amoebobacter purpureus*) was described based on six isolates, which were classified into two groups according to their cell size and substrate utilization (Eichler and Pfennig, 1988). Differences between two groups may be enough to allow the recognition of two different species.

8. *Marichromatium* (*Mcr.*) (Truly marine) (Imhoff *et al.,* 1998b) Imhoff (2015)

Cells are straight to slightly curved rods, 1.0-1.7 μm, motile by means of polar flagella and multiply by binary fission. Cells are single or in pairs and may stick together and form clumps. They are Gram (-) negative, contain vesicular type internal photosynthetic membranes with bacteriochlorophyll *a* and carotenoids of spirilloxanthin and okenone as photosynthetic pigments. Growth occurs photolithoautotrophically under anoxic conditions in the presence of light with sulfide and elemental sulfur as electron donors. During oxidation of sulfide, elemental sulfur (S^0) is formed as an intermediate and stored in the form of highly refractile globules inside the cells **(Fig. 3.58).** The final oxidation product is sulfate. Molecular H_2 may be used as an electron donor. In the presence of sulfide and bicarbonate, organic substrates are photo assimilated. Chemolithotrophic and chemoorganotrophic growth may occur under microoxic conditions in the dark. Optimum growth is observed at incubation temperature 25–35 °C and pH 6.5–7.6 and salt concentrations typically of marine bacteria. Anoxic marine sediments and stagnant waters containing hydrogen sulfide and exposed to the light, marine sponges and other marine invertebrates are the selective habitats for their growth. The mol% G + C of the DNA is 68.9–70.4. *Mcr. gracile* (Imhoff *et al.,* 1998b) is the type species. The characteristic properties for distinguishing *Marichromatium* species are presented in **Table 3.32.**

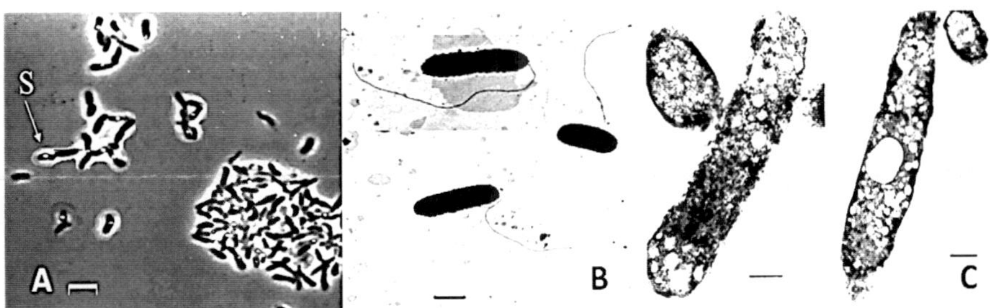

Fig. 3.58. Phase-contrast micrograph of cells of *Mcr. indicum* showing elemental sulfur globules (Bar=5 μm), B. Negatively stained cells showing monotrichous flagellum (Bar=1 μm), C. Ultrathin section of cells showing vesicular type photosynthetic membranes (Bar=0.5 μm).

Table 3.32. Differential Characteristics of *Marichromatium* Species

Characteristics	***Marichromatium gracile***	***Mcr. purpuratum***	***Mcr. indicum***	***Mcr. bheemlicum***	***Mcr. littoris***	***Mcr. chrysaorae***
Cell shape	Rods	Rods	Rods	Rods	Rods	Rods
Cell size (μm)	1.0–1.3 x 2.6–6.0	1.2–1.7 x 3.0–4.0	0.8–1.0 x 2.0–7.0	0.8–1.0 x 2.0–4.0	1.0–1.2 x 2.0–4.0	1.0–1.3 x 2.0–5.0
Motility	+	+	+	+	+	+
Rosette formatium	-	-	-	+	-	-
Color of cell suspension	Brownish red	Purple red	Reddish brown	Reddish brown	Brown	Reddish brown
Carotenoid composition (mol%)	Rp (75), Sp (9), Rv (7), Ahrv (4), Ly (5)	Okenone (100)	Rp (50), Sp (35), Rv (1), Ahrv (13), Ly (1)	Rp (58), Sp (30), Ahrv (10), Ly (1)	Rp (65), Sp (9), Rv (11), Ahrv (5), Ly (10)	Rp (58), Sp (20), Rv (7), Ahrv (6), Ly (4)
Polar lipids	DPG, PG, PE, OL, L1–5	DPG, PG, PE, OL, AL3, PL1,2, L3,6,8	DPG, PG, PE, OL, AL3, PL1,2, L1–6,8	DPG, PG, PE, OL, PL1,2, L4	DPG, PG, PE, OL, AL2,3, PL1, L3,4,7	DPG, PG, PE, OL, AL2,3, PL1–4, L3,4,6,8

Characteristics	*Marichromatium gracile*	*Mcr. purpuratum*	*Mcr. indicum*	*Mcr. bheemlicum*	*Mcr. littoris*	*Mcr. chrysaorae*
Major quinone (ratio)	Q8, MK7 (8:2)	Q8, MK7 (8:2)	Q8, MK7 (8:2)	Q8, MK7 (9:1)	Q8, MK7 (8:2)	Q8, MK7 (8:2)
Growth modes	POH, PLA, COH, CLA, CLH	POH, PLA, COH, CLA, CLH	PLH, PLA, CLH, POH (with cysteine) PLH	PLH, PLA, POH	PLH, POH	PLH, POH, COH, CLH
Absorption maxima of living cells (nm)	ND	370, 400, 524, 586, 805 and 825	372, 462, 489, 522, 591, 798 and 852	374, 488, 590, 797 and 851	377, 461, 488, 590, 800 and 851	377, 461, 488, 590, 800 and 851
Mol% G+C of DNA	68.9-70.4	68.4-68.9	66.7	67.0	68.4	70.5
Growth factors	Not required	Not required	Not required	Pyridoxal phosphate	Not required	Not required
Sulfate assimilation	+	ND	+	+	+	+
pH Optimum (Range)	7.2-7.4	7.2-7.6	6.0-7.5	6.5-8.5	6.5-7.5	7.0-8.0
Temperature Optimum (Range) (◦C)	20-35	20-30	30-35	30-35	25-30	25-35
NaCl range (%)	0.5-8.0	2.0-7.0	0.05-13.0	1.5-11.0	2.0-5.0	2.0-6.0
NaCl optimum)%(	2.0-3.0	5.0	1.0-4.0	1.5-8.5	2.0-4.0	2.0-5.0
Sulfide tolerance (mM)	4.0	4.0	4.0	8.0	6.5	7.0
Carbon sources/ Electrondonors						
Hydrogen	+	ND	ND	ND	ND	ND
Sulfide	+	+	+	+	+	+
Thiosulfate	+	+	+	+	+	+
Sulfur	+	+	-	-	-	-
Sulfite	+	ND	-	+	-	-
Formate	+	-	ND	-	-	-
Acetate	+	+	+	+	+	+
Pyruvate	+	+	+	+	+	+
Propionate	+	+	+	-	+	+
Butyrate	+	+	+	-	(+)	(+)
Lactate	+	+	+	-	+	-

Characteristics	*Marichromatium gracile*	*Mcr. purpuratum*	*Mcr. indicum*	*Mcr. bheemlicum*	*Mcr. littoris*	*Mcr. chrysaorae*
Fumarate	+	+	-	+	+	+
Succinate	+	+	-	+	+	+
Malate	+	+	+	+	+	+
Fructose	-	(+)	+	-	-	-
Glucose	-	-	-	-	-	-
Ethanol	-	-	-	+	-	-
Propanol	ND	-	-	ND	-	-
Glycerol	-	-	-	+	-	-
Glycolate	ND	ND	ND	+	-	-
Crotonate	+	ND	ND	-	-	-
Valerate	ND	+	-	+	(+)	(+)
Casamino acids	+	(+)	-	ND	-	-
Nitrogen sources						
Urea	-	-	+	-	-	-
Glutamine	-	-	-	+	+	+
Glutamate	+	+	-	+	+	-

Symbols: +, positive in most strains/ substrate utilized; -, negative in most strains/ substrate not utilized; ±, variable; (+), weak growth; ND, not determined. CLA, chemolithoautotrophy; Whole Fatty acid composition: +, presence of compound but their percentage is not available, -. not present; CLH, chemolithoheterotrophy; COH, chemo-organoheterotrophy; PLA, photolithoautotrophy; PLH, photolithoheterotrophy; POH, photo-organoheterotrophy; Rp, rhodopin; Sp, spirilloxanthin; Rv, rhodovibrin; Ahrv, anhydrorhodovibrin; Ly, lycopene; Ok, okenone PE, phosphatidyl ethanolamine; DPG, diphosphatidyl glycerol; PG, phosphatidyl glycerol; OL, ornithine lipid; AL1–3 unknown aminolipids; PL1–4, unknown phospholipids; unknown lipids. Numbering of polar lipids is according to the convention adopted by Kampfer *et al.* (2006). Data for *Mcr. gracile, Mcr. purpuratum, Mcr. indicum, Mcr. bheemlicum, Mcr. littoris* and *Mcr. chrysaorae* is taken from Serrano *et al.*(2009), Imhoff and Truper (1979), Arunasri *et al.* (2005), Anil Kumar *et al.*(2007) and Shivali *et al.* (2012) respectively.

Differentiation of *Marichromatium* from other Genera

Species of *Marichromatium* are different from other members of the family *Chromatiaceae* by their specific salt requirement and very high mol% G+C content of their DNA. *Marichromatium* species can be distinguished on the basis of cell suspension colour and pigment content, substrate utilization, as well as differences in cell size and salt response. Differentiating characteristics of the genus *Marichromatium* from other phototrophic members of *Chromatiaceae* are precised in **Table 3.40.** The phylogenetic relationships of *Marichromatium* species to other members of *Chromatiaceae* based on 16S rDNA and 16S rRNA gene sequencing are presented in **Figs. 3.46** and **3.47** .

Taxonomic Comments

Species of this genus have been previously included in the genus *Chromatium* (Pfennig, 1989b), but because of significant phenotypic and genotypic differences from true *Chromatium* species and other members of *Chromatiaceae* a new genus *Marichromatium* (Imhoff *et al.,* 1998b) was created. Five species are validly published such as *Mcr. gracile, Mcr. purpuratum, Mcr. indicum, Mcr. bheemlicum, Mcr. littoris, Mcr. chrysaorae* and *Mcr. fluminis*. Based on distinct phenotypic, chemotaxonomic and molecular differences, *Marichromatium fluminis* JA418T was reclassified into a new genus *Phaeochromatium* as *Phaeochromatium fluminis* (Shivali *et al.,* 2012).

9. *Phaeochromatium* (*Phc.*) (Brown *Chromatium*) (Sucharita *et al.,* 2010) Shivali *et al.* (2012)

Cells are rod-shaped, slightly curved Gram (-) negative bacteria (**Fig. 3.59 A** .) Cells are motile by polar flagella and multiply by binary fission (**Fig. 3.59 B** .) Vesicular type intracytoplasmic membrane structures containing bacterio-chlorophyll *a* and carotenoid of rhodopin are present as photosynthetic pigments (**Fig. 3.59 C**). Growth occurs photolithoheterotrophically. Growth factors are required. $C_{16:0}$, $C_{16:1}\omega6c/C_{16:1}\omega7c$ and $C_{18:1}\omega7c$ are the major fatty acids. Diphos-phatidyl glycerol (DPG), phosphatidyl glycerol (PG), phosphatidyl ethanolamine (PE) and an unidentified amino lipid (AL1, 2) are present as the major polar lipids. They are mesophilic marine phototrophic bacteria. The G + C content of genomic DNA is 68–69 mol% (HPLC). Delineation of the genus is determined by the phylogenetic information from 16S rRNA gene sequences, MLS Aphylogenetic information and MLSA barcoding. The type species is *Phaeochromatium fluminis*. Only a single species is validly published and its characteristics are precised in **Table 3.33** .

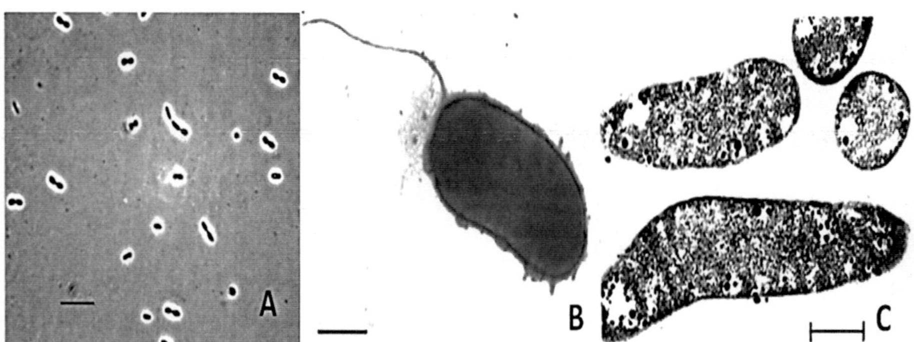

Fig. 3.59. Phase-contrast micrograph of *Phaeochromatium fluminis* showing diplococcus cells (Bar, 16.6 μm), B. Cells showing single polar flagellum (Bar= 16.6 μm), C. Transmission electron photomicrograph showing vesicular type intracyto-plasmic membrane (Bar= 869 nm).

Table 3.33. Differential Characteristics of species of *Phaeochromatium, Phaeobacterium, Rhabdochromatium* and *Thermochromatium*

Characteristics	*Phaeochromatium*	*Phaeobacterium*	*Rhabdochromatium*	*Thermochromatium tepidum*
Cell shape	Slightly curved rods	Rods	Rods	Rods
Cell size (μm)	0.4–0.5 x 4.0–6.0	0.9-1.3 x 2.0-4.0	1.5-1.7 x 16-32	1.0-2.0 x 2.8-3.2
Motility	+	+	+	+
Rosette formatium	-	-	-	
Color of cell suspension	Reddish brown	Reddish brown	Beige, Orange brown	Red
Carotenoids (major component)	Spirilloxanthin series (rhodopin)	Spirilloxanthin series (rhodopin)	Spirilloxanthin or rhodopinal series (lycopene)	Spirilloxanthin series
Polar lipids	DPG, PG, PE, unidentified AL1,2	DPG, PE, PG, L1–8	ND	ND
Major quinone	ND	Q-10	ND	ND
Growth modes	PLH, PLA, POH	PLH, POH	PLA, POH	PLA, POH
Absorption maxima of living cells (nm)	374, 488, 593, 803 and 857	376, 488, 592, 702, 800 and 836	375, 485, 592, 797 and 863	599, 858 and 808
Mol% G+C of DNA	71.4	65.5	60.4	61.5
Growth factors	Not required	Thiamine and riboflavin	Not required	Not required
Sulfate assimilation	+	-	ND	ND
pH Optimum (Range)	6.5-7.5	7.5 (5.0-9.0)	7.2-7.3	7.0
Temperature Optimum (Range) (°C)	30-35 (20-35)	30 (22-37)	30	48-50 (35-60)
NaCl range (%)	0.5-6.5	tolerate upto 7.0	1.0-6.5	Not required
NaCl optimum (%)	1.5-3.0	1.5-3.0	1.5-5.0	-
Sulfide tolerance (mM)	5.0-6.5	1.0	1.25	ND
Carbon sources/Electron donors				
Hydrogen	ND	ND	+	-
Sulfide	+	+	+	+
Thiosulfate	+	+	+	-
Sulfur	+	-	+	+
Sulfite	+	-	ND	-
Formate	+	-	-	-
Acetate	+	+	+	+
Pyruvate	+	+	+	+
Propionate	+	+	+	-
Butyrate	+	-	-	-
Lactate	-	ND	-	ND

Characteristics	*Phaeochromatium*	*Phaeobacterium*	*Rhabdochromatium*	*Thermochromatium tepidum*
Fumarate	+	+	-	-
Succinate	+	+	-	-
Malate	+	+	-	-
Fructose	+	-	-	-
Glucose	-	-	-	-
Ethanol	-	ND	-	ND
Propanol	ND	ND	-	ND
Glycerol	-	-	-	ND
Glycolate	ND	+	-	-
Crotonate	ND	-	-	-
Valerate	+	+	-	-
Casamino acids	ND	-	-	-
Nitrogen sources	Ammonium chloride, Urea, Glutamate	Ammonium chloride, Glutamine, Nitrate and Yeast extract	ND	Ammonia, Urea and Glutamine

Symbols: +, positive in most strains/ substrate utilized; -, negative in most strains/ substrate not utilized; ±, variable; (+), weak growth; ND, not determined. CLA, chemolithoautotrophy; CLH, chemolithoheterotrophy; COH, chemo-organoheterotrophy; PLA, photolithoautotrophy; PLH, photolithoheterotrophy; POH, photo-organoheterotrophy; PE, phosphatidyl ethanolamine; DPG, diphosphatidyl glycerol; PG, phosphatidyl glycerol; OL, ornithine lipid; AL1–2 unknown aminolipids; L, unidentified lipid. Numbering of polar lipids is according to the convention adopted by Kampfer *et al.* (2006). Data for *Phaeochromatium* and *Phaeobacterium* is taken from Shivali *et al.* (2012) and Nupur *et al.* (2015). Data for *Rhabdochromatium* and *Thermochromatium* is taken from Dilling *et al.* (1995) and Madigan (1986), Imhoff *et al.* (1998b).

Differentiation of *Phaeochromatium* from other Genera

Species of *Phaeochromatium* are different from other members of the family *Chromatiaceae* by their specific phenotypic and genotypic characters. *Phaeochromatium* species can be distinguished on the basis of cell size, pigment analysis, substrate utilization and fatty acid composition. Differentiating characteristics of the genus *Phaeochromatium* from other phototrophic members of *Chromatiaceae* are precised in **Tables 3.33** and **3.40.** The phylogenetic relationships of *Phaeochromatium* species to other members of *Chromatiaceae* based on 16S rRNA gene sequencing are presented in **Figs. 3.54** and **3.60** (Shivali *et al.,* 2012).

Taxonomic Comments

Phaeochromatium fluminis previously has been included in the genus *Marichromatium* (Sucharita *et al.,* 2010), but because of significant phenotypic and genotypic differences from true *Chromatium* species and other members of *Chromatiaceae,* a new genus *Phaeochromatium* (Shivali *et al.,* 2012) was created. *Marichromatium fluminis* were slightly curved rods while all the other species of *Marichromatium* wire rods. Based on distinct polar lipid composition, phenotypic properties (cell shape, cell division, cell size and motility), genotypic and phylogenetic analyses separation of *Marichromatium fluminis* strain JA418^T

into a new genus *Phaeochromatium* as *Phaeochromatium fluminis* (Shivali *et al.*, 2012).

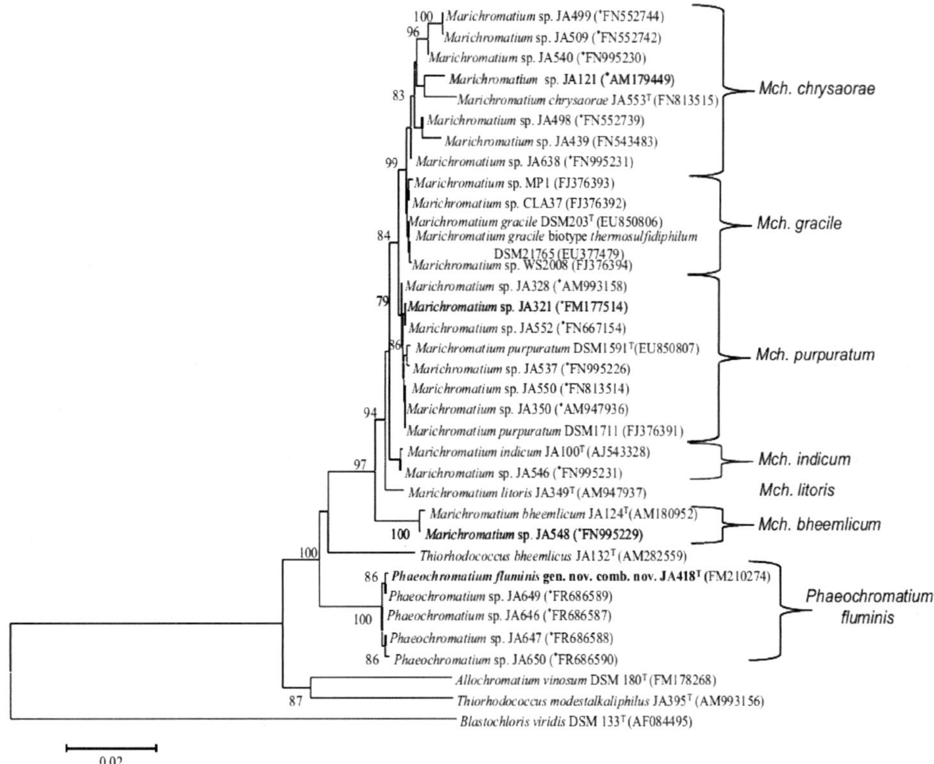

Fig. 3.60. Phylogenetic tree of *Phaeochromatium* species and other genera of *Chromatiaceae* family based on 16S rRNA. The sequence of *Blastochloris viridis* was used as an outgroup to root the tree.

10. *Phaeobacterium* (*Phb.*) (The brown rod) Nupur *et al.* (2015)

Cells are rod-shaped, motile and Gram-negative bacteria (**Fig. 3.61**) (Nupur *et al.,* 2015). They are oxidase-positive and catalase-negative. The phototrophically grown cells are brown colour and anaerobic. Vesicular type internal photosynthetic membrane with bacteriochlorophyll *a* and carotenoids of rhodopin series are present as photosynthetic pigments. The major fatty acids are $C_{12:0}$, $C_{16:0}$, $C_{18:1}\omega7c$ and summed feature 3 ($C_{16:1}\omega7c$ and/or iso-$C_{15:0}$ 2-OH). The predominant polar lipids are diphosphatidyl glycerol (DPG), phosphatidylethanolamine (PE), phosphatidylglycerol (PG) and eight unidentified lipids. Q-10 is the predominant respiratory quinone with 10 isoprene units. Growth occurs photoorganoheterotrophically under anaerobic conditions in the presence of light by utilizing a number of organic substrates. Unable to grow photoautotrophically and chemotrophically in the presence of thiosulfate and sulfide as electron donors and $NaHCO_3$ as carbon source. Thiamine and riboflavin are required for their growth. They are mesophilic, optimum growth occurs at temperature range 30ºC (range 22-37ºC) and optimum pH 7.5 (range 5.0–9.0). Grow under both acidic and alkaline conditions and does not require NaCl for growth, but can tolerate concentrations up to 7% (w/v), with optimum growth at 1.5–3.0% NaCl. The DNA G+C content of the type strain of the type species is 65.5 mol%. The mol% of DNA G+C content is 65.5. Only a single species is validly published i.e., *Phaeobacterium nitratireducens* which is the type species (Nupur *et al.,* 2015). The characteristic feature of this species is presented in **Table 3.33**.

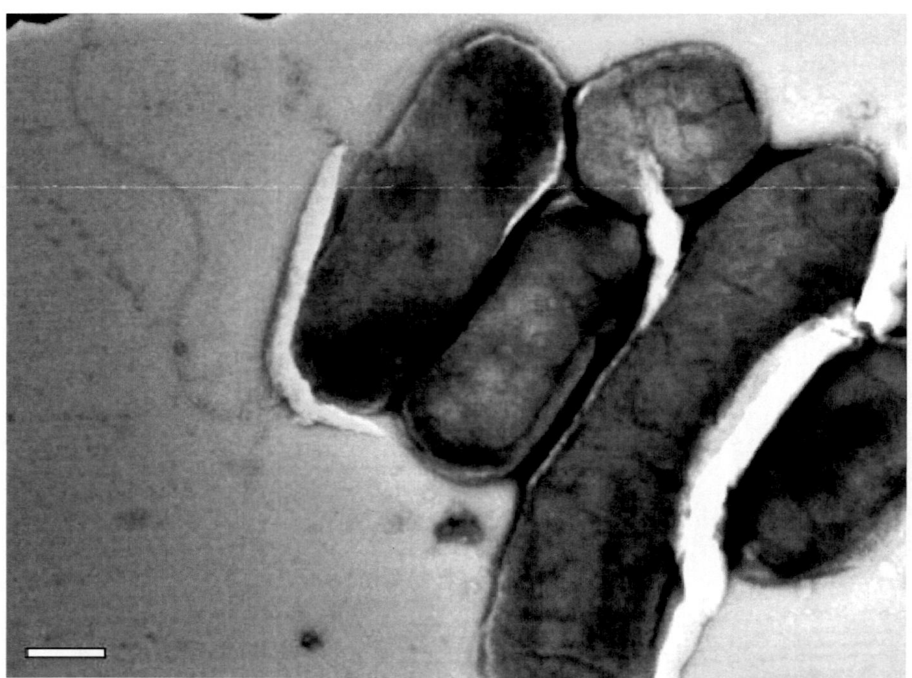

Fig. 3.61. Electron micrograph of negatively stained cells of *Phaeobacterium nitratireducens* (Bar=0.5μm)

Differentiation of *Phaeobacterium* from other Genera

Species of *Phaeobacterium* are different from other members of the family *Chromatiaceae* by their specific phenotypic and genotypic characters. *Phaeobacterium* species can be distinguished based on cell size, presence of sulfur globules, pigment analysis, substrate utilization, polar lipids and fatty acid composition and DNA G+C content. Differentiating characteristics of the genus *Phaeobacterium* from other phototrophic members of *Chromatiaceae* are precised in **Tables 3.33** and **3.40.** The phylogenetic relationships of *Phaeobacterium* species to other members of *Chromatiaceae* based on 16S rRNA gene sequencing is presented in **Fig. 3.62** (Nupur *et al.,* 2015).

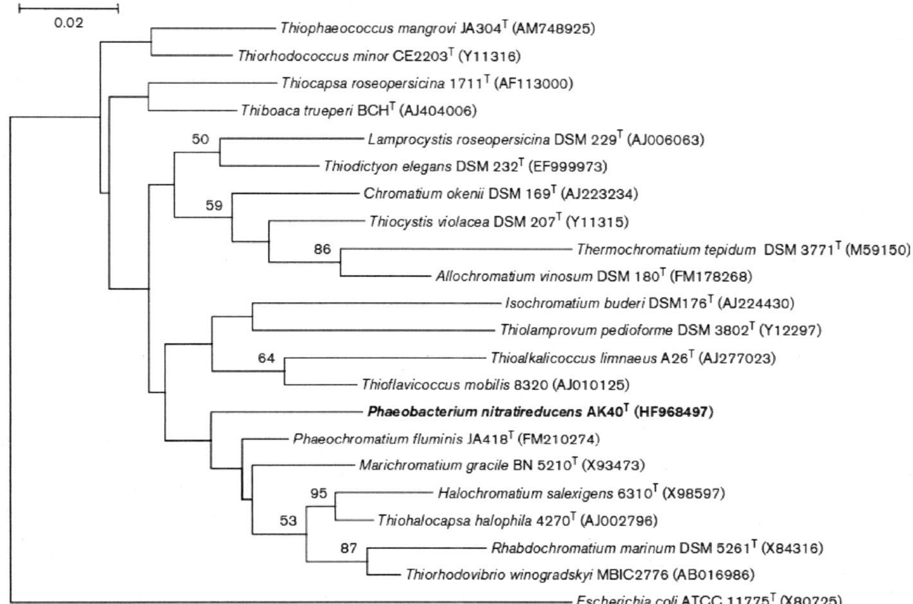

Fig. 3.62. Phylogenetic tree of *Phaeobacterium* species and other genera of *Chromatiaceae* family based on 16S rRNA. The sequence of *E. coli* was used as an outgroup to root the tree.

Taxonomic Comments

Phylogenetic analysis based on the 16S rRNA gene sequence showed that *Phaeobacterium nitratireducens* strain AK40T was closely related to the genus *Phaeochromatium* of the family *Chromatiaceae.* In this study, a polyphasic approach, including genotypic, phenotypic and chemotaxonomic characterizations, was used to determine the taxonomic position of these bacteria. Based on pairwise sequence similarity, *Phaeobacterium nitratireducens* strain AK40T was closely related to *Phaeochromatium fluminis*, with 95.2% similarity and to other members of the family, *Chromatiaceae* is 90.8–94.8%. Based on the sequence comparison data, *Phaeobacterium nitratireducens* was positioned differently outside the group formed by genera of *Marichromatium gracile* , *Halochromatium salexigens* , *Thiohalocapsa halophila, Rhabdochromatium marinum* and *Thiorhodovibrio winogradskyi*(**Fig. 3.62**). Therefore, based on distinct morphological, physiological and genotypic differences, a new genus was classified such as *Phaeobacterium* with species *Phaeobacterium nitratireducens* (Nupur *et al.,* 2015).

11. *Rhabdochromatium* (*Rbc.*) (Coloured rod) (Dilling *et al.,* 1995) Imhoff (2015)

Cells are straight or slightly bent rods, motile by a polar tuft of flagella and multiply by binary fission **(Fig. 3.63 A & B)** (Dilling *et al.,* 1995). Gram-negative and contains vesicular type internal photosynthetic membranes with photosynthetic pigments bacteriochlorophyll *a* and carotenoids of lycopene (**Fig. 3.63 C**). Obligately phototrophic and strictly anaerobic. Photolithoautotrophic growth occurs with hydrogen, sulfide, elemental sulfur (S^0) and thiosulfate as electron donors. During the oxidation of sulfide, sulfur globules are stored inside the cells as an intermediary product. The final oxidation product is sulfate. In the presence of sulfide and bicarbonate, simple organic substrates are photoassimilated. Mesophilic marine bacteria require NaCl for growth, optimum growth at neutral pH and a temperature of 20–35°C. The mol% G + C of the DNA is 60.4. *Rbc. marinum* (Dilling *et al.,* 1996) is the type species. Characteristics of this genus for differentiation from other phototrophic members of the family *Chromatiaceae* are given in **Table 3.33.**

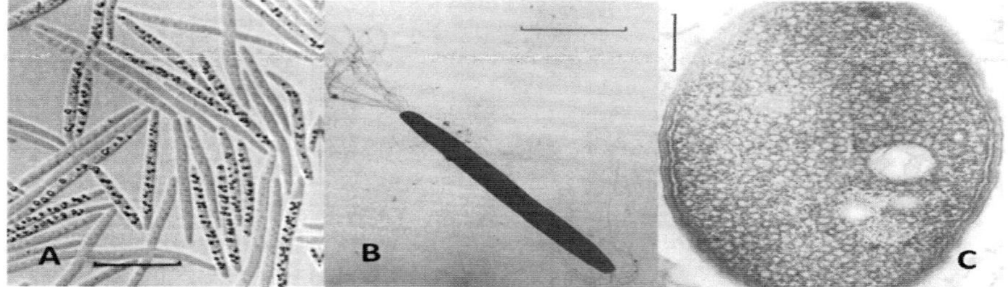

Fig. 3.63. A. Bright-field photomicrograph of *Rhabdochromatium marinum* (Bar=10 μm), B. Negatively stained cells showing bipolar flagellar tufts (Bar= 5 μm), C. Ultrathin section of cells showing vesicular type intracytoplasmic membrane system (Bar=3 μm).

Differentiation of *Rhabdochromatium* from other Genera

Differentiating characteristics of *Rhabdochromatium* from other phototrophic members of *Chromatiaceae* are precised in **Tables 3.33** and **3.40**. The genetic relationship of *Rhabdochromatium marinum* to other members of *Chromatiaceae*, based on 16S rDNA sequence is illustrated in **Fig. 3.46.**

Taxonomic Comments

The genus *Rhabdochromatium* was described by Winogradsky (1888) as he observed long, rod- or spindle-shaped, red sulfur bacteria in natural habitats. He described three species from microscopic observations of reddish cells containing sulfur globules and differing in cell size (*Rbc. roseum* , *Rbc. minus* and *Rbc. fusiforme*). Two more species were added by Bavendamm, (1924). Since none of the described species has so far been isolated in pure culture or found again in samples from nature and enrichment cultures. The genus was not included in the Approved Lists of Bacterial Names, in spite, species of this genus were included in the genus *Chromatium* (Pfennig and Truper, 1974), because long and irregularly swollen cells were found in pure cultures of *Chromatium okenii* , *Chromatium weissei* , *Isochromatium buderi* and *Allochromatium warmingii* , when the mineral salts concentrations of the culture media were inappropriate. *Rhabdochromatium* was regenerate for a long rod-shaped red sulfur bacterium. The morphology of which proved to be stable in pure culture. Only one species is validly published i.e., *Rhabdochromatium marinum* which represents the cluster of truly marine species of *Chromatiaceae* .

12. *Thermochromatium* (*Tch.*) (Hot *Chromatium*) (Imhoff *et al.*, 1998b) Imhoff and Madigan (2015)

Cells are straight to slightly curved rods and motile by means of polar flagella (**Fig. 3.64 B**) (Madigan, 1986). Multiply by binary fission and Gram (-) negative. Vesicular type of internal photosynthetic membranes of with photosynthetic pigments bacteriochlorophyll *a* and carotenoids of spirilloxanthin are located (**Fig. 3.64 C**). Obligately phototrophic, strictly anaerobic and require sulfide-reduced media. Photolithoautotrophic growth occurs under anoxic conditions in the presence of light with sulfide and elemental sulfur (S^0) as electron donors. During oxidation of sulfide, elemental sulfur is transiently stored inside the cells in the form of highly refractile globules (**Fig. 3.64 A**). Sulfate is the final oxidation product. The organic substrates are photoassimilated in the presence of sulfide and bicarbonate. They are thermophilic freshwater bacteria with optimum growth at temperatures 40°C and neutral pH 7.0, without salt requirement. Sulfide-containing hot springs of neutral to alkaline pH having temperatures between 35 and 60°C are the selective habitats for their growth. The mol% G + C of the DNA is 60–63. *Tch. tepidum* (Imhoff *et al.,* 1998b) is the type species. Morphological and physiological characteristics for the differentiation of *Thermochromatium* species from other phototrophic members of *Chromatiaceae* are presented in **Table 3.33** .

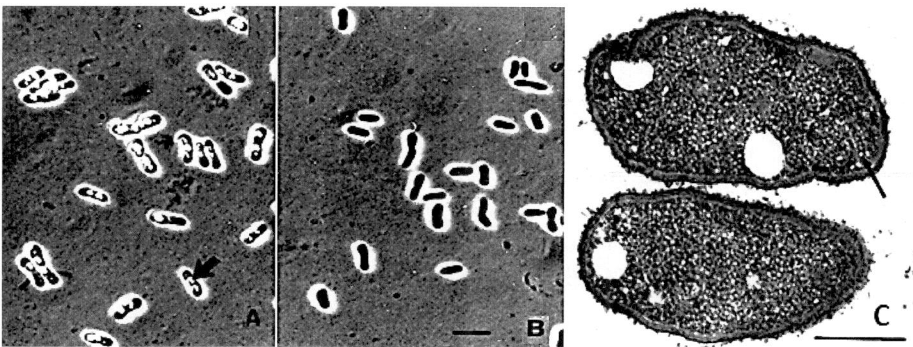

Fig. 3.64. A. Photomicrographs of cells of *Tch. tepidum* from mid-logarithmic growth filled with elemental sulfur (arrow), B. Sulfur-free cells from stationary-phase cultures. The cells were grown photoheterotrophically with sulfide, acetate and CO_2 (Bar = 3 μm), C. Electron micrograph of ultrathin sections of cells showing intracellular membrane of the chromatophore type. A darkly staining outer cell wall layer was seen (Bar = 0.5μm).

Differentiation of *Thermochromatium* from other Genera

Analysis of 16S rDNA sequence reveals that the genus *Thermochromatium* belongs to *Chromatiaceae* containing motile freshwater bacteria, characterized by high-temperature optima above 40ºC (*Tch. tepidum* at 48–50ºC) and incapable of chemotrophic growth. The type strain of *Tch. tepidum* has a distinct and quite characteristic absorption spectrum. Intact cells and photosynthetic membrane preparations showed near-infrared absorption peak at 920 nm (Garcia *et al.,* 1986). This absorption is because of the novel light-harvesting complex and is the longest wavelength-absorbing band recorded for a photosynthetic light-harvesting complex containing bacteriochlorophyll *a*. Whether this different complex is related to the thermophilic lifestyle of *Tch. tepidum* is not clear. Differentiation of morphological and physiological characteristics of species of *Thermochromatium* from other phototrophic *Chromatiaceae* are summarized in **Tables 3.33** and **3.40**. The phylogenetic relationship of *Thermochromatium tepidum* from other members of *Chromatiaceae* based on 16S rDNA sequence is illustrated in **Fig. 3.46** .

Taxonomic Comments

High-temperature requirement range for the growth of *Thermochromatium tepidum* restricts this bacterium to hot springs and other thermal environments with sulfide and light (Castenholz and Pierson, 1995). The successful adaptation to growing at high temperatures was considered to be taxonomically relevant property of *Thermochromatium tepidum*, which is also phylogenetically different from mesophilic freshwater *Chromatiaceae* species. Therefore, based on the above characteristics *Chromatium tepidum* was reclassified into a new genus i.e., *Thermochromatium* (Madigan, 1986; Imhoff *et al.,* 1998b) with single species such as *Thermochromatium tepidum.*

13. *Thioalkalicoccus* (*Tac.*) (Sulfur sphere from soda) (Bryantseva *et al.,* 2000) Imhoff (2015)

Cells are spherical or oval, form diplococcus-shaped cells during cell division and multiply by binary fission (**Fig. 3.65A**) (Bryantseva *et al.,* 2000). They are Gram (-) negative contains tubular-type internal photosynthetic membranes (**Fig. 3.65B**). Photosynthetic pigments are bacteriochlorophyll *b* and carotenoids of spirilloxanthin series with tetrahydrospirilloxanthin as a major component. They are strictly anaerobic and obligately phototrophic. Growth occurs photolithoautotrophically with sulfide as electron donor and globules of elemental sulfur (S^0) are accumulated inside the cytoplasm. The final oxidation product is sulfate. Organic substrates are photoassimilated in the presence of sulfide and bicarbonate. They are mesophilic, an obligate alkaliphilic bacterium that recorded optimum growth at temperatures 20–25ºC. Optimal growth occurs in the presence of low concentrations of sodium salts under alkaline conditions. The surface of sediments rich in organic matter and microbial mats of soda lakes containing hydrogen sulfide and exposed to the light are the ideal substrates for their growth. The mol% G + C of the DNA is 63.6–64.8. *Thioalkalicoccus limnaeus* (Bryantseva *et al.,* 2000) is the type species.

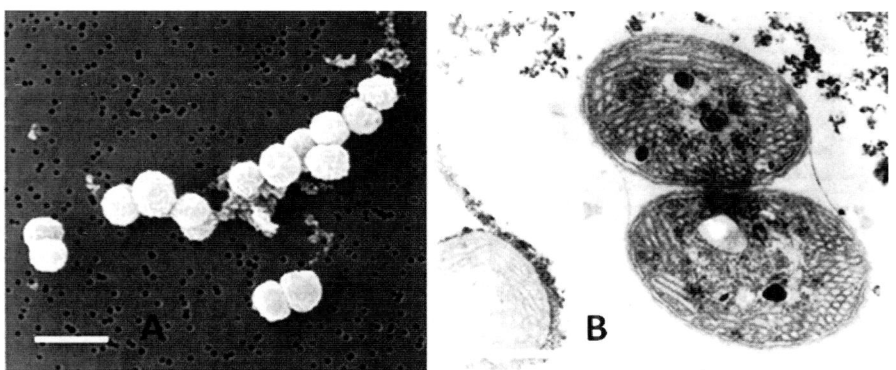

Fig. 3.65. A. Scanning electron photomicrograph of cells of *Tac. limnaeus* showing spherical shape and diplococcus-like division stages (Bar= 2 μm). B. Electron microphotograph of an ultra-thin section of cells of *Tac. limnaeus* grown photoautotrophically showing tubular internal photosynthetic membrane (Bar= 1 μm). Cells were harvested after complete oxidation of elemental sulfur and late stages of cell division are shown with the two cells already completely separated.

Differentiation of *Thioalkalicoccus* from other Genera

Based on the results of 16S rDNA sequence analyses **(Fig. 3.46)** , *Thioalkalicoccus* belongs to *Chromatiaceae* which mainly marine and halophilic bacteria. Differentiation of *Thioalkalicoccus* from other phototrophic *Chromatiaceae* species is summarized in **Tables 3.34** and **3.40** .*Thioalkalicoccus limnaeus* is one of the very few species that contain tubular internal photosynthetic membranes which are also seen in *Thiococcus pfennigii* and *Thioflavicoccus mobilis* .

Table 3.34. Differential Characteristics of *Thioalkalicoccus, Thiococcus, Thiodictyon, Thioflavicoccus* and *Thiohalocapsa* species

Characteristics	*Thioalkalicoccus limnaeus*	*Thiococcus pfennigii*	*Thiodictyon elegans*	*Thiodictyon bacillosum*	*Thiodictyon syntrophicum*	*Thioflavicoccus mobilis*	*Thiohalocapsa halophila*	*Thiohalocapsa marina*
Cell shape	Sphere	Sphere	Rod	Rod	Oval-round/ cocci	Sphere	Sphere	Sphere
Cell size (μm)	1.3-1.8	1.2-1.5	1.5-2.0 x 3.0-8.0	1.5-2.0 x 3.0-6.0	1.4-2.0 x 4.0	0.8-1.0	1.5-2.5	1.5-2.0
Motility	-	-	-	-	-	+	-	-
Color of cell suspension	Yellowish brown, Orange brown	Orange brown	Purple voilet	Purple voilet	Purple-red	Yellowish brown, Orange brown	Purple red	Purple red

Chara-cteristics	*Thioalka-licoccus limnaeus*	*Thiococcus pfennigii*	*Thiodi-ctyon elegans*	*Thiodi-ctyon bacill-osum*	*Thiodi-ctyon syntro-phicum*	*Thioflavi-coccus mobilis*	*Thiohal ocapsa halo-phila*	*Thiohal ocapsa marina*
Carotenoid group	Tetrahydro spirilloxa-nthin	tetrahydro spirilloxa-nthin	Rhod-opin	Rhod-opin	Okenone	tetrahydro spirillo-xanthin	Oke-none	Oke-none
Absortion spectra of living cells (nm)	410, 462, 492, 530, 602, 835 and 1030	368, 397, 527 and 780	ND	ND	374, 528, 582 and 833	410, 462, 492, 530, 602, 835 and 1025	359, 490, 523, 551, 590, 803, 830 and 883	395, 509, 584, 803, 803 and 845
Mol% G+C of DNA	63.6-64.8	69.4-69.9	65.3-66.3	66.3	66.22	66.5	65.9-66.6	64.8
Growth factors	Not required	-	Not required	Not required	Not required	-	Vitamin B12	Not required
Sulfate assimilation	ND	-	ND	ND	+	ND	-	-
Chemolitho trophic growth	+	+	-	+	-	-	-	-
pH optimum (range)	8.8-9.5 (8.0-10)	7.2-7.4	6.7-7.3	6.7-7.3	6.8-7.5	7.2-7.4	7.0 (6.0-8.0)	7.5 (6.5-8.5)
Tempera-ture Optim-um (◦C)	20-25 (15-35)	20-35	20-30	20-30	20-23 (5-25)	25-30	20-30 (25-35)	25-30 (25-35)
NaCl range)%(	0.5-7.0	0-3.0	Not required	Not required	Not required	1.0-3.0	3.0-2.0	1.0-6.0
NaCl opti-mum (%)	1.0-6.0	0.5-2.0	-	-	-	2.0	4.0-8.0	2.0
Carbon sources/Electron donors								
Hydrogen	+	ND	ND	ND	+	ND	+	-
Sulfide	+	+	+	+	+	+	+	-
Thiosulfate	-	-	ND	ND	+	-	+	-
Sulfur	+	+	+	+	-	+	+	-
Sulfite	ND	ND	-	-	-	ND	+	-
Formate	-	-	ND	ND	-	-	-	-
Acetate	+	+	+	+	+	+	+	+
Pyruvate	+	+	+	+	+	+	+	+
Propionate	+	+	ND	ND	-	-	-	-
Butyrate	-	-	ND	ND	-	-	-	-
Lactate	-	+	ND	ND	-	-	+	+
Fumarate	(+)	+	ND	ND	-	-	-	+
Succinate	+	+	ND	ND	-	-	-	+
Malate	+	+	ND	ND	-	-	-	-
Fructose	-	+	ND	ND	+	-	+	-

Chara-cteristics	*Thioalka-licoccus limnaeus*	*Thiococcus pfennigii*	*Thiodi-ctyon elegans*	*Thiodi-ctyon bacill-osum*	*Thiodi-ctyon syntro-phicum*	*Thioflavi-coccus mobilis*	*Thiohal ocapsa halo-phila*	*Thiohal ocapsa marina*
Glucose	-	-	ND	ND	-	-	(+)	+
Ethanol	-	-	ND	ND	-	-	-	-
Propanol	ND	-	ND	ND	-	-	-	-
Glycerol	-	-	ND	ND	-	-	(+)	-
Glycolate	-	-	ND	ND	ND	-	-	-
Crotonate	ND	-	ND	ND	ND	-	-	ND
Valerate	-	-	ND	ND	ND	-	-	-
Casamino acids	-	(+)	ND	ND	ND	-	-	+

Symbols: +, positive in most strains/ substrate utilized; -, negative in most strains / substrates not utilized; ±, variable; (+), weak growth; ND, not determined/no data available; Data for *Thioalkalicoccus limnaeu* and *Thiococcus pfenigii* is taken from Bryantseva *et al.* (2000) and Eimhjellen (1970), Imhoff *et al.* (998b) respectively. Data for *Tdc. elegans, Tdc. bacillosum* and *Tdc. syntrophicum* is taken from Winogradsky (1888), Pfennig and Truper (1971b) and Luedin *et al.* (2018). Data for *Thioflavicoccus mobilis* is taken from Imhoff and Pfennig (2001). Data for *Thiocapsa halophila* and *Thiocapsa marina* is taken from Anil Kumar *et al.* (2009) and Caumette *et al.* (1991).

Taxonomic Comments

Based on 16S rDNA gene sequence analyses *Thioalkalicoccus limnaeus* is closely related to the branch of marine and halophilic species of *Chromatiaceae* family. They show low similarity i.e., 92% with *Thiococcus pfennigii* and most closely related to known purple sulfur bacteria. The successful adaptation to growth at low salinity and alkaline environments are considered to be taxonomically relevant property of *Thioalkalicoccus limnaeus*, which is also phylogenetically different from mesophilic freshwater *Chromatiaceae* species. They also differed in phenotypic properties such as bacteriochlorophyll, carotenoids and growth conditions. Therefore, based on above characteristics a *new* genus was classified i.e., *Thioakalicoccus* with single species such as *Thioalkalicoccus limnaeus* (Bryanteseva *et al.,* 2000).

14. *Thiocapsa* (*Tca.*) (Sulfur box) (Winogradsky, 1888) Imhoff and Caumette (2015)

Cells are spherical, 1.0–3.0 μm in diameter, diplococci before multiplication by binary fission and non-motile **(Fig. 3.66).** Tetrads may be formed after consecutive division in two perpendicular planes. Cells are surrounded by a strong slime capsule and which contains gas vesicles. They are Gram (-) negative contains vesicular internal photosynthetic membranes with photosynthetic pigments bacteriochlorophyll *a* and carotenoids of spirilloxanthin series. Growth occurs photolithotrophic ally under anoxic conditions in the presence of light by utilizing sulfide, thiosulfate and sulfur as electron donors. Hydrogen may also be used. During oxidation of sulfide, elemental sulfur (S^0) is transiently stored inside the cells in the form of highly refractile globules. Sulfate is the final oxidation product. Simple organic substrates are photoassimilated in the presence of sulfide and bicarbonate. Some species may grow chemoautotrophically or chemoheterotrophically under microoxic to oxic conditions in the dark. Storage materials are polysaccharides, poly-β-hydroxybutyrate, elemental sulfur and polyphosphates. Vitamin B_{12} may be required as a growth factor. The mol% G + C of the DNA is 63.3–66.3. *Thiocapsa roseopersicina* (Winogradsky, 1888) is the type species. Differentiation of *Thiocapsa* from other members of *Chromatiaceae* are presented in **Table 3.35.**

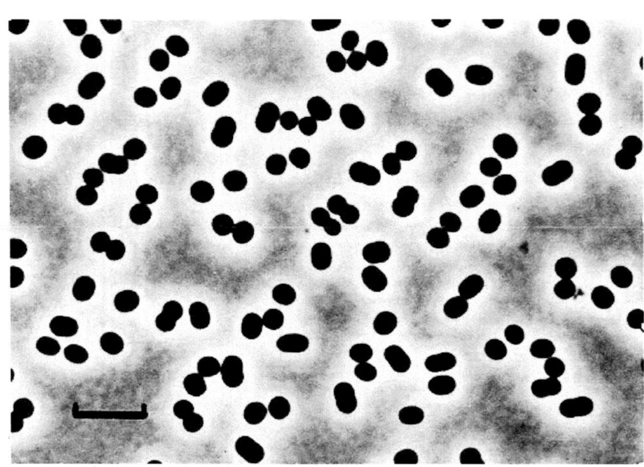

Fig. 3.66. Phase-contrast micrograph of *Thiocapsa roseopersicina* cultured photoautotrophically with sulfide (Bar= 5 μm).

Table 3.35. Differential Characteristics of *Thiocapsa* Species

Characteristics	***Thiocapsa roseopersicina***	***Tca. rosea***	***Tca. pendens***	***Tca. littoralis***	***Tca. marina***	***Tca. imhoffii***
Habitats	Stagnant water, wastewater lagoons; estuaries/salt marshes	Freshwater	Stagnant freshwater and mud	White Sea littoral shore	Coastal brackish to marine lagoons	Soap Lake alkaline water/ sediments
Cell shape	Sphere	Sphere	Sphere	Spherical	Spherical	Spherical
Motility	-	-	-	-	-	-
Cell diameter (μm)	1.2-3.0	2.0-3.0	1.5-2.0	1.5-2.5	1.5-3.0	1.7-2.0
Gas vesicles	-	+	+	-	-	-
Aggregate formation						
Clumps of cells	-	-	-	-	-	-
Irregular aggregates	+	+	-	+	+	+
Irregular clumps	-	-	-	-	-	-
Irregular net like aggregates	-	-	-	-	-	-
Platelets	-	-	-	+	-	-
Tetrads	+	-	-	+	+	+

Characteristics	***Thiocapsa roseopersicina***	***Tca. rosea***	***Tca. pendens***	***Tca. littoralis***	***Tca. marina***	***Tca. imhoffii***
Color of cell suspension	Pink, Rose red	Pink, Rose red	Pink, Rose red	Rose-red	Purple-red	Pink-rose red
Carotenoid group	Spirilloxanthin	Spirillo-xanthin	Spirillo-xanthin	Spirillo-xanthin	Okenone	Spirilloxanthin, anhydrorhodo-vibrin
Absorption spectra of living cells (nm)	484, 512, 549, 589, 800 and 844	ND	ND	375, 484, 516, 552, 590, 800, 855 and 896	521, 579, 799, 828 and 885	486, 514, 553, 603, 612, 775, 802, 847 and 897
Mol% G+C of DNA	63.3-66.3	64.3	65.3	64	62.7-63.2	
B12 requirement	-	+	+	+	-	Not required
Sulfate assimilation	+	-	-	+	-	+
Chemolithotrophic growth	+	+	+	-	+	+
pH optimum	7.3	6.7-7.5	6.7-7.5	6.5	7.2-7.4	8.5 (7.5-9.5)
Temperature optimum (◦C)	20-35	20-35	20-35	30	30-35	32
NaCl optima (%)	none[c]	none	none	1.0	1.0-2.0	0-4.0
Carbon sources/Electron donors						
Hydrogen	+	-	-	ND	+	ND
Sulfide	+	+	+	+	+	+
Thiosulfate	+	+	+	+	+	+
Sulfur	+	+	+	+	+	-
Formate	-	-	-	-	+	-
Acetate	+	+	+	+	+	+
Pyruvate	+	+	+	+	+	+
Propionate	±	+	+	+	+	-
Butyrate	-	-	-	+	-	-
Lactate	±	+	+	+	+	+
Fumarate	+	-	-	+	+	-
Succinate	+	-	-	+	+	+
Malate	+	±	+	+	+	+
Fructose	+	+	-	+	+	-
Glucose	-	-	+	+	+	-
Ethanol	-	-	ND	-	ND	-
Propanol	-	-	ND	-	ND	-
Glycerol	+	-	-	-	+	+
Glycolate	-	ND	ND	ND	-	-
Crotonate	ND	ND	ND	ND	-	-
Valerate	ND	-	-	+	ND	-
Casamino acids	ND	ND	ND	+	+	-

[a]Symbols: +, positive in most strains/ substrate utilized; -, negative in most strains/ substrate not utilized; ±, variable; ND, not determined or no data available. Data for *Thiocapsa roseopersicina, Tca. rosea, Tca. pendens, Tca. littoralis, Tca. marina, Tca. Imhoffii* is taken from Guyoneaud *et al.* (1998); Imhoff and Caumette (2005), Pchkova *et al.* (2000), Caumette *et al.* (2004) and Asao *et al.* (2007)

Differentiation of *Thiocapsa* from other Genera

Characteristics features for differentiation of *Thiocapsa* from other phototrophic members of the family *Chromatiaceae* are listed in **Table 3.40**. Cells of *Thiocapsa* may form tetrads and therefore it can be wrongly identified as *Thiopedia rosea* or *Thiolamprovum pedioforme*. The cells of *Thiopedia* and *Thiolamprovum* species are readily distinguishable from *Thiocapsa* by the formation of platelets, which generally consist of 8, 16 or more regularly arranged cells. The phylogenetic relationship of *Thiocapsa* species from other members of *Chromatiaceae* based on 16S rDNA gene sequence is illustrated in **Fig. 3.46** and **3.67**.

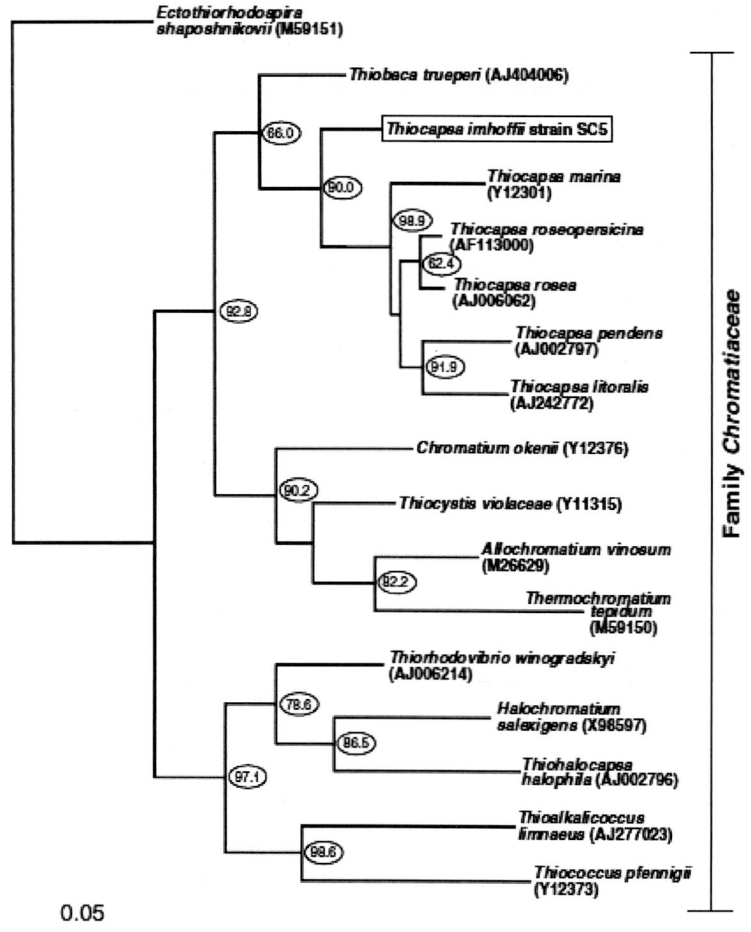

Fig. 3.67. Phylogenetic tree of *Thiocapsa* species and other genera of *Chromatiaceae* family based on 16S rRNA. The sequence of *Ectothiorhodospira* was used as an outgroup to root the tree.

Taxonomic Comments

Earlier, non-motile, spherical, PSB without gas vesicles have been assigned to *Thiocapsa*. Therefore, the results of genetic analyses indicate that two of the three species placed in this genus (the primary is *Thiocapsa halophila* and *Thiocapsa pfennigii*) were phylogenetically different from the type species, *Thiocapsa roseopersicina*. Both species are associated with the cluster containing truly marine *Chromatiaceae* species (**Figs. 3.46, 3.47** and **3.67**). Therefore, they were removed from *Thiocapsa* and placed in *Thiococcus* and *Thiohalocapsa* as *Thiococcus pfennigii* and *Thiohalocapsa halophila*, respectively (Imhoff *et al.*, 1998b). In addition, *Thiocapsa rosea* and *Thiocapsa pendens* were previously known as *Amoebobacter roseus* and *Amoebobacter pendens* (Pfennig, 1989a) are genetically close to *Thiocapsa roseopersicina.* Therefore, they are transferred to the genus *Thiocapsa* (Guyoneaud *et al.,* 1998). The type species of *Amoebobacter* and *Thiocapsa* were merged and decision concerning the genus name is formed.

Thiocapsa is the well-known genus and more suitable name and the *Thiocapsa roseopersicina* is one of the best-studied species of purple sulfur bacteria. Therefore, according to rule 42 and recommendation 42 of the International Code of Nomenclature of Bacteria (Lapage *et al.,* 1992), the two bacteria are included in the genus *Thiocapsa* such as *Thiocapsa rosea* and *Thiocapsa pendens* and *Thiocapsa roseopersicina* retained as the type species (Guyoneaud *et al.,*1998). To date, 6 species are validly published such as *Tca. imhoffi, Tca. roseopersicina, Tca rosea, Tca. pendens, Tca. littoralis* and *Tca. marina* in this genera based on their phenotypic, genotypic and phylogenetic properties (16S rRNA gene sequence) (**Fig. 3.67**) (Asao *et al.,* 2007).

15. *Thiocystis* (*Tcs.*) (Sulfur bag) (Winogradsky, 1888) Imhoff (2015)

Cells are rod-shaped or ovoid to spherical, before cell division from diplococcus-shape. Motile utilizing flagella, occur in single or in pairs, form irregular aggregates surrounded by slime and multiply by binary fission. They are Gram (-) negative and contain vesicular type internal photosynthetic membranes in which the photosynthetic pigments bacteriochlorophyll *a* and carotenoids of rhodopinal are located. Growth occurs photolithographically under anoxic conditions in the presence of light with sulfide and elemental sulfur (S^0) as electron donors. Thiosulfate and hydrogen may also be used. During oxidation of sulfide, S^0 is transiently stored inside the cells in the form of highly refractile globules **(Fig. 3.68)** (Winogradsky, 1888). Sulfate is the final oxidation product. Simple organic substrates are photoassimilated in the presence of sulfide and bicarbonate. Some species grow chemosautotrophically or chemoorganotrophically under microoxic to oxic conditions in the dark. Storage materials are polysaccharides, poly-β-hydroxybutyrate, S^0 and polyphosphates. They are ubiquitous, mesophilic with optimum growth at 25–35°C and pH 6.5–7.6. It can be isolated from freshwater, brackish water and marine environments. Anoxic sulfide-containing water and mud of freshwater ponds and lakes, as well as brackish water and marine environments exposed to light, are the ideal habitats for their growth. The mol% G + C of the DNA is 61.3–67.9. *Tcs. violacea* is the type species. Sic species are validly published and their characteristics properties are precised in **Table. 3.36.** The species of *Thiocystis* can be identified with the help of the key provided.

I. Cells spherical.

A. Sulfur globules appear to be randomly distributed within the cell, cells contain carotenoids of the rhodopinal group and cultures are purple-violet. .... ***Tcs. violacea***

B. Sulfur globules occur only at the inner periphery of the cell, cells contain carotenoids of the okenone group, and cultures are purple-red.. ----***Tcs. gelatinosa***

C. Sulfur globules appear to be randomly distributed within the cell, cells contain carotenoids of the okenone group and cultures are purple-red. .... ***Tcs. chemoclinalis***

II. Cells rod-shaped.

A. Carotenoids of the okenone group are present; only acetate, pyruvate, and glucose are utilized. .... ***Tcs. minor***

B. Carotenoids of the rhodopinal group are present, hydrogen utilized but no sugars are utilized. .... ***Tcs. violascens***

C. Carotenoids of the okenone group are present and only elemental sulfur, acetate, pyruvate, lactate, glucose and ethanol are utilized .... ***Tcs. cadagnonensis***

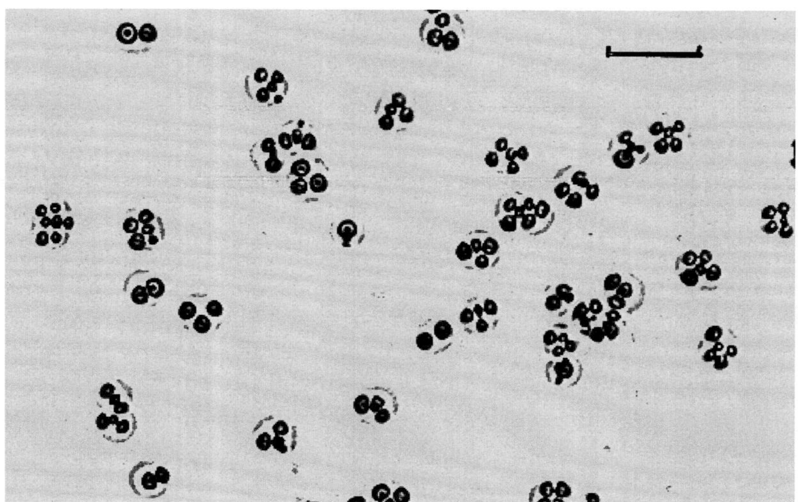

Fig. 3.68. Bright-field micrograph of *Thiocystis violacea* DSM 207 cultured photoautotrophically with sulfide. The cells contain sulfur globules (Bar= 5 μm).

Table 3.36. Differential Characteristics of *Thiocystis* Species

Characteristics	*Thiocystis violacea*	*Thiocystis gelatinosa*	*Thiocystis minor*	*Thiocystis violascens*	*Thiocystis chemoclinalis*	*Thiocystis cadagnonensis*
Cell shape	Sphere	Sphere	Rod	Rod	Oval/sphere	Oval/rod
Cell size (μm)	2.5-3.0	3	2.0 x 2.5-6.0	2.0 x 2.5-6.0	2.3-3.6	2.3-4.7
Motile by flagella	+	+	+	+	+	+
Color of cell suspensions	Purple violet	Purple-red	Purple-red	Purple violet	Purple-red	Purple-red
Carotenoid group	Rhodopin	Okenone	Okenone	Rhodopin	Okenone	Okenone
Mol% G + C of DNA	62.8-67.9	61.3	62.2	61.8-64.3	68.1	69.5
B12 requirement	-	-	-	-	-	-
Sulfate assimilation	±	-	-	+	+	+
Chemotrophic growth	+	+	+	+	+	+
pH optimum (range)	7.0-7.3	7.0-7.3	7.0-7.3	7.0-7.3	7.2 (6.6-7.5)	7.2 (6.8-7.5)
Temperature optimum (range)°C	25-35%	30	30	25-35	20 (5-25)	20 (5-25)
Salt requirement	0-2%	0.-1%	Not required	Not required	Not required	Not required
Carbon sources/ Electron donors						
Hydrogen	+	ND	ND	+	ND	ND
Sulfide	+	+	+	+	+	+
Thiosulfate	+	-	+	+	+	+
Sulfur	+	+	+	+	+	+
Sulfite	±	-	ND	+	-	-
Formate	ND	ND	-	+	-	-
Acetate	+	+	+	+	+	+
Pyruvate	+	+	+	+	+	+
Propionate	±	-	-	+	-	-
Butyrate	ND	ND	-	±	-	-
Lactate	-	-	ND	ND	-	+
Fumarate	+	-	-	-	-	-
Succinate	±	-	-	-	-	-
Malate	ND	ND	-	-	-	-
Fructose	±	-	-	-	+	-
Glucose	±	-	+	-	+	+
Ethanol	-	-	-	-	+	+
Propanol	±	-	-	-	-	-
Glycerol	±	-	-	-	-	-

Symbols: +, positive in most strains/ substrate utilized; -, negative in most strains/ substrate not utilized; ±, variable; ND, not determined/no data available; Marine strains may tolerate low concentrations of NaCl. Data for *Thiocystis violacea, Thiocystis gelatinosa, Thiocystis minor, Thiocystis violascens* is taken from Winogradsky (1888), Pfennig and Truper (1971), Imhoff *et al.* (1998), Perty (1852) respectively. Data for *Thiocystis chemoclinalis* and *Thiocystis cadagnonensis* is taken from Peduzzi *et al.* (2011) respectively.

Differentiation of *Thiocystis* from other Genera

The results of 16S rDNA and 16S rRNA sequence analyses, suggests that *Thiocystis* belongs to a branch of motile freshwater species of *Chromatiaceae.* Differentiating characteristics of *Thiocystis* species from other phototrophic members of *Chromatiaceae* are summarized in **Table 3.40**. The phylogenetic relationship of *Thiocystis* species to other *Chromatiaceae* species, based on 16S rDNA and 16S rRNA sequence is shown in **Fig. 3.46** and **3.69** (Peduzzi *et al.,* 2011).

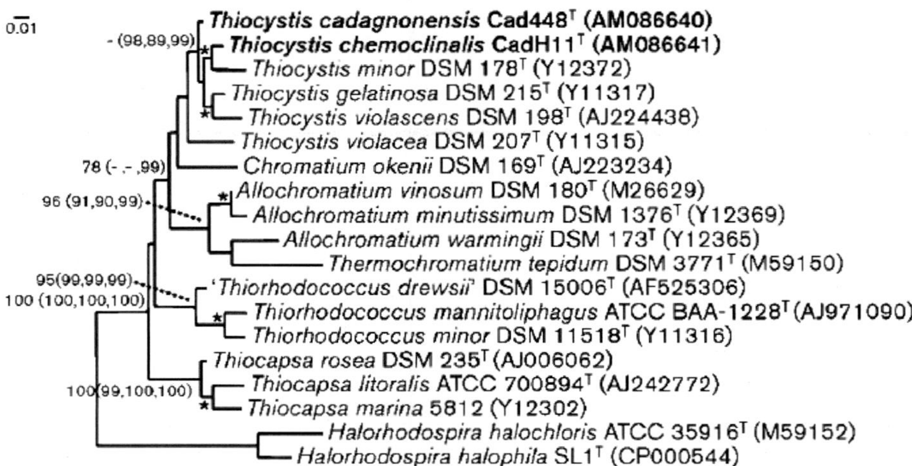

Fig. 3.69. Phylogenetic tree of *Thiocystis* species and other genera of *Chromatiaceae* family based on 16S rRNA. The sequence of *Halorhodospira halochloris* was used as an outgroup to root the tree.

Taxonomic Comments

Earlier, the genus *Thiocystis* was differentiated from other genera of *Chromatiaceae* on the basis of its spherical morphology, flagellar motility and lack of gas vesicles (Pfennig and Truper, 1989). Winogradsky (1888) reported *Thiothece* from *Thiocystis* based on the location of the intracellular sulfur globules. In the monotypic genus *Thiothece*, the sulfur globules are located at the inner periphery of the cells, while in *Thiocystis* the globules are distributed randomly in the cell (Winogradsky, 1888). Pfennig and Truper (1971a) included *Thiothece gelatinosa* in the genus *Thiocystis* as maintenance of *Thiothece* . Furthermore, *Thiocystis* is not justified. Moreover, both the 16S rRNA oligonucleotide catalogs of *Chromatiaceae* species (Fowler *et al.,* 1984) and 16S rDNA sequence analyses (Imhoff *et al.,* 1998b) not only substantiated the close relationship of *Thiocystis gelatinosa* and *Thiocystis violacea* (**Fig.3.46**), but also revealed a high degree of genetic relatedness between these species and *Thiocystis minor* (formerly *Chromatium minus*) and *Thiocystis violascens* (formerly *Chromatium violascens*) (Imhoff *et al.,* 1998b). Based on phenotypic, genotypic and phylogenetic analysis, *Chromatium violascens* and *Chromatium minus* were transferred to the genus *Thiocystis*. At present six species were validly published (**Fig. 3.69**) (Peduzzi *et al.,* 2011).

16. *Thiococcus* (*Tco.*) (Sphere with sulfur) (Imhoff *et al.*, 1998b) Imhoff (2015)

Cells are spherical, with diplococcus-shaped division stages. They are non-motile, multiply by binary fission and Gram (-) negative. Internal photosynthetic membranes consisting of bundles of ribbon-like branched tubes continuous with the cytoplasmic membrane are observed (**Fig. 3.70**) (Eimhjellen *et al.,* 1967). Color of cell suspensions is yellowish to orange-brown. Bacteriochlorophyll *b* and carotenoids of tetrahydrospirilloxanthin are present as photosynthetic pigments. They are obligately phototrophic and strictly anaerobic. Growth occurs photolithoautotrophic under anoxic conditions in the presence of light with sulfide and elemental sulfur (S^0) as electron donors. During oxidation of sulfide, S^0 is transiently stored inside the cells in the form of highly refractile globules. The final oxidation product is sulfate. Simple organic substrates are photoassimilated in the presence of sulfide and bicarbonate. They are mesophilic with optimum growth at temperatures 20–35°C and neutral pH (6.5–7.5). May require salt for optimum growth. The mol% G + C of the DNA is 69.4–69.9. *Thiococcus pfennigii* (Eimhjellen *et al.,* 1967; Imhoff *et al.* (1998b) is the type species. Single species are validly published and their differentiating characteristics from other members of *Chromatiaceae* are presented in Table **3.34.**

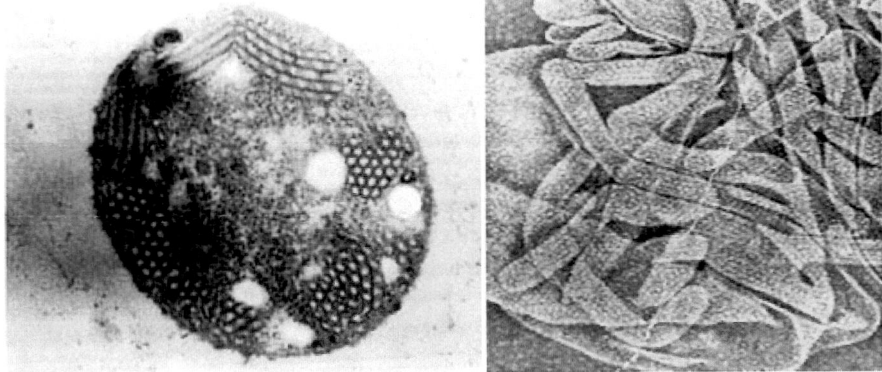

Fig. 3.70. Electron microscopic photomicrographs of ultrathin sections showing internal membrane structures of cells of *Thiococcus pfennigii* with tubular internal membranes

Differentiation of *Thiococcus* from other Genera

Differentiation of *Thiococcus* from other phototrophic members of *Chromatiaceae* are given in **Table 3.34** and **3.40.** The phylogenetic relatedness of

Thiococcus pfennigii to other members of *Chromatiaceae* based on 16S rDNA analysis is illustrated in **Fig. 3.46.**

Taxonomic Comments

Though *Thiococcus* was proposed by Eimhjellen *et al.* (1967), it was validly published only in 1970 as *Thiocapsa pfennigii*. Due to striking genetic and phenotypic difference between *Thiococcus pfennigii* and members of *Thiocapsa* , it was later named as a new genus *Thiococcus* by reclassified (Imhoff *et al.*, 1998b) as a neotype strain of *Thiococcus pfennigii* of true marine species of *Chromatiaceae* (**Fig. 3.46**), which requires at least 1% NaCl for growth. However, the salt dependence and natural distribution of this species are poorly understood. The original type strain which was isolated from a freshwater source (river water), was lost and additional strains were isolated both from freshwater and marine sources. However, chemotaxonomic, genetic analyses and salt dependence have not yet been determined.

17. *Thiodictyon* (*Tdc.*) (Sulfur net) (Winogradsky, 1888) Imhoff (2015)

Cells are rod-shaped with rounded ends, occasionally appear spindle-shaped and multiply by binary fission. They are non-motile under all conditions may form aggregates in which the cells are arranged end-to-end in an irregular net-like structure with variable shape. Clumps are also observed that are more compact or break up into individual cells. Contain large irregularly shaped gas vesicles in the central part of the cell (**Fig. 3.71 A**). They are Gram (-) negative and contain vesicular-type internal photosynthetic membranes with bacteriochlorophyll *a* and carotenoids of rhodopinal as photosynthetic pigments. Obligately phototrophic and anaerobic. Growth occurs photolithoautotrophically under anoxic conditions in the presence of light with sulfide and elemental sulfur (S^0) as electron donors. During sulfide oxidation, elemental sulfur (S^0) is transiently stored in the peripheral part of the cell as sulfur globules that are free of gas vesicles (**Fig. 3.71 B**) (Luedin *et al.*, 2018). The final oxidation product is sulfate. Simple organic substrates are photoassimilated in the presence of sulfide and bicarbonate. Mud and stagnant water of ponds and lakes containing hydrogen sulfide are the ideal substrates for their growth. The mol% G+C of the DNA is 65.3–66.3. *Thiodictyon elegans* (Winogradsky 1888) is the type species. Two species are validly published so far and their characteristics used for differentiation of *Thiodictyon* from other phototrophic members of *Chromatiaceae* are presented in **Table 3.34**. Identification of *Thiodictyon* species can be made with the help of the following key.

I. Cells able to grow in form of typical netlike aggregates under certain cultural conditions. ....***Tdc. elegans***

II. Cells that do not form netlike aggregates, usually growing in the form of free single cells. .... ***Tdc. bacillosum***

III. Cells able to grow single cells, as well as in cell aggregates with upto 100 cells. .... ***Thiodictyon syntrophicum***

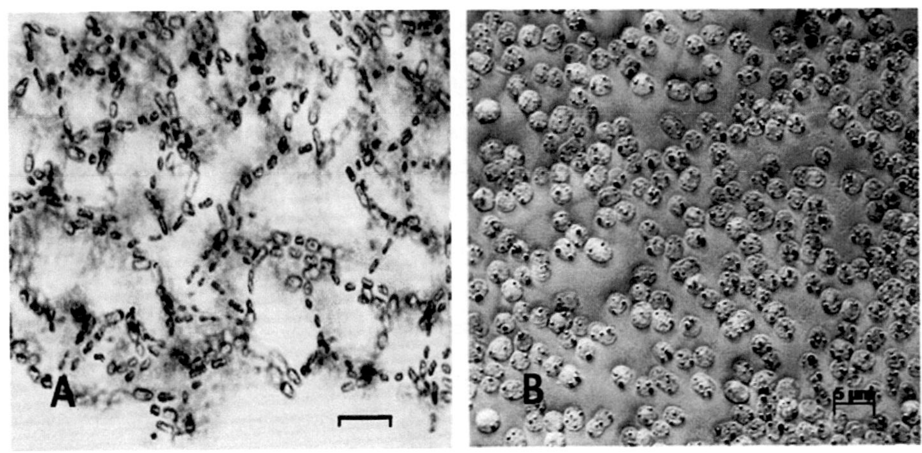

Fig. 3.71. A. Phase-contrast micrograph of cells of *Thiodictyon elegans* DSM 232 showing the typical, irregular netlike arrangement (Bar= 12 µm), B. Phase-contrast micrograph *Thiodictyon syntrophicum* cells showing the elementary sulfur globule inclusions which are visible as black dots within the cells (Bar= 5 µm).

Differentiation of *Thiodictyon* from other Genera

The differentiation of *Thiodictyon* from other phototrophic members of *Chromatiaceae* is given in **Tables 3.34** and **3.40.** The genetic relationship among *Thiodictyon* species and other members of the *Chromatiaceae* based on 16S rDNA sequence comparison is shown in **Fig. 3.47**.

Taxonomic Comments

At present, the genus *Thiodictyon* comprises three species (*Tdc. elegans*, the type species and *Tdc. bacillosum* .)*Tdc. bacillosum* which was originally described by Winogradsky (1888) as *Amoebobacter bacillosus*. But Pfennig and Truper (1971a) have transferred this species to *Thiodictyon*, based on cell shape i.e., rod-shaped cells which are very similar to those of *Thiodictyon elegans*. Furthermore, according to the original description of the type species of *Amoebobacter* (*Amoebobacter roseus*) (Winogradsky, 1888). This genus should comprise only spherical, non-motile species with gas vesicles. Based on 16S rRNA oligonucleotide pattern analysis *Thiodictyon elegans* is related to *Thiocapsa* species (*Thiocapsa pendens* and *Thiocapsa rosea*) (Fowler *et al.,* 1984). 16S rDNA sequence similarities of *Thiodictyon* species are expected to confirm this relationship. However, due to lack of genetic sequence information, species of *Thiodictyon* have not been included in a recent reclassification of *Chromatiaceae* (Imhoff *et al.,* 1998b). Recently *Thiodictyon syntrophicum* was added to the genus based on morphology and 16S rRNA sequence analysis. This species is closely related to *Tdc. bacillosum* and *Tdc. elegans* with 99% sequence identity (partial 16S rRNA sequences) (Luedin *et al.*, 2018).

18. *Thioflavicoccus* (*Tfc.*) (Beige-yellow coccus with sulfur) (Imhoff and Pfennig, 2001) Imhoff (2015)

Cells are spherical with diplococcus-shaped division stages, motile by means of flagella and multiply by binary fission (**Figs. 3.72 A & B**) (Imhoff and Pfennig, 2001). Gram (–) negative and contain tubular type internal photosynthetic membranes (**Fig. 3.72 C**). Bacteriochlorophyll *b* and carotenoids of tetrahydro spirilloxanthin are present as photosynthetic pigments. Obligate phototroph and strictly anaerobe. Growth occurs photolithoautotrophic under anoxic conditions in the presence of light with sulfide and elemental sulfur as electron donors. During oxidation of sulfide, elemental sulfur is transiently stored inside the cells as highly refractile globules. The final oxidation product is sulfate. Simple organic substrates are photo assimilated in the presence of sulfide and bicarbonate. They are mesophilic bacteria that grow well at temperatures 20–30°C and at neutral pH (pH range 6.5–7.5). Sodium chloride is required for optimum growth. The mol% G + C of the DNA is 66.5. *Tfc. mobilis* (Imhoff and Pfennig, 2001) is the type species and it is only one validly published species that contain tubular internal photosynthetic membranes. The characteristics properties of these species are precised in **Table 3.34.**

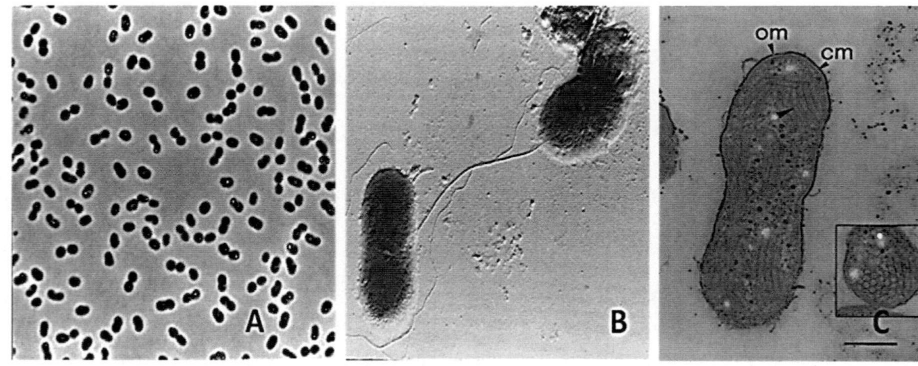

Fig. 3.72. A, Phase-contrast photomicrograph of *Thioflavicoccus mobilis* grown photoautotrophically with sulfide. Different stages of cell division are seen. Some cells contain globules of elemental sulphur. B. Electron photomicrograph of negatively stained cells showing single–monotrichous flagella and pili. (Bar, 0.5 μm), C. Ultrathin section of *Tcf. mobilis* showing tubular internal membrane structures in longitudinal section (t) and in cross-section (inset) (om, Outer membrane; cm, cytoplasmic membrane) (Bar= 0.5 μm).

Differentiation of *Thioflavicoccus* from other Genera

As a result of 16S rDNA sequence analyses of *Thioflavicoccus*(**Fig. 3.46**) it belongs to the branch of *Chromatiaceae* containing marine and halophilic species. Characteristics for differentiation of these members from other phototrophic members of *Chromatiaceae* are given in **Tables 3.34** and **3.40.**

Taxonomic Comments

Thioflavicoccus mobilis is the only one species of the very few species (similar to *Thiococcus pfennigii* and *Thioalkalicoccus limnaeus*) that contains tubular internal photosynthetic membranes. These distinct phenotypic properties and phylogenetic analysis data differentiate these bacteria from other members of *Chromatiaceae*. The phylogenetic relationship of the *Thioflavicoccus* species to other *Chromatiaceae* species based on 16S rDNA sequence is shown in **Fig. 3.46.**

19. *Thiohalocapsa* (*Thc.*) (The sulfur capsule of the salt) (Imhoff *et al.*, 1998b) Imhoff and Caumette (2015)

Cells are spherical, non-motile and multiply by binary fission (**Fig. 3.73 A** .) Individual cells are surrounded by a strong slime capsule and cell aggregates are formed. They are Gram (-) negative and contain vesicular-type internal photosynthetic membranes (**Fig. 3.73 B**) with bacteriochlorophyll *a* and carotenoids as photosynthetic pigments. Growth occurs photolithoautotrophic under anoxic conditions in the presence of light with sulfide, elemental sulfur (S^0) and some other electron donors. During oxidation of sulfide, S^0 is transiently stored inside the cells as highly refractile globules. The final oxidation product is sulfate. Simple organic substrates are photo assimilated in the presence of sulfide and bicarbonate. Chemolithotrophic or chemoorganotrophic growth is possible under microoxic to oxic conditions in the dark. They are mesophilic with optimum growth at temperatures 20–30°C and pH 6.5–7.5. An elevated concentration of sodium chloride is required for growth. Sulfide-containing waters and sediments of saltern and other coastal habitats with elevated salt concentration and exposure to the light are the ideal substrates for their growth. The mol% G + C of the DNA is 65.9–66.6. *Thiohalocapsa halophila* (Caumette *et al.*, 1991) is the type species. Only a single species is validly published and their differential characteristics from other members of *Chromatiaceae* are summarized in **Table 3.34** .

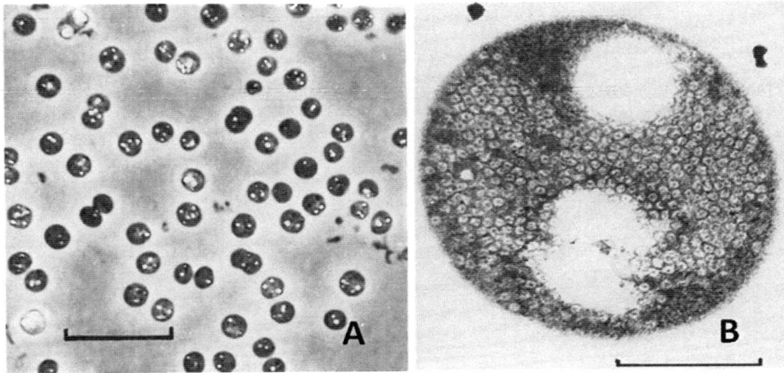

Fig. 3.73. A. Phase contrast microphotograph of *Thiocapsa halophila* cells grown photolithotrophically on sulfide (Bar=10 μm), B. Ultrathin section of cells showing vesicular type internal photosynthetic membrane system (Bar=1 μm).

Differentiation of *Thiohalocapsa* from other Genera

Characteristics for differentiation of *Thiohalocapsa* from other members of *Chromatiaceae* are given in **Tables 3.34** and **3.40.** The genetic relationship of *Thiohalocapsa halophila* to other *Chromatiaceae* species, based on 16S rDNA sequence is shown in **Fig. 3.46.**

Taxonomic Comments

Thiohalocapsa halophila was previously described as a *Thiocapsa* species (Caumette *et al.,* 1991), because of its phenotypic properties which are in agreement with the classification, of Pfennig and Truper (1989) based on spherical and non-motile cells. However, phylogenetic relationships and salt response clearly differentiate *Thiohalocapsa halophila* from species of *Thiocapsa* (Guyoneaud *et al.,* 1998; Imhoff *et al.,* 1998b). Hence, it was transferred to a new genus (*Thiohalocapsa*) by Imhoff *et al.* (1998b).

20. *Thiolamprovum* (*Tlp.*) (Bright egg with sulfur) (Guyoneaud *et al.,* 1998) Caumette *et al.* (2015)

Cells are nearly spherical to oval, 2.0 x 2.0–3.0 μm in size, may form regular platelets of 4–16 cells (**Fig. 3.74 A**) (Eichler and Pfennig, 1986). Non-motile cells and multiply by binary fission. Gram (-) negative and contains vesicular type internal photosynthetic membranes with photosynthetic pigments bacteriochlorophyll *a* and carotenoids of spirilloxanthin (**Fig. 3.74 B**). Growth occurs photolithoautotrophic under anoxic conditions in the presence of light with sulfide, thiosulfate and sulfur as electron donors. During oxidation of sulfide, S^0 is transiently stored inside the cells as highly refractile globules. Sulfate is the final oxidation product. Simple organic substrates can be photo assimilated in the presence of sulfide and bicarbonate. Chemoautotrophic or mixotrophic growth occurs under microoxic to oxic conditions in the dark. The mol% G + C of the DNA is 64.5–66.5 (Bd). *Tlp. pedioforme* (Eichler and Pfennig, 1986; Guyoneaud *et al,* 1998) is the type species. The characteristics of these species is presented in **Table. 3.37** .

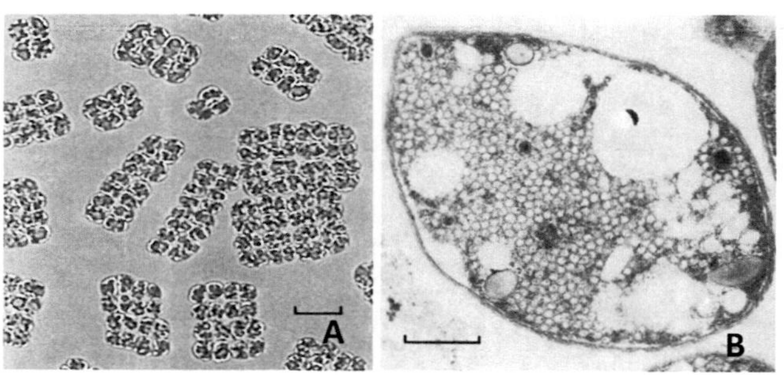

Fig. 3.74. Light field photomicrograph of *Thiolamprovum pedioforme* (Bar= 5μm), B. Ultrathin section of cells showing vesicular type internal photosynthetic membrane (Bar= 0.5 μm).

Differentiation of *Thiolamprovum* from other Genera

The differentiating characteristics of *Thiolamprovum* species from other phototrophic members of *Chromatiaceae* are given in **Tables 3.37** and **3.40**. *Thiolamprovum* is characterized by the formation of platelets of 8–16 or more regularly arranged cells as in *Thiopedia rosea*. These bacteria can be differentiated by their carotenoid composition and colour of cell cultures. *Thiolamprovum* cells contain carotenoids of the spirilloxanthin group with pinkish colour colonies, while *Thiopedia rosea* cells contain okenone carotenoids with purple-colored colonies. The phylogenetic relationship of *Thiolamprovum pedioforme* with other *Chromatiaceae* species based on 16S rDNA sequence is illustrated in **Fig. 3.47.**

Table 3.37. Differential Characteristic of *Thiolamprovum* and *Thiopedia*

Characteristics	*Thiolamprovum pedioforme*	*Thiopedia rosea*
Habitats	Wastewater ponds	Sulfide containing stratified freshwater lakes
Cell shape	Sphere	Sphere
Motility	-	-
Cell diameter (μm)	2.0 x 2.0-3.0	2.0-2.5 x 2.5-3.0
Gas vesicles	+	+
Aggregate formation		
Clumps of cells	-	-
Irregular aggregates	-	-
Irregular clumps	-	-
Irregular net-like aggregates	-	-
Platelets	+	+
Tetrads	-	-
Color of cell suspension	Pinkish- red/ Rose red	Purple-red
Absorption spectra of living cells (nm)	370, 480, 484, 508, 512, 547, 584, 801 and 828	368, 521, 580, 805 and 832
Carotenoid group	Spirilloxanthin anhydrorhodovibrin, rhodopin and OH-spirilloxanthin	Okenone
Mol% G+C of DNA	65.5 ± 1.0	62.5-63.5
Growth factors	Not required	Not required
Sulfate assimilation	-	-
Chemolithotrophic growth	+	-
pH optimum	7.4-7.6	7.3-7.5
Temperature optimum (°C)	37	20
NaCl optima (%)	Not required	Not required
Substrates		
Hydrogen	-	-

Characteristics	*Thiolamprovum pedioforme*	*Thiopedia rosea*
Sulfide	+	-
Thiosulfate	+	+
Sulfur	+	-
Formate	-	ND
Acetate	+	+
Pyruvate	+	+
Propionate	-	-
Butyrate	-	+
Lactate	+	-
Fumarate	-	±
Succinate	-	±
Malate	-	±
Fructose	-	±
Glucose	-	+
Ethanol	-	-
Propanol	-	-
Glycerol	-	-
Glycolate	-	-
Crotonate	-	-
Valerate	-	+
Casamino acids	-	-

Symbols: +, positive in most strains/ Substrate utilized; -, negative in most strains/ Substrate not utilized; ±, variable; (+), weak growth; ND, not determined/no data available; Data for *Thiolamprovum pedioforme* and *Thiopedia rosea* is taken from Eichler and Pfennig (1986) and Eichler and Pfennig (1991).

Taxonomic Comments

Thiolamprovum pedioforme is the only validly published species of the genus and previously known as *Amoebobacter pedioformis* (Eichler and Pfennig, 1986). Sequence analyses of 16S rRNA genes have shown that *Thiolamprovum pedioforme* is genetically different from *Lamprocystis* and *Thiocapsa* species (Guyoneaud *et al.,* 1998). Hence, it is transferred into a new genus *Thiolamprovum* as *Thiolamprovum pedioforme.*

21. *Thiopedia* (*Tpd.*) (Sulfur plain) (Winogradsky, 1888) Imhoff (2015)

Cells are spherical to ovoid, non-motile and multiply by binary fission. Rectangular platelets formed with 4, 8, 16, 32, or more regularly arranged cells (up to 128 or 256 cells may stick together). Gram (-) negative and contain irregularly shaped gas vesicles in the central part of the cell. Vesicular internal photosynthetic membranes with bacteriochlorophyll *a* and carotenoids okenone as photosynthetic pigments. Growth occurs photolithoautotrophic under anoxic conditions in the presence of light with sulfide and elemental (S^0) as electron donors. During oxidation of sulfide, $S^0$is transiently stored in the peripheral part of the cell as highly refractile globules **(Fig. 3.75)** (Eichler and Pfennig, 1991). The final oxidation product is sulfate. Simple organic substrates are photo assimilated in the presence of sulfide and bicarbonate. Reduced sulfur compounds are required as a sulfur source. Mud and stagnant water of ponds and lakes containing hydrogen sulfide are the ideal substrates for their growth. The common planktonic bacterium in the sulfide-containing hypolimnion of freshwater lakes. The mol% G + C of the DNA is 62.5–63.5. *Thiopedia rosea* (Winogradsky, 1888) is the type species. Single species are validly published and their characteristics for differentiating *Thiopedia* species from other spherical and non-motile purple sulfur bacteria of *Chromatiaceae* species are precised in Table **3.37.**

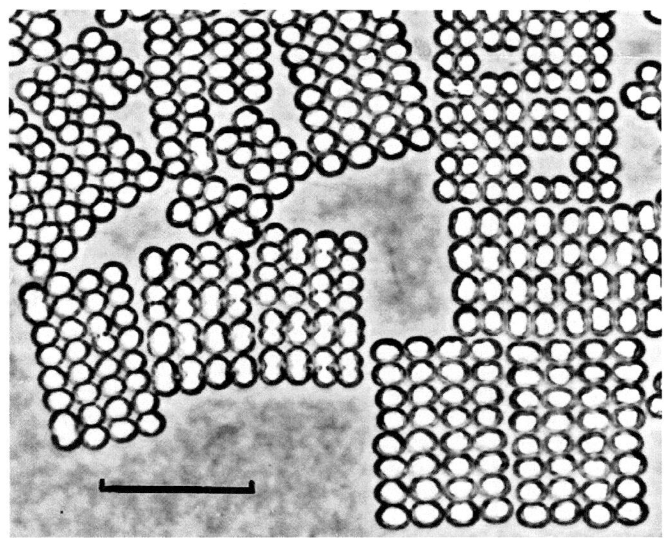

Fig. 3.75. Phase-contrast micrograph of *Thiopedia rosea* strain 4211 cultured photoautotrophically with sulfide and acetate. The light areas inside the cells are the gas vacuoles. The spherical sulfur globules are visible in the peripheral parts of the cells. In some of the platelets, synchronously dividing cells can be recognized (Bar 10 µm).

Differentiation of *Thiopedia* from other Genera

Thiopedia species can be distinguished from other phototrophic members of the family *Chromatiaceae* based on the following characteristics 1. The arrangement of the cells in rectangular platelets, 2. Possession of gas vesicles and 3. The purple-red colour of cultures and colonies, due to the presence of okenone carotenoids. Platelets are similar to those of *Thiopedia* species formed by *Thiolamprovum pedioforme*. Characteristics for differentiation of *Thiopedia* species from other spherical and non-motile purple sulfur bacteria of *Chromatiaceae* species are given in **Table 3.37** and **3.40.**

Taxonomic Comments

Purple sulfur bacteria *Thiopedia rosea* are widely distributed in sulfide-containing ponds and lakes were first described by Eichler and Pfennig (1991). Fine-structure studies of *Thiopedia*-like cell material collected from a purple bacteria bloom in a small forest pond (Michigan, USA) recorded the presence of at least two cell types with distinct internal membrane systems. One type contained a vesicular membrane system and the other type is disk-shaped lamellae present in the membrane system. Ultrathin section photomicrographs of *Thiopedia rosea* (type strain) showed a vesicular membrane system (Remsen, 1978).

Inappropriately, the ability of *Thiopedia* to multiply in rectangular platelets are characteristically found in many natural habitats is gradually lost over a period of time in pure culture. Hence, it is not a taxonomically reliable character and cannot differentiate *Thiopedia* from other *Chromatiaceae* species (Pfennig and Truper, 1989). The taxonomic unreliability of this property raises a question about the existence of *Thiopedia*. Therefore, the studies are needed to show whether *Thiopedia rosea* is a close relative of *Thiolamprovum pedioforme,* which also forms rectangular platelets (Eichler and Pfennig, 1986), or *Thiocapsa* species, form tetrads of cells but does not form large platelets.

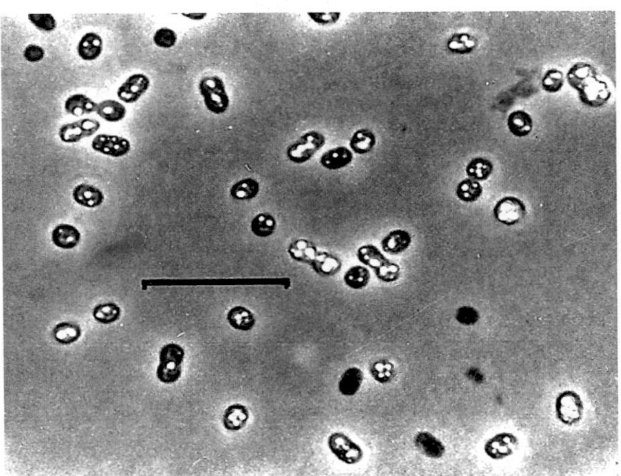

Fig. 3.76. *Thiorhodococcus minor* DSM 11518T grown photolithoautotrophically with sulfide. Phase-contrast microphotograph (Bar=10 μm).

22. *Thiorhodococcus (Trc.)* (Rose sphere with sulfur) Guyoneaud *et al.* (2015)

Cells are spherical to slightly ovoid, diplococcus-shaped before cell division and motile by flagella. Cells are single or in pairs but may form irregular aggregates, multiply by binary fission. Gram (-) negative and contain internal photosynthetic membranes of vesicular type with bacteriochlorophyll *a* and carotenoids as photosynthetic pigments. Growth occurs photolithoautotrophic under anoxic conditions in the presence of light with sulfide and S^0 as electron donors. During oxidation of sulfide, S^0 is transiently stored inside the cells as highly refractile globules **(Fig. 3.76)**. The final oxidation product is sulfate. Simple organic substrates are photo assimilated in the presence of sulfide and bicarbonate. Growth occurs chemotrophically under microoxic conditions in the dark. Water and sediment of marine and brackish environments containing hydrogen sulfide are the ideal substrates for their growth. The mol% G + C of the DNA is 66.8–67.0. *Thiorhodococcus minor* (Guyoneaud *et al.,* 1998) is the type species. Six species are validly published and their characteristics are presented in **Table 3.38.**

Table 3.38. Differential Characteristics of *Thiorhodococcus* Species

Characteristics	***Thiorhodococcus minor***	***Trc. mannitoliphagus***	***Trc. kakainadensis***	***Trc. modestalkaliphilus***	***Trc. fuscus***	***Trc. alkaliphilus***
Habitats	Anoxic sediments from coastal lagoons, marine environments rich in sulfides.	Microbial mats of the estuary	Water of marine aquaculture pond	Sediment of lagoons	Sediments of lagoon	Sediments of salt pan
Cell shape	Sphere	Coccoid	Coccoid	Coccoid	Coccoid	Coccoid
Cell size (μm)	1.0-2.0	1.5-2.0	3.0-5.0	1.5-2.5	2.0-4.0	2.0-4.0
Motility	+	+	+	+	+	+
Color of cell suspension	Brown-orange	Purple–violet	Purple-violet	Brown	Brown	Reddish-brown
Carotenoid group (Major)	Spirilloxanthin	Rhodopinal	Lycopene	Lycopene	Spirilloxanthin (rhodopin)	Spirilloxanthin (rhodopin)
Major quinones	ND	ND	ND	ND	Q-8	Q-8, MK-8
Absorption spectra of living cells (nm)	376, 460, 518, 591, 804, 854 and 875	375, 458, 493, 528, 591, 805, 856 and 888	374, 390, 460, 494, 527, 590, 803 and 857	378, 460, 490, 527, 594, 803 and 858	374, 491, 530, 590, 803 and 857	377, 488, 510, 800 and 854
Mol% G+C of DNA	66.9	61.8	60	57.8	64.0	62.4
Growth factors	-	Vitamin B12	Niacin, Biotin, Pantothenate	Not required	Yeast extract	Not required
Sulfate assimilation	-	+	-	-	-	-
Chemolithotrophic growth	-	-	-	-	-	+
Growth modes	PLA, CL, COH	PLA	PLA, PLH	PLH, PLA, POH	POH, PLA	PLA
pH optimum (range)	7.0-7.2 (6.0-8.0)	7.0-7.5 (7.0-8.5)	7.2 (6.5-7.2)	8.5 (6.5-9.5)	7.5 (6.0-8.5)	10.0-11.0 (6.0-12.0)

Characteristics	*Thiorhod-ococcus minor*	*Trc. man-nitolip-hagus*	*Trc. kaka-inadensis*	*Trc. mod-estalkali-philus*	*Trc. fuscus*	*Trc. alkali-philus*
Temperature Optimum (◦C)	30-35	25-30	25-30 (25-35)	25 (25-30)	30 (25-35)	25 (25-30)
Salinity range)%(	0.5-9.0	0.1-3.0	0.5-5.0	0.5-4.0	0.5-7.0	0-6.0
NaCl optimum)%(	2.0	0.5-2.0	1.0-2.0	1.5	2.0-4.0	1.0-3.0
Carbon sources/Electron donors						
Hydrogen	+	ND	ND	-	ND	ND
Sulfide	+			+	+	+
Thiosulfate	+	+	-	+	+	+
Sulfur	+	+	-	+	+	+
Sulfite	-	+	-	+	+	+
Formate	-	-	-	-	+	-
Acetate	+	+	(+)	+	+	+
Pyruvate	+	+	+	+	+	+
Propionate	+	+	+	-	-	-
Benzoate		-	-	-	+	
Butyrate	-	-	-	-	-	-
Lactate	+	+	(+)	+	+	-
Fumarate	+	+	+	+	-	+
Succinate	+	+	+	+	+	+
Malate	(+)	+	+	+	+	+
Fructose	+	(+)	-	+	+	+
Glucose	-	+	-	+	-	-
Ethanol	+	(+)	-	+	-	-
Propanol	+	(+)	-	+	-	ND
Glycerol	-	-	-	-	-	-
Glycolate	+	(+)	-	+	-	ND
Crotonate	-	-	-	-	-	+
Valerate	-	(+)	-	-	+	+
Casamino acids	-	+	-	+	+	-
Polar lipids	ND	ND	ND	ND	PG, PE, AP, L1-4,	PG, PE, AP, L1-4

Symbols: +, positive growth in most strains/ substrate utilized; -, negative/no growth in most strains/substrate not utilized; ±, variable in strains; (+), weak growth; ND, not determined/ no data available. PG, Phosphatidylglycerol, PE, phosphatidylethanolamine, AL, amino phospholipid and L1-4 four unknown lipids; PLA, photolithoautotrophic; PLH, photolithoheterotrophic; POH, photoorganoheterotrophic, CL, chemolithotrophic, COH, chemo organoheterotrophic. Data of *Trc. minor, Trc. mannitoliphagus, Trc. kakainadensis, Trc. modestalkaliphilus, Trc. fuscus* and *Trc. alkaliphilus* is adopted from Guyoneaud *et al.* (1997), Rabold *et al.* (2006), Anil Kumar *et al.* (2007), Sucharitha *et al.* (2010), Lakshmi *et al.* (2015), Prathyash Ushus *et al.* (2017).

Differentiation of *Thiorhodococcus* from other Genera

Thiorhodococcus can be differentiated from other members of *Chromatiaceae* by phenotypic and genetic properties. The morphological characters such as spherical-shaped cells and motility, of *Thiorhodococcus* are similar to *Thiocystis* species but differ from them by their salt requirement and 16S rDNA sequence composition. *Thiorhodococcus minor* forms a separate lineage not specifically affiliated to other *Chromatiaceae* species as shown in **Figs. 3.46** and **3.77** (Imhoff *et al.,* 1998b). Important characteristics for the differentiation of *Thiorhodococcus* from other members of *Chromatiaceae* are given in **Table 3.40.**

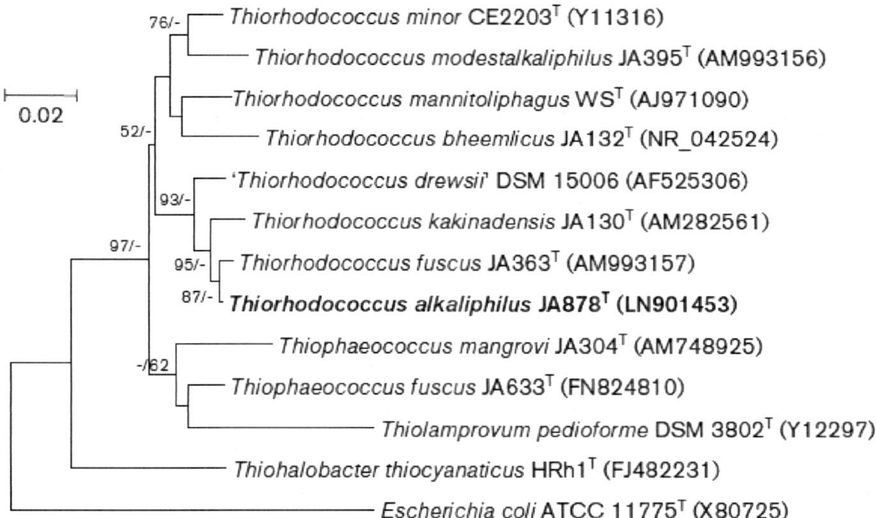

Fig. 3.77. Phylogenetic tree of *Thiorhodococcus* species from other genera of *Chromatiaceae* family based on 16S rRNA gene sequences. The sequence of *E.coli* was used as an outgroup to root the tree.

Taxonomic Comments

Thiorhodococcus forms a separate clade not specifically affiliated to other *Chromatiaceae* species as shown in **Fig. 3.46** (Imhoff *et al.,* 1998b) and **3.77** (Prathyash Ushus *et al.,* 2017). To date, eight species are identified in this genus but only seven species are validly published such as *Thiorhodococcus minor, Trc. modestalkaliphilus, Trc. mannitoliphagus, Trc. fuscus, Trc. alkaliphilus, Trc. kakinadensis* and *Trc. bheemlicus*. Based on phenotypic (colour, growth factors requirement, organic substrate utilization) chemotaxonomic (polar lipids) and phylogenetic analyses, *Trc. bheemlicus* is separated from *Thiorhodococcus* and reclassified into a new genus *Imhoffiella* as *Imhoffiella bheemlicus* (Nupur *et al.,* 2017). *Thiorhodococcus drew*sii (Zaar *et al.,* 2003) is also a well-characterized species but not validated so far which is isolated from anoxic sediment of a fishpond (manmade coastal lagoon), a microbial mat of salt marsh and microbial mat of an estuary. Therefore only six species are retained in this genus which shows distinct characteristics from other genera of *Chromatiaceae* family.

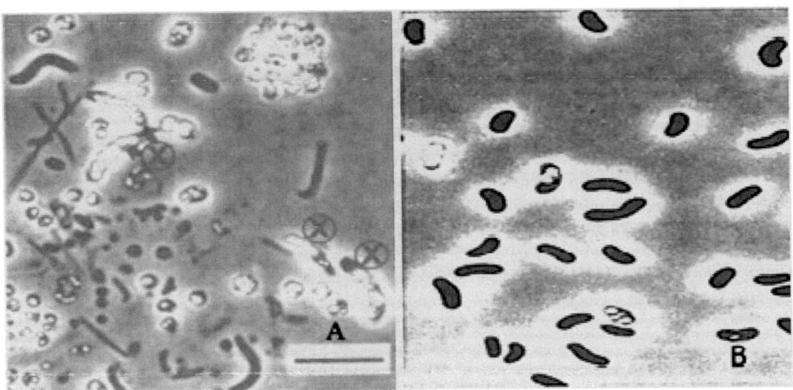

Fig. 3.78. A. Phase contrast photomicrograph vibrioid-shaped bacteria (X) containing small intracellular sulfur globules. B. Pure culture of *Thiorhodovibrio winogradskyi* grown under 100 gmol m $^{-2}$ s^{-1} tungsten light (Bar= 10 μm).

23. *Thiorhodovibrio* (*Trv.*) (Rose vibriod with sulfur) (Overmann *et al.*, 1992) Imhoff and Overmann (2015)

Cells are vibrioid to spirilloid, motile by a single polar flagellum, multiply by binary fission. Gram (-) negative and contain Vesicular type internal membranes with bacteriochlorophyll *a* and carotenoids of spirilloxanthin as photosynthetic pigments. They grow photolithoautotrophically under anoxic conditions in the presence of light with sulfide and elemental sulfur (S^0) as electron donors. During oxidation of sulfide, S^0 is stored inside the cells in the form of highly refractile globules **(Fig. 3.78)** (Overmann *et al.,* 1992). The final oxidation product is sulfate. In the presence of sulfide and bicarbonate, simple organic substrates are photo assimilated. Chemotrophically grow at reduced oxygen partial pressure. Surface layers of marine and Salt Lake sediments and microbial mats containing hydrogen sulfide are the ideal substrates. The mol% G+C of the DNA is 61–63. *Thiorhodovibrio winogradskyi* (Overmann, *et al.*, 1992) is the type species. Only a single species is validly published and its characteristics are presented in **Table 3.39** .

Table 3.39. Differential Characteristics of *Thiovibrio, Thiospirillum, Thiobaca* and *Thiophaeococcus* Species

Characteristics	*Thiorhodovibrio winogradskyi*	*Thiospirillum jenense*	*Thiobaca trueperi*	*Thiophaeoccus fuscus*	*Tpc. mangrovi*
Habitats	Salt lake sediments	Mud, stagnant water ditches, Freshwater ponds	Freshwater sediments	Lagoons	Mud

Characteristics	*Thiorhodovibrio winogradskyi*	*Thiospirillum jenense*	*Thiobaca trueperi*	*Thiophaeococcus fuscus*	*Tpc. mangrovi*
Cell shape	Vibrio-spirilloid	Sphere	Rods	Coccoid-Spherical	Spherical
Cell size (μm)	1.2-1.6 x 2.6-4.0	2.5-4.5 x 30-40	1.5 x 3.0-4.5	2.5-3.0	2.0-2.5
Motility	+	+	+	+	+
Color of cell suspension	Pink, Pink red	Orange brown	Brown	Brown	Sandy brown
Carotenoid group (Major)	Spirilloxa-nthin (Rhodopin)	Spirilloxanthin	Lycopene	Spirilloxanthin	Lycopene
Major quinones	ND	ND	ND	Q-8	ND
Absorption spectra of living cells (nm)	370-372, 483 484, 510, 590-591,794-795 and 867	ND	380, 462, 489, 524, 598, 805 and 875	374, 458, 488, 524, 593, 800 and 860	398, 457, 491, 524, 590, 800 and 857
Mol% G+C of DNA	61.0	45.5	63.0	64.5	68.1-68.9
Growth factors	Not required	Vitamin B12	Not required	Thiamine and para-amino-benzoic acid	Yeast extract
Sulfate assimilation	ND	ND	-	+	-
Chemolithotrophic growth	-	-	-	-	+
Growth modes	PLA	PLA	PLH, POH	POH	PLA
pH optimum (range)	7.0-7.4	7.0 (6.5-7.5)	7.2-7.4	6.5-7.5 (7.0)	7.2
Temperature Optimum (◦C)	33 (14-37)	20-25	25-30 (36)	30 (25-35)	6.5-7.5
Salinity range)%(	0-7.2	Not required	2.5	1.0-4.0	0.5-4.0
NaCl optimum)%(	2.2-3.2	-	0-0.5	2.0	1.0
Polar lipids	ND	ND	ND	DPG, PG, PE, PC, APL, PL1-3	DPG, PG, PE, PC, APL, PL1,
Carbon sources/Electron donors					
Hydrogen	ND	ND	-	-	-
Sulfide	+	+	+	+	-
Thiosulfate	-	-	-	+	+
Sulfur	+	+	-	+	-
Sulfite	ND	-	-	+	-
Formate	-	ND	-	-	-
Acetate	+	+	+	+	-
Pyruvate	+	ND	+	+	+

Characteristics	*Thiorhodovibrio winogradskyi*	*Thiospirillum jenense*	*Thiobaca trueperi*	*Thiophaeococcus fuscus*	*Tpc. mangrovi*
Propionate	+	ND	+	+	-
Benzoate			ND	-	-
Butyrate	-	ND	ND	+	-
Lactate	-	ND	-	-	-
Fumarate	-	ND	+	+	-
Succinate	-	ND	+	-	+
Malate	-	ND	-	+	+
Fructose	-	ND	-	+	+
Glucose	-	ND	-	-	+
Ethanol	-	ND	ND	-	-
Propanol	-	ND	ND	-	-
Glycerol	-	ND	-	-	+
Glycolate	-	ND	ND	-	-
Crotonate	-	ND	ND	-	-
Valerate	-	ND	ND	+	-
Casamino acids	-	ND	ND	-	+

Symbols: +, positive growth in most strains/ substrate utilized; -, negative/no growth in most strains/ substrate not utilized; ±, variable in strains; (+), weak growth; ND, not determined/ no data available. Marine strains may tolerate low concentrations of NaCl. PLA, photolithoautotrophic; PLH, photolithoheterotrophic; POH, photoorganoheterotrophic. DPG, Diphosphatidylglycerol; PG, phosphatidylglycerol, PE; phosphatidylethanolamine; PC, phosphotidylcholine, an unknown APL, aminophospholipid and PL, Polar lipids, were the major polar lipids. Data of Thiorhodovibrio *winogradsky,* is taken from Overmann *et al.* (1992), Winogradsky (1888) (*Thiorhodovibrio winogradsky*), Ehrenberg, (1838) (*Thiospirillum jenses*) Rees *et al.* (2002) (*Thiobaca trueperi*) and Divyasree *et al.* (2014) (*Thiophaeococcus fuscus*), Anil Kumar *et al.* (2008) (*Thiophaeococus mangrovi*)

Differentiation of *Thiorhodovibrio* from other Genera

Characteristics for differentiation of *Thiorhodovibrio* from other *Chromatiaceae* members are given in **Tables 3.39** and **3.40**. The genetic relationship of *Thiorhodovibrio winogradskyi* to other members of *Chromatiaceae*, based on the 16S rDNA gene sequence is depicted in **Fig. 3.46.**

Taxonomic Comments

Thiorhodovibrio winogradskyi was described based on phenotypic differences of the significance of two isolates (Overmann *et al.,* 1992). Both strains may be considered as different species and the description of *Thiorhodovibrio winogradskyi* is that of the type strain (DSM 6702, SSP1). The following properties distinguish strain 06511 from the type strain of *Thiorhodovibrio winogradskyi*. Cells are spirilloid, 0.7–0.9 × 2.5–3.9 μm. Thiosulfate is used as a photosynthetic electron donor and the mol% G+Ccontent of the DNA is 62.4. Propionate does not support the growth of strain 06511. Furthermore, the substrates used by the type strain, succinate, fumarate, malate, tartrate, malonate, glycerol and peptone stimulate growth slightly of strain O6511. (Overmann *et al.,* 1992).

24. *Thiospirillum* (*Tsp.*) (Sulfur spirillum) (Winogradsky, 1888) Imhoff (2015)

Cells are curved rods or spirals (sigmoid), division by binary fission, motile by means of multitrichous polar flagella. Cells may be non-motile and form irregular aggregates surrounded by slime. Gram (-) negative contain vesicular type internal photosynthetic membranes with photosynthetic pigments bacteriochlorophyll *a* and carotenoids of rhodopinal are present. Obligate phototrophs and strictly anaerobes. Growth occurs photolithoautotrophically under anoxic conditions in the presence of light with sulfide and elemental sulfur (S^0) as electron donors. During oxidation of sulfide, S^0 is transiently stored inside the cells as highly refractile globules **(Fig. 3.79).** Sulfate is the final oxidation product. Storage materials are polysaccharides, poly-β-hydroxybutyrate, S^0, and polyphosphate. The mol% G + C of the DNA is 45.5. *Thiospirillum jenense* (Ehrenberg, 1838) is the type species and its characteristic properties are presented in **Table 3.39** .

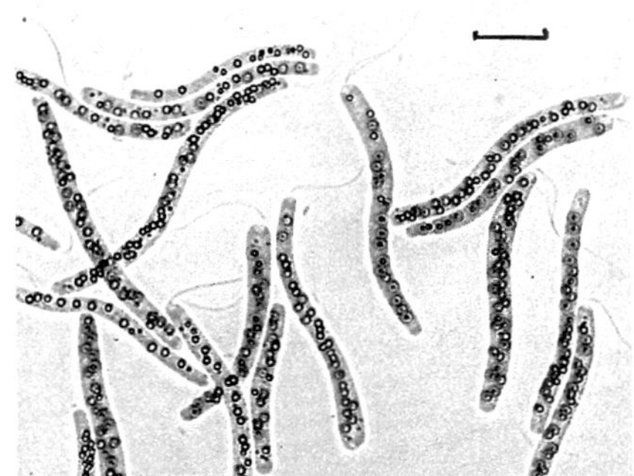

Fig. 3.79. Bright field micrograph of *Thiospirillum jenense* DSM 216 cultured photoautotrophically with sulfide. Intracellular sulfur globules and the short, slightly sigmoidal tufts of polar flagella are observed. (Bar= 10 μm).

Differentiation of *Thiospirillum* from other Genera

Characteristics used for the differentiation of *Thiospirillum* species from other phototrophic members of *Chromatiaceae* are given in **Table 3.39** and **3.40.**

Taxonomic Comments

Along *Thiospirillum jenense*, two other species i.e, *Tsp. sanguineum* and *Tsp. rosenbergii* have been described from observations of water and mud samples. Based on the same size and morphology the former species was described as *Thiospirillum jenense*, with which it could be confused. More recent reports of the existence of *Thiospirillum rosenbergii* are lacking. Winogradsky (1888) originally designated *Thiospirillum sanguineum* as the type species of this genus. This species, in contrast to *Thiospirillum jenense*, has never been obtained in pure culture, therefore, *Thiospirillum jenense* was later recognized as the type species (Pfennig, 1989d). Further detailed studies of strains isolated in pure culture and 16S rDNA sequence analyses are required to determine the relationship of *Thiospirillum* species within the genera and with other members of *Chromatiaceae*. According to 16S rRNA oligonucleotide catalogs, *Thiospirillum jenense* differs from *Chromatium* and *Allochromatium* species (Fowler *et al.,* 1984), though it may be distantly related to these bacteria are a group of motile freshwater *Chromatiaceae* species.

Table 3.40. Differential Characteristics of Genera of *Chromatiaceae*

Characte ristics	***Alc.***	***Chr.***	***Hch.***	***Isc.***	***Imf.***	***Lpb.***	***Lps.***	***Mcr.***	***Phc.***	***Phb.***	***Rbc.***	***Tch.***	***Tac.***
Cell shape	R	CR	R/O	R	C	R-O	S-O	R	R	R	R	R	S-O
Cell size (μm)	1.0-4.0	1.0-3.0	1.0-4.0	3.5-4.5	1.0-7.2	2.0-2.5	2.0-3.5	1.0-1.7	0.4-0.5	2.0-4.0	0.4-0.5	1.2-2.0	1.3-1.8
Motility	+	+	+	+	+	+	+	+	+	+	+	+	-
Colour of cell su-spension	OB	PuR	PR	PV-PuV	Pu	PuR	PV	B-PuR	DB	RB	Beige-OB	ND	YB-OB
Gas vesicles	-	-	-	-	-	+	+	-	ND	-	ND	-	-
Internal photosy-nthetic mem-branes	V	V	V	V	V	V	V	V	V	V	V	V	T
Slime capsule	-	-	-	-	-	-	-	-	ND	ND	ND	ND	-
Bacteriochloro-phyll	*a*	*a*	*a*	*a*	*a*	*a*	*a*	*a*	*a*	*a*	*a*	*a*	*b*
Caroten-oid group	RA	OK	SP	RA	RA	OK	RA	SP, OK	RN	RN	LY	SP	THS
Optimum pH	6.5-7.6	7	7.2-7.6	6.5-7.6	7.5-8.0	7.5	7.0-7.3	6.5-7.6	6.5-7.5	7.5	7	7	808-9.2
Optimum Temp-erature (°C)	25-35	20-35	20-35	25-30	20-37	23-27	20-30	25-37	-	22-37	20-35	48-50	25
Salt req-uirement	-	-	+	+	-	+	-	+	ND	-	+	-	+
Growth modes	PLA	PLA	PLA, CLA, COH	PLA	PLH POH	PLA, CLA	PLA, CLA	PLA CLA COH	PLH	POH	PLA	PLA	PLA
Sulfate assi-milation	+	+	+	+	+	-	NA	+	ND	-	+	+	ND

Characteristics	*Alc.*	*Chr.*	*Hch.*	*Isc.*	*Imf.*	*Lpb.*	*Lps.*	*Mcr.*	*Phc.*	*Phb.*	*Rbc.*	*Tch.*	*Tac.*
Vitamin B_{12} requirement	-	+	±	+	-	+	NA	-	-	-	ND	-	-
DNA G+C content (mol%)	63.1	48.0-50.0	64.6-66.3	62.2-62.8	63.1	60.0-64.0	63.4-64.1	68.9-70.4	68-69	65.5	60.4	60-63	63.6-64.8

continued....

Characteristics	*Tac.*	*Tca.*	*Tcs.*	*Tco.*	*Tdc.*	*Tfc.*	*Thc.*	*Tlp.*	*Tpd.*	*Trc.*	*Trv.*	*Tsp.*	*Tba.*	*Tpc.*
Cell shape	S-O	S	R, S-O	S	R	C	S	S-O	S-O	S-O	Vi-S	R-S	R	S
Cell size (µm)	1.3-1.8	1.2-3.0	2.5-3.0	1.2-1.5	1.5-2.0	0.8-1.0	1.5-2.5	2	2.0-2.5	1.0-2.0	1.2-1.6	2.5-4.0	1.6	2-2.5
Motility	+	+	+	+	+	-	-	-	+	-	-	+	-	-
Colour of cell suspension	YB, OB	P, RR	PuV	OB	PuV	Y beige-OB	PuR	P, RR	PuR	BO	P	Y-OB	B	SB/CB
Gas vesicles	-	±	-	-	+	-	-	+	+	-	ND	ND	-	-
Internal photosynthetic membranes	T	V	V	T	V	T	V	V	V	V	V	V	V	V
Slime capsule	-	-	+	-	ND	-	+	-	-	-	ND	ND	ND	-
Bacteriochlorophyll	*b*	*a*	*a*	*b*	*a*	*b*	*a*	*a*	*a*	*a*	*a*	*a*	*a*	*a*
Carotenoid group	THS	SP	RA	THS	RA	THS	OK	SP	OK	SP	SP	LY, RN	LY	LY
Optimum pH	808-9.2	7.3	6.5-7.6	6.5-7.5	6.7-7.3	6.5-7.5	6.5-7.5	7.4-7.6	7.3-7.5	7.0-7.2	7.0-7.4	7	ND	7.5
Optimum Temperature (°C)	25	20-35	25-35	20-35	20-30	20-30	20-30	37	20	30-35	14-37	20-25	25-30	20-35
Salt requirement	+	-	-	±	-	+	+	-	-	-	+	-	-	+

Charac teristics	Tac.	Tca.	Tcs.	Tco.	Tdc.	Tfc.	Thc.	Tlp.	Tpd.	Trc.	Trv.	Tsp.	Tba.	Tpc.
Growth modes	PLA	PLA, POH, CLA, COH	PLA, CLA, COH	PLA	PLA	PLA	PLA, POH, CLA, COH	PLA, CLA	PLA	PLA, CLA	PLA	PLA	PLH, POH	PLA
Sulfate assim-ilation	ND	+	±	-	ND	+	-	-	-	-	+	+	-	-
Vitamin B_{12} requi-rement	-	-	-	-	ND	-	+	-	-	-	-	+	ND	-
DNA G+C content (mol%)	63.6 - 64.8	63.3-66.3	61.3-67.9	69.4 - 69.9	65.3 - 66.3	66.5	65.9-66.6	64.5 - 66.5	62.5 - 63.5	66.8 - 67.0	61	45.5	6.3	68.5

Symbols: +, Present; -, absent; ±, variable; ND, Not determine/no data available. RA, rhodopinal; R, Rods; CR, Circular rods; O; Ovoid; C, Coccoid; S, Spherical; Vi, Vibrioid; YB, Yellowish brown; OB, Orange brown; P, Pink; RR, Rose Red; PuV, Purple violet; PuR, Purple Red; BO, Brown Orange; Y-OB, Yellow-Orange Brown; B, Brown; SB, Sandy Brown; CB, Choclate Brown; SP, spirilloxanthin; RN, rhodopin; THS, tetrahydro spirilloxanthin; Ok, okenone; LY, lycopenal; T, Tubules; V, Vesicles; PLA, photolithoautotrophic; PLH, photolithoheterotrophic; POH, photo-organoheterotrophic; CLA, chemolithoautotrophic; CLH, chemolithoheterotrophic; COH, chemo-organoheterotrophic. *Alc. Allochromatium* (Imhoff *et al.,* 1998; Imhoff, 2015); *Chr, Chromatium* (Perty, 1852; Imhoff, 2015); *Hch. Halochromatium* (Imhoff *et al.,* 1998b; Imhoff and Caumette, 2015); *Isc. Isochromatium* (Imhoff *et al.,* 1998b; Imhoff, 2015); *Imf. Imhoffiella* (Nurpur *et al.,* 2017); *Lpb. Lamprobacter* (Gorlenko *et al.,* 1988; Gorlenko and Imhoff, 2015); *Lps. Lamprocystis* (Schroeter, 1886; Imhoff, 2015); *Mcr. Marichromatium* (Imhoff *et al.,* 1998b; Imhoff, 2015); *Phc. Phaeochromatium* (Sucharita *et al.,* 2010; Shivali *et al.,* 2012); *Phb. Phaeobacterium* (Nupur *et al,.* 2015); *Rbc. Rhabdochromatium* (Dilling *et al.,* 1995; Imhoff, 2015); *Tch. Thermochromatium* (Imhoff *et al,.* 1998b; Imhoff and Madigan, 2015); *Tac. Thioalkalicoccus* (Bryantseva *et al.,* 2000; Imhoff, 2015); *Tca. thiocapsa* (Winogradsky, 1888; Imhoff and Caumette 2015); *Tcs. Thiocystis* (Winogradsky, 1888; Imhoff and Caumette 2015); *Tco. Thiococcus* (Imhoff *et al.,* 1998b; Imhoff, 2015); *Tdc. Thiodictyon* (Winogradsky, 1888; Imhoff, 2015); *Thc. Thiohalocapsa* (Imhoff and Pfennig, 2001; Imhoff, 2015); *Tlp. Thiolamprovum* (Imhoff *et al.,* 1998b; Imhoff and Caumette, 2015); *Tpd. Thiopedia* (Winogradsky 1888; Imhoff, 2015); *Trc. Thiorhodococcus* (Guyoneaud *et al.,* 2015); *Trv. Thiorhodovibrio* (Overmann *et al.,* 1992; Imhoff and Overmann, 2015); *Tsp. Thiospirillum*) (Winogradsky 1888; Imhoff, 2015); *Tba. Thiobaca* (Rees *et al.,* 2002) and *Tpc. Thiophaeococus* (Anil Kumar *et al.,* 2008)

25. *Thiobaca* (*Tba.*) (Berry with sulfur) Rees *et al.* (2002)

Cells are rod-shaped, occurring as single cells or in pairs, dividing by binary fission. Motile by means of flagella. Gram (-) negative. The division is by binary fission. Vesicular type photosynthetic pigments with bacteriochlorophyll *a* and carotenoids of lycopene are present. Grows to occur photolitho heterotrophically or photoorganoheterotrophically under anoxic conditions with sulfide as an electron donor. Sulfur globules are accumulated transiently inside cells (**Fig. 3.80 A & B**). No growth occurs with hydrogen, thiosulfate and Fe^{2+} Optimum growth occurs at temperatures 25-30ºC (range up to 36ºC). The freshwater sediments are the ideal habitats for their growth. The mol% of the G+C content of DNA is 63. The type species is *Thiobaca trueperi*. The characteristic property of this species is presented in **Table. 3.39** .

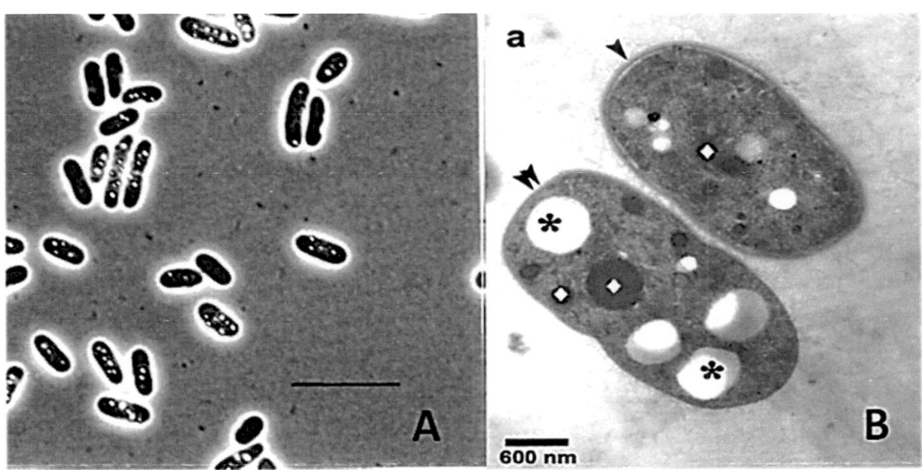

Fig. 3.80. A. Phase-contrast micrograph of cells of *Thiobaca trueperi* strain BCH^T containing refractile sulfur globules (B= 10 μm). B. Ultrathin structure of transversely sectioned cells, which contain electron-translucent poly-3-hydroxybutyrate inclusions (*) and dark inclusions of various sizes (□ .) Double arrows indicate a cell that has lacking S layer, whereas single arrow showing S layer of a second cell.

Differentiation of *Thiobaca* from other Genera

Characteristics for differentiation of *Thiobaca* from other *Chromatiaceae* members are given in **Tables 3.39** and **3.40**. The genetic relationship of *Thiobaca trueperi* to other members of *Chromatiaceae*, based on 16S rDNA gene sequence is depicted in **Figs. 3.46** and **3.81.**

Taxonomic Comments

Only one species is validly published according to Rees *et al.* (2002). The genetic relationship of *Thiobaca trueperi* to other members of *Chromatiaceae*, based on the 16S rRNA gene sequence is depicted in **Fig. 3.81.** They are closely related to *Thiocapsa roseopersicina* , *Thiocapsa rosea* and *Thiocapsa pendens* , *Lamprocystis roseopersicina* ,*Lamprocystis purpurea* and strain Thd2. Based on morphological (carotenoids, cell shape and substrate utilization) and phylogenetic analysis these genera was described. Based on morphological (carotenoids, cell shape and substrate utilization) and phylogenetic analysis, the new genus i.e., *Thiobaca* with species *Thiobaca trueperi* was described.

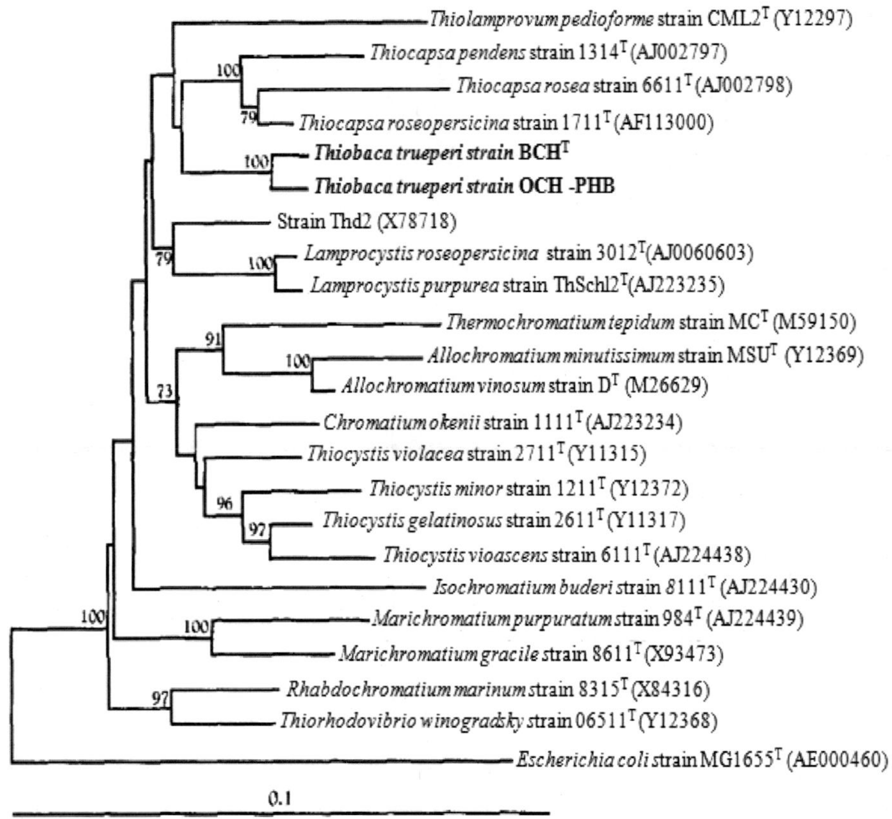

Fig. 3.81. Phylogenetic tree of *Thiobaca trueperi* from other genera of *Chromatiaceae* family based on 16S rRNA gene sequences. The sequence of *E.coli* was used as an outgroup to root the tree.

26. *Thiophaeococcus* (*Tpc.*) (Brown sulfur-reducing coccus) Anil Kumar *et al.* (2008)

Cells are spherical, multiply by binary fission and are motile by means of a polar flagellum. Gram (-) negative, Vesicular type internal photosynthetic membranes with photosynthetic pigments, bacteriochlorophyll *a* and carotenoids are present (**Fig. 3.82**). Obligately phototrophic and strictly anaerobic. During growth on reduced sulfur sources as electron donors, elemental sulfur is intermediately deposited as small granules within the cell. Simple organic substrates are photo assimilated in the presence of sulfide and bicarbonate. Salt and growth factors are required for growth. Anoxic sediment that is rich in organic matter and hydrogen sulfide exposed to light are ideal habitats. The Mol% G+C of DNA is 68–69 (HPLC). Two species are validly published i.e., *Thiophaeococcus mangrove* is the type species and *Thiophaeococcus fuscus*. The characteristic properties of these two species are presented in **Table 3.39** .

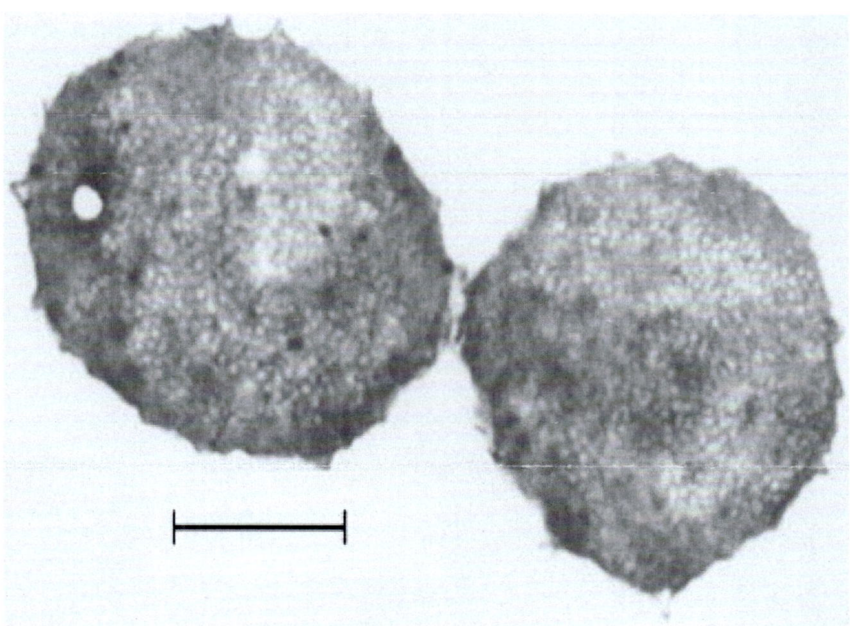

Fig. 3.82. Electron photomicrograph of an ultrathin section of a pair of cells of *Thiophaeococcus mangrove* showing the vesicular type of photosynthetic membranes extending throughout the cell (Bar=500 nm .)

Differentiation of *Thiophaeococcus* from other Genera

Characteristics for differentiation of *Thiophaeococcus* from other *Chromatiaceae* members are given in **Tables 3.39** and **3.40**. The genetic relationship of *Thiophaeococcus* to other members of *Chromatiaceae*, based on 16S rRNA gene sequence is depicted in **Fig. 3.83**.

Taxonomic Comments

The phylogenetic relationship between *Thiophaeococcus mangrovi* strain JA304^T and other purple sulfur bacteria based 16S rRNA gene sequencing revealed that the isolate belongs to the family *Chromatiaceae*(**Fig. 3.83**). The highest sequence similarities were observed between *Thiophaeococcus mangrovi* strain JA304^T and *Thiocystis violacea* (91%) and *Allochromatium vinosum* (90.5%). The similarity to other species of *Thiocystis, Chromatium, Allochromatium* and *Thermochromatium* was 86.7–89.6%. The similarity to other species of *Marichromatium* clade was 84.0–88.5%. The phylogenetic differences of *Thiophaeococcus mangrovi* from its closest relative, *Thiocystis violacea* are requirement for salt for growth spherical shaped cells, colour of the cell suspension, obligately anaerobic metabolism and the requirement for complex growth factors. Therefore, based on both phenotypic and genotypic evidence, this genus was created with *Thiophaeococcus mangrovi* as type strain.

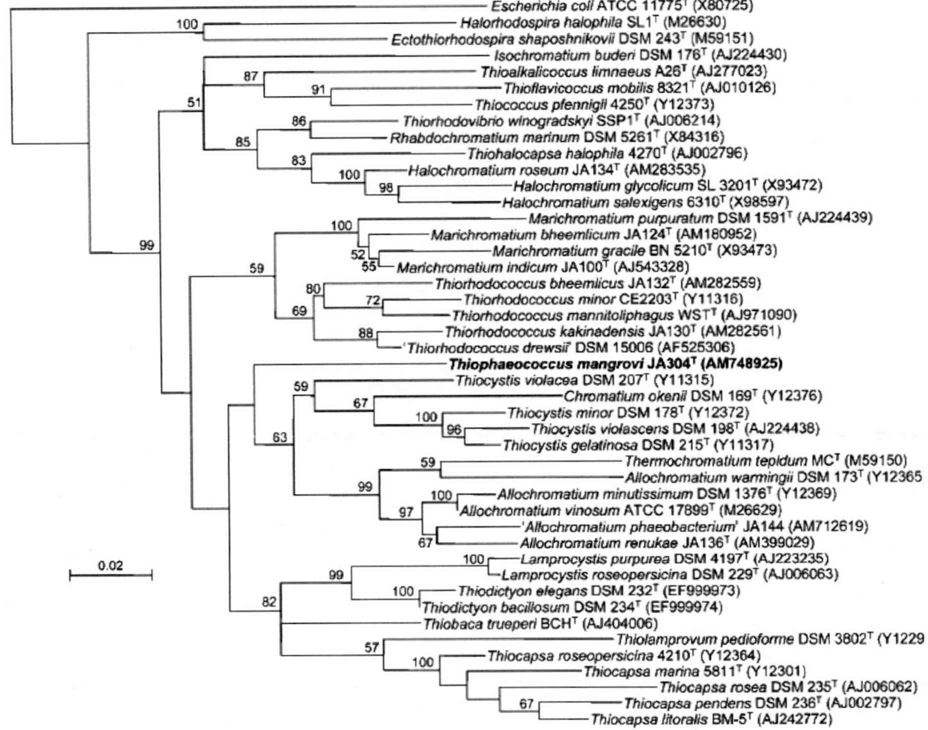

Fig. 3.83. Phylogenetic tree of *Thiophaeococcus* from other genera of *Chromatiaceae* family based on 16S rRNA gene sequences. The sequence of *E. coli* was used as an outgroup to root the tree.

Family 2. *Ectothiorhodospiraceae* (Imhoff, 1984b) Imhoff (2015)

The family is originally separated from *Chromatiaceae* to include phototrophic representatives that contain bacteriochlorophyll *a* or *b* and carotenoids as photosynthetic pigments. They are strictly chemotrophic bacteria. Cells are spiral or vibriod or rod-shaped bacteria, motile by means of polar flagella and divide by binary fission, with or without gas vesicles. Gram (-) negative, contains internal photosynthetic membranes that are continuous with the cytoplasmic membrane. Phototrophic representatives preferably grow anaerobically in the presence of light using reduced sulfur compounds as electron donors, oxidize sulfide to elemental sulfur (S^0), which is deposited outside the cells and eventually in the peripheral periplasmic space of the cell body, forming sulfate as the final oxidation product. Members of the *Ectothiorhodospiraceae* are found in marine to extremely saline environments containing neutral to extremely alkaline pH and containing sulfides such as estuaries, salt flats and soda lakes. Species of this family are the most halophilic eubacteria. The separation of *Ectothiorhodospira* species from *Chromatiaceae* based on oligonucleotide patterns of 16S rRNA molecules was justified.

Growth occur photoautrophically under anaerobic conditions with reduced sulfur compounds or hydrogen as electron donors or photoheterophically with a few simple organic compounds. Some species are able to grow under microaerobic to aerobic conditions in the dark. Growth factors are not required, but vitamin B_{12} enhances the growth of some strains. Storage products are polysaccharides, poly–β–hydroxybutyrate and polyphosphate. Studies on DNA-DNA and r-RNA-DNA hybridization analysis of lipopolysaccharides, DNA restriction patterns as well as lipid, fatty acid and quinone composition among the members of *Ectothiorhodospiraceae* revealed homogeneity in phylogenetic and taxonomic relations. The mol% G+C of the DNA is 50.5–69.7. *Ectothiorhodospira* (Pelsh, 1936) is the type genus.

Based on their phylogenetic analysis, different groups of sulfur or nitrite-oxidizing Chemolithotrophic bacteria and aerobic and facultatively anaerobic chemoheterotrophic bacteria were later classified within this family. Therefore, till date only 4 (*Ectothiorhodospira, Ectothiorhodosinus, Halorhodospira* and *Thiorhodospira*) out of 15 (*Ectothiorhodospira ,Alkalilimnicola, Alkalispirillum, Aquisalimonas, Arhodomonas, Ectothiorhodosinus, Halorhodospira, Methylonatrum, Natronocella, Nitrococcus, Thioalbus, Thioalkalivibrio, Thiohalomonas, Thiohalospira* and *Thiorhodospira*) genera are anoxygenic phototrophs. Common phenotypic properties are observed in many members of the family in their requirement for alkaline or saline or hypersaline conditions for growth.

Differentiation of Ectothiorhodospiraceae

Characteristics for differentiation of *Ectothiorhodospira* , *Ectothiorhodosinus* , *Halorhodospira* and *Thiorhodospira* are summarized in Table **3.41.** Characteristic fatty acid compositions of *Halorhodospira* and *Ectothiorhodospira* species are given in **Table 3.42**. The genetic relationships of *Ectothiorhodospiraceae* based on 16S rDNA sequences are presented in **Fig. 3.84** (Petri and Imhoff, 2000).

Table 3.41. Differential Characteristics of the Genera of the *Ectothiorhodospiraceae*

Characteristics	***Ectothiorhodospira***	***Ectothiorhodosinus***	***Halorhodospira***	***Thiorhodospira***
Gram stain	Negative	Negative	Negative	Negative
Motility	+	-	+	+
Flagellation	A polar tuft of flagella	+	Bipollar flagella	Polar tuft
Internal membranes	Lamellar stacks	Ring-shaped	Lamellar stacks	Lamellar stacks
Cell shape	Vibrioid to spirilloid	Vibriod-semicircular	Vibrioid to spirilloid	Vibrioid to spirilloid
Cell size (μm)	0.7-1.5 x 2.0-3.5	0.3-0.5 x 0.7-1.0	0.5-1.2	3.0-4.0
Bacteriochlorophyll	*a/b*	*a*	*a*	*A*

Characteristics	*Ectothiorho-dospira*	*Ectothiorho-dosinus*	*Halorhodospira*	*Thiorhodospira*
Carotenoids	Spirilloxanthin, dihydroxylycopene diglucoside diesters	Spirilloxanthin	Spirilloxanthin	Spirilloxanthin
Growth medium[c]	EM 5% or mMPM	EM 30%	EM 15-30%	-
Optimum salinity (range)	1.0-5.0	1-7%	15-25%	0.5-1.0%
Optimum pH (range)	7.0-9.5	8.3-9.1 (7.6-10.1)	8.5-9.0 (6.0-9.2)	9.0-9.5 (7.5-10.5)
Reaction to oxygen	Anaerobes, microaerobes	Anaerobes	Anaerobes	Anaerobes
Compatable solutes	Glycine betaine, sucrose	ND	Glycine betaine, ectoine, trehalose	ND
Thiosulfate consumption	-	- /+	+	+
Photoautotrophic growth on sulfide	)+(	+	)+(	+
Photoorganotrophic growth	-	-	+	+
Fatty acid clusters[d]	V-VII	ND	I-IV	ND
Major quinone components[e]	MK7 and Q7 or Q8	MK-4/5, MK-8, Q-8	MK8, Q8 and MK4/5	ND
Mol% G+C of DNA	61.4-68.4	57.5	50.5-69.7	56.0-57.4

Data are taken from Imhoff (2005) (*Ectothirhodospira*), Gorlenko *et al.* (2004) (*Ectothiorhodosinus*,) Imhoff (2005) (*Halorhodospira*) and Bryantseva *et al.* (1999b) (*Thiorhodospira*); ND, not determined/no data available; EM, standard medium for haloalkaliphilic *Ectothiorhodospiraceae* species according to Imhoff (1988a) with varying salt concentrations; mMPM: modified marine Pfennig' s medium (Truper, 1970;) Fatty acid clusters according to Thiemann and Imhoff (1996); Major quinone components according to Imhoff (1984a) and Ventura *et al.* (1993).

Taxonomic Comments

The taxonomic position of bacteria classified in *Ectothiorhodospiraceae* since their discovery by Pelsh (1936). He distinguished these bacteria, which he called *Ectothiorhodaceae*, from those phototrophic purple bacteria with elemental sulfur inside their cells, which he called *Endothiorhodaceae* (Pelsh, 1937). Pelsh's isolates were poorly characterized and are not obtained in pure cultures even lost soon after their isolation. Truper (1968) reisolated *Ectothiorhodospira mobilis* and Pfennig and Truper (1971c) included this genus as an exceptional genus in the *Chromatiaceae* because of its ability to perform an oxidative dissimilatory sulfur metabolism similar to other phototrophic purple sulfur bacteria. Because bacteria known at that time as *Ectothiorhodospira* species were significantly different from all other genera of *Chromatiaceae* and because representatives of the purple non-sulfur bacteria have been found to oxidize sulfide to extracellular elemental sulfur. Pfennig (1977) suggested the removal of *Ectothiorhodospira* from the *Chromatiaceae* and its placement in the purple non-sulfur bacteria. However, based on significant differences in their polar lipid composition, three major groups of phototrophic purple bacteria were distinguished by Imhoff *et al.* (1982) such as *Chromatiaceae*, the bacteria known at that time as *Ectothiorhodospira* and the purple non-sulfur bacteria. A clear differentiation of *Ectothiorhodospira* species from *Chromatiaceae* was later demonstrated by the oligonucleotide patterns of 16S rRNA molecules (Stackebrandt *et al.,* 1984). With the support of this information, *Ectothiorhodospiraceae* family was separated (Imhoff, 1984b). The separation of two groups of species within the family had been recognized not only based on physiological properties and lipid composition (Imhoff, 1984a) but also by similarities of rRNA oligonucleotide patterns (Stackebrandt *et al.,* 1984). However, a formal separation into genera was not proposed till date.

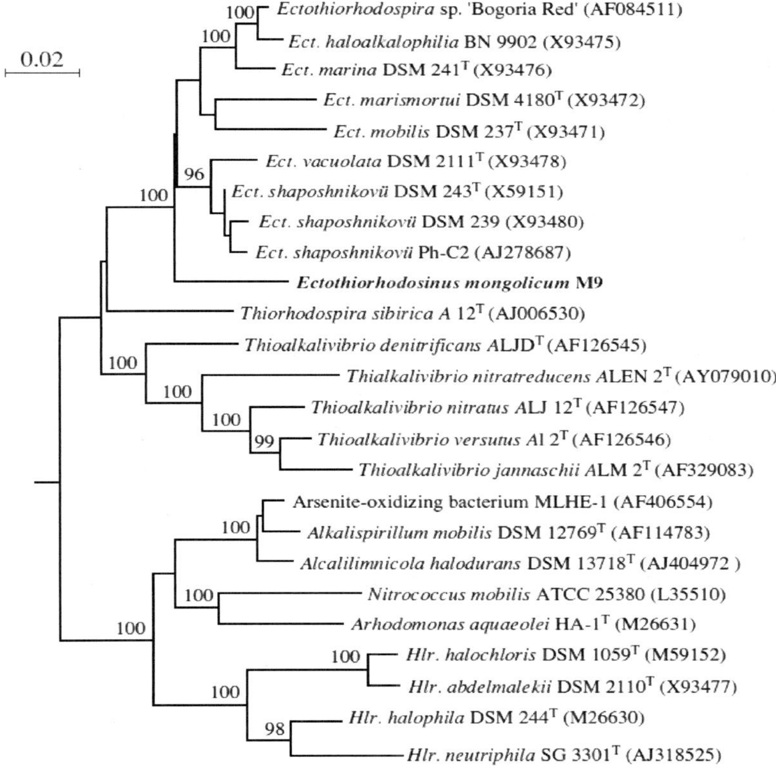

Fig. 3.84. 16S rDNA-derived phylogenetic tree of the family *Ectothiorhodospiraceae* including species of *Ectothiorhodospira ,Ectothiorhodosinus ,Halorhodospira , Thiorhodospira.* The sequence of *E.coli* was used as an outgroup to root the tree.

The complete sequence analysis of the 16S rDNA from a large number of strains, including all available type strains of *Ectothiorhodospiraceae*, not only supported their separation from the *Chromatiaceae*, but also the divergence of the two groups within this family (Imhoff and Suiling, 1996). Because of the large phylogenetic distance between these two groups, which is also reflected in a number of distinctive phenotypic properties, the extremely halophilic species were transferred to the new genus *Halorhodospira* (Imhoff and Suiling, 1996). With this taxonomic reevaluation of the *Ectothiorhodospiraceae*, their taxonomy now follows their phylogenetic relationships, based on 16S rDNA sequences. Two new isolates have expanded our knowledge of the *Ectothiorhodospiraceae* .

A new phototrophic sulfur bacterium isolated from a Siberian soda lake, *Thiorhodospira sibirica*, has been shown to be genetically related to the genus *Ectothiorhodospira*, and many of its properties resemble to those of *Ectothiorhodospira* species (Bryantseva *et al.,* 1999b). This impressively large bacterium is unique in that it forms S^0 globules outside the cells as well as inside. Sulfur globules are located in the cells only in the very peripheral part, which is represented by an large periplasmic space and they are not located in the cell interior. Therefore, the mechanism of S^0 formation and the location of the involved enzymes appear to be the same as in other *Ectothiorhodospiraceae* . However, one has to be cautious to distinguish between internal and external deposition of S^0 .

Another new halophilic and the strictly chemotrophic bacterium has been found to be genetically related to the *Ectothiorhodospiraceae*, in particular *to Halorhodospira* (Imhoff and Suiling, 1996). This bacterium, *Arhodomonas aquaeolei*, is a close, purely chemotrophic relative to the phototrophic *Ectothiorhodospiraceae*. It uses a restricted spectrum of carbon sources for growth. It is considered as a member of the *Ectothiorhodospiraceae* primarily because of its phylogenetic relationship by 16S rDNA sequence.

Key to the genera of the *Ectothiorhodospiraceae*

I. Halophilic *Gammaproteobacteria* that require >10% total salts for optimum growth.

- A. Cells contain photosynthetic pigments typical of phototrophic purple bacteria, perform a strictly phototrophic metabolism and require alkaline pH for optimum growth. .... ***Halorhodospira***

II. Halophilic phototrophic purple sulfur bacteria require alkaline pH and salt concentrations below 10% for optimum growth.

- A. S^0 globules are deposited exclusively outside the cells. .... ***Ectothiorhodospira***
- B. S^0 globules deposited both outside and inside the cells. .... ***Thiorhodospira***

III. Moderately halophilic and alkaliphilic purple sulfur bacteria, non-motile. Elemental sulfur globules are the intermediary products that are deposited outside the cells. .... ***Ectothiorhodosinus***

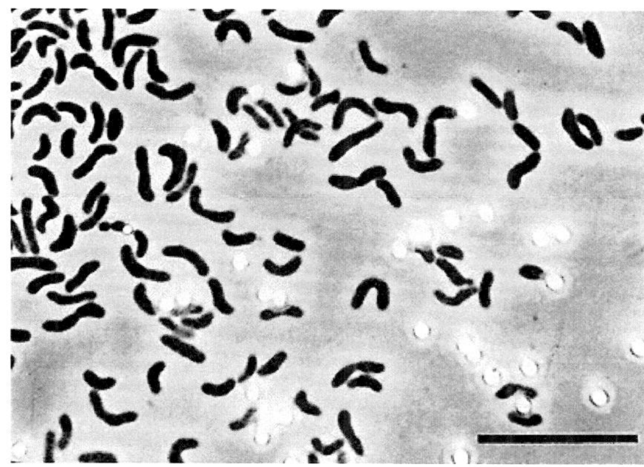

Fig. 3.85. Phase-contrast micrograph of *Ectothiorhodospira mobilis* DSMZ 237 showing extracellular sulfur globules (Bar= 10 μm).

1. *Ectothiorhodospira* (*Ect.*) (Spiral rose with sulfur outside) (Pelsh, 1936) Imhoff (2015)

Cells are rod-shaped or vibrioid, also appearing as true spirals, 0.7– 1.5 μm in diameter, motile by a polar tuft of flagella, multiply by binary fission, and may contain gas vesicles **(Fig. 3.85)** (Truper and Imhoff, 1977). Gram (-) negative, internal photosynthetic membranes are lamellar stacks that are continuous with the cytoplasmic membrane. Photosynthetic pigments are bacteriochlorophyll *a* and carotenoids. Grow photoautotrophically under anoxic conditions with reduced sulfur compounds or hydrogen as electron donors, or photoheterotrophically with a limited number of organic compounds. Sulfide oxidized to sulfate, with S^0, which is deposited outside the cells, as an intermediary product. Some species grow micro aerobically or aerobically in the dark. Sodium chloride is required for growth, which is dependent on saline and alkaline conditions. Compatible solutes include glycine, betaine, sucrose and N-carbamoyl-L-glutamine. Growth factors are not required, but vitamin B_{12} enhances the growth of some strains. Storage products are polysaccharides, poly-β-hydroxybutyrate and polyphosphate. Most species live in marine and saline environments that contain sulfide, have slightly too extremely alkaline pH, and are exposed to light, such as estuaries, salt flats, salt lakes and soda lakes, generally in soil. The mol% G + C of the DNA is 61.4–68.4. *Ectothiorhodospira mobilis* (Pelsh, 1936) is the type species. Six species are validly published and their characteristic features, carbon sources and electron donors used by different species of *Ectothiorhodospira* are summarized in **Table 3.43.** The genetic relationship of *Ectothiorhodospira* species based on 16S rDNA sequences is shown in **Fig. 3.84.**

Differentiation of *Ectothiorhodospira* from other Genera

Differentiation of *Ectothiorhodospira* and *Halorhodospira* species is based on both molecular and physiological properties (**Table 3.41** .) *Ectothiorhodospira* species contain MK-7 and either Q-7 or Q-8 as major components, whereas

Halorhodospira species lack significant proportions of homologs with 7 isoprenoid units, but have MK-8, Q-8, and a short-chain MK component as major components (Ventura *et al.,* 1993). Both genera differ in their fatty acid compositions (**Table 3.42**) (Thiemann and Imhoff, 1996) and 16S rDNA sequences by a number of signature sequences and sequence similarities of 87.2–89.9% (**Table 3.41**). In contrast to *Ectothiorhodospira* species that form sulfur globules outside their cells, *Thiorhodospira* forms sulfur globules enclosed by the outer membrane (Bryantseva *et al.,* 1999b).

Table 3.42. Fatty acid Clusters of *Ectothiorhodospiraceae* and *Halorhodospira* Species

Charact-eristics	*E. mobilis*	*E. haloa-lkalip-hila*	*E. mar-ina*	*E. mari-smor-tui*	*E. shapo-shni-kovii*	*E. vacu-olata*	*H. halop-hila*	*H. halo-phila*	*H. abde-lmal-ekii*	*H. halo-chloris*
Fatty acid cluster	V	VII	VII	V	VI	VI	I	II	III	IV
$C_{12:0}$	0.7-0.8	0.9-1.2	0.9	1.1	1.3-1.7	nd	2.4-3.4	2.8-5.0	2.4	1.1-2.9
$C_{16:0}$	35.1-36.4	29.0	24.8	33.2	244-26.4	21.6	9.2-11.1	10.8-14.5	21.4	21.0-24.0
$C_{16:1\ \omega 7c}$	2.7-3.5	4.8-5.1	5.9	3.8	4.1-4.6	3.6	0.4-0.6	0.6-0.8	1.4	1.2-1.6
$C_{18:0}$	5.7-7.6	6.8-7.1	9.2	5.0	5.6-6.8	5.5	14.2-17.7	11.3-13.2	6.0	6.3-11.6
$C_{18:1\ \omega 7c}$	12.1-16.2	38.2-38.6	29.2	18.5	57.1-59.0	65.3	62.8-65.5	58.6-62.8	62.8	57.7-57.9
$C_{19:0cyc}$	33.5-37.7	15.9-16.3	25.8	34.7	ND	ND	2.5-3.4	3.0-5.8	0.4	1.3-2.0

Symbols, ND, Not determine/No data available, [b]Data from Thiemann and Imhoff (1996)

Metabolic properties such as the utilization of carbon, sulfur, and nitrogen sources are very similar among the species of both genera. Several species of *Ectothiorhodospira* are able to grow under chemotrophic conditions in the dark, while *Halorhodospira* species are obligately phototrophic. *Ectothiorhodospira* species have growth optima at 1–8% total salts, whereas *Halorhodospira* species cannot grow below 10% total salts.

Table 3.43. Differential Characteristics of *Ectothiorhodospira* Species

Characteristics	*Ect. mob-ilis*	*Ect. haloal-kaliph-ila*	*Ect. ma-rina*	*Ect. marism-ortui*	*Ect. shapo-shnik-ovii*	*Ect. vacuo-lata*	*Ect. varia-bilis*	*Ect. salini*	*Ect. magna*
Habitats	Salt flats, Galapa gos Islands	Alkaline salt and soda lakes	Coa-stal marine sedi-ments	Hypersaline sulfur spring, Dead Sea, Israel	Alkaline salt lakes, Jordan, Kenya	Alkaline salt lakes, Jordan, Kenya	Soda lake, Wadi Natrun, Egypt	Solar saltern, India	Soda lake, Trans-baikal region, Russia
Cell diameter (μm)	0.7-1.0	0.7-1.2 x 2.0-3.0	0.8-1.2 x 1.5-4.0	0.9-1.3 x 1.5-3.3	0.8-9.0 x 1.5-2.5	1.5 x 2.0-4.0	0.8-1.2 x 1.2-1.5	1.0-1.5 x 2.0-3.5	2.0-3.2 x 9.6-2.0
Motility	+	+	+	+	+	+	+	+	+

Characteristics	*Ect. mobilis*	*Ect. haloalkaliphila*	*Ect. marina*	*Ect. marismortui*	*Ect. shaposhnikovii*	*Ect. vacuolata*	*Ect. variabilis*	*Ect. salini*	*Ect. magna*
Color of cell suspension	Red	Red	Red	Red	Red	Red	Red-purple	Reddish-brown	Red - brownish red
NaCl optimum)%(	2.0-3.0	5	2.0-6.0	3.0-8.0	3	1.0-6.0	2.0-2.0	5.0	0.5-1.5
Salinity range (%)	1.0-5.0	2.5-15	0.5-10.0	1.0-20	0-7.0	0.5-10.0	5.0-8.0	0.5-12.0	0.0-8.0
pH optimum (range)	7.6-8.0	8.5-10.0	7.5-8.5	7.0-8.0	8.0-8.5	7.5-9.5	9.0-9.5 (7.5-10.0)	7.5 (7.0-10.0)	9.0-10.0
Growth medium	mPF	mPF/EM5%	mPF	mPF	mPF/EM5%	mPF/EM5%	ND	ND	ND
Growth factors	Not required except some strains require B12	Not required	Not required	Not required	Not required	Not required	Not required	Para-aminobenzoate, pantothenate, pyridoxal phosphate	Not required except some strains require B12
Chemolithotrophic growth	-	+	ND	-	+	ND	-	-	-
Sulfate assimilation	+	+	(+)	-	+	-	-	-	ND
Major quinones	MK7/Q8	MK7/Q8	MK7/Q8	MK7/Q8	MK7/Q7	MK7/Q7	ND	MK-7, Q-7	MK-7, Q-7
Mol% G+C of DNA of the species	67.3-68.4	62.2-63.5	62.8	65.0	62.0-64.0	61.4-63.6	62.3-63.3	63	59.2
Mol% G+C of DNA of the type strain	67.3 (Bd)	63.5 (T_m)	62.8 (Bd)	65.0 (T_m)	62.0 (T_m)	63.6 (T_m)	-	-	-
Major fatty acids or a cluster of fatty acidsl	$C_{18:1}$, $C_{16:0}$, $C_{18:0}$ (V)	VII	ND	V	$C_{18:1}$, $C_{16:0}$, $C_{16:1}$ (VI)	VI	ND	$C_{18:1}$ ω7c, $C_{16:0}$, $C_{19:0}$ cyclo ω8c, $C_{16:1}$	$C_{18:1}$, $C_{16:1}$, $C_{16:0}$
Carbon sources/Electron donors									
Hydrogen	+	ND	ND	+	+	+	ND	-	+
Sulfide	+	+	+	+	+	+	+	-	+
Thiosulfate	+	+	+	-	+	+	+	-	-
Sulfur	+	+	+	+	+	+	+	-	+
Sulfite	+	ND	+	ND	+	ND	-	-	-

Characteristics	*Ect. mobilis*	*Ect. haloalkaliphila*	*Ect. marina*	*Ect. marismortui*	*Ect. shaposhnikovii*	*Ect. vacuolata*	*Ect. variabilis*	*Ect. salini*	*Ect. magna*
Acetate	+	+	+	+	+	+	+	+	+
Pyruvate	+	+	+	+	+	+	+	+	(+)
Propionate	±	ND	+	(+)	+	+	+	-	(+)
Butyrate	±	ND	-	-	+	-	-	-	-
Lactate	±	ND	+	(+)	+	-	+	-	-
Fumarate	+	+	+	+	+	+	+	-	+
Succinate	+	+	+	+	+	+	+	-	+
Malate	+	+	+	+	+	+	+	+	+
Fructose	±	-	-	-	+	(+)	-	-	-
Glucose	±	-	-	-	ND	-	-	-	-
Ethanol	-	-	-	-	-	-	-	-	-
Propanol	-	-	-	-	-	-	-	-	-

Symbol: +, characteristic positive or substrate utilized by ost strains; ±, characteristic positive or substarte utilized by some srains; -, negative or not utilized; ND, Not determine/ No data available; (+), weak growth. None of the species uses formate, methanol, ethanol, propanol, citrate or benzoate. MK-7, Menaquinone with 7 isoprene units; Q-7/8, quinone with 7/8 isoprene units. Data from Thiemann and Imhoff (1996) is taken for the composition of Cluster V, VI, and VII. Data is taken from Truper (1968) (*Ect. mobilis*), Imhoff and Suling (1996) (*Ect. haloalkaliphila*), Imhoff and Suling (1996) (*Ect. marina*), Oren *et al.* (1989), Ventura *et al.* (2000) (*Ect. marismortui*), Cherni *et al.* (1969) (*Ect. shaposhnikovii*), Imhoff *et al.*(1981), Ventura *et al.* (2000) (*Ect. vacuolata*), Gorlenko *et al.* (2009) (*Ect. variabilis*), Ramana *et al.*(2010) (*Ect. salini*), Bryantseva *et al.* (2010) (*Ect. magna* .)

Taxonomic Comments

After reevaluation of the taxonomy of *Ectothiorhodospiraceae* on the basis of 16S rDNA sequences, the extreme halophilic species were removed from *Ectothiorhodospira* and placed in a new genus *Halorhodospira* (Imhoff and Suiling, 1996). Analyses of 16S rDNA sequences and fatty acid compositions supported the identification and species assignments of some strains to species, but resulted misclassification of others (Imhoff and Suiling, 1996). A number of strains have been tentatively assigned to *Ect. mobilis*. Strains 8112 (DSMZ 237), 8113 (DSMZ 238), and 8815 (Mandel *et al.,* 1971) were confirmed as strains of *Ect. mobilis.* Strain 8115 (Mandel *et al.,* 1971) was recognized as *Ect. shaposhnikovii*. Strain 51/7 (BN 9903, ATCC 51935) and strain C (BN 9902) (Imhoff *et al.,* 1978) were recognized as a new species *Ect. haloalkaliphila* . Strain BA 1010 (DSMZ 241) (Matheron and Baulaigue, 1972) was found to represent the new species *E. marina*, and strain BA 1011 (Matheron and Baulaigue, 1972) was found to be most closely related to strain BA 1010. The recognition of strain DSMZ 239 as *Ect. shaposhnikovii* was close between *Ect. shaposhnikovii* and *Ect. vacuolata* and that between *E. mobilis* and *E. marismortui* are supported by the DNA–DNA hybridization (Ivanova *et al.,* 1985) and DNA restriction pattern analysis (Ventura *et al.,* 1993). These results were substantiated by DNA–DNA reassociation studies (Ventura *et al.*, 2000) and proposed the transfer of *Ect. vacuolata* to *Ect. shaposhnikovii* and *Ect.*

marismortui to *Ect. mobilis*. Therefore, this proposal cannot be accepted because of clearly distinctive phenotypic properties and ribotyping patterns are not available. In addition In addition, strain BN 9903 (ATCC 51935) has been intensively studied and, though it is known from the literature as *Ect. mobilis* (Zahr *et al.,* 1992; Ventura *et al.,*1993). Recent studies have led to this strain to be as the type strain of *Ect. haloalkaliphila* (Imhoff and Suiling, 1996).

2. *Ectothiorhodosinus* (*Ets.*) (Red curved rod with sulfur outside) Gorlenko *et al.* (2004)

Cells are vibrio- shaped or semi-circular and non-motile (**Fig. 3.86 A**). Gran (-) negative, contain internal photosynthetic membranes have the form of concentric lamellae lining the cell wall (**Figs. 3.86 B & C**). Photosynthetic pigments are bacteriochlorophyll *a* and carotenoids of spirilloxanthin. Growth occurs photolithoautotrophically, photolithoheterotrophically or photoorganoheterotrophically under anaerobic conditions in the presence of light with sulfide as an electron donor. Globules of elemental sulfur are formed as an intermediary product of sulfide oxidation and are deposited outside the cells. Simple organic compounds are photo assimilated in the presence of sulfide and carbonates. Growth depends on intermediate salinity and alkalinity. The mol% G + C of the DNA is 57.5. Till date single species is validly published i.e., *Ectothiorhodosinus mongolicum* (type species) (Gorlenko *et al.,* 2004) and its characteristics are presented in **Table 3.44.**

Differentiation of *Ectothiorhodosinus* from other Genera

Species of *Ectothiorhodosinus* are moderately halophilic and alkaliphilic as they grow with optimum pH 8.3-9.1. This species will grow with 0.5-9.0% salt concentrations. The differential characteristics of *Ectothiorhodosinus* species from other genera of *Ectothiorhodospiraceae* are precised in **Tables 3.41** and **3.44.** The genetic relationship of *Ectothiorhodosinus* species based on 16S rDNA sequences is shown in **Fig. 3.84.**

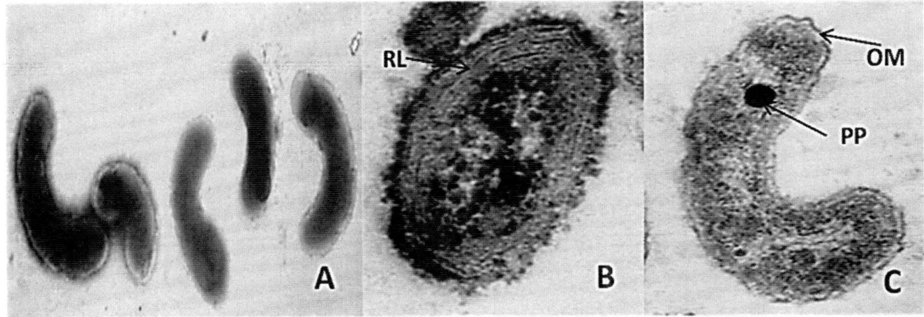

Fig. 3.86. A.Phase-contrast micrograph of vibriod-shaped or semi-circular cells of *Ectothiorhodosinus mongolicum* (Bar=5 μm), B & C. Ultrathin section of internal photosynthetic membrane showing ring-shaped lamellae (RL), Polyphosphates (PP), Outer membrane (OM).

Table. 3.44. Differential Characteristics of *Ectothiorhodosinus, Halorhodospira* and *Thiorhodospira* species

Characteristic	*Ectothiorhodosinus mongolicum*	*Hlr. halophila*	*Hlr. abdelmalekii*	*Hlr. halochloris*	*Hlr. neutrophila*	*Thiorhodospira sibrica*
Cell size (μm)	0.3-0.5 x 0.7-1.0	0.6-0.9 x 2.0-8.0	0.9-1.2 x 4.0-6.0	0.5-0.6 x 2.5-8.0	1.0-1.2 x 2.0-5.0	3.0-4.0
Motility	-	+	+	+	+	+
Color of cell suspension	Pink to brownish red	Red-purple	Pale green-green	Pale green-green	Pinkish-red	Brownish red to red
Bacteriochlorophyll	*a*	*a, c*	*b*	*b*	*a*	*A*
Carotenoids	Spirilloxant hin	Spirillo-xanthin	Dihydro-xylycopene diglucoside diesters	Dihydro-xyly-copene diglucosid e diesters	Spirilloxa nthin	Spirilloxanthi n
NaCl optimum (range) (%)	1.0-7.0 (0.5-9.0)	11.0-32.0 (3.0-30)	12.0-18.0 (5.0-30)	14.0-27.0 (10-35)	6.0-30.0 (9.0-12.0)	25.0-30.0 (5-80)
pH optimum (range)	8.3-9.1 (7.6-10.1)	8.5-9.0	8.0-9.2	8.1-9.1	6.8-7.0 (6.0-8.5)	7.5-10.5 (9.0-9.5)
Temperature optimum (range) °C	30-35	47	30-40	48	30-35	25-30
Sulfate assimilation	ND	-	-	+	-	ND
Major fatty acid cluster	ND	$C_{18:1}$, $C_{19:0\ cycl}$, $C1_{8:0}$, $C_{16:0}$ (II)	$C_{18:1}$, $C_{16:0}$, $C_{19:0\ cycl}$ (III)	$C1_{8:1}$, $C_{16:0}$, $C_{19:0\ cycl}$ (IV)	ND	ND
Major quinones	Q-8	MK8/Q8[a]	MK8/Q8[a]	MK8/Q8[a]	ND	ND
Mol% G+C of DNA of the species	57.5	66.5-69.7	63.3-63.8	50.5–52.9	74.1-74.5	56.0–57.4
Mol% G+C of DNA of the type strain	-	68.4 (Bd)	63.8 (T_m)	52.9 (T_m -	-	)
Carbon sources/Electron donors[b]						
Hydrogen	ND	ND	ND	ND	ND	ND
Sulfide	(+)	+	+	+	(+)	+
Thiosulfate	+	+	-	-	-	-
Acetate	+	+	+	+	+	+

Characteristic	*Ectothiorhodosinus mongolicum*	*Hlr. halophila*	*Hlr. abdelmalekii*	*Hlr. halochloris*	*Hlr. neutrophila*	*Thiorhodospira sibrica*
Pyruvate	+	+	+	+	+	+
Propionate	+	+	+	+	+	+
Lactate	+	±	-	-	-	-
Fumarate	+	+	+	+	+	+
Succinate	+	+	+	+	+	+
Malate	+	+	+	+	-	+

Symbol: +, characteristic positive or substrate utilized by ost strains; ±, characteristic positive or substarte utilized by some srains; -, negative or not utilized; ND, no data available; (+), weak growth. [a] In addition a short chain isoprenoid qinone is present in large amounts. [b] No species uses formate, butyrate, glucose, fructose, methanol, ethanol or propanol. Data for composition of cluster II, III and IV id taken from Thiemann and Imhoff (1996). Data is taken from Raymond and Sistrom (1967, 1969), Imhoff (2005) (*Halorhodospira halophila*), Imhoff (2005), Imhoff and Truper (1981) (*Hlr. abdelmalekii*), Imhoff (2005), Imhoff and Truper (1977) (*Hlr. halochloris*), Hirschler-Rea *et al.* (2003) (*Hlr. neutrophila*), Bryantseva *et al.* (1999) (*Thiorhodospira sibrica)* and Gorlenko *et al.* (2004) (*Ectothiorhodosinus mongolicum* .)

Taxonomic Comments

After reevaluation of the taxonomy of *Ectothiorhodospiraceae* based on a complete 16S rDNA sequence, the moderate halophilic species were removed from *Ectothiorhodospira* and assigned to a new genus *Ectothiorhodosinus* (Gorlenko *et al.,* 2004). Only one species is retained in this genus which is isolated from Steppe soda lake, Mongolia. Based on phylogenetic analysis of 16S rDNA gene sequences this species belongs to the family *Ectothiorhodospiraceae* (**Fig. 3.84**). A new separate branch showed 93.9-96.3% sequence similarity with *Ectothiorhodospira* and 93.2% with separately branched *Thiorhodospira sibirica* . The DNA G+C content of *Ectothiorhodosinus mongolicum* strain M9 was 57.5 mol%. This value is lower than that typical of the family *Ectothiorhodospiraceae*, where it varies within the range of 61.4–74.5 mol(%**Table 3.41**), with the exception of the extremely halophilic species *Halorhodospira halochloris* (50.5-52.9 mol%) and *Thiorhodospira sibirica* (56–57.4 mol%). *Ectothiorhodosinus mongolicum*has differed in the phenotypic characters such as a number of unusual phenotypic features such as morphology; the absence of flagellum assisted motility, ring-shaped lamellae as photosynthetic membranes, the size of the peripheral antenna of pigment-protein complex which is 2 or 3 times smaller in other purple bacteria, poor photoautotrophic growth led to the proposal of a new genus i.e., *Ectothiorhodosinus* and the type species is *Ectothiorhodosinus mongolicum.*

3. *Halorhodospira* (*Hlr.*) (The spiral rose from salt lakes) (Imhoff and Suiling, 1997) Imhoff (2015)

Cells are spirals or rods, 0.5–1.2 μm in diameter, motile by means of bipolar flagella, multiply by binary fission **(Fig. 3.87)**. Gram (-) negative. Internal photosynthetic membranes appear as lamellar stacks that are continuous with the cytoplasmic membrane. Photosynthetic pigments are bacteriochlorophyll *a* or *b* and carotenoids. Grow photoautotrophically under anoxic conditions with reduced sulfur compounds as electron donors, or photoheterotrophically with a limited number of simple organic compounds. Sulfide oxidized to S^0, which is deposited outside the cells and may be further oxidized to sulfate. Growth requires high saline and alkaline conditions. The mother then 10% (w/v) of salt concentration is required by all known species. Some of the species grow in saturated salt solutions. Glycine, betaine, ectoine and trehalose accumulated as compatible solutes in response to salt and osmotic stress. Growth factors are not required. Storage products are polysaccharides, poly-β-hydroxybutyrate and polyphosphate. *Halorhodospira* species are found in hypersaline and extremely saline environments with slight to extreme alkaline pH (up to pH 11–12) that contain sulfide and are exposed to light, such as salt flats, salt lakes and soda lakes. The mol% G + C of the DNA is 50.5–69.7. *Halorhodospira halophile* (Raymond and Sistrom, 1969) is the type species (Imhoff and Suiling, 1997). Distinguishing characters of properties for differentiation of *Halorhodospira* species are summarized in **Table 3.44.**

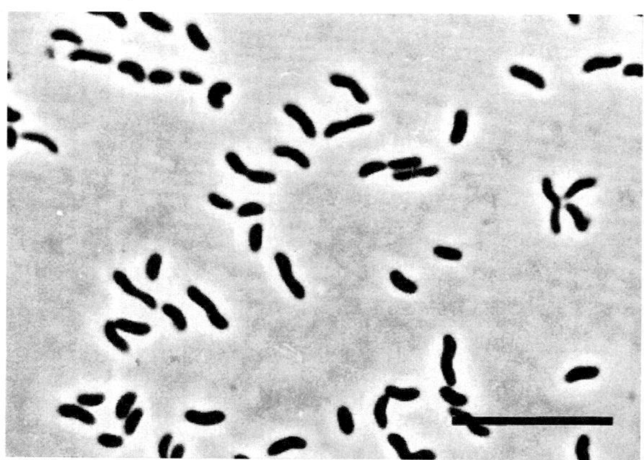

Fig. 3.87. Phase-contrast micrograph of *Halorhodospira halophila* BN 9621 (Bar=10 µm).

Differentiation of *Halorhodospira* from other Genera

Species of *Halorhodospira* represent the most halophilic species and are well adapted to form massive blooms in hypersaline environments and alkaline soda lakes. *Halorhodospira* species will not grow below 10% salt concentrations, whereas *Ectothiorhodospira* species grow well below 10% salt concentration. Species of *Halorhodospira* are easily differentiated from other purple sulfur bacteria of *Chromatiaceae* by the deposition of S^0 outside their cells when grown with sulfide as the photosynthetic electron donor. The differential characteristics of *Halorhodospira* species from other genera are precised in **Tables 3.41** and **3.44.** The genetic relationship of *Halorhodospira* species based on 16S rDNA sequences is shown in **Fig. 3.84.**

Taxonomic Comments

After reevaluation of the taxonomy of *Ectothiorhodospiraceae* based on a complete 16S rDNA sequence, the extremely halophilic species were removed from *Ectothiorhodospira* and assigned to a new genus *Halorhodospira* (Imhoff and Suiling, 1996). Two different but closely related groups of *Hlr. halophila* strains (12 strains included in the study) have been recognized based on fatty acid composition (Thiemann and Imhoff, 1996), 16S rDNA sequences (Imhoff and Suiling, 1996) and DNA restriction pattern analysis (Ventura *et al.,* 1993). This diversity was considered to be intraspecies variation, most likely related to different salt concentration responses. The mol% G+C content of the DNA of *Hlr. halochloris* is surprisingly low when compared to that of two other species of this genus. Such a difference of nearly 20% within the same genus is unusual, particularly when given similar sequences of the 16S rRNA gene. However, the mol% G+C values in **Table 3.44** have been confirmed by two independent analyses of a total of three strains.

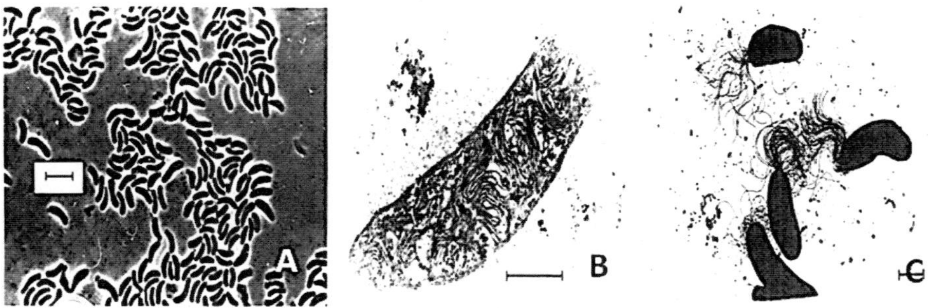

Fig. 3.88. A. Phase contras photomicrograph of cells of *Thiorhodospia sibirica* grown photoautotrophically after consumption of elemental sulfur (Bar = 20 μm). B. Ultrathin section of *Trs. sibirica* showing lamellar type internal membrane structure (Bar = 2 μm), C. Negatively stained cells showing the polar tuft of flagella (Bar = 3 μm).

4. *Thiorhodospira* (*Trs.*) (Spiral rose with sulfur) (Bryantseva *et al.,* 1999) Imhoff and Gorlenko (2015)

Cells are vibrioid or spiral-shaped (**Fig. 3.88 A**), motile by means of a monopolar flagellar tuft and multiply by binary fission (**Fig. 3.88 C**). Gram (-) negative contains internal photosynthetic membranes that are parallel to lamellae penetrating the cytoplasm lengthwise or underlying the cytoplasmic membrane (**Fig. 3.88 B**). Bacteriochlorophyll *a* and carotenoid of spirilloxanthin series are the photosynthetic pigments. Obligate phototrophs and strictly anaerobic. During photolithoautotrophic growth with sulfide as the electron donor, globules of elemental sulfur (S^0) are formed outside the cytoplasm in the medium attached to cells or in the periplasm. The final oxidation product is a sulfate. Simple organic substances may be photo assimilated in the presence of sulfide. Growth will dependent on sodium salt concentrations and alkaline conditions. The surface of sediments rich in organic matter and microbial mats from brackish soda lakes that contain hydrogen sulfide is the ideal substrates for their growth. The mol% G + C of the DNA is 56.0–57.4 (Tm). *Trs. sibirica* (Bryantseva *et al.*, 1999) is the type species. Only one species is validly published in this genus and their characteristics are presented in **Table 3.44.**

Differentiation of *Thiorhodospira* from other Genera

Thiorhodospira species can be distinguished from other purple sulfur bacteria based on the deposition of elemental sulfur (S^0) both inside and outside the cells. It can be differentiated from *Ectothiorhodospira* and *Halorhodospira* on the basis of sulfur deposition both outside and inside the cells. The irregular arrangement of the membrane stacks of internal photosynthetic membranes, the low mol% G+C content and differences in the 16S rDNA gene sequence are precised in **Tables 3.41** and **3.44.**

Taxonomic Comments

Based on 16S rDNA gene analysis analyses *Thiorhodospira* shows 88–92% similarity to those of *Ectothiorhodospira* species. Metabolic characteristics such as the utilization of carbon, sulfur and nitrogen sources are very similar to other members of *Ectothiorhodospiraceae* . *Thiorhodospira* species grow at salt concentrations between 0.5 and 8.0% as same as *Ectothiorhodospira* species. The genetic relationship of *Thiorhodospira* species based on 16S rDNA sequences is shown in **Fig. 3.84.**

3.3. GREEN SULFUR BACTERIA (Imhoff, 2014) Overmann (2015)

The green sulfur bacteria (GSB) are predominantly aquatic bacteria that grow photosynthetically under anoxic conditions. Similar to *Chromatiaceae* (PSB) and *Proteobacteria* (PNSB), the photosynthetic metabolism of GSB varies from that of Cyanobacteria, algae and green plants in which the water cannot serve as an electron donor and molecular oxygen is not generated. The external electron donors for photosynthetic CO_2 assimilation have low redox potentials and comprise reduced sulfur compounds and molecular hydrogen. GSB can be distinguished from the *Chromatiaceae* by their distinct bacteriochlorophyll (Bacteriochlorophyll *a*), the presence of special light-harvesting structures (chlorosomes) and particularly photo lithotrophic metabolism.

The green sulfur bacteria are distantly related to other bacteria and represent the phylum *Chlorobi*, though the known representatives are taxonomically treated as *Chlorobiaceae* with retained genera *Ancalochloris, Chlorobium* , *Chlorobaculum* , *Prosthecochloris, Pelodictyon* and *Chloroher-peton.* The *Chlorobi* share a common root with *Bacteroides* in both of the major reference trees. The phylum contains a single class, order and family. They are Gram (-) negative, spherical, ovoid, straight, or curved rod-shaped bacteria. Obligately phototrophic and strictly anaerobic. Cells grow preferentially by simple organic compounds which are photo assimilated in the presence of light with sulfide and bicarbonates. Some species may utilize sulfide or thiosulfate as an electron donor for CO_2 assimilation. Acetate is incorporated by reductive carboxylation which requires the reducing equivalents from sulfide and involves a reduced ferredoxin. Most strains reach light saturation of growth at 10 $\mu mol.m^{-2}.5^{-1}$ (daylight fluorescence tubes). Light inhibition has been observed at intensities above 200 $\mu mol.m^{-2}.5^{-1}$.

Photolithoautotrophic growth occurs with sulfide and sulfur as electron donors. During oxidation of sulfide, elemental sulfur (S°) globules are accumulated outside the cells. The sulfur may be further oxidized to sulfate. Ammonia and dinitrogen are used as nitrogen sources. Molecular nitrogen is fixed by many strains. Hydrogen sulfide containing mud and freshwater, brackish water and marine environments are the ideal habitats for their growth. The mol% G+C of DNA is 49.0-58.1 (Mandel *et al.,* 1971). *Chlorobium* is the type genus (Nadson 1906).

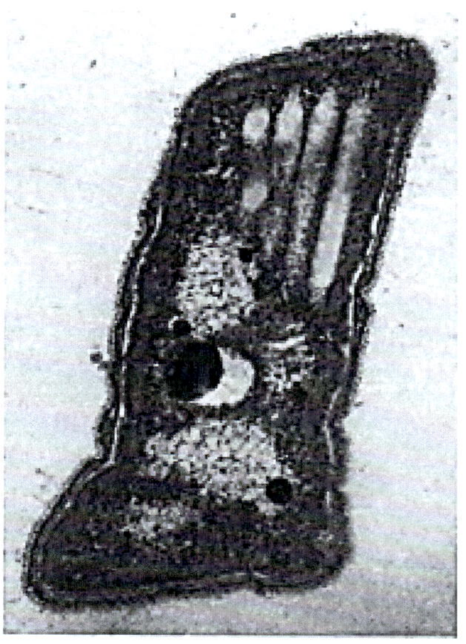

Fig. 3.89. Electron micrograph (X 105,000) of the structure of *Pelodictyon clathratiforme*

Class I. *Chlorobia*, Order I. *Chlorobiales*, Family I. *Chlorobiaceae* (Copeland, 1956) Overmann (2015)

The gram (-) negative cells are spherical, ovoid, straight, or curved rod-shaped, multiply by binary fission or binary plus ternary fission. In *Chloroherpeton*, the cells form long unicellular filaments, highly flexible and motility occurs by gliding movement, while all other genera are non-motile. Six of the fourteen species of this group possess gas vesicles. Photosynthetic pigments are located in the cytoplasmic membrane and in the light-harvesting structures such as chlorosomes (previously *Chlorobium* vesicles). The latter is an ovoid structure that measures 70-180nm long and 30-60 nm wide, that underlie and are attached to the cytoplasmic membrane **(Fig. 3.89)** (Cohen-Bazire *et al.,* 1964). The chlorosomes envelopes with width 2-3 nm consist of monogalactosyl-diglycerides and several proteins. Chlorosomes are also found in green non-sulfur bacteria such as *Chloroflexus* and *Oscillochloris* which are phylogenetically distinct.

The existing species shows one of two distinct colours i.e., either green (grass green) or brown (chocolate-brown). All the strains that appear green in colour are due to the presence of bacteriochlorophyll *c* or *d* **(Table 3.45)** as the major component and small amounts of bacteriochlorophyll *a.* The major carotenoids of the green species are chlorobactene in addition to OH-chlorobactene and γ-carotene. Whereas all the strains that exhibit brown colour are due to the presence of bacteriochlorophyll *e* as the major component as well as small amounts of bacteriochlorophyll *a.* The carotenoids of the brown species are isorenieratene and β-isorenieratene. The members of *Chloroherpeton* contain γ-carotene as the main carotenoid.

Table 3.45. Nomenclature of Bacteriochlorophylls of Green Bacteria

Bacteriochlorophyll	Characteristic Absorption Maxima in Living Cells (nm)
a	375, 590, 805
c	335, 460, 745-755
d	325, 450, 710-740
e	345, 450-460, 700-710

Antenna bacteriochlorophylls are mainly of farnesol esters. In each species, several bacteriochlorophyll homologs are present which varies for aliphatic side chains attached to the tetrapyrroles (methyl, ethyl, n-propyl, isobutyl, or neopentyl). A high degree of alkylation occurs in cells grown at low light intensities (Huster and Smith, 1990). A minor fraction of the bacteriochlorophylls contains distinct esterifying alcohol. In general, upto 15 different bacteriochlorophyll homologs can be observed in one strain (Borrego and Garcia-Gil, 1994). All chlorosomes contain about 10,000 molecules of bacteriochlorophyll *c, d,* or *e* , small amounts of bacteriochlorophyll *a* and three isoprenoid quinones (1'-oxomenaquinone-7, a derivative thereof and menaquinone-7) and is energetically connected to 5-10 reaction centers. The size of the photosynthetic antennae of green sulfur bacteria greater than that of *Chromatiaceae* .

These bacteria are different from all purple bacteria, which carry out autotrophic CO_2 assimilation by the reductive tricarboxylic acid cycle (Fuchs *et al.,* 1980) that requires less ATP per molecule of CO_2 fixed than the Calvin cycle. This may moderately explain why GSB exhibits a much lower maintenance energy requirement than *Chromatiaceae* (Overmann *et al.,* 1992a). The lowest light intensity ($<0.25\ \mu mol.m^{-2}.5^{-1}$) organism supporting the growth of a *Chlorobium phaeobacteroides* MN1 isolated from the chemocline of the Black Sea (Overmann *et al.,* 1992a). Furthermore, to reduced sulfur species, molecular hydrogen is utilized as an electron donor several GSB. Though, due to the lack of assimilatory sulfate reduction, a reduced sulfur source is required during growth conditions. The bond for sulfide of GSB is one degree higher than that of purple sulfur bacteria, while the bond for acetate is one degree lower. Vitamin B_{12} is required several strains of all species. In its absence, the specific bacteriochlorophyll content is highly reduced and cells are inadequate to form chlorosomes (Fuhrmann *et al.,* 1993).

Based on the presence or absence of gas vacuoles and selective enrichment cultures condition, two subgroups can be differentiated based on physiological and ecological properties among the GSB. The first group includes members of *Chlorobium* and *Prosthecochloris* which do not contain gas vacuoles and are selectively grow in liquid cultures with high sulfide concentration (4-8 mM) and light intensities (100 μmol.m^{-2}.5^{-1}). The second subgroup includes the gas vacuole containing species such as *Pelodictyon clathratiforme ,Pld. phaeoclathratiforme* and *Ancalochloris perfilievii* which can grow successfully only at very low sulfide concentrations (0.4-2 mM), low incubation temperatures (10-20°C) and low light intensities (5-20 μmol.m^{-2}.5^{-1}). The mol% G+C of the DNA is 47.8-58.1 (Bd). *Chlorobium* is the type genus. The phylogenetic differentiation of genus *Chlorobium* and other genera are illustrated in **Fig. 3.90.** The major differentiating characteristics of genera of green sulfur bacteria are presented in **Table 3.46** .

Molecular Characterization

To date, green sulfur bacteria have been divided into seven genera and 19 species. Among them, the species *Clathrochloris sulfurica* was eliminated in the first edition of the Bergey's Manual (Pfennig, 1989) because of its morphology (trellis-shaped cell aggregates and formation of gas vesicles) which resembles closely that of *Pelodictyon luteolum.* Based on their physiological capacities are too similar to allow differentiation among the taxa, the six genera *Ancalochloris, Chlorobium, Chlorobaculum, Chloroherpeton, Pelodictyon* and *Prosthecochloris* have been classified according to their morphologies, pigment compositions and formation of gas vesicles.

Table 3.46. Characteristic Properties of Genera and Species of Green Sulfur Bacteria

Genus and Species Name	Motility	Cell size (µm)	Cell shape	Colour of cell suspension	Absorption spectra of living cells	Salt required (%) (range)	Vitamins	Major Bchl	G + C values	Gas vesicles
Chlorobium* - *Chl	–	0.3-1.2 x 0.4-1.3	Ovoid, straight/ curved rods	Green/ Brown	ND	+/-	B_{12}	*c, d, e, (a)*	46.7–58.1	+/–
Chl. limicola	–	0.7-1.1 x 0.9-1.5	Rods	Green	ND	Not required	B_{12}	*c/e, (a)*	51.0–52.5	–

Genus and Species Name	Motility	Cell size (μm)	Cell shape	Colour of cell suspension	Absorption spectra of living cells	Salt required (%) (range)	Vitamins	Major Bchl	G + C values	Gas vesicles
Chl. phaeobacteroides	–	0.6-0.8 x 1.03-2.7	Curved Rods	Yellowish-reddish brown	Longer wavelenght at 725	1.0	B_{12}	*e, (a)*	49.0–50.0	–
Chl. clathratiforme	–	0.7-1.2 x 1.5-3.0	Rods	Green/ Brown	ND	Not required	B_{12}	*c/e, (a)*	48.0–49.0	+
Chl. ferrooxidans	–	0.5 x 1.0-1.5	Rods	Brown	335, 457 & 756	Not required	ND	*c, (a)*	ND	–
Chl. luteolum	–	0.6-0.9 x 1.2-2.0	Curved rods/ vibriods	Yellowish green	ND	> 1.0	B_{12}	*c,d, (a)*	53.5–58.1	+/–
Chl. phaeovibrioides	–	0.3-0.4 x 0.7-1.4	Curved rods/ vibriods	Yellowish-reddish brown/ choclate brown	ND	> 1.0	B_{12}	*c,d.e, (a)*	52.0–53.5	+
Chl. chlorochromatii	–	0.5 x 2.7	Rods	Yellowish green	433, 453, 662 & 748	Growth inhibited at ≥ 0.8	B_{12}	*c, (a)*	46.7	–
Prosthecochloris - Ptc.	–	0.5–0.8	Spherical-ovoid, vibrioid/ rod	Green/ Brown	ND	> 1-	B_{12}	*c,d.e, (a)*	50–56.1	–
Ptc. aestuarii	–	0.5-0.7 x 1.0-1.2	Spherical/ vibrioid	Green	740-750	2.0-5.0 (1-8.0)	B_{12}	*c, (a)*	52–56.1	–
Ptc. vibrioformis	–	0.5-7.0 x 1.0-2.0	Curved rods/ vibriods	Green/ Brown	ND	> 1.0	B_{12}	*c,d,e, (a)*	53.5	+/–
Ptc. indica	–	0.8-1.0 x 1.0-1.2	Spherical-oval	Dark green	335, 433, 458, 463, 490, 746 & 805	2.0-5.0 (0.5-7.0)	YE	*c,(a)*	53.0	–
Pelodictyon-Pld.	–	–	Rod-ovoid	Green/ Brown	ND	–	B_{12}	*c,d or e (a)*	48.5-58.1	–
Pld. phaeum	–	0.6-0.9 x 1.0-2.0	Rods/ curved rods	Choclate brown	ND	3.0	B_{12}	*e*	ND	+
Chlorobaculum - Cba.	–	0.3–1.1	Vibrioid-rod	Green/ Brown	ND	0-3.0	B_{12}	*c,d,e*	54–58.1	–
Chlorobaculum tepidum	–	0.6–0.8 x 1.3-2.6	Straigt/ Curved rods	Dark green	460 & 753	Not required	B_{12}	*c(a)*	56.5-58.2	–
Cba. limnaeum	–	0.6–0.8 x 1.3-2.7	Rods	Brown	ND	Not required	B_{12}	*e, (c, a)*	49.0	–

Genus and Species Name	Motility	Cell size (μm)	Cell shape	Colour of cell suspension	Absorption spectra of living cells	Salt required (%) (range)	Vitamins	Major Bchl	G + C values	Gas vesicles
Cba. thiosulfatophilum	–	0.7–1.1 x 0.9-1.5	Rods	Green	ND	Not required	B_{12}	*c*	58.1	–
Cba. parvum	–	0.5-0.7x 1.0-1.2	Curved rods or vibrioid	Green	ND	> 1.0 (2.0)	B_{12}	*d*	56.1–56.6	–
Cba. chlorovibrioides	–	0.3-0.4 x 0.4-0.8	Curved rods or vibrioid	Green	ND	2.0-3.0	B_{12}	*d*	54.0	–
"Cba. macestae"	-	0.6-0.9 x 1.8-4.2	Straigt/ Curved rods	Green	435, 460, 489, 660 & 736	Not required (tolerate upto 2)%	B_{12}	*c*	58.3	–
Anacalochloris-Anc	-	0.5-1.0	Irregular star	Green/ Yellowish green	ND	ND	B_{12}	*a*	ND	+
Anc.perfilievii	-	0.5-1.0	Irregular star	Green/ Yellowish green	ND	ND	B_{12}	*a*	ND	+
Chloroherpeton - Chp.		8.0-20 x 1.0	Long rods	Green	ND	1.0-2.0	B_{12}	*c*	45.0-48.2	+
Chp. thalassium		8.0-20 x 1.0	Long rods	Green	ND	1.0-2.0	B_{12}	*c*	47.8	+

Continued

Genus and Species Name	Major carotenoids	Thiosulfate	Optimum temperature °C (range)	Optimum pH (range)	Carbon sources/ electron donors utilized	Substrates not utilized	Major fatty acids (% of total)		
							14:0	16:0	16:1
Chlorobium – Chl	Chlorobactene, isorenieatene	+/-	20-32	ND	Simple organic substrates/ Sulfite, Bio-carbonates, Sulfur, Thiosulfate/ Hydrogen	-	10–16	15–29	47–64
Chl. limicola	Chlorobactene, isorenieatene	+/-	25-35	6.5-7.0	Sulfide, Biocarbonates Sulfur, Hydrogen, Acetate, Propionate	Thiosulfate, Ethanol, glutamate, Fructose & higher fatty acids	13	17	57
Chl. phaeobacteroides	Isorenieratene and β-isorenieratene	-	20-7.3	(6.5-7.3)	Sulfide, Biocarbonates Sulfur, Hydrogen, Acetate, Fructose	Thiosulfate, Alcohols, Pyruvate, Propionate & Higher fatty acids	16	15	64

Genus and Species Name	Major carote-noids	Thio-sul-fate	Optimum tempe-rature °C (range)	Optimum pH (range)	Carbon sources/ electron donors utilized	Substrates not utilized	Major fatty acids (% of total)		
							14:0	16:0	16:1
Chl. clathra-tiforme	Chloro-bactene, isoren-ieatene	+/-	25-30	6.7-7.0	Sulfide, Sulfur, Biocarb-onates, Acetate, Fructose	Hydrogen, Propionate, Methane, Pyruvate & Higher fatty acids	ND	ND	ND
Chl. ferro-oxidans	Chlorobac tene	ND	22-25 (18-30)	6.5-6.7 (6.3-7.1)	Hydrogen, Ferrous iron, Acetate	Sulfide, Thisulfae/ elemental sulfur & other organic substrates	ND	ND	ND
Chl. luteolum	Chlorobac tene	-	20-25	(6.5-7.0)	Sulfide, Sulfur, Biocarbonates, Acetate, Propionate & some simple compounds	Thiosulfate, Pyruvate, Succinate, Peptone & Higher fatty acids	14	21	47
Chl. Phaeovi brioides	Chlorobac tene, isorenieat ene	+/-	20-30	(6.5-7.3)	Sulfide, Bicarbonates, Acetate, Propionate	ND	10–12	23–29	51–52
Chl. chloroc hromatii	γ-caro-tene, OH-γ-carotene glucoside laurate	-	25	7.0-7.3	Sulfide, Ferrous iron, Acetate, Peptone	Thiosulfate, Sulfur, Hydr-ogen, Alanine, Benzoate, Butyrate, Caprylate, Citrate, For-mate, Fuma-rate, Fructose, Glucose, Malate, Propi-onate, Pyru-vate, Casa-mino acids & other substrates	ND	ND	ND
Prosthe coch-loris - Ptc.	Chloro-bactene, isorenieat ene	-	-	-	Sulfide, Elemental sulfur, Bio-carbonates & Simple organic substrates	Thiosulfate	ND	ND	ND
Ptc. aestu-arii	Chloro-bactene, rhod-opin, or lycopene or OH deriva-tives.	-	-	6.7-7.0	Sulfide, Elemental sulfur, Biocar-bonates, Acetate, Glutamate & Pyruvate	Thiosulfate & other organic compounds	ND	ND	ND

Genus and Species Name	Major carote-noids	Thio-sul-fate	Optimum tempe-rature °C (range)	Optimum pH (range)	Carbon sources/ electron donors utilized	Substrates not utilized	Major fatty acids (% of total)		
							14:0	16:0	16:1
Ptc. vibrio-formis	Chloro-bactene, isorenieat ene	-	25-35	6.5-7.3 (6.0-7.5)	Sulfide, Sulfur, Hyd-rogen, Bicar-bonates & some simple organic compounds	Thiosulfate, Ethanol, Succinate or Higher fatty acids	ND	ND	ND
Ptc. indica	Chloro-bactene	-	30 (25-35)	6.8-7.2 (6.3-7.7)	Sulfide, Bicarbonates, Acetate, Pyruvate & Glutamate	Thiosulfate, Lactate, Ben-zoate, Fruc-tose, Malate, Mannitol, Citrate, Casa-mino acids, Propionate, Melanoate, Glycolate, Asparagine, Valerate, Cap-roate, Fuma-rate, Succinate, Tratarate, Glu-cose, Glycerol, Ethanol & Methanol	ND	ND	ND
Pelod-ictyon- Pld.	Chlorobac tene, isorenieat ene	-	(15-30)	ND	Sulfide, Sul-fur, Bicarbo-nates & Sim-ple organic compounds	ND	ND	ND	ND
Pld. phaeum	Isoreniera tene	-	25	7.0	Sulfide, Sul-fur, Acetate	ND	ND	ND	ND
Chloro-bacu-lum - Cba.	Chlorobac tene, iso-renieatene	+	ND	ND	Sulfide, Ele-mental sulfur, Thiosulfate, Biocarbonates & Simple organic substrates	ND	21-25	10–25	43–52
Chloro-baculum tepidum	Chlorobac tene	+	47-48 (32-52)	6.8-7.0	Sulfide, Thiosulfate, Acetate, Pyruvate	Sugars and other organic or fatty acids (including propionate).	ND	ND	ND
Cba. limna-eum	Isoreniera tene, β-isoren-ieratene,	-	20-30	6.7-7.0	Sulfide, Sulfur, Hydrogen, Bicarbonates, Acetate, Fructose & some simple organic compounds	Thiosulfate, Propionate, Pyruvate & other organic compounds	ND	ND	ND

Genus and Species Name	Major carote-noids	Thio-sul-fate	Optimum tempe-rature °C (range)	Optimum pH (range)	Carbon sources/ electron donors utilized	Substrates not utilized	Major fatty acids (% of total)		
							14:0	16:0	16:1
Cba. thiosulfatophilum	Chlorobactene	+	20-30	ND	Acetate, Propionate, Pyruvate, Glutamate, Fructose	ND	21	10	43
Cba. parvum	Chlorobactene	+	25-35	ND	Sulfite, Sulfide, Sulfur, Thiosulfate, Hydrogen, Biocarbonates, Acetate, Propionate & some organic compounds	ND	24	23	43
Cba. chlorovibrioides	Chlorobactene	-	20-30	6.8	Sulfide, Sulfur, Bicarbo*-nates, Formate, Aceate & Propionate	Thiosulfate, Methanol, Ethanol & Glutamate	ND	ND	ND
''Cba. macestae''	Chlorobactene	+	25-28	6.8 (6.5-7.2)	Sulfur, Thiosulfate, Elemental sulfur, Acetate	ND	16	25	50
Anacalochloris-Anc	Sphaeroidene	-	18-20	6.8-7.0	Sulfide & some organic compounds	ND	ND	ND	ND
Anc. perfilievii	Sphaeroidene	-	18-20	6.8-7.0	Sulfide & some organic compounds	ND	ND	ND	ND
Chloroherpeton - Chp.	γ-carotene, OH-γ-carotene glucoside laurate	ND	25-30	6.8-7.2	Sulfide, Bicarbonates, Acetate, Propionate, Malate, Succinate or Glutamate	Glucose, Fructose or Casamino acids	ND	ND	ND
Chp. thalassium	γ-carotene, OH-γ-carotene glucoside laurate	ND	25-30	6.8-7.2	Sulfide, Bicarbonates, Acetate, Propionate, Malate, Succinate or Glutamate	Glucose, Fructose or Casamino acids	ND	ND	ND

Data for *Chlorobium* (Nadson 1906) is taken from Nadson (1906) (*Chl. limicola*); Pfennig (1968) (*Chl. Phaeobacteroides & Chl. phaeovibrioides*); Szafer (1911) & Imhoff (2003); (*Chl. clathratiforme*); Schmidle (1901) & Imhoff (2003) (*Chl. luteolum*); Heising *et al.* (1999) (*Chl. ferrooxidans*); Vogl *et al.* (2006) (*Chl. chlorochromatii*). Data for *Anacalochloris* is taken from Gorlenko and Lebedeva (1971) (*Anc. perfilievii.*) Data for *Chloroherpeton* is taken from Gibson *et al.* (1984) (*Chloroherpeton thalassium*). Data for *Prosthecochloris* is taken from (*Pld. phaeum*); Gorlenko (1970) (*Ptc. aestuarii*); Pelsh (1936) (*Ptc. vibrioformis*); Anil Kumar *et al.* (2009) (Ptc. indica). Data for *Chlorobaculum* is taken from Imhoff (2003) (*Chlorobaculum*), Keppen *et al.* (2008) (*Cba. macestae*), Wahlund *et al.* (1996) (*Cba. tepidum*) Pfennig (1968) & Imhoff (2003) (*Cba. limnaeum*), Imhoff (2003) (*Cba. thiosulfatophilum & Cba. parvum*), Gorlenko *et al.* (1974) & Imhoff (2003) (*Cba. chlorovibrioides .*)

A phylogenetic tree of different strains of GSB, representing 19 different species and 6 different genera has been recently constructed based on their 16S rRNA sequences (**Fig. 3.90**) and *fmo A* (Fenna–Matthews–Olson protein) gene sequences (**Fig. 3.91**) including essential signatures of amino acids of the FMO protein and nucleotides (Alexander *et al.,* 2002). This is quite remarkable and strongly supports the 16S rRNA-based phylogeny of these bacteria. The results of these sequence A (Alexanalyses with most of the strains contained in the Pfennig collection clearly revealed that almost similar grouping in constructed trees from 16S rRNA and *fmo A* genes and suggested a highly compatible evolution of both FMO and 16S rRNander *et al.,* 2002). The corresponding phylogenetic relationship found within these two genes gave strong support for the phylogenetic taxonomy of the green sulfur bacteria (Imhoff, 2003). Their phylogenetic relationship as illustrated in **Fig. 3.90** shows the 16S rDNA-based relationships between type strains and a larger number of supplementary cultures of the GSB (Imhoff and Thiele, 2010). Among six genera, *Ancalochloris perfilievii, Chlorobium limicola* (Nadson, 1906), *Pelodictyon phaeum* (Gorlenko 1972), *Prosthecochloris aestuarii* (Gorlenko, 1970), *Chloroherpeton thalassium* (Gibson *et al.,* 1985) and*Chlorobaculum tepidum* (Alexander *et al.,* 2002), six major groups of GSB are representing the six currently recognized genera *Ancalochloris, Chlorobium ,Pelodictyon, Prosthecochloris ,Chloroherpeton* and *Chlorobaculum .Prosthecochloris* and *Chloroherpeton* are true marine bacteria. *Chloroherpeton thalassium* is highly different from other species with <87% similarity based on 16S rRNA gene sequences (Gibson *et al.,* 1984) and represents the earliest divergence in the phylogenetic tree.

The phylogenetic grouping of the GSB is not in agreement with their current classification or with the current designation of species to genera and of strains to species. Based on three examples phylogenetic tree was constructed, (i) The *Chlorobium* species appear in all major groups and those of the genus *Pelodictyon* in two of them, (ii) The strains of different species, including *Chlorobium limicola* , *Chlorobium phaeobacteroides* , *Chlorobium vibrioforme* and *Chlorobium phaeovibrioides*, appear in at least two of the major groups, (iii) *Chlorobium limicola* subsp. *thiosulfatophilum* and *Chlorobium vibrioforme* sub sp. *thiosulfatophilum*, are described by their type strains which are different from their reference species at the genus level.

From the analysis of all available information, it becomes clear that besides 16S rRNA and other gene sequences, the DNA G+C content is appropriate for species identification. Currently, it is difficult to find phenotypic properties that can be determined easily and that are in agreement with the genetic relationship among the green sulfur bacteria. The essential properties required for the taxonomic classification include the salt concentration, the composition of lipids and fatty acid analysis (Imhoff, 1988; Imhoff and Bias-Imhoff, 1995). Therefore, the properties that have been used so far for the classification are pigment composition, presence of gas vesicles, thiosulfate utilization and the formation of cell aggregates are evidently haveless importance in a phylogenetic system of the GSB.

Similarity values of >90.1%. K_{max} values of <0.11 indicate close phylogenetic relatedness among all strains. The 16S rDNA sequence of *Chloroherpeton thalassium* was recently determined and it has been included in the phylogenetic tree presented in **Fig. 3.90**. Within the green sulfur bacterial group, *Chp. thalassium* is the most isolated branch. The GSB represents an isolated group within the eubacterial radiation, confirming earlier results of ribosomal oligonucleotide cataloging (Woese *et al.,* 1985b). Consequently, the family *Chlorobiaceae* should be maintained as a coherent systematic group (Truper and Pfennig, 1971). Strains that have been combined in the species of *Chl. limicola* or *Chl. vibrioforme* clearly represents different taxa. This confirms the genetic heterogenicity of *Chl. limicola*, in which the mol% G+C varies between 49-58.1 and total genomic size ranges from 1435 to 3342 kb (Mendez-Alvarez *et al.,* 1996). Subsets of closely related strains of *Chl. limicola* has been described. Also, the genus *Pelodictyon* will have to be redefined. The DNA base composition of the *Chlorobiaceae* ranges from 45 to 58 mol% G+C of DNA with clearly different ranges in the four genera i.e., in *Chloroherpeton* it is 45.0–48.2 mol%, in *Chlorobium* 47.9–58.1 mol%, in *Chlorobaculum*, 54.0–58.1 mol% and in *Prosthecochloris* 52.0–56.1 mol% (Overmann, 2001).

Based on phylogenetic heterogenicity, a smaller number of strains analyzed so far, it appears premature to establish a new classification system for GSB at this point. Additional nucleic acid analyses are required in order to avoid further rearrangements. The previous arrangement of genera and species of the GSB, which was based on simple phenotypic characteristics (Pfennig, 1989) has therefore been kept in the second edition of Bergey's Manual of Systematic Bacteriology. According to 16S rRNA oligonucleotide catalogs, *Chloroherpeton* is phylogenetically more closely related to *Chlorobium* of the green sulfur bacteria than to the multicellular filamentous gliding bacterium *Chloroflexus* . The phylogenetic relationships of the species and genera of the green sulfur bacteria among themselves are not known. The taxon *Chlorobiaceae* is not used here because an emended description of the family is required. Such a description should be based on nucleic acid analyses in order to avoid further rearrangement. The present arrangement of genera and species of the green sulfur bacteria is based on simple phenotypic characteristics as in the eighth edition of Bergey's manual. Nucleic acid studies for determining phylogenetic relationships are lacking.

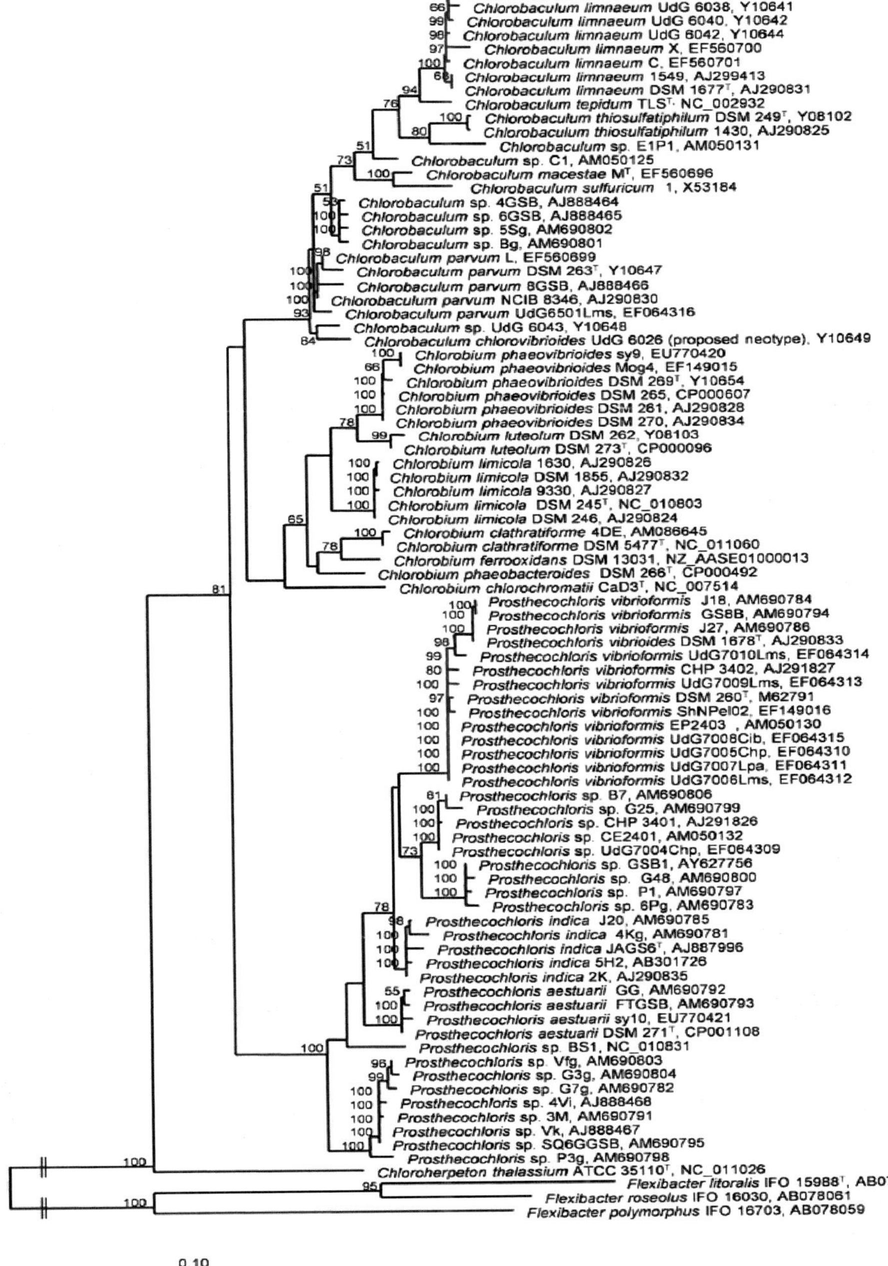

Fig. 3.90. Phylogenetic tree of the green sulfur bacteria based on 16S rRNA gene sequences showing the relationships within the family *Chlorobiaceae* calculated by the maximum likelihood method (Imhoff and Thiel, 2010).

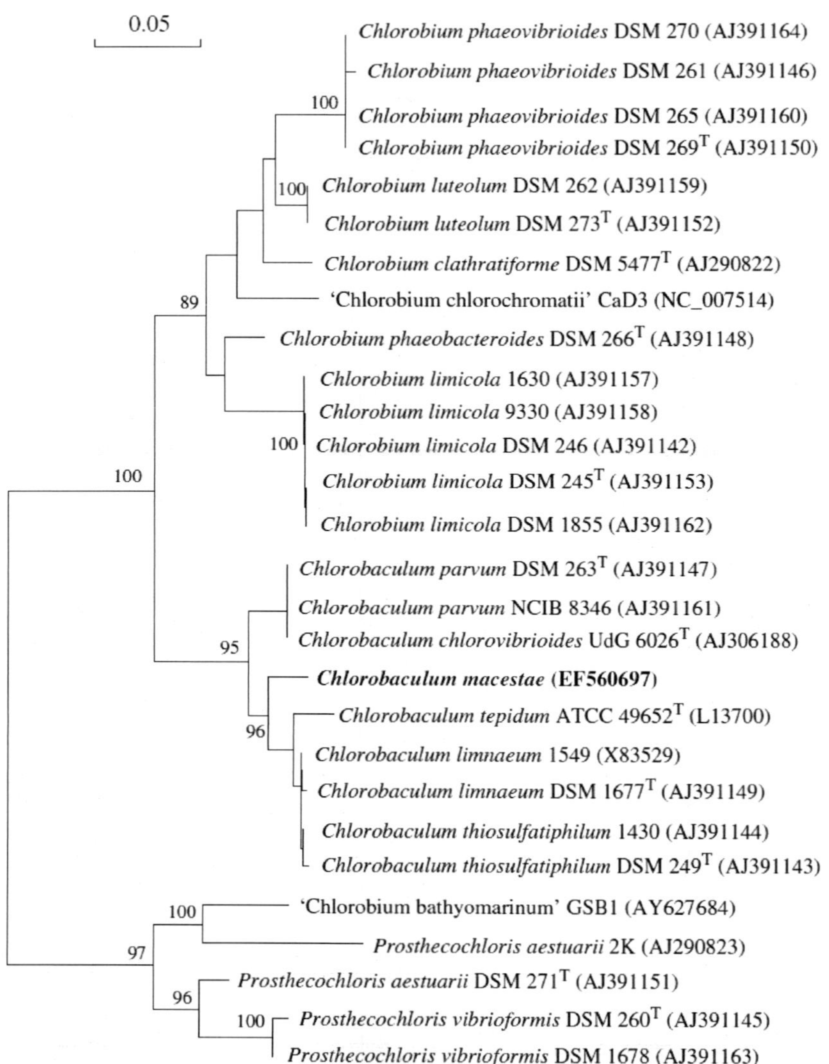

Fig. 3.91. Phylogenetic tree of the green sulfur bacteria based on analyses of the amino acid sequences of FMO protein. A gene sequence showing the relationships within the family *Chlorobiaceae* calculated by the maximum likelihood method (Keppen *et al.*, 2008).

Quinones of Green Sulfur Bacteria

A unique set of menaquinones with seven isoprene units, MK-7 and 1'-oxo-MK-7 also called chlorobiumquinone, are determined in *Chlorobium limicola*. Both quinones were also recorded in strains of *Chlorobium luteolum* strain 2530, *Prosthecochloris vibrioformis* strain 6030 and *Chlorobaculum parvum* strain

8327 appear to be characteristic components of the family *Chlorobiaceae* (see Imhoff and Bias-Imhoff, 1995).

Polar Lipids of Green Sulfur Bacteria

The members of the *Chlorobiaceae* family contain considerable amounts of glycolipids. More than 80% was found in association with the chlorosomes, whereas phospholipids were found to be dominant in the cytoplasmic membrane. At least five different glycolipids were found in *Chlorobaculum parvum* (previously *Chlorobium vibrioforme* f. *thiosulf.* strain 8327=DSM 263), *Chlorobaculum thiosulfatophilum* (strain 6230= DSM 249) and *Chlorobium limicola* (strain 6430). Two of these were similar to monogalacto-syldiacylglycerol (MGDG) and glycolipid II. Another two have been described as lipid 1 and lipid 2 were found in Chlorobaculum *parvum* (strain 8327) and *Chlorobaculum thiosulfatophilum* (strains 6230 and 2230) (Imhoff and Bias-Imhoff, 1995). Lipid 1 appeared to be an amino glycol sphingolipid with carbohydrate components as neuraminic acid and the main fatty acid as myristic acid. These lipids were exclusively found in the cytoplasmic membrane but not in the chlorosomes of *Chlorobium* and *Chlorobaculum* members. It was therefore proposed that the lipid might operate as an anchor for the hydrophilic bacteriochlorophyll *a*-protein, which forms a two-dimensional crystal between the plasma membrane and the chlorosomes. Neither bacteriochlorophyll *a*-protein nor lipid 1 could be detected in *Chloroflexus*, which shows significant differences in the functioning of the photosynthetic apparatus in both groups of bacteria (Imhoff and Bias-Imhoff, 1995). Lipid 2 is a poorly characterized lipid of distinct nature and with atypical staining reactions, for which the presence of mannose and two unidentified sugars was proposed. One or two minor glycolipids with chromatographic properties resembling but not similar to the sulfolipid sulfoquinovosyl diacylglycerol were found in GSB. In fact, the absence of the sulfolipid was confirmed in many *Chlorobiaceae* such as *Chlorobaculum parvum* (strain 8327=DSM 263), *Chlorobaculum thiosulfatophilum* (strain 6230=DSM 249) and *Chlorobium limicola* (strain 6430) by radioactively labelled sulfur isotopes in cysteine and thiosulfate cultured experiments (Imhoff and Bias-Imhoff, 199).

However, cardiolipin (CL) and phosphatidylglycerol (PG) were commonly identified phospholipids of GSB. Phosphatidyl ethanolamine (PE) has been recorded only in few strains (*Chlorobaculum parvum* strain 8327=DSM 263, *Chlorobaculum thiosulfatophilum* strain 6230=DSM 249 and *Chlorobium limicola* strain 6430), but not recorded in other members of *Chlorobiaceae* (Imhoff and Bias-Imhoff, 1995). In the end, the polar lipid composition of GSBis characterized by the presence of PG, CL, mono-galactosyl diacylglycerol (MGDG), glycolipid II, lipid 2 and the amino glycosphingolipid (lipid 1), while the presence of PE may be the main feature for the differentiation among the species and phosphatidylcholine, phosphatidylinositol and sulfolipids are absent in members of *Chlorobiaceae*. Undoubtedly, more studies are needed on the lipid composition of the different species of *Chlorobiaceae* .

Fatty Acids of Green Sulfur Bacteria

Quite characteristic for green sulfur bacteria is the presence of straight-chain, saturated and monounsaturated fatty acids with large portions of C-14:0, C-16:0 and C-16:1 with only small portions to traces of C-18:0, C-14:1 and C-18:1.In particular, the presence of large amounts of myristate C-14:0, a major component of the amino glycosphingolipid in *Chlorobium* is characteristic of green sulfur bacteria. In addition variable amounts of C-17: cy fatty acid and hydroxy fatty acids (3-OH-14:0; 3-OH-16:0) may be present (Imhoff and Bias-Imhoff, 1995; Imhoff, 2014).

Genomic Information and Differentiation from Other Bacteria

The first complete genome sequence of a GSB was published from thermophilic *Chlorobaculum tepidum* which is 2.15Mbpin size comprises a single, circular DNA molecule and contains two rRNA operons, 2288 coding sequences, 50 t-RNAs and two smalls stable RNAs (Eisen *et al.,* 2002). More than half of the encoded proteins (53%) are functional in known ones and the remaining functions are unknown. However, more than 15 genome sequences from at least 10 species of GSB mostly type strains have been sequenced completely and have given information regarding biochemical and physiological features of this group of bacteria (Bryant *et al.,* 2012). The comparison of these genome sequences showed a large number of common genes (approximately 50–90%), which shows similar physiological and biochemical characteristics of all the GSB (Bryant *et al.,* 2012). The formation of groups and distinction of genera within the GSB is mainly based on phylogenetic analyses using 16SrRNA and *fmo A* gene sequences (Imhoff, 2003) highly supported by the comparison of the genomes (Bryant *et al.,* 2012). For instance, the high similarity in gene content between *Chlorobaculum tepidum* and *Chlorobaculum parvum* and between *Chlorobium phaeovibrioides* and *Chlorobium luteolum* are highlighted (Bryant *et al.,* 2012). The separate phylogenetic position of *Chloroherpeton thalassium* is also depicted in the genome which shares <50% of its genes with other *Chlorobiaceae* members.

Furthermore, the members of *Chlorobiaceae* are distinct from the phototrophic bacteria of other phyla by several characteristic properties such as the structure and properties of photosynthetic pigments and photosynthetic apparatus. However, chlorosomes are also present in the family *Chloroflexaceae* and the recently discovered *Chloroacidobacterium thermophilum*, they differ in their structure and function of the different groups (Bryant *et al.,* 2007). Furthermore, different types of photosystems are present. A type-I reaction center involving the reduction of ferredoxin in *Chlorobiaceae* and *Chloroacidobacterium* and a type-II reaction center (RC) involving electron transfer through membrane-bound quinones is present in *Chloroflexaceae*. Although, the biosynthesis of bacteriochlorophyll *c* in *Chloroflexus aurantiacus* and *Chlorobaculum tepidum* appears to be relevant pathways. The FMO protein is lacking in *Chloroflexus* and also the chlorosomes proteins of both types of bacteria share an only allow level of homology (Gregersen *et al.,* 2011). Moreover, *Chloroflexus* and *Chlorobiaceae* have two different pathways for autotrophic CO_2-fixation and several other contrasting physiological properties (Bryant *et al.,* 2012).

For a long time, no chemotrophic relative had been distinguished indicating even slow association with *Chlorobiaceae.* This included first environmental metagenome studies using specific primers for the 16S rRNA genes of green sulfur bacteria (Alexander *et al.,* 2002). Recently, chemotrophic bacteria have been recognized, which are phylogenetically identical to the *Chlorobi,* in all probability at a phylum level. These bacteria form a profound branching new lineage which is recommended to represent a novel phylum, the *Ignavibacteriae* (Kadnikov *et al.,* 2013). *Ignavibacterium album* is a metabolically adaptable chemoorganotrophic bacterium, lacking photosynthesis and sulfur oxidation and its 16S rRNA gene sequence is just 77–83% identical to that of *Chlorobiaceae* (Liu *et al.,* 2012). Additionally, *Melioribacter roseus* is a facultatively anaerobic organotrophic bacterium, lacking photosynthetic capabilities and the autotrophic CO_2 fixation pathway (reductive TCA) of GSB (Kadnikov *et al.,* 2013). The new bacteria from different phylogenetic clusters outside the class *Chlorobia* in which still no chemotrophic representative has been identified. Based on metagenomic and metatranscriptomic studies, evidence has been provided for the existence of a further aerobic bacterium, *Candidatus Thermochlorobacter aerophilum*, which is supposed to be a photoheterotrophic bacterium belonging to the *Chlorobi* phylum (Liu *et al.,* 2012), but to date, it is not available in pure culture.

1. *Chlorobium* (Green life) (Nadson, 1906) Pfennig and Overmann (2015)

Cells are spherical, ovoid, straight or curved, rod-shaped bacteria, 0.3-1.1 μm wide and 0.4-3.0 μm long or sometimes much longer. Cells are also formed in united chains which are similar to Streptococci or filaments. Curved rod-shaped strains may form long spirals. Multiply by binary fission, non-motile and are Gram (-) negative. The cell cultures exhibit one or two distinct colours, either green (grass-green) or brown (chocolate-brown). Photosynthetic pigments are located in the cytoplasmic membrane and the chlorosomes which underlie are attached to the cytoplasmic membrane. Bacteriochlorophyll *c, d,* or *e* occur as major photosynthetic pigments in addition to a small amount of bacterio-chlorophyll *a.* Chlorobactene or isorenieratene are the major carotenoids. Obligately anaerobic and phototrophic. Growth occur photolitho autotrophically in the presence of light with sulfide or sulfur as an electron donor. During sulfide oxidation, globules of elemental sulfur (S°) are formed outside the cells **(Fig. 3.92).** The sulfur may be further oxidized to sulfate. In sulfide-reduced media, thiosulfate may be used as an electron donor. A number of simple organic substances can be photo assimilated in the presence of reduced sulfur compounds and bicarbonate. Ammonia is used as a nitrogen source and molecular nitrogen is fixed by many strains. Growth occurs at optimum temperature is 20°C -35°C (range 5°C – 40°C) except for *Chlorobium tepidum* (temperature range 32-52°C). Storage materials are polyphosphate and polysaccharides. The growth occurs at optimum pH 6.8 (range 6.5-7.2). Vitamin B_{12} is required for its growth. Hydrogen sulfide containing mud and water of freshwater, brackish water and marine environments are ideal substrates for their growth. The mol% of G+C of the DNA is 49.0-58.1 (Bd). A total of eight species is validly published, from which *Chlorobium vibrioforme* (Pelsh, 1936) is transferred to *Prosthecochloris vibrioformis* (Imhoff, 2003), while *Chlorobium chlorovibrioides* (Gorlenko *et al.*, 1974) and *Chl. Tepidum* (Wahlund *et al.*, 1996) were transferred to a new genus *Chlorobaculum chlorovibrioides* (Imhoff, 2003)and *Cba. tepidum* (Imhoff, 2003). *Chlorobium limicola* is the type species (Nadson, 1906). The distinguishing characteristics of the species of *Chlorobium* are presented in **Table 3.46.**

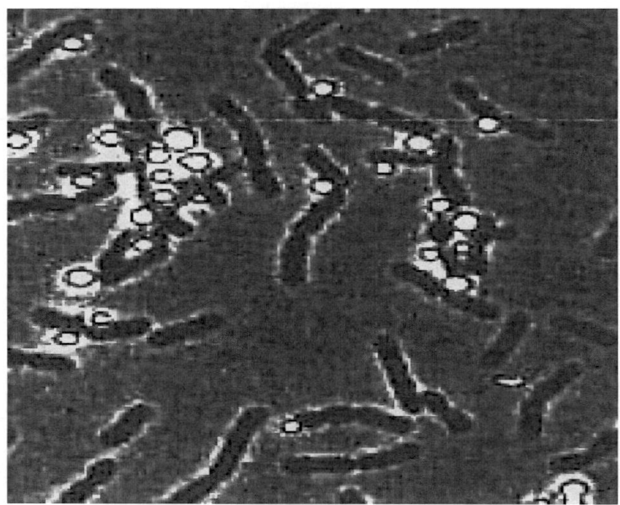

Fig. 3.92. A. Phase-contrast micrograph of *Chlorobium limicola* f. sp. *thiosulfatophilum* strain DSMZ 249 cultured photoautotrophically with sulfide. Globules of elemental sulfur are formed outside of the cells (Bar = 5μm).

Differentiation of *Chlorobium* from other Genera

The differentiation of the *Chlorobium* species from other genera of the green sulfur bacteria is presented in **Table 3.46**. The genus *Chlorobium* is retained with green or brown strains with spherical to straight or curved rod-shaped cells which do not form gas vacuoles under any conditions of culture or storage at 4°C. The cells of *Chlorobium* species that do not form prosthecae are the characteristic feature of the genera *Prosthecochloris* and *Ancalochloris*. For the detection of the prosthecae, high-resolution phase-contrast microscope objectives (magnification at X 100 oil immersion) should be used. Based on the non-motile nature of *Chlorobium* strains under all conditions, the unicellular filamentous gliding green sulfur bacterium *Chloroherpeton* is differentiated. The genetic relationship among *Chlorobium* species and other members of the *Chlorobiaceae* based on 16S rRNA gene sequence comparison is shown in **Fig. 3.90** .

Taxonomic Comments

Based on simple morphological and biochemical characteristics (photosynthetic pigments), the species of *Chlorobium* are assigned to this genus. In the two species *Chl . limicola* and *Chl . vibrioforme* (present *Prosthecochloris vibrioforme*), they forma specialist *thiosulfatophilum* was recognized for the thiosulfate-utilizing strains (Pfennig and Truper, 1971c). These strains contain an additional cytochrome C_{551} which is absent in the non-thiosulfate-using strains of the two species (Meyer *et al.,* 1968; Steinmetz and Fischer, 1982) and which is assumed to be involved in thiosulfate metabolism. However, in the non-thiosulfate-using strains, the first oxidation product of sulfide is elemental sulfur (S^0), whereas the *thiosulfatophilum* strains produce thiosulfate as the first oxidation product and release it into the medium. In a second step, thiosulfate is then further oxidized to elemental sulfur (S^0). Based on these results it is clear that the thiosulfate-utilizing strains of the two species differ significantly from the other strains in their metabolism of sulfide. The wide change of the DNA base ratios within the genus and single species signifies a marked genetic heterogeneity of the physiologically very homogeneous group. Based on 16S rRNA gene sequences and DNA–DNA hybridization studies a meaningful classification is established.

2. *Prosthecochloris* (*Ptc.*) (Green bacteria with appendages) (Gorlenko, 1970) Gorlenko (2015)

Cells are spherical to ovoid (**Fig. 3.93 A** and **B**) (Puchkova and Gorlenko, 1976), 0.3-0.6 x 0.5-0.8 µm, forming non-branching 10-20 prosthecae and multiply by binary fission in various directions. When the separation is incomplete, cells form groups and branched chains. The configuration will depend on the direction of fission. Cells are non-motile and Gram (-) negative. Cell culture appears green or chocolate-brown. Bacteriochlorophyll *c, d* or *e* as major bacteriochlorophyll pigments and carotenoids of chlorobactene and rhodopin or lycopene are present as photosynthetic pigments. The photosynthetic apparatus includes antenna structures, chlorosomes, i.e. elongated ovoid vesicles underlying and attached to the cytoplasmic membrane and fill the prosthecae (**Fig. 3.93C**) (Puchkova, 1984). Cells do not form gas vacuoles. Strictly anaerobic. Growth occurs photolithographically in the light with H_2S and elemental sulfur (S^0) as electron donors. During the oxidation of sulfide, sulfur is formed as an intermediate product. H_2S and elemental sulfur are deposited outside the cells in the form of globules which are partly or fully oxidized to sulfate. Thiosulfate cannot be utilized. Photoautotrophic growth occurs with CO_2 as the sole carbon source and requires vitamin B_{12}. A number of simple organic substances are assimilated in the presence of H_2S and CO_2. Ammonium salts are utilized as a nitrogen source. They are moderately halophilic, growth is possible with optimum NaCl concentration 0.5-2% and the salinity ranges from 0.2-10%. The salt tolerance of *Prosthecochloris* strains correlates with their tolerance of sulfide (maximum up to 2 g/L; optimum 500-700mg) and their ability to live under high light intensities up to 10,000lx. The mol% G+C of the DNA is 50.0-56.1 (T_m, Bd). Three species are validly published (*Ptc. aestuarii* , *Ptc. vibrioformis* and *Ptc. indica* .) *Prosthecochloris aestuarii* is the type species (Gorlenko 1970). The differential characteristics of three species of *Prosthecochloris* are presented in **Table 3.46.**

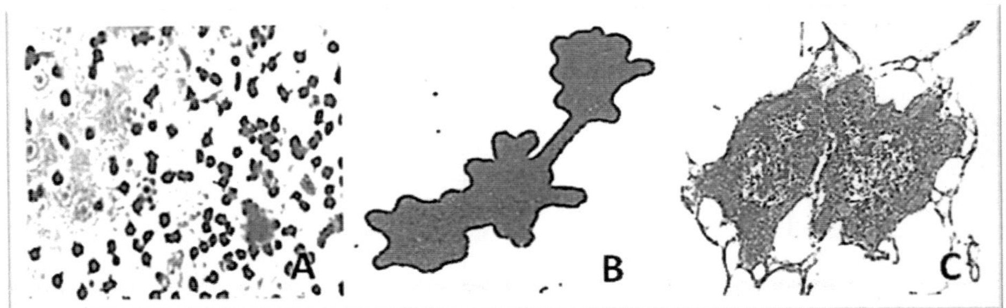

Fig. 3.93. A. Phase-contrast micrograph of cells of *Prosthecochloris aestuarii.* B. Electron micrograph of cells stained with phosphotungstic acid (X 20,000), C. Electron micrograph of ultrathin section of *Ptc. aestuarii* cells fixed with OSO_4 (X 100,000).

Differentiation of the *Prosthecochloris* from other Genera

The genus *Prosthecochloris* morphologically distinct from other genera of GSB in the presence of short prosthecae with thickened ends and absence of gas vacuoles. It can be differentiated from *Chlorobium* ,*Pelodictyon* and *Ancalochloris* which have gas vacuoles. The shape and number of prosthecae are different in *Ancalochloris* and *Prosthecochloris*. The *Ancalochloris* possesses prosthecae, but the cells contain gas vacuoles. The differential characteristics of *Prosthecochloris* from other genera are precised in **Table 3.46.** The genetic relationship among *Prosthecochloris* species and other members of the *Chlorobiaceae* based on 16S rRNA gene sequence comparison is shown in **Fig. 3.90.**

Taxonomic Comments

Ptc. vibrioformis originally assigned to *Chlorobium vibrioforme* (Pelsh, 1936). Based on the analysis of 16S rRNA gene sequences (**Fig. 3.90**) and morphological differences this species is transferred to the genus *Prosthecochloris* as *Ptc. vibrioformis* . *Ptc. aestuarii* strain DSMZ 271^T is phylogenetically closely related to the type strain of *Chlorobium vibrioforme* strain DSMZ 260^T and therefore does not represent an isolated branch within the radiation of GSB. *Prosthecochloris vibrioformis* is a basonym of *Chlorobium vibrioforme* and other strains assigned to this species, including subspecies, must be assigned to different species. Based on sequence similarities of more than 99% for both the 16S rRNA and *fmo A* genes supported the addition of *Chlorobium phaeo vibrioides* subspecies *thiosulfatophilum* (Former *Chlorobium vibrioforme*) and *Chlorobium vibrioforme* strain CHP 3402 to the genus *Prosthecochloris* (Imhoff,

2003) as additional strains. No, further studies on the genetic relationships of the species of *Prosthecochloris* with other GSB have been conducted till date.

3. *Chlorobaculum (Cba.)* (The green rod) Imhoff (2003)

Cells are vibrioid or rod-shaped cells that develop singly, cell size ranges from 0.3-1.1 μm in width **(Fig. 3.94A)**. Bacteriochlorophyll *c* and *d* or bacteriochlorophyll *e*, with carotenoids of the chlorobactene series in green strains and isorenieratene series in brown strains, are present as photosynthetic pigments. Some species and strains may contain gas vesicles. NaCl is required for growth. Vitamin B_{12} is required as a growth factor by most of the species. The major fatty acid composition is characterized by approximately 43% $C_{16:1}$and approximately 21–25% $C_{14:0}$.

The genus assigned with those strains and species that are previously placed in the genus *Chlorobium*, but form phylogenetic group 4 of the GSB. Growth occurs photolithoautotrophically anoxic conditions in the presence of light with reduced sulfur compounds such as sulfide, elemental sulfur (S°) and thiosulfate as electron donors (Alexander *et al.*, 2002). During oxidation of sulfide, S° is temporarily deposited outside cells. The final oxidation product is sulfate. Several simple organic substrates are photo assimilated in the presence of sulfide and bicarbonate. Obligately phototrophic and strictly anaerobic. Chlorosomes of *Chlorobaculum tepidum* cells are shown in **Fig. 3.94 B**. The DNA G+C content is 54.0–58.1 mol% (Bd). *Chlorobaculum tepidum* (Wahlund *et al.,* 1991) is the type species. Five species are validly published and their differential characteristics are given in **Table3.46.**

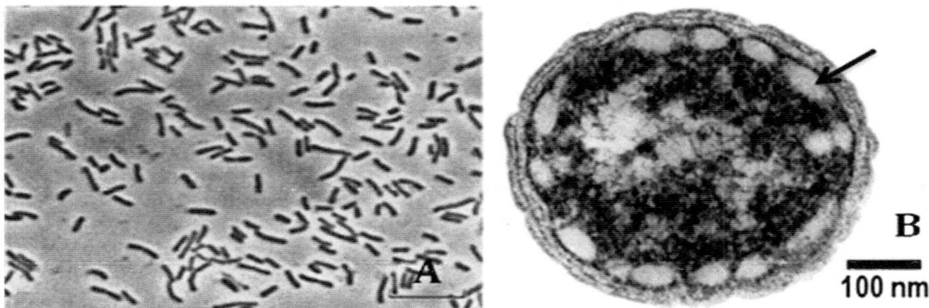

Fig. 3.94. A. Phase-contrast micrograph of cells of *Chlorobaculum tepidum* (Bar=5 μm), B. Electron micrograph of ultrathin section of *Chlorobaculum tepidum* showing chlorosomes (Arrow mark) (Bar 100nm).

Differentiation of *Chlorobaculum* from other Genera

The differential characteristics of *Chlorobaculum* from other genera are precised in **Table 3.46.** The genetic relationship among *Chlorobaculum* species and other members of the *Chlorobiaceae* based on 16S rRNA gene sequence comparison is shown in **Fig. 3.90** .

Taxonomic Comments

Chlorobaculum tepidum formerly called *Chlorobium tepidum,* it is a moderately thermophilic, freshwater species without a salt requirement and with a DNA G+C content of 56.5 mol%. It is phylogenetically different from all other strains belongs to this group. Therefore, based on different morphological and biochemical characteristics the *Chlorobium tepidum* was reclassified into a new genus *Chlorobaculum* with type species as *Chlorobaculum tepidum* (Imhoff, 2003) . The other strains of *Chlorobium* species due to their distant phylogeny, higher DNA G+C content and morphology, are transferred to the genus *Chlorobaculum* as *Chlorobaculum parvum* (formerly *Chlorobium vibrioforme* subsp. *thiosulfatophilum* strain DSM 263), *Chlorobaculum chlorovibrioides* (formerly *Chlorobium vibrioforme* strain UdG 6043) *Chlorobaculum limnaeum* (former *Chlorobium limicola* strain UdG 6038) and *Chlorobaculum thiosulfatophilum* (formerly *Chlorobium limicola* subsp. *thiosulfatophilum*, strain DSM 249 and strain 1430) (Imhoff, 2003). *Chlorobaculum indica* was added to genus based on phylogenetic analysis and phenotypic characteristics (Keppler *et al.,* 2008). It shows the closest similarity with *Chlorobaculum* and *Chlorobium* i.e., 96.8%. According to data recorded, among representatives of the *Chlorobaculum* , *Chlorobaculum indica* strain M is phylogenetically closer to *Cba. parvum* and *Cba. chlorovibrioides*. Although, it shows a difference in phenotypic characteristics and the content of G+C base pairs in the DNA (**Table 3.46** .)

4. *Pelodictyon* (*Pld.*) (Dark coloured net) (Lauterborn, 1913) Pfennig and Overmann (2015)

Cells are rod-shaped to ovoid, form singly or in netlike or more or less spherical aggregates (**Fig. 3.95**). Multiply by binary fission, branching may occur as a result of ternary fission. They are non-motile, Gram (–) negative and contain gas vacuoles. Cultures appear either green or brown. Photosynthetic pigments are located in the cytoplasmic membrane and in the chlorosomes which underlie and are attached to the cytoplasmic membrane. Photosynthetic pigments are bacteriochlorophyll *c, d* or *e* occurs as the major pigments and small amount of bacteriochlorophyll *a*. Chlorobactene or isorenieratene is the major carotenoid pigments. Obligately anaerobic and phototrophic. Growth occurs photoautotrophically with sulfide or sulfur as an electron donor. During sulfide oxidation globules of elemental sulfur are formed outside the cells and the sulfur may be further oxidized to sulfate. A number of simple organic substances may be photo assimilated in the presence of reduced sulfur compounds and bicarbonates. Hydrogen sulfide containing water and mud of freshwater, brackish water and marine environments are suitable environments for their growth. The growth temperature is 18.0-30.0°C. The mol% G+C of the DNA is 48.5-58.1 (Bd). *Pelodictyon clathratiforme* (Szafer 1911; Lauterborn 1913) is the type species. At present only one species is retained in the genus and their differentiating characteristics are listed in **Table 3.46.**

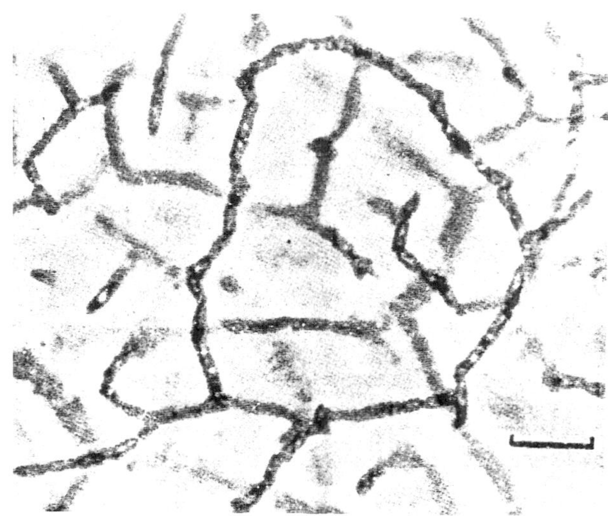

Fig. 3.95. Phase-contrast micrograph of *Pld. clathratiforme* has grown phototrophically with sulfide. Gas vacuole-containing cells are united in three-dimensional net like structures (Bar= 5μm).

Differentiation of the *Pelodictyon* from other Genera

The differentiating phenotypic characteristics used to differentiate the species of the *Pelodictyon* from other genera are given in **Table 3.46**.

Taxonomic Comments

The genus *Pelodictyon* contains straight or curved rod-shaped bacteria with gas vacuoles. They are either green are brown coloured species. According to previous studies, four species are included in *Pelodictyon* such as *Pelodictyon clathratiforme* (Pfennig, 1977), *Pld. phaeoclathratiforme* (Overmann and Pfennig, 1989), *Pld. luteolum* (Pfennig and Truper, 1971c) and *Pld . Phaeum* (Gorlenko, 1972) based on their morphological characteristics, cell culture colour, cell shape, metabolism and presence of gas vacuoles. However, the species of *Chlorobium* and *Pelodictyon* are physiologically and biochemically similar. The gas-vacuolated *Pelodictyon* species are buoyant because of their gas vesicles. Consequently, these species occur predominantly in sulfide-containing stagnant bodies of water, in which they differed significantly from *Chlorobium* species. The 16S rDNA sequences of *Pelodictyon clathratiforme* strain PG and *Pld. phaeoclathratiforme* DSM 5477^T which are almost similar and closely related to those of *Chlorobium ferrooxidans* and *Chlorobium phaeobacteroides* . As already proposed (Imhoff, 1999), both species are transferred to the genus *Chlorobium*. As the *Pld. clathratiforme* is the type species of the *Pelodictyon*, all the members of the genus must be transferred to another genus. Bacterial pigmentation is the main differentiating characteristic between *Pld. clathratiforme* and *Pld. phaeoclathratiforme* that has been disregarded as a species determinative property. Due to 16S rDNA sequence identity and lack of a typical culture of *Pld. clathratiforme*, hence it is proposed to combine both species and recognized this strain DSM 5477^T as the type strain of *Chlorobium clathratiforme*. Therefore, *Pld. clathratiforme, Pld. phaeoclathratiforme* are transferred to the *Chlorobium* as *Chlorobium clathratiforme*, whereas *Pld. luteolum* changed as *Chlorobium luteolum* (Imhoff, 2003). At present only single species, *Pld. phaeum* (Gorlenko, 1972) is retained in the genus which is also lost.

5. *Ancalochloris* (*Anc.*) (Arm producing green microbe) (Gorlenko and Labedeva, 1971) Gorlenko (2015)

Cells are irregular starlike shape (**Fig. 3.96**), 0.5-1.0 μm wide, forming prosthecae of irregular length, wide at the base and pointed towards the end. The length of prosthecae can exceed the cell diameter. These bacteria multiply by unequal fission and form irregular chains of cells and typical perforated microcolonies. They are non-motile, Gram (-) negative and contain bacteriochlorophyll *a* and carotenoids of sphaeroidene series as photosynthetic pigments. Cell cultures appear green or yellowish-green. The photosynthetic apparatus includes antenna structures of the chlorosomes type and ovoid vesicles underlying and attached to the cytoplasmic membrane. Cells contain gas vacuoles that are present in the central parts of the cells and consist of elongated gas vesicles measuring 40-80 × 230-360 nm. Strictly anaerobic, carry the photosynthesis in the presence of hydrogen sulfide and sulfide is deposited extracellularly. These bacteria occur in nature as typical conglomerates consisting of 10-30 cells, less frequently, they have been observed as short chains of cells with long prosthecae. The length of the prosthecae differed due to growth conditions. The longer prosthecae are observed at the low light intensity in the habitat of *Ancalochloris*. The mol% G+C of the DNA is not known. *Ancalochloris perfilievii* (Gorlenko and Labedeva, 1971) is the type species. Anoxic hypolimnion of dimictic and meromictic freshwater lakes with low H_2S content are ideal habitats for their growth. Only one species is retained in the genus and their characteristics are presented in **Table 3.46.**

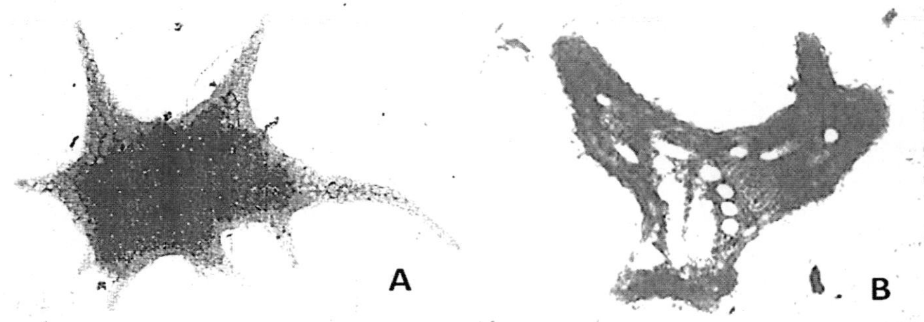

Fig. 3.96. A. Electron micrograph of a single cell stained with phosphotungstic acid (X 30,000), B, Ultrathin section of *Anc. perfilievii.* (X 50,000).

Differentiation of *Ancalochloris* from other Genera

The genus *Ancalochloris* is characterized by the star-like shape of the cells and the possession of gas vesicles. The prosthecae distinctly differ from those of *Prosthecochloris* members by their greater length and tapering ends. The differentiating characteristics of *Anc. perfilievii* from other genera are precised in **Table 3.46.**

Taxonomic Comments

Only one species is retained in *Ancalochloris* ,*Anc. perfilievii*, failed to obtain in pure culture. Therefore, the use of pure cultures is required to clarify the systematic position of *Ancalochloris* and its relationship to other genera of the GSB.

6. *Chloroherpeton* (*Chp.*) (Green creeping organism) (Gibson *et al.,* 1984) Gibson (2015)

Cells are long rods, **(Fig. 3.97A)** (Gibson *et al.,* 1984), 8–20 × 1 µm, separated promptly after division and no septa are seen in each unit. Cells able to grow in clumps and produce some extracellular slime. They are Gram (-) negative and gliding motility is observed. Cells flex through up to 180°. Cell cultures are green in colour. Bacteriochlorophyll *c* and carotenoids of γ-carotene are the main photosynthetic pigments that are located in the cytoplasmic membrane and chlorosomes which underlie or attached to the cytoplasmic membrane (**Fig. 3.97B**). Strictly anaerobic and obligately phototrophic. Sulfide and bicarbonates are required for growth. Good growth occurs by acetate, propionate, malate, succinate, or glutamate, but not on glucose, fructose, or casamino acids. During sulfide oxidation, elemental sulfur (S°) is formed outside the cells and oxidized slowly further to sulfate. Growth occurs at optimum temperature 25.0ºC and pH 6.8-7.2. The mol% G+C of the DNA is 45.0-48.2 (Tm). *Chloroherpeton thalassium* is the type species. Single species is validly published and their characteristic features are presented in **Table 3.46.**

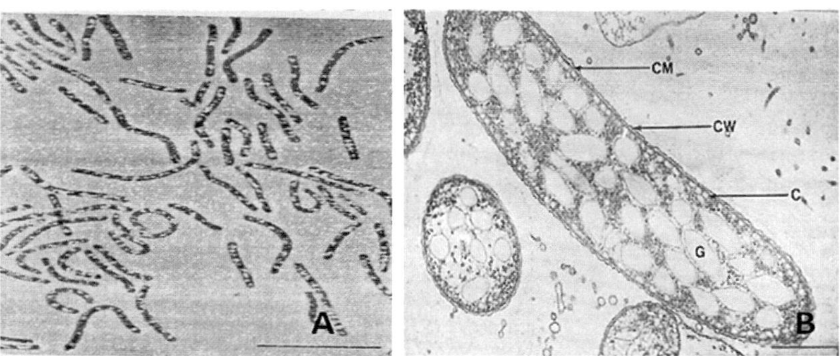

Fig. 3.97. A. Phase-contrast micrograph of *Chloroherpeton thalassium* (Bar = 10 µm), B. Electron micrograph of a longitudinal thin section of a single cell showing G, vesicles; C, chlorosomes; CW, cell wall; CM, cytoplasmic membrane (Bar= 0.5 µm).

Differentiation of the *Chloroherpeton* from other Genera

The differentiating characteristics of *Chloroherpeton* from other genera of green sulfur bacteria are presented in **Table 3.46**. They also differ from the multicellular filamentous green bacteria *Chloronema* (Dubinina and Gorlenko, 1975), *Oscillochloris* (Gorlenko and Pivovarova, 1977) and *Chloroflexus* (Pierson and Castenholz, 1974a) in cell dimensions, in being unicellular rather than filamentous and in photosynthetic pigment composition.

Taxonomic Comments

The members of *Chloroherpeton* show similar characteristics to green species of *Chlorobium* in their physiology and pigmentation. A major differentiating characteristic shows in the flexing and gliding motility, which might suggest affinities with *Chloroflexus*. However, the 16S rRNA gene sequence analysis indicates that *Chloroherpeton* belongs to the green sulfur bacteria (*Chlorobiaceae*) (Gibson, 1980; Gibson *et al.,* 1985b), which determines that motility and flexing are of lesser taxonomic importance compared to other characteristics. Therefore, based on DNA mol% G+C, DNA–DNA similarity and 16S rRNA gene sequence data, they are classified into a new genus *Chloroherpeton* with type species *Chloroherpeton thalassium* .

3.4. GREEN NON-SULFUR BACTERIA (MULTICELLULAR FILAMENTOUS GREEN BACTERIA) (Gupta *et al.,* 2012) Hanada (2014)

Green non-sulfur bacteria (GNSB) belong to the phylum *Chloroflexi,* order *Chloroflexales* and family *Chloroflexaceae*, which contains a highly divergent group of bacteria that includes anoxygenic photoautotrophs, aerobic chemoheterotrophs, thermophilic organisms, as well as anaerobic organisms that obtain energy by reductive dehalogenation of chlorinated organic compounds such as tetrachloroethane and trichloroethane (Garrity and Holt, 2001a; Hugenholtz and Stackebrandt, 2004). The description of the phylum was first defined by Garrity and Holt in Bergey's Manual of Systematic Bacteriology, the second edition, volume one (Garrity and Holt, 2001) was slightly amended by Hugenholtz and Stackebrandt in 2004. Prior to 2001, this phylum was retained with single class *Chloroflexi* which consists of two orders *Chloroflexales* and *Herpetosiphonales* , each consists a single family such as *Chloroflexaceae* and *Herpetosiphonaceae* , together consists of five genera (Garrity and Holt, 2001a; Castenholz, 2001a).

Although, in the past 10 years, this phylum has undergone enormous development both by addition of newly discovered species as well as by joining of species that were previously part of other bacterial phyla (e.g. *Thermomicrobia* , *Actinobacteria*) (Garrity and Holt, 2001b; Hugenholtz and Stackebrandt, 2004). As a result, the phylum is currently made up of six classes such as *Chloroflexi, Thermomicrobia, Dehalococcoidetes, Anaerolineae, Caldilineae* and *Ktedonobacteria* based on 16S rRNA gene sequences (**Fig. 3.98**) (Hugenholtz and Stackebrandt, 2004; Yamada *et al.,* 2006; Moe *et al.,* 2009;

Yarza *et al.,* 2010; Yabe *et al.*, 2010; Euzeby, 2011). Among these six classes, the photosynthetic ability within this phylum is only present within the species belonging to the class *Chloroflexi* (Pierson and Castenholz, 1992; Garrity and Holt, 2001a; Hanada and Pierson, 2006). As this phylum *Chloroflexi* represent one of the five phyla of cultured bacteria that contain anoxygenic phototrophic bacteria are mainly defined on the basis of 16S rRNA sequence analysis and the comparative genomic analyses focusing on several conserved proteins such as GroES (a chaperon unit) and FtsZ (a tubulin homolog), they have been extensively studied (Olson and Pierson, 1987; Blankenship, 1992; Pierson, 1994; Gupta *et al.*, 1999; Xiong *et al,*. 2000; Raymond *et al.,* 2002; Bryant and Frigaard, 2006; Mulkidjanian *et al.,*2006; Blankenship, 2010; Gupta *et al.,* 2012).

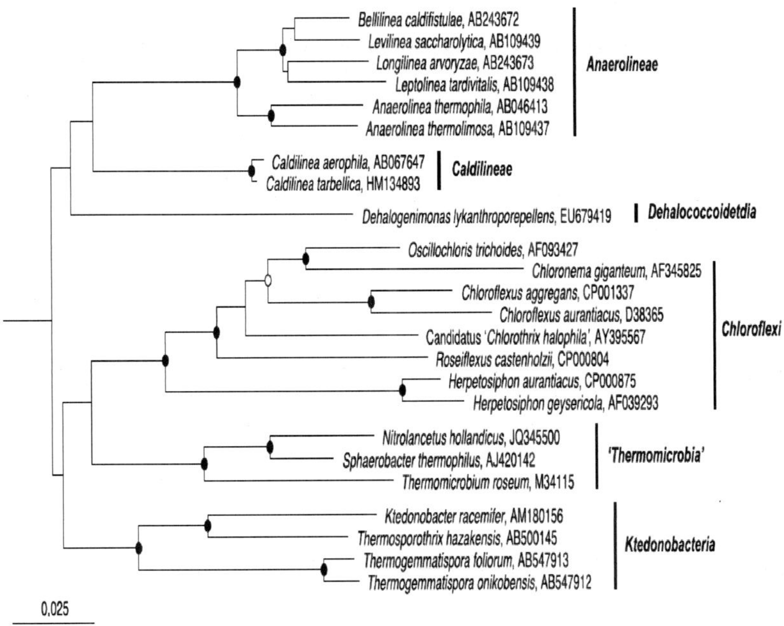

Fig. 3.98. Phylogenetic tree of species of the phylum *Chloroflexi* based onanalyses of 16S rRNA gene sequences obtained from the SILVA database (Hanada, 2014).

In this manual, the phylum basically comprised of the filamentous anoxygenic phototrophic organisms and furthermore contained one non-phototrophic family, *Herpetosiphon.* Although, a related thermophile, *Thermomicrobium roseum*, was separated from the phylum *Chloroflexi* and a determined phylum, *Thermomicrobia*, was authorized for the organism. This phylum was revised based on exact phylogenetic analyses of the 16S rRNA gene sequence (Hugenholtz and Stackebrandt, 2004). *Thermomicrobia* was placed in the phylum as a class and a new number, *Sphaerobacter thermophilus* was renamed into the class *Thermomicrobia.* These two organisms are non-phototrophic thermophiles without appearing filamentous morphology.

The majority of the known microorganisms in this class have multicellular filamentous morphology regardless. The class contains two orders, *Chloroflexales* and *Herpetosiphonales.* The former comprises filamentous anoxygenic phototrophic bacteria regardless. The order *Herpetosiphonales* consist of a single-family *Herpetosiphonaceae .Herpetosiphon aurantiacus* (Holt and Lewin, 1968) is the type genus and species. The non-phototrophic filamentous bacterium demonstrating gliding motility has been separated from slimy covered springs. The order *Chloroflexales* assigned with three families, *Chloroflexaceae* (Gupta *et al.* 2012), *Roseiflexaceae* (Gupta *et al.,* 2012) and *Oscillochloridaceae* (Keppen *et al.,* 2000; Gupta *et al.,* 2012). The former two families contain thermophilic phototrophic bacteria isolated from terrestrial hot springs, whereas the family *Oscillochloridaceae* consists of freshwater phototrophic bacteria. All the phototrophic bacteria are explained in detail in a separate section. They are mostly Gram-negative, filamentous bacteria with unusual cell envelopes. No peptidoglycan or atypical peptidoglycan present in cell walls. Lipopolysaccharide-containing outer membrane not present in studied representatives (Hugenholtz and Stackebrandt, 2004).

Order 1. *Chloroflexales* (Garrity and Holt, 2001a) Gupta *et al.* (2013)

This order is described on the basis of 16S rRNA gene sequences, which include filamentous anoxygenic phototrophic bacteria that exhibit gliding motility. They are facultative aerobic and able to grow photoheterotrophically by utilizing a number of organic substances during their phototrophic or chemotrophic metabolism. The order consists of two suborders *Chloroflexineae* and *Roseiflexineae* and three families *Oscillochloridaceae ,Chloroflexaceae* and *Roseiflexaceae* .

The species from this order are differentiated from all other bacteria by the CSIs in the following proteins, (i) Conserved inserts of 1 aa each in the proteins Lysyl tRNA synthetase, Prolyl tRNA synthetase, signal recognition particle FtsY and precorrin- 4 C11-methyltransferase, (ii) 3 aa deletion in Alanyl-tRNA synthetase,(iii) 6 aa insert in AMP-dependent synthetase/ligase protein, (iv) 2 aa insert in signal recognition particle protein, (v) 2 aa deletion in acetolactate synthase small subunit and (vi)1 aa deletion in the protein magnesium chelatase.

Suborder 1. *Chloroflexineae* Gupta *et al.* (2013)

It is the suborder of the genus *Chloroflexus* assigned with green-coloured bending bacteria. Consists of two families *Oscillochloridaceae* and *Chloroflexaceae* of straight or spiral filamentous phototrophic bacteria showing gliding mobility. Filaments are green in colour. Cells contain chlorosomes and Bchl *c* and additionally, they may also contain either Bchl *a* or Bchl *d* in minor amounts. The major carotenoids are γ- and β-carotenes and the major quinone is MK-10 with 10 isoprene units. The species from suborder are distinguished from other bacteria by the CSIs in the following proteins; (i) 3 aa insert in nucleotide sugar dehydrogenase, (ii) Conserved inserts of 1 aa each in a Phage SPO1 DNApol-like protein, nucleoside diphosphate kinase and the translation initiation factor-2 proteins, (iii) Conserved inserts of 2 aa each in the proteins threonine synthase and ArsA and (iv)1 aa insert in the acetolactate synthase large subunit.

Family 1. *Chloroflexaceae* (Truper, 1976, Gupta *et al.*, 2013) Pierson (2015)

Anoxygenic phototrophic bacteria with no branching filamentous morphology of indefinite length are included in these family members. Cells are filaments of indefinite length, 0.5-1.5 μm wide and motile by gliding. Bacteriochlorophylls *a* and *c* and carotenoids of β- and γ-carotenes and hydroxyl-γ-carotene glucoside esters are present as photosynthetic pigments. They are primarily photoheterotrophic and are able to grow by oxygen respiration under aerobic conditions. This family contains only a single genus *Chloroflexus* and the type species is the *Chloroflexus aurantiacus*. Based on 16S rDNA and 16S rRNA oligonucleotide catalogs, the genus *Chloroflexus* is related neither to the green sulfur bacteria nor to any other eubacteria. However, the evidence of nearest genetic relatives to *Chloroflexus* is, *Oscillochloris, Heliothrix* and chemoorganotrophic flexibacterium *Herpetosiphon*(**Figs. 3.98** and **3.99**) (Keppen *et al.,* 2000). This supports the perspectives of Castenholz (1973) who considers the genera *Chloroflexus* and *Heliothrix* as phototrophic flexibacteria. The sequence of 5S rRNA was the primary proof to support the relationship of *Heliothrix* to *Chloroflexus* and *Herpetosiphon* (Pierson *et al.,* 1985) and it was the principal evidence to demonstrate that *Oscillochloris* was likewise more closely identified with this group of bacteria than some other phototrophic bacteria (Keppen *et al.,* 1994). The 16S rRNA affirmed these relationships. So far, no phylogenetic information is available to support the position of the genus *Chloronema* in this group. In spite of the fact the filamentous phototrophic bacteria are grouped as the most profound division of photosynthetic bacteria, their phylogenetic relationship to one another needs further explanation and it might be premature to build a higher taxonomic order around them.

Proof for the presence of a specific and novel sequence of 16S rRNA in microbial mats has been used to study the dispersion of specific species in the mats and to determine the relative abundance and diversity of known and novel species in mats. Weller *et al.* (1992) recovered two *Chloroflexus* like sequences from the Octopus Spring mat in Yellowstone. Neither was similar with *Cfl. aurantiacus* or *Htr. oregonensis*. In fact, no sequences indistinguishable from *Cfl. aurantiacus* (J-10-fl) or another *Cfl. aurantiacus* strain developed from a similar mat were obtained (Ward *et al.,* 1990; Weller *et al.,* 1992). These investigations demonstrate that the extrapolation from specifically cultured strains to the mass population in the local hot spring mat must be done cautiously. Species diversity inside the mat is more prominent than previously thought (Ward *et al.,* 1990).

The major differentiating characteristics of the *Chloroflexus* are presented in **Table 3.47.** The species from this family or genus form a separate clade in the 16S rRNA gene tree and the tree based upon protein sequences. They are distinguished from all other *Chloroflexales* species by the CSIs in the following proteins; (i) 1 aa deletion in CTP synthetase, (ii) 4 aa insert in the pyruvate flavodoxin/ferredoxin oxidoreductase protein and (iii) 2 aa insert in the magnesium-protoporphyrin IX monomethyl ester cycle protein.

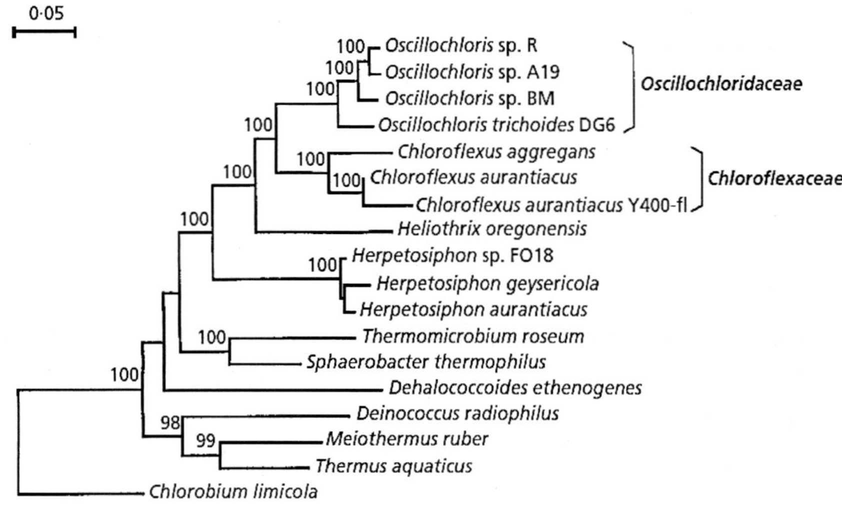

Fig. 3.99. Phylogenetic tree including filamentous phototrophs *Chloroflexus, Oscillochloris* and *Heliothrix* based on 16S rDNA gene sequence.

In light of 16S rRNA sequences, the genus *Chloroflexus* and its related phototrophs are not closely identified with the green sulfur bacteria or with some other photosynthetic bacteria. They belong to one of the most profound phyla called the green non-sulfur bacteria forming the only phototroph-containing branch occurring before the nearly simultaneous radiation of all the various bacterial phyla (Stackebrandt *et al.,* 1996).

Table 3.47. Characeristics Properties of Genus and Species of Green Non-Sulfur Bacteria (Multicellular Filamentous Phototrophic Bacteria)

Characteristics	*Chloroflexus*			*Oscillochloris*		*Chloronema giganteun*		*Roseiflexus castenholzii*	*Heliothrix oregonensis*
	Chloroflexus aurantiacus	*Chloroflexus aggregans*	*Chloroflexus islandicus*	*Osc. chrysea*	*Osc. trichoides*	*Cln. giganteun*	*Cln. spiroideum*		
Habitat	Neutral - alkaline hot springs	Neutral - alkaline hot springs	Iceland	Sulfidic fresh-water mud	Benthic mats, cold and warm springs	Fresh water lakes	Fresh water lakes	Neutral-alkaline hot springs	Neutral-alkaline hot springs
Cell shape	Filaments with indefinite length	Filaments with indefinite length	Unbranched multicellular filaments of varying lengths	Uniseriate multicellular straight filaments	Straight filaments	straight spiral or Cylindrical filaments	Spiral filaments	Spiral filaments	Septate filaments

Characteristics	*Chloroflexus*			*Oscillochloris*		*Chloronema giganteun*		*Roseiflexus castenholzii*	*Heliothrix oregonensis*
	Chloroflexus aurantiacus	*Chloroflexus aggregans*	*Chloroflexus islandicus*	*Osc. chrysea*	*Osc. trichoides*	*Cln. giganteun*	*Cln. spiroideum*		
Cell size (µm)	0.5-1.5 x 2.0-6.0	1.0-1.5 x 2.0-6.0	0.6 x 4.0-7.0	4.5-5.5 x 3.5-7.0	0.8-1.4 x 2.0-5.0	2.0-2.5 x 3.5-4.5	1.5-2.0	0.8-1.0	1.5 x 10.0
Sheath formation	± (thinly sheathed)	-	-	-	±	+	+	-	-
Gliding movement velocity µms¯1	0.01-0.04	10-3.0	ND	≤7.0	0.1-0.2	10	ND	ND	0.1-0.4
Colour of cell suspension	Green, Orange	Green, Greenish-Brown	Yellowish green	Green, Yellow-Green	Green, Yellow-Green	Yellow-Green	Yellow-Green	Red, Reddish-Brown	Bright orange
Bacteriochlorophyll	*a*, c_s	*a*, *c*	*c*,*a*	*c*, *a*	*a*, *c*	*d*, *c* (*a*)	*d*, *c* (*a*)	*a*	*a*
Carotenoids	β-γ-Carotene, OH-γ-carotene glucoside fatty acid ester. Aerobic cells contain echinenone, myxobactone	γ-Carotene, (β-carotene), OH-γ-carotene glucoside fatty acid ester.	β- and γ-Carotene	β-γ-Carotene	β-γ-Carotene	ND	ND	γ-Carotene derivatives, methoxy-keto-myxocoxanthin, methoxy-keto-myxocoxanthin glucoside fatty acid ester	4-Oxo-γ-carotene, 4-oxo-3',4'-didehydro-OH-γ-carotene, myxobactone
Absorption spectra of living cells (nm)	462, 740, 802 & 865	464, 740, 803 & 868	461, 741, 805 & 868	ND	748, 852	720	ND	479, 590, 801, & 878	865, 795
Major cellular fatty acids	$C_{16:0}$, $C_{17:0}$, $C_{18:0}$, $C_{18:1}$, $C_{18:0}$-OH	ND	$C_{16:0}$, $C_{18:0}$, $C_{18:1}\omega9$, $C_{18:0}$-OH	ND	$C_{18:1}$, $C_{16:0}$, $C_{16:1}$	ND	ND	$C_{16:0}$, $C_{14:0}$, $C_{15:0}$	ND
Major quinones	MK-10	MK-10 (& MK-4)	MK-10	ND	MK-10	ND	ND	MK-11	ND
Chlorosomes	-	-	+	+	+	+	+	+	+
Gas vesicles	-	-	+	+	+	+	-	-	-
Growth factors	Thiamine	Thiamine, folic acid	Folic acid	ND	ND	ND	ND	ND	ND

Charac teris-tics	*Chloroflexus*			*Oscillochloris*		*Chloronema giganteun*		*Rosei-flexus casten-holzii*	*Heliothrix oregon-ensis*
	Chloro-flexus aurant-iacus	*Chloro-flexus aggregans*	*Chloro-flexus islandicus*	*Osc. chrysea*	*Osc. trichoides*	*Cln. giganteun*	*Cln. spiroi-deum*		
Metabo lism	Photo-organo-hetero-troph, photolithoa utotroph (H_2S, H_2,) aerobic chemohe-terotroph	Photo-organo-hetero-trophs, aerobic chemoh-eterotrophs	Photohe-terotrophs, aerobic chemohe-terotrophs	Photo-organo-hetero-trophs, aerobic chemohet erotrophs	Photolitho-heterotrphs, photolith-oautotrophs (H_2S, H_2)	Photoorgano-heterotrophs, aerotaxis	Photohe-terotr-ophs, Photoau-totrophs, aerotaxis	Photo-organo-hetero-trophs, aerobic chemohet erotrophs	Photo-organo-hetero-trophs, aeroto-lerant
Opti-mum tempe-rature (°C) (range)	52-60 (40-66)	55	55 (46-59)	10-20	28-30	4.0-15	Mesophil ic	50	40-55
pH opti-mum	7.6-8.4	ND	7.5-7.7	7.5-8.5	8.0-8.5 (7.5-8.5)	ND	ND	7.5-8.0	8.1-8.5
Subs-trates uniti-zed	Acetate, Aspartate, Butyrate, Casamino acids, Glycerol, Glucose, Lactate, Pyruvate, Glutamate, Yeast extract	Aspartate, Butyrate, Casamino acids, Glucose, Glutamate, Lactate, Malate, Succinate, Yeast extract	Acetate, Glucose, Fructose, Glutamate, Aspartate, Succinate, Sucrose, Mannitol, Methanol, Ethanol, Glycyl-glycine, Glycerol, Yeast extract, Casamino acids, Soytone, Casein hydrolysate	ND	Acetate, Casein hydrolysate, Pyruvate, Yeast extract	ND	ND	Casamino acids, Citrate, Glucose, Lactate, Yeast extract	Acetate
G+ C of DNA (mol%)	54.9	56.7-57.0	59.6	ND	59.2	ND	ND	62.0	ND

Symbols: +, Present/utilized by strains, -, Absent or not utilized; ±, strain variability; cs, bacteriochlorophyll *c* esterified with stearol; *a*p, bacteriochorophyll a esterified with phytol; ND, Not determine/No data available. Data for *Chloroflexus* is taken from Pierson and Castenholz (1974) (*Cfl. aurantiacus*), Hanada *et al.* (1995) (*Cfl. aggregans*) and Gaisin *et al.* (2017) (*Cfl. islandicus*). Data for *Oscillochloris* is taken from Gorlenko and Pivovarova (1977), Keppen *et al.* (1993, 1994, 2001) (*Osc. chrysea* and *Osc. trichoides*). Data for *Chloronema* is taken from Dubinina and Gorlenko (1975), Gorlenko and Pierson (2001) (*Cln. giganteun, Cln. spiroideum*). Data for *Roseiflexus* is taken from Hanada *et al.* (2002) (*Roseiflexus castenholzii*). Data for *Heliothrix* is taken from Pierson *et al.* (1984, 1985) (*Heliothrix oregonensis* .)

1. *Chloroflexus (Cfl.)* (Green bending) Pierson and Castenholz (2015)

Cells are filaments of indefinite length (**Fig. 3.100A**), 0.5-1.5 x 2.0-6.0 μm, none differentiated. Cell division by fission, no branching occurs. No internal proliferation of cell membrane, except mesosomes, chlorosomes are present when anaerobically grown. Motile by gliding, no flagella formation and Gram (-)

negative. Thin sheath sometimes present, anaerobic and some are facultatively aerobic. Bacteriochlorophylls *a* and *c* and carotenoids of β- and γ- carotene, hydroxyl-γ-carotene glucoside ester (Haffen *et al.,* 1972) are present as photosynthetic pigments under anaerobic conditions. Primarily growth occurs photoheterotrophically, may also grow photoautotrophically (not all strains) and chemoheterotrophically (not all strains). Acetate, glycerol, glucose, pyruvate and glutamate are used as carbon sources. The optimum growth occurs at the temperature 52-60°C (range of upper limit is 65-70°C and the lower limit is 30-35°C) in the laboratory. At temperature below 40°C, filaments formed aggregates and often adhere to culture vessel walls. The major quinones are menaquinone with 10 isoprene units (MK-10) and the major fatty acids are $C_{16:1}$ (palmitic acid), $C_{18:0}$ (stearic acid) and $C_{18:1}$ (oleic acid). Poly-β-hydroxybutyric acid granules, polyphosphate granules and, possibly, glycogen bodies may be seen with the help of transmission electron microscopy **(Fig. 3.100B).** Polyglucose is present in cells under certain conditions. The mol% G + C of the DNA is 53.1-54.9. *Chloroflexus* the monotypic only single species validly published. *Chloroflexus aurantiacus* is the type genus. The differentiating characteristics of *Chloroflexus* are presented in **Table 3.47.**

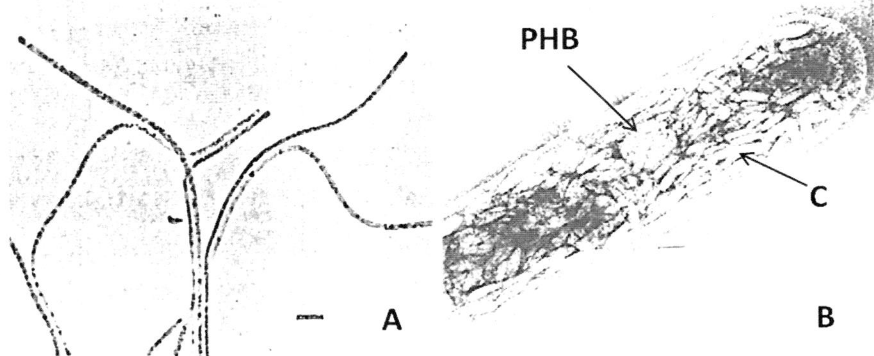

Fig. 3.100. A. Phase-contrast micrograph of *Cfl. aurantiacus*. The transverse septa are not apparent (Bar = 5 μm), B. Cells showing numerous chlorosomes (C) and Poly-β-hydroxybutyric acid (PHB) granules (Bar =0.5 μm).

Differentiation of the *Chloroflexus* from other Genera

The differentiation of *Chloroflexus* from other genera is precised in **Table 3.47** . Ultrastructural and biochemical characteristics are required. All are filamentous. The larger trichome diameter across the abundance of chlorosomes on complete and deficient septa appears to be adequate to separate *Oscillochloris chrysea* (Gorlenko and Pivovarova, 1977) from *Chloroflexus*, despite the fact that this species of *Oscillochloris* has not been isolated in pure culture. The smaller *Oscillochloris trichoides* would be hard to recognize from *Chloroflexus* without the use of electron microscopy and biochemical analysis. The possession of gas vesicles, a trademark so inconsistently spread through the Bacteria and Archaea, ought to most likely not be used as a character to differentiate genera. Characteristics of the genus *Chloronema* (Dubinina and Gorlenko, 1975), i.e., gas vesicles, trichome diameters of 2.0–2.5 μm (*Chloronema giganteum*) or 1.5–2.0 μm (*Chloronema spiroideum*) and the presence of an obvious sheath, are not really sufficient for raising a genus separate from *Chloroflexus*. Again, neither species of *Chloronema* has been developed in culture, other than as enrichments. The likely presence of bacteriochlorophyll *d* rather than of bacteriochlorophyll *c* in *Chloronema* is likewise not a good distinctive characteristic. Since, bacteriochlorophylls *c* and *d* occur in different species of *Chlorobium* and even in a single strain (Broch-Due and Ormerod, 1978) and are deficient for a generic distinction. Recent detailed phylogenetic studies of the green sulfur bacteria affirm that gas vesicles and pigmentation are not helpful characters for making nonexclusive differentiation or generic distinctions (Overmann and Tuschak, 1997). *Heliothrix* is a thermophilic, filamentous bacterium that falls inside the scope of cell diameters for species of *Chloroflexus* (1.5 μm). However, *Heliothrix* has neither chlorosomes nor the accessory bacteriochlorophyll *c* or *d* that would be related to these structures (Pierson *et al.,* 1984a, 1985). *Herpetosiphon* and other aerobic filamentous bacteria are of dimensions and colour similar to orange *Chloroflexus* show ever chlorophylls of any need type (Soriano, 1973). A few carotenoids such as γ-carotene and carotenoid glucosides are like those of *Chloroflexus* (Kleinig and Reichenbach, 1977).

Taxonomic Comments

Based on 16S rRNA gene sequence analyses and DNA-DNA hybridization three species are validly published in *Chloroflexus*. According to the previous classification, two species are included in this. Recently, *Chloroflexus islandicus* , a thermophilic filamentous APB was included in the genus (Gaisin *et al.,* 2017). The phylogenetic analysis of the 16S rRNA gene sequence of this species shows 97.0% and 81.4% similarity with *Chloroflexus aggregans* DSM 9485^T and other strains of *Chloroflexus* respectively (**Fig. 3.101**) (Gaisin *et al.,* 2017). The species of *Chloroflexus* showed the closest relationship with *Roseiflexus* and distantly related to *Oscillochloris* .

The species *Chloroflexus aggregans* has been assigned with an invalid generic name, "*Chlorocrinus aggregans*", in any rate of distribution (Ward *et al.,* 1998). As noted here, *Cfl. aggregans* have been portrayed as a substantial species of the genus *Chloroflexus* (Hanada *et al.,* 1995a). It was initially assigned in a family *Chloroflexaceae* characterized by the possession of chlorosomes, accessory bacteriochlorophylls *c* or *d,* filamentous morphology and gliding motility (Truper, 1976). Pfennig (1989) gathered the clearly related filamentous phototrophic bacteria into a group assigned the multicellular filamentous green bacteria in the previous edition of this Manual. Woese (1987) insisted the phylogenetic group containing *Chloroflexus* in the domain *Bacteria* as the green non-sulfur bacteria while advised that there was little phenotypic attachment to the group. These terms have been generally used. Since sulfide-dependent autotrophy develops in this group. Truper (1987) questioned the green non-sulfur bacteria designation. It is deluding. Truper (1987) further called attention to that if *Heliothrix oregonensis* undoubtedly was phylogenetically identified with *Chloroflexus,* at that point the family definition would need to be improved to evacuate the prerequisite for chlorosomes and accessory bacteriochlorophylls. Pierson and Castenholz (1995) have discussed the taxonomic issues of the filamentous phototrophs completely and prescribed assigning them as a group called the "filamentous anoxygenic phototrophs" or the "photosynthetic flexibacteria" as initially proposed by Castenholz (1973). These terms would precisely describe all the phototrophs as of now found in the same phylogenetic group as several non-phototrophs. Since both 5S (Pierson *et al.,* 1985) and 16S rRNA (Weller *et al.,* 1992) sequences revealed that *Heliothrix oregonensis* is phylogenetically connected with *Chloroflexus,* such acknowledgment of the inappropriateness of the "green" assignment is required.

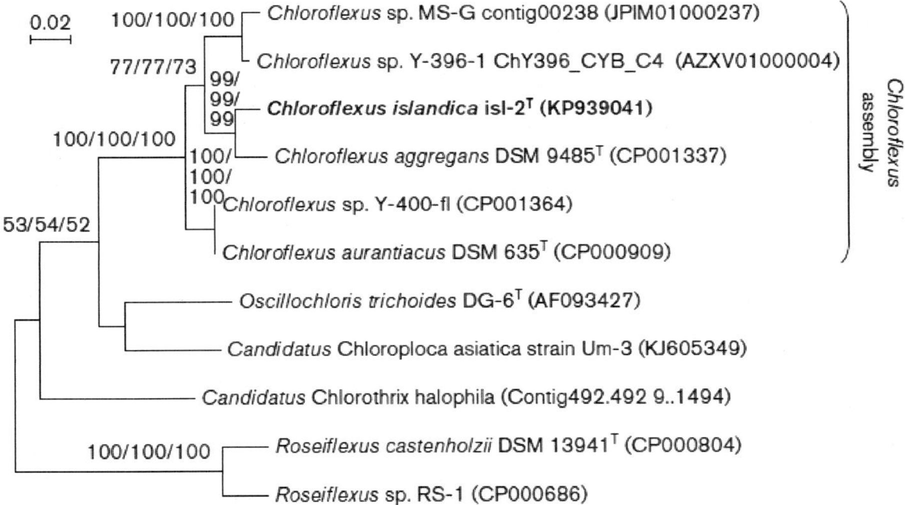

Fig. 3.101. Phylogenetic tree of *Chloroflexus* species and other members of *Chloroflexi* based on 16S rRNA gene sequences.

Family 2. *Oscillochloridaceae* (Keppen, 2000) Gupta *et al.* (2013)

This family was assigned with two genera *Oscillochloris* and *Chloronema* of green-colored filamentous bacteria with gliding motility. The species from this family group together in phylogenetic trees based upon 16S rRNA gene sequences and phenotypic traits such as metabolism, they are mesophilic, contain gas vesicles and an outer sheath. Cells are arranged in multicellular flexible filaments of constant diameter. Shows gliding motility, trichome formation is seen which don not form branches and are uniformly wide throughout their length. Its length is variable with or without sheaths. Multiply by the separation of short segments or single cells from the parental trichome, appear green or yellowish-green. Gas vesicles are present. Bacteriochlorophylls *a* and *c* as well as carotenoids of *β*-carotene and *γ*-carotene are present as photosynthetic pigments. The light-harvesting structures of the photosynthetic apparatus are chlorosomes. Growth occurs photolithoautotrophically or photolithoheterotrophically under anoxic conditions. Some strains are capable of growing under microaerobic conditions. The differences in the sequences of 16S rRNA gene between members of the families *Oscillochloridaceae* and *Chloroflexaceae* are in the range 13.5-17.5%. The type genus of the family is *Oscillochloris* (Gorlenko and Pivovarova, 1977, 1989) and the type species is *Oscillochloris trichoides* (Keppen *et al.,* 2000). Among *Oscillochloridaceae* members, the *Oscillochloris* strains showed only remote relatedness with non-phototrophic members of the phylum (69.2–76.6% sequence similarity). The phototrophic genera *Oscillochloris* ,*Chloroflexus* and *Heliothrix* formed a monophyletic cluster in which the highest level of sequence similarity was observed between branches of *Chloroflexus* and *Oscillochloris* (82.5–86.5% sequence similarity). Although, a similar level of sequence similarity had been recorded for members of related families. The level of sequence divergence of *Osc .trichoides* DG-6T as well as the new isolates and other phototrophic genera of the phylum have a close similarity with one another.

1. *Oscillochloris* (*Osc.*) (Oscillating green bacteria) (Gorlenko and Pivovarova 1989) Keppen *et al.* (2015)

Cells are arranged in uniseriately multicellular flexible filaments, 0.8-1.4µm wide. Shows gliding motility (**Fig. 3.102A**) (Gorlenko and Pivovarova, 1977). Trichomes uniformly wide throughout their length is variable (**Fig. 3.102 B** .) Non-branching, with or without a sheath. Sometimes they are surrounded by a thin layer of slime. Trichomes multiply by hormogonia. Cells contain intracellular gas vesicles are formed along the cell septa. Gram stain may be positive or negative. Trichomes appear green or yellow-green. Bacteriochlorophylls *c* and small amounts of *a* and carotenoids of β- and γ- carotene are present as photosynthetic pigments. The light-harvesting structures of the photosynthetic apparatus are Chlorosomes. Growth occurs in photolithoheterotrophically or Photolithoautotrophically under anaerobic conditions, capable of photosynthesis in the presence of hydrogen sulfide or molecular hydrogen as an electron donor. They deposit elemental sulfur globules outside the cells during the oxidation of sulfide. Capable of chemoorganotrophic growth under aerobic conditions in the dark. Strictly anaerobic and phototrophic. Mesophilic bacteria show growth at optimum temperature 20-30ºC and pH 6.8-9.0. The optimal sulfide ($Na_2S.9H_2O$) concentration is 0.5-0.7 g/L. The mol% G + C of the DNA is 57.3-59.2% (Tm). *Oscillochloris chrysea* is the type species (Gorlenko and Pivovarova 1977, 1989). Two species are validly published and their major differentiating characteristics are given in **Table 3.47** .

Cell walls are formed by diaphragm-like narrowing and closing septa and subsequent formation of the other components of the cell wall. Multiple incomplete septa are typical in *Osc. chrysea* (**Fig. 3.102C**). The formation of porous thick cell septa results in the breakage of the trichome and its multiplication. Short sections of trichomes may separate at the end of a filament, or the entire filaments may break up into segments resembling hormogonia. Long trichomes as well as "hormogonia" possess gliding motility at rates of 0.1-0.2 µm/s in *Osc. trichoides* and up to 7 µm/s in *Osc. chrysea*. The cell wall consists of multiple layers and appears similar to the Gram-negative type in *Osc. trichoides*. However, deep studies are needed.

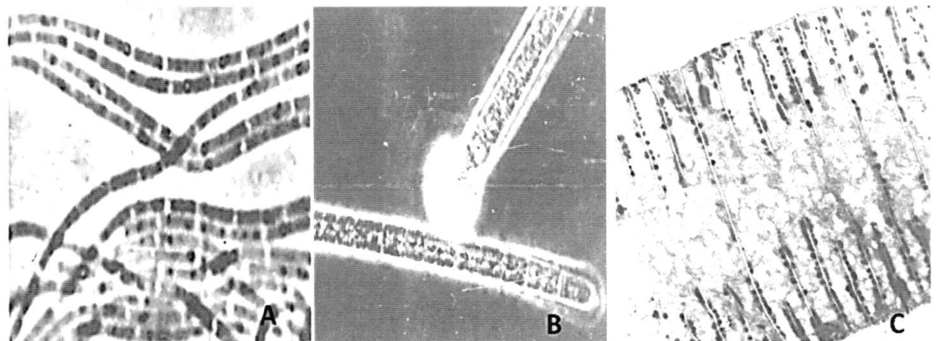

Fig. 3.102. A. Phase-contrast micrograph of *Osc. trichoides* (X 2500) . B. Phase contrast light micrograph of *Osc. chrysea* with India ink (X 2700), C. Electron micrograph of ultrathin section fixed with $KMnO_4$ (X 2000).

Differentiation of the *Oscillochloris* from other Genera

The major differentiating characteristics of *Oscillochloris* from other genera of multicellular filamentous bacteria are given in **Table 3.47.**

Taxonomic Comments

Mesophilic members of the multicellular filamentous bacteria were first studied in pure culture, mixed culture, or in samples from the mass development in nature (Gorlenko and Pivovarova, 1977; Gorlenko and Korotkov, 1979). The primary description of the genus *Oscillochloris* depended on field perceptions of *Osc .chrysea*, which was never in culture. Culture SR-1, based upon the primary description of *Osc .trichoides* it was not a pure culture and was subsequently lost . *Osc .trichoides* strain DG-6 is the first strain to be effectively developed as a pure culture and the first to be exposed to molecular analysis. The genus description provided here incorporates extra information gained from the investigations of the pure culture of *Osc . trichoides*, strain DG-6 (Keppen *et al.,* 1993, 1994; Ivanovsky *et al.,* 1999).

The accessibility of sequence data (5S and 16S rRNA) links *Osc .trichoides* DG-6 to the *Chloroflexus* and *Heliothrix* group of organisms (**Fig. 3.99**) (Keppen *et al.,* 2000). The low DNA hybridization results reported by Keppen *et al.* (1994) with *Chloroflexus*, the higher mol% G+C content in *Oscillochloris*, and the low (82.5–86.5%) sequence comparability in 16S rRNA gene between

Chloroflexus and *Oscillochloris* show that various families are most likely justified for these two genera. However, the fact that original reference material for *Osc .chrysea* and *Osc .trichoides* culture SR-1 is no longer available as they are failed to be obtained in pure culture. Whereas *Osc .trichoides* strain DG-6 is available in pure culture. *Osc .chrysea* has been assigned as the type species for *Oscillochloris*. Culture SR-1 was not validated as the type strain for *Osc . trichoides* and therefore it has been lost. The new strain DG-6 is now the type strain for this species.

2. *Chloronema (Cln.)* (green filament) (Dubinina and Gorlenko, 1975) Gorlenko and Pierson (2015)

Cells are cylindrical, 2.0-2.5 × 3.5-4.5 µm in diameter, combined in a trichome surrounded by a sheath. Trichomes are straight or spiral (75-250 µm long), appear yellow-green (**Fig. 3.103A**). Trichomes are surrounded by a slimy sheath of varying thickness. In iron salts containing, lakes, the sheaths of *Chloronema* develop with iron oxides, sometimes in significant amounts. In this case, the sheath appears brown colour. Cells multiply by the separation of sections of trichomes (hormogonia) of varying lengths. Motile by gliding. The rate of gliding is about 10 µm/s. The major pigment is bacteriochlorophyll *d* and also contains small amounts of bacteriochlorophyll *c* and *a.* The photosynthetic structures are chlorosomes which are found in the periphery of cells but also occur along the septa with the membrane invaginations. The cytoplasmic membrane is mesotype with multiple invaginations. During cell division, the cytoplasmic membrane extends into the inner portion of the cell wall. The central part of each cell is filled with elongated gas vesicles (**Fig. 3.103B**). They are capable of anoxygenic photosynthesis. *Chloronema* requires low light intensity and low temperature (4-15°C). Metalimnion and upper hypolimnion of stratified freshwater lakes with high ferrous iron content and no/or low concentration of H_2S are the selective habitats for their growth. These species develop along with several other species of purple- and green sulfur bacteria. The mol% G + C of the DNA has not been determined. *Chloronema* is monotypic and *Cln. giganteum* is the type species. The characteristic feature of *Chloronema* is presented in **Table 3.47.**

The sequencing of the 16S rRNA gene of *Cln. giganteum* in a mixed culture through molecular techniques, its phylogenetic position has been settled (Gich *et al.*, 2001). The *Chloronema* sp. strain UdG9001 was isolated as a mixed culture containing a non-phototrophic bacterium and its spectroscopic properties were analysed (Gich *et al.*, 2003). The absorption peak of chlorosomes in *Chloronema* sp. strain UdG9001 is recorded at 745 nm and this value is different in previously reported i.e., at 720 nm (Dubinina and Gorlenko 1975).

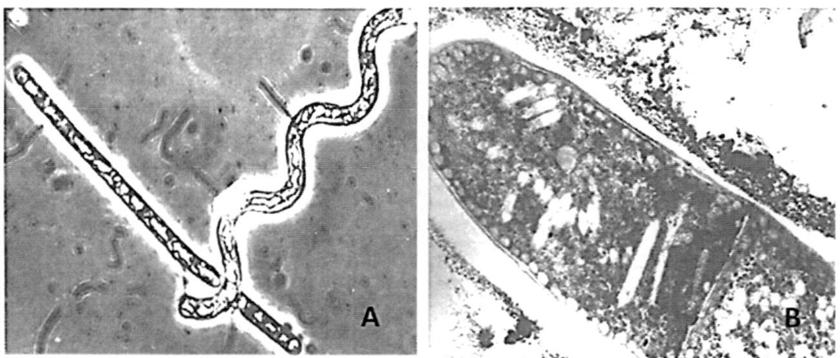

Fig. 3.103. A. Phase-contrast micrograph of *Cln. giganteum* (X 2000); B. Electron micrograph of ultrathin section of *Cln. giganteum* fixed with OsO_4 and glutaraldehyde (X 20,000).

Differentiation of the *Chloronema* from other Genera

The major differentiating characteristics of *Chloronema* from other genera of multicellular filamentous green bacteria are precised in **Table 3.47** .

Taxonomic Comments

The lack of cultures with appropriate physiological information and the lack of any molecular phylogenetic information bargain our understanding of the legitimacy of the genus and species assignments of these life forms. The characters that recognize them from the other chlorosome-containing genera (*Chloroflexus* and *Oscillochloris*) are not many and variable enough to warrant alert in the taxonomy of this group. Its large size, presence of gas vacuoles, bacteriochlorophyll *d*, and planktonic distribution lead to its acknowledgment *in situ*. As in the case of the other conspicuous uncultured filamentous phototrophs, the presence of *Chloronema* is prominent in the habitats where it is found and various examiners have seen it and studied it *in situ* (Steenbergen and Korthals, 1982; Steenbergen *et al.,* 1989; Eichler and Pfennig, 1990). Abella and Garcia-Gil (1992) studied the distribution of the straight and spiral forms in a few lakes documenting the significance of gas vacuoles and morphology to the migration of *Chloronema.* It is a major constituent of the plankton in lakes and subsequently may have a huge in spite of the fact that till now not yet determined environmental impact. As a result of its widespread occurrence and clear significance, the genus should probably remain until culture or molecular information an explanation of its taxonomy and phylogeny. In its present structure, the description allows the recognition of the organisms *in situ* which thusly guarantees their proceeded study.

Family 3. *Roseiflexaceae* (Gupta *et al.,* 2013) Chander and George (2013)

This family consists of two genera, *Roseiflexus* (Hanada *et al.*, 2002) and *Heliothrix* (Pierson *et al.*, 1985) of straight filamentous bacteria showing gliding mobility. Facultative phototrophs with the optimal growth temperature between 45°C and 55°C. Filaments are reddish-brown to reddish-orange in colour. Bacteriochlorophyll *a* and carotenoids, derivatives of γ-carotene and ketomyxocoxanthin are presented as photosynthetic pigments. Chlorosomes are absent. The fatty acids are $C_{16:0}$, $C_{14:0}$, $C_{15:0}$ differ from those of members of *Chloroflexus* and *Oscillochloris* (Hanada and Pierson 2006). The family is named after the genus *Roseiflexus* as species of *Heliothrix* is not available in axenic culture. A number of conserved signature indels (CSIs) in the following proteins are specifically observed in the sequenced *Roseiflexus* spp; (i) 3 aa insert in the DUF59 protein, (ii) conserved inserts of 1 aa each in a PSP1 domain-containing protein, in Ser (Thr)-tRNAhydrolase protein and in the exonuclease ABC subunit C protein. However, sequence data for these CSIs (genes) is not available for *Heliothrix* spp., few of them are expected to provide molecular markers for the entire family and suborder. Type genus is *Roseiflexus* (Hanada *et al.,* 2002).

1. *Roseiflexus* (*Rof.*) (Rose-coloured bending bacteria) (Hanada *et al.,* 2002) Gupta et *al.* (2013)

Cells are unbranched, multicellular filaments 0.8-1.0 showing gliding motility (**Fig. 3.104**). Gram (-) negative. Bchl *a* and γ-carotene derivatives are present. Chlorosomes and Bchl *c are* absent. Cultures are red to reddish-brown in colour. Thermophilic, facultative phototrophs that grow at 45-55°C and pH 7.0-9.0. Aerobic dark growth occurs at full atmospheric oxygen tension. Neither photoautotrophic nor fermentative growth is observed. The major respiratory quinone is MK-11. Cellular fatty acids consist of saturated $C_{14:0}$, $C_{15:0}$ and $C_{16:0}$, unsaturated fatty acids are not detected. DNA G-C content is 62.0 mol% (by HPLC). Type species is *Roseiflexus castenholzii* (Hanada *et al.,* 2002). Single species is validly published and their characteristics are presented in **Table 3.47.**

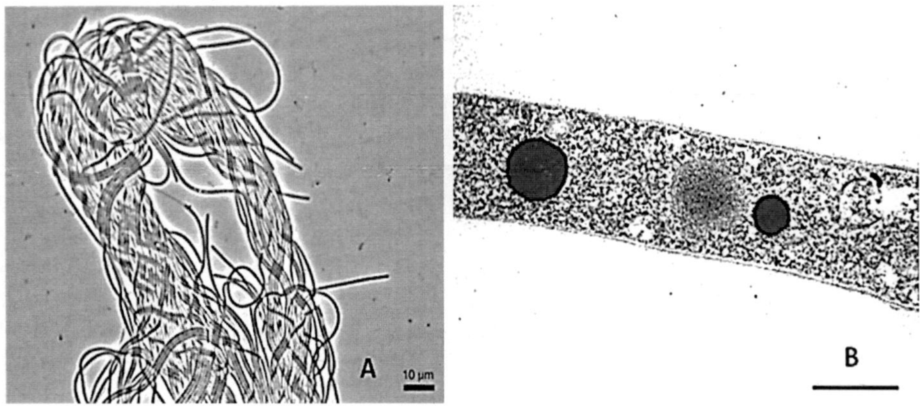

Fig. 3.104. A. Phase-contrast micrograph of *Roseiflexus castenholzii* HLO8^T which shows multicellular filamentous morphology (Bar = 10 µm), B. Ultrathin section of the cells, no evidence of intracellular vesicles or extensive internal membranes (Bar = 0.5 µm).

Differentiation of the *Roseiflexus* from other Genera

The differential characteristic of *Roseiflexus* from other multicellular filamentous green bacteria is precised in **Table 3.47** .

Taxonomic Comments

Phylogenetic analysis based on 16S rDNA sequence showed that *Roseiflexus castenholzii* strain HLO8^{T} was assigned to the anoxygenic filamentous phototrophs (**Fig. 3.105**). The most firmly related grouping to *Roseiflexus castenholzii* strain HLO8T was that of OS type C', which was not a sequence of an isolated strain but rather a 16S rDNA clone recovered from Octopus Spring in Yellowstone (Weller *et al.,* 1992). The sequence comparability was significantly high (95.7%; the compared length was 1087 bp). The position of the isolate was anyway absolutely from different members of this group. The sequence similarities of this species with other related species such as *Chlorflexus aurantiacus* is 83.2% (1343 bp), *Chloroflexus aggregans* is 82.1% (1356 bp), *Osc .trichoides* is 83.8% (1375 bp) and *Htr .oregonensis* is 78.6% (868 bp). The sequence similarities to these species were showing less than 83.2% and these values were low enough for the classification of a new genus in this group (Stackebrandt and Goebel, 1994).

Based on phenotypic properties of *Roseiflexus castenholzii* strain HLO8^{T} , supported the creation of a new taxon. There was evident contrast *in vivo* spectra of the near-IR region between the present isolate and its relatives. The *in vivo* absorption spectrum of the isolate showed different peaks at 878 and 801 nm due to the presence of bacteriochlorophyll *a*, probably bound to membranous proteins. Peaks in the absorption spectrum of *Htr .oregonensis* were marginally blue shifted (peaks at 865 and 795 nm). In *Chloroflexus* spp., peaks were recorded at 868 and 808 nm, with an important peak at 740 nm indicates the presence of bacteriochlorophyll *c* (chlorosomes). In *Osc .trichoides,* absorption peaks were recorded at 852 and 748 nm indicates the presence of bacteriochlorophyll *a* and *c*, respectively. Chlorosomes in the genus *Chloronema* contained bacteriochlorophyll *d* instead of *c* which showed an absorption peak at 720 nm (Dubinina and Gorlenko, 1975; Gorlenko, 1988 .)

The phenotypic data used for the differentiation of *Roseiflexus castenholzii* strain HLO8^{T} from other related species include the absence of chlorosomes, which is a typical feature of *Chloroflexus* spp. and relatives, contains MK-11 as a major quinone and saturated cellular fatty acids, whereas related species contained unsaturated fatty acids. The genomic G+C content of the isolate was 62.0 mol%, which is marginally higher than values for related species (under 60 mol%). In view of these genotypic and phenotypic characteristics, the name *Roseiflexus* was created with type species *Roseiflexus castenholzii* (Hanada *et al.,* 2002).

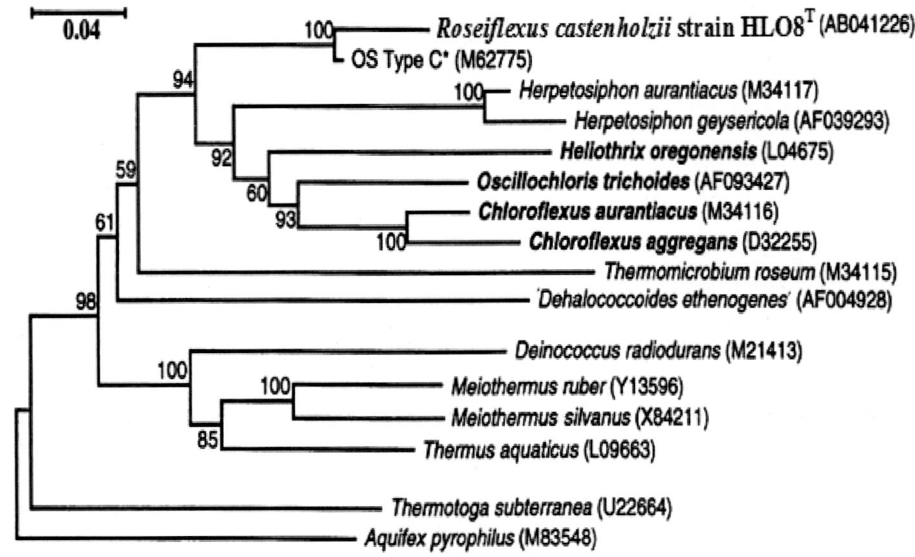

Fig. 3.105. Phylogenetic tree showing the relationship between *Roseiflexus castenholzii* from other members of *chloroflexi* based on 16S rDNA gene sequence analysis.

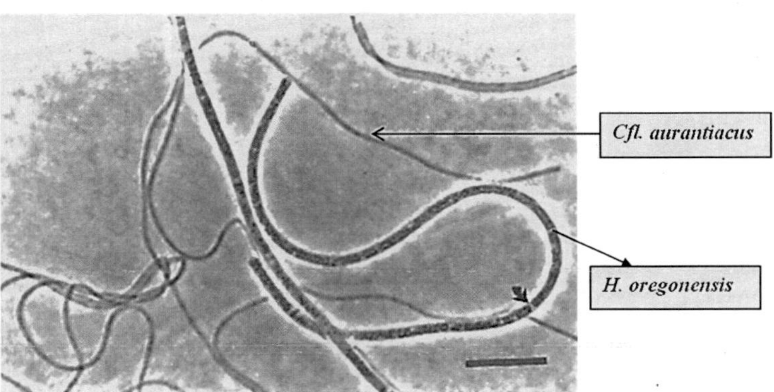

Fig. 3.106. Phase-contrast micrograph of *Htr. oregonensis* (Bar = 10 μm). The narrow filaments of *Cfl. aurantiacus*are added to the collection for comparison. A cross-wall is indicated by an arrow.

2. *Heliothrix (Htr.)* (Sun hair) (Pierson *et al.,* 1986) Pierson and Castenholz (2015)

Cells of these bacterium form septate filaments of indefinite length unbranched and undifferentiated. Cells are about 1.5 μm in diameter, much longer than breath (**Fig. 3.106**). The thin sheath may be present or not. Cell division by fission in one plane only. Gliding motility, no flagella is seen and Gram (-) negative. Cells are granular, poly-β-hydroxybutyric acid inclusion bodies are seen. Gas vacuoles and internal membranous structures are unknown. Bacteriochlorophyll *a* is present, but accessory bacteriochlorophylls or chlorosomes are absent. Filaments appear bright orange due to the presence of carotenoids such as oxygenated derivatives of γ-carotene and glycosides of γ-carotene. Aerotolerant, grow photoheterotrophically under anaerobic conditions in the presence of light with acetate. Growth occurs at optimum temperature 40–50°C (range 35°C–56°C). Top layers over cyanobacterial mats in relatively few hot springs in Oregon (U.S.A) are the ideal habitats for their growth. *Heliothrix oregonensis* (Pierson *et al.,* 1985) is the type species. The characteristic features of this species are presented in **Table 3.47.**

Differentiation of the *Heliothrix* from other Genera

The major differentiating characteristics of *Heliothrix* from other genera were presented in **Table 3.47.** Based on 16S rRNA gene sequence analysis, *Htr. oregonensis* was partially determined (Weller *et al.,* 1992) and phylogenetically distinct from another genus and species in the family, *Roseiflexus castenholzii* (the sequence similarity is 78.6%) (**Fig. 3.105**). However, both the genus share the common phenotypic features such as the absence of chlorosomes, bacteriochlorophyll c and b-carotene and the presence of γ- carotene derivatives (precisely, keto-3', 4'-didehydro-OH- γ-carotene, myxocoxanthin) as major carotenoids (Pierson *et al.,* 1985; Takaichi *et al.*, 2001; Hanada *et al.,* 2002).

Taxonomic Comments

Based on the complete nucleotide sequence of the 5S rRNA of *Htr. oregonensis,* a similarity of 0.767 was recorded compared with that of *Chloroflexus aurantiacus*. However, these two bacteria are not very closely related, they are related to each other than to any other phototrophs (Pierson *et al.,* 1985). The partial sequence of the 16S rRNA of *Heliothrix oregonensis* showed that it was a different genus from *Chloroflexus* and was more closely related to *Chloroflexus* (Ward *et al.,* 1998). *Chloroflexus* specific antibodies that reacted substantially with 13 different strains of *Chloroflexus* isolated from all over the world showed no cross-reactivity with *Htr. oregonensis* (Tayne *et al.,* 1987).

3.5. *FIRMICUTES* (Gibbons and Murray, 1978) Murray (1984)

The *Firmicutes* are a phylum of bacteria that consists of Gram (+) positive cell wall structure. *Megasphaera, Pectinatus, Selenomonas*, and *Zymophilus* are genera that have a porous pseudo-outer membrane that causes them to stain Gram (-) negative. This phylum consists of 3 classes *Bacilli* , *Clostridia* and *Thermolithobacteria.* Scientists once classified the *Firmicutes* to include all Gram (+) positive bacteria, but have recently defined them to be of a core group of related forms called the low G+C group, in contrast to the *Actinobacteria* . They have round cells, called cocci (singular coccus), or rod-like forms (*Bacillus*). Many *Firmicutes* produce endospores, which are resistant to desiccation and can survive extreme conditions. They are found in various environments and the group includes some notable pathogens. Among different families of *Firmicutes* , *Heliobacteriaceae* members are the photosynthetic bacterium, which produces energy through photosynthesis. As the members of *Heliobacteriaceae* belong to anoxygenic phototrophic bacteria, which are included in the class *Clostridia* and order *Clostridiales*. Hence, it is discussed briefly.

Class I. *Clostridia* Rainey (2015)

The class is phenotypically, chemotaxonomically, physiologically and ecologically diverse. Both Gram (+) positive and Gram (-) negative organisms are included in this class. It is a phylogenetically coherent cluster based on the analysis of 16S rRNA gene sequences. The class contains the orders *Clostridiales* ,*Halanaerobiales* and *Thermoanaerobacterales*. The type order is *Clostridiales* (Prevot 1953).

Order I. *Clostridiales* (Prevot, 1953) Rainey (2015)

The order is phenotypically, chemotaxonomically, physiologically and ecologically diverse group of bacteria, contains both Gram (+) positive and Gram (-) negative bacteria. The order forms a phylogenetically relevant cluster based on the analysis of 16S rRNA gene sequences. The order contains the families *Clostridiaceae* ,*Eubacteriaceae* ,*Gracilibacteraceae* ,*Heliobacteriaceae* ,*Lachnospiraceae* , *Peptococcaceae* , *Peptostreptococcaceae* , *Ruminococcaceae* , *Syntrophomonadaceae* and *Veillonellaceae* alongwith nine families *Incertae sedis*. The type genus is *Clostridium* (Prazmowski, 1880). Among these families, *Heliobacteriaceae* includes photosynthetic bacteria. Therefore the characteristics of *Heliobacteriaceae* are briefly described.

Family 1. *Heliobacteriaceae* (Madigan, 2001) Sattely and Madigan (2014)

The family *Heliobacteriaceae* belongs to the order *Clostridiales* of the phylum *Firmicutes* is the only group of phototrophic bacteria included. The closest relative of this family is the *Peptococcaceae*, a family that contains genera such as *Desulfotomaculum, Pelotomaculum* and *Desulfitobacterium*. The latter genus contains species that are closely related to the members of *Heliobacteriaceae* . The family *Heliobacteriaceae* consists of 4 genera such as *Heliobacterium* , *Heliobacillus* ,*Heliophilum* and *Heliorestis* (Asao and Madigan, 2009). A fifth genus was recently added with a novel heliobacterial strain, designated *Candidatus*, *Heliomonas lunata*, which is yet to be obtained in pure culture (Asao *et al.,* 2012). A total of 12 species within the *Heliobacteriacaea* has been described (Asao *et al.*, 2012). *Heliobacterium (Hbt.) chlorum* was the first *Heliobacteria* to be studied (Gest and Favinger 1983). Furthermore, as the cells of *Hbt. chlorum* are susceptible to lysis upon reaching the late exponential phase, few laboratories could culture this species for their research studies. It is a *Heliobacterium modesticaldum,* thermophilic bacteria that has emerged as a model organism for biochemical and genomic studies of heliobacterial physiology and photosynthesis (Sattley and Swingley, 2013).

This family includes all the anoxygenic phototrophic bacteria that synthesize bacteriochlorophyll (Bchl) *g*as the major photosynthetic pigment. In addition, small amounts of 8^1-OH-Chlorophyll (Chl) *a* and Bchl *g* are esterified with the C_{15} alcohol farnesol. *Heliobacteria* members synthesize C_{30} carotenoids either 4, 4'-diaponeurosporene, while in alkaliphilic species, OH-dianeurosporene glucoside esters are seen (Takaichi *et al.,* 2003). Internal photosynthetic membrane systems contain chlorosomes, chromatophores, lamella, or thylakoids. Heliobacterial pigments are confined to cytoplasmic membrane-bound photosynthetic reaction centers (Miller *et al.,* 1986).

Cells are straight, curved, or coiled rods or spirillum-shaped cells varying in length from few micrometers to filaments of 20 μm or more. Heliobacterial members exhibit flagellar motility by means of polar, sub-polar or peritrichous flagella and some are capable of gliding movement. Multiply by binary fission. Like other *Firmicutes* but unlike all other phototrophic bacteria, cells of heliobacteria stain gram (-) negative. However, an outer membrane is lacking. They are strictly anaerobic. Growth occurs photoheterotrophically under anaerobic conditions with a limited number of organic carbon sources. In non-alkaliphilic species growth occur chemoorganotrophically in the dark by fermentation of pyruvate or other simple organic compounds (Kimble *et al.,* 1994). Pyruvate is used by all heliobacteria. Aerobic or microaerobic dark growth has not been observed. In addition, photoautotrophic growth on CO_2, H_2 or on CO_2+ H_2S has not been observed in any species of heliobacteria, and the genome of *Heliobacterium modesticaldum* (Sattley *et al.,* 2008) lacks genes encoding proteins of any known autotrophic pathway.

Heliobacteria vary in their sulfide tolerance, few species are quite a sulfide sensitive and others are quite tolerant. However, sulfide present in the media is oxidized and precipitates in the form of elemental sulfur and deposited outside the cells. Most of the heliobacteria grow at optimal pH 7.0 (range of 5.5-8.0) depending on the species. Species of the *Heliorestis* are alkaliphilic and grow at optimum pH 9.0 (range 8.0-10.0). The optimal temperature is 37-42°C (52°C for thermophilic species) and 30-32°C for alkaliphilic species. Ammonium salts and glutamine are used as nitrogen sources by all heliobacteria. All heliobacteria that have been tested fix nitrogen (Kimble and Madigan, 1992; Madigan, 1995). Biotin is required as a growth factor. Some species require to reduce sulfur compounds such as sulfide, thiosulfate, or cysteine for their growth. Neither poly-β-hydroxybutyrate nor glycogen is produced by these bacteria. Absorption spectra of living cells of heliobacteria show a major peak at 784-792 nm (species dependent) due to absorption by bacteriochlorophyll g and comparatively small peaks are observed at about 575 and 670 nm is due to a small amount of 8^1-OH-chlorophyll present in the reaction center. The cell wall of heliobacteria is fragile and stationary phase cells of many species form sphaero blasts. The peptidoglycan of the heliobacteria is present in small amounts and contains L, Ldiaminopimelic acid instead of meso-diaminopimelic acid as the diamino acid. They are primarily found in soil and in this property, they differ significantly in their ecology from purple- and green anoxygenic phototrophic bacteria. *Heliobacterium modesticaldum* is thermophilic and inhabits neutral to alkaline hot springs worldwide at temperatures below 60ºC.

Phylogenetic studies based on 16S rRNA gene sequence analyses revealed that the species of the family *Heliobacteriaceae* form a monophyletic group inside the *Firmicutes* that bifurcates into two particular clades in neighbor-joining phylogenetic trees. The significant clade contains the genera *Heliobacterium* ,*Heliobacillus* and *Heliophilum*, while the minor clade contains all the assigned species of *Heliorestis* (**Fig. 3.107**) (Sattley and Madigan, 2014). Before the identification of *Candidatus Heliomonas lunata*, an uncommon alkaliphilic heliobacterium was cultured from the sediment-water interface of Soap Lake, Washington, USA (Asao *et al.*, 2012). The two clades of heliobacteria exhibited a clear phylogenetic separation indicating particular ancestors for the neutrophilic *Heliobacteria* and the alkaliphilic *Heliorestis* species. In any case, the grouping of *Candidatus Heliomonas lunata* inside the clade containing neutrophilic heliobacteria obscures the phylogenetic delineation of these physiologically and ecologically particular heliobacteria. Even though they are not closely related to neutrophilic species of heliobacteria i.e., <94% sequence similarity. The addition of *Candidatus Heliomonas lunatic* lose to the major clade proposes a possible alkaliphilic origin for the heliobacteria.

The four described species of *Heliorestis* i.e., *Hrs. daurensis* ,*Hrs. baculata* , *Hrs. convolute* and *Hrs. acidaminivorans* form a tightly packed clade, with 16S rRNA gene sequence similarities ranging from 94% to 98% (**Fig. 3.107**). Species of the major clade made up of the other three built-up genera of heliobacteria demonstrate a high degree of level of separation than within the alkaliphilic clade. The only species of the genus *Heliophilum* is *Heliophilum fasciatum* is the most revolutionary separation of the neutrophilic species showing <93% 16S rRNA gene sequence similarity to any other heliobacterium. This species is additionally recognized from other heliobacteria dependent on several phenotypic distinctions and accordingly, its existence as a novel genus is very much supported (Asao and Madigan, 2010).

Heliobacterium chlorum groups intimately with *Heliobacillus mobilis*, the sole representative of the genus (**Fig. 3.107**). In spite of their taxonomic distinction, the two species have 16S rRNA gene sequences that are about 98% similar. By this standard, *Hbt. chlorum* is more closely related to *Hba. mobilis* than it is to different species of *Heliobacterium* such as *Hbt. gestii, Hbt. modesticaldum, Hbt. sulfidophilum* or *Hbt. undosum* to which it shows <96% 16S rRNA gene sequence similarity. Subsequently, this is a point of taxonomic vulnerability that warrants further investigation. In light of their nearby phylogenetic relationship, as well as several shared phenotypic properties appears to be likely that *Hbt. chlorum* and *Hba. mobilis* are two species of the same genus. Genuinely, the genus designation *Heliobacillus* ought to be relinquished with *Hba. mobilis* being absorbed into the genus *Heliobacterium*, this was suggested by Bryantseva *et al.* (2000b) and Asao and Madigan (2010). In any case, a second probability is that *Heliobacillus* is a valid genus and should be assigned considered in this manner likely bringing the renaming of the *Hbt. chlorum* as a species of *Heliobacillus*. Genomic DNA-DNA hybridization studies have just been used for the correct classification of several species of heliobacteria and the use of such an analysis in this situation would resolvethis issue.

Genomic DNA-DNA hybridization (DDH) studies support taxonomic conclusions for few, yet not all, type strains of heliobacteria. The initial descriptions for most species of *Heliorestis* included DDH analysis. Bryantseva *et al.* (2000a) determined *Hrs. baculata* strain OS-H1T to have a genomic DNA similarity of 43% to *Hrs. daurensis*, the only described species of the genus at the time. Afterward, Asao *et al.* (2006) determined a 48% DNA similarity among *Hrs. convoluta* strain HH^T and *Hrs. daurensis*, its nearest phylogenetic relative based on a 16S rRNA gene sequence identity of 97.8%. DDH analyses for other heliobacteria have been restricted to a solitary report, in which two novel species of *Heliobacterium* , *Hbt. sulfidophilum* and *Hbt. undosum* were described (Bryantseva *et al.,* 2000b). The author's determined *Hbt. sulfidophilum* strain $BR4^T$ showed 25% and 38% DNA similarity to *Hba. mobilis* and *Hbt. chlorum* , respectively. In addition, *Hbt. sulfidophilum* strain $BR4^T$ was found to have a 25% DNA similarity to *Hbt. undosum* strain $BG29^T$, an isolate with which it shared approx. 97.5% 16S rRNA gene sequence similarity (Bryantseva *et al.,* 2000b). DDH studies have not been conducted for the remaining described species of heliobacteria, including *Hph. fasciatum* , *Hbt. chlorum* , *Hbt. modesticaldum* and *Hrs. acidaminivorans* .

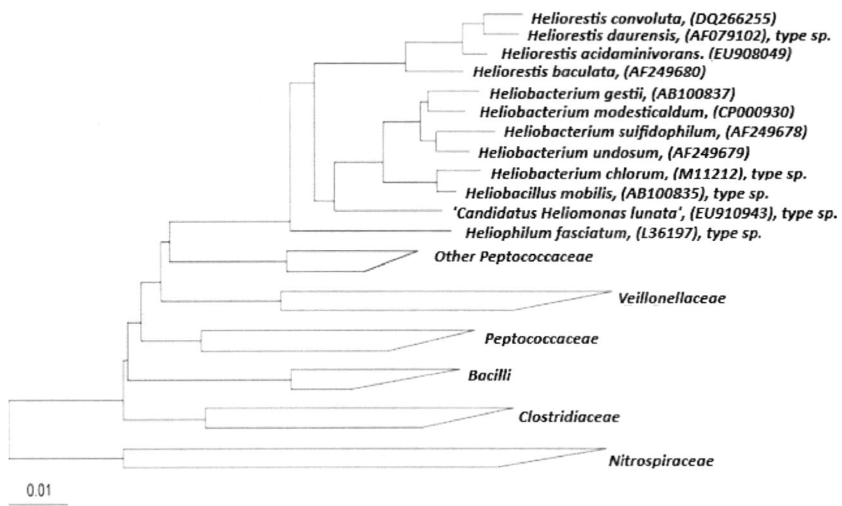

Fig. 3.107. Phylogenetic tree of *Heliobacteriaceae* members and related *Firmicutes* based on 16S rRNA gene sequence analyses.

1. *Heliobacterium* (*Hbt.*) (Sun bacterium) (Gest and Favinger, 1985) Madigan (2015)

Cells are rod-shaped are slightly curved, 0.8-1.0 x 2.5-9.0 µm **(Fig. 3.108 A)** or spirilla, 1.0-1.2 µmin width and are with similar length to rod-shaped cells (**Fig. 3.108B**) (Ormerod *et al.,* 1996). They are Gram (-) negative show gliding motility. Bacteriochlorophyll *g** is the major photosynthetic pigment, intracytoplasmic membranes are extensively developed similar to that of anoxygenic photosynthetic bacteria (**Fig. 3.108 C**) (Kimble *et al.,* 1995). Cells do not contain chlorosomes or gas vacuoles. Mature cells are typical ~1 µm wide and variable in length. The colour of cell culture is brownish-green in young cultures which tends to become emerald-green with age. Growth occurs at optimal temperature 35-42°C (range of 20-45°C). These bacteria can be grown from dry soil under strictly anaerobic conditions using intense light (10,000 lux) at 40ºC. They are strictly anaerobic and growth occurs photoheterotrophically in the light by utilizing few organic compounds. These are nitrogen fixers particularly in rice fields and can be exploited as biofertilizers. Both 16S rRNA analysis and lack of lipopolysaccharide in the cell wall place these organisms with gram (+) positive bacteria and it has been suggested that they are phylogenetically related to *Clostridia*. Further, both the groups produce the same type of endospores. The mol% G + C of the DNA of the type strain is 52 (Tm). *Heliobacterium chlorum* is the type species. The characteristics of five species of *Heliobacteria* are précised in **Table 3.48.**

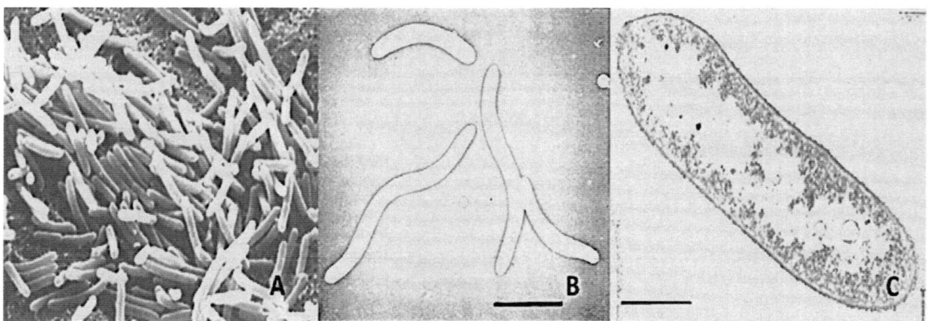

Fig. 3.108. A. Scanning electron micrograph of *Hbt. chlorum* cells (X 5000), B. Cells of *Hbt. gestii* (Bar = 5 µm), C. Transmission electron micrograph of ultrathin section cells of *Hbt. modesticaldum* strain Ice1 (Bar= 1 µm).

Differentiation of the *Heliobacterium* from other Genera

The differentiating characteristics of *Heliobacterium* from other genera are shown in **Table 3.48**. The genus *Heliobacterium* is differentiated from other genera of heliobacteria based on morphological characteristics i.e., motility (gliding or sub-polar flagella in species of *Heliobacterium* while it is peritrichous in *Heliobacillus*) and that cells exist as single entities rather than as cell bundles that move as a unit in *Heliophilum* culture. The *Heliobacterium* is distinguished from the *Heliorestis* by an inability to grow at pH 9.0 and an optimal growth temperature near 40ºC instead of 30ºC. As a group, heliobacteria are differentiated from all other anoxygenic phototrophs due to the presence of bacteriochlorophyll *g*. The genetic relationship among *Heliobacterium* and other members of *Heliobacteriaceae* based on 16S rRNA gene sequence is illustrated in **Fig. 3.107** .

2. *Heliobacillus* (*Hba.*) (Sun rod) (Romero and Gest, 1998) Madigan (2015)

Cells are rod-shaped, motile by means of peritrichous flagella (**Fig. 3.109A** .) Gram (-) negative, multiply by binary fission. Photosynthetic pigments are bacteriochlorophyll *g*, appear green in colour. Strictly anaerobic, growth occurs photoheterotrophically in the presence of light under anaerobic conditions and chemotrophically under dark in the presence of simple organic substrates. The intracytoplasmic membrane system is well developed, chlorosomes are absent (**Fig. 3.109 B**). They are nitrogen fixers. Acetate, pyruvate, lactate, butyrate, CO_2 and yeast extract are utilized photoheterotrophically. The optimal growth was observed at pH 6.5-7.0 and temperature at 38-42ºC. The mol% G + C of the DNA of the type strain is 50.3. A monotypic genus with single-type species *Hba. mobilis* (Romero and Gest, 1998) and their characteristics are presented in **Table. 3.48** .

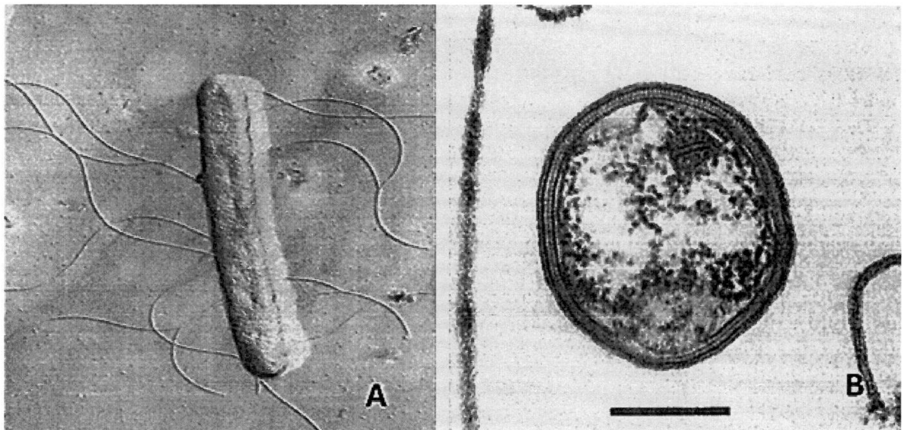

Fig. 3.109. A. Electron micrograph of *Heliobacillus mobilis* showing peritrichous flagella (X 11800), **B.** Ultrathin section of cells showing intracellular photosynthetic membrane (Bar= 1μm).

Differentiation of *Heliobacillus* from other Genera

The differentiating characteristics of *Heliobacillus* from other genera of *Heliobacteriaceae* are presented in **Table 3.48** and the genetic relationship among *Heliobacillus mobilis* and other *Heliobacteriaceae* members is based on 16S rRNA gene sequence is illustrated in **Fig. 3.110.**

Table 3.48. Differentiating Characteristics of Genus and Species of *Heliobacteriaceae*

Characteristic	*Heliobacterium chlorum*	*Hba. gestii*	*Hba. modesticaldum*	*Hba. sulfidophilum*	*Hba. undosum*	*Heliobacillus mobilis*	*Heliophilum fasciatum*	*Heliorestis daurensis*	*Hrs. baculata*	*Hrs. convoluta*	*Hrs. acidaminivorans*	*Candidatus Heliomonas lunata*
Habitat	Temperate soil	Tropical paddy soil	Neutral/ alkaline hot spring and volcanic soils	Alkaline sulfidic hot spring	Alkaline sulfidic hot springs	Dry soil	Tropical paddy soil	Microbial mats of brackish soda lakes	Soda lake shoreline soil	Soda lake shoeline soil	Shoreline sediments of lake	Water or benthic sediments of lake
Morphology	Rod	Spirillum	Rod/ Curved rod	Rod	Rod/ slightly twisted spirillum	Rod	Straight rods with tapered ends grouped in bundles of two to many cells	Coiled to the bent rod or short filament	Rod/ Curved rod	Coil	Rod	Curved rod
Cell size (µm)	1.0 x 7.0-9.0	1.0 - 1.2 x 5.0-10	0.8-1 x 2.5-9	0.6-1 x 4.0-7.0	0.8-1.2 x 7.0-20.0	1.0 x 7.0-10	0.8-1.0 x 5.0-8.0	0.8-1.2 x 10.0-20.0	0.6-1.0 x 6.0-10.0	0.6 x variable lenght	0.6-0.9 x 3.0-12.0	0.6 x 2.0-7.0
Motility	Gliding	Multiple Subpolar flagella	Flagella or none	Peritrichous flagella	Peritrichous flagella	Peritrichous flagella	Polar to Subpolar flagella; cell bundles move as a unit	Peritrichous flagella	Peritrichous flagella	Unknown	Flagella	Flagella
Major carotenoides	4, 4'-diaponeurosporene	4, 4'-diaponeurosporene	4, 4'-diaponeurosporene	Neurosporene	Neurosporene	4, 4'-diaponeurosporene	4, 4'-diaponeurosporene	OH-diaponeurosporene glucoside esters	OH-diaponeurosporene glucoside esters	OH-diaponeurosporene glucoside esters	OH-diaponeurosporene glucoside esters	OH-diaponeurosporene glucoside esters
Absorption spectra of living cells (nm)	375, 575, 670, 718 & 788	375, 415, 480 , 575, 670, 718 & 788	430, 575, 670, 718 & 788	375, 412, 575, 670 & 788	370, 412, 576, 671, 720 & 790	375, 575, 670, 718 & 788	415, 480, 577, 673, 723 & 792	370, 405, 576, 671 & 788	370, 413, 575, 671, 755 & 788	572, 592, 666, 746 & 786	478, 575, 591, 670, 749 &787	451, 480, 576, 592, 670, 744 & 789
Nitrogen fixation	+	+	+	+	ND	+	+	ND	ND	+	+	-
Growth factors	Biotin	Biotin	Biotin	Biotin	Biotin	Biotin	Biotin	Biotin	Biotin	Biotin	Not required	Yeast extract
Endospore formation	-	+	-	+	+	+	-	-	+	+	+	-

Characteristic	*Heliobacterium chlorum*	*Hba. gestii*	*Hba. modesticaldum*	*Hba. sulfidophilum*	*Hba. undosum*	*Heliobacillus mobilis*	*Heliophilum fasciatum*	*Heliorestis daurensis*	*Hrs. baculata*	*Hrs. convoluta*	*Hrs. acidaminivorans*	*Candidatus Heliomonas lunata*
Optimum temperature (° C) (range)	38-42	40 (38-42)	50-52 (25-56)	32	31-36	38-42	37-40	25-30 (25-35)	30.0	25-30	30-37 (18-45)	25-30 (18-37)
Optimum pH (range)	6.2-7.0	7.0	6.0-7.0	7.0-8.0	7.0-8.0	6.2-7.0	7.0	9.0 (7.5-10.0)	8.5-9.5 (8.0-10.2)	9.0	8.0-9.0 (8-10.5)	8.0-9.5
Optimum NaCl %)(	ND	Not required tolerate upto 1.0%	Not required	Not required	Not required	ND	Not required	Not required	0.5-1.0	Not required	0.5-4.0	1.5-3.0
G+C content (mol%)	52	54.5-55.6	54.6-55	51.3	57.2-57.7	50.3	51.8	44.9	45.0	ND	ND	ND
Pyruvate fermentation	+	+	+	+	+	+	+	+	+	ND	-	-
Substrates utilized	Pyruvate, lactate, Yeast extract	Pyruvate, Lactate, Acetate, butyrate, Fructose, Glucose, Ethanol, +(CO_3) Yeast extract	Pyruvate, Lactate, Acetae, Yeast extract	Pyruvate, Lactate, Acetate, Butyrate, (+ CO_2,) Casein hydrolysate, Yeast extract	Pyruvate, Lactate, Acetate, Casein hydrolysate, Propionate, yeast extract	Pyruvate, Lactate, acetate, Yeast extract, Butyrate (+CO_2)	Pyruvate, Lactate, Acetate, Butyrate (+CO_2)	Pyruvate, Acetate, Butyrate, Propionate +(HCO_3^- / CO_3^{2-})	Pyruvate, Lactate, Acetate, (+ CO_3^{2-})	Pyruvate, Acetate, Butyrate, Propionate, (+HCO^{3-} / CO_3^{2-})	Pyruvate, Acetate, Casamino acids, Propionate, Yeast extract, Alanine, Aspartate, Arginine, Glucose, Glutamate, Glutamine, Histidine, Lysine, Methionine, Serine	Pyruvate, Butyrate, Malate, Yeast extract, Alanine, Lysine, Serine, Threonine

Symbols: +, Positive/present; -, Negative / absent; ND, Not determined/No data available. Data for members of *Heliobacteriaceae* is taken from Gest and Favinger (1983) (*Heliobacterium chlorum*), Ormerod *et al.* (1996) (*Hba. gestii & Heliophilum fasciatum*), Kimble *et al.* (1995) (*Hba. modesticaldum*), Bryantseva *et al.* (2000) (*Hba. sulfidophilum & Hba. undosum*), Beer-Romero and Gest (1987) (*Heliobacillus mobilis*,) Bryantseva *et al.* (1999) (*Heliorestis baculata, Hrs. daurensis*), Asao and Jung (2006) (*Hrs. convoluta)* and Asao *et al.* (2012) (*Hrs. acidaminivorans & Candidatus Heliomonas lunata* .)

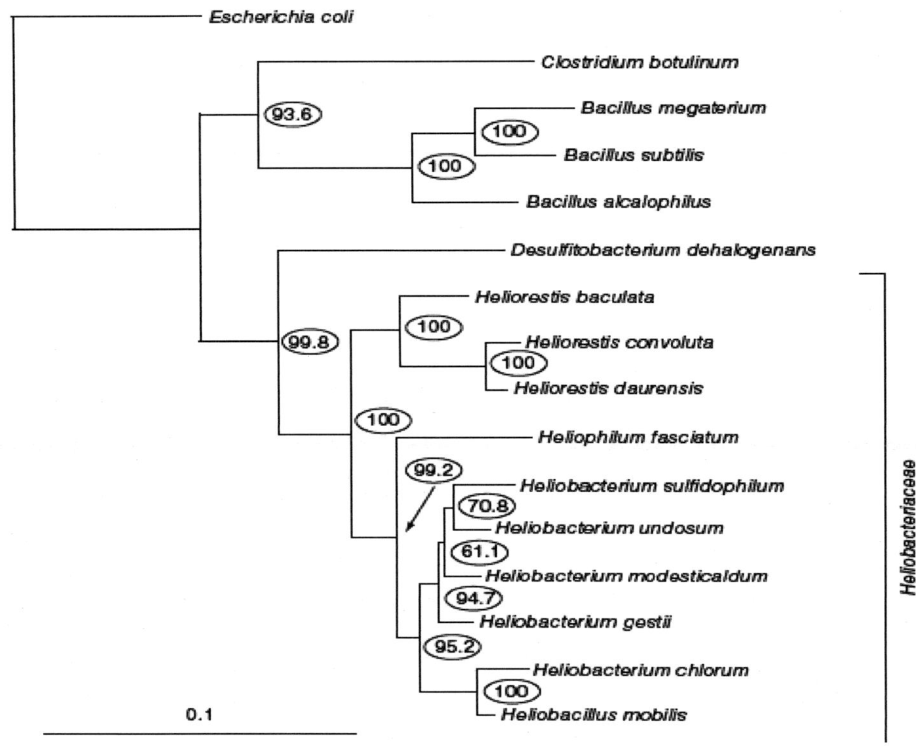

Fig. 3.110. Phylogenetic tree of *Heliobacteriaceae* members and other genera of *Firmicutes* based on 16S rRNA gene sequence analyses. The sequence of *E.coli* was used as an outgroup to root the tree.

3. *Heliophilum* (*Hph.*) (Cell bundles) (Ormerod *et al.,* 1996) Madigan (2015)

Cells are straight rods with tapered ends, grouped in bundles of two to many cells, large, multiply by binary fission (**Fig. 3.111 A**) (Ormerod *et al.,* 1996). They are gram (-) negative. Motile by means of polar to sub-polar flagella, cell bundles move as a unit (**Fig. 3.111 B**). Photosynthetic pigments are bacteriochlorophyll g is present but intracytoplasmic photosynthetic membranes are absent (**Fig. 3.111 C**). Growth occurs both photoheterotrophically in light under anaerobic conditions with acetate, pyruvate, lactate and butyrate and chemotrophically in dark under aerobic conditions. CO_2 assimilation takes. SO_4^{2-}, $S_2O_3^{2-}$ are sulfur sources. The optimum growth occurs at pH 7.0 and temperature of 37-40 °C. Biotin is required for their growth. NaCl is not required and growth inhibits at concentration 1.0%. Endospore formation is seen, which contains dipicolinic acid and calcium. The mol% G + C of the DNA of the type strain is 51.8. Paddy soils are suitable substrates for their growth. Other properties are the same as those given for the family *Heliobacteriaceae* , *Heliophilum fasciatum* is the type species of monotypic genus and characteristic features of this species are precised in **Table 3.48.**

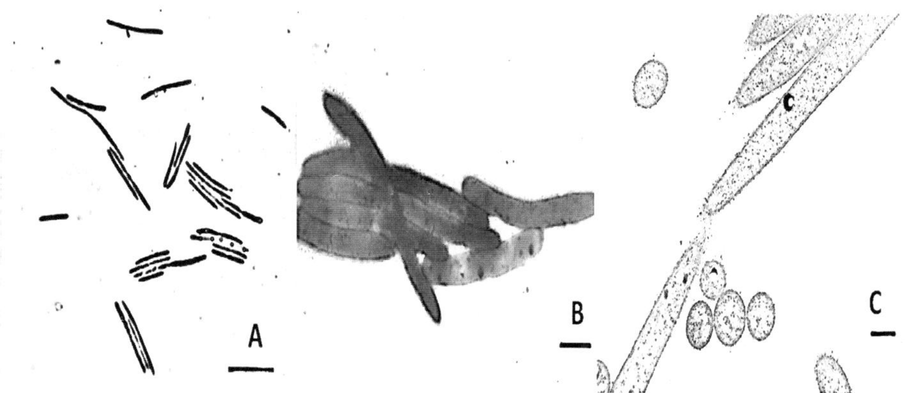

Fig. 3.111. A. Transmission electron micrograph of cells of *Heliophilum fasciatum* (Bars =15 μm), B. Cells showing polar or subpolar flagella (Bars=2μm), C. Ultrathin section of the cells (Bars=1 μm).

Differentiation of *Heliophilum* from other Genera

The differentiating characteristics of *Heliophilum* from other genera of *Heliobacteriaceae* are presented in **Table 3.48** and the genetic relationship among *Heliophilum fasciatum* and other *Heliobacteriaceae* members is based on 16S rRNA gene sequence is illustrated in **Fig. 3.107** and **110.**

4. *Heliorestis (Hrs.)* (Sun rope) (Bryantseva *et al.*,1999) Madigan (2015)

Cells are rod-shaped, form coils or bent filaments of variable length up to 20 μm. Gram (-) negative but lack an outer membrane (**Fig. 3.112 A** and **B**) (Bryantseva *et al.,* 1999). Motile by means of peritrichous flagella (**Fig. 3.112 C**). Bacteriochlorophyll *g* and neurosporene-like carotenoids are present as photosynthetic pigments but lack intracytoplasmic membranes or chlorosomes (**Fig. 3.112 D** .) Biotin and small amounts of yeast extract are used as growth factors (0.025-0.05%). Obligately anaerobic, growth occurs photoheterotrophically in the presence of light under anaerobic conditions with organic compounds. During sulfide oxidation, elemental sulfur and polysulfide are formed. Photoautotrophic growth is not observed. They are alkaliphilic and mesophilic. Optimal growth occurs at pH 8.5-9.5 (range pH 7.5-10.0) and temperature of 25-35°C. Endospore formation is seen. The mol% of G+C content and DNA is 44.9% (Tm). *Heliorestis daurensis* is the type species (Bryantseva *et al.,* 2000). Phylogenetic position within the heliobacterial clade of the low G+C, gram (+) positive bacteria but basal to the genera *Heliobacterium, Heliobacillus* and *Heliophilum* . Alkaline soils or soda lakes are suitable substrates for their growth. Four species are validly published and their characteristic properties were presented in **Table 3.48**.

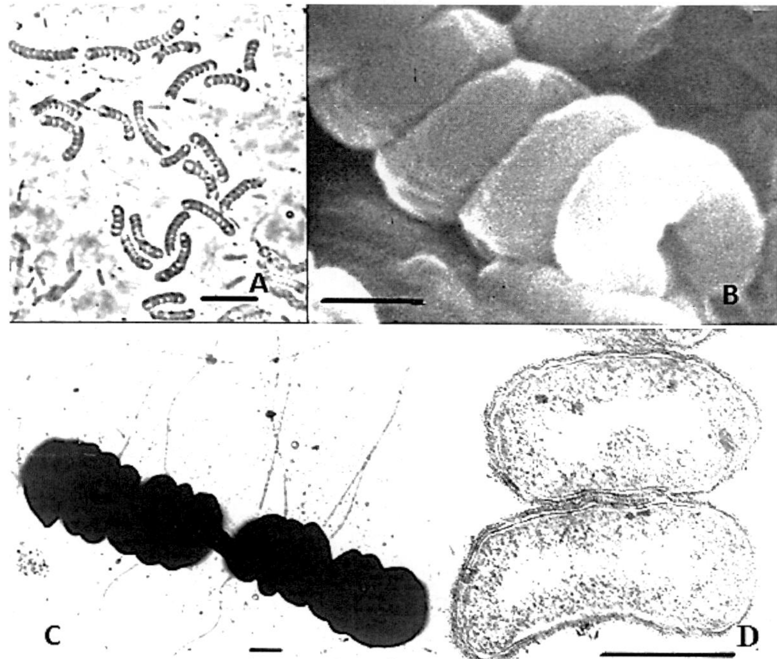

Fig. 3.112. A. Phase-contrast micrograph of cells of *Hrs. daurensis*, (Bar= 20µm), B. Co-culture of *Hrs. daurensis* and the small rod (Bar= 1µm), C. Transmission electron micrograph of coiled cells of *Hrs. daurensis* showing peritrichous flagella (Bar= 1µm), D. Ultrathin section of cells (Bar= 1µm).

Differentiation of *Heliorestis* from other Genera

The differentiating characteristics of *Heliorestis* from other genera of *Heliobacteriaceae* are presented in **Table 3.48** and the genetic relationship among *Heliorestis* and other *Heliobacteriaceae* members based on 16S rRNA gene sequence is illustrated in **Figs. 3.107** and **3.110.**

Taxonomic Comments

Phylogenetic analysis of the 16S rRNA gene sequence of the *Heliorestis* species showed that it groups within the heliobacterial clade (**Fig. 3.107**and **109**.) Although, its branching order was phylogenetically different to all previously investigated species of heliobacteria. The G+C content of the DNA of *Hrs. daurensis* strain BT-H1 (44.9 mol%) was also quite distinct from that of other heliobacteria. This specific assemblage of properties such as its alkaliphilic nature and phylogenetic differences implicates the reclassification of a new genus *Heliorestis* with type species *Hrs. daurensis .Heliorestis daurensis,* named for its unusual morphology ("restis" is Latin for "rope") and for the Daur Steppe in Russia in which these soda lakes are located. Four species are validly published and are assigned to the genus based on their phenotypic and genotypic properties.

5. *Candidatus Heliomonas* (Solar unit of crescent-shaped cells) Asao *et al.* (2012)

Cells are curved rods measuring 0.6 x 2.0–17.0 µm and motile by means of flagella (**Figs. 3.113 A, B** and **C**) (Asao *et al.,* 2012). These are anaerobic anoxygenic phototrophs. Cell cultures appear green and contain bacteriochlorophyll *g* and carotenoids of OH-diaponeurosporene glucosyl esters are present as photosynthetic pigments. Absorption maxima of living cells show a peak at 789 nm (major) and 670 nm (major). Absorption at 670 nm in intact cells kept anoxic during assay is significantly higher than in other species of heliobacteria. Cells lack intracytoplasmic membranes or chlorosomes (**Fig. 3.113 D**). They are alkaliphilic and mesophilic. Optimum growth occurs at pH 8.0–9.5 (range 8.0–9.5) and 25–30°C (range 18–37°C). NaCl is not required, but tolerate up to 1–3% NaCl which is growth stimulatory. Growth occurs photoheterotrophically under anaerobic conditions in the presence of light with pyruvate or malate as carbon sources. Moderate growth is also observed with alanine, butyrate, lysine, serine, threonine, or yeast extract as carbon sources. Photoautotrophy or pyruvate fermentation is not observed. The co-culture containing strain SLH was obtained from 23-m water/benthic sediment of Soap Lake, Washington (USA). *Candidatus Heliomas lunata* is the type species and their characteristics are presented in **Table 3.48** .

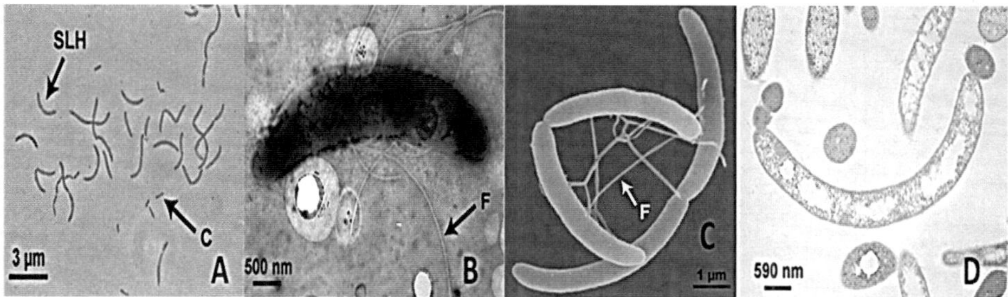

Fig. 3.113. A. Phase-contrast micrograph of curved rod-shaped cells of *Candidatus Heliomonas lunata* strain SLH, B. Negatively stained cells showing flagella (F), C. Scanning electron micrograph of the cells and D. Transmission electron micrograph of ultrathin section of the cells lacking intracytoplasmic membrane or chlorosomes.

Differentiation of *Heliomonas* from other Genera

The differentiating characteristics of *Heliomonas* from other genera of *Heliobacteriaceae* are presented in **Table 3.48** and their genetic relationship among *Heliomonas* and other *Heliobacteriaceae* members based on 16S rRNA gene sequence is illustrated in **Figs. 3.107** and **3.114** .

Taxonomic Comments

Based on phylogenetic analysis of the 16S rRNA gene sequence of the *Heliomonas lunata* strain SLH, it is closely related to the neutrophilic heliobacteria such as species of *Heliobacterium* and *Heliobacillus* (**Fig. 3.107** and **114**). The 16S rRNA gene sequence of *Heliomonas lunata* SLH is highly divergent from that of any cultured heliobacterium. *Hba. mobilis* and *Hbt. gestii* are the closest relatives of *Heliomonas lunata* strain SLH which showed 94.0% sequence similarity, which is highlighting the unique phylogeny of strain SLH of *Heliomonas lunata.* Although, its branching order was phylogenetically different from all previously investigated species of heliobacteria. Due to the non-axenic condition of the strain SLH culture, such a new taxon must have '*Candidatus*' status and thus the name '*Candidatus Heliomonas lunata'* was proposed by Asao *et al.* (2012) to accommodate strain SLH into a taxonomy.

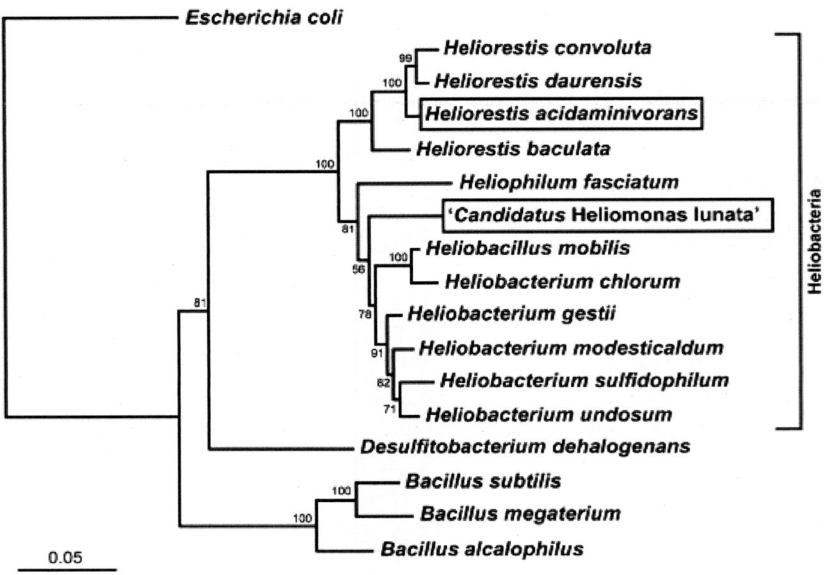

Fig. 3.114. Phylogenetic tree of *Canditatus Heliomonas lunata* and other genera of *Firmicutes* based on 16S rRNA gene sequence analyses. The sequence of *E. coli* was used as an outgroup to root the tree.

■■■■

Chapter 4

PHOTOSYNTHESIS

Photosynthesis is the process that uses energy from the sun and ultimately provides all our food and has provided all of the fossil fuels. This is a process utilizing light energy for biosynthesis. It can be divided into light reactions that convert light energy into biological energy and dark reactions that utilize the biological energy in the biosynthesis of carbohydrates. The time between the years 1771 and 1804 was an exciting period in the history of photosynthesis (Blankenship, 2002). The role of oxygen in photosynthesis was established by Joseph Priestly during the period 1733-1804 and that of light was demonstrated by Jan Ingenhousz between 1730 and 1799. The involvement of carbon dioxide in photosynthesis was shown by Jean Senebier between 1742 and 1809 and Nicolas de Saussure discovered the importance of water during the period 1767-1845. As a result of these studies the basic chemical equation of photosynthesis was established, which could be written as:

Carbon dioxide + Water + Light → Organic matter + Oxygen

It was another sixty years before the careful measurements of the photosynthetic quotient, which is the ratio between the carbon dioxide assimilated and the oxygen produced, was made by a Frenchman and Boussingault in 1864. Photosynthesis can be thought of as the photochemical reduction of CO_2 into organic substances (Blankenship, 2002). The minimally balanced equation for photosynthesis can now be written as:

$$6CO_2 + 6H_2O \xrightarrow{hv \downarrow} C_6H_{12}O_6 + 6O_2 \quad (1)$$

Many years after the overall reaction was established, attention has turned to elucidating the details of the mechanism of the process. The photosynthetic process can be conveniently divided into two parts, the light reactions and the subsequent dark reactions. The light reactions are those where solar energy is used to power the synthesis of the ATP and NADPH. The dark reactions, commonly called the Calvin cycle is where carbohydrate is synthesized from CO_2 in an enzymatic process.

Photosynthetic organisms use light energy to perform several biosynthetic processes. In anoxygenic photosynthesis, organic or sulfur compounds are used as electron donors whereas in oxygenic photosynthesis oxygen is generated where water is used as the electron donor. Plants, algae and cyanobacteria carry out oxygenic photosynthesis, whereas the photosynthetic bacteria obtain energy from anoxygenic photosynthesis. Aerobic anoxygenic phototrophic bacteria use light energy in a similar way as the purple bacteria and are a group of photosynthetic bacteria that grow under aerobic conditions.

Phototrophic organisms have a photosynthetic apparatus consisting of a reaction center intimately associated with antenna molecules (or a light-harvesting complex). The antenna molecules and the reaction center absorb light energy. The energy is concentrated at the reaction center that is activated and initiates light-driven electron transport. Halophilic archea convert light energy through a photophosphorylation process.

Microorganisms utilizing light energy include eukaryotic algae and cyanobacteria, photosynthetic bacteria and aerobic anoxygenic phototrophic bacteria among the prokaryotes. The halophilic archea synthesize ATP through photophosphorylation, but they are not considered to be photosynthetic organisms since they lack photosynthetic pigments.

Algae and cyanobacteria have similar photosynthetic processes, using chlorophyll, as plants. However, cyanobacteria are members of the proteobacteria according to their cell structure and ribosomal RNA sequences. Photosynthetic bacteria are different from these photosynthetic organisms. They have different photosynthetic pigments and do not use water as their electron donor. Some of them can grow chemoorganotrophically in the dark.

Plants are not the only types of organisms that can carry out photosynthesis, but most species of bacteria merely survive by respiration. However, some bacteria are able to use photosynthesis to grow (Blankenship, 2002). In 1941, Cornelis van Neil carried out series of experiments on bacterial photosynthesis in which no oxygen was involved. According to van Neil's findings, the bacteria must have access to a reducing substrate in order to provide hydrogen for the reduction of CO_2. The general equation he proposed was:

$$CO_2 + 2H_2A \xrightarrow{hv} (CH_2O) + H_2O + 2A^- \quad (2)$$

This reaction is a general form of the reaction (1). H_2A may be H_2S, H_2O, or a variety of simple organic molecules. Wherein H_2A is oxygen atom the process is plant photosynthesis, if it is sulfur atom then the process is the photosynthesis performed by sulfur bacteria. Among five major types of photosynthetic bacteria such as purple bacteria, green sulfur bacteria, green non-sulfur bacteria, *Heliobacteria* and *Cyanobacteria* (Madigan *et al.,* 1997), the first four are anoxygenic bacteria, where they do not evolve oxygen. Purple bacteria are one of the main groups of photosynthetic organisms that play an important role in revealing how the process of photosynthesis works. These bacteria have a very simple form of photosynthesis and have been the subject of detailed structural and spectroscopic studies, making them the best understood of all photosynthetic organisms in terms of energy collection and primary electron transfer.

4.1. ANAEROBIC PHOTOSYNTHETIC BACTERIA

Photosynthetic bacteria are grouped according to their photosynthetic pigments and the electron donors used. These are purple bacteria, green bacteria and heliobacteria. Purple- and green bacteria are further divided into purple non-sulfur and purple sulfur bacteria and green sulfur and filamentous anoxygenic phototrophic bacteria (**Table 4.1**). The purple non-sulfur bacteria grow under all electron-accepting conditions (aerobic respiratory, anaerobic respiratory and fermentative conditions) in addition to anaerobic photosynthetic conditions.

Table 4.1. Photosynthetic Pigments and Electron Donors Present in Different Anoxygenic Phototrophic Bacteria

Characters	**Purple bacteria**		**Green bacteria**		***Firmicutes***	**Aerobic APB**
	Non-sulfur	**Sulfur**[b]	**Sulfur**	**Filamentous APB (Non-sulfur)** [c]	***Heliobacteria***	
Bacteriochlorophyll	*a, b*	*a, b*	*a,c,d,e*	*a,c,d*	*g*	*a*
Sulfur granules accumulation	–	Intra-cellular	Extra-cellular	–	–	–
H_2S as e^- donor	±[a]	+	+	+	–	–
H_2 as e^- donor	+	+	–	+	+	+
Organic compounds as e^- donor	+	+	–	+	+	+
Carbon source	CO_2 organics	CO_2 organics	CO_2 organics	CO_2 organics	Organics	Organics
Aerobic dark growth	+	–	–	+	–	+

Symbols: [a], Depending on the strain; [b], Members of *Chromatium* accumulate sulfur granules intracellularly, while the members of *Ectothiorhodospiraceae* accumulate sulfur granules extracellularly; [c], Species of *Chloroflexus* accumulates sulfur granules extracellularly. –, Growth absent; +, Growth present.

4.1.1. Purple Bacteria

Purple bacteria include sulfur and non-sulfur bacteria (Imhoff, 1995) based on their ability or not to use sulfur as the electron donor in the photochemical reaction. There are two families of purple sulfur bacteria, namely, *Chromatiaceae* and *Ectothiorhodospiraceae.* The *Chromatiaceae* deposit globules of elemental sulfur inside the cell (Intracellular), while members of *Ectothiorhodospiraceae* deposit sulfur outside the cell (Extracellular). The *Rhodospirillaceae* family of the purple non-sulfur bacteria is the most diverse among the phototropic purple bacteria (Imhoffand Truper, 1989). The purple sulfur and non-sulfur bacteria have pheophytin-quinone-type reaction centers.

Most of the *Rhodospirillaceae* are able to survive by respiration under aerobic and dark conditions. However, when the growth conditions become anaerobic and light is available the cells quickly change from respiration to photosynthesis (Law and Cogdell, 2008). The natural habitat of these organisms is in the anaerobic layer in ponds and lakes below the chlorophyll-containing oxygenic phototrophs (**Fig 4.1**) (Cogdell *et al.,* 2006). This situation due to the lack of oxygen causes the cytoplasmic membranes, where photosynthesis takes place to become invaginated and extended in area. As a result, intracytoplasmic membranes (ICM) are created into which the cell inserts the necessary complexes to harvest light energy (Drews *et al.,* 1995). These complexes contain all the necessary pigments and proteins required for light harvesting (Law *et al.,* 2008).

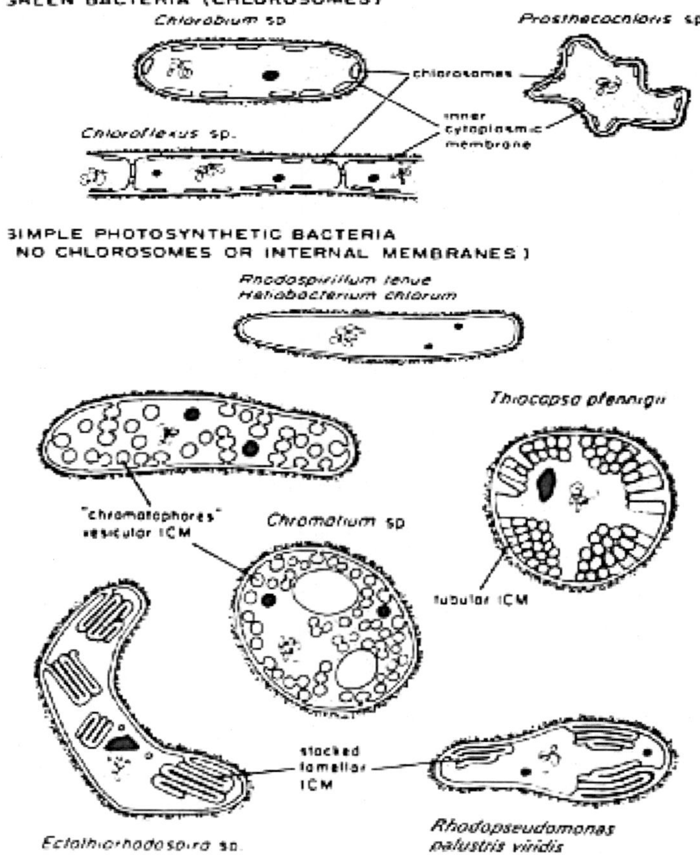

Fig. 4.1. Diagrammatic representation of different structures of intracytoplasmic membranes (ICM) of purple bacteria (Sprague and Varga, 1986).

The spectrum of the solar energy that reaches the purple bacteria has only green and near-infrared light (>750 nm) to pass through. It is only this infrared wavelength of light (>750 nm) that is available for APB. Carotenoid pigments contained in the purple bacteria capture green light, while bacteriochlorophyll (Bchl) captures the far-red light (Cogdell, *et al.,* 2006). Most purple bacteria contain bacteriochlorophyll *a*, whereas a few contain bacteriochlorophyll *b* . These bacteriochlorophyll *a* by themselves are not able to harvest energy in purple bacteria. These pigments can harvest light only when they are oriented to specific and are able to perform photosynthetic process and bound to the reaction center (RC), which participate in electron transfer. If they are bound to the Light-Harvesting (LH) apoproteins, they participate in energy transfer (Cogdell *et al.,* 2006). Under anaerobic conditions, APB develops intracytoplasmic membranes (ICM) which vary from species to species. Most of the purple photosynthetic bacteria have extensive invaginations of the cell membrane that can take the form of stacked lamellae (*Ectothiorhodospira* sp., *Rhodopseudomonas palustris* ,) vesicles (*Rhodospirillum* sp. and *Rhodopseudomonas* sp.), or tubules (*Thiocapsa pfennigii*) **(Fig. 4.1)**. A few photosynthetic bacteria do not have internal membrane or chlorosomes (*Rhodospirillum tenuis, Heliobacterium chlorum* .) However, the shape of these structures is very stable and therefore, they have great taxonomical value.

(i) **Purple Non-Sulfur Bacteria:** The purple non-sulfur bacteria belonging to α- and β-proteobacteria are more diverse. They grow photosynthetically under anaerobic conditions, and many of them can grow chemoorganotrophically under aerobic conditions.

(ii) **The Purple Sulfur Bacteria:** These include members of the *Chromatiaceae* and *Ectothiorhodospiraceae* within the gammaproteobacteria. The former accumulates sulfur granules intracellularly, while the latter extracellularly.

4.1.2. Green Bacteria

These include two physiologically and phylogenetically distinct groups. They are green sulfur bacteria and green non-sulfur bacteria or filamentous anoxygenic phototrophic bacteria. These are strictly anaerobic and obligately photo lithotrophic green sulfur bacteria. The green sulfur bacteria cannot grow heterotrophically, contains iron-sulfur type reaction centers.

4.1.3. The Filamentous Anoxygenic Phototrophic Bacteria

These are facultatively anaerobic. These have the pheophytin-quinone type reaction centers. The filamentous anoxygenic phototrophic bacteria can grow heterotrophically under aerobic dark conditions. The latter, members of the *Chloroflexaceae,* belong to a deep-branching lineage of bacteria. These are also called photosynthetic flexibacteria. They stain Gram-negative but lack lipopolysaccharide.

4.1.4. The Photoheterotrophic *Heliobacteria*

These include three genera i.e., *Heliobacterium, Heliobacillus, Heliophilum* and *Heliorestis.* They do not grow aerobically in the dark and fix N_2. They do not have photosynthetic organelles and photosynthetic pigments but include the unique bacteriochlorophyll *g,* which is located in the cytoplasmic membrane. They have an iron-sulfur-type reaction center. *Heliobacterial* cells have several unusual features. They are extremely fragile and lyses when approaching the stationary phase. They stain Gram (-) negative due to the lack of lipopolysaccharide like the filamentous anoxygenic phototrophic bacteria, they do not fix CO_2 , but phylogenetically they belong to the low G+C, Gram (+) positive bacteria.

4.2. AEROBIC ANOXYGENIC PHOTOTROPHIC BACTERIA (AAPB)

Phototrophic bacteria utilize light energy under anaerobic conditions and their photosynthetic pigments are not synthesized under aerobic conditions. However, many bacteria are known to synthesize bacteriochlorophyll *a* and carotenoids under aerobic conditions. They can harvest light energy and respire oxygen. These are referred to as quasi-photosynthetic bacteria or aerobic anoxygenic phototrophic bacteria. They comprise at least 11% of the total microbial community in the upper open ocean. They inhabit a variety of locations including the extreme environments of acidic mine drainage waters, hot springs and deep ocean hydrothermal vent plumes.

AAPB is found in seawater include species of *Erythrobacter* ,*Roseibium, Roseivivax, Roseobacter, Roseovarius* and *Rubrimonas.* Freshwater is the habitat of other aerobic anoxygenic phototrophic bacteria including species of *Sandaraacinobacter, Erythromonas, Erythromicrobium* ,*Roseococcus, Porphyrobacter* and *Acidiphilium.* These have the pheophytin-quinone-type reaction center. These genera include not only aerobic phototrophic bacteria but also species unable to synthesize bacteriochlorophyll *a.* It is not clear that the ability to synthesize the photosynthetic pigment was lacking or transferred through lateral gene transfer during their evolution. Aerobic anoxygenic photosynthesis is also known in species of *Bradyrhizobium*, syntrophically growing on the stems of tropical legume plants. However, these are obligate aerobes with high carotenoids and low bacteriochlorophyll contents, they are closely related to purple photosynthetic bacteria in many respects.

Some of the AAPB divide in an unusual manner. Budding, in addition to binary division, occurs in *Porphyrobacter neustonensis* and *Erythromonas ursincola.* Ternary fission and branching are exhibited by *Erythromicrobium ramosum* and *Erythromicrobium hydrolyticum* .

4.3. PHOTOSYNTHETIC APPARATUS-LIGHT HARVESTING PIGMENTS

Photosynthetic pigments are divided into two categories. They are reaction center pigments (chlorophylls) and light-harvesting pigments include bacteriochlorophylls, carotenoids and phycobiliproteins. They function to absorb light/ energy and transmit it to the photosynthetic reaction center. The light-harvesting pigments are sometimes called Antennae pigments or Accessory pigments. The light-harvesting pigments are critical to photosynthesis as they absorb light of different wavelengths and supply energy to the reaction center. The groups of pigments that comprise an antenna system are group-specific and their cumulative light absorptive properties determine the wavelengths of light over which the photosynthesis occurs. As a consequence of the variation in the composition of antenna systems, phototrophs are collectively capable of utilizing light as an energy source. All radiant energy falls in the range of visible and near-infrared light.

4.3.1. Bacteriochlorophylls

In purple bacteria, bacteriochlorophyll *a* is the main light-harvesting pigment contains a bacteriochlorin ring and a phytol chain attached to C-17. The bacteriochlorin ring is planar with an Mg^{2+} ion in the center, coordinated to four pyrrole nitrogen atoms.

Bacteriochlorophyll differs from Chl *a* in that the presence of acetyl group at the C-3 position and the presence of double bond between C-7 and C-8 in ring B (**Table 4.2** and **Fig. 4.2 A**). Because of asymmetric conjugated π system, the electronic transition of bacteriochlorophyll has two main transitional dipole moments, Qx and Qy. The Qx absorption band is found in the visible region (700-420 nm) at ~570 nm, the Qy absorption band is in the near-infrared region (NIR >770 nm), a third absorption band Soret band is located in the blue region of the spectrum (~380 nm) **(Fig. 4.3)** and Bacteriopheophytin is similar to bacteriochlorophyll but differs only in the absence of Mg^{2+} ion at the center of the bacteriochlorin ring (**Fig. 4.2 B** .)

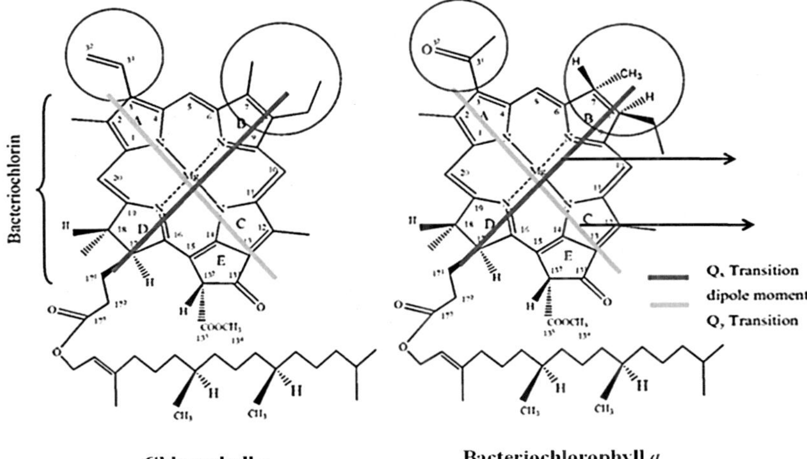

Fig. 4.2. A. Structures of Chl *a* and Bacteriochlorophyll *a*

Table. 4.2. Comparison of Structure of Chlorophyll *a* and Bacteriochlorophyll *a*

Chlorophyll *a*	**Bacteriochlorophyll *a***
At C-3 position	Acetyl group at C-3 position
The double bond in ring B between C-7 and C-7	Single bond in ring B between C-7 and C-7
Chlorophyll contains Mg in the center	Bacteriophaeophytin (Bphe) does not contain Mg

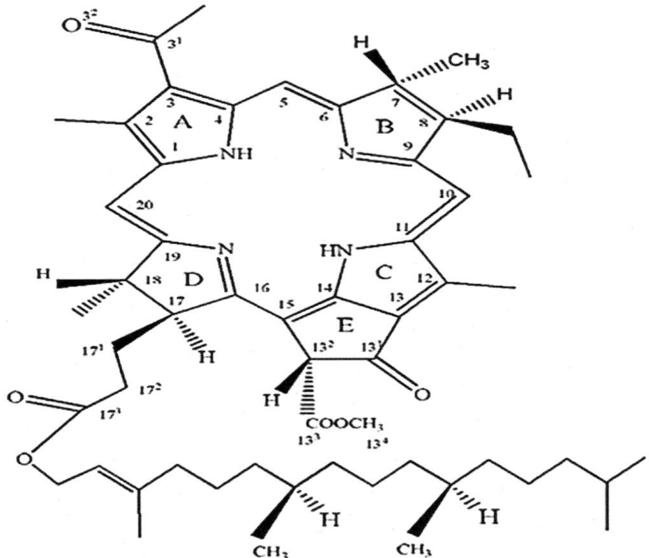

Fig. 4.2B. Structure of Bacteriopheophytin*a*.

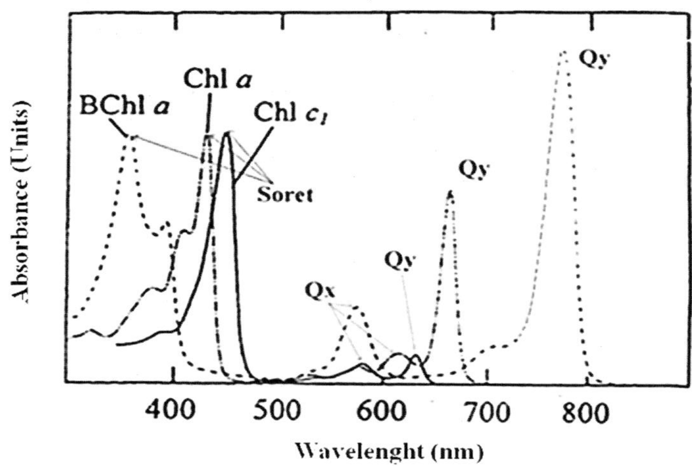

Fig. 4.3. Absorption spectra of bacteriochlorophyll *a* in diethyl ether

Atleast seven kinds of bacteriochlorophylls (Bchl) occur in various groups of phototrophs, absorb light intensely in two regions of spectrum **(Fig. 4.4)**, the violet around 400 nm, and the red or near-infrared, around 600 to 1100 nm. The region of peak absorption within the longer wavelengths varies with the particular chlorophyll types and proteins with which they are associated. All chlorophylls share the same molecular ground plan along with detailed variations in the value of R_1 to R_7 found in different prokaryotes and eukaryotes are given in **Table 4.3.**

Table 4.3. Chemical Differences among Chlorophyll and Bacteriochlorophyll

	R1	R2	R3	R4	R5	R6	R7	Absorption maxima (nm)
Chlorophyll								
a	$-CH\text{-}CH_2$	$-CH_3$	$-C_2H_5$	$-CH_3$	$-C(=O)-O-CH_3$	Phytyl	–H	680
b	$-CH\text{-}CH_2$	$-CH_3$	$-C_2H_5$	$-CH_3$	$-C(=O)-O-CH_3$	Phytyl	-H	675
Bacteriochlorophylls								
a	$-C(=O)-CH_3$	CH_3^{a}	$C_2H_5^{a}$	$-CH_3$	$-C(=O)-O-CH_3$	Phytyl, geranyl-geranyl, or Farnesyl	–H	800-810, 850-910
b	$-C(=O)-CH_3$	$-CH_3^{b}$	$-CH{=}CH_3^{b}$	$-CH_3$	$-C(=O)-O-CH_3$	Phytyl	–H	835-850, 1020-1035
g	$-CH{=}CH_2$	$-CH_3^{b}$	$-CH{=}CH_3^{b}$	$-CH_3$	$-C(=O)-O-CH_3$	Geranyl-geranyl	–H	670, 788
c	$-C(H)(OH)-CH_3$	$-CH_3$	C_2H_5	$\bar{C_2H_5}$	–H	Farnesyl	$-CH_3$	745-760
d	$-C(H)(OH)-CH_3$	$-CH_3$	$-C_2H_5$	$\bar{C_2H_5}$	–H	Farnesyl	–H	725-745
e	$-C(H)(OH)-CH_3$	$-C{=}O$ ‖ O	$-C_2H_5$	$\bar{C_2H_5}$	–H	Farnesyl	$-CH_3$	715-725

[a]No double bond between C –3 and C –4; additional –H atoms at C –3 and C –4;[b]No double bond between C – 3 and C –4; additional –H atoms at C –3.

Bacteriochlorophyll molecules have tetrapyrrole nucleus, containing central metal ion, magnesium (Mg^{2+}) is chelated (**4.2B**). They form a complex with proteins embedded in the membranes of the photosynthetic apparatus. The tetrapyrrolic nucleus (rings I, II, III and IV) has the same derivation as that of the hemes, but it is chelated with magnesium. In bacteriochlorophyll, one or more of the pyrrole rings are reduced; in this diagram, ring IV is shown reduced, as is characteristic of bacteriochlorophyll *a*; R1, R2 and so forth, the designated aliphatic side chains attached to the tetrapyrrolic nucleus. The presence of ring V, the pentanone ring and the substitution of R7 by long-chain alcohol are characteristic features of the bacteriochlorophyll that do not occur in hemes. The different substitutions on the pyrrole rings and the number of double bonds in ring II distinguish the chlorophylls from each other. In bacteria bacteriochlorophylls (Bchl) are classified as *a, b, c, d, e* and *g*. Recently a Zn-bacteriochlorophyll '*a*' which contains Zn^{+2} in place of Mg^{2+} has been discovered in aerobic anoxygenic phototrophic bacteria. It is also observed in an acidophilic photosynthetic bacterium called *Acidiphilium rubrum*. In general, the chlorophylls show one of the side chains of the tetrapyrrole nucleus esterified to phytol tail (alcohol). The absorption spectrum of bacteriochlorophyll of purple bacteria

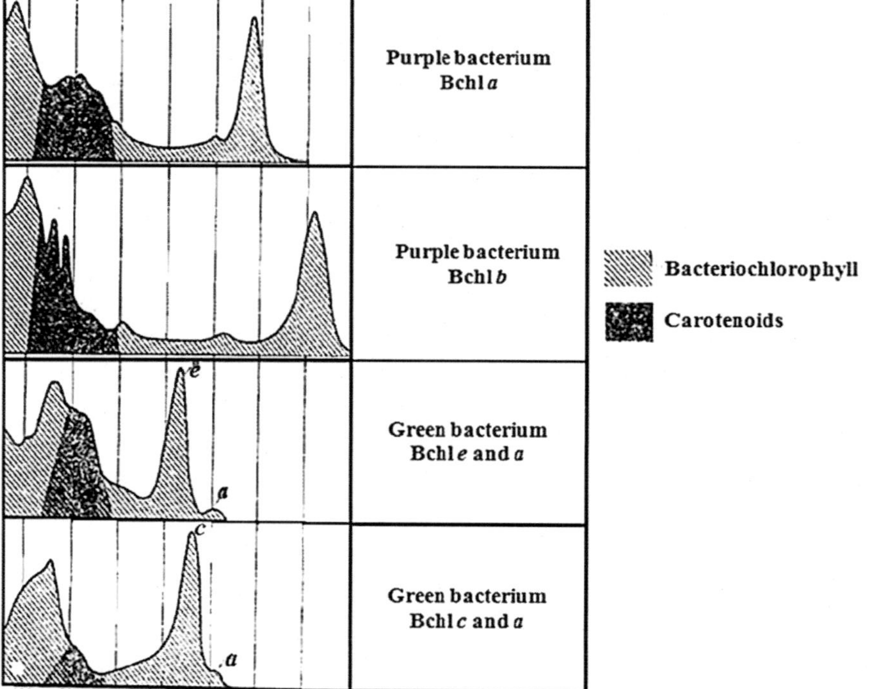

Fig. 4.4. Absorption spectrum of bacteriochlorophyll of purple bacteria (*Rhodopseudomonas palustris*) and green bacteria.

4.3.2. CAROTENOIDS

Carotenoids are long conjugated molecules, which are synthesized by the condensation of molecules of isoprene. They are joined head-to-tail, forming C20, and then tail-to-tail between two C20's, to form C40. This C40 molecule is the precursor of all carotenoids and is called phytene (Gross, 2006). All other carotenoids are formed from this precursor by reactions such as hydroxylation, desaturation and methylation (Rodriguez-Amaya, 2001). Carotenoids are found in all known photosynthetic organisms and in some non-photosynthetic organisms (Frank *et al.,* 2000). They have several important functions in photosynthesis such as accessory light-harvesting, photoprotection through the quenching of Chlorophyll (Bacteriochlorophyll) triplet states, singlet oxygen scavenging, excess energy dissipation and structure stabilization (Frank and Cogdell, 1996). The chemical structure of several carotenoids recorded in purple bacteria is shown in **Fig. 4.5.** Some quinone molecules are also involved in the light reactions of purple bacteria. Ubiquinones are found in all purple bacteria, whereas menaquinones are found in some purple bacteria but found in all green bacteria (Imhoff, 1984a). These molecules are involved in the redox reactions in the electron transfer chain within the RC. The chemical structure of ubiquinone is illustrated in **Fig. 4.6.**

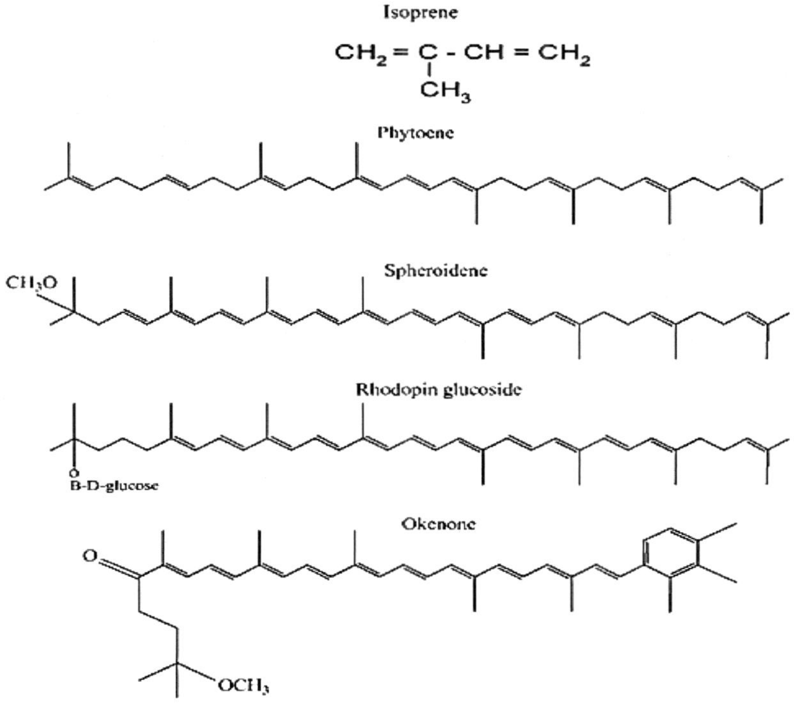

Isoprene

$CH_2 = \underset{|\atop CH_3}{C} - CH = CH_2$

Phytoene

Spheroidene

CH_3O

Rhodopin glucoside

B-D-glucose

Okenone

O

OCH_3

Fig. 4.5. Structure of carotenoids found in purple bacteria

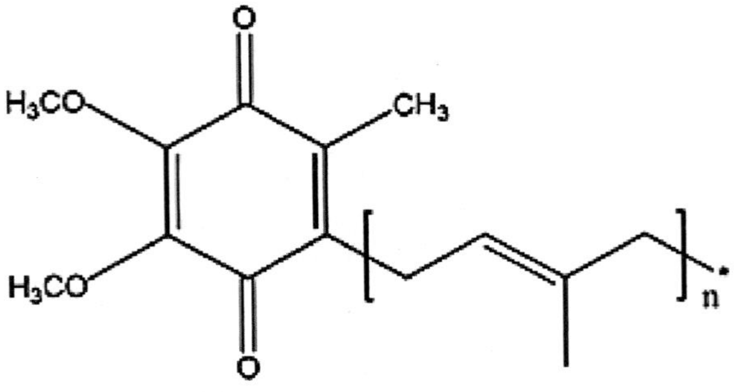

Fig. 4.6. Structure of ubiquinone. n – number of isoprene units

Modified carotenes are called Xanthophylls (e.g., by hydroxylation). Other isoprenoids are the side chains of quinones, the phytol chain in chlorophyll and the retinal chromophore in bacteriorhodopsin, halorhodopsin and rhodopsin. Carotenoids have conjugated double bonds, which account for their absorption spectrum in the visible range at wavelengths poorly absorbed by chlorophylls and make them valuable light-harvesting pigments. The absorption spectrum of different carotenoid pigments is listed in **Table. 4.4.**

Table 4.4. Absorption Spectrum of Different Carotenoids of Phototrophic Bacteria

Different Carotenoids	**Absorption Peak**
Chlorobactene	458
Isorenieratene	517
Lycopene, rhodopin	463, 490, 524
Okenone	521
Rhodopinal	497, 529
Sphaeroidene	450,482,514
Spirilloxanthin	486, 515, 552
β- carotene (βαϒ)	433, 483
Phycobilins	**Absorption peak**
Phycocyanin	620-650
Phycoerythrin	560-566

During photosynthesis under high light intensities and electrons in bacteriochlorophyll can go from the excited singlet state to a lower-energy triplet state. In the triplet state, the spin of the excited electron is changed, so that there are two unpaired electrons as opposed to paired electrons in the singlet state (phosphorescence occurs from the triplet state) (**Fig. 4.7**). When triplet state bacteriochlorophyll reacts with oxygen, the oxygen becomes energized to its first excited state, singlet oxygen, O_2. Singlet oxygen is very reactive and combines with cell molecules such as unsaturated fatty acids, amino acids and purines, thus causing oxidative damage to various cell components including lipids, enzymes and nucleic acids. The carotenoids in the reaction center quench the triplet state in bacteriochlorophyll. Thus, carotenoids prevent photooxidation under high light intensities. When the carotenoid absorbs the energy from the triplet bacteriochlorophyll, the carotenoid itself becomes energized to the triplet state (triplet-triplet energy transfer). However, the energy in triplet-state carotenoid is dissipated harmlessly as heat to the medium. Because, the triplet state energy of oxygen, carotenoids will also quench singlet oxygen. Carotenoids are useful in detecting changes in membrane potential. The energy levels of electrons in the conjugated double bond system of the carotenoids are altered by the electric field of the membrane potential. When the membrane potential changes, the carotenoid undergoes a rapid shift in its absorption spectrum peaks.

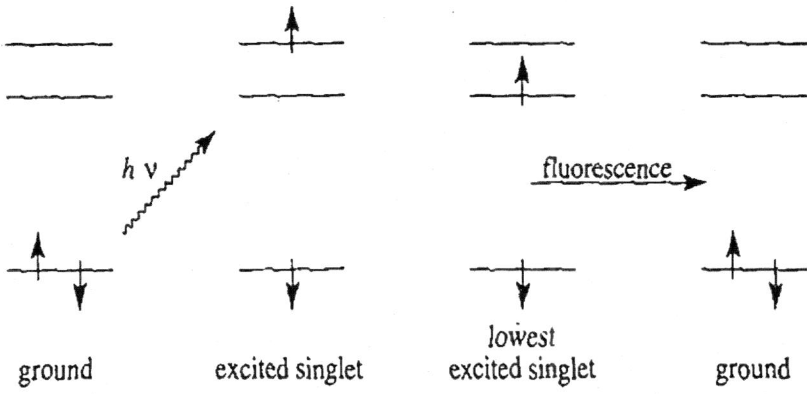

Fig. 4.7. Different states of the electron under the influence of light

4.3.3. Pheophytin

Pheophytin is a chemical compound that serves as the primary electron carrier intermediate in the electron transfer pathway of Photosystem-II (PS-II) in plants, and the type-II photosynthetic reaction center (RC P870) found in purple bacteria. In both PS-II and RC P870, light drives electrons from the reaction center through pheophytin, which that point passes the electrons to a quinone (QA) in RC P870 and RC P680. The general mechanisms, roles and purposes of the pheophytin molecules in the two transport chains are similar to one another. This pigment has the structure of chlorophyll but without Mg^{2+} or Zn^{2+}. Cyanobacteria and photosynthetic eukaryotes possess pheophytin, while bacteripheophytin (**Fig. 4.2**) is found in phototrophic bacteria with the phaeophytin-quinone type reaction center (RC) serving as an electron carrier. This pigment is not found in green sulfur bacteria and *Heliobacteria* that have the iron-sulfur type reaction center. Before its function was established, pheophytin was regarded as a chlorophyll degradation product.

4.3.4. Structure of Chlorosome

The chlorosome is surrounded by a galactolipid layer, which contains some protein and is attached to the reaction center in the cell membrane through a base plate **Fig. 4.8** .

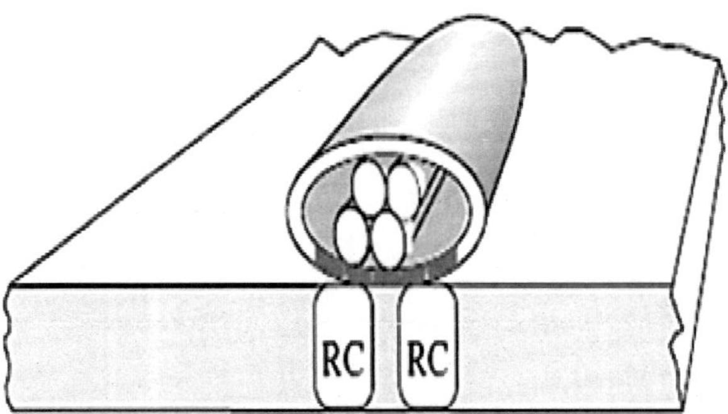

Fig. 4.8. Structure of Chlorosome

There is bacteriochlorophyll *a* in the base plate and in the membrane. One model for the inside of the chlorosomes postulates rod-shaped proteins to which is attached the major light-harvesting pigment, bacteriochlorophyll *c, d*, or *e*. An alternative model is that the bacteriochlorophyll exists as an aggregate, not attached to the protein. It is believed that light is absorbed by the bacteriochlorophyll in the interior of the chlorosome and the energy is transmitted to bacteriochlorophyll *a* in the base plate, then to the bacteriochlorophyll *a* attached to the reaction center. Wide variation exists in photosynthetic membranes of anoxygenic phototrophic bacteria. The green bacteria such as *Chlorobium, Chloroflexus* and *Prosthecochloris* contain chlorosomes attached to the cytoplasmic surface of the cell membrane.

Phototrophic organisms utilize light energy to reduce $NADP^+$ and synthesize ATP through the proton motive force. This energy transduction is facilitated by the photosynthetic pigments and electron carriers arranged in the photosynthetic apparatus. Separate photosynthetic structures are not found in the heliobacteria and aerobic anoxygenic phototrophic bacteria. Their reaction centers and antenna molecules are located at the cytoplasmic membrane. The reaction center is the key component for the primary events in the photochemical conversion of light into biological energy. Coupling to secondary electron donors and acceptors allows the electrons and accompanying protons to be transferred to other components of the photosynthetic apparatus synthesizing ATP or reducing $NADP^+$.

Photosynthetic reaction centers can be classified into two categories based on the nature of the electron acceptors. Those of the purple bacteria, filamentous anoxygenic phototrophic bacteria and photosystem-II of cyanobacteria belong to the 1. Pheophytin-quinone type, 2. Iron-sulfur type is found in green sulfur bacteria, heliobacteria and photosystem-I of cyanobacteria. Pheophytin (or bacteriopheophytin) and iron-sulfur centers participate in the electron transfer reactions of each type of reaction center.

Green bacteria have a photosynthetic apparatus called the chlorosomes on their cytoplasmic membrane. Chlorosomes contain antenna molecules and their baseplates are bound to the reaction center that is part of the cytoplasmic membrane (**Fig. 4.9**). The chlorosome has the structure of a galactosyl diglycerides monolayer membrane filled with rod-shaped antenna molecules. Bacteriochlorophyll *c, d,* or *e* constitutes the antenna molecules together with carotenoids. The base plate contains bacteriochlorophyll that transfers photons to the reaction center. The reaction center contains bacteriochlorophyll *a* and the electron transport chain. In addition to bacteriochlorophylls, Fe-S proteins are found in the reaction centers of green sulfur bacteria and bacteriopheophytin (Bphe) in filamentous anoxygenic phototrophic bacteria.

The green sulfur bacteria such as *Chlorobium tepidum,* thrives in anaerobic aquatic environments where sulfide is available with very dim light. To capture light efficiently this bacterium contains about 200-250 chlorosomes per cell with more than 200 x 103 bacteriochlorophyll *c* molecules per chlorosome. Thus, this bacterium contains up to 50 x 106 bacteriochlorophyll *c* molecules per cell. The monolayer chlorosomes contain antenna molecules and their base plates are bound to the reaction center that is a part of the cytoplasmic membrane.

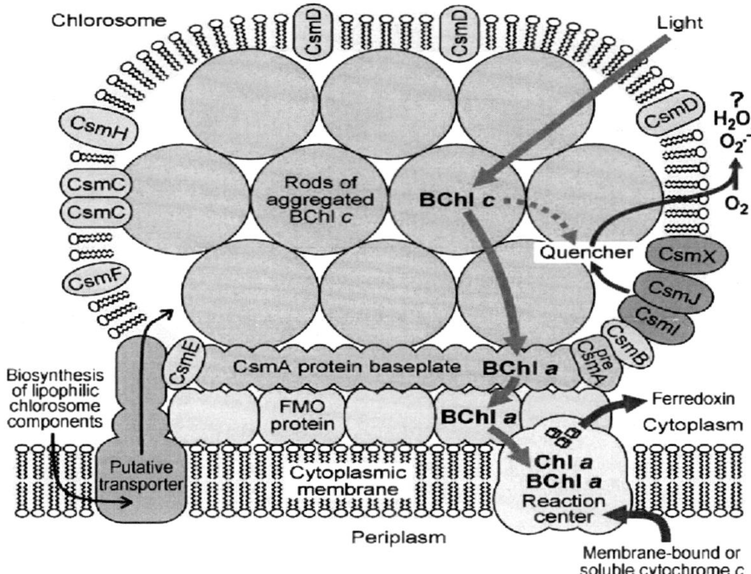

Fig. 4.9. Chlorosome bound to the cytoplasmic membrane in green bacteria (Frigaard and Bryant, 2004).

i. **Purple Bacteria:** The purple bacteria have a less developed photosynthetic structure than cyanobacteria and green bacteria. An intracellular membrane structure contains antenna molecules and reaction centers **(Fig. 4.10).** This is a phospholipids bilayer membrane continuous with the cytoplasmic membrane. The shape of the intracellular membrane structure varies depending on the strain.

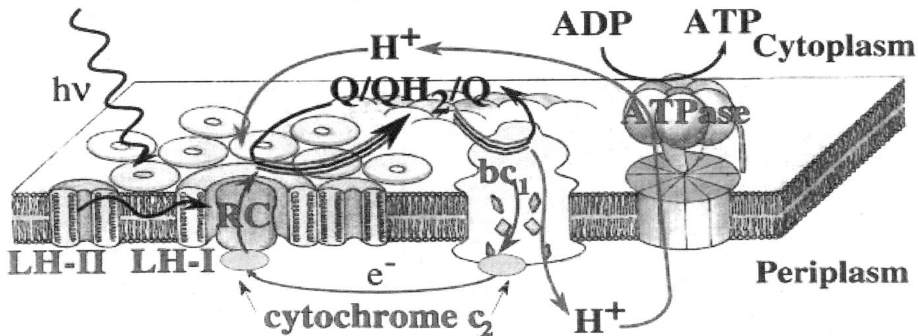

Fig. 4.10. The schematic representation of the photosynthetic apparatus in the intracytoplasmic membrane of purple bacteria (Hu *et al.,* 2002). The reaction center (RC, red) is surrounded by the light-harvesting complex-I (LH-I, green) to form the LH-I-RC complex, which is surrounded by multiple light-harvesting complexes (LH-IIs) (green), forming altogether the photosynthetic unit (PSU). Photons are absorbed by LHs and excitation is transferred to the RC initiating a charge (electron-hole) separation. Electrons are shuttled back by the cytochrome $c_2$complex (blue) from the ubiquinone-cytochrome bc_1complex (yellow) to the RC. The electron transfer across the membrane produces a large proton gradient that drives the synthesis of ATP from ADP by the ATPase (orange). Electron flow is represented in blue, proton flow in red, quinone flow, likely confined to the intramembrane space, in black.

ii. ***Heliobacteria* and Aerobic Anoxygenic Phototrophic Bacteria:** The photosynthetic apparatus in these organisms is not well developed and lack differentiated structures such as chlorosomes or intracytoplasmic membranes. The antenna molecules and reaction centers present within the cytoplasmic membrane.

4.4. PROPERTIES OF LIGHT

Light radiant energy is always transferred in discrete packets known as photons, whose energy content is inversely proportional to wavelength (**Fig. 4.11** .) Wavelength longer than 1200 nm is in the infrared region and beyond have energy content so small that the absorbed energy is converted into heat. It cannot mediate chemical change. Wavelengths less than 200 nm in the range of X-rays, α-particles and cosmic rays are termed as ionizing radiations because their energy content is so high that molecules in their path are immediately ionized. Between these two extremes (200 nm and 1200 nm) lies the region of electromagnetic radiation that is capable of mediating biological reactions. It is this portion of the electromagnetic spectrum (ultraviolet, visible and near-infrared) that can serve the performance of photosynthesis. It constitutes a major portion of the energy content of solar radiation at the earth's surface. The shorter and longer wavelengths of the broad spectrum of electromagnetic radiation from the sun are filtered out during passage through the earth's atmosphere. Indeed, at sea level, three-quarters of the energy content of solar radiation is contained in the light of wavelengths between 400 and 1100 nm, and it is within these limits, pigments are responsible for light capture in photosynthesis. The light absorption spectrum of different photosynthetic pigments is illustrated in **Fig. 4.11.**

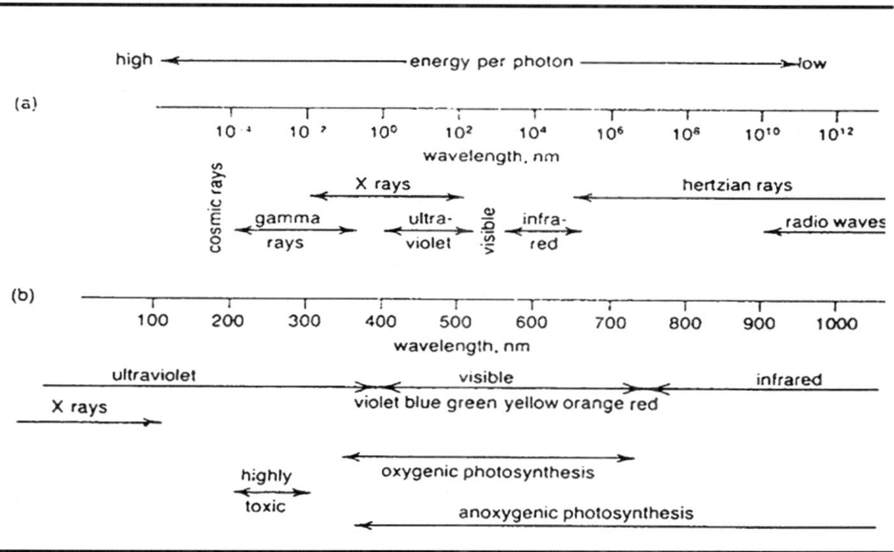

Fig. 4.11. The electromagnetic spectrum of light. (a) Light spectrum scale. (b) The ultraviolet, visible, and near-infrared (enlarged).

When a photon is radiated to a surface, it may be reflected, transmitted, or absorbed. Different pigments absorb photons of different wavelengths depending on the nature of the absorbing pigment. When a photon is absorbed by a pigment, the energy of light is converted to kinetic energy in the form of an excited electron. When a molecule (or atom) absorbs a photon, an electron is boosted to a higher energy level by transferring an electron from the normal shell to the outer shell (**Fig. 4.12**).

4.4.1. Light-Harvesting Complexes

The past several years have witnessed dramatic progress in our understanding of the reactions taking place in the early events of photosynthesis. This has been mainly due to research involving purple phototrophic bacteria. These anaerobic photosynthetic prokaryotes have been and continue to be an excellent model to investigate the basic mechanism of light-harvesting and reaction center (RC) photochemistry.

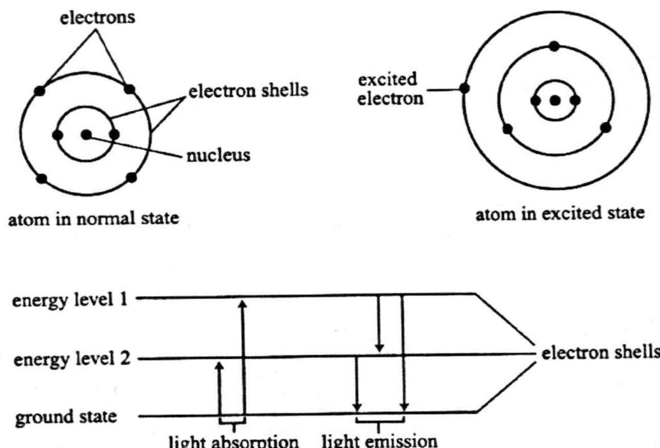

Fig. 4.12. Photon exciting an electron of a light absorbing pigment molecules.

Light-harvesting complexes, also known as antenna complexes, have evolved to increase the effective cross-sectional area for light absorption for each reaction center. Antenna complexes supply's enough photons to each reaction center to allow photosynthesis to occur at reasonable rates over a wide range of incident light intensities **(Fig. 4.13)** (Cogdell *et al.,* 2006). The solubilized photosynthetic membrane was layered onto the top of the sucrose-density gradient and centrifuged overnight resulting in the separation of the two major light-harvesting complexes. These two complexes have different colours as different carotenoids are bound. The LH2 complex has absorption bands at 800 – 850 nm, while the RC-LH1 complex has an absorption band at 875 nm. They also absorb a wider spectral range of light than the RC itself, so that more of the incident solar spectrum can be used efficiently (Cogdell *et al.,* 1999).

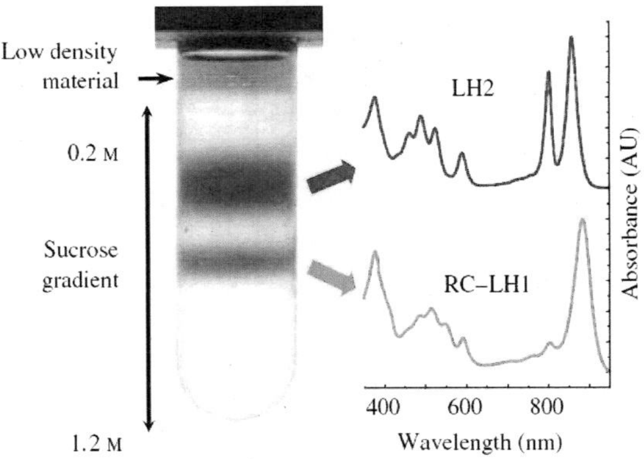

Fig. 4.13. A sucrose gradient is the first step to isolate and purify of LH2 and RC-LH1 complexes from *Rps. acidophila* strain 10050.

The major light-absorbing pigments in purple bacteria are bacteriochlorophyll *a* or *b* and carotenoids. These pigments are non-covalently bound to two types of integral membrane proteins forming either reaction centers or antenna complexes. All species of purple bacteria contain "core" antenna complexes (light-harvesting complex 1 (LH1)), which surround the reaction centers. LH1 complexes typically have a single strong near-infrared absorption band between 870 and 890 nm (Bchl *a*) or at 1,020 nm (Bchl *b*). Many species also contain a second type of antenna complex (LH2), which is arranged more peripherally. LH2 complexes typically have two strong absorption bands in the near-infrared (800 and 820 or 850 nm). The exact ratio of LH2/LH1 complexes present in the photosynthetic membrane is controlled by a variety of environmental factors including light. Therefore, the structure of the PSU is very variable.

Antenna complexes have evolved to increase the effective cross-sectional area for light absorption of each reaction center. Antenna complexes supply enough photons to each reaction center to allow photosynthesis to occur at maximum rates over a wide range of incident light intensities. They also absorb light over a broader spectral range than the reaction centers alone and allow more incident solar spectrum to be used productively.

Prior to the determination of the structure of LH2 complex in 1995, purple bacterial antenna complexes have been extensively studied by a wide range of biochemical and molecular biological techniques. Antenna complexes have been prepared and characterized by a large number of purple bacteria. Each complex is an oligomer of a basic unit which consists of a pair of small, hydrophobic apoproteins (named α- and β). Hydropathy analysis of these apoproteins showed that they all contained polar N and C termini and a central hydrophobic region between 20 and 24 amino acids which were confirmed and led to the idea that these apoproteins span the photosynthetic membrane with the central hydrophobic region folded into a single membrane-spanning α-helix. It was confirmed by experiments with proteases and inside-out and right-side-out membrane vesicles. The energy transfer reactions occurring within isolated antenna complexes and intact photosynthetic membranes were studied by picosecond flash photolysis. The time of energy transfer for the B800 and B850 step in LH2 was put at less than 1 ps, while transfer from LH2 to LH1 was rather multi-exponential with the main time being 3 to 5 ps, but with some slower components are also visible. The energy transfer step from LH1 to the reaction center took 30 to 50 ps. Once the excitation energy reaches the reaction center, the primary photochemical redox reactions are initiated. As a result of this, cyclic electron transport occurs, a transmembrane proton motive force is generated and ATP is synthesized.

4.5. ELECTRON TRANSPORT IN PURPLE BACTERIA

In purple bacteria, the photosynthetic apparatus consists of RC and several LH complexes. The RC can absorb photons directly but usually, energy is transferred to them from the light-harvesting complexes (Cogdell *et al.,* 1999). In all species of purple bacteria, the reaction center is surrounded by a light-harvesting complex called LH1. This complex is therefore called the LH1-RC "core" antenna complex (Cogdell *et al.,* 1996). In some purple bacteria a different type of complex called LH2, is also present and is arranged peripherally around the 'core' complex (Cogdell *et al.,* 1999).

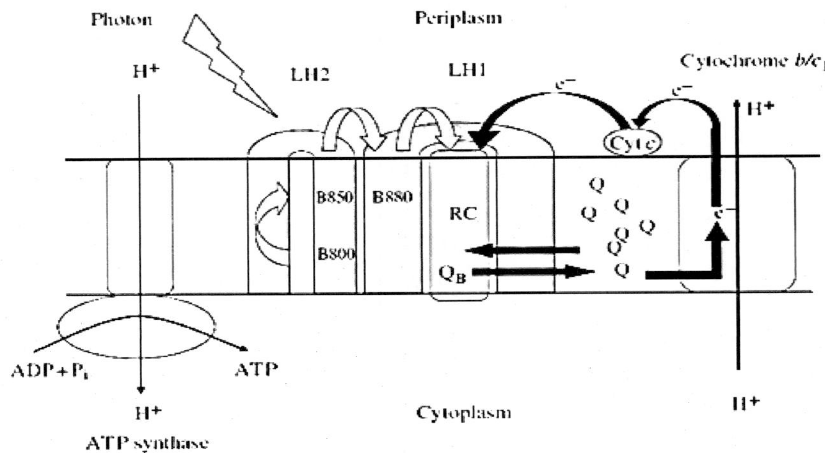

Fig. 4.14. Schematic presentation of the photosynthetic membrane in purple bacteria (Cogdell *et al.,* 2006). Energy transfer is indicated by the grey arrows, whereas the redox reactions involved in cyclic electron transfer are shown by the black arrows. The cyclic movement of electrons results in a proton gradient across the membrane. This proton motive force is used to drive ATP synthesis.

The light reaction processes in the purple bacterial photosynthetic membrane are shown in **Fig. 4.14**. The process begins when a photon absorbed by a bacteriochlorophyll or carotenoid molecule and this energy is transferred to the RC. The RC then catalyzes a series of electron transfer reactions resulting in charge separation across the membrane. This initial charge separation reaction is stabilized by subsequent slower electron transfer reactions where (ubi- or mena) quinone is reduced to (ubi- or mena) quinol. The reduced quinone molecule diffuses out from the RC-LH1 'core' into the lipid bilayer and migrates to its oxidation site on the cytochrome *b/c1* complex. The reduced quinone is oxidized and the released proton is pumped across the membrane. The electrons are used to reduce cytochrome *c*, which in turn reduces the RC to complete the electron transfer cycle. The resulting proton gradient drives the generation of ATP through the action of ATP synthase (Berg *et al.,* 2002). The two well-defined pigment-protein complexes, LH2 and RC-LH1 'core' complex of *Rhodopseudomonas acidophila* 10050 can be separated on a sucrose gradient and be seen by the eye (Cogdell *et al.,* 1983) (**Fig. 4.13**) LH_2 has two strong NIR absorption bands at ~800 and 850 nm, whereas the RC-LH1 complex has a single strong absorption band at ~875 nm. These complexes are also referred to as

B800-850 and B875 respectively (Cogdell *et al.,* 1985). In both complexes, the bacteriochlorophylls and carotenoids in the LH parts are non-covalently bound to two hydrophobic proteins called α- and β .

4.6. THE STRUCTURE OF THE LIGHT HARVESTING COMPLEXES

(i) Structure of LH1: LH1-RC "core" complexes were first visualized in electron microscopy studies on membranes of *Rhodopseudomonas viridis* which has only LH1 complexes and it lacks LH2. Its intracytoplasmic membranes are lamellar and contain large regions of quasi-crystalline 2D arrays of LH1-RC core complexes. These core structures are circular with a diameter of ~120 Å. Image processing of these structures shows that they had six-fold symmetry. Early determination of stoichiometry indicated bacteriochlorophyll *a*/reaction center ratio of 24:1 and the reaction center were surrounded by an $\alpha_{12}\beta_{12}$ LH1 rings (Miller, 1982; Stark *et al.,* 1986). These workers produced 2D crystals of LH1-RC cores from *Rhodobium marinum* with and without the reaction center and images showed that removal of the reaction center resulted in a loss of the electron density in the middle of the LH1 ring (Meckenstock *et al.,* 1994). Karrasch *et al.* (1995) also produced 2D crystals of reconstituted LH1 complexes from *Rsp. rubrum* which was well enough ordered to be studied by electron diffraction and a 2D projection map of LH1 at a resolution of 8.5 Å was produced. This map showed a ring structure consisting of 16 α-β- pairs, very similar to that for LH2. When the data were processed at a lower resolution, it appeared to have six-fold symmetry. However, at a higher resolution, this pseudo- six-fold symmetry broke down to reveal eightfold symmetry. They also showed that with the larger ring structure, the hole in the middle (diameter 68 Å) was just big enough to accommodate the reaction center. The diameter of the LH1 ring seen by Karrasch *et al.* (1995) is very similar to the size of the core complexes seen previously both in *Rps. viridis* (Scheuring *et al.,* 2003) and *Rhodobium marinum* (Qian *et al.,* 2000). This new model of the LH1-reaction center core complex as a 16-meter conflict with the previous measurements of the bacteriochlorophyll *a*/reaction center stoichiometry of 24:1 (Zuber and Cogdell, 1995). The new model would imply a ratio of 32:1 rather than 24:1.

The LH1 structure in some species, such as *Rba. sphaeroides, Rsp. rubrum* or *Blc. viridis,* is rather flexible (Fotiadis *et al.,* 2004). Further, shape complications were found in *Rba. sphaeroides,* which is related to the presence of a protein called PufX. When this protein present in *Rba. sphaeroides* RC-LH1 complex ,they are dimeric, whereas in a PufX -phenotype the RC-LH1 complex is monomeric (Siebert *et al.*, 2004). PufX is a protein that has a role in photosynthetic cyclic electron transport. It was suggested that the presence of this protein is required to allow a sufficiently rapid redox connection between quinol reduced by the RC with the site of its oxidation on the cytochrome *b/c*1complex (Barz *et al.,* 1995). It was further suggested that PufX provides a portal to allow the quinol to escape through the LH1 helices into the photosynthetic membrane **(Fig. 4.15)** (Cogdell *et al.,* 2006).

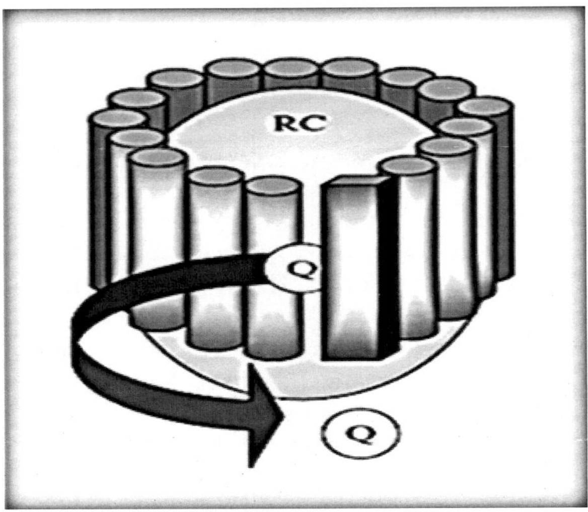

Fig. 4.15. A schematic diagram LH1 that is interrupted by PufX protein.

(ii) Structure of LH2: The 3D crystal structure to 2.5 Å resolution of the LH2 complex from *Rps. acidophila* strain 10050 has allowed the structure and function of LH2 (**Fig. 4.16**) (Cogdell *et al.,* 1999). The overall structure is a circular nanometer composed of nine pairs of α-and β- apoproteins ($\alpha_9\beta_9$ nanometer). It is rather like an elongated, hollow cylinder. The inner walls of the cylinder are formed from the transmembrane α-helices of the α-apoproteins, and the outer walls are formed from the α-helices of the β-apoproteins. The pigments are all located within these walls of protein. The structure is "capped" top and bottom by the N and C termini of the apoproteins which fold over and interact with each other. Overall, the complex consists of 18 polypeptides, 27 bacteriochlorophyll *a* and 9 carotenoid molecules. The complex can be split into nine heterodimers, each containing one α- and one β- apoprotein, one bacteriochlorophyll a molecule that absorbs at 800 nm (B800), two bacteriochlorophyll molecules that absorb at 850 nm (B850) and one carotenoid. The α-apoprotein helices make up the inner wall of the complex, while the β-apoprotein helices form the outer wall. The C-termini of both apoproteins are located at the periplasmic side of the photosynthetic membrane and the N-termini are at the cytoplasmic side. Both of these termini bend over, almost parallel to the membrane surfaces, to enclose the pigments. The eighteen B850 bacteriochlorophyll molecules are situated in between the helices, with the plane of their bacteriochlorin rings perpendicular to the proposed membrane surface. In each α - β-apoprotein pair, one bacteriochlorophyll is bound to the α-polypeptide and one to the β-polypeptide. The nine B800 bacteriochlorophyll molecules are situated between the β-polypeptides. The positions of the B800 bacteriochlorophyll rings are almost parallel to the membrane surface (**Fig. 4.17 and 4.18** .)

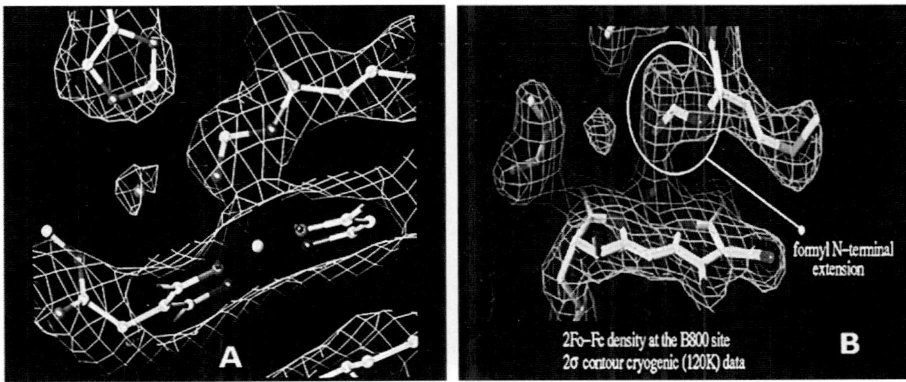

Fig. 4.16. Comparison of the electron density in the region of the B800 Bchl *a* binding pocket of LH2 from *Rhodopseudomonas acidophila* at a resolution of 2.5 Å (A) and 2.0 Å (B). (A) With a resolution of 2.5 (Å) The extension of the N-terminal methionine residue of the α-apoprotein is clearly seen in the electron density map, together with the "modeled" formyl group. (B) At this improved resolution of 2.0 Å, the N-terminal extension is also clearly seen. Now, however, it can be seen to bifurcate. The modeled formyl group no longer gives a satisfactory fit to this higher-resolution data. The electron density is shown as the white or blue cages (Cogdell *et al.,* 1999).

The major carotenoid molecule present in the LH2 of *Rps. acidophila* strain 10050 is rhodopin glycoside, although there are a small number of others (Gardiner *et al.,* 1992). These nine molecules are in Vander Waals contact with the apoproteins and the bacteriochlorophyll molecules. The carotenoid is found between the two B850 bacteriochlorophyll a molecule in the monomer and also comes into close contact with the B800 bacteriochlorophyll a molecule. The carotenoids must be in close contact with the bacteriochlorophyll molecules in order to quench the in-triplet states before they can continue to make singlet oxygen. The RC-LH1 'core' complexes were first visualized in EM pictures of membranes from *Rps. viridis* and referred to as *Blastochloris viridis*, and confirmed to have a hexagonal shape surrounding the RC (Hiraishi, 1997). Since then, many other studies have suggested a variety of shapes for LH1; circular (Gerken *et al.,* 2003), square (Stahlberg *et al.,* 1998), 'S'-shaped (Scheuring *et al.,* 2005), elliptical (Fotiadis *et al.,* 2004) or even just arcs (Bahatyrova *et al.,* 2004). Basically, there are at least two distinct classes of RC-LH1 'core' complexes. In one class they are monomeric, namely one RC surrounded by one LH1 complex. Examples of this class are in the RC-LH1 complex of *Rsp. rubrum* and *Rps. palustris.* The second classes are dimeric, consisting of two RC-LH1 units. This kind of RC-LH1 'core' complex can be found in *Rba. sphaeroides* (Siebert *et al.* 2004).

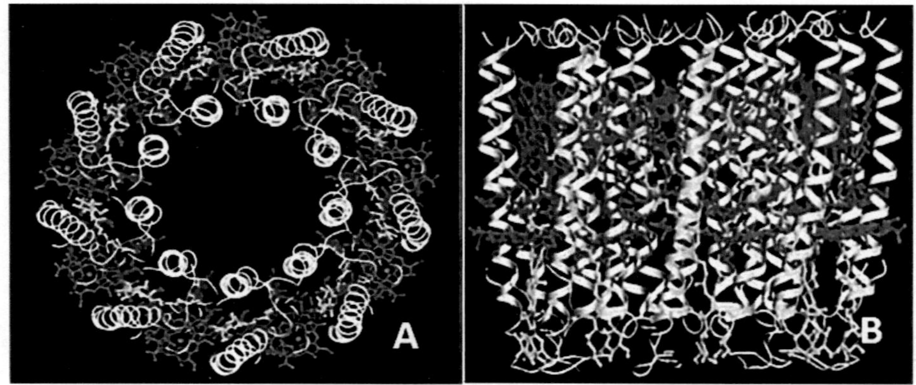

Fig. 4.17. LH2 holocomplex of *Rps. acidophila* (A) The cytoplasmic side of the complex (B) A view of the complex perpendicular to the symmetry axis.

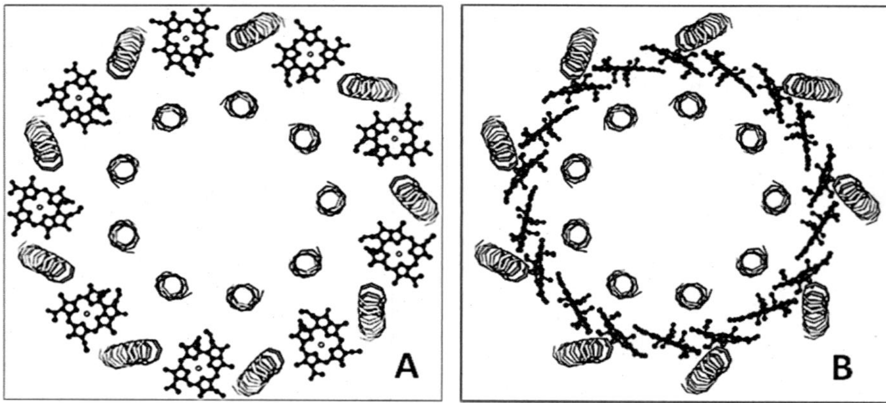

Fig. 4.18. Location and organization of the B800 bacteriochlorophylls (A) and organization of B850 bacteriochlorophylls (B) in theLH2 of *Rhodopseudomonas acidophila* (Cogdell *et al.,* 1999).

4.7. FUNCTION OF LIGHT HARVESTING COMPLEXES

The LH complexes function in a series of photophysical and photochemical processes. These processes are the result of the interaction of light with matter. There are three energy states that play important role in the photophysics of organic molecules, namely, the ground state (S0), the lowest singlet excited state (S1) and the lowest triplet excited state (T1). The absorption of light by a molecule results in the promotion of an electron from the highest occupied molecular orbital (HOMO) to the lowest unoccupied molecular orbital (LUMO) (**Fig. 4.19**) to produce the first excited singlet state, S1. The singlet excited state (S1) can decay to the ground state through several radiative and non-radiative processes. All of these processes are noted in a Jablonski diagram (**Fig. 4.20** .) The vertical scale represents energy and the arrows represent the type of possible transitions between different electronic states. Ground state molecule (S0) is promoted to an excited state (S1 or S2) by absorbing light energy. Non-radioactive decay mechanisms (Internal conversion (IC) and Intersystem crossing (ISC)) return the molecule to the ground state whereas radioactive transitions (Fluorescence (Fl) or Phosphorescence (Ph) also decay excited molecule by emitting light (Garrison, 2005).

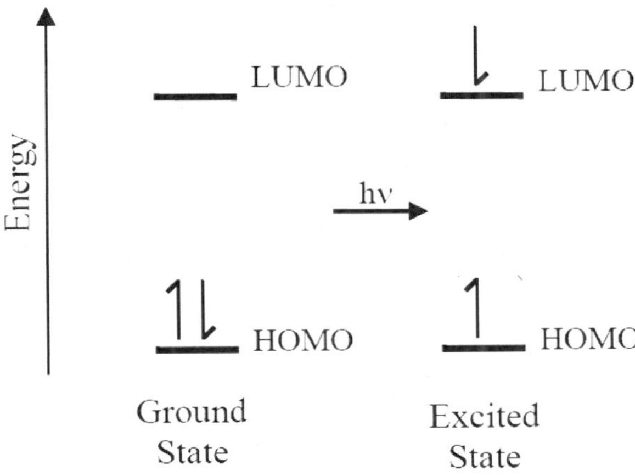

Fig. 4.19. Electronic ground and excited state levels of a typical molecule.

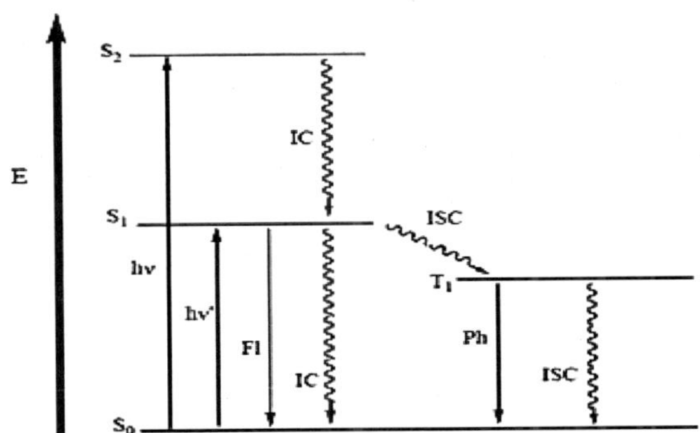

Fig. 4.20. A Jablonski diagram showing the electronic states of a molecule.

Radioactive decay consists of emissive processes, fluorescence (Fl) or phosphorescence (Ph), whereas non-radioactive processes consist of internal conversion (IC) and intersystem crossing (ISC) (Kavarnos and Turro, 1986). Fluorescence ($S_1 \rightarrow S_0 + hv$) is a spin allowed process in which the molecule relaxes to the ground state via the emission of a photon of light. Phosphorescence ($T_1 \rightarrow S_0 + hv$) is a forbidden radioactive process involving emission of light from the triplet state (T_1). IC is non-radioactive decay process where energy is lost through vibrational motion ($S_2 \rightarrow S_1$ + Heat; $S_1 \rightarrow S_0$ + Heat), and involves a transition between states of the same spin. ISC is also a non-radioactive decay process but it performs a forbidden transition that incorporates an electron spin-flip of the electron ($S_1 \rightarrow T_1$; $T_1 \rightarrow S_1$; $T_1 \rightarrow S_0$). The bacteriochlorophyll molecule, for example, when excited may participate in energy transfer or electron transfer processes that compete with the normal modes of relaxation. Energy transfer is the process whereby energy is transferred from an excited molecule to a second ground state molecule. Electron transfer is a process where an electron is transferred from a donor, excited molecule, to an acceptor molecule. When light drives the electron transfer process, it is usually called Photoinduced Electron Transfer (PET). A schematic figure of both the processes is illustrated in **Fig. 4.21**. If a photon is absorbed in the correct wavelength, an electron is excited from the ground state (S0) to the singlet state (S1). (1) Energy transfer occurs if the singlet state electron returns to the ground state by releasing the energy which is then captured by a second ground state molecule. (2) Electron transfer is a process where the singlet state electron is transferred from a donor into an acceptor molecule (Bchl).

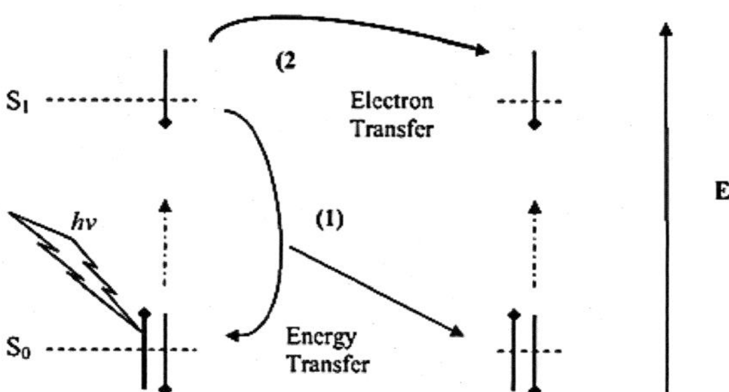

Fig. 4.21. A schematic model of energy and electron transfer.

The light-absorbing pigments are non-covalently bound to two types of integral membrane proteins, forming either reaction center or antenna complexes (Cogdell *et al.,* 1999). The function of the pigments depends on the protein where they are bound. LH polypeptides hold the pigments in the correct orientation so that they are able to perform light harvesting. In this case the protein 'scaffold' allows these pigments to absorb light energy and then efficiently and rapidly transfer that energy to the RC.

The pathway of energy transfer in the purple bacterial antenna complexes is as follows: B800→B800, B800→B850, B850→B850, LH2→LH1 (B850→B875) and LH1→RC. Initially, when a LH2 B800 bacteriochlorophyll molecule absorbs a photon the molecule is excited to the first excited singlet state (B800*). B800→B800 singlet-singlet energy transfer can occur. From B800*, energy can then be transferred to the B850 molecules that are excitonically coupled in a ring (Sundstorm and Van Grondelle, 1995).

Rapid energy migration around the B850 ring is functionally important in the bacterial photosynthetic membrane. The energy migrates around the B850 molecules very rapidly >100-200 fs. The rate of energy transfer from *Rps. sphaeroides* and *Rba. acidophila* LH2 (B850) → LH1 (B875) has been time-resolved. This energy transfer takes place in 2-5 ps. The energy transfer from B875 → RC takes between 20 and 50 ps was illustrated in **Fig. 4.22**. Transfer between B800 takes about 500 fs. Transfer from B800 to B850 takes about 1-2 ps. Transfer between B850 takes about 100-200 fs. Transfer from B850 to B875 (LH1) takes about 2-5 ps. Then the energy migrates to RC in about 20-50 ps. The energy "funnels" its way to the RC, through the LH.

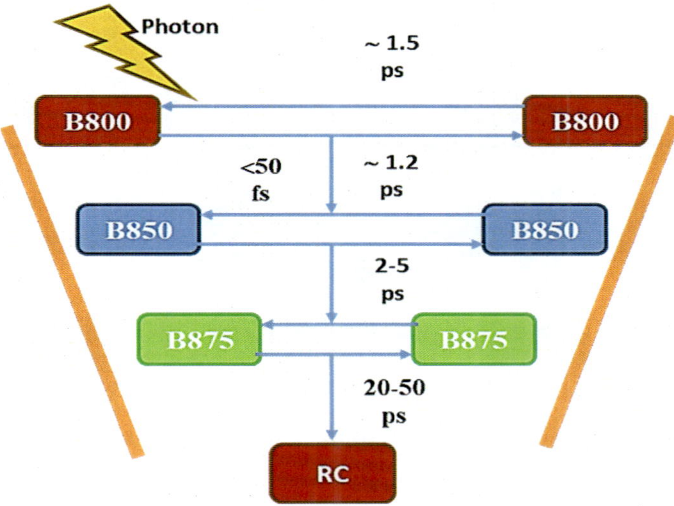

Fig. 4.22. Illustration of energy transfer times in the light-harvesting complex.

Carotenoids function as accessory LH pigments and have a stabilizing role in the complex. However, the most important role of carotenoids in LH complexes is photoprotection by absorbing energy from excited bacteriochlorophyll a can led to the production of bacteriochlorophyll a triplet and on to the production of singlet oxygen. Singlet oxygen is very toxic to the cells. When carotenoids are present, energy transfer between bacteriochlorophyll triplets and carotenoids leading to the production of carotenoid triplets rather than singlet oxygen (Bittl *et al.,* 2001).

4.8. THE REACTION CENTER (RC)

The RC is a pigment-protein complex, in which energy from LH is used to drive the charge-separation reaction across the photosynthetic membrane (Cogdell *et al.,* 2008). The RC consists of both bacteriochlorophylls and other electron transfer cofactors, such as quinones or iron-sulfur centers, along with hydrophobic peptides. The RCs contain a "special pair" of chlorophylls, which have unique properties because of their environment in the RC protein (Blankenship, 2002). Anoxygenic bacteria, such as purple bacteria, have only one type of RC, while oxygenic photosynthetic organisms have two RC which are called Photosystem-I (PSI) and Photosystem-II (PSII). Purple bacteria, on the other hand, contain a pheophytin-quinone type RC similar to PSII.

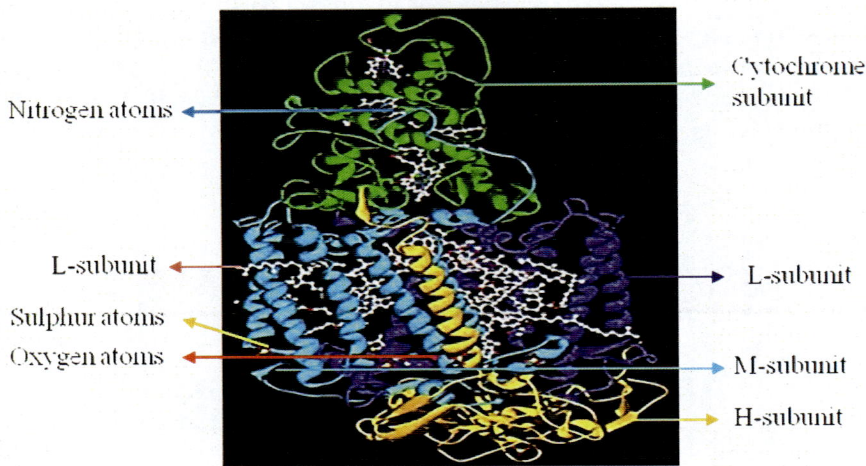

Fig. 4.23. Structure of the *Blc. viridis* reaction center consisting of a cytochrome subunit and the M, L and H subunits. Molecules that carry out the reaction center function, the cofactors, are represented as ball-and-stick models.

4.8.1. Purple Bacterial Reaction Center

Most purple bacterial reaction centers contain three protein subunits; Heavy (H), Medium (M), Light (L) and additional bound cytochrome *c* (C). The RC from *Blc. viridis*, has an additional submit (**Fig. 4.23**) (Deisenhofer, 2009). The L- and M- subunits have five transmembrane helices each, whereas the H-subunit has only one. The L- and M-subunits in all bacterial RC have a similar overall protein fold and are located both at the periplasmic and cytoplasmic sides of the membrane. The H-subunit is located on the cytoplasmic side of the membrane alone. The C- subunit consists of five segments that bind covalently four iron-containing heme groups (**Fig. 4.23**). These are; the N-terminal segment, the first heme-binding segment, a connecting segment, a second heme-binding segment and the C-terminal segment. The two segments that contain binding sites for the heme groups make up the core of the cytochrome, while the other three non-heme binding segments contain little secondary structure. The first heme-binding segment contains the binding sites for heme-1 and heme-2, whereas the second contains the binding sites for heme-3 and heme-4 (Lancaster, 2008).

The RC cofactors are surrounded and held in place by the protein subunits and are arranged into two symmetry-related branches (**Fig. 4.24**). The branch mainly located near the L- subunit is active in electron transfer and is usually called the 'A' branch, whereas the non-active, M-subunit side is called the 'B' branch (Lancaster, 2008). Both branches start from a 'special pair' (P) of bacteriochlorophyll molecules located near the periplasmic side of the membrane. Then each branch contains a single monomeric bacteriochlorophyll, followed by a Bpheo molecule and ends with a quinone, either mena- or ubiquinone. Non-heme iron is located between the two quinones and a carotenoid molecule is in Vander Waals contact with the monomeric bacteriochlorophyll on B branch.

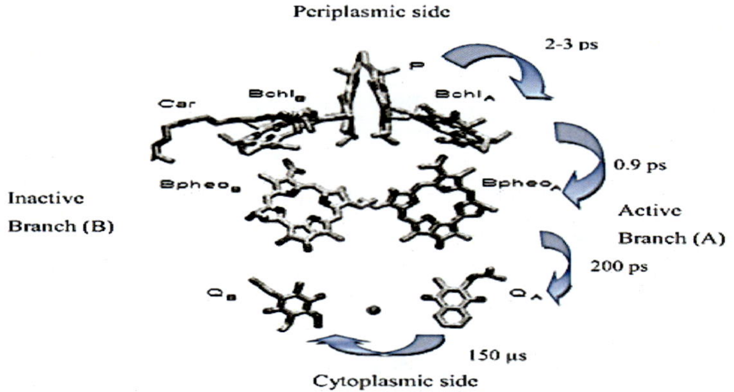

Fig. 4.24. RC cofactor arrangement, Special pair (P), bacteriochlorophyll (BA, BB), Bacteriopheophythin (BpheoA, BpheoB), Quinone (QA, QB). The tails of the cofactor molecules have been removed for clarity. The electron transfer reactions are indicated by the arrows.

4.8.2. Functioning of Reaction Centers

Energy is transferred from the LH system into the RC and is used to excite P to P*. Electron transfer down the A-branch between P* and BA occurs in about 2-3 ps. Next, the electron hops quickly in about 0.9 ps to Bpheo A and is then transferred on in the next step to QA at the cytoplasmic side in about 200 ps. Each of these electron transfers successfully slows down the back reaction by 3 orders of magnitude leaving P^+. The positive charge left on P is then neutralized by electron flow from C. QB then requires a second turnover to take up two H^+ from the cytoplasmic side of the membrane to form QBH2, which diffuses out into the lipid bilayer and away from the RC.

4.9. PHOTOSYNTHETIC UNIT (PSU)

In order to describe the purple bacterial light-harvesting systems, we need to have a complete picture of the organization of the antenna complexes *in vivo* . This is not yet possible. However, if for this purpose we can take the Karrasch *et al.* (1995) structure for LH1 at face value, then it is possible to model a high-resolution generic LH1-RC core complex. Two groups have done this, by using the structures of the two LH complexes and recognizing the strong structural homology of the LH1 and LH2 apoproteins, especially in the transmembrane a-helical regions. The LH1- reaction center core structure can be put together with the known structure of LH2 to produce a model of the whole PSU **(Fig. 4.25)**

(Cogdell *et al.,* 1999). One striking feature of this model is the way in which the macrocycles of tightly coupled rings of bacteriochlorophyll a (the B850 ring in LH2 and the B875 ring in LH1) line up at the same depth in the membrane. This also corresponds very closely to the transmembrane location of the special pair of bacteriochlorophyll a in the reaction center. This minimizes the distance between the rings of bacteriochlorophyll and maximizes the rate of energy transfer from LH2-LH1 and from LH1 to the reaction center. This model is very useful to understand the different types of pigment- pigment interactions that are involved in each of the different energy transfer reactions that occur in the light-harvesting system. The situation *In vivo,* is more heterogeneous than these simple models suggested. The composition of the purple bacterial PSU is dynamic and changes will depend on the growth conditions as described above. Moreover, previous studies on energy migration in the photosynthetic membranes from a range of different species of purple bacteria have shown that the precise supramolecular organization of the LH2 and LH1-RC cores are also species-dependent. There is an urgent need to examine the architecture of the PSU in intact photosynthetic membranes.

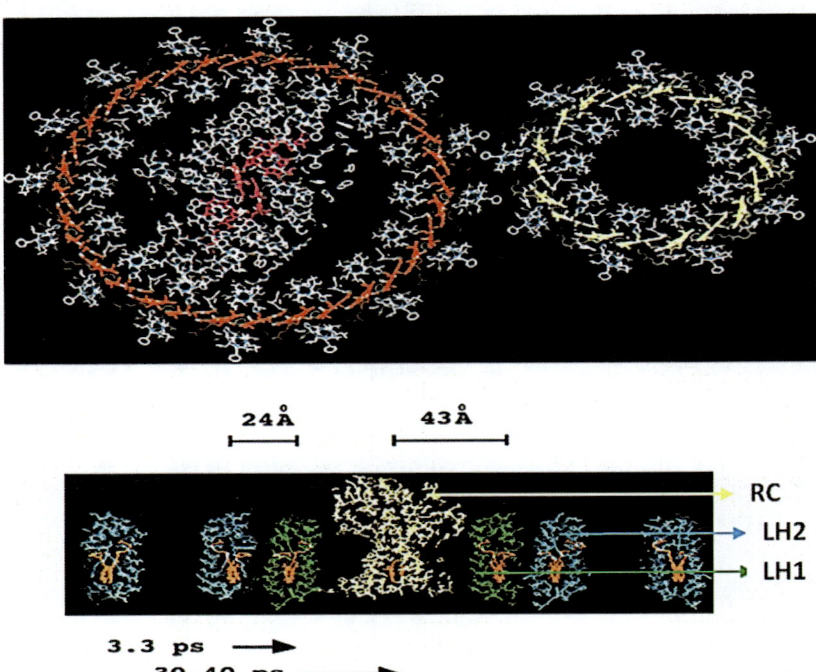

Fig. 4.25. Two views of a model of the purple bacterial PSU. (A) A top view, looking perpendicular to the assumed plane of the membrane. This point in the LH complexes where the tightly coupled rings of bacteriochlorophylls *a* are located .The smaller LH2 sits outside the large LH1 complex. (B) A side view, looking from within the assumed plane of the membrane (blue, LH2; green, LH1; yellow, RC) (Cogdell *et al.,* 1999).

4.9.1. Energy Transfer within the PSU

The purple bacterial PSU has proved to be a very attractive system for physical chemists interested in the study of the light-harvesting process for two reasons 1.There is atomic resolution, structural information and 2. Unlike in plants, there is an excellent spectral separation between different pigment groups in the energy transfer processes. Progress in this area has also been greatly assisted by parallel advances in laser technology which now allow energy transfer events to be probed with red-white), (C) Alignment of the carotenoid to the Qx dipoles, (D) Alignment of the carotenoid with respect to the Qy dipoles. The distances between the center of a B800 *a* and the Femto second-time resolution. **Fig. 4.26** shows the structural context in which the discussion of energy transfer within LH2 should begin. (A) Alignment of the Qx dipoles (yellow-green), (B) Alignment of the Qy dipoles (α-bound and β-bound B850 bacteriochlorophyll a in the same α-β-apoprotein pair are 17.4 and 18.2 Å, respectively. Qx and Qy are labels for the two bacteriochlorophyll a absorption bands at 590 nm and 800 or 850 nm respectively (Cogdell *et al.,* 1999). The transition dipoles, which correspond to these absorption bands, lie within the plane of the bacteriochlorin rings, at right angles to each other, diagonally between the nitrogen atoms, which coordinate the central Mg^{2+}. One of the factors which control the rate of excitation energy transfer between two molecules is the angle between the transition dipoles involved. When the dipoles are parallel, energy transfer is favorable when they are orthogonal, energy transfer is much less favorable.

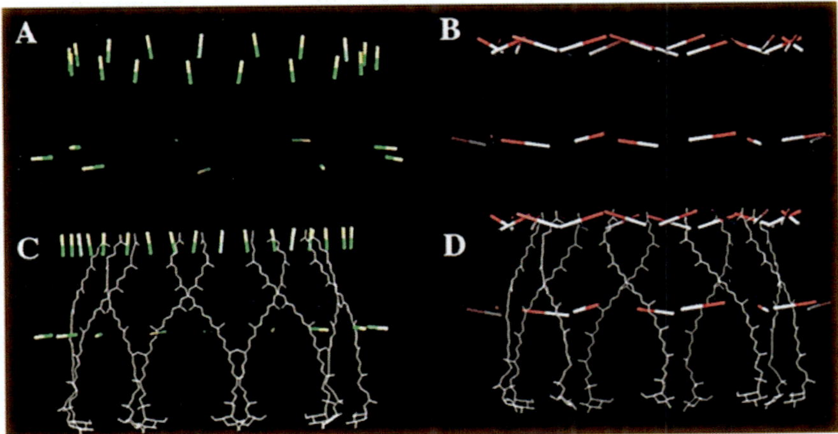

Fig. 4.26. Relative orientation of the Qx and Qy transition dipoles in the B800 and B850 bacteriochlorophyll *a* and the carotenoid in the LH2 complex from *Rhodopseudomonas acidophila* (Cogdell *et al.,* 1999).

The inter pigment distances and the relative orientation of the transition dipole moments of the major electronic states are shown. For many years now, it has been well documented that light absorbed by the light-harvesting carotenoids can be used to drive the light reactions in purple bacterial photosynthesis

(Koyama *et al.,* 1996). The efficiency of carotenoids→ bacteriochlorophyll *a* energy transfer varies from species to species and from LH complex to LH complex, from 30 up to 100% (Frank and Cogdell, 1993). The exact mechanisms involved here have not yet been defined. The basic problem is as follows. For many years, carotenoids were thought to be non-fluorescent (Koyama *et al.,* 1996), implying that their fluorescence lifetimes must be very short. If this was true, then how can there be enough time for such efficient singlet-singlet energy transfer (i.e., light-harvesting) to occur before the carotenoid's excited singlet state is lost by other competing processes. The answer to this question is two-fold. 1. Carotenoids do fluoresce (Koyama *et al.,* 1996), and 2. They have rather unusual photochemistry. In particular, the allowed singlet state transition from the ground state goes to the S2 state rather than the S1 state. The direct one-photon-induced transition from the ground state to the S1 state cannot occur because it is symmetry forbidden. Following excitation, the carotenoid's S2 state lasts only for a few hundred femtoseconds before it relaxes into the S1 state. S1 then decays back to the ground state in a few picoseconds **(Fig. 4.27).**

In the LH2 complex of *Rhodopseudomonas acidophila* 10050, the efficiency of energy transfer between carotenoid to bacteriochlorophyll *a* is singlet-singlet is about 55% (Frank and Cogdell, 1996). We can, therefore, ask two questions: which excited state of the carotenoid is acting as the energy donor and which group of bacteriochlorophyll a (i.e., B800 or B850) receives this energy?. If the carotenoid rhodopin-glucoside is excited with a 60-fs excitation pulse, the measured lifetime of the S_2 state depends upon where that carotenoid is present (Macpherson *et al.,* 1998). In benzyl alcohol, its lifetime is ~130 fs. In LH2, this lifetime is shortened to ~61 fs. This reduction in a lifetime is due to energy transfer to the bacteriochlorophyll *a.* This can be seen directly by exciting the carotenoids and measuring the kinetics of the arrival of that energy in the B800 or B850 manifolds. Measuring at 851 nm the kinetics of the increase in B850 fluorescence is biphasic. About 70% of the fluorescence rises with a time constant of ~63 fs. The other 30% rises with a time constant of ~900 fs. Clearly, the last phase of this energy transfer event corresponds to direct singlet- singlet energy transfer from the S2 state of rhodopin-glucoside to the B850 manifold. However, what is the slower rise due to excitation of LH2 at 800 nm has allowed the time course for the B800-B850 singlet-singlet energy transfer to be determined (Sundstrom *et al.,* 1999). The rate constant for this process is about 900 fs. The slower phase in the carotenoid- to-B850 energy transfer reflects that energy that has gone via the B800 bacteriochlorophylls. The kinetics of the carotenoid to B800 energy transfer also indicates that the S_2 state of rhodopin-glucoside is the major energy donor. On the basis of the energy levels of the carotenoid's two excited singlet states, it has been suggested that energy transfer from S_2 goes by way of the Qx transition (Qx is the bacteriochlorophyll *a* absorption band at 590 nm) of bacteriochlorophyll a and that from S_1 goes by way of the lower-energy Qy transition (Sundstrom *et al.,* 1999). In the case of LH2 from *Rhodopseudomonas acidophila,* it appears that nearly all of the carotenoid to bacteriochlorophyll *a* singlet-singlet energy transfer comes from the S_2 state (Frank and Cogdell, 1996). In contrast, other antenna complexes, such as LH2 from *Rhodobacter sphaeroides* (where the efficiency of carotenoid-to-bacteriochlorophyll *a* energy transfer is nearly 100%) are also able to harvest energy from the carotenoid's S_1 state. Once the excited state has reached the B850 manifold, it is very rapidly depolarized. This means that it hops very rapidly around the B850 ring. This hopping time has been estimated to be on the order of a few tens of femtoseconds. If there is no other LH complexes nearby that can accept the excitation energy, the excited state will decay in 1 ns. Consequently, the excited state visits each B850 bacteriochlorophyll *a* many time during its lifetime and is, therefore, available for energy transfer out of any site in the ring with equal probability. This is very important because it means that there does not have to be a fixed supramolecular arrangement of LH2 and LH1 complexes in the PSU for efficient energy transfer to occur. As long as the next antenna complex is sufficiently close, energy transfer will occur with equally high efficiency from anywhere in the LH2 ring.

Following excitation of LH2, the times of energy transfer from LH2→LH1 and LH3 reaction center steps can also be measured (Sundstrom *et al.,* 1999). The kinetics of energy transfer from LH2 to LH1 phase takes 3 to 4 ps. The slower phases probably arise from a rather heterogeneous arrangement of the LH complexes in the membrane so that there are several LH2→LH2 steps before an LH1 complex is encountered. The final energy transfer step from LH1 to the reaction center is the slowest and takes 30 to 50 ps. This is clearly due to the longer distance involved compared to LH2→LH1 (**Fig. 4.27**). Even though most, if not all, of the times of the energy transfer steps between the absorption of a green photon by LH2 carotenoid and the arrival of that energy at the reaction center, have been resolved, the details of the exact molecular mechanisms involved are reviewed excellently by Sundstrom *et al.* (1999).

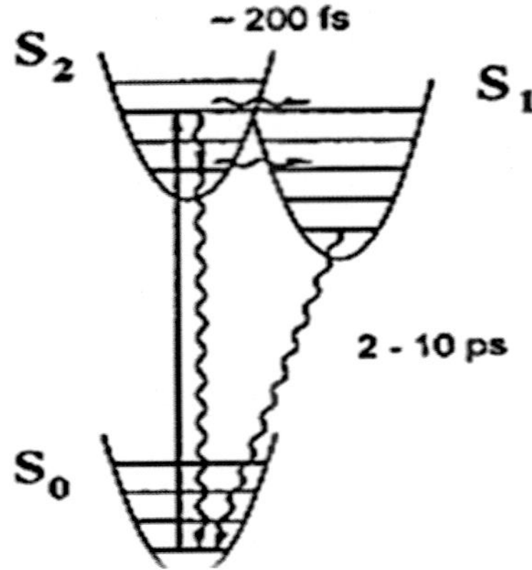

Fig. 4.27. Schematic representation of the two low-lying excited single states of S_1 and S_2 carotenoids. The approximate positions of the S_1 and S_2 excited singlet states are shown. The $S_0 \rightarrow S_2$ represents the optically allowed (One- Photon) transition that gives rise to the carotenoids well-known, strong absorption spectrum. The approximate times for the $S_2 \rightarrow S_1$ and $S_1 \rightarrow S_0$ transitions are also showed in this fig (Cogdell *et al.,* 1999).

4.9.2. Excitation of Antenna Molecules and Resonance Transfer

The excited electron is very unstable and returns to the ground state resulting in one of three reactions. These reactions are (1) Transfer of energy to the adjacent molecule exciting its electron (resonance transfer), (2) Reduction of a second compound (oxidation, photo-induced charge separation), or (3) Emission of fluorescent light. The antenna molecules are excited about absorbing photons. The energy gained by the pigments through such excitation is referred to as the exciton. The exciton is transferred from antenna molecules to the reaction center through resonance transfer and this reaction takes in about 0.1 picoseconds. This energy is referred to as resonance energy. The electron transport reaction at the reaction center is initiated by oxidation with the consumption of resonance energy. Resonance transfer takes place from molecules of a higher exciton to those with a lower exciton. For this reason, photosynthetic pigments in photosynthetic apparatus are arranged to facilitate such resonance transfer.

4.10. ELECTRON TRANSPORT

When the resonance energy excites the reaction center of chlorophyll, its redox potential becomes very low enough to transfer electrons to a lower potential electron carrier. The oxidized chlorophylls are reduced again by either oxidizing externally supplied electron donors in a process known as non-cyclic electron transport or by taking the original electrons through the cyclic electron transport system. Green bacteria have only cyclic electron transport and reduced sulfur or organic compounds serve as the electron donor.

4.10.1. Green Sulfur Bacteria

Green sulfur bacteria utilize light energy through anoxygenic cyclic electron transport similar to that of cyanobacterial photosystem-I with a P840 iron-sulfur type reaction center. Bacteriochlorophyll *a* is the photosynthetic pigment of P840 (**Fig. 4.28**). Antenna molecules on the chlorosomes transfer the exciton to the reaction center on the cytoplasmic membrane. Bacteriochlorophyll *a* mediates exciton transfer through the baseplate. The normal state P840 has a redox potential of +0.3V and this decreases to lower than -1.0V, low enough to reduce the primary acceptor (A_0, bacteriochlorophyll *a*) when excited by photons. Electrons either flow back to P840 generating a proton motive force or are transferred to $NAD(P)^+$. Electrons that are consumed to reduce $NAD(P)^+$ are replaced by oxidizing electron donors such as sulfide. These organisms cannot grow chemoorganotrophically under aerobic conditions. On the other hand, the other members of the filamentous anoxygenic phototrophic bacteria can grow chemoorganotrophically under aerobic conditions. The latter has a different photosynthetic electron transport system from that of the green sulfur bacteria which is similar to that of the purple bacteria with a pheophytin-quinone-type reaction center.

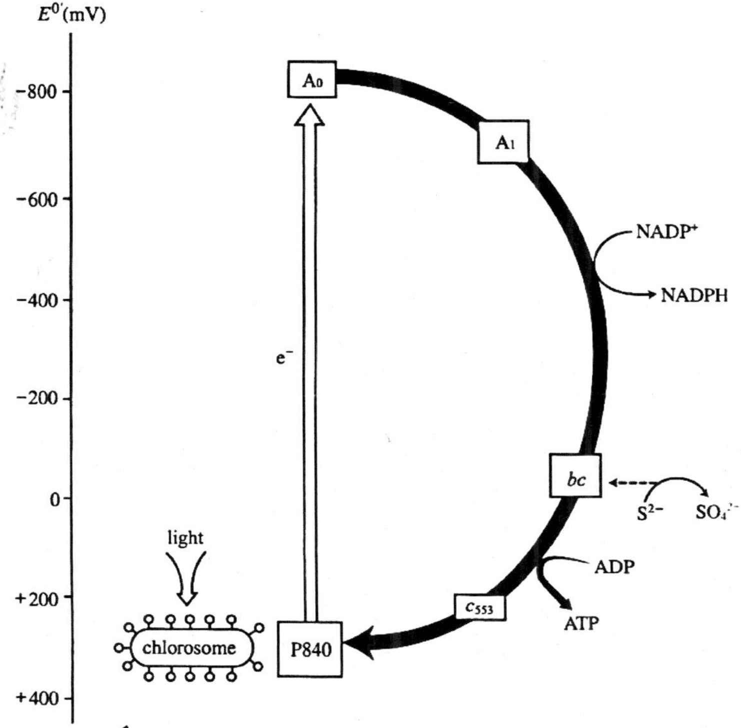

Fig. 4.28 Photosynthetic electron transport in green bacteria. A_0, primary electron transport (bacteriochlorophyll *a*); A_1, a quinone compounds; *bc*, cytochrome *bc* complex; $c_{533,}$ cytochrome $c_{553.}$

4.10.2. *Heliobacteria*

These bacteria have P788 as the reaction center and different from that of green sulfur bacteria, although the photosynthetic mechanisms are similar.

4.10.3. Purple Bacteria

These bacteria utilize light energy to generate a proton motive force which is used to synthesize ATP or reduce NAD(P)+ through reverse electron transport. Purple bacteria have a reaction center with maximum absorption at 870 nm, which is excited by photons with a decrease in redox potential. The excited P870 reduces bacteriopheophytin to begin the cyclic electron transport for the generation of a proton motive force. $NAD(P)^+$ is reduced through reverse electron transport oxidizing quinol. Sulfur and organic compounds are used as electron donors reducing cytochrome c_2 (**Fig. 4.29**). Though purple sulfur bacteria cannot grow under aerobic conditions, purple non-sulfur bacteria can grow chemoorganotrophically under aerobic conditions.

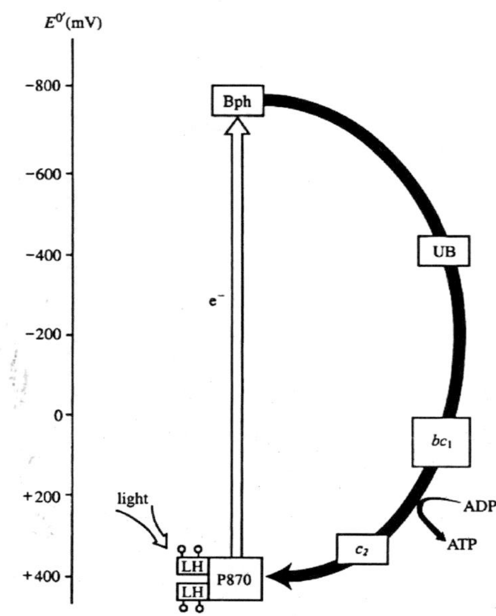

Fig. 4.29. Light reactions in purple bacteria. Bph, bacteriopheophytin; UB, ubiquinone; bc_1, cytochrome bc_1 complex; c_2, cytochrome c_2 .

4.10.4. Aerobic Anoxygenic Phototrophic Bacteria

These bacteria synthesize bacteriochlorophyll *a* and carotenoids only under aerobic conditions. They do not use light energy under anaerobic conditions. They have a pheophytin-quinone-type reaction center. The cytoplasmic membrane accommodates the reaction center and antenna molecules. They have a cyclic electron transport system as in the purple bacteria, but NAD(P)H is supplied from the metabolism of organic compounds.

4.11. PHOTOPHOSPHORYLATION IN HALOPHILIC PHOTOTOPHIC BACTERIA

Halophilic archaea, including species of *Halobacterium*, can grow at a NaCl concentration of over 2.5 M. They swim away from the light when enough electron donors and O_2 are available. With a limited O_2 supply, they move towards the light. They generate a proton motive force transporting H^+ and Cl^- across the membrane using light energy in a process known as photophosphorylation. A group of proteins known as rhodopsins facilitates motility and photophosphorylation.

Sensory rhodopsin II is synthesized for movement away from the light when the dissolved O_2 (DO) concentration is high, and light attracts halophilic archaea under a limited O_2 supply. Sensory rhodopsin I is synthesized with bacteriorhodopsin and halorhodopsin under low dissolved oxygen conditions. With light energy bacteriorhodopsin exports, $H^+$and halorhodopsin imports Cl^- to generate a proton motive force. These rhodopsins are purple proteins bound with the retina (**Fig. 4.30** .)

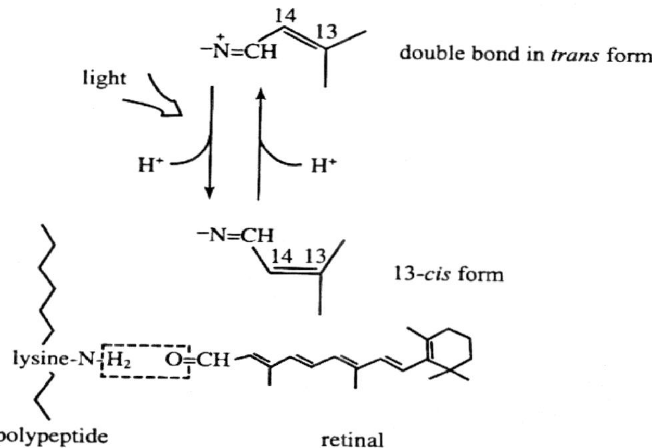

Fig. 4.30. The structure of retinal bound to rhodopsins in species of *Halobacterium*.

Rhodopsins have a similar structure with retinal forming a Schiff's base with a lysine residue of the peptide. The Schiff's base releases H^+ when it receives light. Bacteriorhodopsin exports H^+ and the sensory rhodopsins pass the information to transducer proteins to control phototaxis **(Fig. 4.31).**

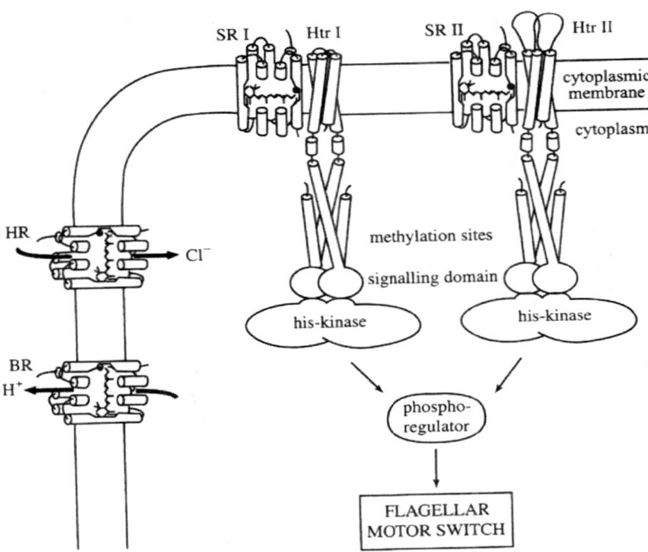

Fig. 4.31. Rhodopsins and their functions in halophilic phototrophic bacteria. Br, Bacteriorhodopsin; Hr, Halorhodopsin; Sr I, Sensory rhodopsins l; Htr I, transducer I; Sr II, sensory rhodopsins II; Htr II, transducer II

4.12. PHOTOCHEMICAL REACTION CENTER

The photochemical reaction centers contain the site where a molecule of chlorophyll becomes photoactivated and oxidized by donating an electron to a carrier molecule. Chlorophyll molecules in the reaction center differ from those in the antenna in two respects. 1. They are associated with certain proteins that interact with them in a manner that decreases the energy required to raise them to the activated state and 2. They are near the carrier molecules that can accept an electron from them when they are activated.

Numerous chlorophyll molecules in the antenna (in the case of most purple bacteria, there are 50 to 500 light-harvesting chlorophyll molecules per reaction center), being unassociated with carrier molecules, do not become oxidized upon activation rather they transfer their absorbed energy by a process termed inductive resonance to an adjacent pigment and eventually to the reaction center. Since the reaction center chlorophylls are more easily activated, this transfer is associated with a slight loss of energy in the form of heat. The energy required to activate a molecule of chlorophyll designated P, in a reaction center can be recovered by the maximum wavelength of a photon that can bring it about. Thus, reaction center bacteriochlorophyll of a purple bacterium that is activated maximally by photons of wave-length of 870 nm is designated as P870.

One of the best-characterized reaction centers is *Rhodobacter sphaeroides.* It contains three polypeptides (21, 24 and 28 kilodaltons), four bacteriochlorophyll molecules, two bacteriopheophytin molecules (bacteriochlorophyll that lack magnesium), two ubiquinone molecules and one iron molecule. It is realized from a variety of experiments that two of the reaction centers of bacteriochlorophylls are in the P870 state, owing to their association with the three polypeptides. When activated, each can donate an electron to an associated molecule of bacteriochlorophyll (B800), which transfers it to a molecule of bacteriopheophytin, and sequentially through two molecules of ubiquinone. The second molecule of ubiquinone seems to act as a gate, accepting two electrons before passing them to the associated electron transport chain.

4.13. PHOTOSYNTHETIC ELECTRON TRANSPORT CHAIN

The photosynthetic electron transport chains located within the photosynthetic membrane are composed of the same sort of carrier molecules-cytochromes, quinines and iron-sulfur centers that are found in respiratory electron transport chains and they appear to function in the same manner. As electron flow through the chain, a proton motive force is generated that is used in part to synthesize ATP by a membrane-located ATP phosphohydrolase.

4.14. PATTERN OF ELECTRON FLOW

The reaction center chlorophyll in its photoactivated and oxidized states serves as electron donor and acceptor for an electron transport chain **(Fig. 4.32 a,b,c).**

Such a system is termed cyclic photophosphorylation. However, electrons cannot be withdrawn from it, so another source of the electron is required to generate reducing power in the form of reduced pyridine nucleotides. Fundamental differences exist in the pattern of electrons depending on the electron donor. It is water, which gets oxidized (removal of electrons) only at a relatively high potential of 0.8V. The product of oxidation is O_2 gas. Thus, those organisms (plant, algae and cyanobacteria) that utilize water are said to carry out oxygenic or plant photosynthesis, while organisms (purple and green bacteria) utilize electron of other sources are said to be anoxygenic or bacterial photosynthesis.

In certain organisms that carry out anoxygenic photosynthesis, reducing power may be generated by reverse electron transport as in chemoautotrophs, while in others, it is generated by a process called non-cyclic photophosphorylation **(Fig. 4.32b).** In this pattern of electron flow, oxidized reaction center bacteriochlorophyll accepts electrons from an electron transport chain fed by an alternate electron source, thus freeing electron emanating from bacteriochlorophyll in the activated state to be used for the reduction of NADP.

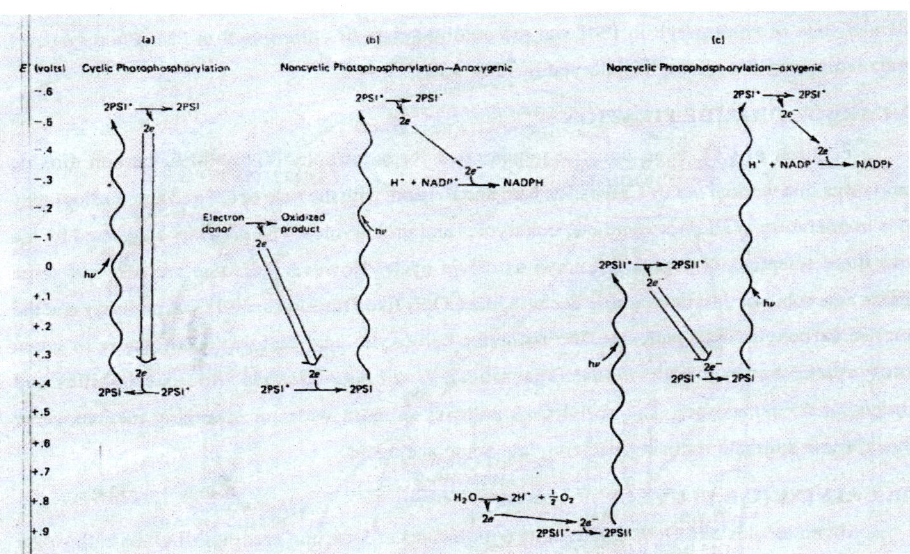

Fig. 4.32 a, b, and c. Patterns of electron flow in different forms of phosphorylation.

The process of non-cyclic photophosphorylation in organisms that carry out oxygenic photosynthesis is more complicated owing to the high potential required to remove electrons from water **(Fig. 4.32c).** Inactivated chlorophyll molecules in the reaction center that we have discussed, collectively termed as photosystem-I (PSI), do not constitute a sufficiently owes' oxidant to accept electrons from water. In oxygenic photosynthesis, photosystem-I (PSI) operates in tandem with another type of reaction center, photosystem-II (PSII) that operates over a much higher range of reduction potential. Photosystem-I and II together can mediate a change in reduction potential above the amount required to oxidize water and reduce NADP. The excess of reducing power is used by a photosynthetic electron transport chain that flows between the activated state of chlorophyll in PSII and the oxidized state of chlorophyll in PSI. Photosystem-I alone mediates cyclic photophosphorylation **(Fig. 4.32a).**

■■■■

Chapter 5

CARBON METABOLISM

According to their carbon source, phototrophs can be classified into photoorganotrophs and photolithotrophs. With a few exceptions, cyanobacteria are photolithotrophs and the green bacteria and purple bacteria can grow photolithotrophically fixing CO_2 as well as photoorganotrophically **(Table 5.1).** Heliobacteria and aerobic anoxygenic photosynthetic bacteria do not fix CO_2 .

Table 5.1. Different Metabolic Pathways in Anoxygenic Phototrophic Bacteria

S.No.	Growth modes	Overall reaction
1.	**Autotrophic**	
a.	Photoautotrophic	$CO_2 + 2H_2 \rightarrow CH_2O + H_2O$
b.	Chemoautotrophic (dark)	
c.	Formate assimilation	$CH_2O + H_2O \rightarrow CO_2 + 2H_2O$
d.	Methanol assimilation	$2CH_3OH + CO_2 \rightarrow 3CH_2O + H_2O$
e.	Carbon monoxide assimilation	$CO + H_2O \rightarrow CO_2 + H_2$
f.	Mixotrophic	
2.	**Heterotrophic (Organic carbon assimilation)**	
a.	**Dark**	
i.	Aerobic	$CH_3COH + O_2 \rightarrow CH_2O + CO_2 + H_2O$
ii.	Anaerobic	$18C_3H_4O_3 + 16H_2O \rightarrow 17CH_3COOH + C_3H_6O_2 + HCOOH + 16CO_2 + 14H_2$
b.	**Light**	
i.	Anaerobic	
a.	Presence of Nitrogen	$C_4H_6O_5 \rightarrow C_3H_6O_3 + CO_2$
b.	Absence of Nitrogen	$CH_3COOH + 2H_2O \rightarrow 2CO_2 + 4H_2$
ii.	Aerobic	$CH_3COOH + O_2 \rightarrow CH_2O + CO_2 + H_2O$

The Calvin cycle is the most common CO_2 fixing mechanism in phototrophs. The key enzymes of the Calvin cycle, phosphoribulokinase and ribulose-1,5-bisphosphate carboxylase are not found in green sulfur bacteria and in one of the filamentous anoxygenic phototrophic bacteria, *Chloroflexus aurantiacus.* Green sulfur bacteria can fix CO_2 into acetyl-CoA through the reductive TCA cycle, and a less common 3-hydroxypropionate pathway in *Chloroflexus aurantiacus.* CO_2 is fixed through the Calvin cycle in purple bacteria and filamentous anoxygenic phototrophic bacteria.

Photosynthetic electron transport pathways are different in anoxygenic and in non-oxygen evolving phototrophs, all contain either a quinone type (type-II) or a Fe-S type (type-I) reaction center. Among six phyla of phototrophic bacteria, anaerobic anoxygenic phototrophic bacteria(ANAPB), green sulfur bacteria (GSB), filamentous anoxygenic phototrophic bacteria (FAPB) or green non-sulfur/gliding bacteria, *Heliobacteria* , *Chloroacidobacteria* and aerobic anoxygenic phototrophic bacteria (AAPB), anaerobic anoxygenic phototrophic bacteria and filamentous anoxygenic phototrophic bacteria contain a type-II reaction center and green sulfur bacteria, *Heliobacteria* and *Chloroacidobacteria* contain type-I reaction center. Phototrophic bacteria with a type-II reaction center operate cyclic electron transport to generate ATP and reverse electron transport to produce NAD(P)H, while, green sulfur bacteria operates both cyclic and non-cyclic electron transport to produce NAD(P)H and ATP, anyhow only cyclic transport is reported in *Heliobacteria* and *Chloroacidobacteria* .

Many phototrophic bacteria play an important role in global carbon, nitrogen and sulfur cycles together are known to assimilate nitrogen or oxidize sulfide in coexistence with their central carbon and energy metabolism. Anaerobic anoxygenic phototrophic bacteria, filamentous anoxygenic phototrophic bacteria and green sulfur bacteria can grow autotrophically through a variety of autotrophic CO_2 assimilation pathways, while aerobic anoxygenic phototrophic bacteria (AAPB) and *Heliobacteria* are obligate photoheterotrophs. AAPB originated from ocean surface waters and certain extreme ecosystems may play an important role in global carbon regulation. *Heliobacteria* have the simplest photosystem, gram (+) ve, form heat resistant endospores, obligate nitrogen fixers, obligate anaerobic photoheterotrophs and hydrogen producers (Madigan, 2006).

APB has a very simple form of photosynthesis and revealed various facts of the process in different organisms. It was generally assumed that photosynthesis in APB is an anaerobic metabolic process. The function of photosynthetic apparatus in these bacteria is the conversion of light energy into an electrochemical gradient of protons across the membrane, which can be used for ATP production, motility, active transport and other energy-consuming processes. However, these bacteria commonly live in approximately deeper layers of water below the planktonic layers **(Fig. 5.1).**

These bacteria carry out anoxygenic photosynthesis with the help of various types of bacteriochlorophylls and carotenoids. In these bacteria, photosynthesis is carried out in anoxic or oxic deficient conditions because the synthesis of photosynthetic pigments is repressed by O_2. These bacteria are incapable of use water as an electron donor, rather use more reduced sulfur compounds such as sulfide, hydrogen and several small organic molecules. They utilize a variety of metabolic pathways for energy generation, carbon assimilation, nitrogen, sulfur and phosphorus metabolism (Madigan and Jung, 2009). Reduced iron as an electron donor has also been reported in some anoxygenic phototrophic purple bacteria (Ehrenreich and Widdel, 1994).

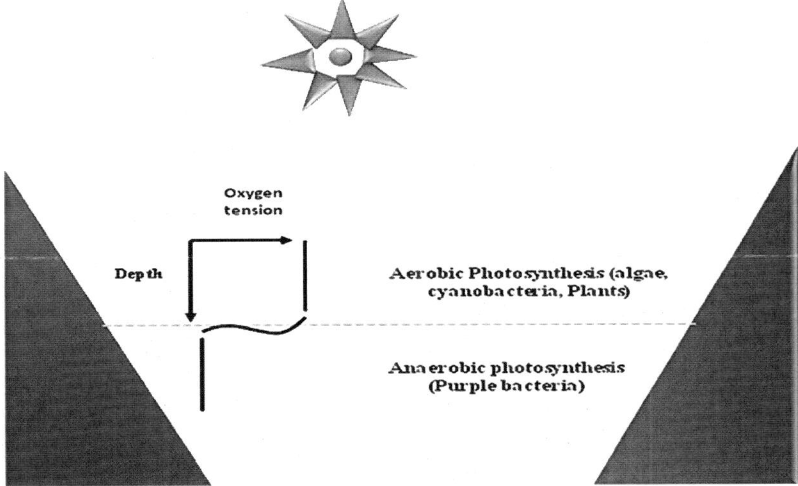

Fig. 5.1. Schematic representation of anoxygenic phototrophic bacteria in their natural habitat.

Among anoxygenic phototrophic bacteria, PSB is a strong photoautotroph that fix CO_2 into cellular biomass and capable of limited photoheterotrophy, which transforms organic matter into cell mass. These are badly equipped for metabolism and growth in the dark. In contrast, PNSB is capable of growing as photoautotrophs, chemoautotrophs, chemoheterotrophs, mixotrophs, carry fermentative and dark metabolism. The various metabolic pathways in anoxygenic phototrophic bacteria are precised in **Table 5.1**. These bacteria have photosystem-Iand don't have photosystem-II. They consume toxic substance H_2S and supply organic matter to anoxic environments by their autotrophic capacities. Primarily they consume organic compounds, which are non-fermentable and offer opportunities to study diverse photosynthetic metabolism.

A number of *Rhodospirillaceae* members are free to grow under aerobic-dark conditions by respiration. Although, under anaerobic light conditions, cells immediately change to phototrophic growth. In anoxic conditions, the cytoplasmic membranes undergo invagination and extend the area of intracytoplasmic membranes (ICM) into the cell which inserts the necessary pigments and proteins to harvest light energy (Law and Cogdell, 2008).

Several metabolites, intermediates, or products produced by these phototrophic bacteria during autotrophic carbon fixation activate cellular material. The Calvin- Benson cycle synthesizes 3-phosphoglycerate (3PG), sugar phosphates such as triose-, tetrose-, pentose-, hexose- and heptose and all other autotrophic carbon fixation pathways produce or restore acetyl- CoA, phosphoenolpyruvate (PEP), pyruvate, oxaloacetate (OAA), glyoxylate, succinyl-CoA and or α-ketoglutarate (**Table 5.2**). All these metabolites are naturally required for the central carbon metabolism of microorganisms.

5.1. ACETATE ASSIMILATION

Many anoxygenic phototrophic bacteria, filamentous anoxygenic phototrophic bacteria, green sulfur bacteria, *Heliobacteria* and aerobic anoxygenic phototrophic bacteria have been reported to grow heterotrophically or mixotrophically (in the presence of CO_2) on acetate with three acetate assimilation mechanisms: 1. The oxidative glyoxylate cycle (**Fig. 5.2**), 2. Pyruvate synthase (**Fig. 5.3**) and 3. The ethylmalonyl-CoA pathway (**Fig. 5.4** .) The glyoxylate cycle and the ethylmalonyl-CoA pathway can restore the metabolites in the TCA cycle. Pyruvate synthase and ethylmalonyl-CoA pathway incorporate both acetate and inorganic carbon (CO_2 /HCO_3^-). All these pathways may directly or indirectly synthesize oxaloacetate. The distribution of the acetate assimilation pathways in anoxygenic phototrophic bacteria is listed in **Table 5.2** .

Table 5.2. Metabolites Produced During CO_2 and Acetate Assimilation by Anoxygenic Phototrophic Bacteria

Pathways	3PG, triose and sugar-phosphate	Acetyl-CoA	Pyru-vate	PEP	OAA	Malate	Glyox ylate	Succinyl-CoA/ Succinate	α-KG
Autotrophic CO_2 fixation pathways	–	–	–	–	–	–	–	–	–
The Calvin-Benson cycle	+	–	–	–	–	–	–	–	–
The reductive TCA cycle	–	+	+	+	+	+	–	+	+
The 3-hydroxy-propionate bi-cycle	–	+	+	–	–	–	+	+	+
Anaplerotic CO_2 assimilation	–	–	+	+	+	+	–	–	–
Acetate Assimilation	+	+	+	+	+	+	+	+	–
The (oxidative) glyoxylate cycle	–	–	–	–	+	+	+	+	–
Pyruvate synthase	+	+	–	–	–	–	–	–	–
The Ethyl-malonyl-CoA pathway	-	+	–	–	+	+	+	+	–

Symbols: +, Positive/Presence of metabolites; –, Negative/absence of metabolite formation during CO_2 and acetate assimilation

5.1.1. *The Oxidative Glyoxylate Cycle*

The oxidative glyoxylate cycle, described by Kornberg and Krebs (1957), initiated four-carbon units (malate, OAA, and succinate) by building two acetyl-CoA molecules (**Fig. 5.2**) (Tang *et al.,* 2011). It is not only important to convert acetate, lipids and some amino acids to carbohydrates, but also other central biosynthetic precursors in the TCA cycle (anaplerotic function). During mixotrophic and heterotrophic metabolism the active glyoxylate cycle has been recommended to channel excess carbon flow away from the branched TCA cycle.

Some APB such as *Rhodopseudomonas palustris* either use an active glyoxylate cycle (McKinlay and Harwood, 2011) or possess genes in the pathway identified in *Rhodospirillum centenum* , *Rhodopseudomonas palustris* and *Rhodobacter capsulatus* (Oda *et al.,* 2008; Strnad *et al.,* 2010; Lu *et al.,* 2010). However, other APB lacks essential genes/enzymes in the pathway identified as in *Rubrivivax gelatinosus* ,*Rhodobacter sphaeroides* and *Rsp. rubrum* (Lim *et al.,* 2009). Genes present in the glyoxylate cycle are also identified in the genome of several filamentous anoxygenic phototrophic bacteria (*Cfl. aurantiacus* , *Cfl. aggregans* , *Chloroflexus* sp. Y-400-fl, *Roseiflexus castenholzii* and *Roseiflexus* sp. RS-1) (Tang *et al.,* 2011). Green sulfur bacteria, aerobic anoxygenic phototrophic bacteria and *Heliobacteria* do not possess an active glyoxylate cycle and employ an alternative acetate assimilation pathway.

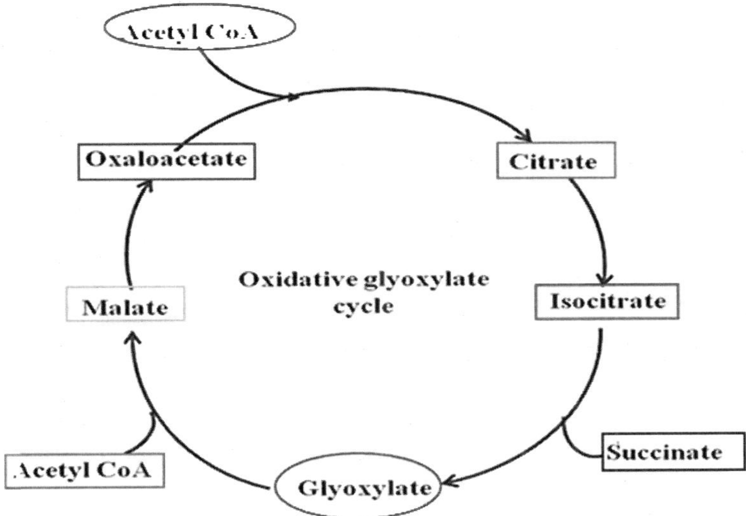

Fig. 5.2. Acetate assimilation by oxidative glyoxylate cycle (modified from Tang *et al.,* 2011).

5.1.2. *Pyruvate Synthase*

In the absence of active glyoxylate cycle reported, green sulfur bacteria and *Heliobacteria* employ pyruvate synthase for acetate assimilation. Pyruvate synthase converts acetyl-CoA to pyruvate and can assimilate both acetate (acetyl-

CoA) and CO_2(**Fig. 5.3**) (Tang *et al.,* 2011). The catalysis requires reduced ferredoxin and thus only anaerobic anoxygenic bacteria can employ the pyruvate synthase pathway. During the mixotrophic growth of green sulfur bacteria, pyruvate synthase regenerates pyruvate from acetyl-CoA which is produced via acetate uptake and the reverse tricarboxylic acid cycle (RTCA) (Tang and Blankenship, 2010). *Heliobacteria* requires acetate as an organic carbon source and CO_2in the medium for its growth (Tang *et al.,* 2010b). The supplementation of CO_2in the medium is required not only for anaplerotic CO_2 assimilation but also produce pyruvate catalyzed by pyruvate synthase (Tang *et al.,* 2010a,b).

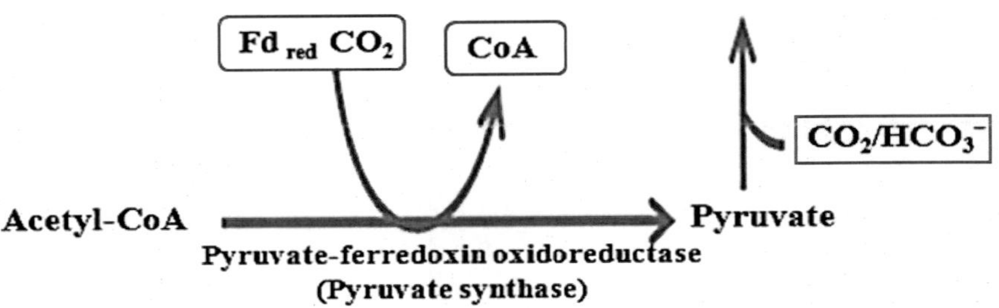

Fig. 5.3. Acetate assimilation by pyruvate synthase pathway (modified from Tang *et al.,* 2011).

5.1.3. *The Ethyl malonyl-CoA Pathway*

The ethyl malonyl-CoA pathway has been identified in the anoxygenic phototrophic bacterium, *Rba. sphaeroides* (Erb *et al.,* 2007) and is also proposed in several APB (Erb *et al.,* 2009). In *Rba. sphaeroides,* glyoxylate gets condensed with acetyl-CoA to generate malonyl CoA and is further hydrolyzed to malate and CoA as this bacterium initiates a different type of ethylmalonyl-CoA pathway (**Fig. 5.4**) (Tang *et al.,* 2011). Similarly, propionyl CoA generated gets condensed with HCO_3 to produce succinate.

Acetate can support the phototrophic and chemotrophic growth of several aerobic anoxygenic phototrophic bacteria (Biebl *et al.,* 2006). The mechanism of acetate assimilation remains to be assumed in these organisms, while some AAPB don't have an active glyoxylate cycle and pyruvate synthase is not identified in all AAPB. Hence, AAPB growing on acetate requires alternative pathways to produce OAA, which is required for producing biomass and building blocks of cells. The ethyl malonyl-CoA pathway employing AAPB is reported to play an important role in CO_2 assimilation (Alber *et al.,* 2006).

5.2. FLOW OF REDUCING EQUIVALENT IN PHOTOAUTOTROPHS AND PHOTOHETEROTROPHS

Green sulfur bacteria cannot grows heterotrophically in dark conditions as these bacteria are obligate photoautotrophs. The reducing equivalents generated from photosynthetic electron transport in the Fe-S type (type-I) reaction center of GSB can be utilized for CO_2 assimilation (through RTCA cycle) and other types of cellular metabolism during phototrophic growth. *Heliobacteria* and other anaerobic anoxygenic phototrophic bacteria with type-I reaction centers do not have an autotrophic CO_2 fixation. The reducing equivalents generated in photosynthetic electron transport are utilized to reduce organic and inorganic molecules in cells. All the heliobacterial species are well-known active nitrogen fixers and hydrogen producers (Madigan, 2006).

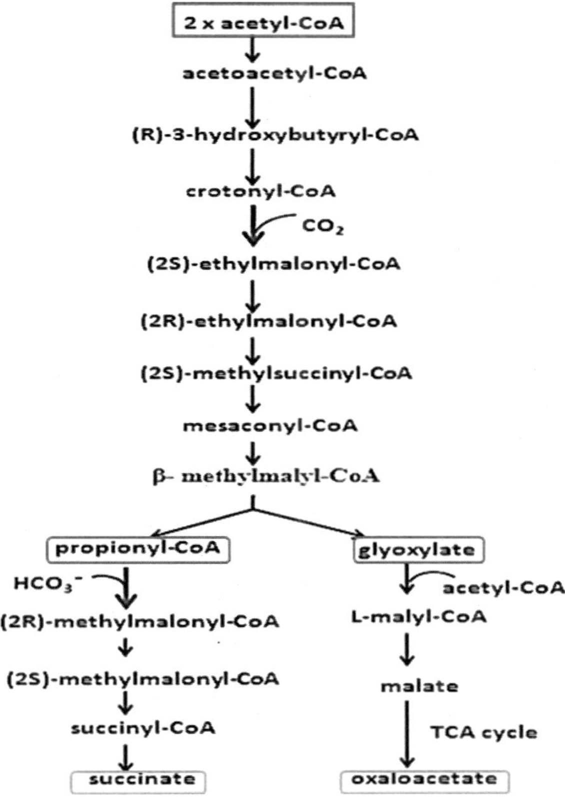

Fig. 5.4. Acetate assimilation by the ethyl malonyl-CoA pathway in *Rba. sphaeroides* (modified from Tang *et al.*, 2011).

In comparison to green sulfur bacteria and *Heliobacteria*, ANAPB can grow both photoautotrophically and photoheterotrophically. Dimethyl sulfoxide is the external electron recipient required by the wild-type, RuBisCO-, and phosphoribulokinase (PRK)-knockout mutant strains of *Rba. sphaeroides* to accept reducing equivalents which are produced during photoheterotrophic growth (Falcone and Tabita, 1991). Similarly, RuBisCO-knockout *Rsp. Rubrum* and *Rba. sphaeroides* produce H_2 by utilizing excess reducing equivalents while growing photoheterotrophically under nitrogen fixation conditions(Joshi and Tabita, 1996).

Metabolic flux analyses on the *Rps. palustris* also suggests that the Cavin-Benson cycle is not only assimilated CO_2 and promotes biomass during photoautotrophic growth but also accepts reducing equivalents during potoheterotrophic growth (McKinlay and Harwood, 2010). *Rps. palustris* uses reducing equivalents to produce H_2 instead of assimilating carbon during photoheterotrophic growth (McKinlay and Harwood, 2011). In *Rba. sphaeroides* during assimilation of acetate using ethylmalonyl-CoA pathway by utilizing reducing equivalents, hydrogen production decreases (Laguna *et al.,* 2011).

These observations are incompatible with findings that RuBisCO-knock out *Rba. sphaeroides* can grow photolithoautotrophically using thiosulfate or sulfide as an electron source and consuming the excess reducing equivalents produced by photosynthesis is less essential(Wang *et al.,*1993). Even though the pathway for assimilating CO_2 in these phototrophic bacteria is not clear, alternative pathways for removing excess reducing equivalents are utilized by certain aerobic anoxygenic phototrophic bacteria during photoheterotrophic growth such as nitrate reduction to ammonium by *Roseobacter denitrificans* (Tang *et al.,* 2009a).

5.3. ANAPLEROTIC CO_2 ASSIMILATION

Anaplerotic reactions restore the intermediates in a metabolic pathway. Anaplerotic CO_2 assimilation is the process that suppliesoxalo acetic acid (OAA) or malate for replenishing the TCA cycle, which is also important to synthesize cellular building blocks. Anaplerotic CO_2 assimilation is unable to serve as an autotrophic CO_2 fixation pathway because they require organic substrate molecules that are not subsequently produced from the incorporation of CO_2 (Tang *et al.,* 2011). Four enzymes catalyzing anaplerotic CO_2 assimilation have been identified, such as 1. Pyruvate carboxylase, 2. Phosphoenolpyruvate (PEP) carboxylase, 3. PEP carboxykinase and 4. The malic enzyme (**Fig. 5.5**) (Tang et al., 2011). Anaplerotic CO_2 assimilations are reported in all types of phototrophic bacteria. Among these, ANAPB and AAPB have an active complete oxidative tricarboxylic acid cycle that can regenerate oxalo acetic acid and can also supplyoxalo acetic acid through anaplerotic CO_2 assimilation (Tang *et al.,* 2009a) and the glyoxylate cycle and/or the ethyl malonyl-CoA pathway to produce oxalo acetic acid (Erb *et al.,* 2007). Anaplerotic CO_2 assimilation is another option, which is essential for the growth of *Heliobacteria* and GSB, as it can only use anaplerotic CO_2 assimilation to produce OAA.

Heliobacteria in contrast to ANAPB, AAPB and FAPB, possess an incomplete OTCA cycle, unable to regenerate oxalo acetic acid and do not possess the glyoxylate cycle and ethyl malonyl-CoA pathway (**Table 5.2** .) Subsequently, the anaplerotic CO_2 assimilation is the only approach to synthesize oxaloacetic acid and is essential for the growth of *Heliobacteria*, compatible with high flux in the anaplerotic CO_2 assimilation (Tang *et al.*, 2010a).

Green sulfur bacteria have active anaplerotic CO_2 assimilation (Tang and Blankenship, 2010), which requires oxalo acetic acid (OAA) to initiate reverse TCA cycle and do not possess the glyoxylate cycle and ethyl malonyl-CoA pathway. However, green sulfur bacteria can produce OAA via ATP citrate lyase (ACL) in the RTC A cycle. Oxalo acetic acid needs to be produced by anaplerotic CO_2 assimilations if large amounts of α-ketoglutarate and other metabolites are removed out of the cycle for nitrogen assimilation and another cellular metabolism. Subsequently, green sulfur bacteria can only produce OAA via anaplerotic CO_2 assimilation.

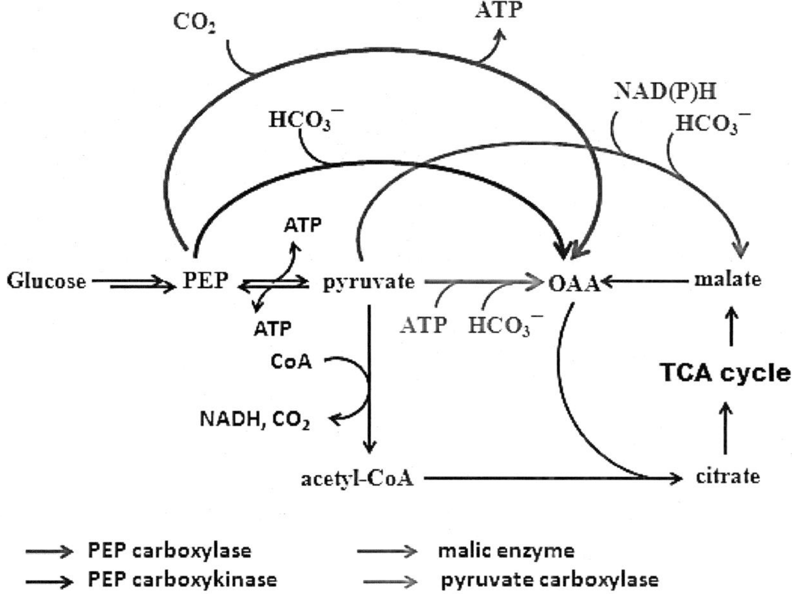

Fig. 5.5. Anaplerotic CO_2 assimilation reactions (the CO_2-anaplerotic reactions catalyzed by pyruvate carboxylase, PEP carboxykinase, PEP carboxylase and malic enzyme are predicted. Participations or consumption of ATP and reducing equivalents are shown) (modified from Tang *et al.*, 2011).

Generally, active anaplerotic reactions are essential not only for photoheterotrophs (*Heliobacteria,*) but also for photoautotrophs (green sulfur bacteria). However, the anaplerotic CO_2 assimilation cannot support in the manner that the autotrophic carbon fixation pathways support photoautotrophic growth as they require pre-existing or endogenous organic compounds that are not produced by the pathway. Anaplerotic CO_2 assimilations are any way important for the cell growth of phototrophic organisms.

5.4. CARBOHYDRATE METABOLISM

A large number of chemoheterotrophic organisms use carbohydrates as a carbon source to construct cellular material and to provide reductants. GSB is the only group of anoxygenic phototrophic bacteria that do not grow on carbohydrates as the sole carbon source (**Table 5.3**). Three carbohydrate catabolic pathways 1. Emden-Meyerhof-Parnas (EMP) pathway (glycolysis),2. Entner-Doudoroff (ED) pathway and3. Phosphogluconate (PP) pathway, which includes the oxidative phosphogluconate (OPP) and non-OPP pathways are in operation in anoxygenic phototrophic organisms. Regulation of EMP and ED pathways provides flexibility for energy generation and bypasses bottlenecks for glucose oxidation. OPP pathway and non-OPP can produce reducing equivalents.

Table. 5.3. Carbohydrate Catabolic Pathways in Anoxygenic Phototrophic Bacteria

	EMP pathway	ED pathway	Oxidative pentose phosphate pathway	Non-oxidative pentose phosphate pathway*
ANAPB	+	+	-	+
AAPB	-	+	-	+
FAPB	+	-	+	+
GSB	-	-	-	+
*Heliobacter*ia	+	-	-	+

*The non-oxidative pentose phosphate pathway is not equal to the reductive pentose phosphate pathway.

5.4.1. *Anaerobic Anoxygenic Phototrophic Bacteria (ANAPB) and Aerobic Anoxygenic Phototrophic Bacteria (AAPB)*

ANAPB is the most metabolically versatile among phototrophic bacteria. However, ANAPB has use either ED or EMP pathway, which depends upon the growth conditions and nutrients available for carbohydrate catabolism, while the OPP pathway is never reported to be active in ANAPB. Applying activity assays, sugar consumption, analysis of labeled amino acids using ^{14}C-labelled sugar in wild-type and Glucose-6-phosphate dehydrogenase deficient mutant, *Rba. capsulatus* and *Rba. sphaeroides* utilize EMP or ED pathway to oxidize fructose during phototrophic or chemotrophic growth and it operates the ED pathway to differentiate glucose during growth in dark conditions (Fuhrer *et al.,* 2005). The genes in the genome of *Rba. capsulatus* and *Rba. sphaeroides* of the ED pathway have been identified, whereas in other ANAPB genes encoding 6-phosphogluconate dehydratase and other genes are missing in the ED pathway.

Genes in the ED pathway are present in the genomes of aerobic anoxygenic phototrophic bacteria (Tang *et al.,* 2009a) and an active ED pathway has been suggested in several aerobic anoxygenic phototrophic bacteria (Yurkov and Beatty, 1998). Recently, ^{13}C-based metabolomic /fluxomic studies, transcriptomics and activity assays indicate that *Roseobacter denitrificans* OCh114 and *Dinoroseobacter shibae* DFL12 mainly use the ED pathway to oxidize glucose. Both aerobic anoxygenic phototrophic bacteria have an inactive EMP pathway (Tang *et al.,* 2009a). However, all of the genes in the EMP pathway have been identified in the *Dinoroseobacter shibae* genome (Furch *et al.,* 2009). Like anoxygenic phototrophic bacteria, the active OPP pathway has not been reported in aerobic anoxygenic phototrophic bacteria (Yurkov and Beatty, 1998).

The marine species of *Roseobacter* which have been sequenced contain genes in the ED pathway in the genome (Brinkhoff *et al.,* 2008). Most of the marine *Roseobacter* species are non-phototrophs, in which some are shown to operate an active ED pathway (Furch *et al.,* 2009). Hence, both phototrophic and non-phototrophic *Roseobacteria* utilize the ED pathway. Many strains of aerobic anoxygenic phototrophic bacteria and anaerobic anoxygenic phototrophic bacteria employ the ED pathway that does not have the *pgd* gene (encoding 6-phosphogluconate dehydrogenase) in the OPP pathway (Strnad *et al.,* 2010) and or the *pfk* gene (encoding6- phosphofructokinase) in the EMP pathway (Tang *et al.,* 2009a .) The particular evolutionary advantages of losing the *pfk* gene and or *pgd* gene and containing an inactive EMP pathway and or the OPP pathway (Tang *et al.,* 2009a)remain unclear.

5.4.2. *Heliobacteria*

D-glucose and D-fructose can support the growth of *Heliobacterium gestii.* Some of the *Heliobacteria* were isolated from tropical paddy soil (Madigan, 2006). *Hbt. modesticaldum* is closely related to *Hbt. gestii* in the area evidence that phylogeny (Bryantseva *et al.,* 1999). All the genes employed in the EMP pathway are present in the genome of *Hbt. modesticaldum* in agreement with genomic annotation and phylogenetic analyses. Recent studies D-fructose, D-glucose, D-ribose and a little amount of yeast extract can support the phototrophic growth of *Hbt. modesticaldum* (Tang *et al.,* 2010b). *Hbt. modesticaldum* use only the EMP pathway for carbohydrate catabolism and H_2 production as it does not possess key genes in the OPP and ED pathways (Sattley *et al.,* 2008) as suggested in certain thermophilic microorganisms (de Vrije *et al.,* 2007). An active ED pathway has not been reported in any *Heliobacteri* in accordant with the fact that very few Gram (+ve)bacteria use the ED pathway.

5.4.3. *Filamentous Anoxygenic Phototrophic Bacteria (FAPB)*

Numerous reports on carbohydrate metabolism of FAPB have made attention to *Cfl. aurantiacus.* It uses the EMP pathway for carbohydrate catabolism during aerobic respiration by utilizing glucose and a number of other sugars. ED pathway is not reported in these bacteria. Gene annotation in *Chloroflexus aurantiacus* specifies that the PP pathway is in complete in accordance with the activities reported for Glucose-6-phosphate dehydrogenase (G6PDH) and 6-phosphogluconate dehydrogenase (6PGDH), which are the two important enzymes in the OPP pathway (Tang *et al.,* 2011). The *fba A* gene encoding FBP aldolase in the EMP/gluconeogenic pathway is missing in the genome of *Chloroflexus* sp. Y-400-fl, *Cfl. aurantiacus* and *Cfl. aggregans* (Tang *et al.,* 2011). If *Cfl. aurantiacus* was unable to synthesize FBP aldolase, an active OPP pathway is used for the interconversion of D-Glucose-6-phosphate and glyceraldehyde-3-phosphate (GAP), so that the sugars could be transformed to pyruvate including other energy-rich substances and vice versa. In *Cfl. aurantiacus* while growth on glucose rather than acetate, showed higher activities of phosphofructokinase and fructose-1,6-bisphosphate (FBP) aldolase have been reported (Hanada and Pierson, 2006). Hence, *Cfl. aurantiacus* and other *Chloroflexi* species may use a novel FBPaldolase. Species of *Roseiflexi* such as *Roseiflexus castenholzii* and *Roseiflexus* sp. RS-1 is closely related to *Chloroflexi* species, has a putative bifunctional FBP aldolase/phosphatase gene identified (Say and Fuchs, 2010). Genes encoding different types of aldolases represent in the genome of *Cfl. aurantiacus* (Tang *et al.,* 2011). However, more detailed investigations are needed to clarify this .

5.5. ALDOLASES EMPLOYED IN THE EMP AND ED PATHWAY

The reactions catalyzed by FBP (Fructose 1, 6- bisphosphate) aldolase and KDPG (2-keto-3-deoxy-6- phosphogluconate) aldolase are reversible and are depicted in **Fig. 5.6** (Tang *et al.,* 2011) . FBP aldolase, catalyzing the interconversion of fructose 1, 6- bisphosphate(FBP) and D-glyceraldehyde-3-phosphate (GAP) or dihydroxy acetonephosphate (DHAP) is widely spread in phototrophs as they employ both EMP pathway and gluconeogenesis .

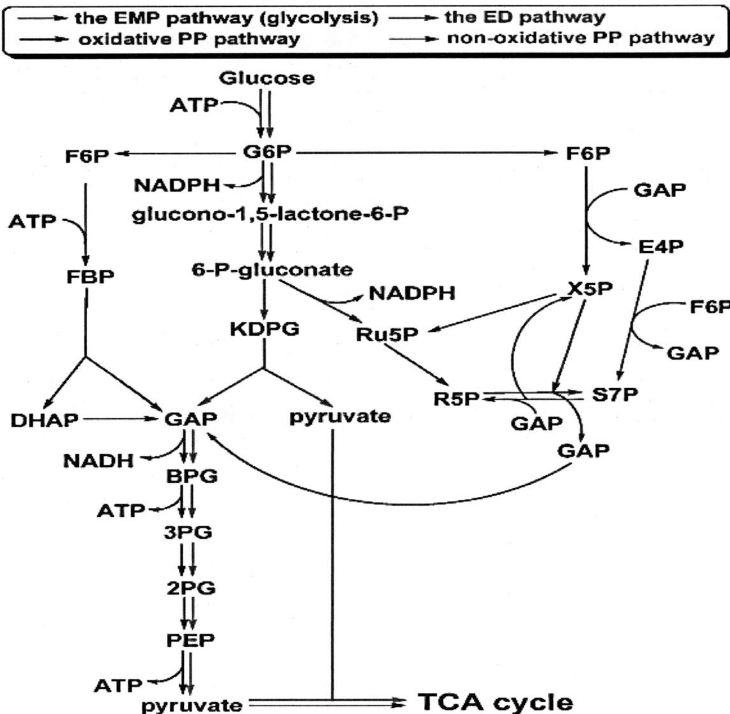

Fig. 5.6. Carbohydrate catabolism, EMP, ED and PP pathways in phototrophic bacteria (Tang *et al.,* 2011).

Fructose 1,6-bisphosphate (FBP) aldolases are identified in archea, bacteria and eukaryotes. Some fructose 1,6-bisphosphate aldolase are also responsible for biological functions. Except for FAPB, all other phototrophic bacteria have *FbaA* gene-encoded Class II FBP aldolase. In the genome of ANAPB, AAPB and *Heliobacteria*, Class I FBP (encoded by the *FbaB* gene) are identified. Further, a bifunctional FBP aldolase or phosphate enzyme is identified in some archea which does not show sequence homology along with the classical fructose 1,6-bisphosphate (FBP) aldolases. The bifunctional enzyme is shown to be an ancestral enzyme in gluconeogenesis. Genes encoding the bifunctional fructose 1,6- bisphosphate aldolase or phosphatase have also been found in the *Roseiflexi* species genome (Say and Fuchs, 2010).

KDPG aldolase, which catalyzes the interconversion of KDPG and GAP or pyruvate (**Fig. 5.7**) (Tang *et al.,* 2011), is one of the two enzymes specialized for ED pathway and the other specialized enzyme is 6-phosphogluconate dehydratase. KDPG aldolase is required in the ED pathway and it is not unique to the ED pathway. Moreover, it is also responsible for arginine, glyoxylate, proline and dicarboxylase metabolism (Conway, 1992). KDPG aldolase has been recognized in ANAPB (Conway, 1992) and AAPB (Tang *et al.,* 2009 a) that employ the ED pathway for carbohydrate oxidation. There are no detailed studies on KDPG aldolases from phototrophic bacteria. The sequence alignments propose that all phototrophic bacterial KDPG aldolases are the Class I aldolases, specifically forms an internal active site lysine-substrate aldimine for catalysis (Fullerton *et al.,* 2006).

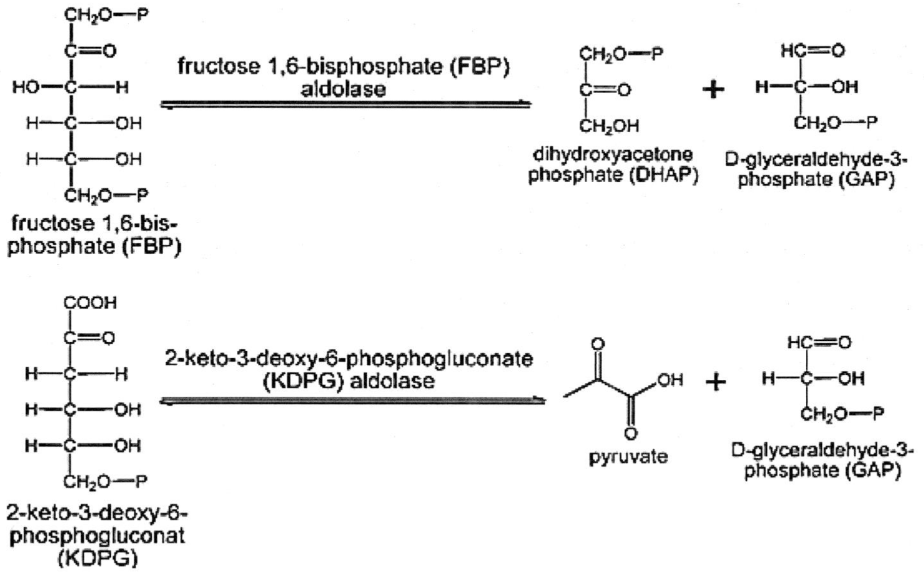

Fig. 5.7. Aldol condensation reactions catalyzed by FBP aldolase in the EMP/ gluconeogenic pathway and KDPG aldolase in the ED pathway (Tang *et al.,* 2011).

5.6. THE TRICARBOXYLIC ACID CYCLES (FORWARD, REVERSE AND BRANCHED)

A large number of metabolites that occur in the TCA cycle are the precursors of amino acids and biomass. At the minimum, three forms of the TCA cycle are identified in phototrophic bacteria: 1. The forward TCA cycle (or OTCA cycle) 2. The reverse TCA cycle (or RTCA cycle) and 3. The branched TCA cycle (or incomplete TCA cycle).

The OTCA cycle or Krebs cycle produces reducing equivalents and ATP and is considered to be the most important central carbon and energy metabolic pathway both in phototrophs and non-phototrophs. ANAPB (Madigan and Jung, 2009) and AAPB (Tang *et al.,* 2009a) use an active OTCA cycle. *Cfl. aurantiacus* operates the OTCA cycle under both aerobic dark or anaerobic light growth conditions, whereas members of filamentous anoxygenic phototrophic bacteria such as *Oscillochloridaceae* have a partial RTCA cycle (Berg *et al.,* 2005).

Green sulfur bacteria transfer the reducing equivalents generated in photosynthetic electron transport to the RTCA cycle, where it functions as an autotrophic CO_2 assimilation pathway for the production of cellular material (Sirevag, 1995). GSB also operates the incomplete OTCA cycle (from citrate to α-ketoglutarate) during mixotrophic growth conditions with acetate (Tang and Blankenship, 2010).

In contrast, to the energy-producing OTCA cycle, metabolic flux in the reverse TCA cycle is energetically unfavorable (Tang *et al.,* 2011). Many reactions in the TCA cycle are reversible. The reductive carboxylation of succinyl-CoA to α-ketoglutarate (catalyzed by α-ketoglutarate synthase) and acetyl-CoA to pyruvate (catalyzed by pyruvate synthase) in the reverse TCA cycle are not thermodynamically successful. The standard Gibbs free energy change of the decarboxylation of pyruvate-by-pyruvate synthase is determined to be –4.6 kcal/mol (Thauer *et al.,* 1977). Hence, GSB and other organisms using the RTCA cycle engage strong reductants such as reduced ferredoxin to initiate the carboxylation reactions catalyzed by pyruvate synthase and α-ketoglutarate synthase to drive the RTCA cycle efficiently.

5.7. ENZYMES INVOLVED IN CITRATE METABOLISMS

Aerobic anoxygenic phototrophic bacteria (AAPB), anaerobic anoxygenic phototrophic bacteria (ANAPB) and filamentous anoxygenic phototrophic bacteria (FAPB) use (*Si*)-citrate synthase to synthesize citrate from acetyl-CoA and oxalo acetic acid for initiating the OTCA cycle. The (*Si*)-citrate synthase (encoded by *gltA* gene) could be identified in the genome of GSB (Davenport *et al.,* 2010). Alternatively, the activity of (*Si*)-citrate synthase is not detected in *Heliobacteria* (Tang *et al.,* 2010b), whereas *gltA* gene is absent in the genome of *Hbt. modesticaldum* (Sattley *et al.,* 2008). Although conducting the enzyme assay under anaerobic conditions, product turnover could be detected in cell-free extracts of *Hbt. modesticaldum* with the addition of acetyl-CoA, oxalo acetic acid and Mn^{2+} metal ions (Tang *et al.,* 2010a). Moreover, the 5[th]carbon position of glucose is found to be ^{13}C-labeled using (1-^{13}C) pyruvate, recommending *Hbt. modesticaldum* synthesizes citrate via (Re)-citrate synthase (Tang *et al.,* 2010a). If citrate is synthesized via (*Si*) - citrate synthase, the primary carbon position of glucose is expected to be labeled using (1-^{13}C) pyruvate. Variations in the labeling patterns of glucose develop from either (*Si*)-citrate synthase or (*Re*-) citrate synthase, attaches the acetyl group at the *pro-S* or *pro-R*arm of OAA respectively. The final product of (*Re*)-citrate synthase and (*Si*)-citrate synthase is the same, hence the citrate is an asymmetric molecule, while the intermediate product of (3*R*)-citryl-CoA and (3*S*)-citryl-CoA is variable (**Fig. 5.8**) (Tang *et al.,* 2011).

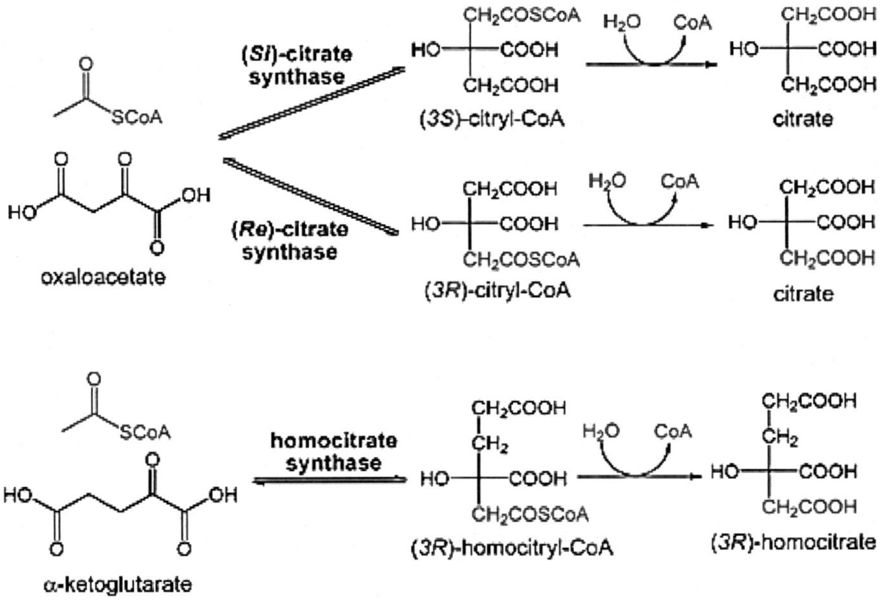

Fig. 5.8. Enzymes are involved in citrate metabolism and central carbon metabolism (Tang *et al.,* 2011).

Compared to (*Si*)- citrate synthase, (*Re*)- citrate synthase is an unconventional enzyme for biosynthesis of citrate and is phylogenetically related to homocitrate synthase (HCS) and 2-isopropyl malate synthase (Li *et al.,* 2007). HCS (EC 2.3.3.14) catalyzes the production of (3*R*)-homocitrate through the Claisen condensation of acetyl-CoA and α-ketoglutarate (Wulandari *et al.,* 2002). Identical sequences and active site structures between 2-isopropyl malate synthase (IPS) and homocitrate synthase (HCS) have also been identified (Qian *et al.,* 2008). Bacteria with the activity of (*Re*)-citrate synthase have also been reported to employ citramalate synthase along with (*Re*)- citrate synthase, homocitrate synthase for isoleucine synthesis (Tang *et al.,* 2010a), 2- isopropyl malate synthase, and citramalate synthase (CMS) (**Fig. 5.9**) (Tang *et al.,* 2011). catalyze Claisen condensation reactions with same α-ketoacid substrates and may have evolved from the same ancestor (**Fig. 5.12**) (Tang *et al.,* 2011). Comparison of sequences between (*Re*)-citrate synthase and (*Si*)-citrate synthase also suggest that two types of citrate synthase are phylogenetically different (Li *et al.,* 2007). As preferable by other enzyme pairs that catalyze the production of stereoisomers (Lamzin *et al.,* 1995). Unidentical active site structures between (*Re*)-citrate synthase and (*Si*)-citrate synthase is supposed to generate unidentical citrate isomers. Furthermore, unlike the potent inhibition of aconitase by(–)-*erythro*-2-fluorocitrate, previous studies have suggested that-)+(*erythro*-2-fluorocitrate synthesized from 2- fluoroacetate and oxalo acetic acid by (*Re*)-citrate synthase is not an inhibitor of aconitase (Tang *et al.,* 2010a). *Hbt. modesticaldum* lack the gene encoding (*Re*)-citrate synthase. Two genes encoding putative homocitrate synthases have been identified in the genome of *Hbt. modesticaldum* (Sattley *et al.,* 2008). Homocitrate serves as the precursor for lysine biosynthesis via the α-aminoadipate pathway.

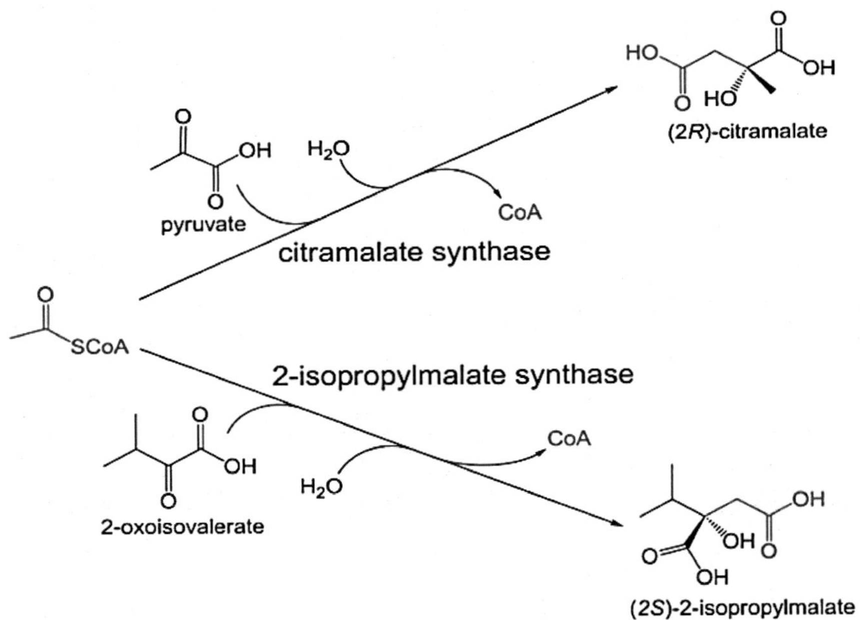

Fig. 5.9. Reactions catalyzed by citramalatesynthase and 2-isopropyl malate synthase (Tang *et al.,* 2011).

Anyhow, except for genes encoding putative homocitrate synthase, other genes in the α-aminoadipate pathway are not identified in the genome of *Hbt. modesticaldum* (**Fig. 5.10**) (Tang *et al.,* 2011), However, the function of the putative homocitrate synthases remains to be verified. Either one of the putative homocitrate synthases or other gene product functions as (*Re*)-citrate synthase in *Heliobacteria* need to be investigated.

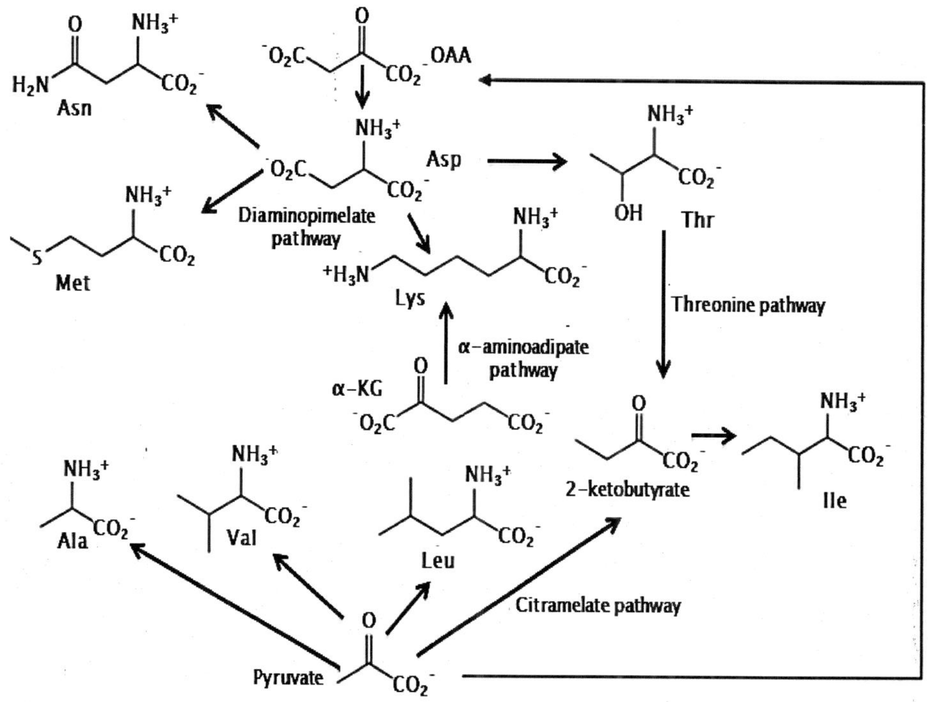

Fig. 5.10. Amino acids biosynthesis in phototrophic and non-phototrophic bacteria (Tang *et al.,* 2011).

Enzymes that alter citrate through the RTCA cycle produce acetyl-CoA and OAA for building cellular material. Citrate cleavage in living organisms is accomplished by three enzymes, 1. Citrate lyase, 2. ATP citrate lyase (ACL), or citryl-CoA synthetase (CCS) and 3. Citryl-CoA lyase (CCL) (Hugler *et al.,* 2007). Citrate lyase cleaves citrate into acetate and oxaloacetate (OAA) and different from ATP citrate lyase and citryl-CoA synthetase/citryl-CoA lyase, which catalyze ATP-dependent cleavage of citrate to oxaloacetate and acetyl-CoA (**Fig. 5.11**) (Tang *et al.,* 2011). The phototrophic green sulfur bacteria and some anaerobic bacteria use ATP citrate lyase to modify citrate produced through the RTCA cycle into acetyl-CoA and oxaloacetate (Voordeckers *et al.,* 2008). The reaction catalyzed by ATP citrate lyase (*Chlorobaculum tepidum*) in *vivo* conditions has been reported to be reversible (Tang and Blankenship, 2010). In some anaerobic chemotrophic bacteria, ATP citrate lyase is responsible for the citrate synthesis and also acetate oxidation in the OTCA cycle (Thauer *et al.,* 1989). In eukaryotic cells, ACL is also responsible for producing cytoplasmic acetyl-CoA through many biological processes, such as lipid production, histone acetylation, a signal for activating glucose catabolism and apoptosis (Zhao *et al.,* 2010).

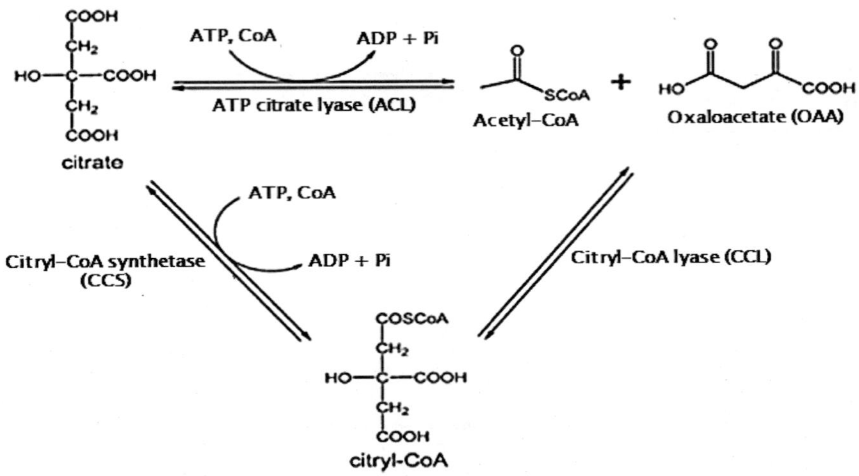

citrate
ATP, CoA
ADP + Pi
ATP citrate lyase (ACL)
Acetyl-CoA
Oxaloacetate (OAA)
ATP, CoA
Citryl-CoA synthetase (CCS)
ADP + Pi
Citryl-CoA lyase (CCL)
citryl-CoA

Fig. 5.11. Retroaldol reactions catalyzed by ATP citratelyase (ACL) and citryl-CoA synthetase (CCS)/citryl- CoAlyase (CCL) (Tang *et al.,* 2011).

5.8. CITRATE METABOLISM AND THE TCA CYCLE

Citrate metabolism occurs in every biological cell and citrate biosynthesis is required for initiating the OTCA cycle and oxidation of citrate via the OTCA cycle produces reducing equivalents and energy. ATP-dependent citrate cleavage generates acetyl-CoA and OAA to build cellular material. (Re)-citrate synthase and (Si)-citrate synthase catalyze citrate biosynthesis and initiate the OTCA cycle and ATP citrate lyase and citryl-CoA synthetase /citryl-CoA lyase operate ATP-dependent citrate cleavage through the RTCA cycle. Sequence alignments suggest that (Si)-citrate synthase and citrate-CoA lyase are phylogenetically related and that ATP citrate lyase is likely a gene fusion product of (Si)-citrate synthase or citryl-CoA lyase and succinyl-CoA synthetase. Sequences of citryl-CoA synthetase and succinyl-CoA synthetase are identical (Aoshima *et al* ,. 2004a). Moreover, slight citryl-CoA lyase activity has been reported in (Si) -citrate synthase (Srere, 1963). Low (*Si*)-citrate synthase activity has been observed in *Hydrogenobacter thermophilus* CCL (Aoshima *et al.,* 2004b). Based on these sequence homology of enzymes involved in citrate metabolism, (*Si* -) citrate synthase may have evolved from CCL and ACL, in turn, may have evolved from CCL or (Si)-citrate synthase ,CCS and succinyl- Co A synthetase (Aoshima *et al.,* 2004b). Some species in *Aquificae* phylum use CCL or CCS and others operate ACL for citrate cleavage (Hugler *et al.,* 2007), recommend that ACL and CCL or CCS may share an identical evolutionary history. ACL, (Si)-citrate synthase and CCL may have evolved from a common ancestor **(Fig. 5.12)** (Tang *et al.,* 2011).2-isopropyl malate synthase, homocitrate synthase, citramalate synthase and (Re)-citrate synthase may have originated from a common ancestor and (Si)-citrate synthase, ATP citrate lyase and citryl-CoA lyase may have originated from another common ancestor. The oxidative TCA cycle understood to

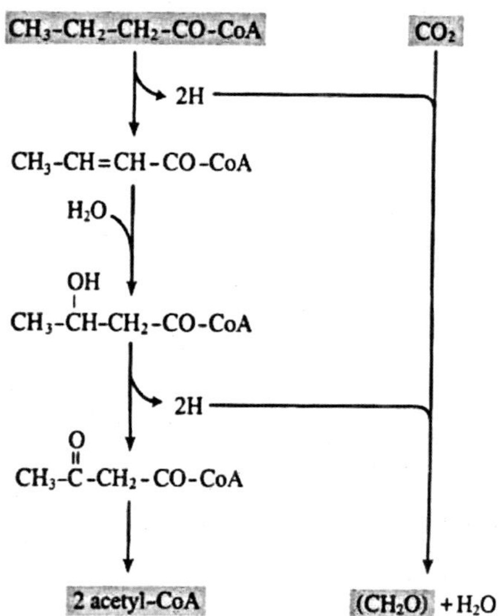

Fig. 5.14. Butyrate metabolism in purple bacteria

5.11. AUTOTROPHIC CO_2 FIXATION (DARK PHASE OF PHOTOSYNTHESIS)

Carbon dioxide is an abundant and available source of carbon, however, it must be reduced to cellular organic carbon at around the level of CH_2O, a cycle that requires reducing power and typically requires energy. Organisms that utilize CO_2 as their essential carbon source are autotrophs and serve as primary producers in ecosystems. Autotrophic carbon dioxide fixation in eukaryotes is represented especially by the Calvin cycle, however, among the prokaryotes, there are several particular pathways of autotrophic CO_2 fixation: (1) The Calvin cycle; (2) The reverse TCA cycle; (3) Acetogenesis and (4) The hydroxyl propionate pathway.

1. The Calvin cycle, which was discovered initially in the green algae, is also found among a large number of the photolithotrophic and chemolithotrophic microbes, where it serves as the essential mechanism for carbon assimilation (**Fig. 5.15**). The primary step of CO_2 fixation is the ribulose 1,5-bisphosphate carboxylase/oxygenase (Rubisco) reaction, resulting in the conversion of one molecule of ribulose-1,5-bisphosphate and CO_2 to two molecules of 3-phosphoglyceric acid. These are then reduced to glyceraldehyde 3-phosphate, followed by a few rearrangements to regenerate ribulose-1,5-bisphosphate. Rubisco from the bacteria has been grouped into two types. Type-I Rubisco is found in plants, cyanobacteria and few different prokaryotes and has a subunit structure comprising of eight large and eight small subunits. Type-II Rubisco is found in some

purple non-sulfur bacteria and a few other organisms and is a dimer of large subunits. The phylogenetic tree for Rubisco shows many branching orders significantly different from that of the 16S rRNA tree and this has been considered as solid proof for uncontrolled genetic transfer of this gene (Delwiche and Palmer, 1996). In addition, genes resembling those encoding Rubisco have been found in archaeal genomes and it has been indicated that the relating proteins have Rubisco activity (Maeda *et al.,* 1999; Watson *et al.,* 1999). These Rubisco homologs form yet another phylogenetic cluster and the metabolic function of these enzymes in their host organisms still can't seem to be resolved on the grounds that most evidence shows that CO_2 in methanogenic archaea is fixed by means of the acetogenic pathway.

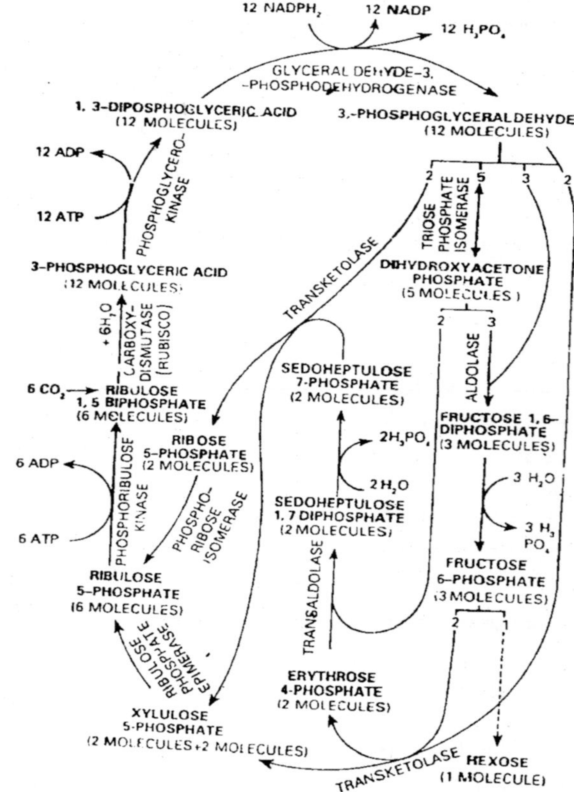

Fig. 5.15. The Calvin cycle (path of carbon in the dark phase of photosynthesis).

2. The *Chlorobiaceae,* referred to conversationally as the "Green Sulfur Bacteria," was indicated not to use the Calvin cycle for CO_2 fixation. Instead, they operate the tricarboxylic acid (TCA) cycle in the reverse, reductive direction, employing it to fix CO_2in the long run into acetyl-CoA and pyruvate, as opposed to the traditional oxidative direction

(**Fig. 5.16**). Most, however not all of the enzymes will catalyze the reactions in the reverse direction. In this manner, the *Chlorobiaceae* have replaced succinic dehydrogenase with fumarate reductase, substituted an α-ketoglutarate dehydrogenase-ferredoxin oxidoreductase for the conventional α-ketoglutarate dehydrogenase complex and replaced the irreversible citrate synthase with an ATP-citrate lyase. Since its revelation in the *Chlorobiaceae,* the reductive TCA cycle has additionally been found in *Desulfobacter hydrogenophilus* in the δ-subphylum of the *Proteobacteria*, in individuals from the *Aquificae*, a profoundly branching hyperthermophilic phylum in the Bacteria and in members of anaerobic *Crenarchaeota*. It was believed to be available in some aerobic *Crenarchaeota*, however recent outcomes show that an alternate pathway functions in those organisms.

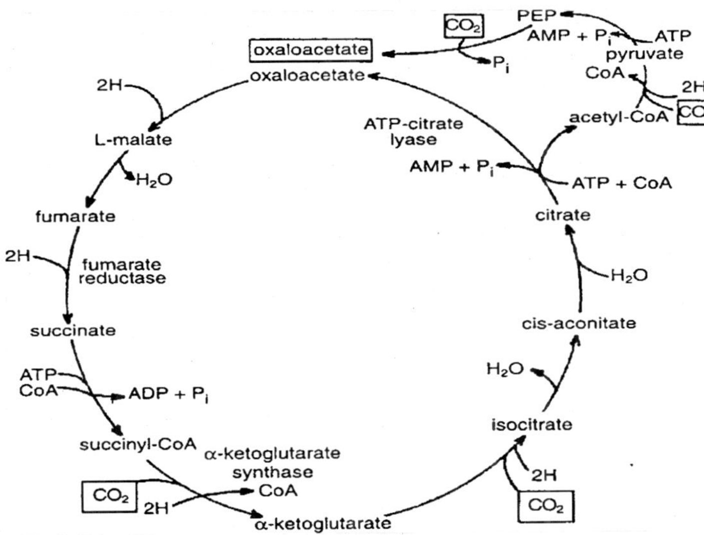

Fig. 5.16. The reductive or reversed tricarboxylic acid cycle (Adapted from Gottschalk, 1986).

3. The acetogenic bacteria can convert $H_2$and CO_2 into acetate, the equivalent of fixing CO_2 into organic matter. To be sure, many acetogenins can grow in the mineral medium using CO_2 as a carbon source and are accordingly autotrophs. The pathway, shown in **Fig. 5.17**, indicates the fixation of two moles of carbon into acetate. The methyl group of acetate is synthesized by reducing one-carbon units to methyltetrahydrofolate. This methyl group is transferred to a corrinoid-containing iron-nickel sulfur enzyme complex called "carbon monoxide dehydrogenase" or "acetate synthetase/ decarbonylase." The methyl group is moved to the cobalt of a corrinoid and a CO_2 is reduced to the equivalent of carbon monoxide. These are then built into an enzyme-bound acetyl group, which is

released as acetyl-coenzyme A (CoA), which can be preserved as ATP in catabolism or can be used for biosynthesis. This pathway costs a cell just a single ATP for every two carbons fixed to acetyl-CoA. This pathway is additionally found in certain sulfate reducers in the δ-subphylum of the *Proteobacteria* (Menendez *et al.,* 1999). A variation of this pathway using methanopterin derivatives instead of folates is found in autotrophic methanogenic Archaea as well as in the sulfate-reducing *Archaeoglobus* (Menendez *et al.,* 1999). Therefore, the acetogenic pathway is responsible for a large part of the CO_2 fixation occurring in anoxic habitats.

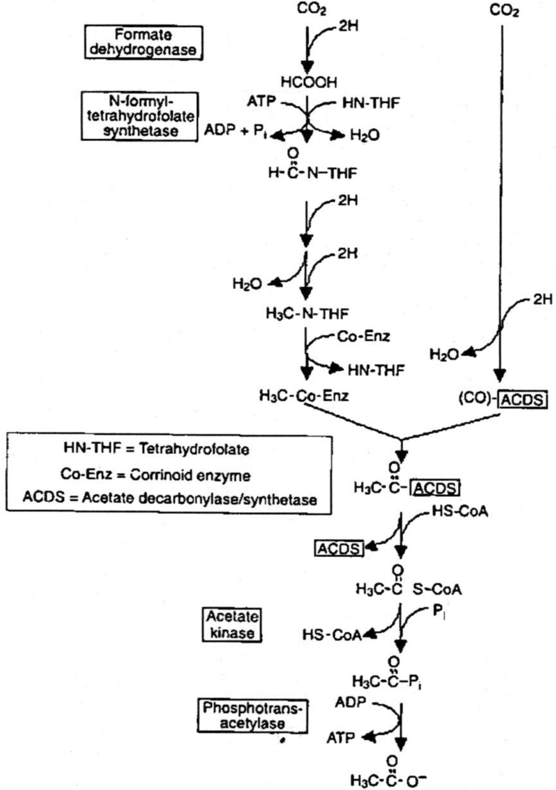

Fig. 5.17. The acetogenic pathway.

4. Lastly, the pathway for CO_2 fixation in the green non-sulfur phototrophic bacterium *Chloroflexus aurantiacus* was not clear until it was illustrated (Strauss and Fuchs, 1993) that a novel pathway dependent on carboxylation of acetyl-CoA and propionyl-CoA (**Fig. 5.18**), once in a while called the "3-hydroxypropionate pathway," is responsible for CO_2 fixation to glyoxalate in these organisms. The

vital enzymes of this pathway are acetyl-CoA carboxylase and propionyl-CoA carboxylase. This pathway has additionally been found in aerobic lithotrophic *Crenarchaeota* such as *Sulfolobus* (Menendez *et al.,* 1999). The distribution of these four autotrophic pathways is summarized in **Table 5.4**

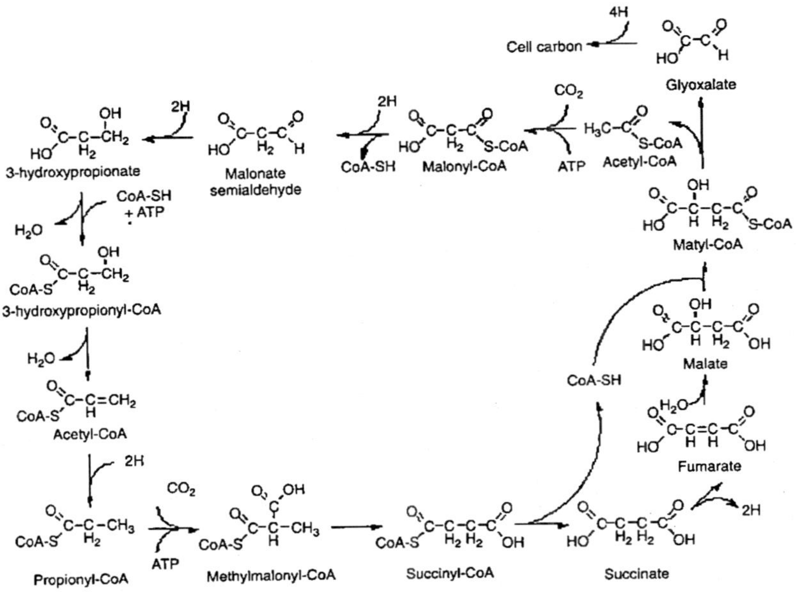

Fig. 5.18. The 3-hydroxypropionate cycle (Menendez *et al.,* 1999)

Table 5.4. Autotrophic CO_2 Fixation Pathways Found in Prokaryotes

Organisms	Pathway	Phylum
Rhodobacter sphaeroides	Calvin cycle	*Proteobacteria*
Thiobacillus ferroxidans		
Chlorobium limicola	Reductive TCA cycle	*Proteobacteria*
Desulfobacterium autotrophicum	Reductive acetyl CoA pathway	*Proteobacteria*
Chloroflexus aurantiacus	Hydroxypropionate cycle	*Chloroflexi*

Abbreviations: TCA, tricarboxylic acid; CoA, coenzyme A

5.12. PHOTOMETABOLISM IN ANOXYGENIC PHOTOTROPHIC BACTERIA

In the light of available evidence an electron transport scheme in APB is outlined in **Fig. 5.19**. This Scheme outlines cyclic photophosphorylation together with photoreduction of NAD^+ via reverse electron flow system. It also includes the possibility of non-cyclic phosphorylation from thiosulfate via ferredoxin with the help of the pyridine nucleotide transhydrogenase. The role of ferredoxin is still disputed but could be mainly reductive CO_2 assimilation. Under aerobic conditions, the electron flows via cytochrome C_2 and cytochrome b to oxygen, whereas NAD reduction occurs via the same reversible electron flow system. The importance of the two NADH dehydrogenases could be for low potential redox systems in the case of heterotrophic growth.

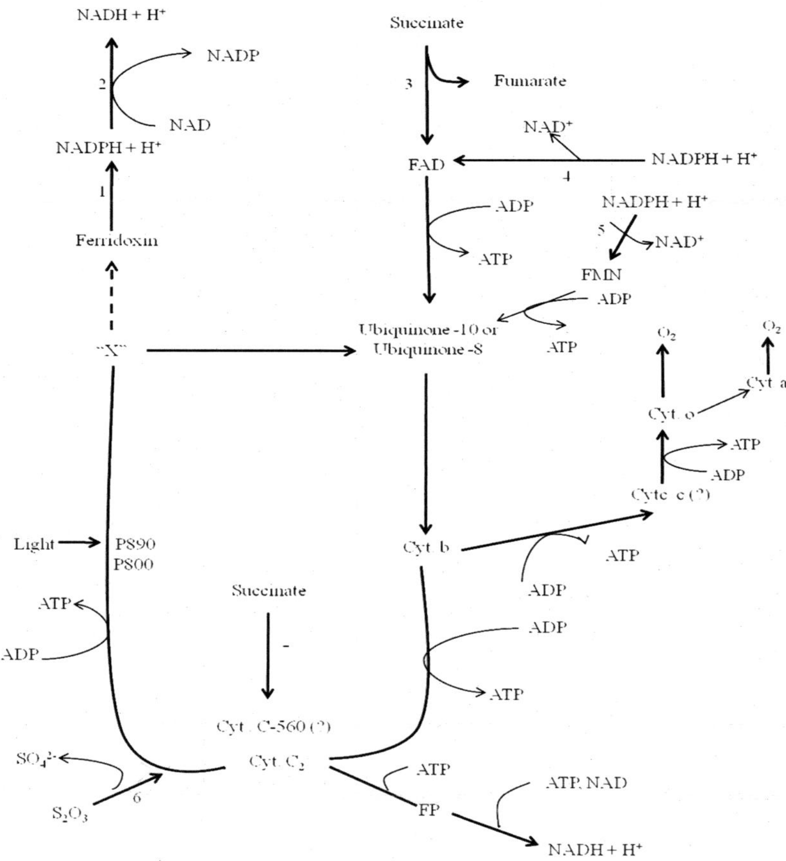

Fig. 5.19. A generalized electron transport scheme in anaerobic photosynthetic bacteria under aerobic and anaerobic conditions. 1. Ferridoxin-NADP reductase, 2. NADPH transhydrogenase, 3. Succinic dehydrogenase, 4. NADH: heme protein oxidoreductase or NADH cyclic reductase, 5. NADH ubiquinone oxidoreductase, 6. Thiosulfate cytochrome c reductase, 7. Succinate: cytochrome *C* reductase.

In the intact cells of *Rhodospirillum rubrum* and *Rhodopseudomonas sphaeroides* rate of decrease of ATP in intact bacteria in the dark under anaerobic conditions is low. On illumination or oxygenation, the formation of ATP increases rapidly from a similar initial rate. However, the total content of ATP is higher during illumination. This illumination, however, inhibits the respiration of bacteria. This may be due to the existence of two independent pathways as both could indirectly compete for the phosphate acceptor. The inhibition of respiration in light may be due to the result of increased ATP formation that causes the depletion of the phosphate. Another reason for the inhibition could be the high ATP/ADP ratio in aerated illuminated cultures. This ratio could function as a regulatory mechanism on the enzymes of the respiratory chain. This hypothesis certainly confirms with observations that the ATP/ADP/AMP form a signal system that can influence the synthesis of certain macromolecules including the bacteriochlorophyll. The possible inhibitory effect of cyclic AMP on NAD^+ photoreduction and energy-linked trans hydrogenation. This inhibition could be an internal effect that occurs only in the actively photosynthesizing cell.

Rhodospirillum rubrum can form pyrophosphate via photosynthetic electron transport. Pyrophosphate is necessary to make the conserved energy available to the cell was the first found example of bacterial inorganic pyrophosphate subject to allosteric inhibition. At low Zn concentration, the enzyme is inhibited by NADH + H^+, NADPH +H^+ and mg-ATP. The effect could be reversed by a high concentration of Zn.

In *Rhodobacter capsulata* and *Rba. sphaeroides* a second type of pyrophosphate appears to exist which exhibits normal Michaelis-Menten reaction kinetics and was stabilized by Ca^{2+} instead of Zn. The pyrophosphate formation is important in so far as the activated phosphate intermediate does not necessarily require ADP but can be transferred to a phosphate molecule. Further, the pyrophosphate drives the synthesis of ATP from ADP and inorganic phosphate with the help of the pyrophosphatase and has a stimulatory effect on CO_2 fixation in a light-independent reaction wherein pyrophosphate could replace ATP to 60% during autotrophic growth and up to 90% during photoheterotrophic growth on acetate. Even AMP, a potent inhibitor of ATP-dependent CO_2 fixation, stimulates the presence of pyrophosphate.

In light-grown cells of *Rsp. rubrum*, the third route of ATP formation was observed. This light causes uptake of protons by applying the chemiosmotic hypothesis for photophosphorylation proposed that light-dependent proton extrusion creates an electrochemical potential that drives the phosphorylation (Edward and Borell, 1969). The electric field was the motive force for ion transfer through the chromatophore membrane against a concentration gradient with NAD+ photoreduction, non-cyclic photophosphorylation and the energy-dependent anion transport, then four sites of energy coupling exist in the chromatophore redox chain.

NADPH→NAD+; NADH→cytochrome b; cytochrome b→cytochrome c and a fourth region after cytochrome C. Each of these coupling sites can provide energy for the generation of membrane potential.

The existence of separate non-cyclic photophosphorylation is still disputed. However, *Rhodospirillum rubrum* would have the cyclic photophosphorylation and NAD reduction, whereas *Rhodopseudomonas* appears to have a second non-cyclic photophosphorylation pathway.

5.13. HYDROGEN AND INORGANIC SULFUR COMPOUNDS

In phototrophic bacteria, two distinct photoreactions can occur involving molecular hydrogen. 1. Photoreduction of carbon dioxide and 2. Photoreduction of hydrogen with subsequent carbon dioxide reduction. Both these reactions are being carried out by *Chlorobium* ,*Chromatium* and *Rhodomicrobium*. However, the significance of these reactions is not fully understood.

If molecular hydrogen serves as an electron donor, it is useful in the reduction of NAD in the presence of ferredoxin-NAD reductase **(Fig. 5.20.)**. This ferredoxin–NAD reaction is called a dark- reaction. The rate of NAD reduction is 3-4 times that of NADP reductase. The enzyme ferredoxin–NAD reductase plays a key role in photochemical NAD reduction.

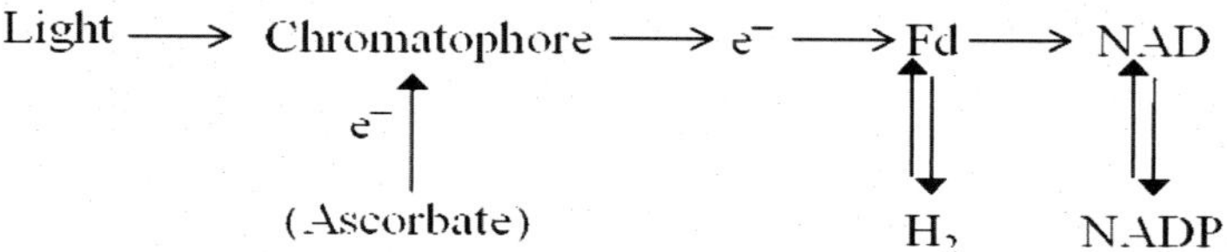

Fig. 5.20. Proposed electron flow in phototrophic bacteria leading to pyridine nucleotide reduction.

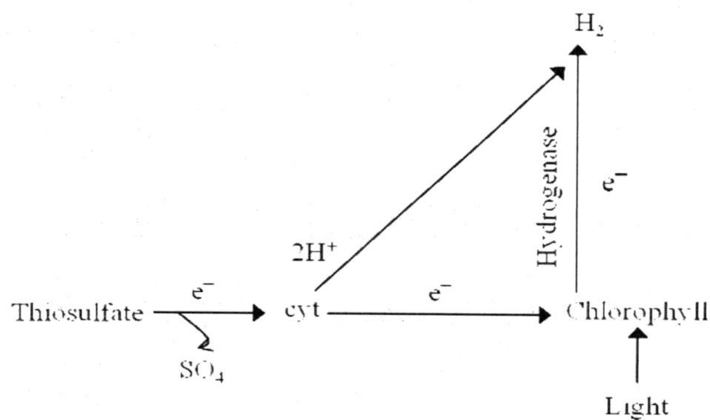

Fig. 5.21. Electron flow for photoproduction of hydrogen from thiosulfate and cytochrome.

The photoproduction of hydrogen requires some inorganic electron donors such as H_2S or thiosulfate. *Chromatium* cytochromes are reduced by thiosulfate and serves as an entry for electrons from the source. In the presence of light and hydrogenase hydrogen is evolved in addition to the reduction of thiosulfate. With one molecule of thiosulfate, two electrons and two protons will combine in the presence of hydrogenase and hydrogen is evolved. This photoproduction of hydrogen can be viewed as a reduction of H^+ by a hydrogenase with the aid of electrons expelled from the excited chlorophyll molecule. The cytochrome system would thus become a gateway for the entry of electrons and their transport to chlorophyll. The electron flow mechanism for the photoproduction of hydrogen from thiosulphate is given in **Fig. 5.21** .

The CO_2 reduction with molecular hydrogen as an electron donor is very much independent of light and substrate concentration. At low light intensities, the rate of CO_2 fixation in the presence of molecular hydrogen and thiosulphate are equal. At higher light intensity, the CO_2 fixation is low with molecular hydrogen.

The hydrogen evolution is inhibited by nitrogen gas and ammonia ions as well as by organic electron donors. However, CO_2 fixation was inhibited either by N_2 or NH_4 as long as thiosulfate was the electron donor which indicates that only the flow of electrons for CO_2 fixation via NAD(P) is inhibited. The sequence of reactions is illustrated below. This evolution of hydrogen is quite common in photosynthetic bacteria **(Fig. 5.22).** *Rhodopeudomonas palustris* is also able to oxidize thiosulfate to tetrathionate.

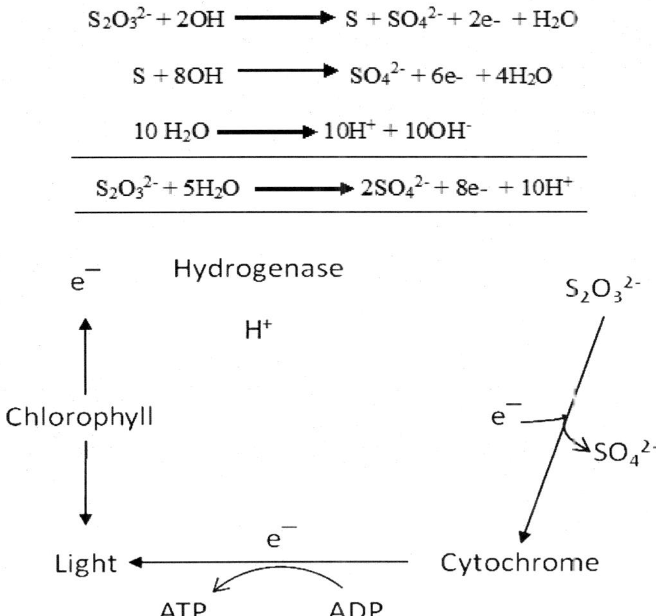

Fig. 5.22. Photoproduction of hydrogen from thiosulfate by purple bacteria.

5.14. ACETATE PHOTOMETABOLISM

Rhodospirillum rubrum can assimilate acetate directly in the absence of CO_2. The acetate is converted to acetyl CoA by an acetate activating enzyme with acetyl adenylate as the anhydride intermediate **(Fig. 5.23)**. The main assimilatory product formed is poly-β-hydroxybutyric acid. As this is a reductive product, some acetate must be oxidized to provide necessary reducing power. The conversion of acetate to the polymer has to complete with CO_2 fixation for limiting the reducing power available in the absence of CO_2. The supply of reducing power for poly-β-hydroxybutyric acid production is an anaerobic process that probably proceeds via the tricarboxylic acid cycle.

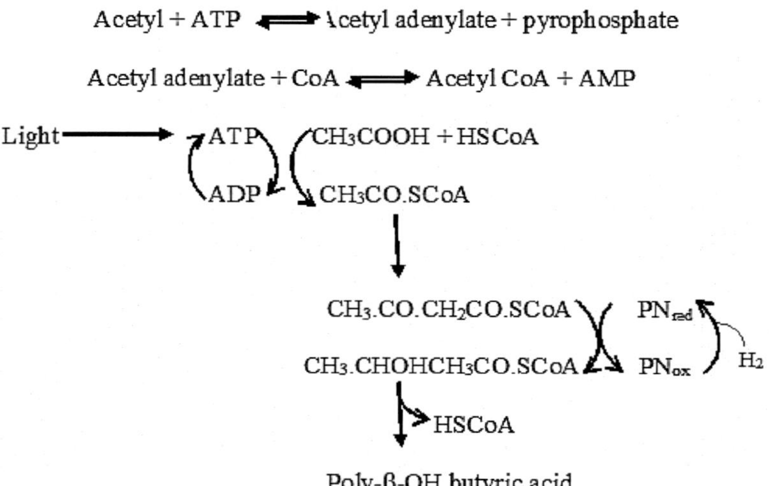

Fig. 5.23. The acetate-hydrogen reaction in purple bacteria.

The acetate-hydrogen reaction in purple bacteria proceeds with the TCA cycle. The formation, economics and usefulness of polymer as a reserve product is remarkable. Informing the polymer the free fatty acids are neutralized and osmotically inert and the cell is able to build up a reserve of reducing power which is used for CO_2 fixation during later stages. *Rhodospirillum rubrum* synthesizes glutamate from acetate and bicarbonate probably via light-dependent anaerobic TCA cycle or reductive CO_2 fixation similar to that of *Chlorobium* . Therefore, *Rhodospirillum* can assimilate acetate in the absence of bicarbonate to produce poly-β-hydroxybutyrate and in the presence of bicarbonate to synthesize glutamate. *Chlorobium thiosulphaticum*, a strict autotroph, assimilate acetate in the presence of bicarbonate and reducing power (*i.e.* Hydrogen gas) with the formation of poly-β-hydroxybutyrate. The capacity of autotrophs to assimilate or metabolize organic compounds resides in their capacity to oxidize and so derive reducing power for biosynthetic reactions leading to cell growth. *Chloropseudomonas ethylicum* probably plays an intermediatory role as this green obligatory photosynthetic bacterium has a nutritional requirement for two carbon compounds and most of the TCA cycle enzymes are present. Polymer synthesis from acetate requires only 1mole of ATP per mole acetate. The function that could be attributed to light is the generation of ATP by cyclic photophosphorylation. 3-hydroxybutyrate dehydrogenase, D-hydroxybutyrate:

NAD oxidoreductases are responsible for the formation and degradation of the polymer and 3-hydroxy acid dehydrogenase which can catalyze the reversible reaction of L(+) 3-hydroxybutyrate into acetoacetate.

$$\text{Acetoacetate} + \text{NADH} + \text{H}^+ \longrightarrow \text{3 - hydroxybutyrate} + \text{NAD}$$

Under anaerobic conditions in the light, *Rhodobacter sphaeroides* could convert acetate to pyruvate via malate which enables the bacterium to generate malate continuously in the presence of malate synthase

$$\text{Acetate} \longrightarrow \text{Acetyl CoA} \longrightarrow \text{Malate} \longrightarrow \text{Pyruvate}$$

The oxidation of malate to pyruvate would then become a site of acetate oxidation and does not pass through the stage of citrate and avoid the use of citrate synthase. The activity of citrate synthase varies with the carbon source and particularly its activity is high in acetate grown cells under aerobic-dark conditions. *Chromatium* and *Rhodospirillum* can transform acetate to citramalate and possibly glutamate.

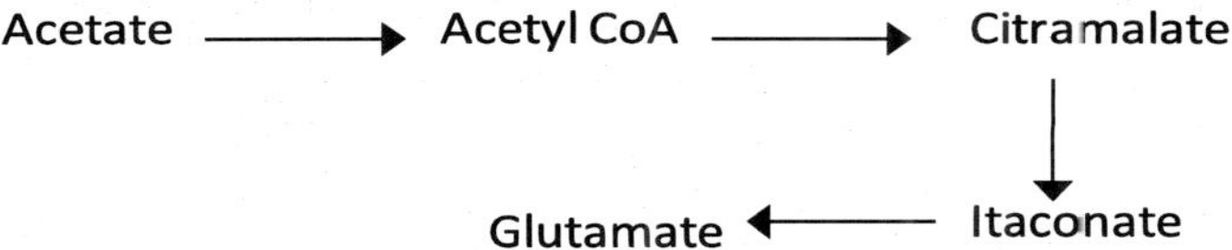

Citramalate and itaconate strongly inhibit the metabolism of acetate and glutamate but not the metabolism of succinate and fumarate, while the addition of C4 dicarboxylic acids releases the inhibitors. *Rhodospirillum rubrum* metabolizes glutamate via the TCA cycle.

5.15. PYRUVATE PHOTOMETABOLISM

It is metabolized in various pathways. *Rsp. rubrum* besides metabolizing pyruvate is also used for the synthesis of cell materials. Under the light, in the presence of the excess amount of pyruvate, storage products such as poly-β-hydroxybutyrate and polysaccharides are formed. Anaerobic light with molecular hydrogen induces the formation of poly-β-hydroxybutyrate, while under anaerobic conditions with molecular nitrogen produces polysaccharides. This is probably regulated by enzyme pyruvate kinase. The presence of CoA-dependent pyruvate carboxylase in *Rhodobacter sphaeroides* indicates that pyruvate can be decarboxylated straight to acetyl CoA and acetate suggesting dismutation reaction with the carboxylation reaction dominating under anaerobic light conditions. The dismutation produces not only poly-β-hydroxybutyrate but also CO_2 and H^+ via the TCA cycle. *Rsp. rubrum* synthesizes poly-β-hydroxybutyrate by fermenting pyruvate with a pyruvate ferredoxin-oxidoreductase and/or pyruvate formate lyase reaction releasing acetate, H_2 and CO_2 or poly-β-hydroxybutyrate. Formation of hydrogen in light-independent during anaerobic dark and light-dependent during anaerobic light conditions. It is also reported that pyruvate produces propionic acid which is inversely related to the production of hydrogen. Thus *Rsp. rubrum* is an ideal organism to differentiate photosynthetic and fermentative metabolism as it can grow equally well in anaerobic light or dark conditions.

5.16. FORMATE PHOTOMETABOLISM

Rhodopseudomonas palustris utilizes formate as an organic carbon source and photoassimilates via CO_2 and the autotrophic CO_2 assimilation in light and anaerobic conditions. The assimilation is mediated by inducible formic hydrogenase which consists two of enzymes soluble formic hydrogenase and particulate hydrogenase. The NAD acts as an electron carrier between two enzymes. As the redox potential of formate/CO_2 is –400mv it does not require light energy and ferredoxin.

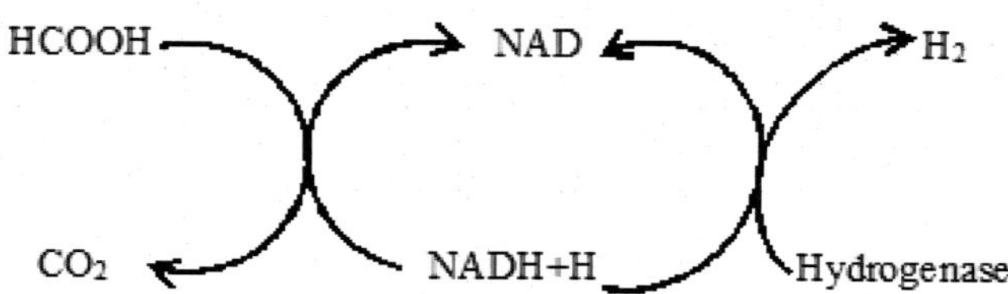

5.17. SUCCINATE PHOTOMETABOLISM

Photometabolism of succinate is slightly more complicated than that of butyrate and acetate. As succinate has lower potential as a hydrogen electron donor which is insufficient to reduce NAD directly, chromatophore of phototrophic bacteria, however, is able to reduce NAD in the presence of light. This reduction is coupled with the oxidation of either succinate or reduced flavin mononucleotide. Since the photochemical reactions by *Rhodospirillum rubrum*, chromatophores are concluded in terms of only one electron transfer system (Pichinoty, 1965), part of the electron transport is operated in the respiratory reaction taking place in the dark with this particular organism. This results in the synthesis of polysaccharides. The substrate is oxidized to pyruvate via fumarate (**Fig. 5.24**) and a continuation of the reverse of the Embden-Meyerhof pathway. A small amount of substrate is also oxidized to acetate and forms the polymer, poly-β-hydroxybutyric acid.

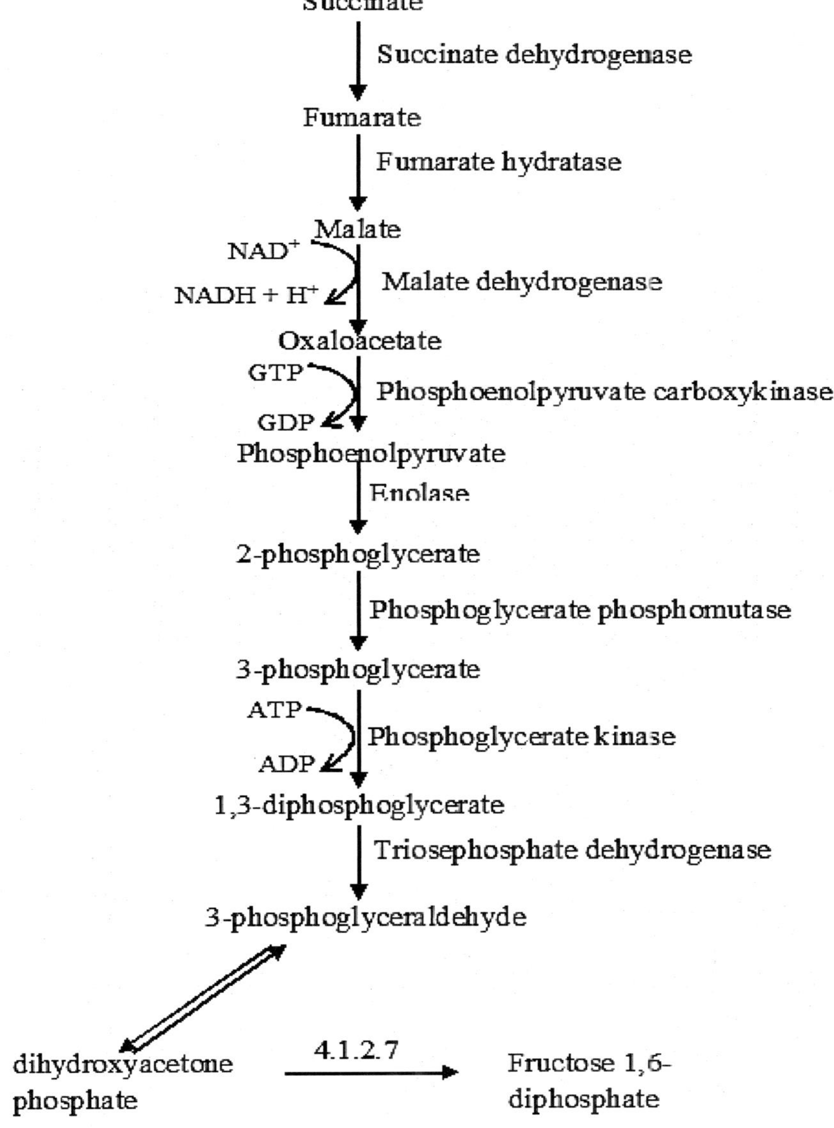

Fig. 5.24. Photometabolism of succinate by *Rhodospirillum rubrum*

A positive correlation was observed between the amount of succinate and NAD reduced. Further, fumarate was the only product of succinate oxidation and succinate the first detectable product of propionate oxidation followed by fumarate and malate. The following pathway is proposed from succinate to hexose formation (**Fig. 5.25**) which is similar to acetate metabolism and carbohydrate cycle in *Chromatium* .

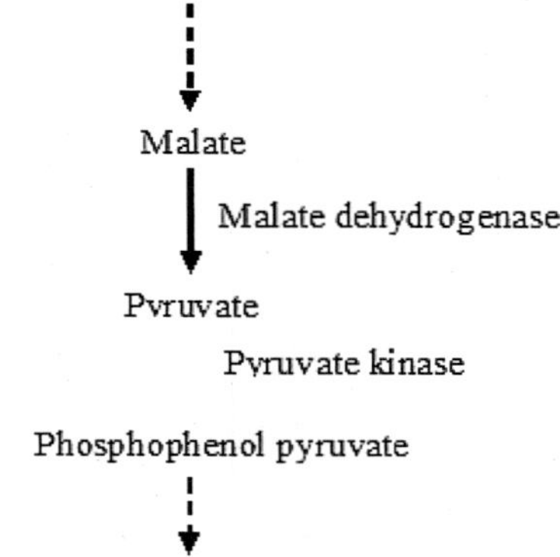

Fig. 5.25. A variation of succinate photometabolism in *Rhodospirillum rubrum.*

Propionate fermentation in *Rsp. rubrum* results in the formation of pyruvate as an intermediate instead of oxaloacetate. The differences in the photometabolism of succinate and propionate is an additional carboxylation step that converts propionate to succinate when propionyl CoA is the substrate for carboxylation and methyl malonyl-CoA is the product of the reaction. If pyruvate is intermediate, the pathway would be as outlined in **Fig. 5.26.**

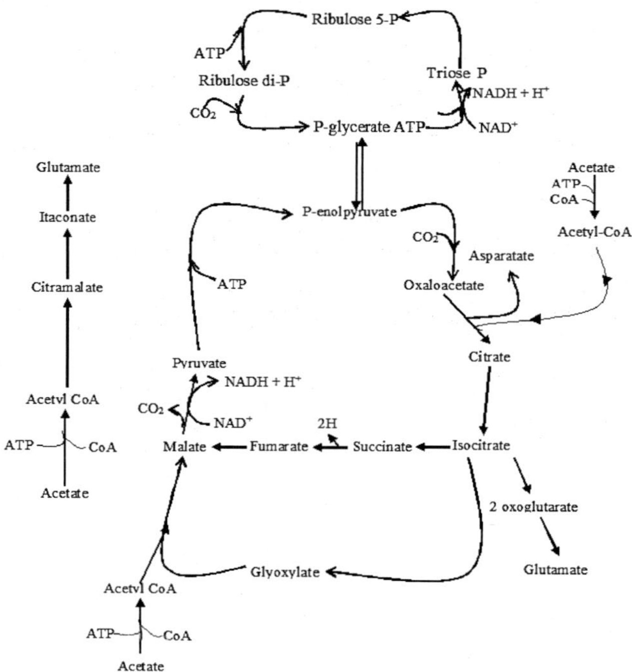

Fig. 5.26. Acetate metabolism and carbohydrate cycle in *Chromatium.*

Malate is converted to pyruvate instead of oxaloacetate in the presence of malate dehydrogenase. Pyruvate thus formed is converted to phosphoenol-pyruvate with the help of pyruvate kinase. Similar metabolism was found in *Chlorobium thiosulfaticum* and *Chromatium*. Therefore, the substrates that are convertible to acetyl units without the formation of pyruvate yield mostly polysaccharides. Synthesis of polysaccharides follows a pathway similar to that of CO_2 fixation which also leads to polysaccharide formation mainly via the Calvin cycle.

5.18. BUTYRATE PHOTOMETABOLISM

Butyrate is assimilated to poly-β-hydroxybutyric acid without CO_2 evolution. The CO_2 that is fixed photosynthetically with molecular hydrogen is converted to polysaccharides.

$$2n\ C_4H_8O_2 \longrightarrow 2\ (C_4H_6O_2)n + 4n\ (H)$$

$$4n\ (H) + nCO_2 \longrightarrow (CH_2O)n + nH_2O$$

$$2n\ (C_4H_8O_2 + nCO_2 \longrightarrow 2\ (C_4H_6O_2)n + (CH_2O)n$$

5.19. ANAEROBIC TRICARBOXYLIC ACID AND REDUCTIVE CO_2 FIXATION

The role of ATP in the photometabolism of organic acids is two folds

1. ATP is required for forming activated substrate to fix CO_2 in the Calvin cycle.
2. ATP activates carbon source such as acetyl CoA from acetate and CoA

The activated compounds are ready to participate in synthetic reactions that are catalyzed by specific enzymes system which function in the dark. Therefore, CO_2 assimilation is a special case of the use of light energy. The photosynthetic transformation of energy is more closely linked with phosphorus than with carbon assimilation.

It is clear from the above discussion that the photometabolism of C_4 dicarboxylic acids such as malate and succinate form significant quantities of CO_2 and H_2. Because of this and the presence of almost all enzymes of the TCA cycle in cell-free extracts, it is now believed that the photosynthetic bacteria use for their assimilation part of the TCA cycle and the glyoxylate cycle.

In *Rhodopseudomonas palustris* α-oxoglutarate is metabolized to cell material via succinate, fumarate, malate and pyruvate anaerobically in light and aerobically in dark.

$$2\ \text{oxoglutarate} + 1.5\ O_2 \longrightarrow 2.5\ (CH_2O) + 2.5\ CO_2 + 0.5\ H_2O\ \text{(light)}$$

2 oxoglutarate + 2.0 O_2 $\longrightarrow$ 2.0 (CH_2O) + 3CO_2 + H_2O (dark)

Succinate + O_2 $\longrightarrow$ 2.5 (CH_2O) + 1.5 CO_2 + 0.5 H_2O (light)

Succinate + 1.5 O_2 $\longrightarrow$ 2.0 (CH_2O) + 2 CO_2 + H_2O (dark)

Fumarate + 0.5 O_2 + 0.5 H_2O $\longrightarrow$ 2.5 (CH_2O) + 1.5 CO_2 (light)

Fumarate + O_2 + 0.5 H_2O $\longrightarrow$ 2.0 (CH_2O) + 2.0 CO_2 (dark)

Malate + 0.5 O_2 $\longrightarrow$ 2.5 (CH_2O) + 1.5 CO_2 + 0.5 H_2O (light)

Malate + O_2 $\longrightarrow$ 2.0 (CH_2O) + 2 CO_2 + H_2O (dark)

Lactate + 0.5 O_2 $\longrightarrow$ 2.5 (CH_2O) + 0.5 CO_2 + 0.5 H_2O (light)

Lactate + O_2 $\longrightarrow$ 2.5 (CH_2O) + CO_2 + H_2O (dark)

Pyruvate + 0.5 O_2 $\longrightarrow$ 2.5 (CH_2O) + 0.5 CO_2 (light)

Pyruvate + 0.5 O_2 $\longrightarrow$ 2.0 (CH_2O) + CO_2 (dark)

The higher rate of oxygen uptake in the dark reaction results in a higher rate of CO_2 formation. The presence of all required enzymes indicates that the TCA cycle may be the pathway for terminal oxidation of the organic acids in a number of purple bacteria such as *Chromatium* and *Rhodospirillum rubrum.* The organic acids in the dark are metabolized via the TCA cycle. Light anaerobic conditions, however, seem to suppress to some extent the activity of the TCA cycle. This suppression could be caused indirectly by stimulation of citrate cleavage reaction which increases the levels of ATP and NADH + H in the cell. The photometabolism of acetate and various dicarboxylic acids in *Rsp. rubrum* and *Chromatium* are illustrated in **Fig. 5.27.**

Oxidation-reduction reactions between carboxylic acids of oxidatively different levels are favored as was indicated when the photometabolism of succinate in *Rsp. rubrum* increased three folds when CO_2 was present suggesting that CO_2 played the role of a hydrogen acceptor. The CO_2 evolution was dependent on the oxidation level of the acids to be metabolized. In the case of *Chromatium* strain D which lacks 2-oxoglutarate dehydrogenase and malate dehydrogenase are substituted by malate dehydrogenase and pyruvate carboxylase are responsible for the modified glyoxylate cycle. *Athiorhodaceae*, however, possess a complete TCA cycle which functions oxidatively in the dark as well as anaerobically in the light.

All the above dicarboxylic acids are members of the TCA cycle are photometabolized without hydrogen production if a suitable electron acceptor is provided or generated by the metabolism of the added substrate. Photoheterotrophically growing purple bacteria possess high levels of ribulose 1,5-

biphosphate carboxylase which is the key enzyme for CO_2 fixation in the Calvin cycle for autotrophs. However, under certain conditions purple bacteria are able to simultaneously fix CO_2 through autotrophic scheme and photoassimilate fragments derived from added organic compounds. In the presence of light, an enzyme source and substrates are more reduced than CO_2, it may be assumed that the major function of the various metabolic cycles of photosynthetic bacteria in the synthesis of new cellular materials rather than the production of energy (Atzpodien and Bode, 1970). *Chromatium* can produce oxaloacetate independently of the TCA cycle. In anaerobes phosphorolytic cleavage of pyruvate to acetyl phosphate, H_2 and CO_2 take place in the presence of coenzyme A, inorganic phosphate and ferredoxin without formate as an intermediate. It was shown that N_2 assimilation takes place at the expense of H_2 evolution and is stimulated by pyruvate (**Fig. 5.28**). Hydrogen evolution was strong in dark, whereas in the light N_2 assimilation was favored.

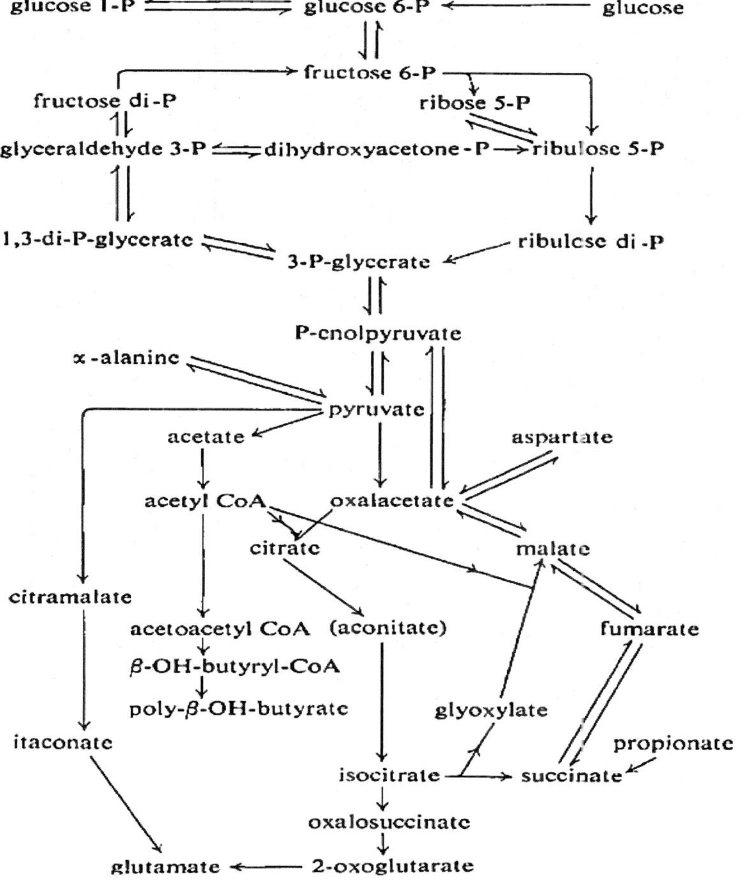

Fig. 5.27. Photometabolism of acetate and dicarboxylic acids in *Rhodospirillum rubrum* and *Chromatium*.

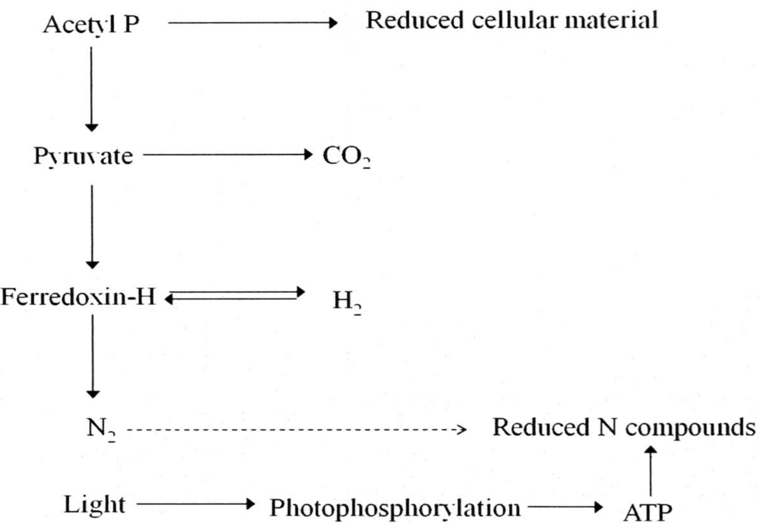

Fig. 5.28. Relationship between N_2 assimilation and hydrogen evolution.

An enzyme pyruvate synthase can use ferredoxin directly as a reductant in carbon assimilation and responsible for the reductive synthesis of pyruvate from CO_2 and acetyl CoA. This mechanism can also function for the assimilation of acetate as well as CO_2. In photosynthetic bacteria, reductive CO_2 assimilation is mediated by ferredoxin which is equal to H_2 in its reducing power.

$$\text{Acetyl CoA} + CO_2 + Fd_{red} \longrightarrow \text{Pyruvate} + \text{CoA} + Fd_{ox}$$

Similarly, 2-oxoglutarate synthase mediates the synthesis of α-ketoglutarate from succinyl-CoA and CO_2

$$\text{Succinyl - CoA} + CO_2 + Fd_{red} \longrightarrow \alpha\text{ - ketoglutarate} + CO_2 + Fd_{ox}$$

Rhodospirillum rubrum possesses both reactions, whereas *Chromatium* and replace the α-ketoglutarate synthase reaction with a third reaction involving the synthesis of phenyl acetyl COA and bicarbonate and reducing the power of ferredoxin and thiamine pyrophosphate.

In *Rhodospirillum rubrum* propionic acid can function as an endogenous CO_2 acceptor forming methylmalonic acid and succinic acid. Similarly, propionyl CoA produces α- ketoglutarate in the presence of 2- oxobutyrate synthase which is dependent on reducing the power of ferredoxin. This reaction is very important as α-ketoglutarate is a precursor in the biosynthesis of isoleucine and γ- amino butyrate. All the above CO_2 assimilation reactions are evidence for the existence of a reversed TCA cycle. This reductive carboxylic acid cycle constitutes a mechanism for utilizing ATP and the reducing power of ferredoxin. The acetyl CoA becomes available for the synthesis of carbohydrates and lipids.

The existence of the carboxylic acid cycle next to the reductive pentose cycle (Calvin cycle) suggests that phototrophic bacteria can switch from one cycle to another depending on the substrate. Lithotrophic photosynthetic bacteria could utilize the reductive pentose cycle, whereas organotrophic photosynthetic bacteria could prefer the reductive carboxylic acid cycle.

Rhodospirillum rubrum photometabolises 1 mole of malate to 3 moles of CO_2 and 3.8 moles of H_2. When H_2 is not produced, malate photo assimilation suggests that this C4-carboxylic acid is converted to CO_2 and a C_3 fragment which is then assimilated to carbohydrates. Hydrogen formation may represent an alternative electron transfer mechanism that comes into play when the metabolic energy balance is such that the ADP level has decreased to a critical level i.e. when the rate of ATP formation has exceeded the rate of utilization. The photo evolution of hydrogen via a terminal hydrogenase system can be interpreted as the manifestation of an alternative pathway of electron transfer which serves as a regulatory function in energy metabolism.

5.20. ACETONE AND ALCOHOL PHOTOMETABOLISM

A number of *Athiorhodaceae* and purple non-sulfur bacteria are able to utilize various alcohols for the photosynthetic reduction of CO_2. The general reactions are

$$\underset{\text{(Ethyl alcohol + Water)}}{CH_3CH_2OH + 3H_2O} \longrightarrow \underset{\text{(Carbon dioxide + Hydrogen)}}{2\,CO_2 + 2H^+}$$

$$\underset{\text{(Carbon dioxide + Hydrogen)}}{3CO_2 + 12H^+} \longrightarrow \underset{\text{(Ethyl alcohol + Water)}}{3(CH_2O) + 3\,H_2O}$$

$$\underset{\text{(Propyl alcohol/Propanol)}}{CH_3CH_2CH_2OH} \longrightarrow \underset{\text{(Carbon dioxide + Hydrogen)}}{3CO_2 + 18H^+}$$

Rhodopseudomonas gelatinosa metabolises acetone to acetoacetate by condensing with CO_2. The enzymes responsible for this reaction are inducible. However, photophosphorylation is strongly inhibited by these lower aliphatic alcohols with chromatophores of *Rhodospirillum rubrum* .

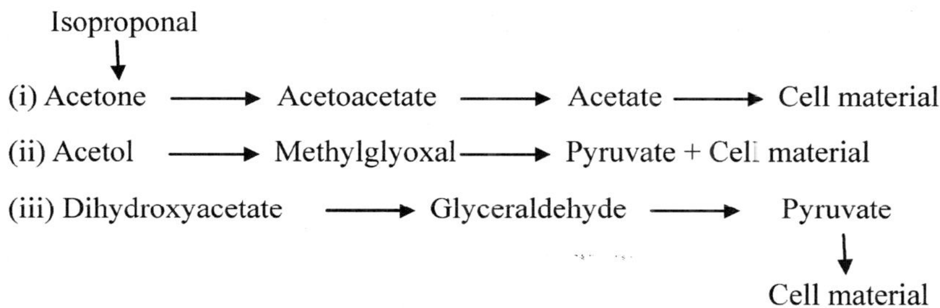

5.21. GLYCINE PHOTOMETABOLISM

Rhodopseudomonas sphaeroides metabolizes glycine to acetyl CoA. However, details of this metabolic pathway is still disputed. It is proved that glycine is exclusively converted to glyoxylate mainly by a transamination reaction wherein pyruvate and oxaloacetate act as amino acceptors. Glyoxylate itself converted to pyruvate via the glyoxylate cycle. It is recently reported that glycine decarboxylase along with methylenetetrahydrofolate metabolises glycine via serine and pyruvate to acetyl CoA. With this system functioning the methylene carbon atom of glycine would be converted to acetyl CoA whereas as the carboxyl carbon atom appears in the newly formed protein. ATP has a marked stimulatory effect on this reaction.

5.22. METHANE PHOTOMETABOLISM

Rhodopseudomonas gelatinosa utilizes methane as the sole electron donor and incorporates methane carbon into cellular material as well as oxidizes methane to CO_2. However, details of this pathway are not yet available.

5.23. AROMATIC COMPOUNDS METABOLISM

Rhodopseudomonas palustris is able to photometabolize aromatic compounds both aerobically and anaerobically. Benzoate supports the growth only under photosynthetic conditions but not aerobically in the dark, whereas other compounds such as β-hydroxybenzoate can be metabolized under both conditions. *Rhodopseudomonas palustris* can utilize *p*-hydroxybenzoate under aerobic dark conditions and converts into protocatechuate 4,5-oxygenase to form γ-carboxy-α-hydroxy malonic semialdehyde. The presence of oxygen was obligatory for this pathway.

The key difference between the aerobic and anaerobic pathways is the reduction of the aromatic nucleus to an aliphatic cyclic acid first before ring fission occurs under anaerobic conditions. Benzoate is first reduced to cyclohexane-1-ene-1-carboxylic acid which is hydrated to the intermediate 2-hydroxy cyclohexane carboxylic acid followed by its dehydrogenation to 2-oxocyclohexane carboxylic acid. Further, hydration of this aliphatic cyclic acid results in the ring fission and formation of pimelic acid (**Fig. 5.29**). It has been suggested that the reduction of aromatic compounds is catalyzed by reductases coupled to ferredoxin.

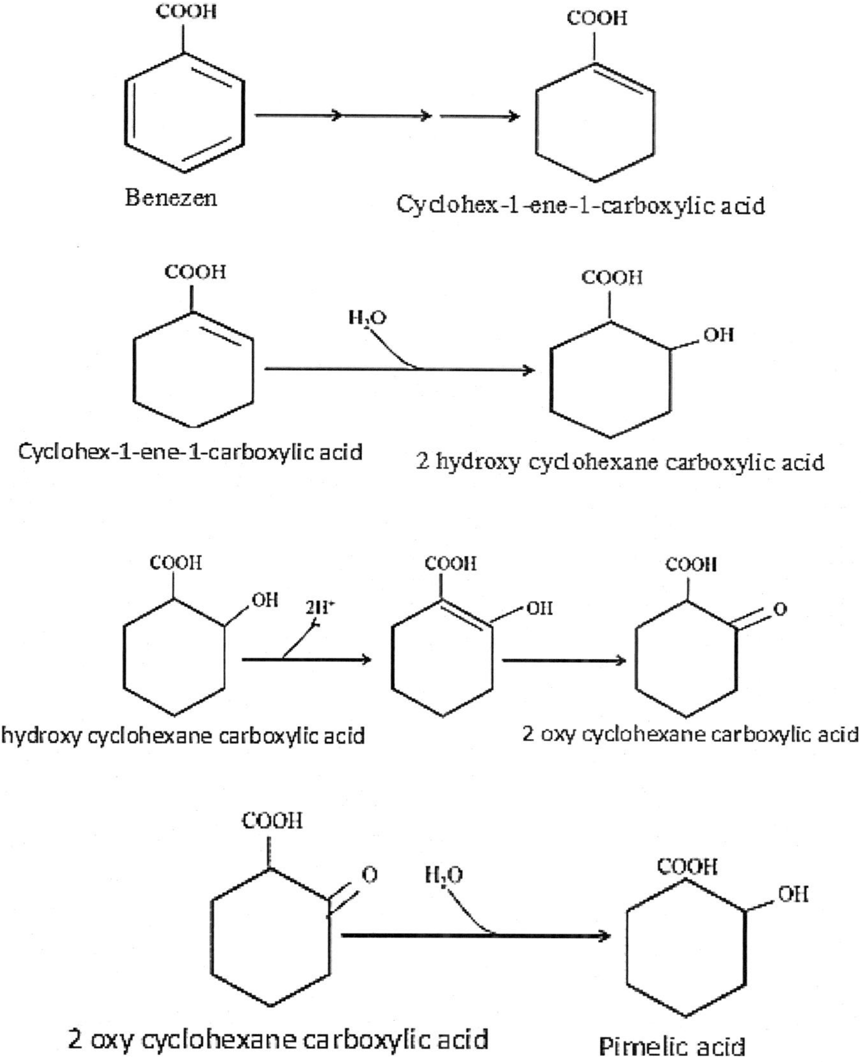

COOH COOH

Benezen → Cyclohex-1-ene-1-carboxylic acid

COOH H_2O COOH OH

Cyclohex-1-ene-1-carboxylic acid → 2 hydroxy cyclohexane carboxylic acid

COOH OH $2H^+$ COOH OH COOH O

2 hydroxy cyclohexane carboxylic acid → 2 oxy cyclohexane carboxylic acid

COOH O H_2O COOH OH

2 oxy cyclohexane carboxylic acid → Pimelic acid

Fig. 5.29. Metabolism of aromatic compounds in anoxygenic phototrophic bacteria.

5.24. PHOTOCHEMICAL NITROGEN FIXATION

Both green and purple bacteria are capable of fixing nitrogen under anaerobic conditions in the light with the help of non-cyclic electron flow in which electrons pass from the substrate via cytochromes to the chlorophyll as was described for hydrogen evolution. The similarity to hydrogen evolution goes even further as nitrogen assimilation was greatly increased by the addition of ATP. The nitrogenase system of *Chromatium*, therefore, appears to have the typical properties of other nitrogenase of heterotrophic bacteria (*Azotobacter* etc.)

as it reduces nitrogen to ammonia and acetylene to ethylene. The energy requirement for the nitrogen fixation is now also known to be qualitative, the same as for CO_2 assimilation suggesting nitrogenase activity is coupled to photochemically generated ATP and reductant. Quantitative measurements of nitrogen fixation in *Rsp. rubrum* revealed that about 6 moles of L-malate, fumarate, or succinate, or 10 moles of pyruvate are consumed per mole of molecular nitrogen. Nitrogen-fixing capability is another way of utilizing the electrons expelled from chlorophyll by light. Such a relationship also exists in cyanobacteria which are known for their photosynthetic nitrogen-fixing ability can be adapted to hydrogen formation.

5.25. THE ELECTRON ACCEPTOR IN PURPLE BACTERIA

Though a lot has been discussed with regard to ATP formation and hydrogen donors in phototrophic bacteria, little is discussed about electron acceptors in these bacteria. It is established that ATP plays a dual role and there exists a relation between CO_2 concentration and H_2 evolution as well as a competition between H_2 evolution and nitrogen fixation. Assuming that the electrons required for the generation of reduced NAD are derived from an accessory inorganic or organic compound which would, therefore, serve as hydrogen donor, the following mechanism is possible (**Fig. 5.30** .)

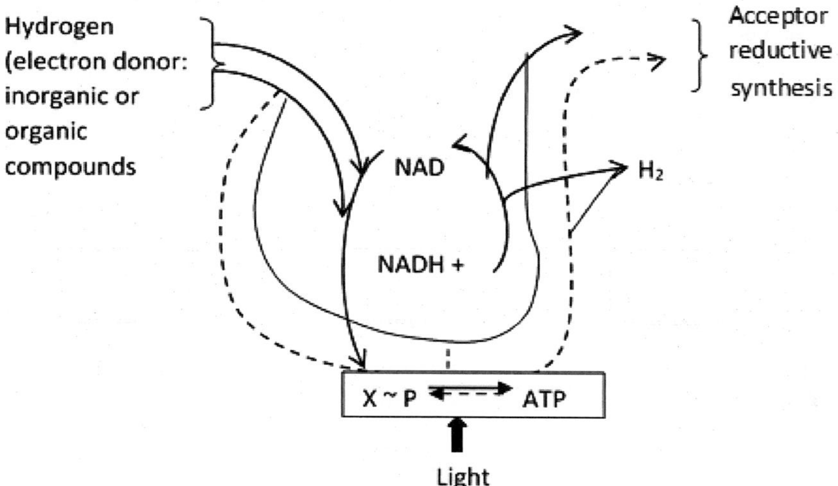

Fig. 5.30. Hydrogen electron flow from donors to acceptors and photoproduction of bacterial photosynthesis.

Depending on the redox potential of the donor or on the steady-state concentration of the reduced and oxidized form of donor and NAD, the formation of reduced NAD may be promoted by energy-rich intermediates created by the action of light on the photochemical apparatus. Reduced NAD would then be available for a great variety of reductive syntheses such as the conversion of CO_2 to cell material, the transformation of C_2 and C_3 compounds to reserve materials, or even reductive aminations to provide the protein synthesis with amino acid precursors. In the case of excess production of ATP and reduced NAD in relation to biosynthetic reaction, reduced NAD can be reoxidised through the liberation of molecular hydrogen by an energy-dependent process. The evolution of H_2 can, therefore, be interpreted as a sort of regulatory device that maintains ATP and reduced NAD at a level consistent with the overall rate of biosynthetic activity.

The only common event left in all types of photosynthesis is the primary photochemical reaction that leads to ATP formation including the transfer of electrons through a closed carrier system coupled to chlorophylls. In green plants, green bacteria and some of the purple sulfur bacteria, major portions of ATP go into CO_2 reduction to drive cell material synthesis. In purple non-sulfur bacteria and some purple sulfur bacteria, it is used to drive cell synthesis from externally provided organic substrates. Therefore, the use of CO_2 as the sole source in photosynthesis is not a fundamental feature of this process but is restricted to those organisms that are called autotrophs.

■■■■

Chapter 6

SULFUR METABOLISM

6.1. INTRODUCTION

Sulfur exhibits high reactivity in reduced form and occurs in several oxidation states. Sulfate or sulfides occurs in soil, water and sulfur dioxide in the atmosphere constitutes the majority of sulfur in nature. Their requirements are small but plays important role in biological systems. Though sulfur has the highest number of allotropes, only a few occur in biological systems. It appears in all organisms in various organic compounds such as amino acids, enzyme cofactors, sulfolipids, vitamins, polypeptides, or carbohydrates. Some of the biological roles of inorganic sulfur compounds are

1. It serves as a source for sulfur dissimilation and incorporation of sulfur into different organic compounds.
2. They serve as donors or acceptors of electrons for energy-generating electron transport. Dissimilatory sulfur-based energy generators move along with mass transformations and exist exclusively among prokaryotes. Although, assimilatory metabolism is not only very frequent in prokaryotes but also obtain in plants, algae and fungi.

In recent times more useful information on genome sequences from green sulfur bacteria and purple sulfur bacteria are available which will throw light on sulfur metabolism also (Frigaard and Bryant, 2008; Sander and Dahl, 2008). Metabolism of organic sulfur compounds in anoxygenic phototrophic bacteria is excellently reviewed by Denger *et al.* (2006) and Baldock *et al.* (2007).

In anoxygenic phototrophic bacteria, reduced sulfur compounds can play an important role as electron donors for photosynthetic CO_2 fixation **(Table 6.1)**. In fact, the utilization of sulfur compounds is common to almost all groups of anoxygenic phototrophic bacteria such as (1) Filamentous anoxygenic phototrophic bacteria (FAPB) (*Chloroflexus* and relatives), (2) Green sulfur bacteria (GSB) (*Chlorobiaceae* members), (3) *Heliobacteria* (*Heliobacteriaceae* which are Gram (+) ve), (4) Purple bacteria (Purple non-sulfur and purple sulfur bacteria) and (5) Phototrophic Acidobacteria *Candidatus Chloroacidobacterium thermophilum* (Bryant *et al.*, 2007). Like green sulfur bacteria, it synthesizes bacteriochlorophyll *a* and *c* in chlorosomes FMO protein and type-I reaction centers, but unlike green sulfur bacteria, it grows under aerobic conditions.

Some of the anaerobic heliobacteria as well as members of the filamentous anoxygenic phototrophic bacteria can oxidize reduced sulfur compounds. Thiosulphate oxidation is widespread among photoheterotrophic aerobic anoxygenic phototrophic bacteria. Most of the purple non-sulfur bacteria can use thiosulfate and or sulfide as electron donors. The utilization of reduced sulfur compounds has been studied intensively both in purple sulfur bacteria and green sulfur bacteria. Details of the metabolism of organo-sulfur compounds have been excellently reviewed by Baldock *et al.* (2007) and Kappler and Schafer (2014).

Table 6.1. Characteristics of Different groups of Anoxygenic Phototrophic Bacteria

Group	**Purple sulfur bacteria (PSB)**	**Purple non-sulfur bacteria**	**Aerobic bacterio-chlorophyll containing (ABC) bacteria**	**Green sulfur bacteria (GSB)**	**Filamentous anoxygenic phototrophic bacteria (FAPB)**	***Helioba-cteria***
Phylogenetic affiliation	*Gammapr oteoba-cteria*	*Alpha* and *Betaproteo bacteria*	*Alpha, Beta* and *Gamma-proteoba-cteria*	*Chlorobia ceae* (phylum *Chlorobi*)	*Chlorofle-xaceae* (phylum *Chloroflexi)*	*Heliobacteri aceae* (phylum *Firmicutes*, order *Clostri-diales*)
Type of bacterio-chlorophyll	BChl *a,b*	BChl *a,b*	BChl *a*, Zn-BChl *a*	BChl *c,d,e* (*a* minor component s in all)	BChl *a*, Some also have BChl *c,d*	BChl *g*
Reaction center type	Pheo-phytin/ quinone (type- II)	Pheo-phytin/ quinone (type-II)	Pheo-phytin/ quinone (type-II)	Fe-S (type-I)	Pheophytin/qu inone (type II)	Fe-S (type-I)
Antenna complexes	LHI and LHII in intracyto-plasmic membr-anes	LHI and LHII in intracyto-plasmic memb-ranes	LHI most have no LHII	FMO protein, Chloro-somes	LH I like , some have also chlorosomes	None[a]
Preferred electron donor fro phototrophic growth	Reduced sulfur comp-ounds, hydrogen (organic substrates)	Organic comp-ounds, hydrogen, (Sulfide, ferrous iron)	None known	Sulfide, hydrogen (Ferrous iron in some)	Organic compounds (Sulfide)	Organic compounds
Chemotro-phic growth	+	+/-	-	+	+	+/-

Group	Purple sulfur bacteria (PSB)	Purple non-sulfur bacteria	Aerobic bacterio-chlorophyll containing (ABC) bacteria	Green sulfur bacteria (GSB)	Filamentous anoxygenic phototrophic bacteria (FAPB)	*Helioba-cteria*
Pathway of auto-trophic CO_2 fixation	Calvin cycle	Calvin cycle	If present: Calvin cycle, No photoautotrophic growth	Reductive TCA cycle	Hydroxypropionate pathway (*Chloro-flexus*), Calvin cycle (*Osci-llochloris*)	None
Major habitat	Anoxic marine and freshwater	Anoxic marine and freshwater	Eutrophic aquatic environ-ments	Anoxic marine and fresh-water	Hot springs, meta and hypolimnion of stratified freshwater pools	Soil

[a]*Heliobacteria* contains no other antenna chlorophyll than those associated with photosynthetic reaction center in the cytoplasmic membrane. Data is taken from Frigaard and Dahl (2008).

The variation in photosynthetic reaction centers (RC), light-harvesting centers and different CO_2 fixation pathways among these groups are quite remarkable (**Table 6.1**). All known photosynthetic RC can be divided into two main groups.

i. Type-I reaction center in which Fe-S cluster acts as a terminal electron acceptor. Photosystem-I of oxygenic phototrophic bacteria, eukaryotes and homodimeric RC of green sulfur bacteria and *Heliobacteria* belong to this group.

ii. Type-II in this type of RC quinone is the terminal electron acceptor. Due to the simpler architecture of photosynthetic apparatus, anoxygenic phototrophic bacteria depend on electron donors with more negative standard potentials than that of water (Overmann, 2008). All APB depends on electron donor with standard potentials are more negative than that of H_2O (reduced sulfur compounds, H_2 and acetate). Probably for this reason these bacteria have narrower ecological niches than those of cyanobacteria.

6.2. TRANSFORMATIONS OF SULFUR COMPOUNDS

Purple sulfur bacteria (PSB) and green sulfur bacteria (GSB) preferentially use reduced sulfur compounds as electron donors during photolithoautotrophic growth and their sulfur-metabolizing capabilities are summarized in **Table 6.2.** Sulfur oxidation capabilities in the aerobic bacteriochlorophyll containing (ABC) bacteria, the filamentous anoxygenic phototrophic bacteria (FAPB) and *Heliobacteria* are limited. Information about the enzymes involved in these bacteria is also not available.

6.2.1. Purple Sulfur Bacteria (PSB)

However, both purple sulfur bacteria and green sulfur bacteria photooxidized sulfide, in general, PSB prefers lower sulfide concentrations than GSB. However, among PSB (*Thiorhodococcus drewsii*) the ability to tolerate sulfide concentration extends up to 11mM (Zaar *et al.,* 2003).

i. ***Chromatiaceae:*** All the members of phototrophic bacteria, *Chromatiaceae* use sulfide and sulfur of the oxidation state zero as photosynthetic electron donors (**Table 6.2**). Unique species such as *Allochromatium warmingii, Isochromatium buderi* and *Thiospirillum jenense* are not able to utilize thiosulfate, H_2 or organic compounds as electron donors (Imhoff *et al.,* 1998). Members such as *Alc. vinosum* and *Tca. roseopersicina* which have been studied for utilization of externally added polysulfides readily use these compounds as photosynthetic electron donors (Visscher *et al.,* 1990). A few organic sulfur compounds can also serve as electron donors for the photosynthetic growth of this group. For instance, *Thiocapsa roseopersicina* splits mercapto malate to fumarate and sulfide. Similarly, mercapto propionate is degraded to acrylate and sulfide and then uses the formed sulfide as an electron donor (Visscher and Taylor, 1993). This organism also oxidizes dimethyl sulfide to dimethyl sulfoxide (Visscher and van Gemerden, 1991). Sulfur compounds such as elemental sulfur can serve as acceptors of electrons which are liberated during the oxidation of stored carbon compounds such as polyhydroxy laconic acid.

Table 6.2. Sulfur Metabolizing Capabilities of Phototrophic Sulfur Bacteria

Genus or Species	Utilization of sulfur substrates	Interme-diates sulfur compounds	End product	Sulfate assi-milation	Chemo-autotrophic growth
Ancalochloris perfilievii[ab]	Sulfide	Sulfur, EC	Sulfate	-	-
Chlorobaculum chlorovibrioides	Sulfide, sulfur	Sulfur, EC	Sulfate	-	-
Chlorobaculum limnaeum	Sulfide, sulfur, thiosulfate	Sulfur, EC	Sulfate	-	-
Chlorobaculum parvum	Sulfide, thiosulfate, sulfur, tetrathionate in some strains)	Sulfur, EC	Sulfate	-	-
Chlorobaculum tepidum	Sulfide, thiosulfate, sulfur	Sulfur, EC	Sulfate	-	-

Genus or Species	Utilization of sulfur substrates	Intermediates sulfur compounds	End product	Sulfate assimilation	Chemo-autotrophic growth
Chlorobaculum thiosulfatiphilum	Sulfide, thiosulfate, sulfur	Sulfur, EC	Sulfate	-	-
Chlorobium chlorochromatii	Sulfide	ND	ND	-	-
Chlorobium clathratiforme	Sulfide, thiosulfate in some strains, sulfur	Sulfur, EC	Sulfate	-	-
Chlorobium ferrooxidans	- +	-	-		-
Chlorobium limicola	Sulfide, thiosu-lfate in some strains, sulfur	Sulfur, EC	Sulfate	-	-
Chlorobium luteolum	Sulfide, Sulfur	Sulfur, EC	Sulfate	- c	-
Chlorobium phaeobacteroides	Sulfide, Sulfur	Sulfur, EC	Sulfate	-	-
Chlorobium phaeovibrioides	Sulfide, sulfur, thiosulfate in some strains	Sulfur, EC	Sulfate	-	-
Chloroherpeton thalassium	Sulfide	Sulfur, EC	Sulfate	-	-
Prosthecochloris aestuarii	Sulfide, Sulfur	Sulfur, EC	Sulfate	-	-
Prosthecochloris vibrioformis	Sulfide, Sulfur	Sulfur, EC	Sulfate	-	-
Purple Sulfur Bacteria (*Chromatiaceae*)					
Allochromatium	Sulfide, sulfur, thiosulfate, sulfite, (latter two not in *Alc. warmingii*)	Sulfur, IC	Sulfate	+ (not in *Alc. warmingii*)	Some species
Chromatium	Sulfide, sulfur	Sulfur, IC	Sulfate	-	-
Halochromatium	Sulfide, thiosulf-ate, sulfur, sulfite	Sulfur, IC	Sulfate	-	(sulfide, thi-osulfate) +
Isochromatium	Sulfide, sulfur	Sulfur, IC	-	-	-
Lamprobacter[a]	Sulfide, thiosulfate, sulfur	Sulfur, IC	Sulfate	-	+
Lamprocystis	Sulfide, thiosulfate, sulfur	Sulfur, IC	S^0 and Sulfate	- /ND	+/-

Genus or Species	Utilization of sulfur substrates	Interme-diates sulfur compounds	End product	Sulfate assi-milation	Chemo-autotrophic growth
Marichromatium	Sulfide, thiosul-fate, sulfur, sulfite (only *Mch. gracile*)	Sulfur, IC	Sulfate	+/-	+/-
Rhabdochromatium	Sulfide, thiosulfate, sulfur	Sulfur, IC	Sulfate	ND	-
Thermochromatium	Sulfide, sulfur	Sulfur, IC	-	ND	-
Thioalkalicoccus	Sulfide, sulfur	Sulfur, IC	Sulfate	ND	-
Thiobaca	Sulfide	Sulfur, IC	Sulfate	ND	ND
Thiocapsa	Sulfide, thiosulfate, sulfur, sulfite (only in *Tca. litoralis* and *Tca. pendens*)	Sulfur, IC	Sulfate	+/-	+/-
Thiococcus	Sulfide, sulfur	Sulfur, IC	Sulfate	-	-
Thiocystis	Sulfide, thiosulfate (not in *Tcs. gelatinosa*,) sulfur, sulfite in some strains	Sulfur, IC	Sulfate	+ In some strains	+
Thiodictyon[a]	Sulfide, sulfur	Sulfur, IC	Sulfate	ND	-
Thioflaviococcus	Sulfide, sulfur	Sulfur, IC	Sulfate	ND	-
Thiohalocapsa	Sulfide, thiosul-fate, sulfur, sulfite	Sulfur, IC	Sulfate	-	+
Thiolamprovum	Sulfide, thiosulfate, sulfur	Sulfur, IC	Sulfate	-	+
Thiopedia[a]	Sulfide, sulfur	Sulfur, IC	Sulfate	-	-
Thiorhodococcus	Sulfide, thiosulfate, sulfur	Sulfur, IC	Sulfate	-	+/-
Thiorhodovibrio	Sulfide, sulfur	Sulfur, IC	Sulfate	ND	+
Thiospirilluma	Sulfide, sulfur	Sulfur, IC	Sulfate	ND	-
Ectothiorhodospiraceae					
Ectothiorhodospira	Sulfide, thiosu-lfate (Except *Ect. marismortu*,) sulfur, sulfite (except some species)	Sulfur, EC	Sulfate	+ In some species	+ In some species

Genus or Species	Utilization of sulfur substrates	Interme-diates sulfur compounds	End product	Sulfate assi-milation	Chemo-autotrophic growth
Ectothiorhodosinus[d]	Sulfide, thiosulfate	Sulfur, EC, Polysulfide, Sulfur, EC	Sulfate	ND	ND
Halorhodospira	Sulfide, thiosul-fate, sulfur only in *Hlr. halophila*	Sulfur, EC	Sulfur or Sulfate	+ in *Hlr. haloch-loris*	-
Thiorhodospira	Sulfide, sulfur	Sulfur, EC & IC	Sulfate	ND	-

Symbols: IC, Intracellular; EC, extracellular; ND, not determined/no data available; [a], 16S rDNA gene sequences are not available for this genera, hence their phylogenetic standings are not clear; [b], Type strains are not available; [c], Sulfate assimilation genes are present but sulfate assimilation has not been investigated in strain DSMZ 273. [d], The genus *Ectothiorhodosinus* has no standing in nomenclature. Data is taken from Frigaard and Dahl (2008); Overmann (2001); Imhoff (2003, 2005a, 2005b, 2008). Additional information is taken from Raymond and Sistrom (1969), Khanna and Nicholas (1982), Rees *et al.* (2002); Zaar *et al.* (2003); Gorlenko *et al.* (2004); Arunasri *et al.* (2005); Vogl *et al.* (2006).

(ii) *Ectothiorhodospiraceae:* Members of *Ectothiorhodospiraceae* grow well on sulfide as electron donors and form extracellular sulfur globules as intermediates (biphasic growth). Beneath the optimal alkaline conditions, *Ectothiorhodospiraceae* species form polysulfides as stable intermediates during oxidation of sulfide to sulfur (Then and Truper, 1984). Species are of *Ectothiorhodospira* are normally able to oxidize sulfide and sulfur to sulfate (**Table 6.2**). However, *Ect. halochloris* grown on elemental sulfur does not oxidize this compound to sulfate but reduces it to sulfide and polysulfide (Then and Truper, 1984). Almost all species of *Ectothiorhodospira* use thiosulfate. On thiosulfate, in all cases, sulfur globules are not detectable as an intermediate. For instance, in *Ect. shaposhnikovii* sulfur is formed only from sulfide but not from thiosulfate as an electron donor (Kusche, 1985). In other circumstances, the ability to form sulfur globules from thiosulfate may or may not occur. It is interesting to note that *Ect. shaposhnikovii* is not stimulated for growth by the thiosulfate but it oxidizes thiosulfate to tetrathionate and later to higher polythionates (Gogtova and Vainstein, 1981). Members of *Ectothiorhodospira* can grow on sulfite which is not seen in all other species (Imhoff, 2005b).

Some species of *Halorhodospira* oxidize sulfide to sulfur that are further oxidized to sulfate. Thiosulfate is used only by *Hlr. halophila* (Raymond and Sistrom, 1969) and poorly used by *Hlr. neutriphila* (Hirschler-Rea *et al.,* 2003). Though, sulfur can be used by *Hlr. halophila* . *Hlr. halochloris* and *Hlr. abdelmalekii,* they have the most restricted sulfur-oxidation capacities and unable to oxidize sulfur to sulfate (Then and Truper, 1984 .) *Thiorhodospira sibirica* utilizes sulfide and sulfur, but not thiosulfate as an electron donor for photoautotrophic growth (Bryantseva *et al.,* 1999b). Though *Ectothiorhodosinus mongolicum* prefers to grow as photo organoheterotroph and it can grow photolithotrophically with sulfide. Thiosulfate is used only in the presence of organic matter and finally oxidizes it to sulfate (Gorlenko *et al.*, 2004).

6.2.2. Green Sulfur Bacteria (GSB)

All members of GSB exhibit little variation in their ability to oxidize sulfur compounds (Vogl *et al.,* 2006). Almost all green sulfur bacteria are capable of oxidizing sulfide and elemental sulfur to sulfate (**Table 6.2**). However, *Chlorobium ferrooxidans* is an exception as it uses only Fe^{2+} and H_2 as photosynthetic electron donors (Heising *et al.,* 1999). Further, it does not grow on sulfide, thiosulfate, or elemental sulfur and that is even inhibited by a 1mM concentration of sulfide when grown on H_2. In general, green sulfur bacteria have a high affinity for sulfide and prefer sulfide over other sulfur substrates (Brune, 1995b). Primarily, sulfide is in completely oxidized to elemental sulfur and get deposited extracellularly as sulfur globules. These sulfur globules are oxidized completely to sulfate when the sulfide has been consumed. Some strains of *Chlorobium* and *Chlorobaculum* oxidize thiosulfate (Imhoff, 2003). Species of *Prosthecochloris cannot* utilize thiosulfate. *Chlorobaculum parvum* NCIB 8346 and *Chlorobium thiosulfatophilum* can use tetrathionate as electron donors (Khanna and Nicholas, 1982). Sulfite utilization by GSB has not yet been reported.

6.2.3. Purple Non-Sulfur Bacteria (PNSB)

Purple non-sulfur bacteria in general are poor in tolerating and using toxic sulfur compounds such as sulfide than purple sulfur bacteria. Although, a number PNSB is able to grow photolithoautotrophically with reduced sulfur compounds (Sander and Dahl, 2008).

Phototrophic *betaproteobacteria* belonging to *Rhodocyclales* and *Burkholderiales* cannot use reduced sulfur compounds as electron donors and sulfide inhibit the growth of these bacteria even at low concentrations (Imhoff *et al* ,. 2005). However, *sox* genes are present in the genome of *Rubrivivax gelatinosus* indicating their potential to oxidize thiosulfate to sulfate (Sander and Dahl, 2008).

Though all three taxonomic groups of the *Alphaproteobacteria*, species of PNSB such as *Rhodospirillales, Rhizobiales* and *Rhodobacterales* use reduced sulfur compounds, they differ considerably with respect to intermediates and final products of the oxidized sulfur compound. In *Rhodobacter capsulatus* and *Rhodospirillum rubrum*, sulfur is formed as the end product of sulfide oxidation is deposited outside the cells, while in others such as species of *Rhodovulum, Rhodopseudomonas palustris,* or *Blastochloris sulfoviridis*, sulfate is the end product of sulfide oxidation (Sander and Dahl, 2008 .)*Rhodomicrobium vanniellii* oxidizes sulfide to tetrathionate, whereas *Rhodobium marinum* oxidizes it to sulfur and thiosulfate when grown photomixotrophically (Imhoff, 1983). Thiosulfate is also used by many species including *Rhodopila globiformis* (Then and Truper, 1981) and oxidizes either to tetrathionate or completely to sulfate by

Rhodovulum species (Imhoff *et al.,* 2005). Still some species, such as *Rhodoplanes roseus* and *Rpl. elegans* use only thiosulfate but not sulfide as an electron donor (Hiraishi and Ueda, 1994b).

Sulfur is also used as a substrate by some species such as *Rhodopseudomonas julia* and most species of *Rhodovulum* (Imhoff *et al.,* 2005). None of these bacteria are able to use sulfite as the sole electron donor. The sulfur oxidation in purple bacteria is excellently reviewed by Sander and Dahl (2008).

6.2.4. Aerobic Bacteriochlorophyll-Containing (ABC) Bacteria

Generally, ABC bacteria unable to grow photolithoautotrophically on reduced sulfur compounds. Although, various representatives of this group can oxidize inorganic sulfur compounds. For instance, *Roseinatronobacter thiooxidans*, a strictly aerobic, obligately heterotrophic alkaliphile that can oxidize sulfide, thiosulfate, sulfite and elemental sulfur to sulfate in the presence of organic compounds. This bacterium appears to be a sulfur-oxidizing litho heterotroph as the efficiency of organic carbon utilization increases notably in the presence of thiosulfate (Sorokin *et al.,* 2000). Thiosulfate oxidizing activity was also shown by *Erythromicrobium hydrolyticum* strain E4 (1) and *Roseococcus thiosulfatophilus,* strain RB-7, in the presence of organic carbon and oxygen (Yurkov *et al.,* 1994). The genomes of ABC bacteria contain the genes *soxB* , *soxAX, soxYZ* and *soxCD* encoding for a multienzyme complex that has been well characterized (Friedrich *et al.,* 2005). In *Pcs. pantotrophus* the encoded Sox proteins catalyze the oxidation of thiosulfate to sulfate. However, little is known about the metabolism of inorganic sulfur compounds in these organisms. Traditionally, members of *Roseobacter* can degrade dimethyl-sulfoniopropionate (DMSP), which is produced in plenty by marine algae. Some isolates have been reported to carry out the major DMSP transformations in natural bacterial communities. Among these two competing pathways for DMSP degradation and the pathway for the incorporation of the sulfur moiety of DMSP into bacterial protein have been reported by Gonzalez *et al.* (2003) and Moran *et al.* (2003). In contrast to sulfide and thiosulfate, DMSP and dimethyl sulfide are more in aerobic marine environments (Liss, 1999).

Sulfur-dependent oxidative sulfur metabolism and chemolithotrophy have most excellently described in *Acidiphilium acidophilum* (Meulenberg *et al.,* 1992b). In ABC bacteria, the utilization of thiosulfate is initiated by the oxidative condensation of two molecules of thiosulfate producing tetrathionate. This step is catalyzed by the periplasmic thiosulfate, cytochrome *c* oxidoreductase (Meulenberg *et al.,* 1993). However, information was obtained that tetrathionate oxidation also takes place in the periplasm (Meulenberg *et al.,* 1993). Further, a tetrathionate hydrolase (de Jong *et al.,* 1997), trithionate hydrolase (Meulenberg *et al.,* 1992a) and sulfite cytochrome *c* oxido reductase (de Jong *et al.,* 2000) have been characterized from these ABC bacteria.

6.2.5. Filamentous Anoxygenic Phototrophic Bacteria (FAPB)

In some strains of *Chloroflexus aurantiacus* (Holo and Sirevag, 1986) and *Oscillochloris trichoides* (Keppen *et al.*, 2000), slow photoautotrophic growth with hydrogen or sulfide as electron donor has been reported but not in *Chloroflexus aggregans* (Hanada *et al.*, 1995). When sulfide is oxidized by cultures of *Chloroflexus* or *Oscillochloris*, elemental sulfur is deposited outside the cells, often affixed to the cells. In the general population of *Chloroflexus* , sulfide-dependent photoautotrophic growth has also been reported (Ward *et al.*, 1998). Four strains of *Oscillochloris trichoides* prefer to grow as photolithoheterotroph, although they grow photoautotrophically with sulfide as electron donors (Keppen *et al.*, 2000). All strains of *Oscillochloris trichoides* tolerate sulfide up to 4mM concentration and oxidize it to elemental sulfur. However, sulfide is not the preferred substrate for photosynthetic cultivation of filamentous anoxygenic phototrophic bacteria under laboratory conditions, sulfide oxidation could be important in certain natural environments. The natural habitats of the species of *Chloronema, Heliothrix* and *Roseiflexus* contain none or very low concentrations of sulfide (Garrity and Holt, 2001b).

6.2.6. *Heliobacteria*

Heliobacteria unable to grow phototrophically using sulfide as the sole electron donor, but sulfide is oxidized to elemental sulfur when added to the medium. *Heliobacterium sulfidophilum* and *H. undosum* are tolerant to sulfide (up to 2 mM concentration at pH 7.5) (Bryantseva *et al.*, 2000). However, the sulfate-fixing ability of *Heliophilum fasciatum* is inhibited by sulfide at concentrations above 0.1 mM (Ormerod *et al.*, 1996). Some strains perform assimilatory sulfate reduction, whereas others require a reduced sulfur compound such as thiosulfate, sulfide, methionine, or cysteine for biosynthetic purposes. Genome sequence information is available only for *Heliobacterium modesticaldum* Ice1 (ATCC 51547T), which can not grow on sulfate as the sole sulfur source but requires thiosulfate, sulfide, methionine, or cysteine (Kimble *et al.*, 1995).

6.3. OXIDATIVE SULFUR METABOLISM

The potentials of sulfur oxidation by different groups of anoxygenic phototrophic bacteria are briefly discussed.

6.3.1. Sulfur Oxidation Pathways

Sulfur oxidation is best described in the purple sulfur bacterium (*Allochromatium vinosum*) and the green sulfur bacterium (*Chlorobaculum tepidum*). The proposed pathway of sulfur oxidation in *Alc. vinosum* is illustrated in **Fig. 6.1** . Several enzymes involved in sulfur metabolism can readily be identified in genome sequences by sequence homology with well-known enzymes. The genome sequences of 15 strains of GSB are available and the occurrence of genes related to sulfur oxidation in these organisms are presented in **Table 6.3** (Venceslau *et al.*, 2014).

Table 6.3. Genome Sequenced Purple Sulfur Bacteria

Organism	S^{2-}	S^0	$S_2O_3^{2-}$	Reference
Chromatiaceae				
Allochromatium vinosum DSM 180^T	+	+	+	Weissgerber *et al.* (2011)
Thiorhodovibrio sp. 970	+	+	-	Unpublished
Lamprocystis purpurea DSM 4197	+	+	+	Imhoff (2001)
Thiocapsa marina 5811 DSM 5653^T	+	+	+	Caumette *et al.* (2004)
Thiocapsa sp. KS1	+	+	+	Unpublished
Thiohalocapsa ML1	+	+	+	Hamilton *et al.* (2014)
Thiorhodococcus sp. AK35	+	+	+	Unpublished
Thiorhodococcus drewsii AZ1 DSM 15006^T	+	+	+	Zaar *et al.* (2003)
Thiocystis violascens DSM 198^T	+	+	+	Imhoff *et al.* (1998)
Marichromatium purpuratum 984 DSM 1591T	+	+	+	Imhoff *et al.* (1998)
Thioflaviococcus mobilis DSM 8321^T	+	+	-	Imhoff and Pfennig (2001)
Ectothiorhodospiraceae				
Halorhodospira halophila SL1 DSM 244^T	+	+	+	Challancombe *et al.* (2013)
*Halorhodospira halochloris*A DSM 1059^T	+	-	+	Singh *et al.* (2014)
Thiorhodospira sibirica ATCC 700588^T	+	+	-	Bryantseva *et al.* (1999)
Ectothiorhodospira haloalkaliphila ATCC 51935^T	+	+	+	Imhoff and Suiling (1996)
Ectothiorhodospira sp. PHS-1	ND	ND	ND	Kulp *et al.* (2008)

Symbols: +, present, -, absent; ND, Not determined; Data is taken from Frigaard and Dahl (2008).

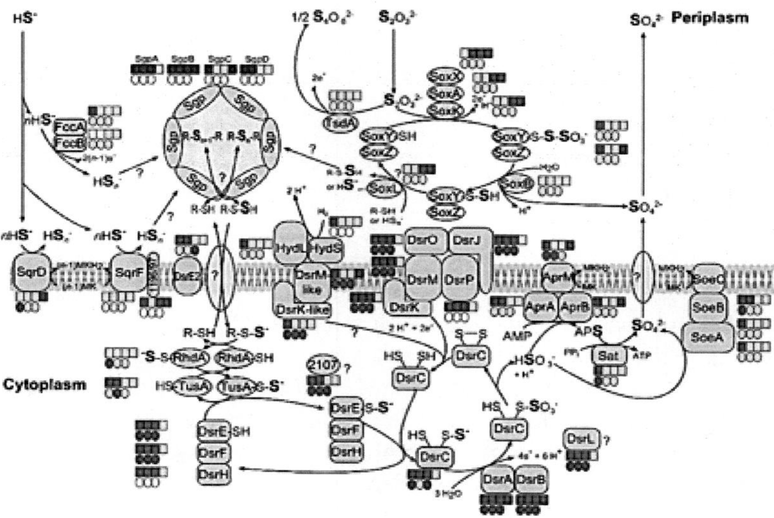

Fig. 6.1. Sulfur oxidation in *Allochromatium vinosum* (Weissgerber *et al.,* 2014a) .

The proteomic profiles (○) and transcriptomic profiles (□) are figured next to the respective proteins. Relative fold changes in mRNA levels above 2.0 (green) were considered as a significant improvement. Relative changes smaller than 0.5 (RED) were considered to deliberate significant decreases in mRNA levels. Relative fold changes between 0.5 and 2.0 (yellow) suggest unchanged mRNA levels. The same colour coding is used to changes the protein level. Here, values above 1.5 (green) and below 0.67 (red) were studied significantly. Those cases where transcriptomic data were not available or the respective protein was not observed in the proteomic approaches that are indicated by white squares or circles. Changes are illustrated that occurred upon a switch from photoorganoheterotrophic growth on malate to photolithoautotrophic growth on sulfide, thiosulfate, elemental sulfur, and sulfite (from left to right). Changes in sulfite were not determined on the proteome level.

i. Oxidation of Thiosulfate: Thiosulfate ($S_2O_3^{2-}$) is oxidized by a large number of photo- and chemotrophic sulfur-oxidizing bacteria. Two completely different pathways of thiosulfate oxidation and are being carried out. Initially, two thiosulfate anions are oxidized to tetrathionate, followed by the activation of the periplasmic *Sox* multienzyme system (Dahl *et al.*, 2008a). The multiple steps lead to complete oxidation of thiosulfate to sulfate. A few bacteria including *Alc. vinosum* both pathways coexist (Hensen *et al.*, 2006). The occurrence of genes related to the two pathways of thiosulfate oxidation in purple sulfur bacteria is summarized in **Table 6.4** .

ii. Oxidation of Thiosulfate to Tetrathionate: Simple oxidation of two molecules of thiosulfate to tetrathionate is widespread among prokaryotes

$$S_2O_3^{2-} \rightarrow S_4O_6^{2-}+2e-.$$

The reaction is not only well established as an intermediate step in the oxidation of reduced sulfur compounds to sulfate in various obligately chemolithoautotrophic bacteria (Muller *et al.*, 2004) and also studied in few PNSB (*Rhodomicrobium vannieli* and *Rhodopila globiformis*) and PSB along with *Alc. vinosum* (Frigaard and Dahl, 2009). Although, the well-described importance of tetrathionate formation in aquatic and terrestrial habitats (Sorokin, 2003), the membrane-bound *doxDA* encoding thiosulfate/quinone oxidoreductase from the thermo acidophilic archaeon *Acidianus ambivalens* was the only tetrathionate forming enzyme characterized on a molecular level for a long period (Muller *et al.*, 2004).

Genes homologous to *doxDA* do not occur in phototrophic prokaryotes. By contrast, a gene *tsd A* encoding a novel periplasmic 27.2 kDa diheme cytochrome *c* thiosulfate dehydrogenase was identified in *Alc. vinosum* (Denkmann *et al* ,. 2012). The crystal structure of the enzyme revealed the two typical class-I *c*-type cytochrome domains wrapped around two hemes. Heme 1 exhibits His/Cys iron coordination and constitutes the active site of the enzyme (Brito *et al.*, 2015). His/ Cys heme iron ligation is rare among prokaryotes, usually leads to a low redox potential of the corresponding heme (Bradley *et al.,* 2012) which has special significance in sulfur-based energy metabolism. In the oxidized state, Heme 2 iron is axially ligated by a histidine and a lysine residue (**Fig. 6.2** .) During reduction, a switch occurs at this heme from Lys to Met axial ligation. This change possibly affects the redox potential of Heme 2 and an important step during the reaction cycle (Brito *et al.*, 2015).

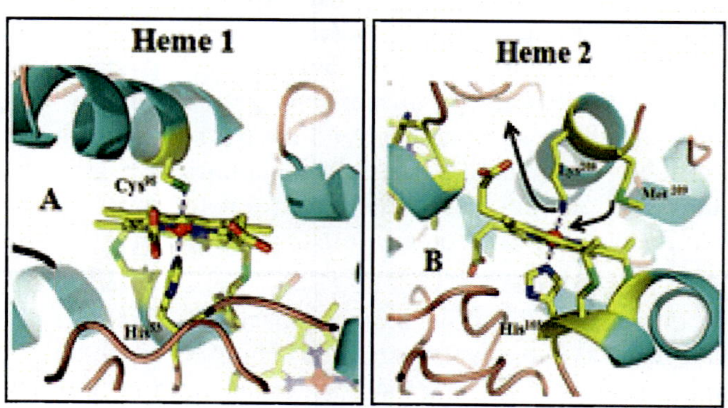

Fig. 6.2. Heme coordination of *Alc. vinosum* TsdA. A. Heme 1 is coordinated by His^{53} and Cys^{96}. B. Heme 2 is coordinated by His^{165} and Lys^{208}. Upon reduction, a ligand switch from Lys^{208} to Met^{209} occurs. Sulfur atoms are illustrated in green (Brito *et al.,* 2015).

Table 6.4. Genes Related to Thiosulfate Oxidation in Genome-Sequenced Purple Sulfur Bacteria

Organism	***Sox BXAKL***	***Sox YZ***	***Tsd BA***
Chromatiaceae			
Allochromatium vinosum DSM 180^T	Alvin_2167 –2171	Alvin_2111/12	Alvin_0091 (only *tsdA*)
Thiorhodovibrio sp. 970	-	Thi970DRAFT_01660	Thi970DRAFT_ 02035/36
Lamprocystis purpurea DSM 4197	WP_026199596, WP_020506632 –36	WP_020506368/67	-
Thiocapsa marina 5811 DSM 5653^T	ThimaDRAFT_ 4579–75	ThimaDRAFT_0728/ 29, 3536–37	-
Thiocapsa sp. KS1	THIOKS 1620009–14	THIOKS11770013/12	-
Thiohalocapsa ML1	WP_058554073 –79	WP_058556710/09	-
Thiorhodococcus sp. AK35	D779_4156–52	D779_3946–47	D779_1816 (*tsdBA* fusion)
Thiorhodococcus drewsii AZ1 DSM 15006^T	ThidrDRAFT_ 2415–19	ThidrDRAFT_2534/35	ThidrDRAFT_3922 (*tsdBA* fusion)

Organism	Sox BXAKL	Sox YZ	Tsd BA
Thiocystis violascens DSM 198^T	Thivi_2200, Thivi_3804–01	Thivi_3138/39	Thivi_3993 (*tsdBA* fusion)
Marichromatium purpuratum 984 DSM 1591^T	MARPU_ 05475–55	MARPU_13720/15	MARPU_02550 (*tsdBA* fusion)
Thioflaviococcus mobilis DSM 8321^T	-	-	Thimo_0460 (*tsdBA* fusion)
Ectothiorhodospiraceae			
Halorhodospira halophila SL1 DSM 244^T	Hhal_1939, Hhal_1948 (*soxXA* fusion), no *soxKL*	Hhal_1941/42	-
Halorhodospira halochloris A DSM 1059^T	M911_11275 (*soxB*), no *soxXAKL*	M911_11265/70	-
Thiorhodospira sibirica ATCC 700588^T	-	ThisiDRAFT_0337/36	-
Ectothiorhodospira haloalkaliphilic ATCC 51935^T	WP_025282121 (*soxB*), no *soxXAKL*	No *soxY,* WP_ 025282120 (*soxZ*)	-
Ectothiorhodospira sp. PHS-1	ECTPHS_10791, (*soxB*), no *soxXAKL*	ECTPHS_10801/796	-

Baits: *SoxBXAKL* (ADC63088-ADC63092), *SoxYZ* (*SoxY,* ADC63033; *SoxZ,* ADC63034) and *Tsd A* (ADC61061) from *Alc. vinosum, TsdB* from *Thiomonas intermedia* (D5WYQ6); Data is taken from Frigaard and Dahl (2008).

TsdA enzymes of different organisms exhibits various catalytic bias (Kurth *et al.* 2015), while the enzyme from *Alc. vinosum* is strongly biased toward activating thiosulfate oxidation (Brito *et al.*, 2015). Tsd A from *Campylobacter jejuni* acts initially as a tetrathionate reductase and allows the organism to utilize tetrathionate as an alternative electron acceptor for anaerobic respiration (Liu *et al.*, 2013). At present, it is largely not clear, whatever the redox carriers mediate the flow of electrons emerging from thiosulfate oxidation into respiratory or photosynthetic electron chains. In many organisms including *Thiomonas intermedia* , *Sideroxydans lithotrophicus*, and *Pseudomonas stutzeri* , *tsdA* is immediately prepared the way to a gene encoding another diheme cytochrome, TsdB (Denkmann *et al.*, 2012). Tsd B is not itself reactive with thiosulfate but accepts electrons from TsdA indeed when TsdA and TsdB do not emerge from the same source (Denkmann *et al.*, 2012).

In purple sulfur bacterium, *Marichromatium purpuratum*, TsdA and TsdB form a fusion protein with TsdB constituting the amino-terminal domain. Tsd BA fusion proteins that are also encoded in other members of the *Chromatiaceae* such as *Thiorhodococcus* sp. AK35, *Thiocystis violascens* , *Thiorhodococcus drewsii* and *Thioflaviococcus mobilis*(**Table 6.4**). Anyways, TsdBA fusions are an uncommon trait in purple sulfur bacteria. In *Alc. vinosum*, a *tsdB* gene is not present (Denkmann *et al.*, 2012). In *Alc. vinosum*, the protein with the closest relationship to *Thiomonas intermedia* or *P. stutzeri* TsdB is Alvin_2879. This cytochrome c_4 (previously cytochrome $C^{-}_{553(550)}$) is membrane-bound probably through the hydrophobic protein Alvin_2880 and has a positive redox potential of +330 mV (Cusanovich and Bartsch, 1969).

Another protein for accepting electrons from TsdA in purple APB is the high potential iron-sulfur protein (HiPIP). *Alc. vinosum* and *Marichromatium purpuratum* produce HiPIP and as this protein has a quite positive reduction potential (+350 mV) it is most suitable as an electron acceptor for TsdA (Bartsch, 1978).

iii. Oxidation of Thiosulfate to Sulfate: The Sox pathway of thiosulfate oxidation is a primary example for the oxidation of protein-bound sulfur atoms in the bacterial periplasm (Zander *et al.*, 2010). Among the organisms carrying this pathway, some store sulfur globules as intermediates such as *Alc. vinosum*, while others do not deposit sulfur such as *Paracoccus pantotrophus*. The Sox pathway in these two physiological groups appears to have one fundamental difference the involvement of the hemomolybdo protein SoxCD (**Fig. 6.3**). In non-sulfur-storing organisms, the recommended mechanism for sulfur oxidation requires four different proteins: SoxB, SoxXA, SoxYZ, and SoxCD (Friedrich *et al* ,. 2001). The heterodimeric SoxYZ protein acts as the central player and carries a pathway between covalently bound to a cysteine residue located near the carboxy- terminus of the SoxY subunit (Sauvé *et al.*, 2007). The c-type cytochrome SoxXA(K) catalyzes the oxidative formation of a disulfide linkage between the sulfane sulfur of thiosulfate and the cysteine of SoxY (Ogawa *et al* ,. 2008). The sulfone group is released hydrolytically as sulfate in a reaction synthesized by SoxB (Sauvé *et al.*, 2009). The following step is the oxidation of the SoxY-bound sulfane sulfur to a sulfone by the hemomolybdoprotein SoxCD and again the hydrolytic release of sulfate (Zander *et al.*, 2010).

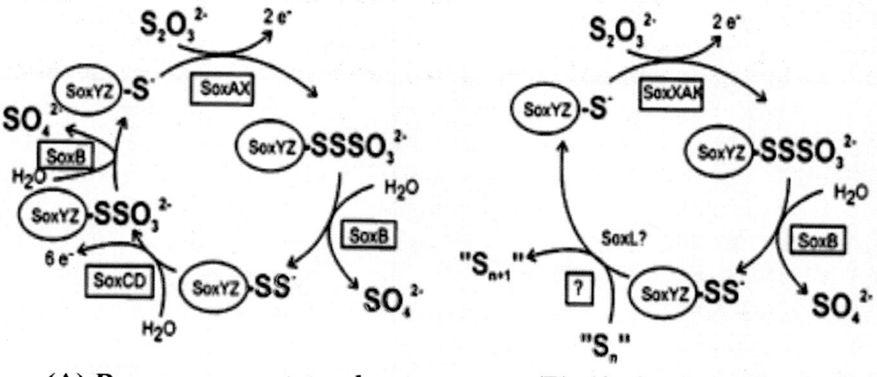

Fig. 6.3. Model of Sox-mediated thiosulfate oxidation in the periplasm of *Paracoccus pantotrophus* (A) and *Alc. vinosum* (B) (Sander and Dahl 2009).

In particular, organisms, forming sulfur as an intermediate, SoxCD is not present and the SoxY-bound sulfane sulfur is transferred to zero-valent sulfur stored in sulfur globules that exist in the periplasm by an unknown mechanism, probably by involving the rhodanese-like protein SoxL (Welte *et al.*, 2009). In *Alc. vinosum* ,*sox* genes are present in two clusters (*soxBXAK*L, Alvin_2167 to 2171 and *sox YZ*, Alvin_2111 and 2112) with *soxBXA* and *sox YZ* being essential for thiosulfate oxidation (Hensen *et al.*, 2006). The protein encoded by *sox K* gene has been identified as a subunit of a SoxXAK complex in the *Chlorobaculum tepidum* a green sulfur bacteria (Ogawa *et al.*, 2008) and possibly fulfils a similar function in purple sulfur bacteria.

iv. Oxidation of Sulfide: Sulfide oxidation is mediated by sulfide/quinone oxidoreductases (SQR) and a flavocytochrome *c* sulfide dehydrogenase (FccAB). In *Rhodovulum sulfidophilum*, a member of the *Rhodobacteraceae*, the Sox system is not only crucial for thiosulfate oxidation but also necessary for the oxidation of sulfide. In *Alc. vinosum* mutants deficient in either flavocytochrome *c* ,*sox* genes, or both sulfide oxidation proceeds with wild-type rates indicating the major role of SQR (Hensen *et al.,* 2006). All characterized SQRs are single-subunit flavoproteins associated with the cytoplasmic membrane (Marcia *et al* ,. 2009, 2010b). Based on the protein structure, six distinct SQR types were identified (Marcia *et al.*, 2010a). The nomenclature suggested by Frigaard and coworkers is followed (Gregersen *et al.*, 2011) to identify the multiple types of *sqr* genes also found in the similar organism.

Members of types SqrA, SqrB, SqrC, SqrE, and SqrF have been biochemically characterized (Shuman and Hanson, 2016). The SqrA is a functionally well-characterized enzyme from the purple non-sulfur bacterium *Rhodobacter capsulatus* (Gregersen *et al.*, 2011) which does not occur in green-and purple sulfur bacteria (**Table 6.5).** The same carry for Sqr E. The Sqr D and SqrF types appear to be widespread significantly in the family *Chromatiaceae*, whereas all the members of the *Ectothiorhodospiraceae* contain a gene encoding SqrB. The SqrF-type enzyme from *Chlorobaculum tepidum* recently has been shown to have a low affinity for sulfide and a high enzymatic turnover rate compatible with a function as a high sulfide habituated SQR (Eddie and Hanson, 2013). The primary reaction product of the SQR reaction is soluble polysulfide (Griesbeck *et al.*, 2002).

Table 6.5. Genes Related to Sulfide Oxidation in Genome-Sequenced Purple Sulfur Bacteria

Organism	sqr B	sqr C	sqr D	sqr F	Alvin_1196/97	fcc BA
Chromatiaceae						
Allochromatium vinosum DSM 180^T	-	-	Alvin_2145	Alvin_119 5	Alvin_1196/97	Alvin_1092/ 93
Thiorhodovibrio sp. 970	-	-	Thi970DR-AFT_02097	Thi970DR AFT_ 01366	Thi970DRAFT _01366	Thi970DRA FT_02375/76

Organism	sqr B	sqr C	sqr D	sqr F	Alvin_1196/97	fcc BA
Lamprocystis purpurea DSM 4197	-	-	WP_020502-950	WP_020508508	-	WP_020504674/73
Thiocapsa marina 5811 DSM 5653^T	-	-	ThimaDRAFT_0111	ThimaDRAFT_0710	-	ThimaDRAFT_3288/87 ThimaDRAFT_4140/39
Thiocapsa sp. KS1	-	THIOKS1270020	THIOKS12830013	THIOKS11770025	-	THIOKS11060009/07 THIOKS12860008/09
Thiohalocapsa ML1	-	-	WP_058557406	WP_058558070	-	WP_058554707/06
Thiorhodococcus sp. AK35	-	-	D779_2797	D779_4227	D779_0811/12	D779_2091/90
Thiorhodococcus drewsii AZ1 DSM 15006^T	-	-	ThidrDRAFT_1076	ThidrDRAFT_2526	ThidrDRAFT_3416/17	-
Thiocystis violascens DSM 198^T	-	-	Thivi_0965	Thivi_3129	-	Thivi_4480/81, Thivi_1815/16
Marichromatium purpuratum 984 DSM 1591^T	-	MARPU_16030	MARPU_12080	MARPU_13750	-	MARPU_11175/70
Thioflaviococcus mobilis DSM 8321^T	-	-	Thimo_0569	Thimo_1567	-	Thimo_2236/35
Ectothiorhodospiraceae						
Halorhodospira halophila SL1 DSM 244^T	Hhal_1665	-	-	-	-	Hhal_1331, Hhal_1163, no *fccA*
*Halorhodospira halochloris*A DSM 1059^T	M911_12365	-	-	-	-	M911_16440 no *fccA*
Thiorhodospira sibirica ATCC 700588^T	ThisiDRAFT_1596	-	ThisiDRAFT_0159	-	-	ThisiDRAFT_025283024, no *fccA*
Ectothiorhodospira haloalkaliphila ATCC 51935^T	WP_025282304	-	-	-	-	WP_025283024, no *fccA*
Ectothiorhodospira sp. PHS-1	ECTPHS_10089	-	-	-	-	ECTPHS_10469, no *fccA*

Baits: FccAB from *Alc. vinosum* (AAA23316-AAB86576), SqrA from *Aquifex aeolicus* (NP_214500), SqrB from *Halorhodospira halophila* (WP_011814451), SqrC from *Chlorobaculum tepidum* (NP_661023), SqrE from *Chlorobaculum tepidum* (NP_661769), SqrF from *Aquifex aeolicus* (NP_213539). Data is taken from Frigaard and Dahl (2008).

In several sulfide-oxidizing species, flavocytochrome *c* is present as a soluble protein in the periplasm or as a membrane-bound enzyme (Kostanjevecki *et al.*, 2000). This protein consists of a larger flavoprotein (Fcc B) and a smaller hemoprotein (Fcc A) subunit. It shows sulfide/cytochrome *c* activity *In vitro* (Bosshard *et al.*, 1986). FccAB is recorded in many PSB and GSB, though some species lack it (Frigaard and Dahl, 2009; Sander and Dahl, 2009). FccAB is beneficial as it has a high affinity for sulfide oxidation mainlyat very low sulfide concentration (Brune, 1995b).

v. Oxidation of Polysulfides: Polysulfides occur as the initial reaction product of the oxidation of sulfide in PSB (Franz *et al.*, 2009) and GSB (**Fig. 6.1**). However, the mechanism of conversion of polysulfides into sulfur globules is not clear (Marnocha *et al.*, 2016).

vi. Oxidation of External Sulfur: Several green- and purple sulfur bacteria are capable to oxidize externally supplied elemental sulfur. The sulfur of oxidation state zero mainly consists of S_8 rings and chain-like polymeric sulfur. Traces of S_7 rings are also observed. Elemental sulfur (S°) is insoluble in water and the mechanism of its binding activation, which takes up this substrate, is not clear. *Alc. vinosum* uses only the polymeric sulfur proportion of commercially available sulfur (Franz *et al.*, 2007). Soluble intermediates like sulfide, polysulfides, or polythionates do not appear to be formed. It is, therefore, not likely to mobilize elemental sulfur by purple sulfur bacteria include the excretion of soluble sulfur-containing substances that can be able to act on substrate far away from the cells (Franz *et al.*, 2009). Rather, direct cell-sulfur contact seems to be necessary for the uptake of elemental sulfur by *Alc. vinosum* (Franz *et al.*, 2007).

vii. Oxidation of Stored Sulfur to Sulfite: The oxidative degradation of sulfur deposits in phototrophic sulfur bacteria is still a major subject of research. Besides the comparatively well-characterized Dsr (dissimilatory sulfite reductase) system, a completely new pathway of sulfur oxidation involving a heterodisulfide reductase-like enzyme system is currently emerging (Dahl 2015; Venceslau *et al.*, 2014) and appears to be implemented in several phototrophic members of the family *Ectothiorhodospiraceae*(**Table 6.6** .)

Table 6.6. Genes Related to Stored Sulfur Oxidation in Genome-Sequenced Purple Sulfur Bacteria

Organism	*rhd-tusA-dsrE2*	*Dsr*	*hdr*-like	*lbpA* and biosynthesis
Chromatiaceae	-	-		
Allochromatium vinosum DSM 180^T	Alvin_2599-2601	Alvin_ 1251-1265	-	-
Thiorhodovibrio sp. 970	Thi970DRAFT_ 01020-22	Thi970DRAFT _01389-75	-	-
Lamprocystis purpurea DSM 4197	WP_040527681 (*rhd*), WP_02050658081	WP_020504936 -22	-	-
Thiocapsa marina 5811 DSM 5653^T	ThimaDRAFT_ 2004-2002	ThimaDRAFT _2915-2901	-	-
Thiocapsa sp. KS1	THIOKS11130019-17	THIOKS 12910007-23	-	-
Thiohalocapsa ML1	WP_058553476-74	WP_058553439-426	-	-
Thiorhodococcus	D779_3058-56	D779_2143-2157	-	-

Organism	*rhd-tusA-dsrE2*	*Dsr*	*hdr*-like	*lbpA* and biosynthesis
sp. AK35				
Thiorhodococcus drewsii AZ1 DSM 15006^T	ThidrDRAFT_0819-817	ThidrDRAFT _2036-2022	-	-
Thiocystis violascens DSM 198^T	Thivi_2161-2159	Thivi_0544-0556, 0030/29 (*dsrRS*)	-	-
Marichromatium purpuratum 984 DSM 1591^T	MARPU_12295-12,285	MARPU_10915-10,850, no *dsrS*	-	-
Thioflaviococcus mobilis DSM 8321^T	Thimo_2639-2637	Thimo_0143-0156, 1749 (*dsrS*)		
Ectothiorhodospiraceae				
Halorhodospira halophila SL1 DSM 244^T	no *rhd*, Hhal_1937 (*tusA*) no *dsrE2*	Hhal_1951–1963, no dsrRS	-	-
Halorhodospira halochloris A DSM 1059^T	M911_16670 (rhd) M911_11250–11,245	-	M911_11240-215	M911_11210—M911_ 11180
Thiorhodospira sibirica ATCC 700588^T	ThisiDRAFT_1542–1540	-	ThisiDRAFT_ 1539–1534	ThisiDRAFT_1533, (lbpA1), 2311–12, (dsrE2-lbpA2), 0082 (radSAM1), 0513 (lplA), 1283 (ger), 1859 (radSAM2)
Ectothiorhodospira haloalkaliphila ATCC 51935^T	no rhd, WP_025282116/15	-	WP_025282114–111 (hdrC1-hyp), WP_025280421 (hdrC2), WP_025282109 (hdrB2)	WP_025282108–105 (lbpA1- rad SAM1), WP_026623531 (lplA), WP_025282103/02 (ger-radSAM2)
Ectothiorhodospira sp. PHS-1	ECTPHS_01344 (rhd), 10,821,10,826	-	ECTPHS_10831–10,856	EctPHS_10861–10,891

The following abbreviations designate occurrence of a core set of several genes: *dsr*, dsrABCEFHMKLJOPNRS; *hdr*-like, hdrC1B1AhypC2B2; *lbpA* and biosynthesis, lbpA1-dsrE3-lpA2-radSAM1-lplA-ger-radSAM2. Baits: Rhd (Alvin_2599), TusA (Alvin_2600), DsrE2 (Alvin_2601), and Dsr proteins (U84760) from *Alc. vinosum.* Hdr-like proteins from *Acidithiobacillus caldus* SM-1: HdrC1, Atc_2352; HdrB1, Atc_2351; HdrA, Atc_2350; Hyp, Atc_2349; HdrC2, Atc_2348; HdrB2, Atc_2347; lipoate-binding protein LbpA and its biosynthesis, LbpA1 (Atc_2346), DsrE3 (Atc_2345), and LbpA2 (Atc_2344); radical SAM protein 1(Atc_2343); single-domain LplA (Atc_2342); geranyl geranyl reductase-like (Atc_2341); and radical SAM protein 2 (Atc_2340) from *Acidithiobacillus caldus* SM-1. Data is taken from Frigaard and Dahl (2008).

a. The Dsr System of Sulfur Oxidation: The best studied of the sulfur oxidation pathways operating in the cytoplasm is Dsr pathway (**Fig. 6.4**) involving the enzyme reverse dissimilatory sulfite reductase (DsrAB) (Dahl *et al.*, 2005). Low-molecular-weight organic persulfides such as glutathione amide persulfide have been proposed as carrier molecules transferring sulfur from the periplasmic or extracellular sulfur deposits in the cytoplasm (Frigaard and Dahl, 2009). It is not yet known how exactly the proposed persulfide carrier molecules are generated and whether specific enzymes are involved in this process or have transporters for such molecules be characterized from any sulfur-oxidizing prokaryote. An extensive Cys-SSH-based sulfur relay system exists in *Alc. vinosum*(**Figs. 6.1** and **6.4**) that traffics sulfur atoms stemming ultimately from sulfur stored in sulfur globules through a cascade of protein persulfide intermediates hosted on a rhodanese, *Tus A*, possibly *DsrE2A, DsrE*, and *DsrC* (Stockdreher *et al.*, 2012) to the active site of the enzyme sulfite reductase (Cort *et al.*, 2008; Dahl, 2015), the enzyme that catalyzes the formation of sulfite.

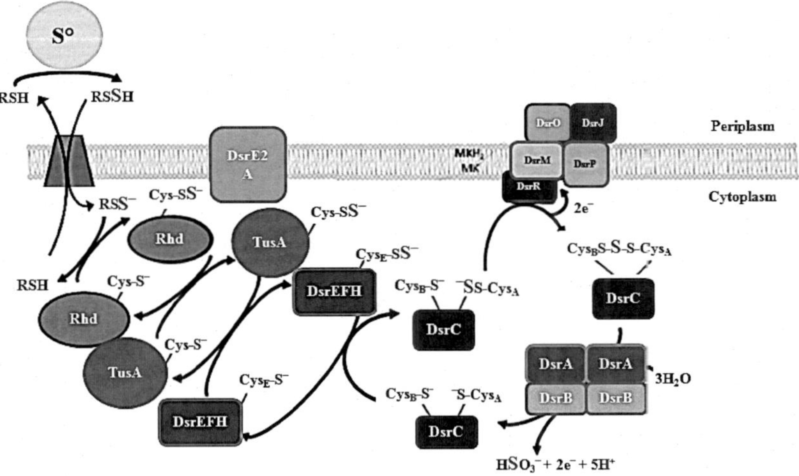

Fig. 6.4. Model of Dsr-mediated sulfane sulfur oxidation in *Allochromatium vinosum* integrating a sulfur-mobilizing function for Rhd, sulfur transfer functions for TusA and DsrEFH, and a substrate-donating function for DsrC.

A *rhd-tusA-dsrE2* or at least a *tusA-dsrE2* alignment occurs in all currently genome-sequenced sulfur oxidizers that possess the Dsr system (Venceslau *et al*,. 2014) (**Table 6.6**). In *Alc. vinosum,* the *tus A* , *rhd* and *dsr E2* genes follow a similar pattern of transcription was detected for the established cytoplasmic sulfane sulfur-oxidizing proteins (the Dsr system) (Stockdreher *et al.*, 2014). A *rhd-tusA-dsrE2*-deficient *Alc. vinosum* mutant strain failed to oxidize sulfur and grow in liquid or solid culture. Tus A is one of the major proteins in *Alc. vinosum,* gene *rhd* and also the *dsrE2A* encoded protein were identified as starting points for sulfur transportation to this protein (Stockdreher *et al.*, 2014). The rhodanese-like Rhdoprotein (Alvin_2599) catalyzes sulfur transport from thiosulfate or glutathione persulfide (GSSH) to cyanide *in vitro* and protein TusA was recognized as a protein accepting sulfane sulfur from the *Alc. vinosum* rhodanese. The DsrE2A protein is less well characterized and its role is not clear. It is firmly recognized that *Alc. vinosum* Tus A is an interaction partner of

DsrEFH, a hexameric protein arranged in a α2β2γ2 structure (Dahl *et al.*, 2008c). Sulfur transfer between Tus A and DsrEFH is reversible in *vitro* conditions (Stockdreher *et al.*, 2014). From DsrEFH sulfur is transferred to DsrC (Stockdreher *et al.*, 2012).

Dsr C protein works eminently as the physiological associate of the DsrAB sulfite reductase not only in sulfur-oxidizing but also in sulfate-reducing prokaryotes (Venceslau *et al.,* 2014). Dsr C is a member of the superfamily (DsrC/TusE/RpsA) and contains two strictly protect redox-active cysteines in a flexible carboxy-terminal arm (Cort *et al.*, 2008), Cys_A is the penultimate residue at the C-terminus and Cys_B is located 10 residues upstream (Venceslau *et al.,* 2014). When they combined in solution in their native, non-persulfurated state, DsrEFH and DsrC form a tight complex (Stockdreher *et al.,* 2012) and each $DsrE_2F_2H_2$ heterohexamer associates with either one or two DsrC molecules. Interaction of DsrEFH with DsrC is strictly dependent on the presence of DsrE-Cys_{78} and DsrC-Cys_A (Stockdreher *et al.,* 2012).

In **Fig. 6.1**, the concept is accomplished that persulfurated DsrC serves as the substrate for DsrAB and oxidation of DsrC-Cys_A-S^- by this enzyme is an idea to result in persulfonated DsrC (DsrC-CysA-SO_3^-) from which sulfite may be released by the formation of a disulfide bridge between Cys_A and Cys_B (Venceslau *et al.,* 2014). Furthermore, it is challenged by the very recent finding that a DsrC trisulfide in which a sulfur atom is bridging the two conserved cysteine residues are released as the product of the reverse reaction i.e., sulfite reduction upon catalysis by DsrAB from a sulfate reducer (Santos *et al.,* 2015). Another model is represented in **Fig. 6.4** synthesize the formation of a DsrC trisulfide probably by the action of the membrane-bound DsrMKJOP electron-transporting complex which contains the heterodisulfide reductase-like subunit DsrK that could well characterize the suggested reaction (Grein *et al.,* 2013). Electrons released during oxidation of the DsrC trisulfide to DsrC and sulfite possibly transferred to the protein Dsr L. This iron-sulfur flavoprotein is essential for sulfur oxidation in *Alc. vinosum* (Dahl *et al.*, 2005; Lübbe *et al.*, 2006). It contains striking sequence similarity to the electron-bifurcating subunit of the NfnAB complex from *Thermotoga maritima* (Demmer *et al.,* 2015), which have the theoretical capacity for reduction of NAD^+. However, experimental evidence is lacking. Knowing the exact mechanistic details of the interaction of DsrC, DsrAB and the other Dsr proteins is in fact that one of the most challenging points in research on sulfur-oxidizing prokaryotes.

b. The Hdr-Like System of Sulfur Oxidation: The *rhd-tusA-dsrE2* alignment does not occur only in all current genome- sequenced sulfur oxidizers that possess the Dsr system but also in a wide array of chemo- and phototrophic sulfur oxidizers which do not contain the Dsr pathway (Venceslau *et al.,* 2014). In these sulfur oxidizers, a gene cluster *hdrC1B1AhyphdrC2B2* encoding an array of proteins appears to be differential subunits of archeal heterodisulfide reductases is consequently present (Venceslau *et al.* 2014). As precised in **Table 6.6**, genes encoding a putative *hdr*-like complex occur in several members of

Ectothiorhodospiraceae. The typical organization of the *hdr*-like gene cluster is illustrated in **Fig. 6.5** for chemotrophic sulfur oxidizer such as *Acidithiobacillus caldus* and the phototrophic *Halorhodospira halochloris*. In almost all cases, the *hdr*-like gene set is instantly linked with *rhd-tusA-dsrE2* or *rhd-dsrE2* alignment promoting the notion that an Hdr-like protein complex is involved in the generation of sulfite from disulfide or even more likely protein-bound persulfide intermediates generated during sulfur oxidation. Heterodisulfide reductases (Hdr) are enzymes present in methanogenic archaea and catalyze the reduction of heterodisulfide, CoM-S-S-CoB formed in the last step of methanogenesis (Thauer *et al.,* 2008). The general thought of involvement of a Hdr-like complex and also specialized sulfur transferases (Rhd, DsrE, Tus A) in sulfite formation was first put forwarded by Quatrini *et al.* (2009) on the basis of microarray transcriptome profiling and quantitative RT-PCR analyses performed with *Acidithiobacillus ferrooxidans* ATCC 23270.

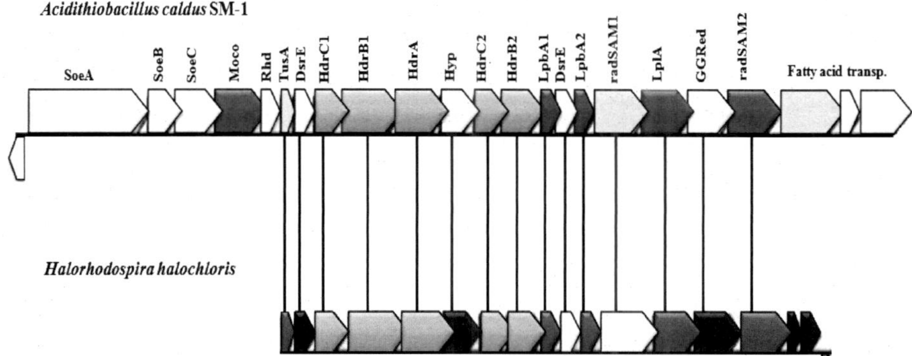

Fig. 6.5. Comparison of the *hdr*-like gene cluster in *Acidithiobacillus caldus* and *Halorhodospira halophila* .

In **Fig. 6.5**, the *soeABC* genes encode a membrane-bound cytoplasmically oriented sulfite-oxidizing enzyme. Rhd, Tus A, and DsrE are sulfur-mobilizing and sulfur-transferring proteins respectively. HdrB1, HdrB2, HdrC1, HdrC2, and HdrA possess similarities to the HdrABC subunits of soluble hetero disulfide reductases from methanogens. Hyp specifies a gene for a hypothetical protein. LbpA1 and LbpA2, lipoate-binding proteins and LplA, single-domain protein lipoate ligase or may be octanoylate transferase (Christensen *et al.* 2011). radSAM1 and radSAM2 are elucidated as radical SAM proteins and could insert sulfur into octanoylated LbpA. Many of the *hdr*-like gene clusters in sulfur oxidizers encode a protein commonly involved in fatty acid transportation, which has been played a role in the construction of lipoate precursors GGred, similarity to geranyl reductase, have been involved in the modification of constructed fatty acids before they are transferred into the specific lipoylation pathway.

In this context, proteomic studies showed high levels of Hdr-like proteins in the presence of reduced sulfur compounds (Ouyang *et al.,* 2013). In various reported studies upregulation in the presence of reduced inorganic sulfur compounds affected the *hdr*-like genes, as well as the sulfurtrans ferase genes, was commonly observed. Tight functional interaction of the encoded proteins is further recorded by the observation that genes *dsrE* to *hdrB2* constitute a single, different transcriptional unit in *Acidithiobacillus ferrooxidans* ATCC 16786 (Ehrenfeld *et al.,* 2013). Recent purification of a *Hdr*-like complex from membranes of *Aquifex aeolicus* (Boughanemi *et al.,* 2016) also confirmed this concept.

The presence of a *gvcH*-like gene encoding a lipoate-binding protein was first reported by Ehrenfeld *et al.* (2013). On the basis of striking sequence identity and the presence of a strictly preserve lysine residue is known to be required for lipoate attachment (Spalding and Prigge 2010). The name LbpA (lipoate-binding protein A) is recommended for this single lipoyl domain protein. Several sulfur oxidizers carry two copies of the gene indicating a functional dimer. However, genes encoding another DsrE-like sulfurtransferase and proteins that are likely to act in the biosynthesis of protein-bound lipoic acid are naturally found in organisms containing *hdr*-like genes but not in sulfur oxidizers following the Dsr pathway (**Fig. 6.5, Table 6.6**). In many cases, these genes are immediately linked with the *hdr* genes as shown in **Fig. 6.5**, and in few cases, they are observed at other places in the genome of the purple sulfur bacterium such as *Thiorhodospira sibirica* (**Table 6.6** .)

In brief Hdr-like enzyme system including dedicated sulfur transferases (Rhd, DsrE, Tus A) and also a dedicated lipoate-binding protein is a central and important element in the bioenergetics of sulfur-oxidizing prokaryotes lacking the Dsr system (Venceslau *et al*., 2014; Dahl, 2015). Although, genetic studies providing omics-derived concepts and biochemical studies that give reaction mechanisms are lacking. Hence, it is premature to suggest a more detailed model of the Hdr-like mechanism.

vi. Oxidation of Sulfite to Sulfate: The final step in the oxidation of reduced sulfur compounds is the oxidation of sulfite to sulfate. Sulfate formation from sulfite is energetically favorable and carried out by a wide range of organisms (Simon and Kroneck, 2013). In addition, several PSB can even use externally available sulfite as a photosynthetic electron donor. Two different pathways for sulfite oxidation have been well established in anoxygenic phototrophic sulfur bacteria (1) Direct oxidation and (2) Indirectoxidation (AMP-dependent oxidation via the intermediate adenylylsulfate) (adenosine-5′-phosphosulfate).

A. Oxidation of Sulfite in the Periplasm: Several sulfite-oxidizing enzymes catalyzing direct oxidation of sulfite are located outside the cytoplasmic membrane. The best-characterized enzyme belonging to this group, Sor AB, molybdopyranoprotein (Mo-PPT) cofactor-carrying subunit (SorA) and a monoheme cytochrome *c* (SorB) (Kappler and Bailey 2005). The second choice for the oxidation of sulfite in the periplasm is the Sox system. It has been shown that sulfite is accepted *in vitro* as a substrate of the reformed Sox system from the chemotrophic *Paracoccus denitrificans* (Sander and Dahl, 2009). Particularly,

Friedrich and co-workers proved this reaction to be independent of the presence of SoxCD, a molybdo haemoprotein catalysing the six-electron oxidation of SoxY-cysteine-bound persulfide to sulfone sulfur. Purple bacteria that form sulfur globules during thiosulfate oxidation contain the Sox system without the SoxCD proteins (Frigaard and Dahl, 2009). The presence of SoxB and SoxXA is not essential for sulfite oxidation in *Alc. vinosum* (Hensen *et al.*, 2006). However, the periplasmic sulfur substrate-binding protein SoxYZ is needed (Dahl *et al.*, 2013). Genes for this protein are present in PSB irrespective of the organisms' substratum, while the presence of SoxXA(K) and SoxB appears to be strictly linked to the ability of the cells to utilize thiosulfate (**Table 6.4** .)

B. Oxidation of sulfite in the Cytoplasm: Oxidation of sulfite to sulfate takes place in either of two pathways.

a. Indirect Pathway (via Adenosine 5′-Phosphosulfate): It is firmly established that a number of purple- and green sulfur bacteria oxidize sulfite in the cytoplasm using an indirect pathway through adenosine-5′-phosphosulfate (APS) catalyzed by APS reductase (AprBA) and ATP sulfurylase (Parey *et al.*, 2013) (**Fig. 6.1** .) An important function of AprM as a membrane anchor which allows the spatial and functional association of oxidative APS reductase with the membrane has been proposed. It has been recommended that AprM serves as an entry point into the membrane for the electrons released during the formation of APS from sulfite and AMP (Meyer and Kuever 2007). In the available complete genome sequences of members of the family *Chromatiaceae*, the same gene arrangements are seen in *Thiorhodovibrio* sp. 970 (**Table 6.7**). In *Thiocapsa marina* 5811, *Thiorhodococcus drewsii* AZ1, *Thiocystis violascens* DSM 198T, and *Thioflaviococcus mobilis* DSM 8321T, *sat* and *aprMBA* are not linked on the chromosome (**Table 6.7**). The occurrence of *aprMBA* has also been reported for *Thiococcus pfennigii* 4520. The *qmoABC* genes are not only present in sulfate-reducing prokaryotes (Ramos *et al.*, 2012) but also occur in many chemotrophic sulfur-oxidizing bacteria, GSB (Rodriguez *et al.* 2011) and in PSB (*Thiodictyon* sp. Cad16) (Gregersen *et al.*, 2011). In sulfur oxidizers, QmoABC acts as an electron acceptor for the electrons released during the generation of APS, which has a function analogous to that of Apr M. Hence, it is convincing to state that the electrons generated by the oxidative formation of APS from sulfite and AMP are delivered into the photosynthetic electron transport chain on the level of menaquinone either by AprM or by the well-characterized QmoABC complex (Grein *et al.*, 2013). It may be chiefly advantageous to be equipped with the Qmo-related electron-accepting unit for APS reductase. The presence of the HdrA- like QmoA in the Qmo complex opens the possibility that in reverse to the mechanism proposed for sulfate reducers .(Grein *et al.*, 2013). When electron bifurcation occurs that could result in the reduction of low potential electron acceptors like ferredoxin or NAD^+ simultaneously. Such a process could have a significant energetic advantage exclusively for chemolithoautotrophic growth that would result in lower energy demand for reverse electron flow.

Table 6.7 Genes Related to Sulfite Oxidation in Genome-Sequenced Purple Sulfur Bacteria

Organism	sat	aprBA	aprM	qmoABC	soeABC
Chromatiaceae					
Allochromatium vinosum DSM 180^T	Alvin_1118	Alvin_1120/21	Alvin_1119	-	Alvin_2491/90/89
Thiorhodovibrio sp. 970	Thi970DRA-FT_00961	Thi970DRA-FT_00963/64	Thi970DRAFT_00962	-	Thi970DRAFT_00955/56/57
Lamprocystis purpurea DSM 4197	WP_02619-9081	WP_020504060/59	-	WP_020504182/83/84	WP_020508252/50/49
Thiocapsa marina 5811 DSM 5653^T	ThimaDRAFT_0331	ThimaDRAFT_4551/52	ThimaDRAFT_4550	-	ThimaDRAFT_0331/30/29
Thiocapsa sp. KS1	THIOKS 1630011	-	-	THIOKS11840023/24/26	THIOKS12550011/12/13
Thiohalocapsa ML1	WP_058556167	WP_058557029/30	WP_058557048	-	WP_058553485/83/82
Thiorhodococcus sp. AK35	D779_2633	D779_0177/78	D779_0176	-	D779_1687/86/85
Thiorhodococcus drewsii AZ1 DSM 15006^T	ThidrDRAFT_3161	ThidrDRAFT_1495/96	ThidrDRAFT_1494	-	ThidrDRAFT_2883/82/81
Thiocystis violascens DSM 198^T	Thivi_0893	Thivi_3300/299	-	Thivi_3114/13/12	Thivi_4531/32/33
Marichromatium purpuratum 984 DSM 1591^T	-	-	-	-	MARPU_14905/00/14895
Thioflaviococcus mobilis DSM 8321^T	Thimo_1948	Thimo_1220/19	Thimo_1221	-	Thimo_1580/81/82
Ectothiorhodospiraceae					
Halorhodospira halophila SL1 DSM 244^T	-	-	-	-	Hhal_1934/35/36
Halorhodospira halochloris A DSM 1059^T	-	-	-	-	M911_11365/01475/01495
Thiorhodospira sibirica ATCC 700588^T	-	-	-	-	ThisiDRAFT_1377/0834/2148
Ectothiorhodospira haloalkaliphila ATCC 51935^T	-	-	-	-	WP_025282138, WP_025281124, WP_025280412
Ectothiorhodospira sp. PHS-1	-	-	-	-	ECTPHS_02816/02811/02806

Baits: APS reductase (AprMBA) from *Alc. vinosum* (U84759), Qmo proteins from *C. tepidum* (QmoA, CT0866; QmoB, CT0867;QmoC, CT0868), SoeABC from *Alc. vinosum* (SoeA, ADC63403; SoeB, ADC63402; SoeC, ADC63401). Data is taken from Frigaard and Dahl (2008).

b. Direct Pathway via Soe ABC: APS reductase pathway is not generally present in PSB nor it is essential in *Alc. vinosum* (Sanchez *et al.,* 2001) (**Table 6.7**). Currently, the membrane-bound iron-sulfur molybdoprotein SoeABC was identified as a key enzyme catalysing direct oxidation of sulfite to sulfate in the cytoplasm of *Alc. vinosum* (Dahl *et al.,* 2013). The function of SoeABC was demonstrated by strongly reduced specific oxidation rates for externally supplied sulfite and by vast excretion of sulfite into the medium throughout the oxidation of sulfide in *Alc. vinosum* SoeABC-deficient strains. The crude extract of a SoeABC-deficient *Alc. vinosum* lacked AMP-independent sulfite-oxidizing activity. However, information for the involvement of SoeABC in dissimilatory sulfur oxidation in *Alc. vinosum* was gathered in recent genome-wide transcriptional profiling (Weissgerber *et al.* 2013). Relative transcription of all three *Alc. vinosumsoe* genes are found to be increased about threefold when photolithoautotrophic growth occurs on sulfide or thiosulfate than photoorganoheterotrophic growth on malate and on thiosulfate, for *soe A, soe B*, and *soe C* , respectively. A modification in the same range was recorded for the genes encoding the enzymes of the APS reductase pathway while thiosulfate was replaced by malate. Meanwhile, relative transcript levels for the *sat-aprMBA* genes were 7.6–9.7 fold higher in the presence of sulfide compared to the presence of malate.

In *Alc. vinosum*, SoeABC is encoded by genes Alvin_2491 (*soeA* ,) Alvin_2490 (*soe B*), and Alvin_2489 (*soe C*). The protein consists of the 108.95 kDa molybdoprotein SoeA carrying one Fe_4S_4 cluster at the N-terminus, the 26.995 kDa iron-sulfur protein SoeB, that is expressed to bind four Fe_4S_4 clusters and a 35.715 kDa NrfD-/PsrC-like membrane protein (Simon and Kern 2008) along with eight transmembrane helices (Jormakka *et al.*, 2008). The AvSoeA and AvSoeB or SoeA or SoeB proteins from purple sulfur bacteria are not synthesized with cleavable TAT signal peptides which are usually present on the active site subunits of the biochemically strongly characterized periplasmic sulfur-metabolizing complex iron-sulfur molybdoproteins (**Table 6.7**), i.e., polysulfide and sulfur reductase (PsrABC, SreABC), thiosulfate reductase (PhsABC) or tetrathionate reductase (TtrABC) (Laska *et al.*, 2003). SoeA and SoeB are thus located in the cytoplasm and attached to the cytoplasmic membrane by interaction with SoeC. The holoprotein is therefore appropriate for the oxidation of sulfite accomplish in the cytoplasm. As it could be recognized that SoeABC and periplasmic Sor-type sulfite dehydrogenases belong to completely different families of molybdoenzymes.

Genes encoding proteins relevant to SoeABC are present in PSBandGSB (Frigaard and Dahl 2009) (**Table 6.7**). Especially, *soeABC*- like genes co-localize with *dsr* genes in many GSB and *Halorhodospira halophila* (Frigaard and Dahl 2009). The possession of the APS reductase pathway in addition to or instead of SoeABC could be beneficial as additional energy is gained by substrate phosphorylation in the ATP sulfurylase catalysed step by transporting the AMP moiety of APS onto pyrophosphate (Parey *et al.,* 2013).

6.4. SULFATE ASSIMILATION

Sulfate assimilation byAPB has been excellently reviewed by Frigaard and Dahl (2009). A few APB are very much specialized for living in habitats containing reduced sulfur compounds as they lack a sulfate reduction pathway. On the other hand, many versatile PSB and PNSB and some green bacteria are capable to assimilate and reduce sulfate in the absence of a reduced sulfur source. Among the FAPB, the ability to assimilate sulfate may or may not be present.

Assimilatory sulfate reduction pathway in *Alc. vinosum* is illustrated in **Fig. 6.6**. The pathway initiates with the uptake of sulfate through the membrane-bound components of a periplasmic substrate-binding transport system identical as in *E. coli*. Once sulfate is inside the cell, which gets activated to adenosine-5′-phosphosulfate by the enzyme ATP sulfurylase. Assimilatory ATP sulfurylases occur in two different forms,i) Heterodimeric CysDN type as in *E. coli* (Leyh, 1993) and ii) Homo-oligomeric Sat-related type as found in other organisms (MacRae *et al.,* 2001). Both types occur in anoxygenic phototrophic bacteria (Sander and Dahl 2009). The sulfate reduction pathway in *Alc. vinosum* do not involve in the formation of phosphoadenosine- 5′-phosphosulfate (Neumann *et al.,* 2000). Instead, a CysH-type iron-sulfur cluster binding APS reductase induces reductive cleavage of APS produces sulfite and AMP. Finally, sulfite is reduced to sulfide by an assimilatory sulfite reductase. Regarding *Alc. vinosum* , this enzyme is a ferredoxin- dependent CysI-type siroheme-(4Fe-4S) cluster-containing protein, which also occurs in other organisms (Dhillon *et al.,* 2005). This type of enzyme is frequent in APB (Frigaard and Dahl, 2009). Biosynthesis of cysteine requires the synthesis of *O*-acetyl-L-serine, which is again transformed to cysteine synthesized by cysteine synthase B (CysM) in a reaction that is dependent on the availability of sulfide.

Glutathione, thioredoxins, or glutaredoxins are reported to be reductants in this reaction (Nakatani *et al.,* 2012). The same reaction sequence is also possible for the assimilation of thiosulfate in *Alc. vinosum*(**Fig. 6.6**). Thiosulfate was previously detected intracellularly in *Alc. vinosum* (Franz *et al.,* 2009). As photoorganoheterotrophic growth of *Alc. vinosum* on organic acids like malate for the biosynthesis of sulfur-containing cell constituents is provided by the assimilatory sulfate reduction pathway in an energy-consuming process (Neumann *et al.*, 2000). At the same time sulfide is readily available without any input of energy under sulfur-oxidizing conditions. The presence of reduced sulfur compounds results in high levels of mRNA and protein for genes/proteins of central oxidative sulfur metabolism, while transcript and protein levels for genes/proteins involved in assimilatory sulfate reduction are negatively affected (Weissgerber *et al.,* 2013, 2014a). These responses are positively correlated to changes in metabolic pathways (Weissgerber *et al.,* 2014b). The same process may interplay between the processes of dissimilatory sulfur oxidation and assimilatory sulfate reduction which is operated in other APB also.

Assimilatory reduction of sulfate in a phototrophic organism such as plants and algae has been studied mostly from a nutritional point of view. The pathway of sulfate assimilation taking place in the following steps. 1. Activation of sulfate to form a sulfonucleotide (adenosine-5'-phosphosulfate), 2. Second activation step with ATP to form3'-phospho-adenosine-5'-phosphosulfate (PAPS), 3. A thioredoxin-dependent PAPS reductase then forms free sulfite, 4. sulfite reductase uses electrons from NAD(P)H or ferredoxin respectively and free sulfide is released and 5. Finally it is integrated into cysteine. It is distinguished that activated sulfate in the form of PAPS is very essential for sulfate transport reactions to numerous substances synthesize an array of sulfonated compounds such as sulfolipids (Negishi *et al.,* 2001).

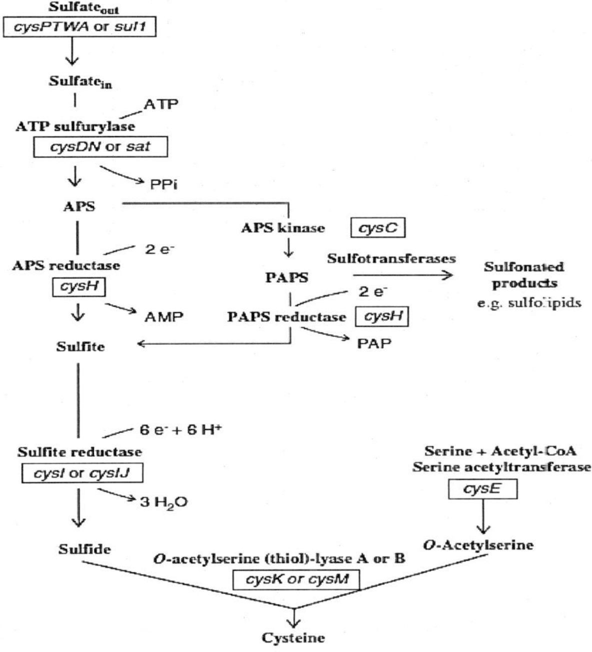

Fig. 6.6. Assimilatory sulfate reduction and cysteine biosynthesis.

The genes related to assimilatory sulfate reduction in the genomes of APB currently available are precised in **Table 6.8**. The enzymes involved in sulfate assimilation in these organisms vary even within the same family. Some anoxygenic phototrophic bacteria such as green sulfur bacteria are very much specialized for living in habitats with reduced sulfur compounds and completely lack a sulfate reduction pathway. Several members of the *Chlorobiaceae* lack genes for the enzymes catalyzing sulfate assimilation, while the specialized members of *Chromatiaceae* fossil fuels (REF) and *Ectothiorhodospiraceae* are also unable to grow photoorganotrophically and could not assimilate sulfate as sulfur source (Imhoff, 2005). On the other hand, many purple and some green bacteria are able to assimilate and reduce sulfate in the absence of reduced sulfur. *Chlorobium ferroxidans* DSMZ 13031^T is one such example of GSB that grows with sulfate as the sole sulfur source, which cannot utilize sulfide, thiosulfate, or elemental sulfur as electron donors (Heising *et al.*, 1999). In accordance with these findings, the *Chlorobium ferroxidans* genome encodes a single gene cluster which includes the assimilatory sulfate reduction genes *cysIHDNCG* and the sulfate permease genes *cysPTWA* that are transcribed in opposite directions. Assimilatory sulfate reduction genes contribute a high degree of sequence similarity with those of the *Clostridia* such as *Clostridium thermocellum* and *Desulfitobacterium hafniense* .

Both the green sulfur bacteria such as *Chlorobium phaeovibriodes* DSMZ265 and *Ptc. aestuarii* DSMZ 271^T) contain *cysC* and *cysN* homologs. However, in these two organisms, the genes occur in a cluster that appears to be involved in another aspect of sulfur metabolism. This cluster also contains a *cysQ* homolog that may encode a phosphatase acting on APS or PAPS (Neuwald *et al* ,. 1992). The homolog of the *ArsB/NhaD* superfamily of permeases that translocates Na^+ and different anions such as sulfate across the cytoplasmic membrane respectively in *Chl. Phaeovibriodes* DSMZ 265. It is further possible that *Chlorobium phaeovibriodes* DSMZ 265 and *Prosthecochloris aestuarii* DSMZ 271^T possess a system that processes APS or sulfite (or both) differently than in other green sulfur bacteria that actively excrete sulfate.

The ability to assimilate sulfate may or may not be present among filamentous anoxygenic phototrophic bacteria. However, *Chloroflexus aurantiacus* harbors a set of genes, the products of which can synthesize the complete activation and reduction of sulfate, *Chloroflexus aggregans* and *Roseiflexus* strains are expected to capable to restart and activate, but not capable of reducing sulfate (**Table 6.8**). Only one genome sequenced representative of the *Heliobacteria* ,*Heliobacterium modesticaldum* cannot grow on sulfate as the sole sulfur source. Consequently, genes for proteins involved in the reduction of (P)APS and sulfite are not present.

Table 6.8 Occurrence of Genes Related to Assimilatory Sulfate Reduction in Anoxygenic Phototropic Bacteria

Organism	**cysPTWA or *sull***	*sat*	*cysDN*	*cysH* [4Fe-4S]$^{2+}$	*cysH* [4Fe-4S]$^{2+}$	*cysJ*	*cysI*
Alphaproteobacteria							
Bradyrhizobiaceae							
Rhodopseudomonas palustris BisA53	*cysPTWA*	—	2 copies (both+APS kinase)	+	—	—	+
Rhodopseudomonas palustris BisB18	*cysPTWA*	—	2 copies (both+APS kinase)	+	—	—	+

Organism	cysPTWA or *sulI*	*sat*	*cysDN*	*cysH* [4Fe-4S]$^{2+}$	*cysH* [4Fe-4S]$^{2+}$	*cysJ*	*cysI*
Rhodopseudomonas palustris BisB5	*cysPTWA*	—	(+ APS kinase)	+	—	—	+
Rhodopseudornonas palustris CGA009 (ATCC BAA-98)	*cysPTWA*	—	(+ APS kinase)	+	—	—	+
Rhodopseudornonas palustris HaA2 (ATCC BAA-1122)	*cysPTWA*	—	(+ APS kinase)	+	—	—	+
Rhodobacteraceae							
Roseovarius sp. 217[a]	*sulI*	(+ APS kinase)	—	+	—	—	+
Roseovarius nubinhibens ISM (DSMZ 15170T)a	no *cys P*, *sulI*	(+ APS kinase)	—	+	—	—	+
Roseobacter sp. MED193[a]	*sulI*	(+ APS kinase)	—	+	—	—	+
Roseobacter denitrificans OCh 114	*sulI*	—	—	+	—	—	+
Rhodobacter sphaeroides ATCC 17025	3 copies *cysPTWA*	(+ APS kinase)	—	+	—	—	+
Rhodobacter sphaeroides ATCC 17029	2 copies *cysPTWA*	(+ APS kinase)	—	—	—	—	+
Rhodobacter sphae-roides(ATCC 55304)	*cysPTWA*	(+ APS kinase)	—	+	—	—	+
Rhodobacterales bacterium HTCC2654	*sulI*	(+ APS kinase)	—	+	—	—	+
Dinoroseobacter shibae DFL_12	*sulI*	+	—	+	—	—	+
Acctobacteraceue							
Acidiphillum cryptum JF-5[a]	—	—	(+ APS kinase)	+	—	+	+
Rhodospirillaceae							
Rhodospirillum rubrum ATCC 11170T	*cysPTWA*	—	(+ APS kinase)	+	—	—	+
Rhodocista centenaria ATCC 51521[a]	*cysPTWA*	—	+	+	—	(+)	+
Betaproteobacteria							
Comamonadaceaee							
Rubrivivax gelatinosus PM1	*cysPTWA*	—	+	+	—	—	+
Rubrivivax ferrireducens T118 (DSMZ 15236 [T])	*cysPTWA*	—	+	+	—	—	+
Gammaproteobacteria							
Congregibacter litoralis KT71[a]	*sulI*	—	—	+	—	—	+

Organism	cysPTWA or *sulI*	*sat*	*cysDN*	*cysH* [4Fe-4S]$^{2+}$	*cysH* [4Fe-4S]$^{2+}$	*cysJ*	*cysI*
Ectothiorhodospiraceae							
Alkalilimnicola ehrlichei MLHE-1 (DSMZ 17681^{T})	2 copies *sulI*	—	2 copies (1 copy + APS kinase)	+	+	—	+
Halorhociospira halophila SL1 DSMZ 254^{T})	*sulI*	—	(+ APS kinase)	+	—	—	—
Chromatiaceae							
Allochrornatium vinosum DSMZ 180^{Tb}		(dissimilatory)	+	+	—	—	+
Thiocapsa roseopersicina MI^{c}		unknown	unknown	—	+	+	+
Green Sulfur Bacteria *(Chlorobiaceae)*							
Chlorobium ferrooxidans DSMZ 13031^{T}	*cysPTWA*	—	(+ APS kinase)	+	—	—	+
Chlorobaculum tepidum TLS (DSMZ 12025^{T})	*cysPTWA*	—	+	+	—	—	+
FAP Bacteria *(Chloroflexaceae)*	—	(dissimilatory)	—	—	—	—	—
Chlorofiexus aggregans DSMZ 9485^{T}	*sulI*	2 copies (one + APS kinase)	—	—	—	—	—
Chloroflexus aurantiacus J-10-fl (DSMZ 35^{T})	*sulI*	+ (+ APS kinase)	—	+	—	—	+
Roseiflexus castenholzii DSMZ 13941^{Ta}	*sulI*	+ (+ APS kinase)	—	+	—	—	—
Roseiflexus sp. RS-1	*sulI*	+ (+ APS kinase)	—	+	—	—	—
Heliobacteria							
Heliobacterium modesticaldum ATCC51547^{Ta}	*cysPTWA*	—	+	—	—	—	—

This data is collected based on complete or draft genome sequence data, except for *Alc. vinosum* and *Tca. roseopersicina* for which only limited sequence data is available. The following purple bacterial genera are ABC bacteria: *Roseovarius, Roseobacter, Dinoroseobacter, Acidiphilium* and *Congregibacter. Cba. tepdium* is shown as an example of those GSB that are incapable of assimilatory sulfate reduction. [a], Genomes were analyzed by BLAST searches using the resources provided by Integrated Microbial Genomes (DOE Joint Genomes Institute, http://img.jgi.doe.gov) or by the Phototrophic Prokaryotes Sequencing Project (http://genomes.tgen.org). Data is taken from Frigaard and Dahl (2008), [a], Neumann *et al.* (2000), [b], Haverkamp and Schwenn (1999).

6.5. UPTAKE OF SULFATE

Assimilatory sulfate reduction initiates with the uptake of extracellular sulfate (**Fig. 6.6**). Most APB encodes genes homologous to *cysPTWA* indicating that they use a sulfate transport system which is similar to that of *Enterobacteria*

(**Table 6.8**). In some APB including *Halorhodospira halophila*, similar genes do not appear to be present. Alternatively, a gene encoding for a potential sulfate permease related to the high-affinity sulfate transporter of *Saccharomyces crevisiae* is present. While in *Halorhodospira halophila, Rhodobacterales* bacterium HTCC2654 and *Roseovarius* sp. 217, this gene are located in the instant vicinity of clusters of genes involved in oxidative sulfur metabolism (Sander and Dahl, 2008). As a result, the encoded transporter may also be involved in the export of sulfate produced in the cytoplasm as an end product of the oxidation of reduced sulfur compounds.

6.5.1. Activation of Sulfate

Once inside of the cell, the low reactivity of sulfate should be activated initially by the formation of a phosphate-sulfate anhydride bond in the compound adenosine-5'-phosphosulfate (adenylylsulfate, APS). This reaction is catalyzed by ATP sulfurylase. Because of the instability of APS and the unfavorable reaction equilibrium ($\Delta G' = +45kJ/mol$), the backward reaction was assumed to be terminated by cleavage of pyrophosphate by high pyrophosphatase activity and additional activation by APS kinase to PAPs (Leustek *et al.*, 2000). Thus, PAPS is strictly very important for sulfate transport reactions to numerous substrates (Negishi *et al.*, 2001).

Assimilatory ATP sulfurylases occur in two different forms: a heterodimeric cysDN type from *E.coli* (Leyh, 1993) and adistinct homooligomeric type found in bacteria, plants and fungi (MacRae *et al.*, 2001). The following polypeptides are related to the sat-encoded dissimilatory ATP sulfurylases of sulfate-reducing and chemotrophic sulfur-oxidizing prokaryotes (Beynon *et al.*, 2001). In both, heterodimeric and homooligomeric, ATP sulfurylase carboxy-terminal APS kinase domains may be present. In APB both types of ATP sulfurylases can be present. Genes encoding the Sat-type of ATP sulfurylase are recorded in members of *Rhodobacteriaceae* and FAPB, whereas other genomes contain *cysDN* genes. In maximum cases, *cysN* is fused with an APS kinase gene.

6.5.2. Reduction to Sulfite

During the incorporation of sulfur into biomolecules such as amino acids, sulfate in APS is reduced to sulfite and finally into sulfide. Depending on the organism this process may occur differently (**Fig 6.6**). At first, APS be phosphorylated by an APS kinase (*cysC* gene) using ATP to produce PAPS and ADP. The PAPS reductase (*cysH* gene) generates free sulfite using reduced thioredoxin as an electron donor. The other pathway involves the direct reduction of APS by APS reductase (*cysH* gene) (Bick *et al.,* 2000).

Assimilatory APS reductases and PAPS reductases are relevant on a sequence level. The name of the plant, algae and cyanobacteria type APS reductase was originally seen in a two-domain structure consisting of a reductase domain with a $(4Fe\text{-}4S)^{2+}$ cluster and a C-terminal thioredoxin/ glutaredoxin domain using glutathione as an electron donor. Distinctly, PAPS reductase was trusted to have only the reductase domain without an iron-sulfur cluster and to use thioredoxin as an electron donor. Almost all of the *cysH* genes present in APB encode proteins predicted to bind an iron-sulfur cluster (**Table 6.8**). Based on the above consideration it appears that these proteins reduce APS instead of PAPS.

6.5.3. Reduction of Sulfite and Sulfide

Sulfite is finally reduced to sulfide by an assimilatory sulfite reductase. The data precised in **Table 6.8** represent the ferredoxin-dependent sulfite reductase which is more common in purple bacteria than the NADPH-dependent enzyme containing two distinct polypeptides. At present *Thiocapsa roseopersicina* strain, M1 is the only purple bacterium for which the existence of a *cysIJ*-encode sulfide reductase has been established (Haverkamp and Schwenn, 1999).

6.5.4. Incorporation of Sulfide

The last step is cysteine biosynthesis which is catalyzed by either *O*-acetylserine (thiol)-lyase A or *O*-acetylserine (thiol)-lyase B encoded by the genes *cysK* and *cusM* respectively (Kredich, 1992, 1996). CysK and CysM synthesized proteins from *E.coli* are 43% identical. CysK synthesizes cysteine from *O*-acetylserine and sulfide. Whereas the CysM protein differs in that it can utilize thiosulfate in addition to sulfide. The reaction among *O*-acetylserine and thiosulfate produces *S*-sulfocysteine and is converted into cysteine by an undifferentiated mechanism. Both functionsare found in many species of *Chromatiaceae* and *Ectothiorhodospiraceae* and in PNSB, belongs to *β-proteobacteria* such as *Rhodocyclus tenius* and *Rubrivivax gelatinosus*. The tested PNSB belongs to *α- proteobacteria* exhibited only *O*-acetylserine (thiol)-lyase A activity which is failed to synthesize the reaction of *O*-acetylserine with thiosulfate. The two different *O*-acetylserine (thiol)-lyases were purified from *Alc. vinosum* (Hensel and Truper, 1981).

6.6. CONCLUSION

From the above discussion on oxidation of sulfur compounds in APB, it can be concluded that sulfide-quinone oxidoreductase (encoded by *sqr* gene) is considered to be important for the oxidation of sulfide, while flavocytochrome *c* is less widespread. Sox genes (*SoxXABYZ*) present both in green sulfur and purple sulfur bacteria are responsible for the oxidation of thiosulfate. Elemental sulfur is considered as an intermediate of sulfide and thiosulfate oxidation. All the thiosulfate utilizing green sulfur bacteria have a similar *sox* gene cluster (*soxJXYZAKBW*). The *soxCD* genes observed in other thiosulfate-utilizing organisms like *Paracoccus pantotrophus* are absent in green sulfur bacteria and purple sulfur bacteria studied. An assumed complex characterizes SoyYZ related thiosulphate binding SoxYZ complex is involved in the processing of an uncharacterized sulfur compound in GSB (Frigaard and Bryant, 2008a).

Siroamide-sulfite reductase (DsrAB) and proteins encoded by various *dsr* genes are essential for the oxidation of sulfur formed as an intermediate during sulfide or thiosulfate oxidation. Mostly green sulfur bacteria and purple sulfur bacteria use sulfide or elemental sulfur to contain dissimilatory sulfite reductase (*dsrNCBLEFHTMKJOP*) genes. However, *Chloroherpeton thalassium* appears to have an alternative enzyme system for elemental sulfur oxidation, the *dsr* genes appear to be involved in elemental sulfur utilization resulting in sulfite production in all other cases. As the APS reductase pathway is restricted only to some members of *Chromatiaceae* and *Chlorobiaceae,* sulfite oxidation in purple sulfur bacteria will require special attention. It has been shown that this pathway is also not essential to *Alc. vinosum*. In addition genes for a classical sulfite, acceptor oxidoreductase is present either in green sulfur bacteria or in purple sulfur bacteria.

Although phylogenies of some enzymes like DsrA protein are in agreement with the organismal phylogeny at the phylum level. The phylogeny of SQR protein is not congruent. Although the green sulfur bacteria are closely related, genome analyses clearly show that gene elimination and LGT substantially influence the distribution of sulfur metabolism genes both within GSB and among prokaryotes from other phyla.

Although assimilatory sulfur reduction occurs in many APB, two different types of sulfate transporters appear to be used. Sulfite is activated to adenosine 5-phosphosulfate through heterodimeric *cysDN* encoded ATP sulfurylase in members of the *Bradyrhizobiaceae, Acetobacteriaceae* and *Rhodospirillaceae* , while in some green sulfur bacteria and *Heliobacterium modesticladium* . The homo-oligomeric *sat*-encoded type of this enzyme occurs in *Rhodobacteraceae* and also present in the purple sulfur bacteria such as *Alc. vinosum* as well as filamentous anoxygenic phototrophic bacteria. In most of the sulfate assimilating APB, APS appears to be next reduced without further phosphorylation to PAPS. The generated sulfite is reduced to sulfide by assimilatory sulfite reductase which appears to be ferredoxin-dependent. A flavoprotein subunit is capable to use NADPH as an electron donor is not encoded in the sequenced genomes of APB.

■■■■

Chapter 7

ECOLOGY

Anoxygenic phototrophic bacteria are universally distributed in nature and are recorded in all kinds of stagnant water bodies. Ecological niches such as anoxic parts of waters and sediments receive light of sufficient quantity and quality which allows the phototrophic development (Imhoff, 2006c). The coloured blooms frequently formed by purple sulfur bacteria can be seen with the naked eye and naturally have attracted many scientists. Along with these macroscopically visible blooms, purple sulfur bacteria develop at a level not visible to the naked eye, as is indicated by positive enrichment cultures that were frequently obtained from sediments or water samples containing hydrogen sulfide, but not showing the visible mass of coloured bacteria (Imhoff, 2001a). Winogradsky (1888) gave a broad description of different morphological forms of PSB that he observed in samples from nature and raw cultures. Later, many investigators followed Winogradsky's descriptions for the identification of PSB in visible enrichments and natural habitats. Bavendamm (1924) gave the first detailed account of their ecology including his own findings in his monograph on colourless and PSB. Many reviews appeared on ecological aspects of the distribution of PSB in nature (Madigan, 1988; Imhoff, 1992; Imhoff, 2001a; Van Gemerden and Mas, 1995). Based on the availability of light, the light intensity and the concentrations of sulfide and 5 oxygen which form countercurrent gradients in most of the natural habitats of purple sulfur bacteria are the most essential properties which determine the development and distribution of phototrophic bacteria in nature. The depth at which the phototrophic sulfur bacteria develop is largely confined to the concomitant availability of light and sulfide. A few microbial activities which determine the slope and position of these gradients, such as oxygen production by algae and cyanobacteria and sulfide oxidation by phototrophic bacteria, show diurnal light-dependent fluctuations. Rests, such as sulfide production by sulfate-reducing bacteria and O_2 consumption during respiration of bacteria, the sulfide horizon rises at night and falls during the day. The motile phototrophic purple bacteria are able to follow the moving sulfide horizon. GSB form separate layers underneath layers of phototrophic purple bacteria and of algae and cyanobacteria under favorable conditions. However, a mixed population of purple- and green sulfur bacteria is observed frequently. Purple non-sulfur bacteria regularly accompany the mass development of phototrophic purple sulfur bacteria (Imhoff, 2006c) .

7.1. THE ROLE OF HYDROGEN SULFIDE

The ability of anoxygenic phototrophic sulfur bacteria to oxidize sulfide and other reduced sulfur compounds under anoxic conditions is one of their most characteristic and ecologically important properties (Imhoff, 2006c). One of the feasible final stages of anaerobic decomposition of organic matter is accomplished by sulfate-reducing bacteria. However the activity of sulfate-reducing bacteria in sediment is enough to raise the sulfide horizon into the photic zone, development of phototrophic sulfur bacteria is probable. Concentrations of sulfide and elemental sulfur and the relation of various phototrophic bacteria to these compounds are essential factors in determining the patterns of natural dominance and gainful competition. The specific importance of the bacterial affinities for these two sulfur compounds, their oxidation rates, the potentiality to utilize external elemental sulfur and to store elemental sulfur inside or outside the cells. The intracellular storage of elemental sulfur by *Chromatiaceae* gives them an advantage over the green GSB and PSB, which store elemental sulfur outside the cells. Elemental sulfur formed by *Chromatiaceae* is not accessible to other bacteria. On the other hand, the extracellularly formed elemental sulfur of green sulfur bacteria is further available free to all bacteria including *Chromatiaceae* .

7.2. THE RELATION TO OXYGEN

However, *Chromatiaceae* are anaerobic and accomplish photosynthetic metabolism under anoxygenic conditions in the presence of light. Some of the species are considered tolerant to oxygen. Several purple sulfur bacteria capable to carry out respiratory energy transformations by using oxygen (Kämpf and Pfennig, 1980). At least two strategies of these bacteria including *Chromatiaceae* enable them to grow, reproduce and successfully compete in the oxic environment are ecologically essential. They are (1) True adaptation by metabolic flexibility in energy conservation specifically use both photosynthetic and respiratory mechanisms and (2) The progress in the effective use of anoxic niches in an apparent oxic environment. The primary strategy facilitates the facultative respiring phototrophic bacteria to grow under diurnal oscillating conditions as part-time phototrophs and chemotrophs at the oxic/anoxic chemocline in common gradient systems (Imhoff, 2006c).

Specifically, *Thiocapsa roseopersicina* has high metabolic flexibility and is well equipped to take the advantage of changing conditions from oxic/dark to anoxic/light conditions and even drives respiration and photosynthesis simultaneously (Schaub and van Gemerden, 1994). Respiratory capabilities are not only favorable in environments with steep chemical and physical gradients, but also in situations of extended darkness where only respiration is possible. *Thiocapsa roseopersicina* is a good candidate to grow under chemolithotrophic conditions during prolonged dark periods (Imhoff, 2001a). The fascinating example of the second strategy is the strictly phototrophic PSB, *Marichromatium purpuratum* , which is adapted to anoxic niches in the bulk oxic environment and even achieved in niches of the open ocean. It was isolated from a strictly oxygen-demanding marine sponge that contained vast cell densities of this species (Imhoff and Truper, 1980). Subsequently, a *Didemnum* species show similar mass accumulations of this species (Imhoff, 1992) and recently, *Mcr. purpuratum* was isolated from the stomach of marine copepods which have a translucent chitin mantle and grow in oxic ocean waters (Proctor, 1997).

7.3. THE ROLE OF LIGHT

Both the quantity and quality of light play a major role in the development of APB depending on their pigment composition and light, which acts as a critical environmental factor. The light quality required by an APB is reflected in the absorption spectra of the photosynthetic pigment-protein complexes of a particular species or group of APB (Imhoff, 2006c). Most of the phototrophic purple bacteria have bacteriochlorophyll α, which have longer wavelength absorption maxima between 800 and 900 nm. However, only a few species have bacteriochlorophyll β, which have absorption maxima at 980–1035 nm. Light absorption in water masses and sediments follows completely distinct scenarios (Imhoff, 1992). In water, the probable light penetration is many meters, whereas in sediments due to low light absorption the enrichment of APB is restricted to the uppermost few millimeters (Taylor, 1964). Therefore, the availability of light severely limits the development of APB in sediments and often they form thin, coloured layers below a surface layer of cyanobacteria (Imhoff, 1992). Bacteriochlorophyll absorption is very essential in sediments, particularly sandy sediments, where infrared radiation deeply penetrates (Hoffmann, 1949) and favors the growth of bacteria that use photosynthetic pigments such as bacteriochlorophyll β with absorption maxima in the far-infrared region. Naturally, bacteria (*Thiococcus pfennigii*) with this pigment are particularly well adapted to sediments never covered by water or only by a thin layer as they occur in many coastal habitats. These bacteria are frequent to shallow coastal habitats and most of them are isolated from such locations, but they are rarely detected in deeper water bodies.

In deeper layers of water, the use of bacteriochlorophylls for light-harvesting is reserved by the significant absorption of infrared radiation (above 800 nm) by water. Hence, the role of carotenoids with absorption maxima at 450–550 nm, light-harvesting increases with the water depth. Okenone as a carotenoid containing purple bacteria is beneficial in deeper water layers due to the efficient light absorption capability by this pigment.

7.4. HABITATS OF PURPLE SULFUR BACTERIA

In addition to the essential and particular environmental factors described above physical and chemical properties such as temperature and salt concentration are also essential environmental factors suitable for the natural distribution of these bacteria. *Chromatiaceae* commonly habituated to moderate temperatures but also found in thermal springs (Madigan, 1986). Even there is evidence of their presence in sea ice (Petri and Imhoff, 2001). Several species of *Chromatiaceae* inhabit various types of freshwater, marine and hypersaline habitats. It should be noted that several investigators have recorded PSB in natural samples. But the identification of genus and species could not be determined and reported.

i. Freshwater Habitats: Visible mass of phototrophic sulfur bacteria observed temporarily in the anoxic parts of different kinds of freshwater habitats such as lakes, shallow ditches, ponds, stagnant water bodies in forests and botanical gardens and mountain lakes. In such habitats, growth usually starts during summer or fall, when a large amount of hydrogen sulfide is formed by sulfate-reducing bacteria from decaying plant material or organic pollution. Pink to purple-red blooms of *Chromatiaceae* is also detected inside the water or in the form of more-or-fewer rich patches that cover the upper or lower side of dead leaves or the mud surface.

Freshwater environments such as lakes are considered as largest and most important freshwater environments for the distribution of PSB. These habitats are more constant over long periods of time than others, thus favor more stable phototrophic communities. The blooms of purple- and green sulfur bacteria are generally observed several meters deep at the chemocline and are not visible on the water surface. Based on the hydrographic conditions, major types of lakes can be differentiated (Pfennig and Truper, 1992).

ii. Meromictic Lakes: Forever stratified meromictic lakes that have an anaerobic sulfide containing hypolimnion, which often consists of saline water supports the growth of APB at a more or less fixed depth for longer periods. However, fluctuations in the population density were observed in response to seasonal variation in sunlight intensity and temperature and also associated consequences containing a concentration of sulfide. A few meromictic lakes such as Ritomsee and Lake Beloved with well-developed blooms of purple-red layers of Switzerland (Duggeli, 1924; Kusnetzov, 1970; Sorokin, 1970) and Lake Cadagno of Switzerland (Bosshard *et al.,* 2000) support the blooms containing *Chromatium okenii.* Mixed populations of purple- and green sulfur bacteria are recorded in Suigetsu Lake and Kisaratsu Reservoir (Takahashi and Ichimura, 1968).

iii. Holomictic lakes: In these lakes that undergo mixing in spring and fall, the anaerobic and sulfide-containing hypolimnion attain the highest level and illumination increases during summer when blooms of phototrophic bacteria occur in the uppermost part of the hypolimnion and form coloured water layers or plates. In few lakes, this layer is dominated by GSB and in others, the green- and purple bacteria are observed either in mixed populations such as *Ancalochloris* ,*Pelochromatium* and *Lamprocystis* as in the Fluss-See or develop in individual layers, as in Wintergreen Lake. GSB containing gas vesicles thrive below layers of *Thiopedia* and *Thiocystis* species in holomictic lakes. The Lunzer Obersee, the Rotsee and Lake are examples of holomictic lakes, in which members of *Chromatium okenii* , *Thiopedia rosea* and *Lamprocystis roseopersicina* dominant in purple-red layers. In the Banyoles karstic area of northeastern Spain, a number of small lakes with mass development of purple sulfur bacteria occur. Lake Cisó was dominated by blooms of *Chromatium minus* (Pedros-Alio and Guerrero, 1991).

iv. Shallow Water Lakes: In shallow lakes with a maximum depth of 15 m, the anaerobic, sulfide-containing zone is confined to the mud sediment and does not extend significantly into the water layers and mass development of PSB are restricted to the sediments. Kolksee and Edebergsee are the best examples, in which blooms of *Thiopedia* and *Pelochromatium* are dominated. However, in lakes like Lunzer Mittersee, the population of *Chromatium okenii* and *Lamprocystis* are dominated. Similarly, in Mulizne Lake (Poland) with a *Thiopedia* bloom (Czeczuga, 1968), the monomictic Zaca Lake with a *Thiopedia* bloom (Folt *et al.*, 1989) and in Haruna Lake with a *Chromatium* bloom dominate in the waters (Takahashi and Ichimura, 1968).

v. Sulfur Springs: Such springs with constant sulfide supply and exposed to the light are suitable habitats for phototrophic sulfur bacteria (Winogradsky, 1888). Mass accumulation of green- and purple sulfur bacteria were recorded in the effluents of sulfur springs in Poland. Green sulfur bacteria occur predominantly in water containing high sulfide concentration and in warm sulfur springs (35–44°C) in Japan. At temperatures up to 43°C, growth of *Allochromatium vinosum* strains and at a temperature of 48–50°C growth of *Thermochromatium tepidum* occurs which is not common. The moderately thermophilic purple sulfur bacterium was isolated from the Stygian Springs of Yellowstone Park (Madgan, 1986).

vi. Wastewater Ponds: Anaerobic stabilized ponds of wastewater plants are ideal habitats for PSB. Purple sulfur bacteria readily develop in waste treatment lagoons with a high organic load treated in shallow ponds (1–2 m in depth) support the development of anoxic conditions and sulfide formation by sulfate-reducing bacteria are ideal for blooms of purple sulfur bacteria in spring and fall. On the other hand, during summer the ponds usually turn green by unicellular algae. The most generally recorded purple sulfur bacteria in waste treatment systems are *Thiocapsa roseopersicina* ,*Thiocapsa rosea* ,*Thiocapsa pendens* and *Thiolamprovum pedioforme* (Eichler and Pfennig, 1986). In addition to these, *Thiocystis violacea* and *Allochromatium* species are also recorded in small numbers.

vii. Marine Habitats: The ocean shorelines with the numerous ecological niches can be treated as the most ideal places for the development of *Chromatiaceae* in the marine environment and a large number of visible phototrophic bacteria were recorded (Imhoff, 2006c). The development of purple sulfur bacteria is observed wherever little water is present such as small splash water ponds along rocky shores, seawater pools and small puddles of salt marshes, closed bays and large lagoons, sediments of sandy beaches, salt marshes and tidal flats and in muddy masses of decaying algae or seaweeds. In these locations, bacterial sulfate reduction is progressed by the degradation of decaying organic materials and the presence of large amounts of sulfate (Truper, 1980). Such habitats are also named "beach sulfurate" (Suckow, 1966) because they build a complete sulfur cycle including bacterial sulfate reduction and sulfide oxidation by phototrophic bacteria and chemotrophic sulfur bacteria. *Chromatiaceae* are found in these habitats. Mass development of purple sulfur bacteria and were recorded on the coasts of Denmark along the ocean coasts all over the world (Imhoff, 2001a).

It is supposed that the metabolic variability is a property to explain their wide occurrence in highly dynamic and unstable marine coastal habitats, which are selectively advantageous. The most flexible phototrophic purple bacteria frequently dominate here (Imhoff, 2001a). This flexibility of phototrophic bacteria may be due to (1) The ability to conserve energy and provide photosynthetic electron donors especially sulfide and thiosulfate, (2) The high potential of photoheterotrophic growth together with the assimilation of sulfate as sole sulfur source and (3) The ability to grow chemotrophically under oxic conditions in the dark, either autotrophically or heterotrophically. *Allochromatium vinosum, Marichromatium gracile, Thiocystis violacea* and *Thiocapsa roseopersicina* were regularly recorded in these environments. *Allochromatium warmingii, Thiocystis violascens, Allochromatium minutissimum, Thiocystis minor ,Thiocapsarosea, Marichromatium purpuratum* and *Isochromatium buderi* are the other bacteria recorded in these environments (Imhoff, 1992; Imhoff, 2001a).

viii. Sediments at Marine Shorelines: As the process of sulfate reduction is common in the sediments, phototrophic sulfur bacteria are abundant in these sediments. However, the production of sulfide is not sufficient to penetrate into the water, the growth of these bacteria is confined to the sediment and form thin coloured layers within the top millimeters of the sediments. Sandy beaches with such coloured layers are called the "Farbstreifensandwatt". Microbial mats of phototrophic bacteria including layers of PSB are common in sheltered areas of the Waddensea and of salt marshes.

An excellent example is observed in the microbial mats of the Great Sippewissett Salt Marsh (Cape Cod, MA, United States), where laminated microbial mats of uncommon thickness frequently occur during summer (Nicholson *et al.*, 1987). The mats about 10 mm thick and contain four to five distinctly coloured layers are formed in these sandy intertidal sediments. Purple sulfur bacteria of the central pink layer and the directly underlying peach-colored layer were identified. This pink layer is dominated by *Thiocapsa roseopersicina* and the peach layer is dominated by *Thiococcus pfennigii* based on the presence of bacteriochlorophyll β and the bundles of tubular intracellular membranes in the coccoid cells. A second bacterium, *Thioflavicoccus mobilis*, resembles *Thiococcus pfennigii* in many aspects but motile and contains bacteriochlorophyll β was isolated from this layer (Imhoff and Pfennig, 2001b). Several other species of purple non-sulfur bacteria were recorded in this water. In the peach-colored layer of the mats, small spirilloid bacteria *Rhodospira trueperi* containing bacteriochlorophyll β was recorded (Pfennig *et al.,* 1997). Another purple non-sulfur bacterium, *Roseospirillum parvum* was also isolated from this habitat (Glaeser and Overmann, 1999). Similarly, a large number of purple- and green sulfur bacteria are reported to originate from this salt marsh.

ix. Coastal Waters, Lagoons and Fjords: There are many reports were reported on the development of vast blooms of phototrophic bacteria called "red waters" in shallow coastal waters such as lagoons saline lakes and anoxic basin. Lake Faro in Sicily (Truper and Genovese, 1968), Veliko and Male Jezero on the Dalmatian island of Mljet (Cviic, 1960), the Bay of Tunis and Bietri Bay of the Ebrie Lagoon, Ivory Coast (Stirn, 1971; Caumette, 1984), the Etang du Prevost near Montpellier, France (Caumette, 1986) lagoons at Messolonghi and Aitolikon in Greece (Hatzikakidis, 1953) and the Solar Lake on the Sinai Peninsula (Cohen *et al.,* 1977), which harbor a variety of phototrophic bacteria are good examples of this category (Gorlenko *et al.,* 1978; Imhoff, 2006c).

x. The Black Sea: The oxygenated waters of the open ocean are devoid of phototrophic sulfur bacteria. However, stratified marine meromictic enclosed basin habitat in the Black Sea is not a part of the open ocean. However, the anoxic layer did not reach the photic zone was observed to contain anoxygenic phototrophic bacteria such as species of *Chromatiaceae* and green sulfur bacteria in the bottom sediments at 600 and 2240 m depth (Kriss and Rukina, 1953). Approximately 10 years later, the situation turned to be distinct when the chemocline reached horizons (68 to 90 m) receiving low-intensity light, brown-colored GSB were detected (Overmann *et al.,* 1992a). However, light intensities at the depth of the chemocline were not sufficient to support the growth of PSB.

xi. Sea Ice: Sea ice that is oxic does not have anoxic niches may not support the growth of anoxygenic phototrophic bacteria. But 16S rDNA analyses revealed the presence of species of *Chromatiaceae* as a major component in a distinct layer suggesting that sea ice may contain potential niches for anoxygenic phototrophic bacteria. This may be attributed to their capacity to tolerate oxygen and carry respiration to live in such a habitat (Petri and Imhoff, 2001).

xii. Salt and Soda Lakes: Several habitats of phototrophic bacteria in the coastal zone contain brackish waters. Shallow waters of splash water ponds, coastal lagoons and closed basins receive deep illumination from the sun resembling evaporation of water causes an increase in inorganic salts and organic matter where anoxic conditions appear and encourage the development of anoxygenic phototrophic bacteria. Many isolates from such places are tolerant to a wide range of salt concentrations and some require or tolerate salt concentrations above seawater salinity (Imhoff, 2001a, 2006c). Splash water ponds along rocky coasts, where salt concentration fluctuates were recorded to contain phototrophic sulfur bacteria. Thalassohaline waters which resulted from evaporation pools of marine waters made marine salterns. A few prominent examples of such habitats are Great Salt Lake (Utah), Dead Sea (Israel), and soda lakes of the Wadi Natrun in Egypt reported to have ecological niches for halophilic phototrophic purple bacteria (Imhoff, 2001a). Numerous species of phototrophic purple sulfur bacteria have been described from marine salterns (Caumette, 1993; Caumette, 1994), alkaline soda lakes in the Egyptian Wadi Natrun (Imhoff and Truper, 1981), Russian soda lakes in Siberia and Mongolia (Bryantseva *et al.,* 1999; Bryantseva *et al.,* 2000) and the Solar Lake (Caumette *et al.,* 1997).

The simple architecture of the photosynthetic apparatus of anoxygenic phototrophic bacteria depends on electron donors with standard potentials more negative than that of water results in comparatively narrow, ecological niches in the extant ecosystem.

7.5. HABITATS OF PURPLE NON-SULFUR BACTERIA

Purple non-sulfur bacteria are widely distributed in nature such as in ponds, marine coastal sediments, sewage, swine, waste lagoons and even in earthworm dropping (Larimer *et al.,* 2004).

Members of *Rhodospirillaceae* occupy a wide range of habitats such as oil-contaminated soil, discarded load and fermentative tank. Some of the purple sulfur bacteria occur in aquatic environments such as fresh stagnant water and anoxic, acids or saline, petroleum-contaminated seawater, ocean and saline soil. *Rhodospirillum* and *Pararhodospirillum* species are recorded in stagnant and anoxic freshwater exposed to light (Kumar *et al.,* 2008 and Lakshmi *et al.,* 2014), while *Rsp. rubrum* was isolated from freshwater habitats (Lakshmi *et al.,* 2014). On the other hand, species of *Pararhodospirillum*(*P. photometricum* and *P. sulfurexigens*) have been isolated from freshwater, while *P. oryza*e was recorded in the soil of paddy fields (Lakshmi *et al.,* 2014). Recent researches have shown that anoxic zones of saline or hypersaline environments such as seawater, brines and salt lakes exposed to light are the preferred habitats of species of *Rhodovibrio* (Blazejak *et al.,* 2006; Maturrano *et al.,* 2006; Boutaiba *et al.,* 2011; Atanasova *et al.,* 2012; Liu and Liu 2013; Schneider *et al.,* 2013 and Ntougias 2014). *Rhodovibrio salinarum* and *Rhv. sodomense* has been recorded in saltern lakes (Imhoff *et al.,* 1998).

Members of *Phaeospirillum* are recorded in freshwater, mud, rhizosphere soil, top layers of aquatic sediments and salt marsh sediments. Different species of *Phaeospirillum* such as *Phs. fulvum* and *Phs. molischianum* (Imhoff *et al.,* 1998) were isolated from mud ditches and sewage ponds. *Phs. chandramohanii* was recorded in a freshwater reservoir in Mudasarlova, India (Kumar *et al.,* 2009). Water or mud from Nelapattu Bird Sanctuary, India contained *Phs. tilakii* (Raj *et al.,* 2012). Species of halophilic *Roseospira* were found to grow in anoxic sediments such as brackish lagoons and saline springs (Guyoneaud *et al.*, 2002). *Ros. marina, Ros. thiosulfatophila* and *Ros. navarrensi* were isolated from anoxic sediments exposed to light and microbial mats of coastal and marine environments. Other species such as *Ros. mediosalina, Ros. visakhapatnamensis* and *Ros. goensis* were isolated from mari Chakravarthy *et al.,* 2007; Guyoneaud *et al.*, 2002). *Rhodospira trueperi* was isolated from a microbial mat in a salt marsh in Massachusetts, USA (Pfennig *et al.,* 1997). *Phaeovibrio sulfidiphilus* only validly published species was isolated from brackish water at Nagapattinam, India (Lakshmi *et al.,* 2011). Species of *Rhodocista* were recorded in freshwater ecosystems. *Rhodocista hanoiens* ne water (Kalyan *is* was isolated from a wastewater pond in Vietnam (Do *et al.,* 2007), while *Rcs. centenaria* was isolated from hot spring (55 °C) of Wyoming USA (Favinger *et al.,* 1989). *Rcs. pekingensis* was isolated from activated sludge of water treatment plant (Zhang *et al.,* 2003). *Rhodopila globiformis* was isolated from the acidic water and sludge sample from a warm sulfur spring at the Gibbon River, Yellowstone Park, Wyoming, USA (Imhoff *et al.,* 1984). *Rhodovastum atsumiense* has been isolated from the paddy soil of Japan (Okamura *et al.,* 2009).

Species of *Rhodomicrobium, Blastochloris* and *Rhodoplanes* belonging to the family *Hyphomicrobiaceae* are recorded in mud, water ponds, lakes and Wastewater. *Rhodomicrobium vannielii* was isolated from a Hot Spring (50-58°C) at Gadek, Malacca, Malaysia (Hamzah *et al.,* 2006), while *Rhodomicrobium udaipurense* was isolated from a freshwater stream sample collected from Udaipur, India (Venkata Ramana *et al.,* 2013). *Blastochloris viridis* was isolated from freshwater, while *Blc. sulfoviridis* and *Blc. gulmargensis* were isolated from cold sulfur spring biofilm (Hiraishi, 1997; Venkata Ramana *et al.,* 2011). *Rhodoplanes roseus* and *Rpl. elegans* have been recorded in freshwater lake sediment and activated sludge respectively (Hiraishi and Ueda, 1994). *Rpl. serenus* was recorded in the pond water at the University of Tokyo, Tokyo, Japan (Okamura *et al.,* 2009), while *Rpl. piscinae* was isolated from a fish pond water (Chakravarthy *et al.,* 2012) and *Rpl. pokkaliisoli* was isolated from the waterlogged brackish paddy soil of Kerala (Lakshmi *et al.,* 2009). *Rpl. oryzae* was isolated from rhizosphere soil of paddy near Rameswaram, Tamilnadu, India (Srinivas *et al.,* 2014).

Rhodoblastus acidophilus and *Rbl. sphangicola* which can fix atmospheric nitrogen has been isolated from acidic (pH 3.5-5.5) Sphagnum peat bog (Imhoff, 2001; Kulichevskaya *et al.* 2006). Species of *Rhodopseudomonas* have been isolated from different ecological niches of India such as soil, pond water, activated sludge, wastewater and saltran sediments etc. *Rps. palustris* recorded in soil or sludge or pond water (Venkata Ramana *et al.,* 2012), while *Rps. parapalustris ,Rps .pseudopalustris* and *Rps. harwoodiae* were recorded in soil, pond water and freshwater sediment samples respectively (Venkata Ramana *et al.,* 2012). *Rps. faecalis* was recorded in chicken feces (Zhang *et al.,* 2002), while *Rps. rhenobacensis* a nitrate reducer was recorded in eutrophic pond (Hougardy *et al.,* 2000). *Rps. thermotolerans* and *Rps. pentothenatexigens* were isolated from paddy soils of Nandyala and Vijayawada (Kumar *et al.,* 2013). *Afifella, Rhodobium ,Roseospirillum* and members of *Rhodobiaceae* have a marine origin (Urdiain *et al.,* 2008; Srinivas *et al.,* 2007b). *Roseospirillum parvum* was isolated from Great Sippewissett salt marsh, USA (Glaeser and Overmann, 1999).

Rhodobacter, Rhodobacca, Rhodovulum, Rubribacterium and *Rhodothalassium* of *Rhodobacteriaceae* have wide ecological distribution covering freshwater and terrestrial environments. *Rhodobacter vinaykumarii* was recorded in the seashore of Visakhapatnam, India (Srinivas *et al.,* 2007d), while *Rba. johrii* was isolated from the rhizosphere soil of Hyderabad (Girija *et al.,* 2010). *Rba. changlensis* a psychrotolerant phototrophic bacteria were recorded in the snow sample of Changla Pass in the Himalayas (Anil Kumar *et al.,* 2007d), while *Rba. ovatus* was recorded in sediments of the polluted pond in Hyderabad (Srinivas *et al.,* 2008). *Rba. azotoformans* was recorded in soil (Hiraishi *et al.,*1997), while *Rba. blasticus* was recorded in a freshwater pond of England (Kawasaki *et al.* 1993). Similarly, *Rba. tardus* (Xian *et al.,* 2020) and *Rba. ruber* (Chen *et al.,* 2021) were isolatd from freshwater pond in Taiwan. In general species of *Rhodobacter* are reported from a variety of habitats of different parts of the world (Hansen and Imhoff, 1985; Arunasri *et al.,* 2008; Venkata Ramana *et al.,* 2008; Shaleem Raj *et al.,* 2013). *Rba. alkalitolerans* isolated from an alkaline brown pond (Suresh *et al.,* 2018), while *Rba. thermarum* (Khan *et al.,* 2018) and *Rba. flagellatus* (Xian *et al.,* 2020) are thermophilic bacterium isolated from a hot spring, Tibet, PR China.

Rhodovulum steppense was recorded in saline soda lakes of Asia (Kompantseva *et al.,* 2010), while *Rdv. lacipunicei* was recorded in purple saline pond water of Orissa (Chakravarthy *et al.,* 2009). *Rdv. kholense* was recorded in mud samples of mangrove forest of Khole, India (Anil Kumar *et al.,* 2008b), whereas *Rdv. strictum* was isolated from tidal and seawater pools of Japan (Hiraishi and Ueda, 1995). *Rdv. tesquicola* haloalkaliphilic bacterium was isolated from brackish steppe soda lakes of southern Siberia (Kompantseva *et al.,* 2012). *Rdv. phaeolacus* (Lakshmi *et al.,* 2011), *Rdv. iodosum* and *Rdv. robiginosu* (Straub *et al.,* 1999) and *Rdv. sulfidophilum, Rdv. euryhalinum* and *Rdv. adriaticum* (Hiraishi and Ueda 1994) have been reported from the brown pond and marine water respectively. Srinivas *et al.* (2006, 2007a,c, 2012 and 2014) have recorded different species of *Rhodovulum* from a variety of coastal substrates of India.

Rhodobaca bogoriensis and *Rca. barguzinensis* were recorded from soda lakes (Milford *et al.,* 2000; Boldareva *et al.,* 2008), while *Rhodobaculum claviforme* was recorded in saline steppe lakes of Siberia (Bryantseva *et al.,* 2015). *Rubribacterium polymorphum* was recorded from soda lakes (Boldareva *et al.,* 2010), while *Rhodoferax saidenbachensis*, a psychrotolerant, is recorded from sediments of a drinking water reservoir of Germany (Kaden *et al.,* 2014). *Rfx. antarcticus* was isolated from a microbial mat of Antarctica (Madigan *et al.,* 2000), while *Rfx. ferrireducens* was recorded in subsurface sediments in Oyster Bay, USA (Finneran *et al.,* 2003). *Rfx. fermentans* were isolated from ditch water and activated sludge (Hiraishi *et al.,* 1991). *Rubrivivax gelatinosus* was recorded in mud (Willems *et al.* 1991), while *Rvi. benzoatilyticus* was recorded from paddy soil of Andhra Pradesh (Ramana *et al.,* 2006). *Rhodocyclus purpureus* was isolated from the dirty-red colored swine-waste lagoon (Pfennig, 1978b).

7.6. HABITATS OF GREEN BACTERIA

GSB is commonly found in anoxic and sulfide-rich fresh water and estuarine environment either in the water columns in sediments or within microbial mats. Recently they have also been observed in an anoxic zone 100 m below the surface of the black sea (Manske *et al.,* 2005) on hydrothermal vents in the Pacific Ocean (Beatty *et al.,* 2005) and in the microbial mats of octopus and mushrooms springs in Yellowstone national park (Ward *et al.,* 1998). The ecological niche of GSB in natural ecosystems is precisely determined. They compete strongly with sulfur purple bacteria, for which photolithoautotrophic and photolithoheterotrophic metabolisms are also basic. However, in stratified lakes green sulfur bacteria dominate more often (Gorlenko *et al.,* 1983).

Thermophilic phototrophic bacteria are the major components of laminated microbial mats that develop in the outflow waters of hot springs worldwide (Brock 1978). In oxygenic microbial mats, either unicellular or filamentous cyanobacteria are present in the upper mat layers in association with the filamentous green bacterium *Chloroflexus aurantiacus* (Bauld and Brock 1973; Pierson and Castenholz 1974a) and some mats, with the filamentous bacteriochlorophyll *a* containing phototrophic bacterium *Heliothrix oregonensis* (Pierson *et al.,*1984a; 1985). Thermophilic purple non-sulfur bacteria are also present in certain oxygenic microbial mats (Resnick and Madigan, 1989). In contrast, anoxygenic microbial mats have been described from certain springs in which relatively high sulfide concentrations prevent the growth of cyanobacteria, hence, oxygen production (Giovannoni *et al.*, 1987; Ward *et al.,* 1989). Such springs are typically pH neutral and average 30-60 mM in sulfide (Ward *et al.,* 1989). Anoxygenic mats can contain nearly pure stands of *Chloroflexus* (Giovannoni *et al.*, 1987) or mixtures of *Chloroflexus* and the thermophilic purple sulfur bacterium *Chromatium tepidum* (Madigan, 1986). Elemental sulfur is abundant in anoxygenic mats due to phototrophic oxidation of sulfide and globules of elemental sulfur can be found attached to the outside of cells of *Chloroflexus* (Giovannoni *et al.,* 1987) or within cells of *Chromatium* (Madigan, 1986). Strains of *Chloroflexus* isolated from anoxygenic mats are dark-green in colour, contain bacteriochlorophyll *c* and are obligately phototrophic, differing in the latter connection from strains of *Chl. aurantiacus* isolated from oxygenic mats, all of which retain the ability to grow aerobically in darkness (Pierson and Castenholz, 1974b). Nevertheless, GCF strains of *Chloroflexus* are filamentous bacteria, morphologically similar to *Chloroflexus* strains isolated from oxygenic mats (Giovannoni *et al.,* 1987). In the central volcanic area of North Island, New Zealand, unlamented anoxygenic mats containing a rod-shaped green bacterium have been found in the outflow of certain hot springs (Castenholz *et al.,* 1990). Springs in which these, mats develop are characteristically acidic (pH 4.5- 6.0), contain high sulfide levels (0.2-2.0 mM) and have temperatures in the range of 45-55°C (Castenholz *et al.,* 1990).

The various photosynthetic tetrapyrrole and carotenoid pigments in phototrophic bacteria contribute to a distinct coloration of water ranging from green, brown and red to purple. These phototrophic sulfur bacteria comprise the PSB and the GSB whose ecology and sulfur metabolism share many similarities (van Gemerden and Mas, 1995; Overmann, 2008), although their phylogeny is different.

In natural environments, light and sulfide occur in opposing gradients, hence the growth of phototrophic sulfur bacteria is restricted to a narrow zone. In open waters, like lakes or lagoons, stratification of oxic and anoxic water layers is maintained by density differences. In benthic environments, gradients of sulfide are much steeper due to the higher rates of sulfate reduction and a lower frequency of turbulent mixing (Overmann, 2008). In some habitats of phototrophic sulfur bacteria, particularly in intertidal sediments, redox conditions change rapidly within hours. Certain species of PSB have adapted to these conditions and can switch to an aerobic chemolithotrophic growth mode and oxidize sulfide or thiosulfate or both with molecular oxygen. Under these conditions, the synthesis of pigments and pigment-binding proteins is repressed and the cells become colorless. In contrast, all GSB are obligately anaerobic and obligately phototrophic. These properties enforce a greater constraint on the possible niches available to GSB.

Phototrophic sulfur bacteria typically develop in non-thermal aquatic ecosystems. However, a few moderately thermophilic members of this group have been described from hot spring mats such as *Chlorobaculum tepidum* and *Thermochromatium tepidum*. Among the PSB, most members of the *Chromatiaceae* are typically found in freshwater and marine environments, whereas *Ectothiorhodospiraceae* inhabit hypersaline waters. Some members of GSB have also been isolated from a hypersaline Athallassohaline lake in Spain (Vila *et al.,* 2002), in coastal and most lacustrine waters, metabolically specialized *Chromatiaceae* dominate (van Gemerden and Mas, 1995). These organisms are obligately photolithotrophic, lack assimilatory sulfate reduction, cannot reduce nitrate and assimilate only a few carbon sources. This is not unexpected as metabolically versatile species of *Chromatiaceae* have no strong selective advantage in most pelagic habitats (Overmann, 2008). In some lakes, the absorption of ultraviolet and blue light by humic substances is so significant that green-colored species of GSB have a selective advantage over their brown-colored counterparts and PSB (Parkin and Brock, 1980).

In most of the phototrophic organisms such as purple bacteria under low light intensities, photosynthetic unit increases in size to accommodate photosynthetic pigments (Overmann, 2008). The specialized photosynthetic antenna organelles of GSB, the chlorosomes, are significantly large and contain about 5000–8000 BChl *c* molecules per RC (Frigaard *et al.*, 2003). In general GSB exhibit, lower maintenance energy requirements and higher sulfide tolerance than PSB and they are well adapted to low-light habitats (Overmann, 2008).

7.7. HABITATS OF HELIOBACTERIACEAE

The distribution of *Heliobacteria* in nature has occurred only from successful isolations. *Heliobacteria chlorum* was first isolated from garden soil by Gest and Favinger (1983) and the following isolation of *Heliobacter mobilis* (Beer-Romero and Gest, 1987) and *Heliobacterium gestii* and *Heliophilum fasciatum* (Ormerod *et al.,* 1996) from paddy soils and may have evolved a mutualistic relationship with plants, particularly, rice plants. These bacteria are strictly anaerobic anoxygenic phototrophic bacteria that synthesize bacteriochlorophyll *g* a unique pigment to species and differ from all other APB. Stevenson *et al.,* (1997), studied the 29 to 159 soil, mud, or water samples, which are found to be positive for *Heliobacteria.* From his findings, he considered that paddy soil is the major habitats for the isolation of *Heliobacteria.* Considering that all other APB are primarily, there is a fundamental difference in the ecological roles of *Heliobacteria* and green- and purple bacteria in nature. Isolation studies illustrated that paddy soil is an excellent and stable source of *Heliobacteria* . Other agricultural soils are rarely yielded *Heliobacteria.* Dry paddy soils are proven to be particularly good sources for their isolation, it is due to a mutualistic relationship between *Heliobacteria* and rice plants and in fact that the endospores of these organisms produce likely at the end of the growing season, would allow these bacteria to tolerate exposure to oxygen even when the paddy soils become desiccated in the dry season. Endospores represent a novel survival strategy for APB and impart special ecological importance to these bacteria. Generally, aquatic habitats such as sewage, pond and lake waters, and sediments, were uniformly negative in isolating enrichment cultures for *Heliobacteria.* However, as previously noted, sediments from Soap Lake yielded a novel strain of *Heliobacteria* strain SLH as having hot spring microbial mats (Bryantseva *et al.* , 2000b; Kimble *et al.,* 1995), and further, it appears that some species of *Heliobacteria* inhabit specialized aquatic habitats.

The alkaliphilic *Heliobacteria* such as *Heliorestis daurensis* (Bryantseva *et al.,* 1999) and *Heliorestis baculata* (Bryantseva *et al.,* 2000) were found to be isolated from microbial mats and soil along the shores of Lake Barun Torey and a soda lake located in the South China region of Siberia (Russia). This lake is unique due to its alkalinity (~pH 9.0) and low salinity. *Heliorestis daurensis* could be recorded from both cyanobacterial mats surrounding the lake and from soils of dried lagoons adjacent to the lake. Subsequent isolation of *Heliorestis baculata*, an alkaliphilic *Heliobacterium* phylogenetically related to *Heliorestis daurensis*, was isolated from soda lake soil adjacent to the mats (Bryantseva *et al.,* 2000). The findings of *H. modesticaldum* and the soda lake *Heliobacteria* thus leave open the possibility that freshwater or marine microbial mats are also considered to be habitats for *Heliobacteria* .

■■■■

Chapter 8

HYDROGEN - NON-CONVENTIONAL ENERGY

The dwindling natural resources, ever-increasing human population and the consequent increase in the demand for goods and services that are beneficial to human welfare have made it necessary to look into newer, safer, cleaner and renewable energy to cater to human needs. Hence, it has become necessary to find an intermediary form of energy that can be produced using the non-conventional primary energy source. Energy is vital to global prosperity, yet dependence on fossil fuels as a primary energy source contributes to global warming and environmental degradation. Hydrogen has the highest gravimetric energy density of any known fuel and is compatible with electrochemical and combustion processes for energy conversion without producing carbon-based emissions that contribute to environmental pollution and climate change. The advantage of biological hydrogen production would mitigate global warming and favorably alter the equation on global energy supply and demand. Further, hydrogen fuel cells and related hydrogen technologies provide the essential link between renewable energy sources and sustainable energy services. The utility and hydrogen production can be represented as in **Fig. 8.1.**

Hydrogen is a clean-burning alternative to fossil fuel. Its energy content per unit masses higher than any other conventional fuel except for nuclear power. Widespread hydrogen utilization will require technological advances in storage fuel cell electrodes and sustainable production. Although 90% of H_2 we use is currently derived from fossil fuels (Hankamer *et al.,* 2007), it can be produced by a variety of sustainable processes including biological (Turner *et al.,* 2008).

Hydrogen produced from primary sources can be stored directly converted to electricity in fuel cells efficiently when needed. Man exploits photobiological and photobiomimetic production of hydrogen. Biological hydrogen production processes offer the prospect of producing H_2 from renewable sources. Species of *Rhodobacter* are photosynthetic purple non-sulfur bacteria that can produce H_2 from small-chain organic acids derived from biomass at the expense of light energy.

Since the fuel consumption rate is very high, biohydrogen needs to be produced at a much faster rate with new strategies basing on energy bionics.

These processes should provide a net energy gain, be economically competitive and producible in large quantities without reducing the food supplies. Further research is needed in all fields to be competitive with conventional technologies. H_2- powered fuel cell electric vehicle option is a clue for 21st century's people on how to change their consumption habits for a sustainable future.

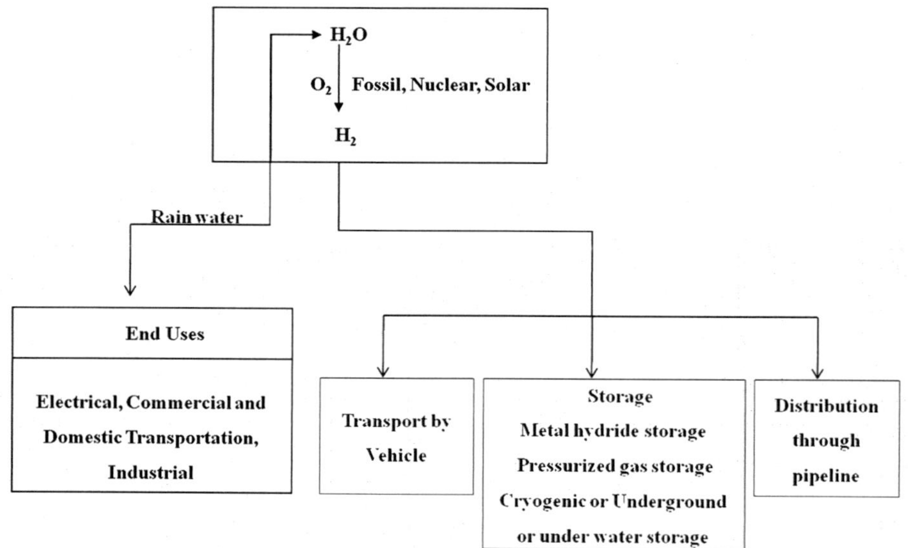

Fig. 8.1. Storage and Utilization of Hydrogen.

However, various bacteria such as aerobic heterotrophic bacteria and thermophilic bacteria (*Thermoanaerobacter, Thermobacterioides* and *Vellenella*) are reported to produce hydrogen, phototrophic bacteria that too anoxygenic phototrophic bacteria such as purple sulfur bacteria (*Chromatium vinosum, C. minutissimum*), Purple non-sulfur bacteria (*Rhodospirillum rubrum, Rsp. molish-ianum, Rhodobacter sphaeroides, Rba. capsulatus, Rba. sulfidophilus, Rba. marinus, Rhodopseudomonas palustris, Rps. acidophila, Rps. rutila, Rhodomi-crobium vannielli* and *Rhodocyclus gelatinosus*) and Green sulfur bacteria (*Chlorobium limicola* f.sp. *sulfatophilum, Pelodictyon*) proved to be efficient in hydrogen production in an economical way.

Many scientists and engineers feel that hydrogen energy could be an ideal alternative as it is the most economical, cleanest, most efficient, renewable, storable, transportable and recyclable fuel which can be used when fossil fuels are getting exhausted. Exceptionally low density, rapid diffusivity in air, high value for upper flammability limit in air, low ignition energy and high flame velocity makes the hydrogen a unique fuel both from a utilization and safety point of view and may become a major source of fuel of tomorrow.

Further, the hydrogen-producing system could be operated at low physiological temperatures (10-40°C). The only input requirement in hydrogen donor possibly industrial and agricultural wastes and solar energy and the equipment required is simple and it is pollution-free. The system is of multiple utilities such as food chemicals, waste treatment and biofertilizers.

Hydrogen holds the promise of being highly efficient energy that can be used in a situation where the transfer of energy as electricity is inefficient, impractical, or impossible. It is the least capital intensive and most inflation fighting. The hydrogen economy a phrase often quoted in the literature refers eventual adoption of hydrogen as the base fuel of the world economy (Veziroglu, 1987). Intensive researches are being carried out all over the world and significant achievements have been made in overcoming the bottle necks for hydrogen utilization as energy for day to day.

8.1. WASTEWATER TREATMENT AND HYDROGEN PRODUCTION

Out of four ways of hydrogen production, the wastewater treatment process is more advantageous as this process produces energy from industrial wastes thereby reducing the pollution load also (Zhu *et al.,* 2002; Shin *et al.,* 2004; Kapdan and Kargi, 2006; Venkatat Mohan *et al.,* 2007; Sangyoka *et al.,* 2007; Yamaoka *et al.,* 2008 and Zhang *et al.*, 2010). Microbial hydrogen production can be fermentative under dark conditions (Tanisho and Isiwata 1994; Yokoi *et al.,* 1998; Kumar and Das 2000; Nakashimada *et al.,* 2002; Oh *et al.,* 2008; Lo *et al.,* 2008), photoheterotrophic conditions under light (Ramchander *et al.*, 2010; Rajyalaxmi *et al.,* 2018) or photo fermentation under dark conditions (Ghirardi *et al.,* 2005; Li and Fang 2008) or using mixed cultures of hetero and/or phototrophs (Fang and Liu 2002; Chiemchaisri *et al.,* 2007; Wang *et al.,* 2008; Vatsala *et al.,* 2008; Manisha *et al.*, 2015).

Hydrogen production using various types of reactors such as batch culture (O-Thong *et al.,* 2007), semi-continuous culture (Vijayaraghavan and Ahmad, 2006; Gavala *et al.,* 2006), or continuous modes of operation (Chang and Lin 2004; Kotsopoulos *et al.,* 2006; Zhang *et al.,* 2007; Younesi*et al.,* 2008; Tolvanen *et al.,* 2008). Immobilization of microbes for hydrogen production has been reported to be advantageous (Wu and Chang 2007; Wu *et al.,* 2006). Hydrogen production from whey (Venkataramana and Vatsala 1990), distillery effluents (Vatsala *et al.,* 2008) and oil mill wastes (Eroglu *et al.*, 2008) using phototrophs and co-cultures have been carried out in reactors of various capacities. Suwansaard *et al.* (2009) could develop two-stage biohydrogen production processes from palm oil mill effluents with the help of anoxygenic phototrophic bacteria. Hay *et al.* (2013). Reported the biohydrogen production through photo-fermentation or dark-fermentation using waste as a substrate. Laurinavichene *et al.* (2018) reported the utilization of distillery wastewater for hydrogen production in one-stage and two-stage processes involving photo fermentation.

Development of new technologies allowing the application of up-to-date clean and neat sources of high energy is one of the main aims of biotechnology

(Teplyakov *et al.,* 2002). Varying amounts of hydrogen production were reported when they have used different technologies. Immobilization of active bacterial cells for prolonged production of hydrogen proved to be most successful (Tsygakov *et al.,* 1994). However, low porosity, diffusion problem and high cost are some of the constraints for this technology which stimulated the development of new approaches to the immobilization of phototrophic cells and the use of other material as cell carriers. Lozinsky *et al,* (2003) have successfully used Cryogel polyvinyl alcohol (PVA) as an immobilizing agent. This process proved to be more superior in view of its mechanical strength and manufacturability which are important for the realization of the microbial process of hydrogen production in practice.

The hydrogen produced by photosynthesis-based systems may not only meet the goal of providing enough hydrogen but also useful for some other applications. Further, continued research will no doubt result in significant improvement in the technology of increased hydrogen production. At present, biohydrogen production may meet the requirement of hydrogen as met by fuel cells but the requirement of huge bioreactors is a main limiting factor.

Kondo *et al.* (2002) demonstrated increased production of hydrogen by combining two types of photosynthetic bacteria with different characteristics. Lee *et al.* (2002) reported good production of hydrogen by purple non-sulfur bacteria with hydrogen fermentation reactor effluents. Ko and Noike (2002) have developed a method by which they suppressed the growth of algae in hydrogen-producing anoxygenic cultures of *Rba. sphaeroides.* APB can use the energy of the sun to produce hydrogen, by consuming a small number of organic acids, while dark anaerobic fermentative bacteria generate hydrogen by the small number of organic acids by consuming carbohydrates. However, efforts to integrate these individual processes for increased efficiency and greater yield of hydrogen are the need of the hour. Biological hydrogen production based on green algae driven by a visible portion of the solar spectrum, coupled with purple photosynthetic bacteria, driven by a near-infrared portion of the solar spectrum, will be more rewarding. Co-cultivation of unicellular green algae and purple phototrophic bacteria will enable us to better utilization of solar radiations (visible and infrared) and integration of nutrient utilization for cost-effective production of a substantial amount of hydrogen gas. It could provide a boost to the economy of the world especially that of underdeveloped and developing nations. As commercial exploitation of green algae and bacteria would be a low-tech business, a method has the potential of generating both biomass (for food) and fuel.

Design of integrated bioprocess in which oxygenic and anoxygenic phototrophs are employed in tandem to harvest near-infrared energy of the sun to convert it to hydrogen energy. Hydrogen can be collected, while biomass from this integrated process can be converted through the use of industrial enzymes into cellulolytic material composed of hydrolysates of poly glucose and protein which can feed directly into dark anaerobic bacteria fermentations. Such non-photosynthetic anaerobic bacteria fermentations would generate hydrogen and a variety of small organic acids which can feed back into anoxygenic phototrophic bacterial hydrogen production. Melis (2005) has elegantly discussed the feasibility of this integrated photoproduction of hydrogen as illustrated in (**Fig. 8.2** .)

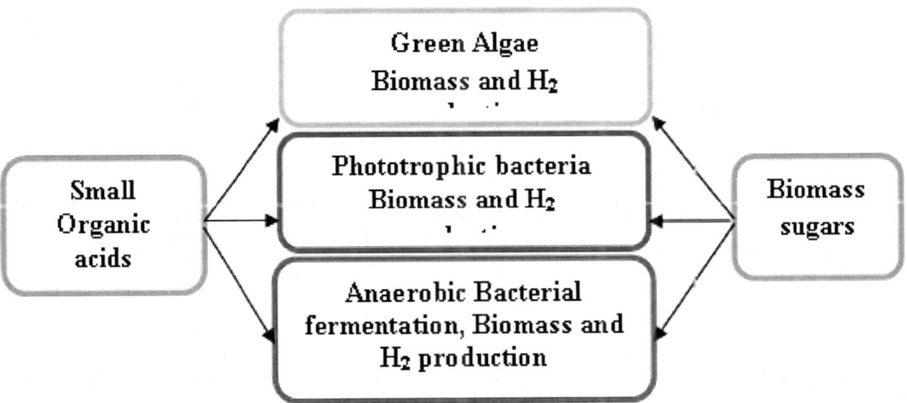

Fig. 8.2. Integrated Three Organism System for Commercial Hydrogen Production.

Efforts to improve phototrophic bacteria for hydrogen production include the elimination of competing organism, such as microalgae using light filters (Ko and Noike, 2002), two-phase fermentation systems in which photoheterotrophic bacteria utilize substrates produced by anaerobic bacteria during the dark fermentation process (Lee *et al.,* 2002), novel photobioreactor designs (Kondo *et al.,* 2002) and use of specific waste streams as substrates for photo-fermentative, substantial improvement in hydrogen production by APB by rapid removal of gas and separation bioreactor design and genetic modifications in microorganisms deserves increased attention.

Gains in hydrogen production may be achieved through genetic modification of hydrogen-producing bacteria. Genetic modifications of bacteria can be achieved by several ways to increase hydrogen synthesis including overexpression of cellulases, hemicellulases and ligases that can maximize substrate availability, elimination of uptake hydrogenases, overexpression of hydrogen evolving hydrogenases that have themselves been modified to be O_2 tolerant and delimitation of metabolic pathways that compete for reducing equivalents required for hydrogen synthesis.

Optimization of bioreactor designs, rapid removal and purification of gases and genetic modifications of enzyme pathways that compete with hydrogen-producing enzyme systems offer exciting prospects for Biohydrogen systems. Even a 10 fold increase in the rate of hydrogen synthesis by some dark fermentation systems would reduce bioreactor size dramatically. This would greatly facilitate overcoming the engineering challenges of scale-up and create new opportunities for practical applications.

Co-cultivation of hydrogen production of green algae and purple bacteria for broad-spectrum integrated photobiological hydrogen production as well as estimates of yield from the commercial exploitation of this integrated process is to be assessed. The advent of biological hydrogen production would mitigate global warming and favorably alter the equation on global energy supply and demand.

Prospects of biological hydrogen production by APB are being investigated all over the world as a potential source of renewable and pollution-free fuel involving fewer capital inputs and simple facilities proved to be promising for a large-scale generation (Carlozzi *et al.*, 2010). Some APB which occur in specialized ecological niches and reported to produce hydrogen efficiently are listed in **Table 8.1.**

Table 8.1. Anoxygenic Phototrophic Bacteria Capable of Producing Hydrogen

Microorganism	**Source of isolation**	**Growth temperature**	**Reference**
Allochromatium sp. GSKRLMBKU-01	Marine water	Mesophilic	Rajyalaxmi *et al.* (2018)
Caldanaerobacter subterraneus subsp. *pacificus*	Marine hot vent	Thermophilic	Sokolova *et al.* (2001)
Carboxydocella themoautotrophica	Hot spring	Thermophilic	Sokolova *et al.* (2002)
Carboxydothermus hydrogenoformans	Hot swamp	Thermophilic	Gerhardt *et al.* (1991)
*Citrobacter*sp. Y9	Wastewater sludge digester	Mesophilic	Oh *et al.* (2003)
Rhodopseudomonas palustris P4	Wastewater sludge digester	Mesophilic	Jung *et al.* (1991)
Rhodopseudomonas rutila	Papermill effluents	Mesophilic	Vasavi *et al.* (2014)
Rhodobacter capsulatus KU002	Leather industry effluents	Mesophilic	Ramchander *et al.* (2010)
Rhodobacter sphaeroides	Sewage water	Mesophilic	Swetha *et al.* (2019)
Rhodobacter sphaeroides CNT 2A	Pond sediments	Mesophilic	Subudhi *et al.* (2016)
Rsp. rubrum	Dead mouse residue	Mesophilic	Zurrer and Bachofen (1979), Gest (1995)
Rubrivivax gelatinosus	Lake sediments	Mesophilic	Uffen (1976)
Rubrivivax gelatinosus L31	Reservoir sediments	Mesophilic	Li and Fang, 2008
Thermincola carboxydiphila	Hot spring	Thermophilic	Sokolova *et al.* (2005)

Microorganism	Source of isolation	Growth temperature	Reference
Thermococcus strain AM4	Hydrothermal vent	Thermophilic	Sokolova *et al.* (2004a)
Thermolithobacter carboxidivorans	Hot spring	Thermophilic	Sokolova *et al.* (2007)
Thermosinus carboxidivorans	Hot spring	Thermophilic	Sokolova *et al.* (2004b)

Photohydrogen production processes with anoxygenic photosynthetic bacteria have been regarded as an efficient system for energy recovery from organic wastewater (Kawasaki *et al.,* 2010). This method proved to be more economical as it would reduce wastewater treatment costs (Lu *et al.,* 2010; Madukasi *et al.,* 2011). This can be easily achieved by improving the efficiency of light energy utilization in photohydrogen fermentation bioreactor (Zhou *et al* ,. 2015 and Zhou *et al.*, 2014). Enhancing hydrogen production could be achieved by a biofilm of anoxygenic photosynthetic bacteria induced by phototaxis (Wang *et al*., 2015).

Hydrogenases catalyze hydrogen metabolism in microorganisms (both chemo- and phototrophs) involving hydrogen production or hydrogen uptake which can be summarized by the following reactions.

1. In the light

$$H_2O \longrightarrow H_2 + \frac{1}{2} O_2$$

$$RH_2 \longrightarrow H_2 + R_2$$

R represents metal hydride

2. Carbondioxide reduction (Photoreduction)

$$2H_2 + CO_2 \xrightarrow{\text{nATP}} (CH_2O) + H_2O$$

$$2H_2 + CO_2 \longrightarrow (CH_2O) + H_2O$$

3. In the dark – hydrogen reduction

$$RH_2 \longrightarrow H_2 + R$$

4. Oxyhydrogen reaction

$$H_2 + \frac{1}{2} O_2 \longrightarrow H_2O$$

5. Hydrogen uptake

$$R + H_2 \longrightarrow RH_2$$

In addition to these reactions, nitrogenase during nitrogen fixation produces hydrogen. Nitrogenase in the absence or the limited availability of its substrate (nitrogen) produces hydrogen from NADH + H. It produces hydrogen according to the following equation (Stevens *et al.,* 1986).

Hydrogen production by *Rhodospirillum rubrum* and other phototrophic bacteria is primarily caused by nitrogenase. Therefore, hydrogen production is induced under nitrogen limitation. If N_2 is eliminated, the nitrogenase produces exclusively hydrogen and thus hydrogen productivity is increased. In a N_2 limited batch culture of *Rsp. rubrum* produces a continuously 0.48 L/day hydrogen per gram of cell dry weight with lactate as an electron donor.

Hydrogen production can also be achieved by co-culturing of green sulfur bacteria *Chlorobium vibrioform* and the chemotrophic sulfur-reducing bacterium, *Desulforomonas acetoxidans* where chemotrophic partner oxidizes acetate and reduces sulfur to H_2S and the phototrophic partner oxidizes H_2S back to sulfur and produces hydrogen using nitrogenase up to 1.3L/day per gram cell dry weight with acetate as an electron donor (Warthmam*et al.,* 1992).

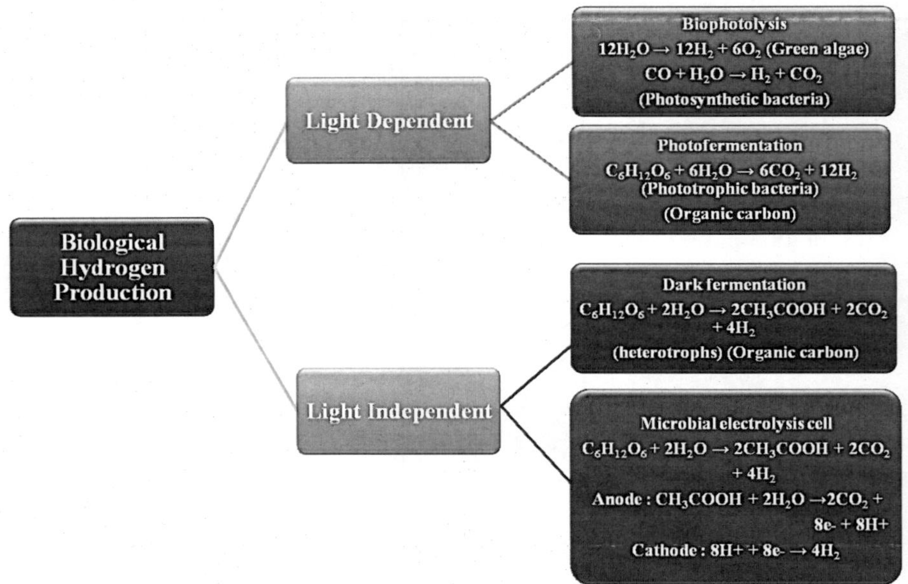

Fig. 8.3. Methods of Biological Hydrogen Production.

Hydrogen can be produced from biological processes that are less energy-intensive and more environmentally friendly in terms of the global reduction of CO_2. These renewable biohydrogen producing technologies have the potential to become cost-competitive as they can use low-value waste biomass as feedstock e.g. municipal, agricultural and industrial organic waste and wastewater (Kotay and Das, 2008). Biohydrogen can be produced by both autotrophic and heterotrophic microorganisms (Das and Veziroglu, 2008). In autotrophic conversions (also known as direct or indirect biophotolysis), solar energy is directly converted to hydrogen via photosynthetic reactions mediated by photosynthetic microorganisms, i.e. microalgae, protists and photosynthetic bacteria. Under heterotrophic conditions, the organic substrates are transformed into simpler organic compounds with simultaneous production of molecular hydrogen (Li and Fang, 2008). There are two types of heterotrophic conversions, photo-fermentation carried out by photosynthetic bacteria and dark fermentation (DF) carried out by anaerobic bacteria that convert carbohydrates into biohydrogen (**Fig. 8.3** .)

Hydrogen production by APB has several advantages over hydrogen production through traditional processes and other microorganisms. Biological hydrogen production requires the use of a simple solar reactor such as a transparent closed box, with low energy requirements while electrochemical hydrogen production via solar battery-based water splitting requires high energy. Low conversion efficiencies of biological systems can be compensated by low energy requirements and reduced initial investment costs. Biological methods are known to be less energy-intensive than chemical or electrochemical processes since they can be carried out at ambient temperature and pressure. Biological hydrogen production can be generated by the transfer of electrons from water to protons by a series of oxidoreduction reactions. Biological hydrogen production represents a potentially revolutionary technology to harvest solar energy (Ghirardi *et al.,* 2005). Advances in biological hydrogen production processes are reviewed by Levin *et al.,* (2004); Das and Veziroglu (2008); Das *et al.* (2008) and Lee *et al.*, (2010). All five types of processes have been developed for biological hydrogen productions.

i. Light-Driven Hydrogen Production
    i. Direct Biophotolysis
    ii. Indirect Biophotolysis,
2. Photo-Fermentation,
3. Dark Fermentation.
4. Hybrid Reactor System.
5. Water–Gas Shift Reaction of Photoheterotrophic Bacteria.

8.2. LIGHT DRIVEN HYDROGEN PRODUCTION

i. Direct Biophotolysis: Hydrogen can be produced directly from the photolysis of water in membranes by solar radiation in the presence of microalgae. It uses the same process found in plants and algal photosynthesis but adapts them for hydrogen gas generation instead of carbon-containing biomass (Das and Veziroglu, 2001) (**Fig. 8.4**). One of the key benefits of this method is that infinite solar energy can be used to convert the most plentiful available resource water to oxygen and unlimited hydrogen production from the earth (Hallenbeck and Benemenn, 2002). Several researchers have reported algal photolysis using various species of green algae (Das and Veziroglu, 2008; Chader *et al.,* 2009), for hydrogen production from water. Unfortunately, the disadvantage of this method is that the hydrogenase enzyme involved in hydrogen metabolism is extremely oxygen-sensitive so light-dependent hydrogen production can only be sustained for some times of minutes because of the oxygen produced in photosynthesis. To overcome this problem, major bottleneck, a process has been developed to sweep out the oxygen during hydrogen production (Benemenn, 1996), using both reversible and irreversible O_2 absorbers and O_2 tolerant hydrogenase enzymes (Ghirardi *et al.,* 1997). Both of these attempts have met with limited success. A photobioreactor is required to carry out the entire process, which can produce and capture hydrogen or O_2 but it is not economically viable to handle the hydrogen or O_2 mixtures in large volumes (Hallenbeck and Benemann, 2002). To overcome such limitations the future studies, have to focus on sweeping out of O_2 from the system. The development of an oxygen insensitive hydrogenase and the solar energy capture area should be within the photobioreactor, which is a great engineering challenge (Nath and Das, 2004).

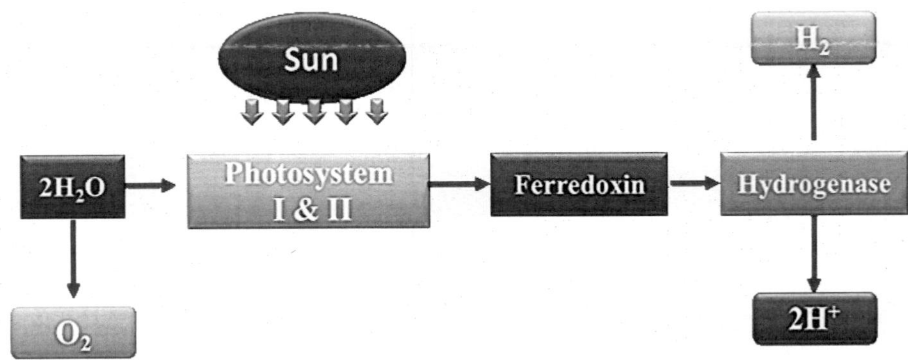

Fig. 8.4. Direct Biophotolysis.

(ii) Indirect Biophotolysis: Production of hydrogen by biophotolysis of water by members of Cyanobacteria and diverse groups of photoautotrophic microorganisms is called indirect biophotolysis (**Fig. 8.5**). Cyanobacteria and microalgae are capable of oxygenic photosynthesis.

$$6CO_2 + 12\ H_2O + \text{light energy} \rightarrow 6\ [CH_2O] + 6O_2$$

$$C_6H_{12}O_6 + 12\ H_2O + \text{light energy} \rightarrow 6CO_2 + 12\ H_2$$

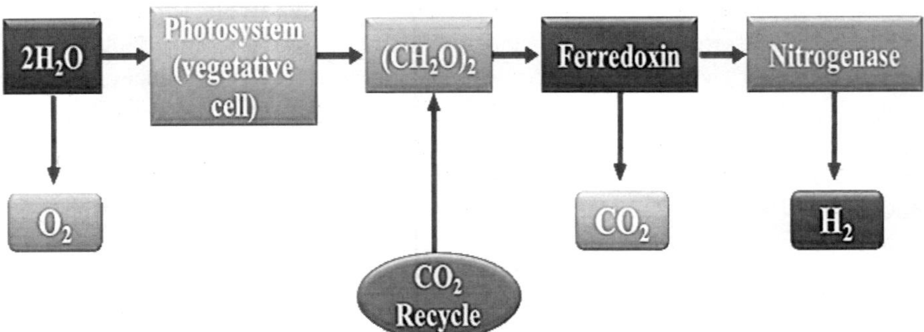

Fig. 8.5. Indirect Biophotolysis.

In this process energy released from the sunlight is utilized by photosystem-II to release two electrons from the water molecule. The electrons thus released from water are transported to the Fe-S protein, ferredoxin, on the reducing side of photosystem-I. The hydrogenase present in the stroma of the algal chloroplast accepts electrons from the reduced ferredoxin and donates two protons to generate one hydrogen molecule (Das and Veziroglu, 2001). The light intensity, nitrogenase, oxygen sensitivity and lower hydrogen evolution vs acetylene reduction are some of the factors which influence biophotolysis (Benemann, 1997). Androga *et al.* (2014) studied the optimization of temperature and light intensity for improved photof ermentative hydrogen production by using *Rhodobacter capsulatus* DSM 1710. Nitrogenase in heterocyst provides an O_2 free environment and fix nitrogen in air (Pinto *et al.,* 2002) and catalyses the production of molecular hydrogen as a by-product, whereas hydrogenase enzyme catalyzes the oxidation of molecular hydrogen synthesized by nitrogenase and bi-directional hydrogenase, which can oxidize and synthesize hydrogen (Levin *et al.,* 2004). Genetic modification of strains to eliminate uptake hydrogenase and even sustain hydrogen production in the presence of O_2 was suggested by Akkerman *et al.* (2002) and Nath and Das (2004).

8.3. PHOTO-FERMENTATION

The efficiency of APB for the production of hydrogen by utilizing light energy from sunlight is theoretically much higher when compared with cyanobacteria because these APB absorb light at longer wavelengths than cyanobacteria, yielding less energy for photosynthesis. They utilize reduced sulfur compounds (H_2S), organic acids and carbohydrates as electron donors (Wakayama and Miyake, 2001). The process is mainly mediated by the nitrogenase enzyme, which catalyzes the reduction of N_2 to NH_3. H_2 production is an inherent activity of the nitrogenase enzyme, which forms 1 mole of H_2 per mole of N_2 fixed (**Fig. 8.6** .)

$$N_2 + 8H^+ + 8e\text{-}\ 16ATP \rightarrow NH_3 + H_2 + 16ADP + 16Pi$$

However, under a limited nitrogen source, the enzyme functions as hydrogenase and catalyzes the reduction of protons to form molecular hydrogen with the expense of 4 moles of ATP. Hence, under limited N_2, 4 times more hydrogen can be produced.

$$2H^+ + 2e\text{-} + 4ATP \rightarrow H_2 + 4ADP + 4Pi$$

Photo fermentative hydrogen production is a microbial process in which electrons and protons generated through oxidation of organic compounds are used to produce molecular hydrogen under anaerobic, N_2-limited conditions utilizing light as an energy source.

The production of hydrogen by phototrophic bacteria takes place in light in the anaerobic atmosphere such as argon or helium from the breakdown of organic substrates like malate and lactate (**Fig 8.6**). Apart from different organic acids and carbohydrates, the use of waste as substrate by PNSB to produce hydrogen has been reported. Mixed cultures of many bacteria (Wang *et al.,* 2008), *Clostridium butyricum* (Fang *et al.,* 2006), *Lactobacillus* (Asada *et al.,* 2006) and *Enterobacter cloace* (Nath and Das, 2009) along with PNSB consortium are also being utilized for hydrogen production (Jahnavi *et al.,* 2014; Raju *et al.,* 2015; Manisha *et al.,* 2015). Poor yield (less than 20%-30%) (De Vrije *et al.,* 2007) and sensitivity to end product accumulation (van Neil *et al.,* 2003) and economically nonviable (Hallenback, 2005) are associated with this process. However, Wang *et al.* (2011) have developed a method by inducing biofilm formation by anoxygenic photosynthetic bacteria by phototaxis and enhanced hydrogen production. Basak *et al.* (2014) studied the photo fermentative molecular biohydrogen production by purple-non-sulfur (PNS) bacteria in various modes. Lazaro *et al.* (2015) studied the effect of inoculum concentration, pH, light intensity and lighting regime on hydrogen production by using a phototrophic microbial consortium. Similarly, Guevara-Lopez and Buitron (2015) evaluated the different supporting materials used with a photo-fermentative consortium mainly *Rhodopseudomonas palustris* for hydrogen production. Photo-fermentative hydrogen production from enzymatic hydrolysate of corn stalk pith using photosynthetic consortium was studied by Jiang *et al.* (2016).

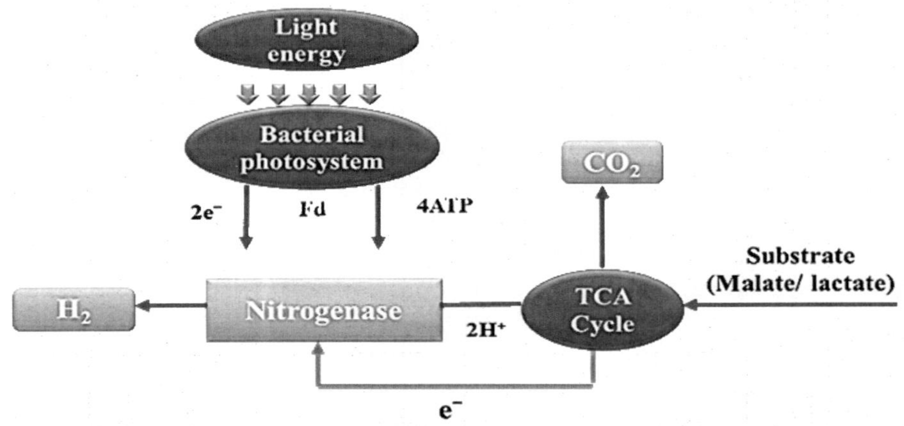

Fig. 8.6. Photofermentative Hydrogen Production.

PNSB plays a significant role in the removal of organic pollutants in water. *Rps. palustris* (Oh *et al.,* 2004; Ramchander *et al.*, 2012c), *Rsp. rubrum* (Najafpour *et al.,* 2004), *Rps. feacalis* RLD53 (Ren *et al.,* 2008), *Rubrivivax gelatinosus* (Mahakhan *et al.,* 2005; Li and Fang, 2008), *Rhodobacter sphaeroides* OU.001 (Nath *et al.,* 2005), *Rhodobacter sphaeroides* (Akroum-Amrouche *et al.*, 2019), *Rhodobacter capsulatus* B10 (Tatyana *et al.*, 2008), *Rba. capsulatus* KU002 (Ramchander *et al.,* 2010), *Rhodobacter capsulatus* (Hongyu Ma *et al.,* 2020), *Rhodopseudomonas rutila* (Vasavi *et al.,* 2014), *Rhodocyclus tenuis* KU 017 (Srinivas *et al.,* 2014), *Allochromatium* sp. GSKRLMBKU-01 (Rajyalaxmi *et al.,* 2018) is the most promising APB for hydrogen production compared to other organisms.

The advantages of this include

i. High theoretical conversion yields

ii. Lack of O_2 evolving activity which otherwise causes O_2 inactivation problems in different biological systems

iii. Ability to use wide spectral light energy

iv. Ability to consume a wide variety of organic substrates derived from the wastes in association with wastewater treatment .

The photosystem of PNSB is not powerful enough to split water, hence, no O_2 evolves, which makes it very suitable for biohydrogen production. Electrons that are liberated through oxidation of organic carbon are funneled through a series of electron carriers, during which protons are pumped through the membrane. This leads to the development of a proton gradient across the membrane, which drives ATP production by ATP synthase. The electrons are either used for replenishment of quinone pool or donated to ferredoxin, which delivers electrons to nitrogenase enzyme to reduce molecular nitrogen to ammonia. When molecular nitrogen is not available, nitrogenase functions as hydrogenase and catalyzes the proton reduction with the electrons derived from ferredoxin (**Fig. 8.7**). In this way, electrons from organic compounds are stored in the form of H_2, by using light energy.

Bacteria produce hydrogen when there is an excess of reducing power mainly to maintain cellular redox balance. Three metabolic pathways compete for electrons.

1. CO_2 fixation,
2. N_2 fixation/H_2 production and
3. Polyhydroxybutyrate (PHB) biosynthesis.

H_2 production is an inherent activity of nitrogenase enzyme functions as hydrogenase and catalyzes the proton which produces 1 mol of H_2 for 1 mole of N_2 fixed. When N_2 is not present, the enzyme functions as hydrogenase and catalyzes the proton reduction to form molecular hydrogen (H_2), with 4 time's higher efficiency (4 ATP is utilized per mole of H_2 produced in the absence of N_2). This energy-intensive process takes place under light metabolic control that is regulated mainly by cellular N_2 status. Nitrogenase activity is tightly regulated by several environmental factors, including NH_3, O_2 and light. (Ni/Fe)-hydrogenases tend to be involved in H_2 consumption, while (Fe/Fe)-hydrogenases are usually involved in H_2 production (Ghirardi *et al.,* 2009; Ramchander *et al.,* 2012). For this reason, hydrogen production is carried out under anaerobic, nitrogen-limited conditions in the presence of light. Liu *et al.* (2009) and Hakobyan *et al.* (2012) studied the effect of Ni^2 , Fe^2 and Mg^2 concentration as factors for enhancing photo-hydrogen production by *Rhodobacter sphaeroides* and *Rhodopseudomonas faecalis* RLD-53 respectively. Hakobyan *et al* . (2012) reported the bio-hydrogen production and the F0F1-ATPase activity of *Rhodobacter sphaeroides*: effects of various heavy metal ions.

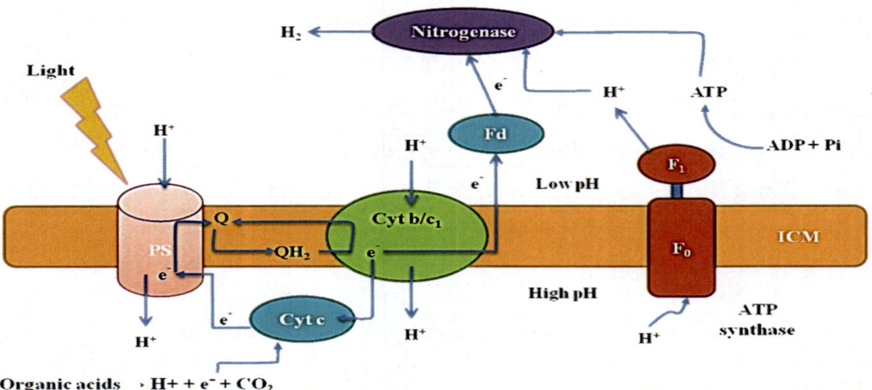

Fig. 8.7. Photofermentative Hydrogen Production by Purple Non-Sulfur Bacteria (Oxidation of organic acids produces electrons, which are transported to cytochrome c and pass through several electron transport proteins and carried to ferredoxin. During this process, protons are pumped through the membranes forming a proton gradient. This proton motive force produces ATP by ATP synthase. Ferredoxin pumps electrons to nitrogenase, which catalyzes the reduction of protons to molecular hydrogen using ATP).

In open systems, solar light intensity and temperature affect biomass production and hydrogen parameters affecting photo fermentable hydrogen production can be represented as in **Fig. 8.5.** Similarly, the diurnal light/dark cycle adversely affects photo fermentative hydrogen production. Fed-batch fermentation systems were found ideal for attaining high hydrogen rates and yields controlled feeding and sequential product removal in fed-batch systems, facilitates continuous photo fermentative hydrogen production and allows feed media optimization performance of photobioreactors (PBRs) is highly dependent on the amount of sunlight received. Therefore, the location and orientation of the PBR in outdoor environments are critical. Most PNSB produce hydrogen optimally between 30 and 35°C. Fluctuating day and night temperatures significantly affect bacterial growth and hydrogen production outdoor. The performances of different outdoor operated PBR are precised in **Table 8.4** .

Biological hydrogen production is environmentally friendly and requires less energy input compared to chemical processes (Ismail *et al.,* 2008). *Rhodospirillum rubrum, Rhodopseudomonas palustris* P4, *Citrobacter* sp., and *Rubrivivax gelatinosus* CBS are photoautotrophic and chemoheterotrophic microorganisms capable of performing the gas-water shift (GWS) reaction (Ismail *et al.,* 2008). Two enzymes that mainly contribute to the GWS reaction in these organisms are CODH and hydrogenase. The former catalyzes the oxidation of CO, while hydrogenase mediates the reduction of protons to hydrogen (Maness *et al.,* 2005). Biological GWS is thermodynamically favorable at room temperature ($CO + H_2O \rightarrow + CO_2$,ΔG = -20 kJ/mol) and atmospheric pressure, compared to the chemical catalysis where a two-stage process is required as well as high temperature (>200 °C) (Benemann, 1999). Therefore, minimum energy requirements and low process costs are expected (Ismail *et al.*, 2008). In a study by Do *et al.* (2007), *Rsp. rubrum* growing on artificial syngas and "real" producer syngas was compared. It was recorded that producer syngas had no negative effect on growth rates, biomass production, hydrogen production, or carbon monoxide consumption (Do *et al.*, 2007).

Table 8.4. Comparison of photo fermentative hydrogen production in outdoor photobioreactors (PBRs)

S. N.	Reactor type and volume	Mode of operation	Microorganism	Carbon and Nitrogen sources	Hydrogen Productivity	Hydrogen Yield	Substrate Conversion Efficiency (%)	Light Conversion Efficiency (%)	Reference
1	Flate plate (33L)	Batch	*Rps. sphaeroides* B5/B	53 mM lactate, 5 mM glutamate	0.18 mol $H_2/m^3/h$	-	78.1	-	Claassen *et al.* (2010)
2	Flate plate (6L)	Semi-continuous	*Rps. sphaeroides* B6	60 mM lactate, 6 mM glutamate or and 150 mM lactate, 30 mM glutamate	0.84 mol $H_2/m^3/h$	-	62.5	-	Kim *et al.* (1982)
3	Flat plate (6.5 L)	Batch	*Rba. sphaeroides* O.U. 001 (DSM 5864)	15 mM acetate, 2 mM glutamate	0.45 mol $H_2/m^3/h$	4.6 mol H_2/mol substrate	-	-	Eroglu *et al.* (2008)
				30 mM malate, 2 mM glutamate	0.01 mol $H_2/m^3/h$	0.6 mol H_2/mol substrate	-	-	Eroglu *et al.* (2008)
				Olive mill wastewater	0.13 mol $H_2/m^3/h$	-	-	-	Eroglu *et al.* (2008)
4	Glass bottle (0.55 L)	Batch	*Rba. capsulatus* DSM 1710	30 mM acetate, 7.5 mM lactate, 2mM glutamate	0.14 mol $H_2/m^3/h$	0.027 g H_2/g substrate	19	0.79	Ozgur *et al.* (2010)
5	Tubular nearly horizontal (80 L)	Fed-batch	*Rba. capsulatus* DSM 1710	40 mM acetate, 2 mM glutamate	0.74 mol $H_2/m^3/h$	0.60 mol H_2/mol acetate	16	1	Boran *et al.* (2010)
6	Flat plate (4 x25 L)	Fed-batch	*Rba. capsulatus* DSM 155	Acetate, lactate, glutamate	0.94 mol $H_2/m^3/h$	-	-	-	Arai *et al.* (1998)
7	Flat plate (8 L)	Fed-batch	*Rba. capsulatus* YO3 (Hup-)	40 mM acetate, 4 mM glutamate	0.30 mol $H_2/m^3/h$	-	44	-	Androga *et al.* (2011)

S. N.	Reactor type and volume	Mode of operation	Micro-organism	Carbon and Nitrogen sources	Hydrogen Productivity	Hydrogen Yield	Substrate Conversion Efficiency (%)	Light Conversion Efficiency (%)	Reference
8	Tubular nearly horizontal (50L)	-	*Rps. palustris* strain 40OL	Malate, glutamate	1.21 mol $H_2/m^3/h$	-	49.7	0.92	Ren *et al.* (2009)
9	Flat plate (4L)	Fed-batch	*Rba. capsulatus* YO3 (Hup-)	Thick juice dark fermenter effluent	1.12 mol $H_2/m^3/h$	-	77	-	Ozkan *et al.* (2012)

Data was taken from Androga *et al.* (2012)

Rsp. rubrum grows quickly and reaches high cell density that uptake CO more rapidly than other similar organisms capable of performing the GWS reaction (Najafpour *et al.,* 2003, 2004). It also tolerates small amounts of O_2 and sulfur often present in syngas (Klasson *et al.,* 1992b). As a result, this strain is the favorite organism for studies investigating biohydrogen production from syngas. *Rsp. rubrum* requires a light source for growth. However, hydrogen production is independent of light intensity (Najafpour *et al.,* 2004). An organic carbon source is needed for this organism to efficiently consume CO. The highest CO consumption (90–97%) has been observed to occur when *Rsp. rubrum* is provided with acetate as the organic carbon substrate (1–2 g/L), resulting in a 98% hydrogen production yield (Najafpour and Younesi, 2007). Investigations involving CSTR of *Rsp. rubrum* and with continuous CO flow resulted in hydrogen yields and CO conversions of 87% and 95% theoretical values respectively (Younesi *et al.,* 2008). This type of bioreactor was stable for continuous operation for 27 days (Ismail *et al.,* 2008). A mixture of pure gases is normally present in synthesis gas at a fixed concentration (e.g., 56.0% N_2, 17.2% CO, 16.3% CO_2 and 8.8% H_2 .)

Scale-up of the PBR system faces several problems including the geometry of PBR, self-shading of the cell which increases with reactor size and cell concentration negatively affect the cell growth and hydrogen production. Mixing becomes a problem in large-scale systems. Similarly, sterilization of PBRs and feed media also leaves a problem in large-scale operations. However, further studies on sterilizing to avoid contamination in the scaled-up PBR system remain to be done.

Improvement of photobioreactor design such as geometry, mixing and mode of operation, immobilization of bacterial cells, adjusting feed media with suitable C/N ratio and minor nutrients and optimum pH may help to maximize the hydrogen production. Genetic modification of the microorganisms in respect of inhibition of hydrogen utilization, optimize the flow of reducing equivalents nitrogenase by inhibition of poly-hydroxybutyrate (PHB) and CO_2 fixation, eliminate decreasing the effect of environmental factors (NH_3, O_2, light, temperature) by reducing the size of antenna pigments to increased light utilization efficiency are some of the strategies to improve hydrogen production by the organism. Kim *et al.* (2008) developed strain produced hydrogen during dark fermentative growth and photo fermentative hydrogen production increased by two folds.

Scientific project, the market strategy is essential to develop a biological hydrogen process. Although considerable progress has been made, still many more basic questions remain to be answered and new questions are being developed. Therefore, more fundamental research is necessary besides the need to test several developed techniques at the lab- and pilot-scale both as standalone and combined.

Since the bioreactor design and materials used are still in the research stage, it would not be useful to comment on a factor, improvement. Kim *et al.* (2008) developed a recombinant strain of *Rhodobacter sphaeroides* produced hydrogen during dark fermentative growth and photo fermentative hydrogen produced increased 2 fold. For biological hydrogen production to be practical and economically viable, biocatalysts must be long-lasting and make the most efficient use of electrons and light sources. Advances in these areas are being made through genetic engineering with the increasing use of systems biology approaches to understand the processes being engineered. There is ample opportunity to study and optimize productive co-cultures in concert with new bioreactor designs and operations. Although, scaling up photobiological processes in an economical manner is a major challenge. However, current laboratory-scale hydrogen production values are encouraging.

Fe-Fe hydrogenases are generally monomeric and consist of the catalytic subunit of Ca 45-48 kDa which bidirectionally catalyze H_2 production (Vignais and Billoud, 2007). The direction of the reaction is determined by the redox status of the cell. Hence anaerobiosis is a prerequisite for hydrogen production by *Rsp. rubrum*. There is also a cascade of cellular reactions controlling the synthesis of hydrogenases. CO-induced [NiFe]-hydrogenase (CooLH), which, together with CO-dehydrogenase, oxidizes CO to CO_2 with the concomitant production of H_2. This allows *Rsp. rubrum* to grow in the dark with CO as the sole energy source.

Photobioreactors which can be classified as open (raceway ponds, lagoons and lakes) or closed (flat plate, tubular) are systems designed to environmental conditions **(Fig. 8.8)** (Tredici, 2004; Androga *et al.*, 2012). Open systems are most suited to biomass production since they cannot provide the anaerobic conditions required for hydrogen production, while the closed systems allow better control of these parameters and results in higher biomass production and bio-hydrogen production (Dasgupta *et al.*, 2010).

Fig. 8.8. (A) Tubular Photobioreactor (B) Flate Plate (Androga *et al.*, 2012).

In open systems, solar light intensity and temperature affect biomass production and hydrogen parameters affecting photofermentable hydrogen production can represent as in **Fig. 8.9.** Similarly, the diurnal light/dark cycle adversely affects photo fermentative hydrogen production. Fed-batch fermentation system was found ideal for attaining high hydrogen rates and yields. Controlled feeding and sequential product removal in fed-batch systems, facilitates continuous photo fermentative hydrogen production. Feed media optimization performance of PBRs is highly dependent on the amount of sunlight received. Therefore, the location and orientation of the PBR in outdoor environments are critical. Similarly, the light distribution effects due to light path length, biomass concentration or type and composition of the feed media which affect the hydrogen production.

Advances in metabolic and regulatory aspects behind the photobiological hydrogen production are needed genetic engineering, co-culturing and bioreactor designs and making use of immobilized cells have the potential to increase conversion efficiencies of light energy to H_2 and to decrease the land area needed for photobiological hydrogen productions (Androga *et al.,* 2012). There is a need to design innovative bioreactor designs that increase the illuminated surface area to volume ratio and a bioreactor in which selected wavelength of light are delivered to layers of cells that have different light absorption maxima. Though laboratory-scale hydrogen production values are encouraging, scaling up photobiological processes in an economical manner is a major challenge (Mc Kinlay *et al.,* 2010). Basak and Das (2009) reported the photo fermentative hydrogen production using purple non-sulfur bacteria *Rhodobacter sphaeroides* O.U.001 in an annular photobioreactor. Liu *et al.* (2015) reported the photo fermentation hydrogen production by *Rhodopseudomonas* sp. nov. strain A7 isolated from the sludge in a bioreactor. Assawamongkholsiri and Reungsang (2015) studied the photo-fermentation hydrogen production of *Rhodobacter* sp. KKU-PS1 isolated from a UASB reactor. Sagir *et al.* (2018) reported the biological hydrogen production from sugar beet molasses by agar immobilized *Rba. capsulatus* in a panel photobioreactor.

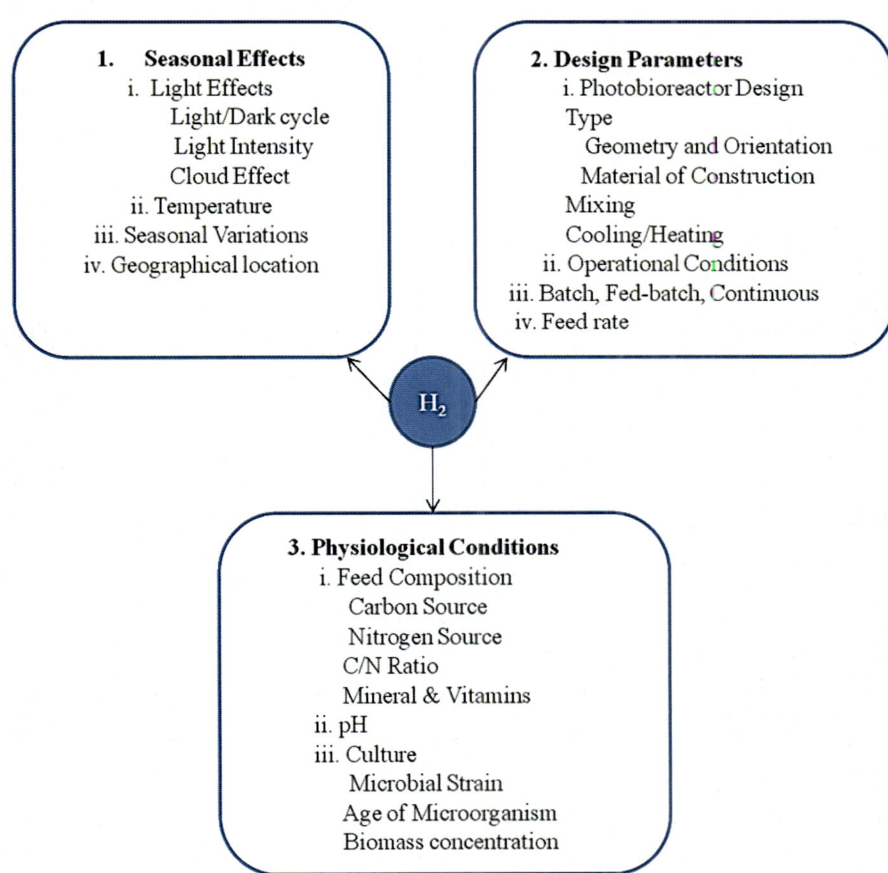

Fig. 8.9. Factors influencing Photo fermentative Hydrogen Production.

8.4. DARK FERMENTATION

The hydrogen production by dark fermentation in the anaerobic metabolism of pyruvate, formed by glycolysis. During the catabolism of various organic compounds such as glucose oxidized to acetyl CoA, which gets converted to acetyl phosphate resulting in the generation of ATP and acetate. Pyruvate oxidation to acetyl CoA requires ferredoxin (Fd) reduction. Reduced ferredoxin is oxidized by hydrogenase, which generates ferredoxin and releases electrons as molecular hydrogen.

$$\text{Pyruvate} + \text{CoA} + 2\ \text{Fd (ox)} \rightarrow \text{Acetyl CoA} + 2\text{Fd (red)} + CO_2$$

$$2\text{Fd (red)} \rightarrow 2\ \text{Fd (ox)} + H_2$$

During this process despite having a higher evolution rate of hydrogen, it is not economically viable in its present form. The fermentations have been optimized for growth naturally by evolution, thus a portion of the substrate (pyruvate) is used in both cases to produce ATP giving a by-product acetate, which is excreted. However, in many organisms, the actual yield of hydrogen is reduced by hydrogen recycling owing to the presence of one or more hydrogenase uptake, which consumes a part of hydrogen produced (Hallenbeck and Benemann, 2002). The metabolic pathway for hydrogen production by dark fermentation is shown in **Fig. 8.9** (Valdez-Vazquez and Poggi-Varaldo, 2009).

8.5. HYBRID REACTOR SYSTEM

In the biological system, the conversion of light energy to hydrogen by using an appropriate carbon source is one of the key factors for hydrogen production. The anaerobic bacteria decompose carbohydrates into organic acids to obtain both energy and electron. Further, the breakdown of organic acids was achieved by phototrophic bacteria to yield maximum hydrogen. In this hybrid reactor system, the combination of photosynthetic bacteria and anaerobic bacteria is used for the maximum yield of hydrogen (Yokai *et al.,* 1998). In this process, the anaerobic bacterium produces low molecular weight intermediates such as organic substances of wastewater from organic acids, which are then converted into hydrogen by phototrophic bacteria in a photobioreactor (Lee *et al.,* 2002). Thus a combination of dark fermentation and photofermentation improves hydrogen production from food processing wastewater and sewage sludge (Kim *et al.,* 2001). The complete degradation of glucose to hydrogen and carbon dioxide is virtually impossible by anaerobic digestion but phototrophic bacteria can accomplish it (Das and Veziroglu, 2001). Ozgur *et al.* (2010) studied the biohydrogen production from beet molasses by sequential dark- and photo-fermentation. Cheng *et al.* (2011) reported the microwave-assisted alkali pretreatment of rice straw to promote enzymatic hydrolysis and hydrogen production in dark- and photofermentation by using mixed photosynthetic bacteria, primarily *Rps. palustris* .

8.6. WATER-GAS SHIFT REACTION OF PHOTOHETEROTROPHIC BACTERIA

The members of PNSB can grow in dark utilizing carbon monoxide as the sole carbon source for ATP production. The oxidation of carbon monoxide to carbon dioxide takes place in the presence of an enzyme carbon monoxide dehydrogenase (CODH) and releases the hydrogen via water- gas shift reaction. The reaction takes place at ambient pressure and temperature. These organisms utilize both organic substrates and carbon monoxide (Levin *et al.,* 2004). However, it is a relatively new approach and its sustainability is yet to be experienced.

$$CO + H_2O \leftrightarrow CO_2 + H_2$$

Table 8.5. Advantages, Major Limitations and Approaches to Overcome in Hydrogen Production by Phototrophic Bacteria

Process	Classification and General reactions of microorg-anisms used	Advantages	Major limitations	Approaches to overcome	Reference
I. Light Dependent Processes **i. Direct Biopho-tolysis**	Micro-algae $2 H_2O$ + light = $2H_2 + O_2$	(i) It can produce H_2 directly from H_2O and sunlight. (ii) Solar conversion energy is increased by 10 folds as compared to trees and crops.	(a) Oxygen sensitivity of hydrogenase enzyme (b) Low light conversion efficiency	(a) (i) Use of O_2 absorbers, both reversible (hemoglobin) and irreversible (glucose-oxidase, dithionite) (ii) Use of O_2 tolerant uptake hydrogenase (b) (i) Genetic manipulation of light-gathering antenna (ii) Optimization of light input into the photobioreactor	Hallenbeck and Benemann (2002) Mc Tavish*et al.* (1995) Polle *et al.* (2002) Gordon (2002)
ii. Indirect Biopho-tolysis	Micro- algae, Cyanobacteria $6 H_2O + 6 CO_2$ + light = $C_6H_{12}O_6 + 6 O_2$ $C_6H_{12}O_6$ + 6 $2H_2O$ = 2 CH_3COOH + $2CO_2$ Overall reaction: $12 H_2O$ + light = $12 H_2 + 6O_2$	(i) It can produce H_2 from H_2O. (ii) It can fix N_2 from the atmosphere	(a) Enzyme inhibition by O_2 (b) H_2 consu-mption by an uptake hydro-genase (c) Overall low production rate	(a) To achieve O_2 tolerant hydrogenase activity by classical mutagenesis (b) Genetic modification of strains to eliminate uptake hydrogenase (c) Genetic modification to increase levels of bidirectional hydrogenase activity	Mc Tavish *et al.* (1995) Levin *et al.*(2004) Carrasco *et al.* (1995)
2. Photo-Fermen-tation	Purple bacteria, microalgae CH_3COOH + $2H_2O$ + light $\rightarrow$ $4H_2 + 2CO_2$	(i) a wide spectral light energy can be used by these bacteria (ii) utilizes different waste materials like a distillery and industrial effluents, waste, etc	(a) Light conversion efficiency is very low (15%) (b) Inhomog-encity of light distribution (c) O_2 is a strong inhibitor of hydrogenase	(a) (i) elimination of competing microorg-anisms (e.g micro-algae) using light filters (ii) Co-cultures of photo heterotrophic bacteria with different light utilization characteristics (iii) Control of photosynthetic protein expression to allow efficient absorption of light energy (b) Improvement in photoreactor designing with light diffuser.	Ko and Noike, (2002) Kondo *et al.* (2002) Hallenbeck and Benemann (2002)

Process	Classification and General reactions of microorg-anisms used	Advantages	Major limitations	Approaches to overcome	Reference
3. Dark Fermen-tation	Fermentative bacteria $C_6H_{12}O_6 + 6 H_2O = 12 H_2 + 6 CO_2$	(i) It can produce H_2 all day long without light. (ii) A variety of carbon sources can be used as substrates (iii) It produces valuable metabolites such as butyric, lactic and acetic acids as by-products. (iv) It is an anaerobic process, so there is no O_2 limitation problem.	(a) Relatively lower achievable yields of H_2 (b) As yields increase, H_2 fermentation becomes thermodynami cally unfavorable (c) Product gas mixture contains CO_2 which has to be separate	(a) Metabolic shift of biochemical pathways to arrest the formation of alcohol and acids. (b) Maintaining low partial pressure of H_2 (c) Efficient removal of gases.	Kumar *et al.* (2001) Mizuno *et al.*(2000) Tanisho *et al.* (1998)
4. Hybrid Reactor System (Combined dark and Photo fermen-tation)	Fermentative bacteria followed by anoxygenic phototrophic bacteria (APB) **Stage I:** $C_6H_{12}O_6 + 2H_2O \rightarrow 4H_2 + 2\ CH_3COOH + 2CO_2$ **Stage II:** $CH_3COOH + 2H_2O + light \rightarrow 4H_2 + 2CO_2$	Two-stage fermentation can improve the overall yield of hydrogen	A relatively newer approach, technological and economic feasibility is yet to be studied	--	Lee *et al.* (2002), Cheng *et al.* (2011)

Rcy. gelatinosus CBS is another promising strain for its use in the syngas to hydrogen conversion. In the presence of CO, this organism carries out the WGS reaction in both light and dark conditions (Maness and Weaver, 2002). These cells are capable of converting 100% of CO in the gas phase to hydrogen in the dark (Markov and Weaver, 2008). Its oxygen tolerance (Maness and Weaver, 2002) and its capacity to use CO as the sole carbon and energy source (Maness *et al.,* 2005; Markov and Weaver, 2008) make it an attractive biocatalyst. In brief, the advantages, major limitations and approaches to overcome limitations in hydrogen production by phototrophic bacteria are precised in **Table 8.5.**

■■■■

Chapter 9

SINGLE CELL PROTEIN

Proteins which are compounds of carbon, hydrogen, oxygen, nitrogen and may also contain sulfur have an essential role in the life processes. They occupy the second position in our diet after carbohydrates. However, there are certain proteins which we directly obtain from food and carry out the different function in our body. The majority of the proteins are synthesized from unbounded amino acids. Still some proteins are formed from the regrouping of amino acids resulting from the digestion of proteins in the cell of a body for the development of cells and tissues such as skin, muscle, blood and bones. Some of the important functions performed by protein are précised in **Table 9.1.**

Table 9.1. Important Functions of Proteins

Type of proteins	Functions
Enzymes	Catalyze or help in biochemical reactions
Transport proteins	Carry different substances in the blood to different tissues
Contractile proteins	Help contraction of muscles and other cells
Hormones	Regulate many body functions
Structural proteins	Form structural elements of cell and tissues
Protective proteins	Help to fight infections

Hence, protein is an important component of our diet. The deficiency of these components causes a disease called 'Kwashiorkor' (A rejected child). The symptoms of the diseases include skin cracks and become scaly. Abdomen swells with thin legs and hands, hair becomes reddish. Life becomes miserable. The cure is simple and provides a sufficient amount of protein in the form of grains, peanuts, animal protein.

The conventional sources of protein such as groundnuts, beans, pulses and cereals (rice, wheat and maize, etc.) from plants and meat, fish, eggs and cheese from the animals, are produced on a very large scale to feed the increasing population. However, there is a growing demand for more and more protein for human nutrition because of the increasing human population. As a result, protein deficiency disease is prevalent in many parts of the world, especially in developing countries. Hence, the search for a non-conventional source of protein began. It is Wilson (1966) of the Massachusetts Institute of Technology, Boston, USA considered that single-cell organisms like bacteria, yeasts and algae can serve as protein and can supplement human food and animal feed and coined the term Single Cell Protein (SCP). Roth (1982) defined single-cell protein as the quick growth of selected microorganisms that are rich in protein and are referred to as a single-cell protein. But the international union of pure and applied chemistry has taken serious protests against microbial cells contain other than protein also. Further, it is suggested that it could be best called Single Cell Biomass (SCB) instead of SCP. However, SCP is termed differently in different parts of the world such as Medical Academy of France prefers to call it proteins debiosynthese. In Japan, it is called petro protein and in Italy, it is called Bio protein.

Microbes are considered to be live-stock of the future era because the microorganisms are good transformers of organic matter as domestic animals. Hence, microbes grow fast and utilize simple and cheap food and produce quality protein. Comparative efficiency of microbial protein production over animal protein production is given in **Table 9.2.**

Table 9.2. The efficiency of Microbial Protein Production over Animal Protein Production (1000gm feed)

Source	Protein production (dry wt. in g)	Doubling time (in weeks)
Chicken	49.0	3.0
Pigs	42.0	8.0
Bullock	14.0	25.0
Fungi	136.0	0.04

Single-cell protein is the total protein extracted from the bacterial cultures. Apart from the nutritional benefits (Adedayo *et al.,* 2011) of single-cell protein, single-cell protein technology is sustainable technology. SCP refers to total proteins extracted from pure microbial cell culture and is produced using a number of different microorganisms including bacterium, fungus and algae. The use of natural cheap substrates and waste industrial products for cultivating microorganisms appear to be a general trend. Microorganisms have a much higher protein content of 30–70% in the dry mass and in particular, PNSB has more than 50% protein content (Kurbanoulu, 2001). The amino acid profiles of many SCP microorganisms often have excellent nutritional quality (Adedayo *et al.,* 2011). Large quantities of single-cell protein animal feed can be produced using bacteria like *Acromobacter delvaevate, Acinetobacter calcoacenticus, Aeromonas hydrophilla, Brevibacterium* (Adedayo *et al*., 2011), *Bacillus megaterium, Bacillus subtilis* (Gomashe *et al*., 2014), *Flavobacterium* species *Lactobacillus* species, *Cellulomonas* species (Rodrigues and Gallardo, 1992), *Methylophilus methylitropous* (Piper, 2004), *Pseudomonas fluorescens* (Rodrigues and Gallardo, 1992), *Thermomonospora fusca* (Dhanasekaran *et al*,. 2011). Single Cell Protein (SCP) is the best way of solving the global food problem (Najafpur, 2007). If waste materials are used as a substrate for SCP production, waste management can also be solved. Proteins not only function as nutrients but also perform other functions (Mahajan and Dua, 1995). The quality of SCP depends on the nature of the substrate and also on the organism used. Microbial cells comprise proteins, fats, carbohydrates, ash ingredients, water, and other elements such as phosphorus and Potassium (Asad *et al.,* 2000 and Nasseri *et al.,* 2011). In many instances, these raw materials have been hydrolyzed by physical, chemical and enzymatic methods before use as substrates for microbial growth (Rhishipinal and Rosamma 1998 and Jhojaosadati *et al.*, 1999). Various hydrocarbons, nitrogenous compounds, polysaccharides and agricultural wastes are used for the production of SCP (Atalo and Gashe 1993 and Ashok *et al* ,. 2000). Microbial protein productions are being carried out several natural range products such as grape juice byproduct, date extract, cashew, apple juice have been used (Kuzmanove *et al.,* 1989, Osho, 1995 and Haider and EL-Hassy, 2000). Several investigations were carried out using cellulose and hemicellulose waste as a suitable substrate for increasing SCP production (Zubi, 2005). Several parts of the crop, such as stems, leaves and roots are not edible but they can be used for the growth of single-cell microorganisms and do not pose a problem and digestible. Microorganisms are ideal for SCP production for the following reasons.

9. 1. MICROORGANISMS

Microorganisms are more suited for developing the non-conventional source of protein as food and feed because of the following reasons.

1. Conventional sources are not adequate to meet the requirement of the growing population.
2. Microorganisms have a very fast growth rate. A protein doubling time is as short as two hours has been recorded in *Neurospora crassa.*
3. They can easily be grown on a variety of substrates including wastes of agriculture, urban and industrial.
4. They can be genetically modified easily for growth on particular substratum under particular cultural conditions.
5. Their protein content is quite high, varying from 35 to 80% on a dry-weight basis.
6. They can be grown in slurry or solid-state fermentation.
7. Their nutritional values are as good as those of other conventional foods rich in proteins.
8. The processes occupy the little land area.
9. Production is independent of seasonal and climatic variation.
10. Consistent product quality.

However, when microorganism has to be used as human food and animal feed, it would have to meet certain minimum requirements in order to become popular and also compete with a conventional protein source. Some of which are:

1. It should have satisfactory biological value (BU) as demonstrated in animal feeding studies and human clinical studies.
2. It should have satisfactory nitrogen, protein content, amino acid profile, lipid contents, minerals and vitamins.
3. SCP product must be free from toxic substances in the cells themselves or introduced into the cells through residues in the substratum such as polyaromatic hydrocarbons and from contamination by pathogenic microorganisms or their toxins.
4. It should have a satisfactory flavor, aroma, colour and texture.
5. Methionine supplementation is required to obtain satisfactory feeding performance in non-ruminants.
6. It should have high protein efficiency ratio (PER).
7. Toxic residues should not accumulate in the tissues of animals to be subsequently consumed by man.
8. It should not show an adverse effect upon reproductive performance in the animal.
9. Should have acceptable physical characteristics such as formulation, blending and storage stability.

Microorganisms selected for SCP should possess the following physiological and other characteristic features.

i. Physiological

1. It should be able to grow fast.
2. It should be able to grow on simple media with no expensive growth requirements.
3. It should have a high affinity and yield coefficient on a carbon substratum.
4. It should have the ability to utilize complex substrates when required.
5. It should have the ability to grow at high cell densities.
6. It should have resistance to a substrate or product toxicity.
7. It should grow stably in continuous culture.
8. It should have a high optimum temperature.
9. It should be resistant to common contaminants.
10. It should be feasible to genetic modification.

ii. Other Characteristics:

1. Protein, fat and carbohydrates of high quality.
2. Nucleic acid content should be low.
3. High nutrient content
4. High digestibility
5. It should not have toxicity
6. Good taste and ease of recovery
7. It should be susceptible to further processing (drying without changes in colour, taste and smell, etc.).

Some of the microorganisms employed in SCP production are precised in **Table 9.3**. Advantages and disadvantages of employing different groups of microorganisms are given below:

Table 9.3. Single Cell Protein Producing Microorganisms

Major group	Genus
Bacteria	
Phototrophs	*Rhodosprillum, Rhodopseudomonas, Rhodobacter, Rhodocyclus, Rubrivivax, Rhodovulum Affifella*
Gram (-ve),aerobes	*Alcaligenes, Pseudomonas*
Gram (-ve), facultative anaerobes	*Aeromonas,Chromobacterium, Flavobacterium, Klebsiella, Hydrogenomonas, Methanomonas*
Coryneform bacteria	*Arthrobacterium, Corynebacterium, Microbacterium*
Actinomycetes	*Actinomyces, Mycobacterium, Nocardia, Streptomyces*
Yeasts	*Rhodotorula, Saccharomyces, Candida, Rhodosporidium, Debaromyces, Endomyces, Saccharomycopsis, Torulopsis*and *Sporobolomyces*
Filamentous fungi	*Cunninghamella, Mortierella, Mucor, Chaetomium Aspergillus, Aureobasidium, Botrytis, Trichoderma* and *Fusarium*
Algae	
Green algae	*Chlorella, Scenedesumus, Euglena* and *Microactinum*
Blue-green algae	*Spirulina*

(i) Algae: These algae such as *Spirulina, Chlorella* ,*Microclanu* ,*Euglena* and *Seredesmus* are reported to yield a good amount of protein at a cheaper cost in view of their phototrophic nature. However, these group of organisms has limitations like

1. The algal growth rate is slower than that of fungal growth.
2. The biological values of algal proteins are less than that of bacteria and yeasts.
3. Algal proteins are deficient in S – amino acids.

4. Poor mammalian digestibility of algal protein except for ruminant of processing of algae as for SCP production is précised as follows
5. The relatively high concentration of unsaturated fatty acids, which causes an adverse effect in animals.

(ii) Fungi: Fungi in view of their universal nature, fast rate of growth on a variety of waste substrates, are receiving greater attention in the exploitation as a source of SCP. At present mycoproteins of human food are being met from mushrooms, *Penicillium notatum, Fusarium* and *Trichoderma harzianum*. Some of the mentioned demerits of fungi as SCP organisms. However, they have disadvantages such as

1. The growth rate of fungi is slower than those of yeasts and bacteria.
2. The protein content is lower.
3. Fungal proteins are deficient in sulfur-containing amino acids.
4. The digestibility of fungal protein is problematic.
5. Fungi are unknown with regard to human nutrition and toxicology.

(iii) Yeasts: These are other groups of microorganisms that are in use. Infact, the concept of SCP is developed from yeasts. They are not only ubiquitous and most familiar to man as a component of beverage or bakery products. Three types of food yeasts are used extensively such as brewer yeasts (*Saccharomyces sp.* ,) baker's yeast (*S. cervisiae*) and dried yeast (*Candida utilis* and *Torula thermophila*). Dried yeast is a byproduct of is being used (*Candida utilis*). It was employed as food during World War I and II by Germans as a meat substitute for the army, prisoner of war and the civilian population. They have several good qualities to select as SCP.

1. Grown on any type of carbohydrate.
2. Sufficient quantities of essential amino acids (Lysine, tryptophan and threonine)
3. Good source of vitamin B, D and E.
4. Low nucleic acid content.
5. The larger size makes the SCP separation easy.
6. They are with lower toxicity potential.
7. Greater acceptance by consumers.

However, these organisms are reported to contain a low concentration of sulfur-containing amino acids.

(iv) Bacteria: Which are microscopic, fast-growing and versatile in nutrient requirement are rich in proteins. Hence, many bacteria such as species of *Bacillus* and *Proteus* are being exploited for SCP. Though they have several advantages such as:

1. Efficient in the conversion of substrate carbon to cell mass,

2. Higher protein conversion efficiency,
3. Lower oxygen requirement,
4. Lower amount and good lipid profile
5. The proteins of bacteria have a fairly good amino acid profile. Similarly, they are rich in sulfur containing amino acids.

They are the best sources with some limitations, including

1. Only a limited number of bacterial species are exploited for food purposes.
2. The small size of the bacteria makes the separation difficult.
3. In view of their growth under limited oxygen condition, they emit an unpleasant odour which may make it unacceptable and
4. Contamination with pathogenic species.

On the other hand, anoxygenic phototrophic bacteria proved to be efficient in the production of SCP as they grow as photoautotrophs or photoheterotrophs under aerobic and anaerobic conditions by utilizing a variety of carbon sources. They are fast in their growth rate, rich in protein, carotenoids and have highly competitive survival besides having several biotechnological applications. Some of the anoxygenic phototrophic bacteria are being exploited are listed in **Table 9.4** .

Table 9.4. Purple Non-Sulfur Bacteria Used in wastewater treatment and SCP Production

Name of the Bacterium	Source of Organism	Protein Production (%, mg/g dry wt)	Application	Reference
Rps. acidophila	Synthetic medium	23	Waste water remediation	Pfennig (1969)
Rhodobacter sphaeroides P47	From pineapple waste	66.6	Feed supplement	Noparatnaraporn and Nagai (1986)
Rba. sphaeroides P47	Dehydrated medium from pineapple peel waste	66.6	Waste water remediation	Noparatnaraporn and Nagai (1986)
Rba. sphaeroides	From cassava waste	Essential amino acids	Fish meal	Noparatnaraporn *et al.* (1986)
Rhodopseudomonas sp.,*Rsp. fulvum*	sugar refinery wastewater	58	Waste water remediation	Balloni *et al.* (1987)
Rhodocyclus gelatinosus R7	Tuna condensate	56	Waste water remediation	Prasertsan *et al.* (1997)
Rps. palustris	Sludge and sago starch processing	72–74	Aquaculture feed	Geetha (1998)

Name of the Bacterium	Source of Organism	Protein Production (%, mg/g dry wt)	Application	Reference
Rhodopseudomonas DK6	Wastewater from a latex rubber sheet processing	65.2-66.7	Used as animal feed	Kantachote *et al.* (2005)
Rps. palustris	Simulated wastewater	45	Waste water remediation	Honda *et al.* (2006)
Rba. capsulatus	Synthetic medium	45	Waste water remediation	Alexandre *et al* . (2009)
Rba. sphaeroides D-8	Medium with poultry dung	58.2	Used as SCP	Paronyan and Gasparyan (2009)
Rba. sphaeroides Z08	Medium withartificial soya bean waste	52	Treatment of soybean waste water	He *et al.* (2010)
Rhodovulvum sulphidophilum	Glutamate Malate Medium	15.6	Fish feed supplement	Suwansaard, (2010)
Rhodopseudomonas sp.	Synthetic 112 medium	11	Waste water remediation	Soon *et al.* (2014)
Rps. palustris P1	Fermented pineapple extract	64.7	Treatment of latex rubber sheet waste water	Kornochalert *et al.* (2014)
Rps. palustris, Rba. blasticus	Wastewater from noodle production	50	Waste water remediation	Worowit *et al.* (2016)
Rhodopseudomonas sp. CSK01	Municipal wastewater	60.1	Waste water remediation	Saejung and Thammarataa (2016)
*Rps. capsulatu*s	Cow dung	Essential sulfur amino acids	Fish feed	Gupta *et al.* (2016)
*Rps. faecalis*PA2	Ponds of sugar company (Thailand)	50	Recycling of sugar industry waste water	Saejung and Salasook (2018)
Rba. sphaeroides SS15	Basic isolation medium	54.0	Used as SCP	Chumpol *et al.* (2018)
Afifella marina STW181	Glutamate acetate medium	46.4	Used as SCP	Chumpol *et al.* (2018)
Rps. faecalis PA2	Waste of animal agriculture	62.7	Animal feed	Patthawaro and Saejung (2019)
Rhodocyclus gelatinosus	Miso-like effluent medium	63	Waste water remediation	Sasaki *et al.* (1981)
	Cassava waste	56	Waste water remediation	Noparatnaraporn *et al.* (1987)

Name of the Bacterium	Source of Organism	Protein Production (%, mg/g dry wt)	Application	Reference
	Seafood processing wastewater	50	Waste water remediation	Prasertsan *et al.* (1993 a, b)
	Poultry slaughterhouse wastewater	67.6	Poultry feed	Ponsano *et al.* (2002 & 2003)

The application of phototrophic bacteria to wastewater treatment is not only attractive methods since the microorganisms are widely available in nature and capable of removing organic substances even under high organic load especially in sunlight intensive areas like tropical countries. The treatment will require minimum additional energy apart from natural sunlight and the treatment cost can be substantially reduced. They apply to organic compounds and wastewater from various industries. Moreover the by-product of the treatment, the phototrophic bacterial cells has high nutrition value such as protein and vitamin. Purple non-sulfur bacteria can be divided into 9 families and 28 genera. They are gram-negative bacteria with varied shapes, including spherical, rod, ovoid, arc, or spiral.

The propagation method of purple non-sulfur bacteria is either budding or by binary split breeding. Bright colours such as purple, purple-red, pink, red and red-brown of the bacterial suspension are another important feature for purple sulfur bacteria (Imhoff, 1992). They are of a special microbial category because all the purple sulfur bacteria have in-vivo bacteriochlorophyll and carotenoids that carry out photosynthesis without oxygen production, which differentiate them from the cyanobacteria and green plants (Howard, 1987). Their photosynthesis will depend on the external electron donor, such as sulfide, molecular hydrogen, or organic substances. PSB has become popular for wastewater purification in recent times (Kosama and Obst, 2009) since there are two flexible metabolic pathways for them, i.e. the respiratory and the fermentative pathways (Siefert *et al.*, 1978), which enable them to adapt to varied environmental condition. Some PSB species have high toxicity resistance levels and even survive in toxic-ridden wastewater by which it utilizes the toxic substances as a carbon source for growth and reproduction (Madukasi *et al*,. 2010). Phototrophic bacteria grown in such substrates are known to produce biomass rich in protein with good contents of essential amino acids, as well as carbohydrates, lipids, minerals, vitamins and carotenoids, which suggests the potential use of the product in animal feed. Phototrophic bacteria lack polyunsaturated fatty acids (PUFA) and highly unsaturated fatty acids (HUFA), which are essential to the survival of larval fish (Loo, 2012). Koi (*Cyprinus carpio*) larvae are unable to survive long when fed with freeze-dried *Rhodovulum sulfidophilum* grown in a synthetic culture medium, but a diet containing the bacteria supplemented with the yolk granules of hard-boiled 'Omega Plus' hen egg significantly improved the survival and growth of Koi larvae (Neik, 2006).

Poor dietary essential fatty acids in fish give rise to abnormal pigmentation, fatty liver, poor swimming ability and raised basal cortisol levels.

The culture of phototrophic bacteria in wastewater rich in lipids may improve their content of PUFA and HUFA by reducing the high organic loadings present in agro-industrial wastewater. Agro-industrial wastewater such as palm oil mill effluents (POME), neutral lipids complex along with high concentrations of carbohydrate, protein, minerals and nitrogenous compounds (Phang, 1990). Kim and Lee (2000) estimated USD 79 per kg production of phototrophic bacteria but can be reduced by the use of agro-industrial wastewater as substrate. Cellulose has emerged as an attractive substrate for SCP and the development of low-cost, high-quality protein feed that is crucial for the future success of the aquaculture industry (Mondal *et al.,* 2012). As single-cell proteins (SCP), microalgae have been used as an essential food for the larval stages of fish and shellfish (Benemann, 1992). The amount of fish meal can be reduced if fish have grown with cheaper alternative protein sources, such as phototrophic bacteria (Bender and Phillips, 2004).

The beneficial effects of these bacteria are known in various areas especially as feed supplements because they are rich in protein, lipid, minerals, carotenoids and other biological cofactors. *Rhodocyclus gelatinosus*, a PNSB, is commonly recorded in many industrial effluents, where it grows well as photoautotrophs because of its ability to hydrolyze starch and gelatin due to hydrolyzing enzymes, particularly it produces. It also produces carotenoid pigments of the spirilloxanthin, an alternative series, which confers red colour to its cultures and may find its application in animal feed due to the possibility of intensifying animal tissue colour. Anoxygenic phototrophic bacteria, especially purple non-sulfur phototrophic bacteria (PNSB) are widely distributed in soil and water (Hoogewerf *et al.*, 2003). PNSB are versatile organisms as they can grow as both photoautotrophs and photoheterotrophs under anaerobic-light or microaerobic-light conditions. Purple non-sulfur bacteria (PNSB) are widely distributed in nature (Okubo *et al.*, 2006). They also can grow anaerobically in the dark using fermentation and many can grow aerobically in darkness using respiration. Anupama and Ravindra (2000) suggested bacterial production of SCP has more advantages than other groups.

Prototrophs have high protein content with essential amino acids and fewer amounts of nucleic acids and recommended purple non-sulfur bacteria as a source of single-cell protein for the fish and poultry industry (Salma *et al.,* 2007). Purple non-sulfur bacteria can use various substrates as sources of carbon and energy with ammonium and/or nitrate. Because of these properties, they are potential for treating various sources of wastewater. In addition, purple non-sulfur bacterial biomass is rich in protein with good quantities of essential amino acids, vitamins and carotenoids. Therefore, single-cell protein (SCP) may be a by-product of wastewater treatment and be used for animal feed (Ponsano *et al* . 2002). Cellulose from agricultural waste sources constitutes the most abundant renewable resource on the planet as potential substrates for SCP production.

Phototrophic bacteria contain 70-72% of crude protein (Kim and Lee, 2000). The mass production of phototrophic bacteria as SCP is a solution to the growing problem of protein-rich supplements for aquaculture .

Kim and Lee (2000) studied the use of *Rhodopseudomonas palustris* as the diet for aquaculture. The composition of essential amino acids of these bacteria is similar to that of soybean or egg proteins (Ponsanio *et al.*, 2002). They can grow both under light in anaerobic or microaerobic conditions as photoautotrophs and photoheterotrophs (Kaewsuk *et al.*, 2010). The application of PNSB has been used to convert to hydrogen gas (Liao *et al.*, 2010) or by-products such as single-cell proteins. PNSB biomass is useful for fish feeding as it is rich in proteins (Honda *et al.*, 2006; Chiemchaisri *et al.*, 2007). The factors which can be optimized for PNSB cultivation are anaerobic condition is light (in lux), pH, carbon, nitrogen, agitation and growth factors (Chitapornpan *et al.*, 2012). The photosynthetic treatment process has been applied to treat the wastewater and cultivate biomass with single-cell protein production. Balloni *et al.* (1986) used a photosynthetic consortium consisting of *Rhodocylcus gelatinosus* with pig farm waste to produce biomass with 7.5% moisture, 50.6% protein, 12% carbohydrates, 22.5% lipids and 15.5% ash. Further, Balloni *et al.* (1987) cultivated a mixed culture of *Rhodopseudomonas* and *Rsp. fulvum* in sugar refinery wastewater and observed biomass with 58% protein, 16.7% carbohydrates, 13.5% lipids and 12.8% ash. Others such as Sasaki *et al.* (1981) cultivated two strains of *Rhodocyclus gelatinosus* in a *miso*-like effluent medium and found 62% and 63% protein in the bacterial biomass. Noparatnaraporn and Nagai (1986) cultivated *Rba. sphaeroides* P47 in pineapple peel waste and obtained biomass with 66.6% protein, while Kobayashi and Kurata (1978) cultivated *Rhodopseudomonas capsulatus* in a synthetic medium and found 60.9% crude protein, 9.9% lipids, 20.8% soluble carbohydrates, 2.9% crude fiber and 5.3% ash in the biomass. Methionine and phenylalanine were in slightly lower concentrations (Noparatnaraporn and Nagai, 1986) in *Rcy. gelatinosus* R1 biomass when compared to other similar findings of other authors. *Rcy. gelatinosus* R1 biomass and amino acid composition were the same as that of algae (Prasertsan *et al.*, 1993a), yeast, animal and plant proteins. Prasertsan *et al.* (1993) cultivated *Rcy. gelatinosus* from seafood processing wastewater and found biomass with 50% protein and COD removal of 86%. Imhoff (2006a) felt that the use of phototrophic purple bacteria is prominent and can be applied in sewage treatment processes as they consume toxic substances and organic compounds present in the wastewater. He *et al.* (2010) reported producing SCP under natural conditions while treatment of soybean wastewater by a wild strain of *Rhodobacter sphaeroides.*

9.2. RAW MATERIALS

A variety of substances (in physical and chemical characteristics) are used as raw materials for the production of SCP. Some of these include wastes from all the three major sources like urban, agricultural and industrial. This will serve two purposes i.e. production of value-added protein and disposal of waste and pollutant materials. Some of the raw materials being exploited for the production of SCP are given in **Table 9.5.**

Table 9.5. List of Raw Materials for SCP Production

A. Agricultural Wastes	Straw (Rye, wheat, barley, rice, oat and grass)
	Stover (Sorghum, Soyabean, Corn)
	Sugarcane bagasse
	Cotton gin
	Husks from grains
	Vegetable and fruit peelings
B. Industrial Wastes	Paper Mill sludge
	Lignocellulose based industrial by-products and wastes
C. Municipal waste	Methane, ethane, propane and butane
D. Hydrocarbons	Liquide hydrocarbons (Gas oil , paraffins, Petroleum, Kerosine) and Alcohols (Methanol, ethanol)

The selection of the microorganisms is mainly dependent upon the available raw materials. Agricultural wastes are rich in lignocellulosic substances. Hence, filamentous fungi which are capable of producing ligases and cellulases are employed for SCP production. However, when the raw material is in a complex state, it is subjected to either acid or enzymatic hydrolysis before being used as a substrate for SCP production. Whenever, suitable inorganic substances are available at a cheaper rate and sunlight is available in plenty, algae and photosynthetic bacteria appear to be more promising SCP microorganisms than others due to many reasons such as

1. They utilize solar energy
2. Make use of cheap inorganic substances
3. SCP from this source is superior to another type in that it is free from toxic substances
4. They purify the environment

9.3. MIXED CULTURES

Mixed cultures proved to be more efficient in SCP production than monoculture in view of the following reasons.

1. Better utilization of a complex carbon source or of carbon source.
2. Removal of organic carbon excreted by one component of microorganisms.
    a. Removal of toxic compounds affecting one component of microorganisms.

b. Increase of the total biomass yield from the primary carbon source.

c. Elimination of foam.

d. Increase the growth rate of one component microorganism.

e. Enable water recycling.

f. Resistance to perturbation in the cultural environment.

g. Resistance to contamination by bacteria, yeast, fungi and phages.

h. Lower maintenance energy requirement.

3. Attainment of desired protein composition.

However, in choosing a substrate, consideration must be given to

1. The cost of the substrate
2. Biomass yield generation
3. Oxygen requirement during fermentation
4. The heat produced and level of fermenter cooling requires
5. Downstream processing costs

9. 4. STEPS IN SCP PRODUCTION

1. Selection of SCP microorganism
2. Substrate
3. Process
4. Product recovery
5. Purification
6. Drying
7. Supplementation of nutrients
8. Packing
9. Marketing

For producing quality SCP one has to take care in every step. The first and foremost important is the selection of microorganisms. The selected microorganism should be fast-growing, wide substrate utilizing capacity, should be non-pathogenic and do not produce any type of toxin, should produce protein not only of high quality but also in good quantity and contain a limited quantity of nucleic acid, etc. The next step would be a selection of substratum. The substratum is not only cheaper but readily utilizable. The organisms employed do not contribute to the liberation of toxic substances. Then the growing condition of the organism should be ambient and do not require any additional requirements. The incubation period should be short and the downstream process should be simple and, in a position, to yield pure protein with a minimum number of steps. The purification and drying of protein should be simple and should not affect the quality of protein. If protein is deficient in some respects that can be improved by supplementation of required substances. Thus, the post-harvest stage of SCP should do simple and amenable for processing. This will help to get a quality SCP and will be useful as a feed additive and for other purposes.

9.5. FERMENTATION PROCESSING FOR SCP PRODUCTION

Fermentation processing for SCP production should be simple and economical. Therefore following aspects should be critically assessed before taking up physiological and engineering aspects for SCP production.

i. Batch and continuation cultures of SCP production
ii. Improvement of media for biomass production
iii. Growth yields of microorganisms
iv. Physiological problems associated with scale-up
v. Sterilization
vi. Structured mixed culture
vii. Recycling
viii. Recovery of SCP products

Some of the economic consideration to be taken in SCP production includes

i. Capital costs: Cost of land, construction of plant and machinery
ii. Site suitability: Availability of transport system, raw materials and utilities size and proximity of markets and safety factors.
iii. Manufacturing (operating costs) cost of raw materials, energy, water supply, waste treatment, labour maintenance and depreciation.

In modern times loop reaction fermenters are being employed as they are more precise and economical. The details of conventional stirred vessels and modern loop reactors are précised in **Table 9.6.**

Table 9.6. Comparison of Fermenter for SCP Production

	Conventional stirred vessels	Modern loop reactors
1	High energy costs	Low energy costs
2	10 KW/m^3	KW/m^3
3	The less defined fluid flow pattern	The quantified fluid flow pattern
4	Intra fermentation cooling (Jacket) complicated heat removal	Extra fermentation cooling (Plate exchanger) Easy removal of the fermentation heat load
5	Limited scale-up	Less limited scale-up
6	Indefinite mixing irregular residence times	More defined mixing regular residence times
7	Cell yield on the oxygen of 10-20%	Cell yield on the oxygen of 40-50%

	Conventional stirred vessels	Modern loop reactors
8	Requiring high aeration rate	Requiring less aeration
9	Batch and continuous operation	Continuous operation
10	Low yield and productivity	High yield and production
11	Limited control	Efficient control
12	Moving part s(easily get contaminated)	No moving parts
13	Expensive capital and running costs	Cheapest installation run

9.6. NUTRITIONAL STATUS OF SCP

Microbial biomass which is harvested as a part of SCP is reported to have both inorganic and organic nutrients that are needed for the healthy growth of man and animals. It is generally fibrous, low in fats and carbohydrates and contains a variety of vitamins and minerals, which is desirable from the health point of view. Chemical analysis of protein SCP from *Pseudomonas* revealed that it contains most of the B-complex vitamins, vitamin A, inositol and choline, etc., in more than the required quantity (**Table 9.7**). The mineral analysis also revealed the presence of macroelements (**Table 9.8**) and microelements in the required quantity and at the same time, some of the heavy metals such as As, Hg, Sb, Pb, Cd and fluorine are below the toxic limits (**Table 9.9**). Varieties of fats are also present both qualitatively and quantitatively in a desirable quantity (**Table 9.10** .) The protein value is on par with protein-rich foods such as soybean, meat, etc., (**Table 9.11**). The quality of protein produced is also good in terms of the essential amino acid index and finally nutrition index. The amino acid composition of SCP is also most satisfactory (**Table 9.12** .)

Table 9.7. Vitamins Content in SCP

Vitamin	Production (mg/kg)
Ca-d pantothenate	1.58
Folic acid	8.74
Nicotinic acid	58.0
Choline	10.0
Biotin	3.12
Inositol	50.0
A-0.00	B_6-0.93
B_1-5.75	B_{12}-23.0
B_2-74.30	E- 28.0

Table 9.8. Macroelements content of SCP

Ash Analyses	(% of Dry wt)
P_2O_5	6.58
MgO	0.53
K_2O	0.30
Na_2O	1.53
CaO	0.04
Fe_2O_3	0.12
Total Ash	9.00

Table 9.9. Microelements Content of SCP

Trace ions (ppm)	Potential ions	Harmful (ppm)
Mn	AS	0.01
Zn	Hg	0.01
Cu	Sb	0.1
Se	Pb	0.05
Co	Cd	0.1
Cl	F	10.0

Table 9.10. Fat Content of SCP

Phospholipid composition	(% of total lipid P)
Phosphatidylethanolamine PEA	76.2
Phosphatidylglycerol PG	12.8
Phosphatidyl serine	8.8
Diphosphatidyl glycerol	2.2
Phosphatidyl choline	0.0
PEA/PG	6.0

Table 9.11. Protein Content of Different Foods alongwith SCP Organisms

Organism	Dry weight (in%)
Phototrophic bacteria	45-50
Mushroom	19-25
Rice	7.3
Wheat	13.2
Soybean	39.1
Milk	25.2

Table 9.12. Comparison of Nutritive Value of Phototrophic bacteria with various foods

EAA index	Amino acids score	Nutrition index
100 pork, chicken	100 pork	59 chicken
Beef	98 chicken, beef	43 beef
99 milk	91 milk	35 pork
98 mushroom	89 mushroom	28 mushroom
91 potato	63 cabbage	28 spinach
kidney beans	59 potatoes	26 spinach
88 corn	53 peanuts	25 milk
86 cucumber		
79 peanuts	50 corn	21 kidney beans
76 spinach	46 kidney beans	20 peanuts
72 mushroom	42 cucumbers	17 cabbage
cabbage	33 turnips	14 cucumbers
68 turnips	32 mushrooms	11 corn
53 carrots	31 carrots	10 turnips
44 tomatoes	28 spinach	9 potatoes
	23 soybean	8 tomatoes
	18 tomatoes	6 carrots
		5 mushroom

9.7. TOXICOLOGICAL STATUS OF SCP

The major drawback with SCP is its comparatively more amount of nucleic acid. Though ruminants (herbivores) possess the metabolism of nucleic acid which are oxidized to uric acid which in turn oxidized to allantoic acid (which is water-soluble) and get excreted in urine and never reach the toxic level. On the other hand, carnivorous animals and man have no mechanism of metabolizing uric acid to allantoic acid. As a result, Arisaema /gout disease is caused besides having interference to native nucleic acids. Risk of the presence of mycotoxins if the SCP organism happens to be a mycotoxin-producing fungus. Some of the SCP bacteria are known to produce endotoxins which may cause health hazards. If the organism selected for SCP produces fatty acids of uneven numbered carbon chain or relatively high content of palmitic acid is likely to have ill effects. The presence of unused carbons may turn out to be carcinogenic. If the SCP produced contains D-amino acids it will not be of much use as these amino acids will not be absorbed by human beings.

9.8. NUCLEIC ACID STATUS

The main hurdle in the utilization of SCP as food and feed is its comparatively high nucleic acid content. However, an attempt is being made to remove excess nucleic acids by different physicochemical methods **(Table 9.13)**. If these methods are successful without affecting nutritive value, SCP can be a good substitute for natural proteins. At present, the SCP is extensively used as a feed supplement to livestock.

Table 9.13. Methods for RNA Reduction in the SCP Product

Method	Protocol	Organism	Initial and final% RNA (of dry wt)
Base-catalyzed	0.12 N NaOH for 30 min at 50°C	*P. variotii*	9% to 3%
Chemical extraction	5% NaCl at 120°C	Baker's Yeast	Complete removal
Cell disruption	Disintegration	Different Microorganisms	Final % RNA =23
Physical properties	glass beads		
a) Enzymatic treatment	Malt sprout nuclease	Yeast	Final % RNA<3
b) Chemical	Succinic anhydride pH 4.2-4.4	*S. carlbergensis*	6.8% to 1.8%
c) Chromatography	Cellex E-Ecteola cellulose	*Candida utilis* *S. cereuisiae* *S. mobilis*	Less than 3% of initial content
Exogenous RNAase	Bovine pancreatic Ribonuclease 55-56°C	*C. utilis* *C. intermedia*	Final % RNA=1.5
Endogenous RNAase	90°C, pH 2.0, 20 min	*R. glutinis*	6.5% to 1.2%
Heat shock	45°C, pH 5.8-6.9, 2h	*C. utilis*	6.8% to 2.0%
	68°C for a few sec	*C. utilis*	7.8% to 1.2%
	50°C 1h, 55°C 1h	*S. cervisiae*	7.8% to 1.4%
	50-60°C + NaCl		

9.9. IMPROVEMENT OF QUALITY OF SCP

The nutritive quality of SCP can be improved by the application of genetic manipulation techniques including rDNA technology. Significant improvement in several aspects of SCP production can be also achieved.

1. Hydrolysis of sugar polymers and utilization of the resultant sugar monomers
2. Utilization of all constituents of multiple carbon sources by one organism.
3. Saving on the energy required for SCP production
4. Increasing the growth rate of the desired organism

5. Isolation of mutants that are able to grow at elevated temperature
6. Changing the morphological properties of the cells to improve the recovery process.
7. Alteration in the amino acid composition of the SCP product.

9.10. DOWNSTREAM PROCESS

At the end of fermentation, the microorganism is harvested by a suitable method depending on the microorganism employed. After harvest care should be taken to remove excess water by employing one of the methods as detailed in **Table 9.13**. The dehydrated biomass is subjected to cell breakage by one of the methods listed in **Table 9.13**. In the end, separation and purification of protein should be taken to make sure it is free from water by one of the suitable methods listed in **Table 9.13**. The procedure by this method will make the protein of good quality and whatever nutrient is deficient can be supplemented before the final packing. For instance, BP yeast Taphrina G protein is improved by supplementing with methionine.

9.11. UTILIZATION OF SCP

Supplementation of feed with anoxygenic phototrophic bacteria was found to be beneficial in both pisciculture and poultry. The food value of anoxygenic phototrophic bacterial cells has also been demonstrated in the culture of brine shrimp (*Artemia salina*) and plankton production (Kobayashi and Yoshida, 1983). Although the toxicity was not observed in the short-term studies, long-term toxicity tests need to be performed before anoxygenic phototrophic bacteria can be commercially used as SCP.

(i) Pisciculture: There is wide variation in the measured content of essential amino acids of different species of fish varied (Cowey, 1994), anoxygenic phototrophic bacteria can give to a large extent. The survival rate of cary fry was significantly enhanced when the feed was supplemented with anoxygenic phototrophic bacteria (Kobayashi and Yoshida, 1983). However, in pisciculture anoxygenic phototrophic bacteria should not be considered solely for young carry fry. The growth of goldfish (*Carassis auratus*) fed on *Rcy. gelatinosus* and yield cell mass (50%, w/w, the dry basis of fish meal) for a 122 day feeding period was significantly superior in terms of specific growth rate, percentage weight gain and fecundity with no mortality or over toxicity (Noparatnaraporn *et al.,* 1987b).

(ii) Poultry: The effect of anoxygenic phototrophic bacteria on egg-laying hen proved to be commercially important. It was shown that by supplementing feed with anoxygenic phototrophic bacteria, the quality, quantity and duration of egg-laying by a hen was improved (Hirayama and Ishigaki, 1988), aging speed of hens was prevented and the gill corruption disease of lobster and Marek's disease of egg-laying hen could be suppressed (Kobayashi and Kondo, 1984)

Anoxygenic phototrophic bacteria can be grown under both anaerobic light and aerobic dark conditions. Noparatnaraporn *et al.* (1983) found that while the growth rate was high with a relatively short time for maximum biomass for *Rcy. gelatinosus* under aerobic dark conditions, cell productivities are higher under anaerobic light conditions (g cell produced per gram starch consumed). They concluded that aerobic dark conditions can be proper for SCP production by *Rcy. gelatinosus* in cassava complex medium containing soybean meal at 40-45 °C. However, if either the time required for the maximum biomass could be minimized or the rate of growth could be improved, an outdoor culture in an area where there is abundant sunlight sufficient for the growth at a high cell mass concentration might be applicable for SCP production from cassava starch.

One of the major problems for the growth of anoxygenic phototrophic bacteria in a photobioreactor is the growth of oxygen-producing algae and sulfate-reducing bacteria. Despite the slow growth rates of the algae, they develop in the reactor liquid and also adhere firmly to the reactor walls. van Niel (1944) suggested the use of short IR illumination (800-1000 nm) as a source of radiant energy to eliminate algal growth. However, since one has to filter the shorter sun radiance, this may not be practical for outdoor photobioreactors. A more practical method is the use of growth inhibitors. The commonly used algal growth inhibitors, like copper sulfate, cycloheximide and nystatin have shown growth toxicity for many anoxygenic phototrophic bacteria as well (Segers and Verstraete, 1985). Chloroxuron was very effective against algal growth with no apparent growth inhibition on APB and was successfully used in photo-bioreactors at a concentration of 5-10 ppm (Segers and Verstraete, 1985; Driessens *et al.,* 1987). The other major contaminating organism is the sulfate-reducing bacteria which can be controlled by using molybdate (Na_2MoO_4 at 0.25 g/liter), an inhibitor of sulfate reduction (Smith and Klug, 1981).

Agglomeration and electrocoagulation were found to be the best methods for harvesting these bacteria on a large scale (Faust and Prave, 1979). Kobayashi (1976) used precipitants like "chitosan" for harvesting anoxygenic phototrophic bacteria after waste treatment. Synthetic cationic polymer like "praestol" which helps the biomass recovery by flocculation was also suggested to be the best method for the recovery of anoxygenic phototrophic bacterial biomass on a large scale (Benjamin *et al.,* 1993). About 1 mg/lit of praestol showed 86% biomass recovery of *Rps. palustris*. Furthermore, there was no inhibitory effect on the growth of the cultures when the medium was recycled in the pond after flocculation. Process economics will be improved when some of the current engineering problems are solved for SCP production.

9.12. PROSPECTS

In view of the above facts, SCP has the greatest potential to meet the increasing shortage of protein to explosive population growth. However, much is to be improved in order to harvest maximum fruits. Some of the important areas of this potential source include

1. Construction and operation costs have to be minimized
2. Use of cheaper raw materials such as hydrocarbons
3. Photosynthetic bacteria are such as preferable anoxygenic phototrophic bacteria which are autotrophic anaerobic require minimum investment can be used for this purpose
4. There is a need to develop agro-based SCP processes
5. Extensive and coordinated research in respect of microorganisms, substrate and processes, recovery, purification, etc.
6. The education of consumers is of prime importance.

■■■■

Chapter 10

BIOREMEDIATION OF WASTEWATER

The fermentation plant utilizes raw materials which are converted to a variety of products. Depending on the fermentative process varying amounts of a range of waste materials are produced. These waste materials or effluents generally include unconsumed inorganic and organic media components, microbial cells and other suspended solids, filter aids, waste wash water from cleaning operations, cooling water, water containing traces of solvents, acids, alkalies, human sewage, etc. In ancient times wastewater used to be disposed of off directly to a convenient area of land or into a nearby watercourse. This cheap and simple method of disposal is not environmentally desirable. The disposal of untreated effluents to municipal sewage treatment works (STW) will all depend on the composition, strength and volumetric flow rate of the effluents. If the discharge from an industrial process is large in volume and intermittently produced, it may be necessary to install storage tanks on-site to regulate the effluent flow. In some locations, municipal sewers are not available or the effluents may be such a composition that wastewater treatment company or regulatory authority requires pre-treatment of effluents may be necessary before discharge to its sewers. In these cases, an effluent – treatment plant will have to be installed at the factory site. Whatever the pollutant load of the liquid effluent its discharge to a sewer will be a cost-centered activity and will incur expenditure from the treatment company.

Fermentation effluents normally may not contain toxic materials which directly affect aquatic flora and fauna but do contain high levels of organic matter which is readily oxidized by the microbial attack and so drastically deplete oxygen levels as a result some species will die off in specific stretches of receiving effluents, while in other regions a different population capable of growth at lower oxygen levels will develop.

The pollution due to effluents may be treated in a variety of ways and it may be possible to recover waste organic materials as a solid and sell it as a by-product which may be used as an animal feed supplement or nutrient to use in fermentation media. The marketable by-product helps to offset the cost of the treatment process. It is now realized water is no longer cheap raw material and needs to be used economically and if the need arises recycled. The introduction of home-keeping will lead to reductions in the volume of water used and the volume of effluent for treatment and final discharge. Recycling and reuse of materials, waste minimization, waste reduction at source and integrated pollution control are now very important factors to consider in the design and operation of any manufacturing facility.

10.1. DISSOLVED OXYGEN CONCENTRATION AS AN INDICATOR OF WATER QUALITY

Since oxygen is essential for the survival of most microorganisms it is important to ensure that there are adequate levels of dissolved oxygen in rivers, lakes and reservoirs. Ideally, the oxygen concentration should be atleast 90% of the saturated concentration at the ambient temperature and salinity of the water. The measure of biologically degradable organic matter present in Wastewater by microorganisms present in a given time interval at a given temperature is called Biological oxygen demand or biochemical oxygen demand (BOD). The oxygen concentration of the effluent or a dilution of it is determined titrimetrically before and after incubation in the dark for 5 days at 20°C and the results are expressed as mg of oxygen consumed per dm^3 of the sample. Mineral nutrients and a suitable bacterial inoculum are usually added to the initial sample to ensure optimal growth conditions. However, by this method recalcitrant or inhibitory compounds cannot be assessed. A more quick method is to determine the quality of wastewater is the determination of chemical oxygen demand which is based on treating effluent water sample with a known amount of boiling acidic potassium dichromate solution for 2.5 to 4 hrs and then titrating the excess dichromate with ferrous sulfate or ferrous ammonium sulfate. The oxidized organic matter present in water is taken as proportional to the potassium dichromate utilized. The COD test is suitable even to measure the quantity of recalcitrant and toxic substances present in Wastewater. The BOD : COD ratios for wastewater are normally between 0.2:1.0 and 0.5:1.0. Very low BOD : COD ratios will indicate a high concentration of non-biodegradable organic matter and consequently biological effluent treatment processes may be ineffective. In such situations, a number of alternative tests are available such as total organic carbon (TOC) and permanganate value to determine the oxygen demand of wastewater.

10.2. SITE SURVEYS

A comprehensive survey of industrial operations is essential for any individual site before an economical, waste treatment program can be planned. It is desirable to divide the facility into as many units as possible as knowledge of the various material streams may show unexpected losses of the finished product, solvent wastage and excessive use of water which might be recycled, recovered or reused within the site. The factors and concentrations ought to be known at all production rates under which an individual unit may operate in a representative time period.

The survey may indicate a need for better control of water usage and should identify sources of uncontaminated and contaminated water that might be reused. It is more economical to treat concentrate effluents rather than a large volume of dilute effluents because of the saving on pumps and settling tank capacities provided that concentrations do not reach toxic or inhibitory levels in biological treatment processes.

The various wastes may be tested in a laboratory and on a pilot scale to assess the best potential methods of chemical and biological treatment. The commonly used biological tests include respirometry aeration-flask tests and continuous-culture experiments. Small flask respirometer and oxygen electrodes are used initially to establish the conditions to use in bio-oxidation of the affluent and to test for the presence of toxic materials. Large respirometers are useful for predicting effluent treatment rates and oxygen requirements. The residues in the flask can be analyzed for the presence of recalcitrants present if any. The use of laboratory continuous-culture vessels fitted with sludge-return pumps and settling tanks can provide detailed information. These tests may help in the design of a full-scale plant.

With the help of the above comprehensive survey, it is possible to establish a wastewater treatment program for a site to establish.

1. Water sources which can be combined or reuse
2. Concentrated waste streams contain valuable wastes to be recovered as food, animal feed, fertilizer, or fuel
3. Toxic effluents needing special treatment or acids or alkalis needing neutralization.
4. The effluent loading is expected under maximum production conditions.
5. The effluents might be discharged directly without treatment on to land or to the watercourse and not cause any pollution.
6. The effluents might be discharged into municipal sewers.

Based on the above information, it is possible to predict the size and type of wastewater treatment plant required and its capital and operational costs. This can be compared with water company charges to treat the waste at a sewage treatment works (STW) with and without on-site treatment.

10.3. THE STRENGTH OF FERMENTATION EFFLUENTS

The presence of high levels of particulate or soluble organic matter in water will result in potential high BODs. This generally occurs in large-scale fermentation processes. The medium rich in the inorganic matter is converted to biomass, primary or secondary metabolites and product which often represents a small portion of the raw material. The spent wastes remaining after the distillation of whisky may account for 90% of the initial raw organic materials, while in an antibiotic fermentation the effluents may represent more than 95%. The data of different industrial fermentations are precised in **Table 10.1.**

Table 10.1. BOD Strength of Wastewater (mg dm^{-3}) of different Industries

Wastewater	BOD
Domestic sewage	350
Paper mill	20,000 – 40,000
Beer	15,000
Brewery effluents	14,000 – 18,000
Alcohol stillage	10,000 – 25,000
Yeast production	3000 – 14,000
Antibiotic	5000 – 30,000 solvents upto $2x10^5$
Penicillin	40,000 – 70,000
Streptomycin	2450 – 5900
Aureomycin	4000 – 7000

Data in **Table 10.1** reveals that fermentation effluents may pose serious potential pollution problems and may be expensive to dispose of unless well-planned processes are used. A number of steps may be taken to reduce BOD in a process. Careful selection of raw materials may have a significant effect on the type and quantities of effluents being produced. The cheap raw material may not be ideal if product yield, recovery cost, effluent disposal cost and possible by-product value are considered together. The liquid fraction of industrial alcohol and distillery stillages can be concentrated to get a dried soluble fraction which can be sold. Some of the components of effluents are highly toxic and need to be detoxified or destructed before disposal. This will add to the overall cost of the process. For example, Vermectin which is a toxic metabolite of *Streptomyces auermitilis* fermentations is collected and degraded chemically .

10.4. WASTE DISPOSAL AND TREATMENT

The effluent disposal procedure adapted by the manufacturer varies and is determined by various factors and concern authorities will determine on the basis of quality and quantity of waste discharge. Some of the methods of waste disposal are

i. Effluents are discharged to land, river, or sea in an untreated state

ii. Effluent is partially removed and disposed of in a landfill site or incinerated.

iii. The effluents are partially treated on the site (eg. by lagoons), before further treatment or disposal by one of the other routes indicated.

iv. Part of the effluent is untreated and discharged as in i or ii, the remainder is treated at a sewage work or the site before discharge.

v. All of the effluents are sent to the sewage works for treatment, although there might be reluctance by the sewage works to accept it, possibly resulting in some preliminary on-site treatment being required and discharge rates and effluent composition defined.

vi. All the effluent is treated at the factory before discharge.

10.4.1. Disposal

i. **Sea and Rivers:** The simplest way of effluent disposal will be on a sea coast or in a large estuary through a pipeline extending below the low watermark. In this, no dilution is required and effluents get diluted by seawater. For the adoption of this method, the effluents must meet the requirement of a local river or drainage authorities. Sometimes stricter standards are often applied depending on the use of the receiving water. In addition, levels of ammonical nitrogen may be stipulated. There are often stringent upper limits for toxic metals and chemicals which might kill the fauna and flora. In recent times pretreatment of industrial wastes is obligatory.

ii. **Lagoons (Oxidation Pond):** Lagoons that are holding ponds may be used by the fermentation industry if the land is available at a reasonable cost. Lagoons are typically 1-2 m deep, simple to build and operate. The effluents are left in a lagoon for a couple of months during which aerobic bacteria digest the organic matter resulting in an anaerobic condition which in turn encourages anaerobic bacteria which break down undigested organic matter releasing water, soluble compounds and gasses **(Fig. 10.1)**. They may be used as the sole method of treatment incorporating both physical (sedimentation) and biological processes but the effluent produced may not reach locally acceptable standards. However, they can provide an initial pretreatment or can be used to polish effluents from the secondary treatment process.

iii. **Spray Irrigation:** Liquid wastes from fermentation can be directly applied to land as irrigation water and fertilizer when they are claimed to have a number of beneficial effects on the soil and plants. If this method of disposal is to be used then it is necessary to have a large area of land near to a fermentation unit in an area of low to medium rainfall. Pipelines costs will often restrict the use of this technique. The wastes of antibiotics and steroid wastes with BODs of 5000 to 20,000 mg dm^3 are disposed of by this method. These wastes were initially chlorinated to lower the BOD and reduce unpleasant odors and then sprayed onto the land until the equivalent of 38 mm rainfall reached. This process was repeated at monthly intervals and improved plant growth. The concentration of heavy metals and certain organic components will require careful monitoring and control to safeguard the environment and public health.

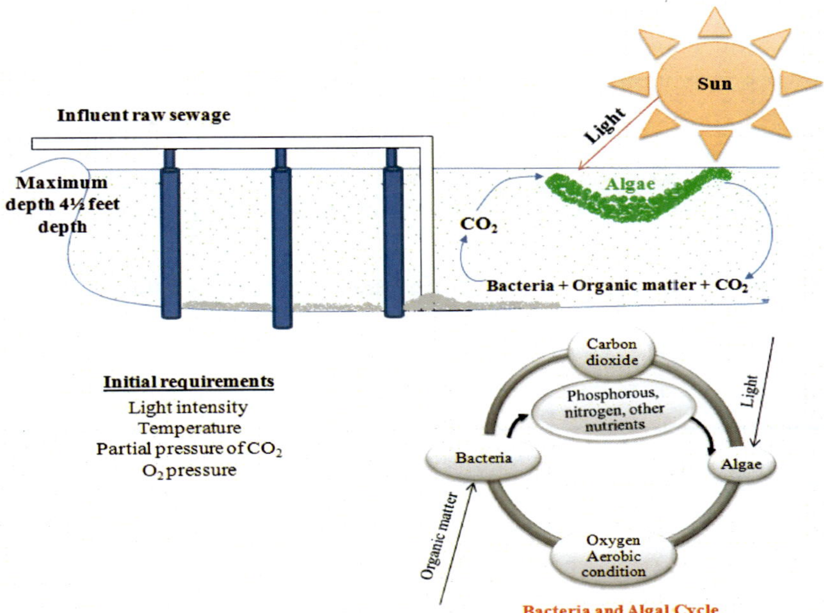

Fig. 10.1. Sewage Disposal by the Waste-Stabilization Pond Method

iv. **Well Disposal:** Disused wells, boreholes or mine shafts may provide an ideal, cheap method for effluents disposal when the volume of waste is limited, underground strata are suitable and the chances of contamination of water supplies are negligible. Careful hydrogeological surveys will be needed to prove that waste disposal in wells will not cause pollution of aquifers and threaten groundwater supplies.

v. **Land Filling:** It is a disposal method for municipal solid waste and industrial wastes. It utilizes natural or manmade voids (Ex. disused clay pits) into which the waste is deposited. Both solid and liquid wastes can be deposited depending on restrictions imposed by the site license. Strict controls exist on the amount of liquid and toxic materials which can be accepted because of the threat of groundwater pollution if leachates escaped from the site. Leachate is generated from liquid deposited in the site, water entering the site naturally via precipitation or surface runoff and by anaerobic microbial action as organic matter in the landfill is degraded by microbial action similar to that in anaerobic digesters which leads to the production of landfill gas (LPG) which being 50-60% methane, if collected efficiently provide a useful source of energy.

vi. **Incineration:** A number of designs exist for the incineration of solid and/or liquid wastes either on-site or at a commercial incinerator including rotary kilns, fluidized beds and multiple earth furnaces. Combustion temperature needs to be controlled carefully to destroy and prevent the formation of dioxins and furans, the formation of which occurs at between 300-800 and destruction occurs at a temperature above 1000 with a retention time of 1 second. Flue gases from the incinerator require cleaning to remove particulates, acid vapors, etc., using electrostatic precipitators, cyclones and wet scrubbers to comply with local environmental protection standards. Waste disposal by incineration is currently significantly more expensive than landfilling.

vii. Disposal of Effluents to Sewers: Municipal authorities and water treatment companies that accept the effluents into their sewage system insist for

1. The sewage works have the capacity to cope with the estimated volume of effluents
2. The effluent will not interfere with the treatment processes used at the sewage works.
3. There are no compounds present in the effluent which will pass through the sewage works unchanged and then cause problems when discharged into a watercourse
4. Generally, local authorities insist on preliminary on-site pretreatment before discharged into sewage to minimize the adverse effects of industrial wastes. Depending on the nature of effluents, pretreatment may include simple sedimentation to complex chemical or biological processes.

10.4.2. Treatment

The fermentation wastes may be treated on-site or at an STW by any or all the three following methods.

1. Physical, 2. Chemical and 3. Biological methods.

The final choice of treatment and disposal processes used in each individual factory will depend on local conditions.

The effluent treatment may also be described in the following manner

i. Primary Treatment: It includes physical and chemical methods (Ex. sedimentation, coagulation, etc.)

ii. Secondary Treatment: Biological methods (activated sludge) conducted after primary treatment.

iii. Tertiary Treatment: Physical, chemical and biological methods (Ex. micro strainer, sand filters and grass plot irrigation) used to improve the quality of liquor from previous stages.

iv. Sludge Conditioning and Disposal: Physical, chemical and biological methods Anaerobic digestion is often used to condition the sludge produced in previous stages. Following dewatering, the sludge can then be disposed of by incineration, landfilling etc.

10.4.2.1. *Physical Processes*

The removal of suspended solids by physical methods before subsequent biological treatment will considerably reduce the BOD of the resulting effluent. In nearly all fermentation processes the cells are separated from the liquid fraction in recovery processes. In effluents with cells and filter aid, the recovered materials may be dealt with in two basic ways.

1. The waste is disposed of without any further treatment.
2. The waste bulk is reduced by mechanical dewatering with a filter press, centrifuge, rotary vacuum filter, or belt press. The compressed waste is incinerated or disposed of in a landfill site.

Solid wastes are produced in some processes such as malting before inoculation. In breweries using malted grain, coarse screens or whirlpool centrifuges may be used to remove spent grain from the wort after it is mashed. Similarly, hops will be recovered on screens in a hop back. Both the spent grain and hop wastes may then be mechanically dewatered before being sold or dumped. Stillage removed from distilleries is mechanically dewatered and dried in rotary driers to yield a potentially marketable residue known as distillers grains. In Scotland about half of the whisky distilleries wastes are being evaporated to a waste syrup containing 45% solids mixing with spent grain, drying and selling the final product as Distillers Dark Grains as low – grade cattle food. Physical processes installed for primary effluents treatment may include the following stages

i. **Screens:** To remove large suspended and floating matter.

ii. **Combinators:** To reduce particle size

iii. **Constant Velocity Channels:** (~0.3 ms^{-1}) for grit removal to prevent damage to the plant in latter processes.

iv. **Sedimentation Tanks:** For removal of finer suspended matter. These are generally circular or rectangular continuous flow tanks operating at retention times of 6-15 hours with a facility for the continuous removal of settled sludge. A sedimentation tank can remove 70% of incoming suspended solids and depend on the nature of waste up to 40% of its BOD load. They can be operated with or without prior chemical coagulation/flocculation. Similar sedimentation processes are also conducted after secondary (biological) treatment.

Physical processes used in tertiary treatment to produce an effluent of better quality than 30:20 standards include micro strainer, slow sand filters, up flow sand filters and rapid gravity sand filters. Suspended solids removal is generally 50-70 and BOD removal around 30-50% depending on the technique used.

10.4.2.2. *Chemical Processes*

Fine suspended particles in an effluent may be removed by coagulation and/or flocculation. Coagulation is essentially instantaneous, while flocculation requires some more time and gentle agitation to achieve aggregation of the particles. Ferrous or ferric sulfate, aluminum sulfate (alum), calcium hydroxide (lime) and polyelectrolytes are often used as chemical coagulants. A solution of coagulant of appropriate strength for effective treatment is added to the effluent in a vigorously mixed tank, a precipitate or floc forms almost immediately and carries down the suspended solids to form sludge. This sludge may be drawn off, mechanically dewatered and subjected to further treatment. The flocs formed on coagulation may be small and will therefore require an extended period to settle and as a consequence for a given throughput of effluent a large sedimentation tank will be needed. Increasing the particle diameter by encouraging small flocs to coalesce (flocculation) increases the rate of sedimentation and thus for a given throughput, a smaller vessel can be operated. Polyelectrolytes are commonly used as flocculants and following addition, the effluent is gently mixed (turbulent mixing would break up the flocs) by passage through sinuous flocculation channels, hydrodynamic flocculation, or mechanically mixed flocculation.

10.4.2.3. *Biological Processes*

Human activities have resulted in the release of a wide variety of pollutants that have a negative impact on the environment in general and human health in particular.

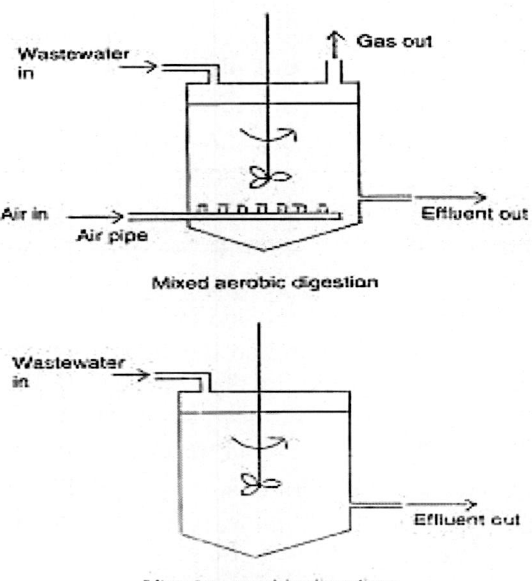

Fig. 10.2. Basic Biological Treatment Processes

Generally, all organic materials are degraded biologically either by aerobic or anaerobic microorganisms in a number of ways. The most widely used aerobic activated sludge processes are trickling filters, rotating disc contractors, activated sludge processes and their modification. The anaerobic processes (digestion, filtration and sludge blanket) are used both in the treatment of specific wastewater and in sludge conditioning. Some of the basic biological treatment processes are illustrated **(Fig. 10.2)**.

The biological treatment of wastewater is to optimize a natural process. For this purpose variety of microorganisms such as heterotrophic bacteria (species of *Pseudomonas* , *Alcaligenes, Citrobacter, Flavobacterium, Nitrobacter, Nitrosomonas, Nitrococcus* and *Bacillus*), anoxygenic phototrophic bacteria (purple non-sulfur bacteria and purple sulfur bacteria) and fungi (species of yeast) are to be optimized. Out of all the above organisms, anoxygenic phototrophic bacteria are more useful and in view of their diversity, wide adaptability and ease of their maintenance. Further, phototrophic bacteria need a large amount of nitrogen and phosphorus for their growth. They will be ideal organisms to decrease COD due to nitrogen, phosphorus and different heavy metals (Takuchi *et al.*, 2016).

Bioremediation as a green process offers a better solution to the conventional method in the treatment of such pollutants. Most of these pollutants can be effectively removed from the environment by the use of phototrophic bacteria. The use of anoxygenic phototrophic bacteria in the process further reduces the environmental impact due to their ability to use CO_2, while bioremediation using some microorganisms produces substances that sometimes are most toxic than the original pollutants. On the other hand, non-oxygenic phototrophic bacteria mostly produce highly valuable substances. Therefore, the application of this group of bacteria is much more economical compared to the use of other groups of microorganisms. Some of the advantages of their use are as follows:

1. They are diverse in their metabolism and can assimilate easily a variety of substrates.
2. They grow fast and build up their population in a minimum time.
3. The treatment process with these organisms can be operated at low physiological temperatures (10-40°C).
4. They grow under a variety of conditions such as photoautotrophy, chemoautotrophy, heterotrophy and fermentation.
5. They require only hydrogen (H_2) donors, possibly H_2O, industrial and agricultural wastes and solar energy.
6. As they can grow under anaerobic conditions, dilution of wastewater is not needed.
7. The use of these organisms does not result in sludge formation and the sludge treatment process can be avoided.

8. They can be easily manipulated for the requirement of the remediation process.
9. These organisms can be grown in a mixed population.
10. The equipment needed for this process is simple and most economical.
11. The treatment system is of multiple utilities such as food chemicals, waste treatment and biofertilizers.
12. The area for the treatment facilities is much less.
13. Photosynthetic treatment of waste produces bacterial biomass rich in protein, many vitamins and carotenoids.
14. The sludge formed during treatment with APB, neither autoclaving nor processing are needed. Further sludge on combustion yields only a small amount of ash remains and low-temperature combustion (700-800°C) does not produce undesirable gases but rather generates hydrogen which augments the combustion process.

The only requirement for the use of these organisms is to maintain temperature, pH, moisture level, aeration, minimize inhibitory metals and minimum nutrients. These organisms will be of much help in the removal of aromatic hydrocarbons (petroleum products) and sulfurous substrates from Wastewater.

A. Wastewater Treatment: APB which are anaerobic phototrophs can be grown at high BOD, needs no dilution of wastewater, require small treatment space, no sludge disposal problem, easy maintenance of treatment space, tolerance of cold, the feasibility of nitrogen, grease removal and biomass rich in protein (60%) and vitamin (B_{12}) are ideal organisms for wastewater treatment. Costa *et al.* (2017) reported the use of *Rba. capsulatus* as a bioremediation agent for biological wastewater treatment. *Rhodopseudomonas palustris* (Perrotta and Harwood, 1994; van der Woude *et al.,* 1994) was reported to metabolize benzoate and useful in the taxonomic identification of the species. Photocatabolism of benzoate by *Rhodobacter capsulatus* (Blasco and Castillo, 1992) has been studied extensively. Frolund *et al.* (1996) and Sheng *et al.* (2010) have excellently reviewed the use of EPS of phototrophic bacteria in biological wastewater systems. Similarly, Judd and Judd (2011) have discussed the principles and applications of membrane bioreactors in wastewater treatment. Therefore, photosynthetic bacteria have an appearance as an engineering alternative for conventional wastewater treatment methods (Talaiekhozani and Rezania, 2017). However, the photosynthetic and non-photosynthetic bacterial consortium can be more effective.

The APB plays a crucial role in biogeochemical cycles by anaerobic catabolism of aromatic compounds. The mineralization of aromatic compounds by facultative or obligate anaerobic bacteria can be coupled to anaerobic respiration with a variety of electron acceptors as well as fermentation and anoxygenic photosynthesis (Carmona *et al.,* 2006). Most of the PNSB plays a vital role in the anaerobic degradation of various organic substrates as a carbon source for their growth. Particularly polymers (Poly-3-hydroxy alkanoates and starch), one-carbon compound, alcohols, aliphatic hydrocarbons and higher fatty acids (Sasikala and Ramana, 1998) proved to be good substrates for the growth of anoxygenic phototrophic organisms. Because of these properties, APB is being exploited for treating wastewater. Species of *Rhodopseudomonas, Rhodobacter* and *Rhodospirillum* are being employed in the degradation of various pollutants (Tadesse *et al.,* 2003; Kim *et al.,* 2004). Zhang *et al.* (2014) recorded the improvement of water quality and change of microbial community with the addition of *Rhodopseudomonas palustris* .

The PNSB was employed in the treatment of odorous swine wastewater (Kim *et al.,* 2004; Shan Wen *et al.,* 2016). Mixed cultures of *Rba. sphaeroides, Chlorella sorokiniana* and *Spirulina platensis* proved to be better for treating organic wastewater (Ogbonna *et al.,* 2000). Treatment of domestic wastewater by using microaerobic membrane bioreactors (Libing *et al.,* 2006) proved to be efficient. Nagadomi *et al.* (2000) have successfully reduced phosphates, nitrates and H_2S and removed COD in synthetic sewage water using porous ceramic immobilized photosynthetic bacteria. Removal of heavy metals and sodium from contaminated shrimp ponds (Saijai *et al.,* 2010), treatment of pharmaceutical wastewater (Madukasi *et al.,* 2010), microaerobic biodegradation of high organic load of wastewater (Madukasi and Zhang 2010) could be successfully removed by exploiting APB. Luo *et al.* (2012) could improve the quality of wastewater from shrimp farms by using a combination of fish, photosynthetic bacteria and vegetation.

APB is efficient in the degradation of organic wastes mostly in agricultural and food industries such as molasses wastes (Sasikala *et al.,* 1992), citric acid fermentation wastes (Yu *et al.,* 1991), refinery, lactic acid wastes (Sasikala *et al.,* 1995), high strength organic wastewater (Ogbonna *et al.,*2000) and oil containing sewage water (Takeno *et al.,* 2005). The denitrification and deammonification capabilities of APB are of special interest because in industrial effluents the ammonia and nitrate concentration is often much higher, which causes many problems in discharging such Wastewater into rivers and lakes (Bashan and Bashan, 2004). *Rhodovulum sulfidophilum* is being used for the treatment and utilization of sardine processing wastewater (Azad *et al.,* 2003). Ponsano *et al.* (2008) and He *et al.* (2010) could successfully treat the wastewater with the help of *Rubrivivax gelatinosus* and *Rba. sphaeroides* respectively. Kaewsuk *et al* . (2010) developed and evaluated membrane sequencing batch reactor (MSBR) with mixed cultured photosynthetic bacteria for dairy wastewater treatment. Feng (2016) employed *Rhodopseudomonas* sp. strain Hy21 for removal of high concentration of organic wastes from soybean Wastewater, citric acid and distillery Wastewater. Amezaga *et al.* (2014) have demonstrated the reduction of monosodium glutamate most effectively photosynthetic bacteria.

PSB has been applied in wastewater treatment for purifying wastewater effectively and recovering valuable bioresource (Kobayashi and Kurata, 1978). Kobayashi (1977) observed that the self-purification process of wastewater

(Kobayashi, 1977), which could degrade the organic matter of Wastewater. Therefore, PSB has been used in the treatment of many types of wastewater such as aquarium water (Nagadomi *et al.*, 1999), sewage wastewater (Nagadomi *et al* ,. 2000), olive oil mill wastewater (Eroglu *et al.*, 2008; Carlozzi *et al.,* 2015), shrimp farms wastewater (Luo *et al.*, 2012), poultry slaughterhouse wastewater (Ponsano *et al.*, 2008), dairy wastewater (Kaewsuk *et al.*, 2010), palm oil mill effluents (Budiman *et al.,* 2014), soybean wastewater (Zhang *et al.*, 2010; Lu *et al.*, 2010; Lu *et al.*, 2011), beer wastewater (Madukasi *et al.*, 2010), latex rubber sheet (Kantachote *et al.*, 2005). Wastewater treatment using PSB with different approaches such as adsorption (Douglas *et al.*, 2016), coagulation (Zhu *et al* ,. 2016), photocatalyst oxidation (Chong *et al.*, 2010) and biodegradation (Liu *et al.*, 2011 and Li *et al.*, 2016) proved to have limitation in applications such as energy, efficiency, stability and economy.

Since, purple sulfur bacteria that are widely distributed in lakes, ocean, soil and other natural environments and carry out photosynthesis under anaerobic conditions using light as an energy source and have diverse metabolic pathways are efficiently degraded organic wastes and have excellent resistance to a variety of toxicants such as chlorines, salts, cyanides and phenols and are capable of photosynthesizing cell material.

The APB is reported to degrade aromatic hydrocarbons (Rajasekhar *et al.,* 2000), benzoates and their derivatives (Wright and Madigan, 1991) and amines (Cabello *et al.,* 2004). Degradation of homocyclic and heterocyclic aromatic compounds by *Rps. palustris* strain has been reported by Cetinkaya and Ozturk (1999). The utilization of aromatic compounds by APB has been reported by several investigators (Shoreit and Shabeb, 1994; Imhoff, 1995). Photoassimilation of heterocyclic aromatic compounds (Rahalkar *et al.,* 1993 and Sasikala *et al.,* 1994), thiols (Visscher and Taylor, 1993) and phenols (Blasco and Castello, 1992) by *Rps. palustris* has also been reported. Phototrophic bacteria are reported to degrade efficiently the aromatic compounds such as benzoates, benzoate derivatives (Wright and Madigan, 1991). Mixed culture of *Rba. sphaeroides* , *Chlorella sorokiniana* and *Spirulina platensis* proved to be better for treating organic wastewater (Ogbonna *et al.,* 2000).

The practical application of APB in biological treatment and bioremediation of organic waste has been reported by Kobayashi *et al*. (1995). *Rba. sphaeroides* , *Rba. capsulatus* , *Rcy. gelatinosus* and *Rps. palustris* are widely used in the remediation of wastewater (Sasikala and Ramana, 1995b). David and Ensign (2005) have isolated *Rps. palustris* capable of growing on acrylamide under photoheterotrophic conditions. The potential of *Rba. sphaeroides* for bioremediation of heavy metal contaminated environments were assessed by Livia *et al* . (2006). Light-dependent transformation of aniline to indole esters was mediated by *Rba. sphaeroide*s (Vijay *et al.*, 2006). Photosynthetic bacteria have been recognized for wastewater purification (Kantachote *et al.,* 2005; Chae *et al.,* 2006; Kosama and Obst, 2009; Shan Wen *et al.*, 2016; Costa *et al.,* 2017). Biofilm formation of anoxygenic phototrophic bacteria induced by phototoxic for enhancing hydrogen production. Frolund *et al.* (1996) and Sheng *et al.* (2010) excellently reviewed the use of extracellular polymeric substances of phototrophic bacteria in biological waste treatment systems. Judd and Judd (2011) have discussed the principles and applications of membrane bioreactors in water and wastewater treatment.

The treatment of wastes by APB may be coupled with biomass and hydrogen production in the following processes (i) One-stage process- a given species of APB is selectively used on a given waste (Mitsui *et al.,* 1985b). (ii) Two-stage process in which the APB phase either precedes cyanobacterial (Balloni *et al.,* 1980) or the algal stage (Kobayashi, 1976) or the heterotrophic bacterial stage (Sawada *et al.,* 1977). The APB stage is followed by the activated sludge process after dilution (Sasaki *et al.,* 1981). (iii) A multistep process involving sequential growth of microorganisms (Kobayashi, 1976). When compared with the single-stage, the use of phototrophic bacteria in the second stage provided increased COD reduction and improved biomass production (Sawada *et al.,* 1977). Two-stage continuous culture with a longer residence time in the second stage was advantageous in effecting assimilation of volatile fatty acids by mixed culture of *Rba. capsulatus* and *Klebsiella* (Sawada and Rogers, 1977b).

The combination of biodegradation and nanotechnology has been suggested to be potentially effective, low-cost and ecofriendly (Huang *et al.*, 2014; Oh *et al.*, 2016). Therefore, the application of phototrophic bacteria such as *Rhodopseudomonas sp.* strain A7 (Liu *et al.*, 2015) proved to be effective in wastewater treatment. Among different phototrophic bacteria, PSB proved to be efficient and eco-friendly that can simultaneously remove carbon, nitrogen and phosphorus (Nagadomi *et al.*, 2000; Lu *et al.*, 2010; Idi *et al.*, 2015). However, biological wastewater treatment has some limitations as such biodegradation processes are usually slow.

Since purple sulfur bacteria (PSB) are a kind of photoheterotrophic bacteria that are widely distributed in lakes, oceans, soil and other natural environments and carry out photosynthesis under anaerobic conditions using light as the energy source and diverse metabolic pathways are efficiently degraded organic waste. Phototrophic bacteria have strong abilities of decomposing and transforming a variety of organic compounds, excellent resistance to toxicants such as chlorine, salts, cyanides and phenol and are capable of synthesizing cell materials (Zhou *et al.*, 2015). He *et al.* (2017) reported the combination of nanoparticles and microbes such as photosynthetic bacteria for wastewater treatment. The efficiency of wastewater pollutants removal by this PSB/Fe_3O_4/biochar agent removed 83.1% of chemical oxygen demand, 87.5% of NH_4^+ and 92.1% of PO_4^{3-} from the Wastewater.

i. BOD/COD/TOC Reduction: A reduction in BOD/COD level of 20-99% has been reported depending upon the nature of wastewater and the organism used. A maximum of 80% reduction in COD was obtained for soybean cooked drain waste using *Rcy. gelatinosus* (Sasaki *et al.*, 1981). BOD reduction of 99% in food processing unit waste solution was reported by Kobayashi and Yoshida (1983). Sewage waste was first biodegraded by *Rba. capsulatus*, which reduced the BOD content of the waste from 5,800 to 2,000 mg/L in 4 days. A reduction of about 90% in COD levels was also recorded by treating waste solutions from swinery (Kobayashi, 1976). Not only the BOD/COD/TOC reduction but also the anoxygenic phototrophic bacteria may also aid in treating the wastes of the harmful microbes since they are known to produce antiviral substances (Hirotani *et al.,* 1989), bacteriocins (Guest, 1974) and antibiotics (Burgess *et al.,* 1991). *Rba. capsulatus* capable of producing an antiviral substance and found effective in removing 97% of coliphages from a pig pen (Hirotani *et al.,* 1989). The reduction of BOD and COD in percentage during wastewater treatment using APB are listed in **Table 10.2**. By wastewater treatment, one can get H_2 which is an eco-friendly energy source.

Table 10.2. Reduction of BOD/COD/TOC during Waste Treatment using APB

Organism	**Type of waste**	**Reduction in BOD/COD/TOC (%)**	**Reference**
Rba. sphaeroides	Night soil	BOD (90)	Kanamori and Kobayashi (1982)
	NO_2 waste	Gas (100)	Kobayashi (1975b)
	HCHO, phenol and cresol	TOC (99)	Kobayashi and Kenkyusho (1982)
Rba. capsulatus	Hydrocarbon Waste	BOD (90) COD (84)	Mochida *et al.* (1975)
Rps. palustris	Sulfate waste	SO_4^{2-} (84)	Gruenwald and Grau (1980)
Rsp. rubrum	Cane molasses	COD (22)	Vatsala and Ramaswamy (1989)
Rc. gelatinosus	Soybean cooked drain	COD (80)	Sasaki *et al.* (1981)
Rps. acidophila	Composting	BOD (99) COD (99)	Mitsubishi Heavy Industries Ltd., Japan (1982)
Phototrophic Organism	Noodle plant	BOD (95)	Mitsubishi (1985)
Rhodopseudomona sp.	Orange processing plant waste	TOD (750 ppm)	Mitsui *et al.* (1985b)
Rba. sphaeroides S , *Rba. sphaeroides* NR-3	Synthetic sewage wastewater	COD (89) NO_3^-(99) Phosphates (77)	Nagadomi *et al.* (2000)
Rcy. gelatinosus	Latex processing wastewater	COD (57)	Choorit *et al.* (2002)

Organism	Type of waste	Reduction in BOD/COD/TOC (%)	Reference
Rps. palustris	Swine wastewater	COD (90)	Kim *et al.* (2004)
Rba. sulfidophilum	Sardine processing wastewater	COD (85)	Azad *et al.* (2004)
Rhodopseudomonas sp. DK6	Latex rubber sheet wastewater treatment plant	COD & BOD (90)	Kantachote *et al.* (2005)
Rcy. gelatinosus	Poultry slaughterhouse wastewater	COD (91)	Ponsano *et al.* (2008)
Rps. palustris & *Rba. blastica*	Noodle processing wastewater	COD (90)	Chiemchaisri *et al.* (2008)
Rps. palustris P1	Latex rubber sheet wastewater	COD (92), SS (87)	Kantachote *et al.* (2010)
Rba. sphaeroides	Pharmaceutical waste water	COD (80)	Madukasi *et al.* (2010)
Rba. sphaeroides	Olive mill wastewater	COD (31-33)	Eroglu *et al.* (2010)
Rba. sphaeroides Z08	Soybean wastewater	COD (90-96)	He *et al.* (2010)
Rvi. gelatinosus	Tilapia fish processing wastewater	COD (43)	de Lima *et al.* (2011)
Rps. palustris	Food processing wastewater	BOD (51) COD (58) SO_4^{2-} (64)	Chitapornpan *et al.* (2013)
Rps. palustris RSOU555, *Rps. thermotolerance*	Polluted waters of Saroornagar lake	BOD (50, 68.16) COD (40, 56.66) TS (53.57, 54.46) TDS (52.28, 54.09) SO_4^{2-} (42.11, 55.26) NO_3^-(56.61, 33.96)	Rajani *et al.* (2016)
Rhodopseudomonas sp. CSK01	Municipal Wastewater	COD (85)	Saejung and Thammaratana (2016)
Rps. faecalis	Sugar industry wastewater	COD (80)	Saejung and Salasook (2018)

Abbreviations: BOD, Biological oxygen demand; COD, Chemical oxygen demand; SO_4^{2-}, sulfates; TOD, Total oxygen dissolved; TOC, Total organic carbon; TS, Total solids; SS, suspended solids; TDS, Total dissolved solids.

ii. Removal of Carbon Monoxide (CO): Carbon monoxide is a natural by-product in a variety of biological and chemical reactions that occur in nature can be oxidized by the enzyme Co-dehydrogenase produced by *Rhodocyclus gelatinosus* (Champine and Uffen, 1987) and *Rsp. rubrum* (Kerby *et al.,* 1992). The waste biomass converted into producer gas ($CO + H_2$) is used for the growth of PNSB and the production of bio polyesters (poly-hydroxybutyric acid) (Maness and Weaver, 1994).

iii. Denitrification and Deammonification: The denitrification and deammonification capabilities of APB are of special interest because in industrial effluents the ammonia and nitrate concentration is often much higher causing many problems if such wastewater is discharged into rivers and lakes (Wiesmann, 1994). Denitrification is a common property of *Rba. capsulatus* (Martinez- Luque *et al.,* 1991) with an enzyme nitrate reductase. The immobilized denitrifying *Rps. palustris* is being exploited effectively to remove NO_3^- nitrogen from the wastewater (Shen and Hirayama, 1993). APB proved to be efficient in the denitrification and deammonification of Industrial Wastewater. Zhang *et al.* (2014) reported that the addition of photosynthetic bacteria decreased the nitrite-reducing bacteria significantly.

Many by-products such as SCP, vitamin B_{12}, hydrogen, 5-aminolevulinic acid (Herbicide) and biomass for pisciculture, poultry industry and biofertilizers during wastewater treatment process by APB are useful in a variety of fields. The biomass of some of these bacteria (*Chromatium vinosum*) is used in the removal of heavy metals like iron and other metals.

iv. Degradation of Heavy Metals: The efficiency of anoxygenic phototrophic bacteria can be improved by the process of immobilization of their cells. The efficiency of *Rhodobacter sphaeroides* in the removal of phosphates and nitrates was improved by immobilizing their cells in a porous ceramic (Nagadomi *et al.,* 2000). Sawayama *et al.* (1998) recorded the stimulation of *Rhodopseudomonas capsulatus* when its cells are immobilized with cellulose beads and resulted in increased removal of phosphates, organic carbon and ammonium from polluted water. *Rba. sphaeroides* were utilized in the removal of phosphorus from oyster farm mud sediments (Takeno *et al.,* 1999).

Pollution of heavy metals in agricultural soils and industrial Wastewater is of increasing concern due to food safety issues and potential health risks because they cannot be biodegraded and may be leached to surface water runoff, groundwater storages, plant absorption and get accumulated in agriculturally important crops and pose not only plant health problems but also pose a health hazard to animals and humans (Yap *et al.,* 2004; Mohktar *et al.,* 2009). Therefore, biosafety of decontaminated soil/water has to be tested for their safety. Varieties of microorganisms are being exploited for the removal of heavy metals from soil/water. However, the use of anoxygenic phototrophic bacteria proved to be more efficient. Many members of *Rhodospirillacaea* show high resistance to heavy metal oxides and oxyanions (Moore and Kaplan 1992). Cadmium and lead ions were absorbed by *Rhodobacter sphaeroides* (Seki *et al.,*

1998). Selenium toxicity has been reduced to a significant level by *Rhodopseudomonas rubrum* (Kessi *et al.,* 1999). Nepple *et al.* (2000) could reduce chromate in industrial effluents with the help of *Rhodobacter sphaeroides*. They could characterize the enzyme chromate reductase responsible for the reduction of toxicity. *Rhodobacter sphaeroides* and *Rhodospirillacaea* could remove cadmium in a batch cultures system (Watanbe *et al.,* 2003). Livia *et al.* (2006) reported good bioremediation of heavy metals by *Rhodobacter sphaeroides* . Panwichian *et al.* (2012) have undertaken detailed studies on the purple non-sulfur bacteria in the remediation of heavy metals and oil-contaminated shrimp ponds.

Two major mechanisms used by purple phototrophic bacteria for the removal of heavy metal ions are bio absorption and the bioremediation process. The difference between the two processes is in requiring energy, the latter requires more energy. The type of mechanism to be used by the organism is dependent on the properties of the metal ions that are exposed. Some PPB may use both the mechanisms and likely to be more efficient when compared to the one that uses only one mechanism.

Microbial reduction of toxic heavy metals to less soluble forms represents a useful detoxification process. Phototrophic purple bacteria are particularly resistant to various heavy metals including Cd, Zn, Al, Fe and As. Kobayashi and Kobayashi (1995) have successfully used anoxygenic phototrophic bacteria for treating industrial wastewater rich in heavy metals .*Rhodospirillum rubrum* has been used for cadmium bioaccumulation (Smiejan *et al.,* 2003) and inactivated *Rhodobacter sphaeroides* for cadmium biosorption (Watanabe *et al.,* 2003). Testing of *Rba. sphaeroides* as *a* heavy metal removal tool has been carried out by Buccolieri *et al.* (2006) and recorded its tolerance to Ni, Co, Cr and Mo at high concentrations. *Rba. capsulatus* was also tolerant to Cd (El-Rab *et al.,* 2006). Hydrogenases isolated from *Thiocapsa roseopersiciana* could reduce the metal ions and oxidize the metals (Zadvorny *et al.,* 2006). Chromate reduction by purple non-sulfur phototrophic bacterium *Rhodopseudomonas palustris* KU003 isolated from tannery effluent has been reported by Ramchander *et al.* (2013). Takaiekhozani and Rezania (2017) have excellently reviewed the use of photosynthetic bacteria in the removal of heavy metal from polluted water. Chromate reduction by *Rhodobacter* sp. GSKRLMBKU-03 isolated from pond water was reported by Rajyalaxmi *et al.* (2017).

With the increased demand for shrimp in the world market, many countries are encouraging to shrimp farming both in coastal areas and freshwater resulting in water pollution including heavy metals. In addition, water removed during shrimp pond drainage during harvesting is often directly discharged into canals and flown into the cultivated areas together with illegal disposal of shrimp pond sediments which can cause serious soil pollution in agricultural areas especially rice-fields and vegetable crops resulting in an accumulation of heavy metals salts and chemical substances. Gazso (2001) elegantly discussed the role of microbes in the remediation of water polluted with toxic metals. Removal of heavy metals from polluted water may be attributed to biopolymer secreted by purple non-sulfur bacteria which binds metals strongly (Watanbe *et al.,* 2003; Iyer *et al.,* 2005; Panwichian *et al.,* 2011).

Accumulation of cadmium in rice may cause some illnesses such as itai itai (Shimbo *et al.,* 2001). Contamination of heavy metals in the water where *Ipomea* is a perennial semi-aquatic or aquatic plant of the tropics and subtropics may cause poisoning to consumers (Gothberg *et al.,* 2002). *Rhodovulum* sp. PS88 and *Rba. sphaeroides* could remove the metal cadmium both under heterotrophic (aerobic-dark) and photoheterotrophic (anaerobic-light) (Watanabe *et al.,* 2003). Feng *et al.* (2007) reported that *Rba. capsulatus* adsorbed metal Trivalent arium. Panwichian *et al.* (2010) isolated purple non-sulfur bacteria for remediation of polluted water of shrimp pond water containing heavy metals such as cadmium, copper, lead and zinc by two purple non-sulfur bacteria. Panwichian *et al.* (2011) reported efficient removal of heavy metals in contaminated shrimp pond under aerobic dark and microaerobic light conditions by *Rba. sphaeroides* KMS24 and *Rhodobium marinum* NW16. The efficiency of removal of metals was in the order of Cu >Zn >Cd which they attributed to the exopolysaccharide present on their surface.

Further, these grow better in the presence of these metals at a higher concentration than that of contaminated shrimp ponds. *Rhodobacter sphaeroides* KMS24 and *Rhodobium maximum* NW16 could remove heavy metals content of shrimp pond water by the process of biosorption (Panwichian *et al.,* 2012). The cadmium bioremediation efficiency of *Rhodobacter sphaeroides* isolated from cd-contaminated soil was studied by Fan *et al.* (2012). In this study, cadmium did not reduce but the oxidation-reductions potential was decreased and it was converted to more stable, less plant-available forms. Similarly, Hua *et al.* (2014) studied arsenic (As) removal from rhizosphere soil by 24% using *Rps. palustris* . On the other hand Khuong *et al.* (2017) studied the remediation of Al^{3+} and Fe^{2+} by using *Rhodopseudomonas* spp. VNW64 and VNS89. These two strains remediate Al^{3+} 42–63% and a mixture Al^{3+} and Fe^{2+} 59–88% through biosorption, showing great potential as acid sulfate soil bioremediation agents. Sakpirom *et al.* (2017) studied the bioremediation of Cd and Zn by *Rps. palustris* TN110 and *Rubrivivax gelatinosus* TN414 isolated from Cd and Zn contaminate paddy fields. In his studies, he observed that *Rps. palustris* TN110 removed respectively 84% and 55% of Cd and Zn (initial concentration of 262 mg l^{-1}), whereas, for *Rubrivivax gelatinosus* TN414, the removals reached to the values of 72% and 74% respectively (initial concentration of 23 mg l^{-1} .)

v. Degradation of Toxic Metals or Radioisotopes: Though many microorganisms either live or dead and free or immobilized are being extensively used for the removal of heavy metals (Dixit *et al.,* 2015). Anoxygenic purple nonanoxygenic phototrophic bacteria proved to be more efficient (Idi *et al.,* 2015). *Rhodobacter sphaeroides* could accumulate large amounts of heavy metals like cadmium, nickel and lead (Buccolieri *et al.,* 2006). Maggini *et al.* (2012) reported the highest amount of zinc removal so far reported by utilizing microorganisms by bio absorption. Sasaki *et al.* (2013) could successfully remove radioisotopic elements like cesium and strontium by *Rba. sphaeroides* . The removal of these metals has been attributed to the negative charges of exopolysaccharides on the surface of the bacterial cell. Batool *et al.* (2017) and Nookongbut *et al.* (2018) reported the reduction in arsenic toxicity and uptake in rice by As-resistant purple non-sulfur bacteria.

APB is known to degrade and remove some toxic compounds present in wastewater such as CH_4, C_2H_6, C_3H_8 and higher liquid paraffinic hydrocarbons (Donald, 1970), putresine (tetramethylene diamine), cadverine (pentamethylene diamine) and other amines (Kobayashi, 1975a), mercaptan (Kobayashi, 1976), fatty acids like butyrate and valerate (Toerien, 1776), volatile organic acids (Kobayashi, 1977a,b; Swada and Rogers, 1977b), dimethyl nitrosamine (Kobayashi *et al.,* 1977), sulfate (Gruenwald and Grau, 1980), cyanide (Balloni *et al.,* 1980), HCHO, phenol and cresol having biocidal effects (Kobayashi Kankyo Kagaku Kenkyusho, 1982), NH_3 (Suhaimi *et al.,* 1988), organo-halo compounds (Kobayashi and Miyazawa, 1990), sulfide (Takeshi and Gotaro, 1991), noxious gases such as CO_x, NO_x and SO_x (Sublette, 1991), NO (Sublette, 1991) and NO_3 (Sasaki *et al.,* 1991a; Shen and Hiryama, 1993)are removed efficiently by APB.

Many of the toxic compounds such as nitrosamines, which are carcinogenic, mutagenic and teratogenic substances (Lijinsky and Epstein, 1970) are formed in nature through microbial activities that can be effectively removed using a purple non-sulfur bacterium, *Rba. capsulatus* (Kobayashi *et al.,* 1977).

vi. Degradation of Recalcitrant Dyes: Azo dyes which are the most recalcitrant xenobiotics generally are degraded under anaerobic conditions (Pandey *et al.,* 2007). Anoxygenic purple non-sulfur bacteria like *Rhodobacter adriaticus, Rba. blasticus, Rba. capsulatus* and *Rhodovulum strictum* could decolorize azo dyes within 2 days of illuminated and anaerobic conditions (Kim *et al.,* 2003). Many purple non-sulfur bacteria which are reported to produce azoreductase break azo bonds and are efficient degraders of azodyes (Yan Bin *et al.,* 2004; Liu *et al.,* 2006). Similarly, Liu *et al.* (2007) reported degradation of azodye by using *Rba. sphaeroides* . *Rhodopseudomonas palustris* has been reported to decolorize azo dyes to the extent of 1 g/L dye (Wang *et al.,* 2008). Tran *et al.* (2012) have reported the discoloration of azodyes by some purple sulfur bacteria and can be exploited in the discoloration of water. The degradation is most efficient under anaerobic and light conditions. Celik *et al* . (2012) reported degrading Reactive Red 195 azodye by using *Rps. palustris .Rba. sphaeroides* ASI 1737 secreted a good amount of azoreductase which functions as nitroreductase and mononucleotide reductase and useful in wastewater treatment. Wang *et al.* (2015) during their investigation on degraded azo dyes observed a clear succession of anoxygenic phototrophic bacteria where purple non-sulfur bacteria are gradually replaced by green sulfur bacteria, which degrade azodyes more efficiently. Srisuwun *et al.* (2018) could decolorize reactive red 159 by a consortium of photosynthetic bacteria using an anaerobic sequence batch reactor (AnSBR).

vii. Degradation of Pesticide: Many purple phototrophic bacteria are capable of utilizing aromatic compounds which are mostly derivatives of benzene consisting of negative resonance energy and stability. These organisms utilize these compounds as a sole source of energy in their catabolism, unlike hetero-trophic organisms. These features make them unique and suitable for pesticide degradation. These bacteria degrade most of these toxic substances under light under aerobic conditions. Kamal and Wyndham (1990) reported the degradation of 3-chlorobenzoate to CO_2 and biomass by *Rhodopseudomonas palustris* in the presence of light and anaerobic conditions. The same 3-chlorobenzoate was degraded by *Rps. palustris* DCP3 under phototrophic and low oxygen conditions (Krooneman *et al.,* 1999). *Rhodopsirillum photometricum, Rps. palustris* and *Rsp. rubrum* is able to grow on 2 and 3 carbon halocarboxylic acid in the presence of CO_2 by reductive dehalogenation and subsequent assimilation of resulting acids (Mc Grath and Harfoot, 1997).

The degradation of pesticides is dependent on the catabolic ability of the bacteria, the chemical structure of the pesticide, physicochemical conditions of the environment in addition to anaerobic and light conditions. The anaerobic condition promotes a favorable environment for the proliferation of these bacteria (Kusha Latha, 2010). *Rhodospirillum* and *Rhodopseudomonas* belonging to purple bacteria have also been reported to degrade halogenated aromatic pollutants like a 3-chlorobenzoate (Kushlatha *et al.,* 2010; Idi *et al.,* 2015).

viii. Degradation of Hydrogen Sulfide from Gas Streams: Hydrogen sulfide is present in small amounts in biogas (usually 0.1–2.0%) and off-gases from wastewater treatment plants. Further, it is toxic and malodorous. H_2S in biogas causes corrosion and poisoning for equipment using the biogas. Therefore, a number of chemical technologies are currently used to remove H_2S from biogas. At present removal of H_2S commercially is removed by chemotrophic bacteria in a fixed-film bioreactor under controlled oxygen conditions (Syed *et al.,* 2006). But the use of fixed-film or suspended-growth of phototrophic bacteria is a good alternative that is cost-effective and without requiring biomass separation steps and sustained growth under high and variable sulfide loads (Holkenbrink *et al.,* 2011). *Rhodopseudomonas palustris* could degrade tributyl phosphate (TBP) from 2.0-0.6 mM within 3 weeks (Berne *et al.,* 2005).

ix. Degradation of Hydrogen Sulfide from Wastewater: Sulfide which is a malodorous, corrosive and toxic compound liberated from organically bound sulfur often present in wastewater streams is readily removed by ecofriendly anoxygenic phototrophic bacteria (Hurse *et al.,* 2008). In general, sulfide is oxidized to elemental sulfur which is further oxidized to sulfate which is soluble in water. On the other hand, oxidation of hydrogen sulfide gets oxidized to elemental sulfur and get deposited due to lack of required oxygen. Anoxygenic phototrophic bacteria which under anaerobic conditions carry out partial oxidation resulting in the formation of sulfur which gets deposited in granular form by the process of sedimentation which reduces downstream problems and also has commercial value as fertilizers and chemical feedstock. Further, the generation of elemental sulfur minimizes pH changes and corrosion and minimizes light requirement. By the use of phototrophic bacteria these conditions such as flow rate and light intensity can be carefully controlled to minimize secondary effects. Similarly, the efficiency of *Rhodobacter sphaeroides* in removing hydrogen sulfide has been enhanced by immobilizing its cells with porous ceramics (Nagadomi *et al.,* 2000). Removal of sulfide from wastewater by green sulfur bacteria have the following advantages

1. Green sulfur bacteria deposit sulfur extracellularly which facilitates recovery of sulfur by sedimentation.
2. GSB can oxidize more sulfide with an increase in the intensity of light in contrast to purple bacteria and is economical.
3. GSB has a higher tolerance for sulfide and more affinity for sulfide uptake than purple bacteria.
4. The rate of sulfide uptake by GSB remains the same even in the presence of organic nutrients but purple bacteria get inhibited. Henshaw and Zu (2001) could remove sulfide efficiently by employing *Chorobium limicola* strain DSM257, while the chemotrophic bacteria are incapable of this process .

x. Degradation of Aromatic Compounds: Apart from the degradation of aliphatic compounds, the capability of anoxygenic phototrophic bacteria to degrade aromatic homocyclic and heterocyclic compounds under anaerobic conditions makes them worth exploiting because degradation of many recalcitrant pesticides/insecticides and wastes from the petroleum industry which cause great environmental pollution is rewarding. Microorganisms in the environment adapt to the use of xenobiotic chemicals as novel growth and energy substrate with specialized enzyme systems and metabolic pathways (van der Meer *et al.,* 1992) offer a special advantage in aromatic waste biodegradation. APB capable of metabolizing aromatic compounds reported so far all belong to the purple non-sulfur group. The aromatic compounds such as benzoates and their derivatives (Ramana *et al.,* 2006; Wright and Madigan, 1991), amides (Cabello *et al.*, 2004), thiols (Visscher and Taylor, 1993), phenols (Blasco and Castillo, 1993) and acrylamides (David and Ensign, 2005) are being degraded by these bacteria. Berne *et al.* (2005) have utilized *Rhodopseudomonas palustris* in the degradation of tributyl phosphate from nuclear fuel processing and chemicals. *Rhodopseudomonas palustris* (Harwood and Gibson, 1988; Sasikala *et al.,* 1994a,b; Perrotta and Harwood, 1994; van der Woude *et al.,* 1994), *Rsp. fulvum* (Pfennig *et al.,* 1965) and *Rcy. purpureus* (Pfennig, 1978) are known to metabolize benzoate. However, aromatic photo catabolism has also been reported in a few other species such as *Rmi. vannielii* (Wright and Madigan, 1991), *Rcy. gelatinosus* (Dutton and Evans, 1978; Rahalkar *et al.,* 1991), *Rba. capsulatus* (Blasco and Castillo, 1992) and *Rps. acidophila* (Yamanaka *et al.,* 1983).

Benzoic acid is the most widely used homocyclic aromatic acid is degraded by many purple non-sulfur bacteria, apart from this compound, methoxylated benzoates such as vanillate, a syringe (Wright and Madigan, 1991), hydroxybenzoates (Harwood and Gibson, 1988), p-aminobenzoic acid and 4-amino phenol (Khanna *et al.,* 1992), aniline and nitrobenezene (Sasikala *et al.,* 1994a,b) and halogenated aromatic compound such as 3-chlorobenzoate and 3-bromobenzoate (Kamal and Wyndham, 1990; van der Woude *et al.,* 1994) are metabolized by this group of microorganisms. A comprehensive understanding of the physiology and biochemistry of degradation of aromatic compounds by various anaerobic bacteria is discussed excellently by Berry *et al.* (1987), Elder and Kelly (1994) and Fuchs *et al.* (1994).

Compared to the homocyclic compounds of aromatic catabolism, the utilization of heterocyclic aromatic compounds for growth and their catabolic pathways have been less studied in this group of microorganisms. Pyrimidines (Kaspari, 1979), purines (Kaspari and Werner, 1986) and pyridine derivatives and pyrazines (Sasikala *et al.,* 1994a) are heterocyclic aromatic compounds reported to be photometabolized for the growth as either sole nitrogen or carbon source. Although thiophene-2-carboxylate could not be photometabolized by *Rps. palustris,* it could transform this compound to (+)-3-hydroxytetrahydro-thiopene-2-carboxylae and tetrahydrothiopene-2-carboxylate (Tanaka *et al.,* 1982). Methylbenzimidazole-2-ylcarbam (carbendazin), a systemic fungicide was used as a carbon and nitrogen source for the growth of *Rps. palustris* (Rajkumar and Lalithakumari, 1992). Most of the photocatabolic pathways involved in the anaerobic degradation of aromatic compounds by this group of microorganisms are still not known. Yoshida and Kobayashi (1992) used APB along with other light- depending microorganisms for purifying ponds and lakes. A 10-hectare pond can treat 100 tons of effluents (30,000 PPm BOD) per day. Simultaneously, it can produce a large amount of H_2 gas and additional methane.

xi. Degradation of Organic Waste: These organisms are reported to be efficient in the treatment process of molasses (Sasikala *et al.,* 1992), citric acid fermentation waste (Yu *et al.,* 1991), refinery and lactic acid waste (Sasikala *et al.,* 1995), high strength organic wastewater (Ogbonna*et al.,* 2000) and oil containing sewage water (Takeno *et al.,* 2005). Hassan *et al.* (1997) have successfully employed *Rhodobacter sphaeroides* in the treatment of oil spills. Similarly, Takeno *et al.* (2005) utilized immobilized photosynthetic bacteria in the treatment of oil-containing wastewater. Ibrahim *et al.* (2006) discussed the utility of *Rhodopseudomonas palustris* in the degradation of sago effluents under different conditions. Madukasi *et al.* (2010) have discussed the potentials of phototrophic bacteria in treating pharmaceutical Wastewater. Madukasi *et al*. (2011) have isolated a wild strain of photosynthetic bacterium which is useful in waste management. Hulsen *et al.* (2014) have successfully used phototrophic bacteria for the recovery of nutrients from domestic wastewater. Feng (2016) employed the photosynthetic bacterium *Rhodopseudomonas* sp. strain HY21 for the removal of three high concentration organic wastewater such as soybean

Wastewater, citric-acid wastewater and distillery wastewater. Amezaga *et al* . (2014) have demonstrated the reduction of monosodium glutamate most effectively in a rotating bioreactor using immobilized photosynthetic bacteria **(Table 10.3).**

Table 10.3. Purple Non-Sulfur Bacteria used in wastewater treatment

Name of the Bacterium	Source of Organism	Application	Reference
Rsp. fulvum	Sugar refinery wastewater	Metabolize benzoate	Pfennig *et al.* (1965)
Rhodocyclus purpureus	-	Metabolize benzoate	Pfennig (1978)
Rps. gelatinosa	Miso-like effluent medium	Treatment of soybean waste	Sasaki *et al.* (1981)
Rmi. vannielii	Freshwater/Pond water	Aromatic photocata-bolism	Wright and Madigan (1991)
Rps. palustris	-	Photocatabolism of benzoate	Blasco and Castilo (1992)
Rba. capsulatus E1F1	-	Degradation of nitrophenols	Blasco and Castillo (1992)
Rcy. gelatinosus	Seafood processing wastewater	Recycling of agro-industrial wastes Treatment of tuna condensate	Prasertsan *et al.* (1997)
Rps. palustris	Simulated wastewater	Metabolize benzoate	Perrotta and Harwood, (1994); Vander woude *et al.* (1994)
Rhodospirillum rubrum	-	Removal of Selenium	Kessi *et al.* (1999)
Rps. palustris	-	Degradation of Tributyl phosphate	Berne *et al.* (2005)
Rhodopseudomonas sp.DK6	the wastewater treatment system of rubber sheet	Treatment of latex rubber sheet wastewater	Kantachote *et al.* (2005)
Rba. sphaeroides S, *Rba. sphaeroides* NR-3	-	Removal of oil from sewage wastewater	Takeno *et al.* (2005) Yamaoka *et al.* (2008)
Rubrivivax benzoatilyticus	Flooded paddy soil	Aromatic hydrocarbon-degrading	Ramana *et al.* (2006)
Rhodobacter sphaeroides	-	Treatment of oil spill	Mohd. Ali Hassan *et al.* (1997)
		Removal of phosphor-ous by strain IL 106	Takeno *et al.* (1999)
		Removal of heavy metals	Buccolieri *et al.* (2006)
		Removal of cadmium and lead	Watanbe *et al.* (2003)
		Chromate reduction	Nepple *et al.* (2000)
Rba. capsulatus	-	Bio reduction of trivalent aurum	Feng *et al.* (2007)

Name of the Bacterium	Source of Organism	Application	Reference
Rba. sphaeroides Z08	From soybean wastewater	Treatment of soybean wastewater	He *et al*. (2010)
Rps. palustris P1	Fermented pineapple extract	Treatment of latex rubber sheet wastewater	Kantachote *et al*. (2010) Kornochalert *et al*. (2014)
Rba. capsulatus KU002	From tannery effluents	Chromate reduction	Ramchander *et al*. (2011)
Rhodobacter sphaeroides KMS24, *Rhodobium maximum* NW16	Pond water	Removal of heavy metals	Panwichian *et al*. (2012)
Rhodobacter sphaeroides SSI	-	Removal of cesium and strontium	Sasaki *et al*. (2013)
Rps. palustris KU003	Tannery effluents	Chromate reduction	Ramchander *et al*. (2013)
Rhodopseudomonas sp. strain HY21	-	Removal of soybean, citric acid and distillery wastewater	Amezaga *et al*. (2014)
Rhodopseudomonas sp.	Synthetic 112 medium	-	Tan (2014)
Rba. sphaeroides SMR 009	-	Bioremediation of wastewater	Ramchander *et al*. (2014)
Allochromatium sp. GSKRLMBKU-01	Marine water	Removal of phosphorous	Rajyalaxmi *et al*. (2015)
Rhodobacter sp. GSKRLMBKU-02	Papermill waste	Bioremediation of slaughterhouse Wastewater	Rajyalaxmi *et al*. (2015)
Rps. palustris RSOU555, *Rps. themotolerance* RSOU555	Polluted water	Treatment of polluted lake water	Rajani *et al*. (2016)
Rba. blasticus *Rba. capsulatus*	-	Treatment of swine wastewater	Shan *et al*. (2016)
Rhodopseudomonas sp. CSK01	Facultative pond	Treatment of municipal wastewater	Saejung and Thammaratana (2016)
Rba. capsulatus	-	Wastewater treatment	Costa *et al*. (2017)
Rhodobacter sp. GSKRLMBKU-03	Pond water	Chromate reduction	Rajyalaxmi *et al*. (2018)
Rhodopseudomonas faecalis	wastewater	Sugar refinery wastewater	Saejung and Salasook (2018)
Rhodospeudomonas palustris TN110		Remediate cadmium through CdS nanoparticles	Sakpirom *et al*. (2019)
Rhodopseudomonas sphaeroides	-	Carbaryl wastewater	Wu *et al*. (2019)

Name of the Bacterium	Source of Organism	Application	Reference
Rhodopseudomonas capsulata	-	Removal of propyzamide from soybean processing wastewater and Biodegradation of carbaryl in soil	Wu *et al.* (2019 a,b)
Rhodopseudomonas sphaeroides	-	Treating mesosulfuron-methyl Wastewater	Wu *et al.* (2020)

B. Aerobic Processes

i. Trickling Filters: It is a misnomer as it is not a filtration but it is a fixed-film – bioreactor. The effluents to be treated are passed down through a packed bed counter-current to a flow of air. The components of the bed are large enough so that air penetrates to the bottom but small enough to maximize the surface area available for microbial activity. Microorganisms adhering to the packed matrix absorb oxygen from the up flowing air and organic matter from the downflowing effluent. The organic matter of effluents is metabolized by adhering to microorganisms and reduce BOD.

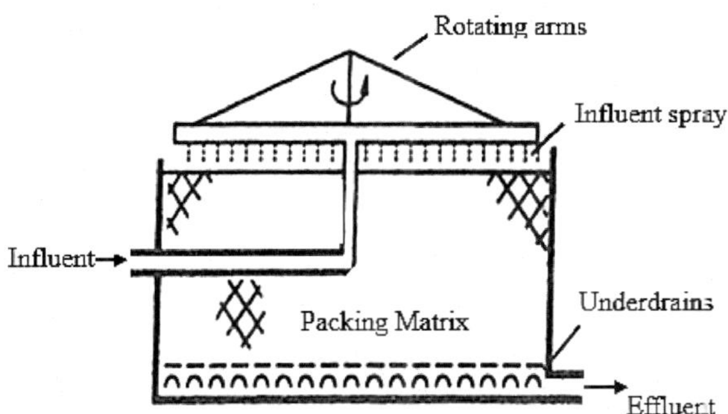

Fig. 10.3. Schematic Diagram of Trickling Filter

The packed bed usually consists of a cylindrical concrete tank 2 to 3 m in depth and 8 to 16 m in diameter (**Fig. 10.3**). Some filters are rectangular but a rotary system allows more uniform hydraulic loading. The tank is packed with a bed of stone (usual granite) or special plastic packings to give more air space the bed being underlaid with drains. Synthetic packing material has a higher surface area and voidage (% air space of total bed volume) allowing higher treatment rates per unit volume of bed and reducing the likelihood of blockages. The trickling filter is always followed by a secondary sedimentation tank or humus tank to remove suspended matter from the treated effluents. The effluent trickles gradually through the bed and a slime layer of biologically active material (bacteria, fungi, algae, actinomycetes, protozoa and nematodes) forms on the surface of the support material. The large surface area created in the bed permits close contact between the air flowing upwards through the bed, the descending effluent and the biologically active growth, complex organic materials are broken down and utilized by biofilm bacteria, nitrogenous matter and ammonia are oxidized to nitrates and sulfides and other compounds are also oxidized in the same manner. A simple filter is inefficient when operated at abnormally high organic loading rates. Initially, there is a very rapid buildup of bacteria, fungi and algae at the top of the filter which cannot be controlled by the resident population of worms and larvae. The voids, therefore, block up resulting in ponding (untreated effluents accumulating on the surface of the filter bed). Film growth can be limited by reducing the dosing frequency, which gives better liquid distribution deeper into the bed. A trickling filter bed should remove 70-95% of the BOD and 90-95% of the suspended solids at organic loading rates of 0.06-0.12 kg BOD m^{-3} day for conditional trickling filtration.

Trickling filters can be modified to increase the capacity for organic loading by use of two sets of filters and settling tanks in series. This is known as alternating double filtration (ADF). The effluent is applied to the first filter at a high hydraulic and organic loading rate, it passes from this filter through the first settling tank and then on to a second filter and settler. After a period of one or two weeks, the sequence of the filters is reversed and the second filter receives the higher loading. In this way, heavy film growth is promoted in the first filter to receive effluents but when the filter sequence is reversed it becomes nutrient-limited, encouraging excess film removal.

Deep beds of 3.5 to 5.5m depth may also be used in which air is blown through the beds by fans. Loading rates up to 12 times that of ordinary filters have also been achieved. The starving of the trickling filter beyond 48 hrs resulted in the near failure of the filter bed due to the death of supporting bacteria. Therefore, supplement or artificially load the filter to maintain viable biomass in the system is needed to maintain viable biomass in the system.

British Chemical Company ICI: Bayer has constructed a bioreactor that is, 30 m and has height circular settling basins are located around the top rim. It allows considerably better oxygen transfer efficiency (**Fig. 10.4** .)

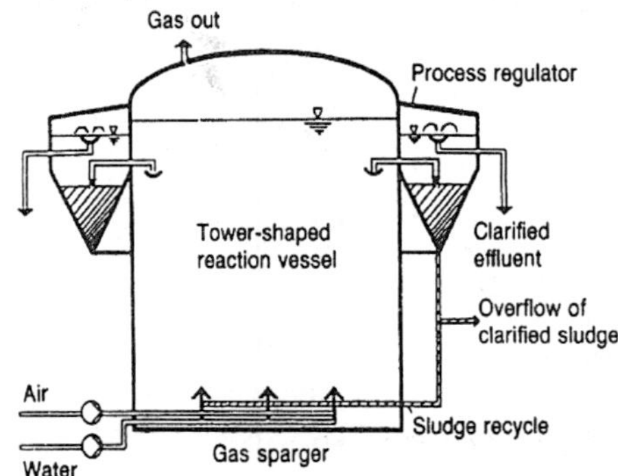

Fig. 10.4. Bayer Biotower

In the Bayer Bio tower process, two-component nozzles have been developed as injectors by means of which the kinetic energy of the liquid stream is utilized to disperse air in small gas bubbles in a manner similar to water spray pump (**Fig. 10.5**). These injectors are installed in groups of four (injector clusters) above the bottom of the lower reactor.

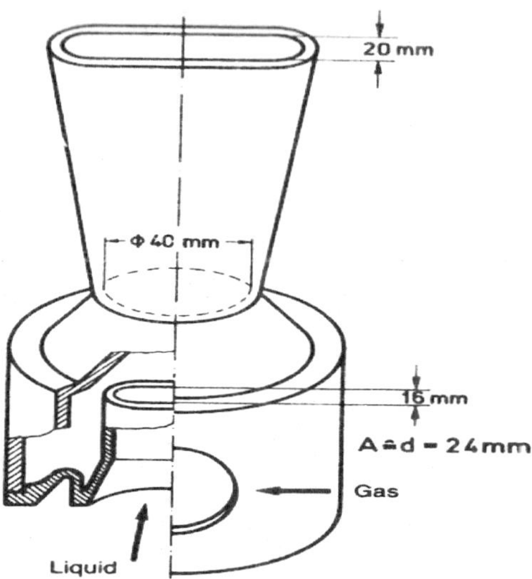

Fig. 10.5. Bayer Injector

ii. Towers: Trickling filters are limited use to treat large volumes of effluents as they do not have a high specific area and a high voidage. Large areas of land would be required for the exclusive and expensive traditional filter beds. Towers of 6-9m in height packed with lightweight (40-80 kg m^{-3}) plastic multifaceted modules or small random packing units have been provided to be a space-saving and relatively inexpensive solution to the problem. These packings have relatively inexpensive solutions to the problem. These packings have a relatively open structure for oxygen transfer and voidage but are expensive when compared with conventional filter packing. They are capable of coping with high BOD loadings. The biological film is similar to that formed on the conventional packing and scouring is due to the hydraulic load applied rather than the predation of higher organisms.

iii. Biologically Aerated Filters (BAFs): These are relatively recent developments based on the trickling filters. They consist of a packed bed that provides sites for trickling filters. The reactor volume is flooded with the effluents to be treated which are passed upwards or downwards through the reactor (i.e. co– or counter–currently to the air supply) depending on the design. The packing matrix may be natural (Ex. pumice) or synthetic (i.e. polyethylene) and may be either a fixed structure or randomly packed **(Fig. 10.6)** .

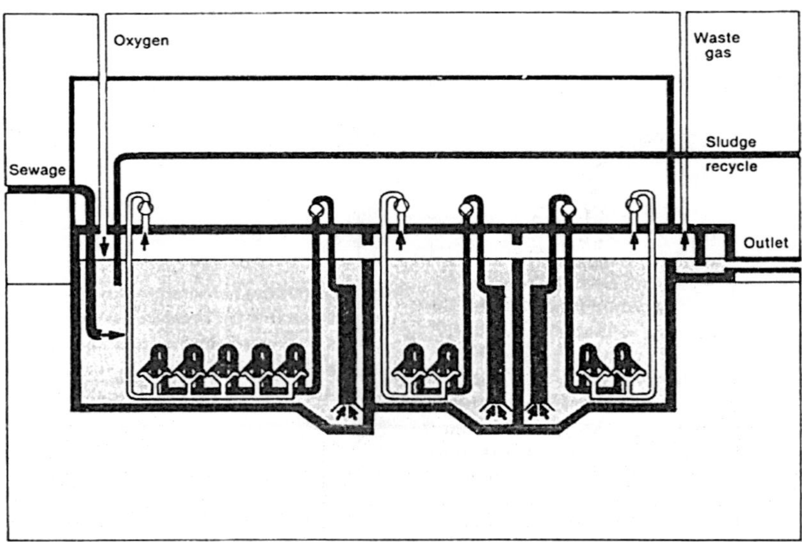

Fig. 10.6. Diagram of Activated Sludge System using Pure Oxygen.

The combination of aeration and filtration allows high rates of BOD and ammonia removal together with solids capture so that sedimentation tanks may not be required. However, regular backwashing is essential to remove filtered solids and excess biomass. They are versatile treatment systems and of that current operation, design capacities vary between 600-700 m^{-3} day^{-1}. In addition to their use as a secondary treatment process, they can also be utilized for tertiary treatment or modified to allow denitrification in a manner similar to that of activated sludge systems.

iv. Rotating Biological Contactors (Rotating Disc Contractors): In this treatment method **(Fig. 10.7)** a unit composed of closely spaced discs (2 to 3m diameter with 1 to 2 cm spacing between discs), on a central drive shaft are rotated slowly (0.5 to 15 ppm) through the effluents so that 40 to 50% of the disc surfaces are submerged. The discs usually made from synthetic materials (Ex. Polyvinyl chloride, polystyrene) are arranged in stages or groups separated by baffles to minimize short-circuiting or surging and to enhance specific treatment requirements such as nitrification. The discs may be flat or corrugated to increase surface area. A microbial film forms on the discs. This is aerated during the exposed part of the cycle and absorbs nutrients during the submerged part. Shear forces produced as the discs rotate through the liquid control the thickness of the biofilm with excess biofilm being sloughed from the discs. A sedimentation tank following the biological stage is, therefore, required to remove biological solids. Rotating biological contactors are compact easily covered for health and aesthetic reasons available as packaged units simpler to operate under varying loads than trickling filters and are easily added onto existing treatment processes. As such they can provide cost effective method of on-site treatment.

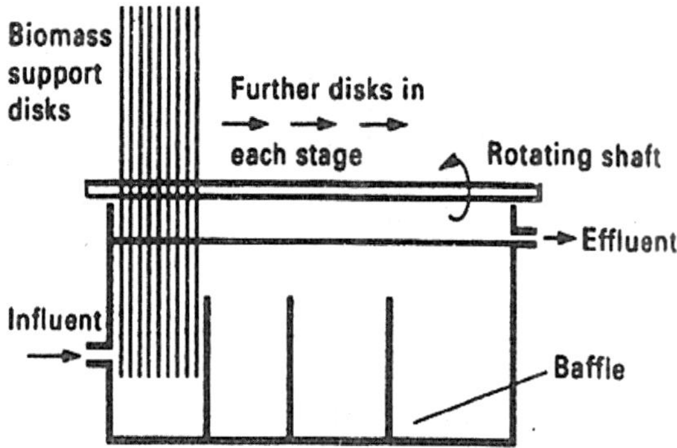

Fig. 10.7. Schematic Diagram of Rotating Biological Contractors.

v. Rotating Drums: Large rotating drums packed with random plastic packing materials or spheres have been manufactured as the develop-ment of rotating discs. Loading rates for the random packing were similar to those of rotating discs, while plastic spheres used in partial treatment could cope with loads of 6kg BOD m^{-2} day^{-1} .

vi. Sequencing Batch Reactor: A sequencing batch reactor (SBR) is a biological reactor quite similar to activated sludge treatment except that all events (aerobic treatment and solids separation) take place in a single tank (i.e no separate clarifier). SBRs are operated in a cyclic batch mode (each cycle consists of 5 discrete events). Fill (add wastewater to a tank), react (aeration of the wastewater with stirring to permit thorough the mixing of the settled microorganisms sludge with the Wastewater), settle (an anoxic period in which the mixed liquor and suspended solids (biomass) are allowed to separate by gravity sedimentation), draw (removal of the clarified supernatant water from the tank) and idle (a period of resting until the next batch of waste-water is ready for treatment). The entire process is illustrated in **Fig. 10.8**. It is one of the innovative and alternative systems for wastewater treatment assessed by EPA and the office of the technology assessment.

vii. Fluidized Bed Systems: These can be operated aerobically, anaerobically, or anoxically for denitrification. It is a recently developed system. It consists of a support matrix (sand, anthracite, reticulated foam) that has a large surface area on which biofilm adheres and thus they are able to operate at high biomass concentrations with high rates of treatment and allows strong Wastewater to be treated in small reactors. These are also useful to Wastewater with variable loadings. The support matrix is fluidized by the upflow of effluent through the reactor and the degree of bed expansion is controlled by the flow rate of Wastewater. The treated effluent can thus be decanted off without loss of the support matrix and with the careful operation, a secondary sedimentation tank may not be needed. The support matrix is regularly withdrawn to remove excess biomass.

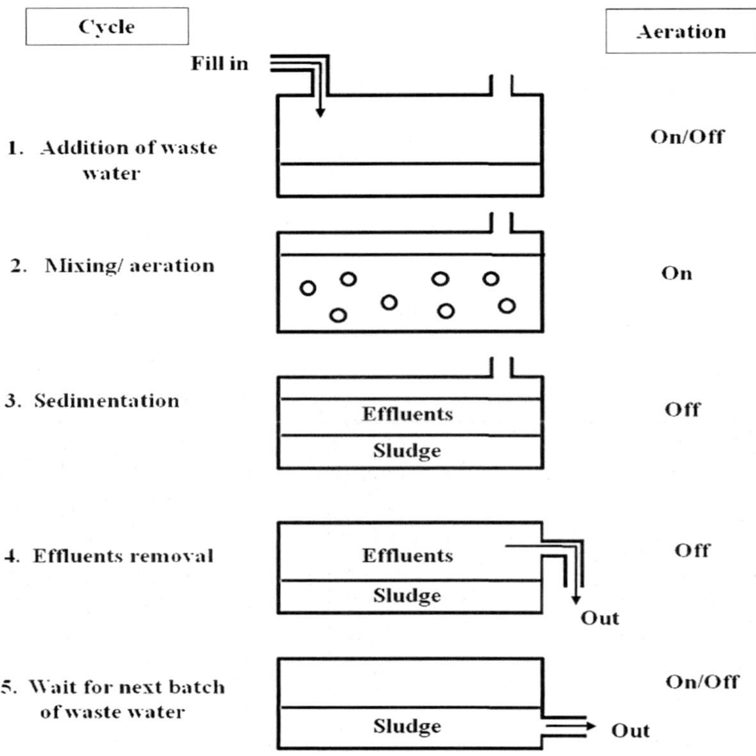

Fig. 10.8. The five cycles of sequencing batch reactor (SBR)

viii. Activated Sludge Process: It is aerobic secondary treatment. It is formed from suspended particles contained in sewage, adsorption of colloids and multiplication of microorganisms on this substance. The main components of activated sludge are bacteria, one of activated sludge contains about 1×10^{-2} bacteria whose overall surface area is 1200 sq. meters. The bacteria one *alpha* and *beta- meso*-saprobic group. Their composition depends on the nature and properties of pollutants. It also contains nitrifying bacteria (3×10^7) per g of activated sludge (unicellular and filamentous bacteria, yeasts and hyphae of molds). Microfauna of activated sludge is represented by mainly unicellular protozoa, rotifers and round warms which are involved in the mineralization of organic substances of sludge. The quality of activated sludge is characterized by the sludge index which is a volume in milliliter occupied by damp sludge after 30 minutes settling with one gram of dry substances.

$$\text{Sludge index} = V/g$$

Where

V= Volume of sludge in mL

g = The dry weight of the sludge in grams

Based on the values of the sludge index, the activated sludges are characterized as follows

1. 60 mL/g is the well-precipitating sludge
2. 80 – 90 mL/g is the looser sludge
3. 300 mL/g is the sludge unfit for treatment

The process of activated sludge digestion is depicted in **Fig. 10.9a** and **b**. It consists of aerating and agitating the effluent in the presence of flocculated suspension of microorganisms on particulate organic matter-the activated sludge.

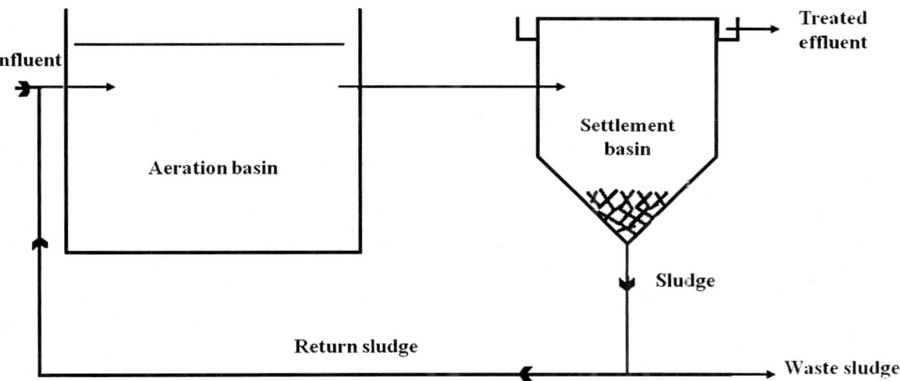

Fig. 10.9a. Simplified Cross Section of an Activated Sludge Process.

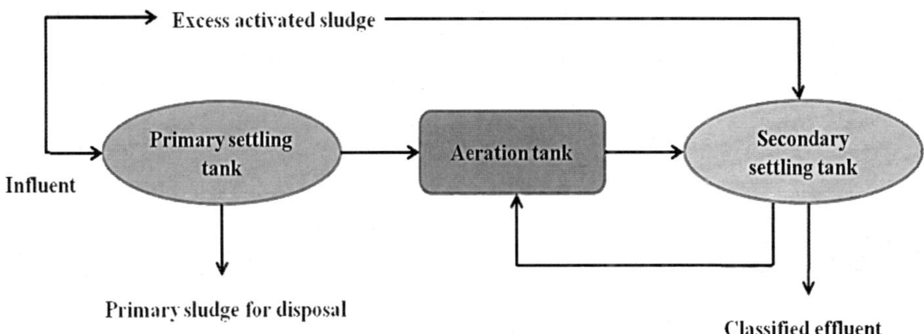

Fig. 10.9b. Process of Activated Sludge Digestion.

After the removal of grit and heavy solids from the raw sewage, the effluents flow into the settling tank, in which more than half of the suspended solids get settle down. The solids are removed into a digestion chamber, while the liquid portion is inoculated with ripe or activated sludge from the previously treated batch of sewage. The energized sewage is allowed into an aeration tank where mechanical agitation and compressed air keep the mixture in a constant motion, promoting aeration. This results in the flocculation of aerobactreria. These flocculent masses of aerobic bacteria provide an ideal environment for rapid decomposition within 6-8 hrs of exposure. Most of the colloidal and dissolved organic materials are oxidized and removed. Finally, the liquid material containing flocculent particles is pumped into the final settling tank or clarified where the flocs of sludge settle down (**Fig. 10.9b** .)

A number of modifications of the basic process have been developed to improve treatment efficiency or for a specific purpose such as denitrification. The tapered aeration and stepped feed aeration are used to balance oxygen demand (which is the greatest at the point of wastewater entry to the aeration basin) with the amount of oxygen supplied. The contact stabilization exploits biosorption processes and thereby allows a considerable reduction in basin capacity for a given wastewater throughput. Denitrification can be accomplished in an activated – sludge plant when the first part of the basin is not aerated.

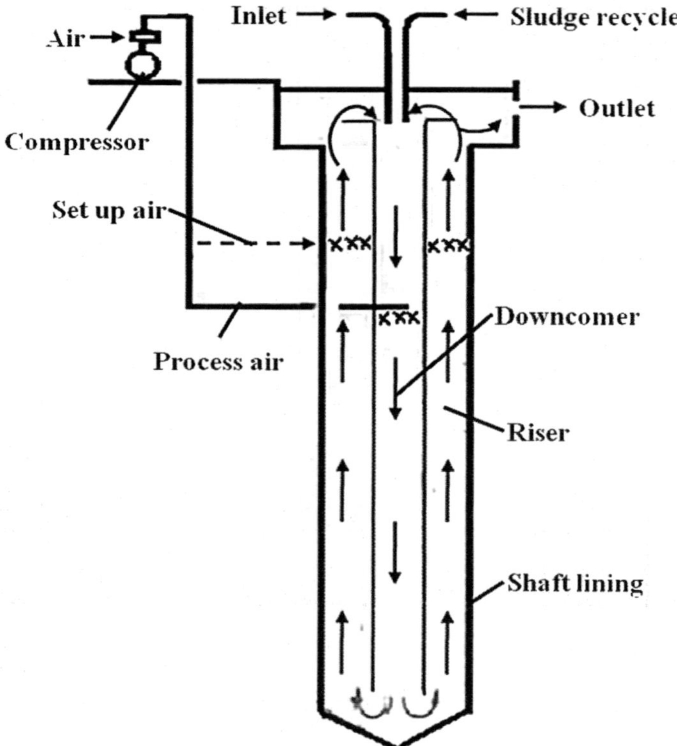

Fig. 10.10. Deep Shaft Effluent Treatment Plant.

Deep shaft, a special type of vessel, is another advanced activated–sludge system in which the amount of available oxygen is increased to improve the treatment rate (**Fig. 10.10).** It is developed by the ICI Plc SCP Process. The deep shaft consists of a shaft 50 -150m deep, separated into a downflow section (down-comer) and an up-flow (Riser) section. The shaft may be 0.5 to 10m in diameter depending on capacity. Fresh effluent is fed at the top of the Deep Shaft and the air is injected into the down–flow section at a sufficient depth to make the liquid circulate at 1 to 2 m s^{-1}. The driving force for circulation is created by the difference in density (due to air bubble volume) between the riser and downflow sections. For starting up, the circulation of liquid is stimulated by injecting air at the same depth in the riser. Air injection is then gradually all transferred to the air injection point in the downcomer, because of the pressure created in the downcomer, oxygen transfer rates of 10 kg $O_2m^{-3}h^{-1}$ can be achieved and bubble contact times of 3 to 5 minutes are possible instead of 15 in differed air systems. BOD removal rates of 90% are achievable at organic loadings of 3.7–66 kg BOD m^{-3} day at hydraulic retention times of 1.17–1.75 hrs. Sludge production was much less than that for original conventional wastewater treatment.

Two types of pure oxygen systems have also been developed to increase the rate of oxygen transfer

i. A closed system that operates in an oxygen-rich atmosphere

ii. An open system employing fine bubble diffusers

a. Closed Systems that Operate in Oxygen–Rich Atmosphere: An example of a closed system is the UNOX process developed by the union carbide corporation in the USA and marketed in the UK. The enclosed oxygenation tank is compartmentalized by baffles. Settled wastewater and returned sludge are fed into the first stage and oxygen pumped into the headspace. The oxygen and wastewater move sequentially through the compartments and the oxygen concentration in the gas phase decreases at each stage as it is consumed by micro-organisms. As the nutrient concentration also falls stage by stage oxygen supply and demand are balanced. Organic loading rates are 3-4 rates higher than those of aerated systems at 2.5–4.0 kg BOD m-3 day^{-1}. These systems also have shorter residence times and produce less sludge with better settling systems and produce less sludge with better settling qualities than conventional systems. The UNOX process has been successfully used for the treatment of brewery wastewater containing 20000 mg BOD dm^{-3} at flow rates of 2269m^{-3} day^{-1} with a 6.6 retention time.

The UNOX systems employ pure oxygen **(Fig. 10.11)**. The first aeration slope consists of two channels each of which is connected in a cascade arrangement to three bioreactors with capacities of 1500, 750 and 730 m^3. In the first bioreactor (activated sludge basin) pure oxygen from an air separation plant is inforced in via injectors. In this stage, 90-93% of the organic decomposable material is eliminated. The co-produced is collected alone with the residual oxygen and treated at 100°C. In the second treated stage (Settling basin), the accumulating sludge is pumped back into the activated sludge basin. The wastewater flows through this second treatment stage to remove material that is not easily broken down. In the final third treatment stage, the clarified wastewater proceeds to the receiving system.

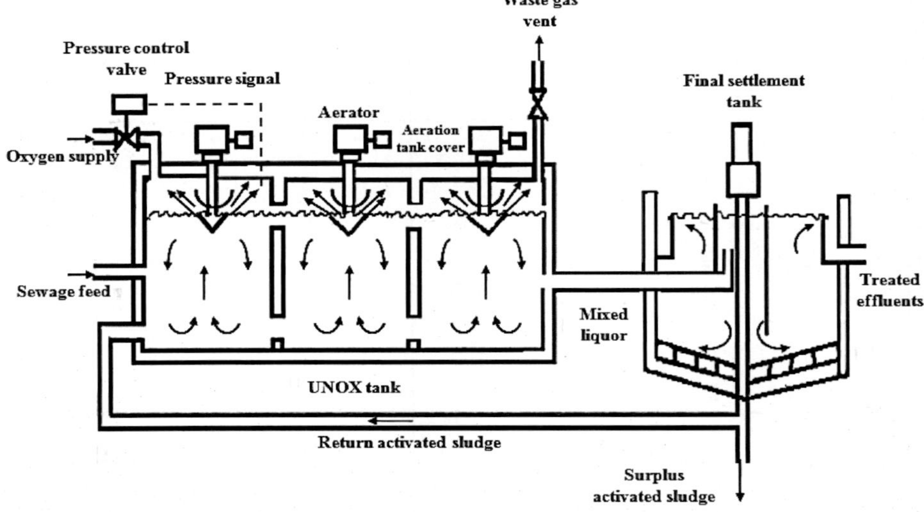

Fig. 10.11. Schematic Diagram of Multistage UNOX System.

b. Open Systems Employing Fine Bubble Diffusers: British Oxygen Company developed a system called VITOX open tank oxygenation system **(Fig. 10.12)**. The main advantage is that it can be used in existing aeration tanks either to replace or upgrade conventional aeration without the need to replace or modify existing units. Oxygenation is achieved by pumping settled effluent water at high pressure through a venturi injector where oxygen is injected. Turbulence and high pressures thus created ensure high levels of oxygen dissolution. The flexibility of this system is particularly useful for the treatment of high-strength intermittently produced Wastewater. Such as those generated by food processing industries.

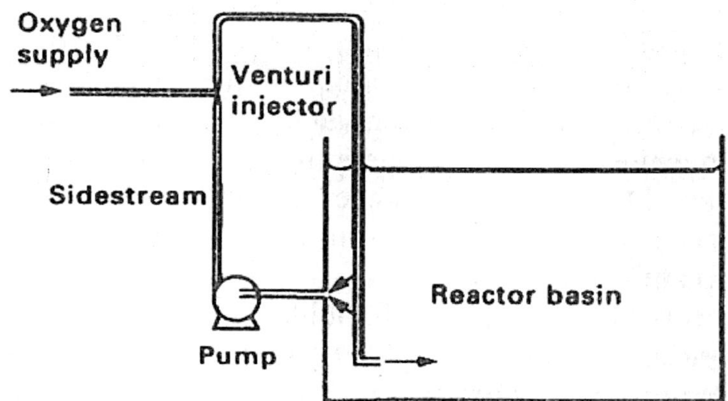

Fig. 10.12. Simplified VITOX Side Stream Aeration System.

C. Anaerobic Processes

Anaerobic treatment processes are based on microbial biodegradation of organic waste in the absence of oxygen. The basic differences between the aerobic and anaerobic treatment of wastewater are precised in **Table 10.4** .

Table 10.4. Comparison of Aerobic and Anaerobic Treatment of Wastewater

Aerobic	Anaerobic
Approximately 50% of biodegradable carbon is converted into new cells	5-10% biodegraded carbon is converted into new cells
Approximately 50% of biodegraded carbon is converted into CO_2	90-95% biodegraded carbon is converted into methane and CO_2
Fast-growing (minutes to hours)	Slow growing (days)
No energy production	The energy is produced in the form of methane/hydrogen.

In these processes, various bacterial species such as the fermentative, acetogen and methanogen bacteria closely interact to effect total mineralization of organic compounds as illustrated in **Fig. 10.13** .

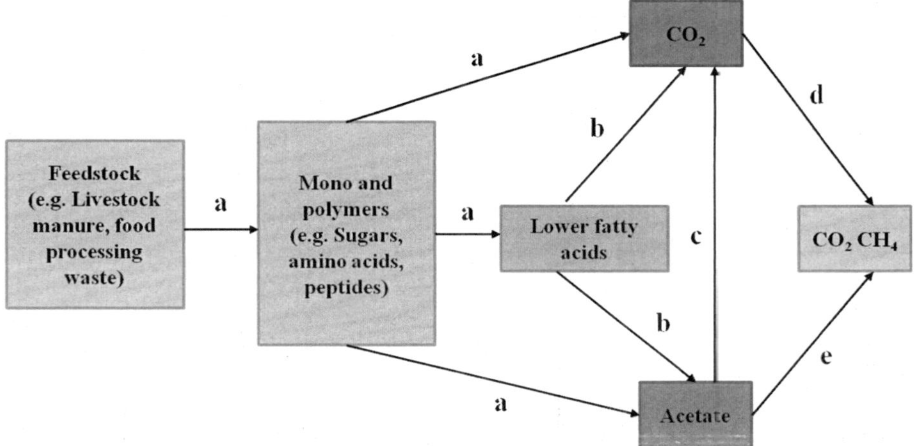

Fig. 10.13. Degradation of Organic Waste by Anaerobic Biological Treatment Processes.

1. **Fermentative/Hydrolytic Bacteria (Group 1):** A group of facultative and obligate bacteria is able to secrete extracellular enzymes that hydrolyze complex polymers into simpler substances such as volatile fatty acids along with CO_2, hydrogen, methanol and trace compounds.
2. **Acetogenic Bacteria (Group 2):** These include species of *Syntrophomonas* and *Syntrophobacter.* These organisms metabolize end products from the fresh group, microorganisms produce mainly acetic acid, CO_2 and hydrogen.

3. **Methanogenic Bacteria (Group 3):** Mainly strict obligate anaerobes associated with methane production. These include acetotrophs such as *Methanosarcina* breaking acetic acid into methane and CO_2 and hydrogenotrophic which reduce CO_2 with simultaneous oxidation of hydrogen to form methane and water. For example, *Methanobacterium thermoautotraphicum* synthesizes methane from CO_2 and hydrogen in seven steps by catalyzing these reactions in several ways depending on environmental conditions.

Traditional anaerobic treatments of effluents are being replaced by a variety of high–rate digesters. Some of the advantages of anaerobic processes in the effluents treatment are

1. Higher loading rates can be achieved than are possible for aerobic treatment techniques
2. Lower power requirements may be needed per unit of BOD treated
3. Useful end products such as digested sludge and/or combustible gases may be produced.
4. Organic matter is metabolized to a stable form.
5. There is the attention of water-binding characteristics to permit rapid sludge dewatering.
6. The reduced amount of microbial biomass leads to easier handling of sludge
7. Low levels of microbial growth will decrease the possible need for supplementary nutrients with nutritionally unbalanced wastes.

i. Anaerobic Digestion: Large volumes of wet sludge which are produced in primary and secondary sedimentation tanks may have to be reduced in volume before disposal. This volume of sludge can be reduced by anaerobic digestion. In sludges containing 20,000-60,000 mg dm^{-1} solid matter, 80% of the degradable matter may be digested which will reduce the solids content by 50%. During anaerobic digestion acid fermenting bacteria degrade the waste to free volatile fatty acids, mainly acetic acid and propionic acid which are then converted to methane (~60%) and carbon dioxide (~40%). The gas produced (biogas) is useful by-products and can be used as biogas fed to engines to generate electricity or used as a vehicle fuel. Anaerobic digesters are also used directly in the treatment of many high-strength wastewater such as food and agricultural industries. A number of anaerobic processes have been developed.

Completely mixed reactors with (often these are simply described as anaerobic digesters) or with sludge recycle, anaerobic filters, upflow anaerobic sludge blankets (UASB) and anaerobic fluidized beds are some of the anaerobic processes developed for anaerobic digestion of fermentation effluents. Advantages of this process of digestion are improved, reliable higher BOD/COD reduction and a large amount of methane production.

ii. Anaerobic Digesters: The anaerobic digestion tanks may be up to 14000 m^3 volume and equipped with heating coils for accurate temperature control normally in the mesophilic range (28-32°C) to increase the rate of digestion which is further improved by mechanical agitation (**Fig. 10.14**). Some digesters are operated thermophilically (50-70°C usually 55-63°C) to increase the rate of degradation and biogas production. Though much of the extra gas produced will be consumed in maintaining the digester temperature. A thermophilic anaerobic digester is generally used for effluents of high strength (16-25g COD dm^{-3} low pH (-3.8) and high-temperature winery effluents. Most digesters operate on a semi-continuous basis with the appropriate volume of digested material being removed and replaced with fresh material on a daily basis.

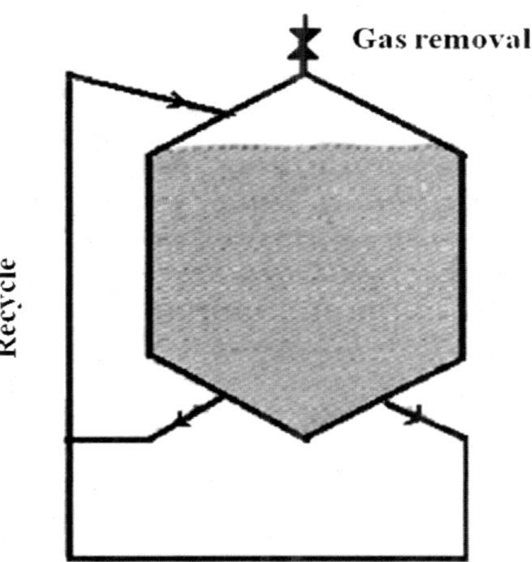

Fig. 10.14. Sludge Digester

iii. Anaerobic Filters: In anaerobic filters as with aerobic filters there is a microbial film growing on an inert support. Anaerobic filters may be operated in an upflow or downflow mode and a wide variety of packings are used both natural and synthetic. Antibiotic wastes, citric acid fermentation wastes, yeast production Wastewater, brewery and winery wastewater and molasses distillery slops and pharmaceutical Wastewater are amenable treatments using the anaerobic filter system.

iv. Up flow Anaerobic Sludge Blankets (USAB): In this system, high levels of active biomass are retained in the reactor by flocculation. No support media is added to the reactor instead the flocculated sludge develops in the reactor and acts as a fluidized bed. The feed is pumped through the bed (the sludge blanket), above which fine particles flocculants and settle back to the blanket as sludge thus preventing washout of organisms. Anaerobic sludge blankets are effective in the treatment of many Wastewater including sugar-beet wastes, domestic sewage, slaughterhouse wastes, agricultural wastes, brewery wastes, winery wastes and distillery wastes.

Two upflow anaerobic reactors are used to achieve the reduction of COD in the leachates. The waste effluent is pumped upward through a dense anaerobic bacterial suspension (the sludge blanket) is biodegraded in the process. The upward flow anaerobic sludge blanket (USAB) process is an anaerobic bioreactor in which wastewater is introduced at the bottom of the reactor and forward to pass through a blanket of sludge (microorganisms). The organics in the wastewater are remediated by the microorganisms into carbon dioxide, water, methane and hydrogen (**Fig. 10.15**).

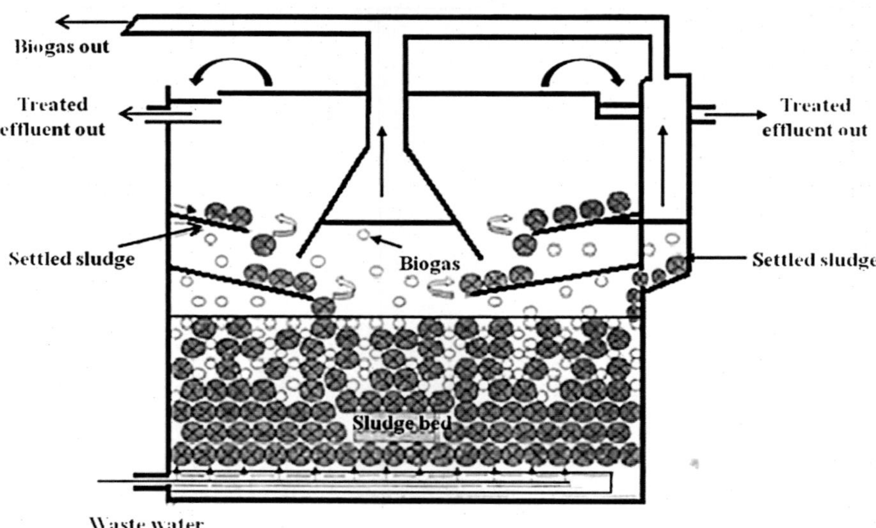

Fig. 10.15. Upward Flow Anaerobic Sludge Blanket Process.

The USAB reactor is designed as a gravity settler to maintain high concentrations of flocculent bacteria inside them. The treated liquid flows over and overflows whether at the top of the reactor and is then discharged or recycled for more treatment. The effective volume of each reactor amounts to 135 m^3. Reactor contents are recirculated through a draw-off piping network to provide mixing and to maintain the temperature of the reactor at 35°C.

■■■■

Chapter 11

BIOTECHNOLOGICAL APPLICATIONS

Increasing global demand for environmentally friendly products has driven several industries to undertake bioprocess technologies rather than process employing harmful chemicals. The acquisition of biomass to byproducts business is inevitable for many industrial sectors to develop new products, potential customers and create market niches. Among the microbes, phototrophic bacteria have the ability to produce a spectrum of bio-based products and energy from renewable feedstocks owing to their versatile unusual metabolism. In addition to biorefineries and biofuels, phototrophic bacteria have also been exploited for various applications such as eco-friendly paper making, natural dye extraction, organic farming and augmentation of bio methanation. This bacterial group can be an efficient system for an entrepreneur who wants to have a compact system for producing numerous value-added products.

Anoxygenic phototrophic bacteria (APB) are ubiquitous in nature and found in soil, water bodies and also under extreme conditions of hot springs and in Antarctica. They produce various carotenoid pigments when grown anaerobically under light or dark conditions. They are nitrogen fixers due to their nitrogenase activity and hence are good biofertilizers. They produce interesting carotenoids and rich in protein which makes them suitable candidates for the food and feed industry. *Rhodospirillum rubrum* produces an attractive pinkish-red pigment called spirilloxanthin which is being used in the food industry and in poultry feed wherein the eggs, other tissues have rich carotenoid content giving a red color. They also produce plant growth hormones and are used in the cultivation of tomatoes and oranges.

Anoxygenic Phototrophic Bacteria (APB) with diverse metabolic versatility grows by various modes of growth such as photoautotrophy, chemoautotrophy, heterotrophy and fermentation coupled to high growth rates and easy manipulation are potential microorganisms for biotechnological applications. The potentials of APB could be realized only recently and needs to be studied extensively for exploitation to the fullest extent to human needs (Burgess *et al.*, 1993). Since light is the source of energy, these bacteria proved to be highly economic and beneficial in different fields of biotechnology. Some of the areas where these can serve as benevolent organisms are precised in **Table 11.1.**

Table11.1. Various Biotechnological Applications of Anoxygenic Phototrophic Bacteria

Microbial product	Applications	Reference
S- Adenosyl- L-cystine	Precursor useful in therapeutic marketing	Yamada *et al.* (1986)
5- Aminolevulinic acid	Herbicide	Sasaki *et al.* (2002) Suwansaard (2010)
Augmentation of CH_4	Fuel	Vatsala and Balaji (1987) Kantha *et al.* (2015) Sakpirom *et al.* (2017)
Biodegradation of organic and inorganic compounds	wastewater treatment	James *et al.* (2000) Nagadomi *et al.* (2000) Ramana and Sasikala (2002) Takeno *et al.* (2005) Koh and Song (2007) Lee *et al.* (2008) Merugu *et al.* (2008) Vasavi *et al.* (2008) Rajyalaxmi *et al.* (2015, 2017) Rana *et al.* (2016) Wu *et al.* (2019 a,b, 2020) Batool *et al.* (2017)
Cytokinins	Plant growth hormone	Serdyuk *et al.* (1993) Scavetta *et al.* (2000)
Carotenoid	Natural dye	Nopartnarporn *et al.* (1986) Frigaard *et al.* (2004) Kuo *et al.* (2012) Wang *et al.* (2012) Englund *et al.* (2015) Saejung and Apaiwong (2015)
Enzymes	Food preparations and Biotechnology	Scavetta *et al.* (2000) Srinivas *et al.* (2007) Ramchander *et al.* (2008, 2010)
Hydrogen	Fuel	Sasaki *et al.* (1991) Fedorov *et al.* (1998) Mahakhan *et al.* (2005) Zabut *et al.* (2005) Merugu *et al.* (2013) Rajyalaxmi *et al.* (2018) Swetha *et al.* (2019) Akroum-Amrouche *et al.* (2019) Hongyu Ma *et al.* (2020)

Microbial product	Applications	Reference
Hopanoids	Therapeutic	Nagumo *et al.* (1991) Garcia Costas *et al.* (2012) Sessions *et al.* (2013) Tushar *et al.* (2014) Tank and Bryant (2015)
Metal ion uptake	Metal recovery	Sanaa *et al.* (2006)
NH_3	Fertilizer	Sasikala and Ramana (1990) Kantha *et al.* (2010, 2012) Gamal-Eldin and Elbanna (2011) Wong *et al.* (2014) Kantha *et al.* (2015) Lee *et al.* (2016) Zarezadeh *et al.* (2019)
Poly-β-hydroxy Alkanoates (PHA) and Poly-β-hydroxy butyrate (PHB)	Biodegradable plastics	Suzuki *et al.* (1995) Kranz *et al.* (1997) Noparatnaraporn (2001) Lorrungruang (2006) Mahuya *et al.* (2005) Sangkharak and Prasertsan (2008) Waligorska *et al.* (2009) Melnicki *et al.* (2009) Ramachander *et al.* (2010a 2010b) Wu *et al.* (2012) Takeuchi and Numata (2019)
Rhodethrin	Cytotoxic and Phytohormonal activity	Ranjith *et al.* (2007)
Rhodestrin	Phytohormonal Activity	Sunayana *et al.* (2005)
Phenols and esters	Biological activity	Ranjith *et al.* (2010)
Rubrivivaxin	Cytotoxic and Cyclooxygenase-i inhibitory metabolite	Ranjith *et al.* (2011)
Single cell protein	Protein source	Sasaki *et al.* (1998) Kantachote *et al.* (2005) Paronyan and Gasparyan (2009) Suwansaard (2010) Gupta *et al.* (2016) Chumpol *et al.* (2018) Saejung and Salasook (2018) Patthawarao and Saejung (2019)
Ubiquinone	Clinical Medicine	Sasaki *et al.* (2002)
Vitamine B_{12}	Vitamin	Goldrick (2003)
Cell biomass	Hypocholesterolemic	Deng (2009) Ramchander *et al.* (2011)

11.1. BIOSYNTHESIS OF CAROTENOIDS AND OTHER TERPENOIDS

Carotenoids or terpenoids are derived from one or more isoprene (C5) units and found in all phototrophic organisms and some chemotropic bacteria, archaea and fungi. These are compounds of yellow, orange and red substances with broad application in food, feed, nutraceuticals, cosmetics and pharmaceutical industries because of their vibrant colours and health-promoting activities (Kirti *et al.,* 2014). Though beta-carotene and astaxanthin are commercially produced by genetically modified *E. coli* and yeasts (Sun *et al.,* 2016), okenone and lycopene derivatives are produced with the help of anoxygenic phototrophic bacteria (Takaichi *et al.,* 1999) **(Fig. 11.1).** These carotenoids are reported to protect chlorophyll molecules from sunlight and prevent the formation of singlet oxygen by quenching. This excitation is more under aerobic conditions than in anaerobic conditions. This means carotenoids cannot be completely removed from phototrophic organisms growing under aerobic conditions. From this, it is clear that carotenoids are not required for anoxygenic phototrophic bacteria which grow under anaerobic conditions and these are available for biosynthetic engineering.

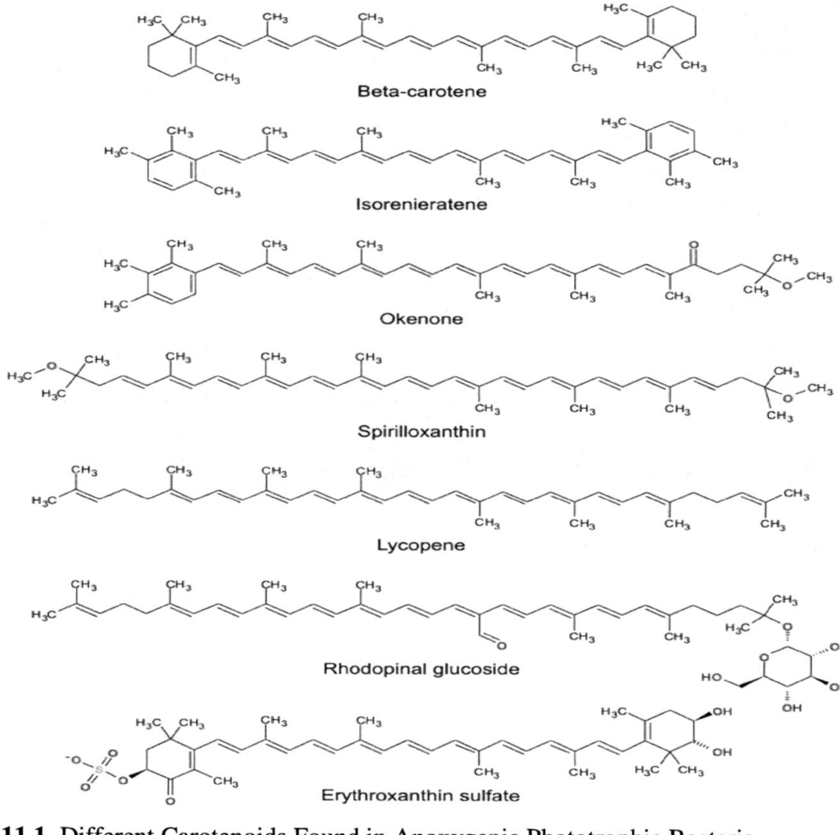

Fig.11.1. Different Carotenoids Found in Anoxygenic Phototrophic Bacteria.

Lycopene's are synthesized commercially by genetic engineering in *Rhodospirillum rubrum* (Wang *et al.,* 2012). Similarly, lycopene and zeto-carotene are being produced by the green sulfur bacterium, *Chlorobaculum tepidum* (Frigaard *et al.,* 2004). Carotenoids are key non-enzymatic plant antioxidants (Ashraf, 2009). The effect of purple non-sulfur bacteria on carotenoids varies, ranging from up to 138% increase in spinach provided with dried purple non-sulfur bacterial cells (Kondo *et al.,* 2008). As carotenoids are not required for anaerobic phototrophic bacteria, the isoprene molecules can be redirected to an isoprenoid compound such as valuable plant-type terpenoids (Englund *et al.,* 2015). Kuo *et al.* (2012) reported the effect of a light source on the growth and carotenoid content of *Rhodopseudomonas palustris*. Saejung and Apaiwong (2015) reported the enhancement of carotenoid production in the new carotenoid-producing *Rps. faecalis* PA2.

11.2. PRODUCTION OF FUNCTIONAL MEMBRANE PROTEINS

The structure and function of proteins can also be studied if it is useful to overexpress the proteins in a foreign host organism and obtain amounts sufficient for experimentation. Further, membrane protein poses a challenge as they get denatured in the absence of a suitable membrane environment. For example, the membrane heterologous proteins that are over-expressed in *E. coli* have much lower titers than heterologous soluble proteins and form inclusion bodies. On the other hand, *Rhodobacter sphaeroides*, purple non-sulfur bacterium, proved to be a good host for overexpression of functional proteins (Laible *et al.,* 2009). This system has an advantage as it contains a large amount of intracellular membrane which can hold the membrane-bound antennae and enzymes of photochemical machinery. In this bacterium these internal membranes form intracytoplasmic (ICM) vesicles that separate newly synthesized foreign proteins and enable easy purification following cell lysis. Using this system challenging membrane proteins in a functional form have recently been prepared for human aquaporin 9 (hAQP 9), human light function protein occludin (OCC), *Rba. sphaeroides* cellulase synthase enzyme complex (BcsAB) and *Rba. capsulatus* cytochrome Cy (Erbakan *et al.*, 2015).

11.3. UBIQUINONES (CO ENZYMES Q)

Ubiquinones (**Fig.11.2**) are the co-enzyme Q, cover a family of 5,6-dimethoxytoluquinones with polyisoprenoid side chains of various lengths at 6-position of the benzene ring of various co-enzymes (Q_6, Q_7, Q_8, Q_9 and Q_{10}) with a different number of isoprene units. Q_{10} proved to be effective in treating advanced cardiomyopathy (Mortensen *et al.,* 1990), granulomatous fibrosing lung disease (Beate *et al.,* 1991) and for protecting fats, fat-containing foods, fat-containing cosmetics and pharmaceutical products by oxidation. The role of ubiquinones as a drug (Baum, 1990) and antioxidants (Robert and Lars, 1990) has been well established. APB are good sources of ubiquinone production in high amounts when compared to other organisms are précised in **Table11.2.**

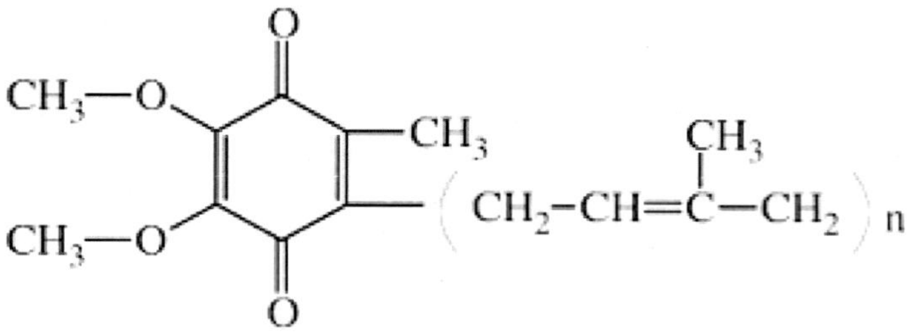

Fig. 11.2. Structure of Ubiquinone

Table 11.2. Production of Ubiquinone by Anoxygenic Phototrophic Bacteria

Organism	Growth Conditions	Ubiquinone Production (µmol/g Dry wt)
Rba. sphaeroides	Aerobic dark	1.51
	Anaerobic light	5.3
Rba. capsulatus	Aerobic dark	2.68
	Anaerobic light	5.3
Rba. sulfidophilus	Anaerobic light	4.2
Rps. palustris	Aerobic dark	0.2
	Anaerobic light	4.5
Rps. viridis	Anaerobic light	3.0
Rsp. fulvum	Anaerobic light	3.8
Rsp. molischianum	Anaerobic light	2.7
Rsp. rubrum	Aerobic dark	1.31
	Anaerobic light	6.3
Rcy. gelatinosus	Aerobic dark	1.71
	Anaerobic light	2.78
Rmi. vannielli	Aerobic dark	0.0
	Anaerobic light	3.0
Chromatium sp.	Anaerobic light	2.7
Heliobacteriumchlorum	Anaerobic light	0.35

Data was taken from Bishop *et al.* (1962); Carr and Exell (1965), Maroc *et al.* (1968), Uragami and Yoshida (1986 a, b) and Sasikala and Ramana (1995a).

Though Q_{10} can be produced by many aerobic and anaerobic microorganisms (Matsumura *et al.,* 1983), anoxygenic phototrophic bacteria are a good source for ubiquinone production as they produce high amounts in comparison to other microorganisms. Quinone composition of different species of anoxygenic phototrophic bacteria is listed in **Table 11.3**. Rhodoquinone (RQ), which is a derivative of ubiquinone in which one of the methoxyl groups is replaced by an amino group (Moore and Folkers, 1966), plays a significant role in the modern chemotaxonomic classification of this group of microorganisms

(Hiraishi and Hoshino, 1984) along with ubiquinones (Hiraishi *et al.,* 1984). Oelze *et al.* (1975) reported the formation of ubiquinone 10 and bacteriochlorophyll (*b*chl) was determined in *Rhodospirillum rubrum* grown under different culture conditions. Some of the purple non-sulfur bacteria have Q_{10} as the sole quinine components, while others have Q_9 or Q_8 (Hiraishi and Hoshino, 1984), together with menaquinone (MK) of the same isoprenoid chain lengths as the major components (Imhoff, 1984a).

Table 11.3. Quinone System Among Anoxygenic Purple Bacteria

Organisms	**Presence of**			**Major Homologue (S)**
	Q	**RQ**	**MK**	
Rhodospirillum				
Rsp. rubrum	+	+	–	10
Rsp. photometricum	+	+	–	8
Rsp. fulvum	+	–	+	9
Rsp. molischianum	+	–	+	9
Rsp. salexigens	+	–	+	10
Rsp. salinarum	+	–	+	10
Rhodobacter				
Rba. capsulatus	+	–	–	10
Rba. veldkampii	+	–	–	10
Rba. sphaeroides	+	–	–	10
Rhodovulum				
Rv. sulfidophilus	+	–	–	10
Rv. adriaticus	+	–	–	10
Rhodopseudomonas				
Rps. palustris	+	–	–	10
Rps. rutile	+	–	–	10
Rps. blastica	+	–	–	10
Rps. acidophila	+	+	+	10
Rps. viridis	+		+	9
Rps. marina	+	–	+	10
Rps. sulfoviridus	+	–	+	8 + 10
Rhodomicrobium				
Rm. vannielli	+	+	–	10
Rhodopila	+			
Rbi. globiformis	+	+	+	9 + 10

Organisms	Presence of			Major Homologue (S)
	Q	RQ	MK	
Rhodocyclus				
Rcy. purpureus	+	–	+	8
Rcy. tenuis	+	–	+	8
Rcy. gelatinosus	+	–	+	8
Rhodoferax				
Rfx. fermentans	+	+	–	8
Chromatium	+			7
Chr. vinosum	+	–	+	8
Chr. warmingii	+	–	+	8
Thiocapsa roseopersicina	+	–	+	8
Thiocystis gelatinosa	+	–	+	8
Ectothiorhodospira				
Ect. mobilis	+	–	+	7
Ect. shaposhnikovii	+	–	+	7
Ect. vacuolata	+	–	+	7
Ect. halophila	+	–	+	8
Ect. halochloris	+	–	+	8
Ect. abdelmalekii	+	–	+	8
Chlorobium				
C. thiosulfatophilum	–	–	+	Different types
C. limicola	–	–	+	7
Chloroflexus				
Clf. aurantiacus	–	–	+	10
Heliobacterium				
H. chlorum	–	–	+	7–10

Symbols: +, present; –, absent, Q, ubiquinone; RQ, rhodoquinone; MK, menaquinone.Data was taken from information from Hiraishi (1989), Hiraishi *et al.* (1984), Hiraishi and Hoshino (1984). Imhoff (1984a). Maroc *et al.* (1968). Hale *et al.* (1983), Frydman and Rapoport, (1963), Redfearn and Powls, (1968) and Sasikala and Ramana, (1995a).

11.4. VITAMINS

Production of vitamins, particularly complex vitamins like cyanocobalamin (B_{12}) by microbiological processes proved to be superior to a variety of chemical processes. Though APB is known to accumulate a variety of vitamins intracellularly, they are yet to be exploited. Currently, only B_{12} production is being studied for possible commercial exploitation using purple non-sulfur bacteria. Some of the APB-producing vitamins are summarized in **Table 11.4.**

Table 11.4. Vitamin Production by Anoxygenic Phototrophic Bacteria

Organism	Carbon Source	Vitamin (μg/g cells)	Reference
Rba. capsulatus	Wastes	Vit B_{12} (35)	Kobayashi *et al.* (1970)
Rba. sphaeroides + *Corynebacterium*	*n*- alkanes	Vit B_{12} (2300)	Florent and Ninet (1979)
Rps protamicus (fused cell hybrid)	KCN	Vit B_{12} (10mg/L)	Takeuchi and Hamada (1985)
Rba. sphearoides	Pineapple wastes	Vit B_{12} (7.9μg/L), Vit B_2 Vit E tocopherol (21)	Noparatnaraporn and Nagai (1986)
Rps. gelatinosa	Malate Cassava Waste	Vit B_{12} (38) Vit C tocopherol (210) Vit B_2 riboflavin (33.2)	Noparatnaraporn *et al.* (1986) Noparatnaraporn and Nagai (1986)
Rcy. gelatinosus + *Rba. sphaeroides*	Cassava waste	Vit B_{12} (44.2)	Noparatnaraporn *et al.* (1987a)
Rsp. rubrum	Succinate	Vitamin B_{12} (69)	Hirayama and Katsuta (1988)

Data was modified from Sasikala and Ramana (1995a)

PNSB can accumulate more intracellular than extracellular vitamin B_{12}. Further, its production by APB could be enhanced 2-3 folds by various manipulations in growth conditions (i) By changing the concentration of a metal ion in the culture medium (Hirayama and Katsuta, 1988), (ii) Stepwise change of culture conditions from aerobic to microaerobic conditions, (iii) Using of *n*-alkanes as the carbon source instead of the regular usage of carbohydrates,(iv) Using mixed cultures, (v) Optimizing pH, temperature and age of culture and (vi) Using a fused cell hybrid obtained by fusing cells of *Rba. sphaeroides* with a cell of a methanol assimilating bacterium *Protaminobacter ruber* (Takenchi and Hamada, 1985). The mixed culture of *Rhodopseudomonas* and *Corynebacterium* could produce increased production of vitamin B_{12}. Noparatnaraporn *et al.* (1986) reported the vitamin B_{12} formation by *Rps. gelatinosa* under microaerobic conditions.

11.5. HORMONES

Although a few PNSB are reported to produce plant hormones, it is still underexploited. *Rhodospirillum rubrum* (Serdyuk *et al.,* 1993) is reported to produce 3 cytokinins, one of which was identified as Zeatinriboside. Kinetin and Zeatin have been isolated from the culture of *Rba. sphaeroides* is grown on organic waste as a substratum (Kuroda *et al.,* 1990). Serdyuk *et al.* (1993) reported 1.0mg/g cytokinin from wet biomass of *Rsp. rubrum* which is higher than that produced by diazotrophic bacteria. Indole Acetic Acid (IAA) is one of the most physiologically active auxins, is a common product of tryptophan metabolism of several microorganisms (Charlie *et al.,* 2013). Auxins are essential for plant development and have a key role in regulating many growths and behavioral processes in the plant's life cycle. They are responsible for plant cell division, extension and specialization (Tsavkelova *et al.,* 2006). Specifically, IAA plays an essential role in the activation of the cell root and the plant mineral uptake (Sakpirom *et al.,* 2017). These plant growth promoters stimulate seed germination, root formation, enhances vegetative growth, fructification, improves photosynthesis, biosynthesis of pigments and metabolites and is responsible for managing the plant growth under stress conditions (Kazan, 2013; Wani *et al.,* 2016). Research has shown that IAA aids in the plant adaptation to salinity stress, and enhances the root and shoot growth under salinity and heavy metal stress (Egamberdieva, 2009; Fahad *et al.,* 2015).

Multiple bacteria have been reported to produce IAA through different biosynthesis pathways. The indole-3-pyruvate and tryptamine pathways identified in *Rhodopseudomonas* sp., while the produced concentrations differ (Spaepen *et al.,* 2007). Sakpirom *et al.* (2017) studied the extracellular production of IAA in four PNSB species under micro-aerobic light conditions yielding 0.65–3.6 mg $_{IAA}$ l^{-1}. This variation in produced IAA concentration may lead to a variety of outcomes, ranging from phytostimulation to pathogenesis. For instance, the addition of auxin to roots at low concentration only promotes the growth (0.02–0.18 lg $_{IAA}$ l^{-1}), while using at higher concentrations is inhibitory to growth (Davies, 1995). This can mainly inhibit root growth, which may result in poor plant development, as the root system plays a major role in water and nutrient uptake. Furthermore, IAA synthesis from PNSB was studied from species isolated from paddy fields and river sediments, as well as insecticide-tolerant species. The species isolated from paddy fields produced IAA in the range 0.65–3.6 mg$_{IAA}$ l^{-1} in their cultivation medium supplemented with 1 mM IAA precursor, tryptophan (Sakpirom *et al.,* 2017). Similarly, *Rps. palustris* GJ-22 produced IAA with the concentration of30 mg $_{IAA}$ l^{-1} when the 3 mM tryptophan was added to the culture medium, while it is 15 mg $_{IAA}$ l^{-1} produced in culture medium without tryptophan (Su *et al.*, 2017). *Rhodopseudomonas* KL9 isolated from river sediments produced the highest concentration of IAA is about 52 mg $_{IAA}$ l^{-1}in a culture medium supplemented with 3 mM tryptophan (Koh and Song, 2007).

Production of IAA was reported by many workers including Rajasekhar *et al.* (1999), Srinivas *et al.* (2002), Ramchander *et al.* (2011) and Srinivas *et al*. (2009) by these PNSB. Indole-3-acetic acid (IAA) was obtained from sawdust manure using *Rba. sphaeroides* (Yokoyama, 1990). IAA and free amino acids production by three anoxygenic phototrophic bacteria was also reported (Rajyalaxmi and Girisham, 2015, Rajyalaxmi *et al.,* 2015). Photo biotransformation of IAA by some of the PNSB plays a major role in the production of phytohormones (Sasikala and Ramana, 1995). A new phytohormone, Rhodestrin was isolated as a metabolite of anthranilate photo biotransformation by *Rba. sphaeroides* O.U 5 which initiates profuse and early rooting in tissue cultured plants than other auxins (Sunayana *et al.,* 2005). Ranjith *et al.* (2007a) have reported the production of novel indole terpenoid ether, Rhodethrin by *Rhodobacter sphaeroides* which has cytotoxic and phytohormonal activities. Mujahid *et al.* (2010) reported the production of Indole-3-acetic acid by *Rubrivivax benzoatiliticus,* JA2, Indole-3-aldehyde by *Rvi. benzoatiliticus* and *Rba. sphaeroides* DSM 158, Indole-3-ethanol by *Rsp. rubrum* ATCC 1170 and anthranilic acid by *Rba. sphaeroides* DSM 158 by aniline-induced tryptophan production. Production of new phenol terpenoid ester, Rubrivivaxin, having cytotoxic and cyclooxygenase-I inhibitory activity by *Rubrivivax benzoatilyticus* JA2 was reported by Ranjith *et al.* (2011).

11.6. ENZYMES

Interest in the production of industrial enzymes by this group of organisms has increased in recent years because of the exogenously produced amylases by *Chloroflexus aurantiacus* release maltotetraose and maltotriose instead of maltose as in the case of most microorganisms (Ratnakhanokchai *et al.,* 1992). Further, amylases produced by APB could hydrolyze oyster glycogen, amylose and pullulan apart from soluble starch (Buranakarl *et al.,* 1988). The amylase produced by *Cfl. aurantiacus* proved to be thermostable showing activity at 71 °C with pH optima 7.5. Protease produced by *Cfl. aurantiacus* are of two types and type-I protease is stable at 70 °C and pH between 4.0 and 11.0. The cell homogenates of *Rhodocyclus gelatinosus* contain a novel protein modifying enzyme with a pH optimum of 3.2 and can be used for food preparations (Isaki and Kamio, 1992). Catalase peroxidase produced by *Rba. capsulatus*, cellulase produced by genetically altered *Rba. sphaeroides*, L-asparaginase (antitumor) produced by *Rba. capsulatus* and L(+)- tartarate dehydrogenase produced by *Rba. sphaeroides* (Roeder *et al.,* 1983) are some of the enzymes which have commercial value in different areas of public interest. Srinivas *et al.* (2004) reported the production of amylase by *Rcy. gelatinosus*. The production of L-asparaginase by *Rba. capsulatus, Rps. palustris, Rps. acidophila* and *Rcy. gelatinosus* isolated from leather industry effluents was reported by Ramchander *et al.* (2008). Selecting an anoxygenic phototrophic bacterium as a source of enzyme should fulfill the following criteria (**Table 11.5** .)

Table 11.5. Critical Parameters for Selecting an Industrial Enzymes

1	**Enzyme activity**
	Specific activity (units/mg), Application dosage
2	**Stability during**
	Storage, application and immobilization for multiple uses
3	**pH range for activity and stability**
	A broad range of pH and a broad range of process conditions
4	**Safety (allergenicity)**
	Non-toxic to man and environment and Nonallergenicity
5	**High yield production**
	High yield expression in a bacterial host, secretion for high yield and easy purification

Some of these organisms serving as a source for industrial enzyme production are listed in **Table 11.6.**

Table 11.6. Bacterial Enzymes and Their Field of Applications

Enzymes	Starch	Deter-gents	Food	Tex-tiles	Fine Chemicals	Brewing and Juices	Paper and Pulp	Feed
Amylases	+5	+2	+2	+2	-	-	-	-
Proteases	-	+5	+2	+1	+1	+1	-	-
Lipases	-	+2	+1	-	-	-	-	-
Esterases	-	-	-		-	-	-	-
Cellulases	-	+2	+1	+3	-	-	+	-
Glucanase	-	-	+1	-	+3	+3	-	-
Xylanases	-	-	+1	-	+2	+2	+2	+1
Glucose isomerase	+4	-	-	-	-	-	-	-
β-lactam acylase	-	-	-	-	-	-	-	-
Phytases	-	-	-	-	-	-	-	+2

Symbols: +++++ to +, the importance of the enzyme class to the specific use is graded on the basis of the number of enzymes produced and its economic value and -, enzyme of no importance to this use.

11.7. ANTIBIOTICS/ANTIVIRAL SUBSTANCES

Plant diseases because of viral or microbial can cause serious damage to crops. The exposure to specific boosts can improves the disease resistance potential and plants that have been exposed to these biotic or abiotic stimuli in the past can quickly react to exposure to a virus by setting off vigorous defense responses. This induced resistance can be synergist for the plant's survival, particularly when triggered before infection (Choudhary *et al.,* 2007). In this context, APB produces IAA, ALA and siderophores, which can induce systemic resistance against viruses (Pieterse *et al.,* 2014). Foliar fertilization using purple non-sulfur bacteria has been reported to be efficient in suppressing plant viral diseases such as tobacco mosaic virus (TMV) (Su *et al.,* 2017). Though these organisms are reported to be poor producers of antimicrobial substances, several antiviral substances have been isolated from several PNSB. *Rba. capsulatus* produced an antiviral substance that could effectively inactivate *Poliovirus, Sindbis* Virus, a couple of fish viruses and *Coliphage* without causing harm to the host (Kobayashi and Hirotani, 1987). The antiviral substances of *Rba. capsulatus* were identified to be a fatty acid, cis-vaccenic acid, and has been exploited for the waste treatment as it could remove *Coliphages* up to 97% from wastewater of pig pen (Hirotani *et al.,* 1991). Antiviral substances from *Rsp. rubrum* and *Chr. vinosum* were also isolated. *Rhodobacter capsulatus* produced cyclic triterpenes which possessed a variety of biological activities (Loeschcke *et al.,* 2017).

11.8. AMMONIA (NH_3)

The phototrophic bacterium is able to leach out ammonia from fixed nitrogen into the medium. Ammonia represents the primary source of nitrogen fertilizers presently used in agriculture. Microbial ammonia production which occurs at ambient temperature and pressures is expected to be a good alternative safe process for commercial production compared to the expensive Haber-Bosch process. Many phototrophic bacteria are reported to leach out ammonia during their growth on amino acids. Ammonia leaching was observed in resting cells of *Rba. sphaeroides* O.U.001 in the absence of MSX under various gas phases (Sasikala and Ramana, 1990). Hiroo (2004) used a mixed culture of photosynthetic bacteria for ammonia leaching. The major group of microorganisms existing in paddy soils contribute significantly to soil fertility (Hanble and Alexander, 1980). Chalam *et al.* (1997) have reported the adverse effects of pesticides on diazotrophic growth and nitrogenase activity of some APB isolated from paddy soils.

11.9. NITROGEN FIXATION

In view of the increased cost of nitrogen fertilizers and the need of more energy and environmental consideration, the interest in biological nitrogen fixation increased considerably. Inspite of exploiting different microorganisms for nitrogen fixation, the need for more nitrogen for increased crop production has become inevitable. In view of the unique ability of anoxygenic phototrophic bacteria to fix dinitrogen, increased attention is being paid to search for different anoxygenic bacteria such as *Chloroflexus aurantiacus* (Gogotov, 1985).

Rhodobacter sphaeroides, Rba. capsulatus, Rhodopseudomonas palustris, Rps. sulfoviridis, Rps. viridis, Rps. blastica, Rps. acidophila, Rhodocyclus gelatinosus, Rhodospirillum rubrum, Rcy. tenuis, Rsp. molischianum, Rsp. photometricum, Rhodothalassium salexigens and *Rhodomicrobium vannielli* are reported to exhibit high nitrogenase activity and fix atmospheric nitrogen efficiently (Sasikala and Ramana, 1995b). Extensive research has been carried out on different aspects of nitrogen fixation by *Rba. capsulatus* (Avtges *et al.*, 1985; Klipp *et al.*, 1988; Kranz and Haelkorn, 1986) which is a testimony for its ability to fix atmospheric nitrogen. Anoxygenic phototrophic bacteria are of particular importance in the rhizosphere/rhizoplane ecological niche, where one of the most significant processes is the hydrogen transformation since these bacteria are capable of nitrogen evolution and hence it is recycling. Thus their presence in natural systems particularly hydrogen-rich sites such as legume fields is of considerable importance in terms of the contribution to nitrogen economy and hydrogen conservation (Sasikala and Ramana, 1995b).

The microbes associated with nitrogen fixation can roughly be categorized into two groups, symbiotic and free-living, while APB belonging to the latter group (Herridge *et al.*, 2008). N_2-fixing microorganisms transform the biological plant unavailable atmospheric molecular nitrogen (N_2), into ammonia/ammonium

(NH_3/NH_4^+), a plant-available nitrogen form (Franche *et al.,* 2009; Olivares *et al.,* 2013). The enzyme that catalyzes nitrogen fixation is nitrogenase, which is present in all nitrogen-fixing bacteria. There are three types of nitrogenase isozymes, namely molybdenum-iron (Mo-Fe), vanadium-iron (V-Fe) and iron-iron (Fe-Fe) nitrogenases (Sakpirom *et al.,* 2017). According to Sakpirom *et al.* (2017), among 235 tested purple non-sulfur bacterial isolates, *Rhodopseudomonas palustris* strain TN110 possessed all three different nitrogenase genes and offered the best nitrogen fixation ability. Nitrogen fixation by APB is most efficient under strict anaerobic conditions, as oxygen inhibits this process (Masepohl and Kranz, 1978). However, some PNSB species of *Rhodopseudomonas* and *Rhodobacter* is very tolerant to oxygen and even fix nitrogen under micro-aerobic conditions (Larimer *et al.,* 2004; Hoffmann *et al.,* 2014). This largely increases the potentials of PNSB in plant production, such as paddy fields, which are micro-aerobic. The N_2 fixation of heterotrophic bacteria can be photo-dependent or photo-independent (Pfennig and Truper, 1989). Although APB fixes nitrogen in considerable amounts under anaerobic conditions only, mixed cultures in symbiotic association with aerobic heterotrophs such as *Bacillus subtilis* (Kobayshi and Haque, 1971*), Azobacter agilis* (Kobayashi and Ye, 1986) and other heterotrophs (Katayama *et al.,* 1965) and the nitrogen-fixing ability of *Rba. capsulatus* increased considerably. *Rhodobacter capsulatus* was cultured together with heterotrophs like *Candida guilliermondi, Lactobacillus planatarum, Rhizobium trifolii* and *Rhizobium japonica* (Kobayashi *et al.,* 1981). *Rhodobacter capsulatus* in association with *Azotobacter vinelandii* fixed atmospheric nitrogen 10 times more than individual capacity. Kobayashi *et al.* (1967) could observe such symbiotic association not only under laboratory conditions but also in nature. The mixed culture of *Rhodobacter capsulatus* and *Rhizobium* was found to be advantageous over *Rhizobium* alone (Hiroto, 1981).

The nitrogen fixation ability of PNSB is higher under light conditions than under dark conditions (Nunkaew *et al.,* 2014a; Wong *et al.,* 2014). Hence, PNSB supplies nitrogen more efficiently to the light habitat zones, where nitrogenase activity is not limited by their chemoheterotrophic metabolism as is the case under dark conditions (Harada *et al.,* 2005). Purple non-sulfur bacteria are distributed in various habitats which are exposed to sunlight and atmospheric N_2 (including paddy fields) and their support to rendering nitrogen bioavailable to plants has been determined. For example, when inoculated in a hydroponic nutrient solution lacking a nitrogen source, ammonium was detected in the medium ($< 10\ \mu M$) (Maudinas *et al.,* 1981). Therefore, PNSB present as one of the main dominant species in paddy fields (Elbadry *et al.,* 1999a,b) which are significantly contribute to nitrogen fertility, where rice yields can reach values up to 2.0–3.5 ton of nitrogen per hectare by utilizing the only nitrogen originating from soil organic matter and biological nitrogen fixation (Kundu and Ladha, 1997).

Gamal-Eldin and Elbanna (2011) studied an improved usage efficiency of synthetic nitrogen fertilizer (a mixture of synthetic fertilizer and *Rhodobacter capsulatus*) were used together. At the same time, the use of the mixture resulted in higher grain yield than that product obtained from synthetic fertilizer or *Rhodobacter capsulatus* when used separately. Biological nitrogen fixation, not only supplies the essential nutrients to plants but also enhances their nitrogen uptake efficiency (Adesemoye *et al.,* 2008; Wong *et al.,* 2014). For example, this was validated during rice cultivation in a hydroponic solution inoculated with *Rhodobacter capsulatus* DSM 155. Particularly, in the nitrogen-deficit medium, the nitrogen content of the root increased by 50–65%, proving the nitrogen-fixing abilities of purple non-sulfur bacteria. In contrast, when the hydroponic solution contained nitrogen, the effect was less considerable (1.3–14% increase) (Elbadry and Elbanna, 1999). Similarly, the nitrogen content of rice straw rose by 9.2% through seed coating with *Rhodobacter capsulatus* DSM 155 was also studied (Gamal-Eldin and Elbanna, 2011).

The nitrogen-fixing abilities of *Rba. capsulatus* DSM 155 were also observed when this PNSB strain was inoculated on the roots of rice seedlings, initiating an increase of rice grain nitrogen content by 7.1% (Elbadry *et al.,* 1999a,b). In this case, the rice straw yield improved by 8.6–24%, with a shrinking effect as the nitrogen content in the fertilizer increased about 0–95 kg N ha^{-1}. Moreover, the establishment of half the amount of suggested synthetic nitrogen fertilizer in combination with PNSB-inoculated seeds resulted in rice grain yields statistically equivalent to the addition of the full amount of synthetic nitrogen fertilizer (Gamal-Eldin and Elbanna, 2011). Therefore, the inoculation of PNSB could considerable as cost savings due to the less use of chemical fertilizers, depending on the required dosage and related production cost for PNSB biomass. A roughly doubled agronomic nitrogen use efficiency (dry yield per unit of nitrogen supplied) was also studied by inoculation with *Rps. palustris* on pak choi (Wong *et al.,* 2014). Similarly, Hsu *et al.* (2015) observed the increased nitrogen efficiency of lettuce and pak choi to 17 and 22-44% respectively with th in *Rhodopseudomon* e inoculation of *Rps. palustris* PS3. Arashida *et al.* (2018) reported the nitrogen fixation *as palustris* co-cultured with *Bacillus subtilis* in the presence of air. Sakpirom *et al.* (2019) reported the abilities of a *Rhodopseudomonas palustris* TN110 to remediate cadmium through the biosynthesis of CdS nanoparticles and to fix nitrogen simultaneously. Therefore, the supply of purple anoxygenic phototrophic bacteria can enhance nitrogen efficiency, potentially contributing to more sustainable agriculture.

11.10. BACTERIOCINS

Bacteriocins are ribosomally synthesized antimicrobial peptides or proteins. The important role of bacteriocins is to inhibit or kill bacteria that are usually closely related to the producer strain (Sprules *et al.,* 2004). Bactericidal effects of bacteriocins have been recognized as an important subject in food fermentation and preservation. Those produced by lactic acid bacteria have been largely studied as potentially safe food-grade preservatives of biological origin (O'Sullivan *et al.,* 2002). The class IIa bacteriocins produced by lactic acid bacteria kill some pathogenic bacteria (Ex. *Listeria*) with high efficiency (Drider *et al.,* 2006). Most research has been dedicated to lactic acid bacteria for a combination of bacteriocin with antimicrobial agents, membrane permeability and lipopolysaccharide (LPS) affinity (Drider *et al.,* 2006). However, bactericidal peptides are produced by several other classes of bacteria, such as obligate and facultative anaerobes. Some facultative anaerobic purple non-sulfur bacteria such as *Rhodobacter capsulatus, Rhodobacter sphaeroides* and *Rhodopseudomonas palustris* have been reported to be good producers of bacteriocins (Wall *et al.,* 1975; Willison *et al.,* 1987). Antiviral substances from *Rba. capsulatus* , *Rsp. rubrum* and *Chr. vinosum* were isolated (Kobayashi *et al.,* 1975) and this substance produced from *Rhodobacter capsulatus* inactivates the poliovirus, send is a virus, a couple of fish viruses and coliphages without causing any damage to the host cells (Kobayashi and Hirotani, 1987; Hirotani *et al.,* 1987). *Rhodobacter capsulatus* ATCC 17016 is also reported to produce bacteriocins (Lee *et al.,* 2009). These bacteria are also reported to produce antibiotics (Burgess *et al* ,. 1992) and antiviral substances (Hirotomi *et al.*, 1991). *Rhodobacter capsulatus* produce cyclic triterpenes which possessed a variety of biological activities (Loeschcke *et al.*, 2017).

11.11. DENITRIFICATION

The denitrification abilities of anoxygenic phototrophic bacteria are of unique interest in light of the fact that in industrial effluents the ammonia and nitrate concentration is regularly a lot higher, causing numerous issues if such wastewater are released into rivers and lakes (Wiesmann, 1994). A few strains of purple non-sulfur bacteria can use nitrate either as an electron acceptor in the anaerobic respiratory process or as a nitrogen hotspot for phototrophic bacterial development. Denitrification is a typical property among the species *Rba. sphaeroides* ,*Rba. capsulatus* and *Rps. palustris* (De Bont *et al.,* 1981; Imhoff, 1989b) and the mechanism of denitrification in these bacteria is entrenched with the enzyme nitrate reductase (Klemme *et al.,* 1980; Kerber and Cardenas, 1982; McEwan *et al.,* 1987; Martinez-Luque *et al.,*1991; Sabaty *et al.,* 1993) and nitrous oxide reductase (Michalski *et al.,* 1986). The immobilized denitrifying cells of *Rps. palustris* was used to adequately remove NO_3-N from the wastewater (Sasaki *et al.,* 1991; Shen and Hirayama, 1993).

11.12. HOPANOIDS

Hopanoids are pentacyclic triterpenoids biosynthesized by bacteria, ferns, fungi, plants, mosses, protists and lichens (Ourisson *et al.*, 1987). They represent the oldest chemo fossils or molecular fossils dating back to late archean times (2.77 billion tears Ga) (Blumenberg *et al.,* 2006), to which a discrete biological source can be attributed (Summons *et al.,* 1999). More than 200 hopanoids are identified from both biological and geological sources (Ourisson and Albrech 1992). Bio hopanoids are buried in the phospholipid bilayer of bacteria and function similarly to steroids of eukaryotic cells playing a key role in regulating and rigidifying the membranes (Talbot *et al.*, 2007). Hopanoids were initially considered to be components of membranes of aerobic bacteria. However, hopanogenes it was independent of oxygen (Blumenberg *et al.*, 2006) and was discovered among many anaerobic bacteria also (Talbot *et al.*, 2007; Garcia Costas *et al.*, 2012; Tushar *et al.*, 2014; Tank and Bryant 2015). Even though hopanoids were reported from some phylogenetically related members of cyanobacteria, purple non-sulfur bacteria, acetic acid bacteria, *Alicyclobacillus* sp., nitrogen-fixing bacteria, methylotrophs, methanotrophs and sulfate-reducing bacteria (SRB), no definitive taxonomic conclusions were drawn from their distribution (Hartner *et al.*, 2005; Talbot *et al.*, 2007; Blumenberg *et al.*, 2009; Tushar *et al.*, 2014). There are no reports of hopanoids from Archaea. Hopanoids are one of the most important lipid biomarkers to record biological activity in sedimentary rocks for environmental diagnostics (Ourisson and Albrech, 1992) and are gaining importance in the chemotaxonomy of bacteria (Vishnuvardhan *et al.*, 2013; Tank and Bryant, 2015).

Among anoxygenic phototrophic bacteria, hopanoids are well characterized from *Rhodopseudomonas palustris* , *Rhodospirillum rubrum* , *Chloracidobacterium thermophilum* ,*Rhodoblastus acidophilus* ,*Rhodomicrobium vannielii* and *Rhodomicrobium udaipurense* (Tushar *et al.*, 2014; Tank and Bryant, 2015). Hopanoids production by *Rhodoplanes* spp. was reported by Tushar *et al.* (2015). Hopanoid which has toxicity against mouse leukemia cells were isolated from *Rhodopseudomonas palustris* (Nagumo *et al.*, 1991) (**Fig. 11.3**). Different hopanoids reported in various anoxygenic bacteria are listed in **Table 11.7.**

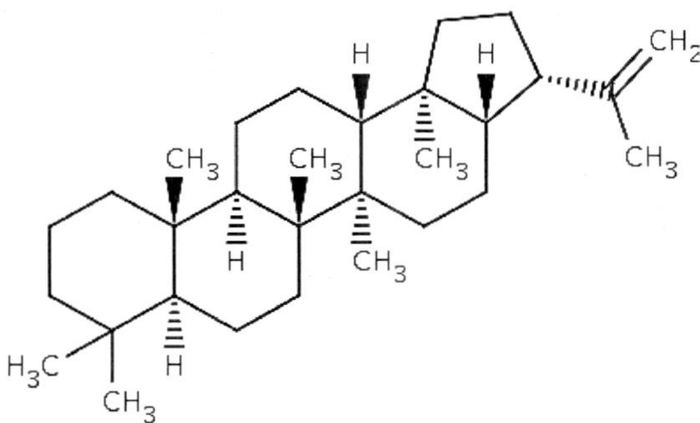

Fig. 11.3. Hopanoid

Table 11.7. Hopanoids of Anoxygenic Phototrophic Bacteria

Name	Hopanoids	Reference
Chloroacidobacterium thermophilum	Diploptene, bacteriohopanetetrol, bacteriohopane cyclitol ether	Tank and Bryant (2015) Garcia Costas *et al.* (2012)
Rhodoblastus acidophilus	Diploptene, diplopterol, uncharacterized bacteriohopane polyols	Neunlist and Rohmer (1985), Neunlist *et al.* (1988)

Name	Hopanoids	Reference
Rhodopseuodomonas Palustris	Diploptene, 2-methyldiploptene, diplopterol, 2-methyldiplopterol, tetrahymanol, 2-methyltetrahymanol, adenosylhopane, ribosylhopane, bacteriohopanetetrol, 2-methylbacteriohopanetetrol, aminobacteriohopanetriol, anhydrobacteriohopanetetrol, BHP-508, BHP-550, BHP-492, BHP-570	Sessions *et al.* (2013), Neunlist *et al.* (1988)
Rhodomicrobium udaipurense	Diploptene, 2-methyldiploptene, diplopterol, 2-methyldiplopterol, tetrahymanol, 2-methyltetrahymanol, bacteriohopanetetrol, 2-methylbacteriohopanetetrol, aminobacteriohopanetriol, N-tryptophanyl-35-aminobacteriohopanetriol	Tushar *et al.* (2014)
Rhodomicrobium vannielii	Diploptene, diplopterol, N-tryptophanyl-35-aminobacteriohopanetriol, N-ornithinylaminobacteriohopanetriol, 3β-hydroxy-17-methylhopane and 29-hydroxy-3, 17-dimethylhopane	Neunlist *et al.* (1985)
Rhodospirillum rubrum	Diploptene, diplopterol, uncharacterized bacteriohopane polyols	Rohmer *et al.* (1984)

11.13. BIOFERTILIZERS

APB is not only having a unique ability to fix dinitrogen but also proved to bethe highest N_2 fixers as evidenced by their nitrogenase activity (**Table11.8** .)

Table 11.8. Nitrogenase Activity of Different PNSB

Organism	Nitrogenase Activity (nmol C_2H_4 formed/ mg dry wt/hr)
Rhodobacter capsulatus	4900
Rhodocyclus gelatinosus	3490
Rba. sphaeroides	2755
Rhodopseudomonas viridis	2510
Rhodomicrobium vannielli	1570
Rhodospirillum tenue	1400
Rhodopseudomonas blastica	1230
Rhodospirillum rubrum	1175

Organism	Nitrogenase Activity (nmol C_2H_4 formed/ mg dry wt/hr)
Rhodospirillum molischianum	1030
Rhodospirillum fulvum	930
Rhodopseudomonas acidophila	880
Rhodopseudomonas palustris	700
Rhodopseudomonas sulfoviridis	520
Rhodospirillum photometricum	755
Rhodospirillum salexigens	630
Rhodobila globiformis	515

Data was taken from Madigan and Cox (1982) and Madigan *et al.* (1984)

APB can be used as a biofertilizer individually or in mixed cultures. Members of purple bacteria (PNSB and PSB) and *Heliobacteriaceae* are good nitrogen fixers for low land rice was established and can be exploited as a natural source of nitrogen fertilizers. The rice fields are ideal places for the growth of APB because of the availability of light as an energy source, organic or inorganic reduced sulfur compounds as carbon source, electron donors and anaerobic conditions are congenial for N_2 fixation. Isolation and characterization of several purple non sulfur bacteria from paddy soils was reported by Montano *et al.* (2009). It is also observed that the Free-air CO_2 enrichment (FACE) enhances the biodiversity of Phototrophic purple bacteria in flooded paddy soil (Feng *et al.*, 2009).

The presence of diazotrophs like *Azotobacter vinelandii* and *Rhodobacter capsulatus* can benefit rice plants for nitrogen nutrition up to the ear stage in the absence of combined nitrogen. Elbadry and Elbanna (1999) have demonstrated the beneficial effects of *Rba. capsulatus* on four rice varieties in hydroponic cultures. Harada *et al.* (2005) reported the inoculation of phototrophic purple bacteria increased grain yield of rice and nitrogenase activity of paddy soil. Application of freeze-dried *Rba. sphaeroides* resulted in an improvement in the fruit quality of tomatoes (Kondo *et al.*, 2006). Koh and Song (2007) and Lee *et al.* (2008) have recorded increased seed germination and growth of tomato with the application of *Rhodopseudomonas* sp. Gamal-Eldin and Elbana (2011) reported the potential of *Rhodobacter capsulatus* as a biofertilizer for flooded rice (*Oryza sativa* L.) under field conditions and minimize the environmental impact of intensive use of chemical fertilizers. Ramchander *et al.* (2012) have reported the stimulatory effect of bio inoculants *Rhodobacter capsulatus* KU002 on two cultivars of paddy. Vareeket and Saytong, (2015) have recorded the application of photosynthetic bacteria to rice fields enhanced plant height, number of tillers and dry weight of rice plant.

The use of APB as a biofertilizer have several advantages over *Cyanobacteria*

1. Grows under subsurface soil which is anaerobic due to waterlogging conditions, on the other hand, it is unsuitable for the growth of *Cyanobacteria* .
2. They grow fast and exist in large numbers.
3. Fix a high amount of nitrogen.
4. Grow even under dark conditions.
5. Can fix nitrogen under dark conditions. Thus N_2 fixation can take place for 24 hrs.
6. Ability to utilize far red infrared light which is not used by rice plants.
7. As APBexists in rhizoplane which makes transport of fixed nitrogen to rice plant easy and fast.
8. These can be cultured easily at a low cost and applied easily.

Similarly, the use of APB is advantageous over free-living nitrogen-fixing organisms

1. They can carry anaerobic nitrogen fixation.
2. They derive energy from light at a low cost.
3. Utilize a variety of substrates.

It has been observed that the anoxygenic phototrophic bacteria release different substances including amino acids, growth substances and nucleic acids either during their growth or after death which promotes the rice plant growth and reproduction contributing to the increased grain yield. Similarly, these organisms are insensitive to herbicides applied to rice plants. It is also observed that the anoxygenic phototrophic bacteria degrade fungicides like carbon and oxycarboxin helping rice plants to protect nitrogen-fixing organisms around them.

The presence of low levels of combined nitrogen in nature may be advantageous to nitrogen-fixing organisms by allowing more healthy growth and thus more vigorous nitrogen fixation than could be achieved with elemental nitrogen as sole nitrogen source using of mixed culture of *Rhodospirillum* , *Rhodopseudomonas* , *Rhodomicrobium* , *Chromatium* and *Ectothiorhodospira* with a highly hygroscopic polymer such as polyacrylic acid, PVA, CMC and an organic/inorganic material was not only a very good soil improver but also a plant growth stimulator (Hessbe and Mitumasu, 1986). Similarly, *Chromatium* and *Chlorobium* cultured in petroleum wastewater pelleted by centrifugation and mixed (1:10) with soil and dried used as a fertilizer in paddy field which increased the crop yield significantly. The mixed cultures of *Rhizobium japonicum* and *Rhodobacter capsulatum* grown in mineral medium and collected by centrifugation, air-dried and applied in the ratio of 1:1 increased the efficiency of nitrogen fixation ability of *Rhizobium japonicum* significantly. Nitrogen fixation activity of *Rba. capsulatum* increased when cultured with heterotrophic bacteria like *Bacillus subtilis* , *B. megaterium* , *Azotobacter agilis* and other heterotrophs even under aerobic conditions. Similarly the nitrogen-fixing ability of *Rba. capsulatus* increased significantly when it was cultured along with *Candida guiliimondia* and *Lactobacillus plantarum*. Association of *Azotobacter vinelandii* with *Rba. capsulatus* and *Rhizobium* is found to be advantageous over *Rhizobium* alone (Hiroto, 1981). Anoxygenic phototrophic bacteria are being utilized as biofertilizers for rice production (Habte and Alexander, 1980), leguminous crops (Kobayashi *et al.,* 1981), tomatoes (Kobayashi, 1982), citrus (Kobayashi, 1976) and persimmon (Kobayashi, 1982). The use of anoxygenic phototrophic bacteria either individually or in mixed cultureshas become a common practice are listed in **Table 11.9** .

Table 11.9. Some Anoxygenic Phototrophic Bacteria used as Biofertilizers

Organisms	**References**
Chromatium and *Chlorobium* cultured in petroleum wastewater, pelleted by centrifugation and mixed (1:10) with soil dried in air and used as fertilizer in a paddy field.	Kobayashi (1970)
Rhizobium japonicum and Rhodobacter capsulatus (in 1:1 ratio) grown on mineral medium collected by centrifugation and sir dried and containing viable cells of 80% *Rhizobium japonicum.*	Hiroto (1981)
Rhodospirillum, Rhodopseudomonas, Rhodomicrobium, Chromatium, Ectothiorhodospira in polyacrylic acid, PVA and CMC.	Hassebe and Mitsumsan (1986)
Rhodopseudomonas sp., used as biofertilizer for the enhancement of growth and yield of tomato under greenhouse conditions.	Lee *et al.* (2008)
Rhodopseudomonas palustris tk123 and lactic acid bacteria strain L35 co-culture in rice straw medium will be used as a tool for producing biofertilizers for organic saline paddy fields.	Kantha *et al.* (2010, 2012)
Rhodopeudomonas palustris PS3 isolate used as bioinoculant on *Brassica rapachinensis* (Chinese cabbage) for integrated nutrient management in agriculture.	Wong *et al.* (2014)
Rhodopseudomonas palustris potentially used as biofertilizers to assist rice *(Oryza sativa L. subsp. indica)* growth under salt stress and to reduce greenhouse gas emissions.	Kantha *et al.* (2015)
Rhodopseudomonas palustris PS3 in alginate, polyethylene glycol (PEG), polyvinylpyrrolidone-40 (PVP), glycerol, glucose, and horticultural oil are used as bio inoculants.	Lee *et al.* (2016)
Rhodopseudomonas palustris strains are used as biofertilizers to enhance rice yields and reduce methane emissions.	Kantha *et al.* (2016)
A dried mass of microalgae including *Chlorella* sp. and *Scenedesmus* sp. and dried mass of *Rhodopseudomonas* sp., used as biofertilizer	Zarezadeh *et al.* (2019)

11.14. BIODEGRADABLE POLYESTERS

A number of APB are reported to produce a considerable amount of poly-β - hydroxyl alkanoates which form several types of polyesters. PHA as polyester for the utilization of either biodegradable or biocompatible plastics for a wide scope of applications. HB/HV copolymer has shown great promise as a biocompatible biodegradable material for medical applications. PHAs have attracted increasing attention due to their biomass-based origin and biodegradability (Numata *et al* ,. 2009). Marchessault *et al.* (1988) have reported the potential role of PHAs as so-called biomass transducers. PHA production by anoxygenic phototrophic bacteria *Rhodospirillum rubrum* is also reported (Brandl *et al.,* 1989; Clemente *et al.,* 2000). Similarly, Kranz *et al.* (1997) reported the PHA production by *Rba. capsulatus*. Brandl *et al.* (2001) reported that *Rba. sphaeroides* produced PHA as the major component (97%) and a small amount of PHV (3%) under anaerobic light conditions. Boaree *et al.* (1993) have developed a biodegradable composite material composing of PHB with hydroxyapatite from anoxygenic phototrophic bacteria for their industrial use. Yamane (1992) has excellently reviewed the biodegradable polyester metabolism. Noparatnaraporn *et al.* (2001) reported the PHA production by *Rba. sphaeroides* and *Rba. sphaeroides* IL 106 from acetic acid. The PHA production by photosynthetic bacteria at 37-40 °C in an aerobic dark condition was reported by Lorrungruang (2006). Marine purple sulfur bacteria proved to be ideal microorganisms for polyhydroxyalkanoates (Takeuchi *et al*., 2016). Similarly, PHA production by marine purple non-sulfur bacteria has been reported (Xiao and Jiao, 2011).

Poly-β-hydroxybutyrate (PHB) is the best-known member of the PHA series of polyesters and it is known to be produced by intracellular energy and carbon reserve. *Rhodospirillum rubrum* is an interesting organism that can accumulate both PHB and glycogen as storage reserves depending on the nature of the carbon source and environmental conditions. When substrates are photometabolized via acetyl CoA without the formation of pyruvate, PHB accumulates. In the absence of CO_2, practically all the acetate assimilated is converted to PHB. On the other hand, when the substrates are metabolized via pyruvate, glycogen is formed. The pathway leading to PHB synthesis from acetate by *Rsp. rubrum* under light can be presented as

Acetate→ Acetyl CoA→ Aceto acetyl CoA → L-(+) β-hydroxy butryl CoA → Crotonyl CoA→D(-) Hydroxy-butryl CoA → Poly-β-hydroxybutyrate.

A part of acetate may undergo oxidation through Kreb's cycle according to the reaction.

$$9N\ \text{Acetate} \xrightarrow{\text{Light}} 4(C_4H_6O_2)\,n + 2N\,CO_2 + 6N\,H_2O$$

However, in the presence of molecular hydrogen, utilization of acetate by *Rsp. rubrum* is very past without partial oxidation resulting 86% of acetate is converted PHB.

$$2n\ Acetate + nH_2 \xrightarrow{Light} (C_4H_6O_2)n + 2nH_2O$$

The reducibility of butyric acid is higher than that of PHB. Therefore, during the polymerization of butyric acid, hydrogen combines with CO_2 to form a polysaccharide.

$$2n\ (C_4H_6O_2) \rightarrow 2(C_4H_6O_2)n + 4n\ (H)$$

$$4n\ (H) + CO_2 \rightarrow (CH_2O)n + nH_2O$$

The whole process can be represented as

$$2n\ (C_4H_6O_2) + nCO_2 \rightarrow 2(C_4H_6O_2)n + (CH_2O)n$$

Rhodocyclus gelatinosus metabolizes acetate phototrophically or chemotrophically to acetoacetate which gets converted to PHB

$$CH_3COCH_3 + CO_2 \rightarrow CH_3COCH_3COOH$$

Chromatium sp. under dark conditions metabolizes internal storage or externally supplied sugar into PHB. Elemental sulfur acts as the electron acceptor.

$$(C_6H_{10}O_5)n + nH_2O \rightarrow (C_4H_6O_2)n + 2n\ CO_2 + 6n\ (H)$$

$$3n\ S + 6n\ (H) \rightarrow 3n\ H_2S$$

$$(C_6H_{10}O_5)n + nH_2O + 3nS \rightarrow (C_4H_6O_2)n + 2n\ CO_2 + 3n\ H_2S$$

Chromatium vinosum and *Rba. capsulatus* form PHB with the help of stored carbon or externally supplied sugars as follows.

Reserve or exogenously supplied sugars → Glucose-1-Phosphate → 2 glyceraldehyde-3-Phosphate → 2 1,3-bi phosphoglycerate → 2 3-phosphoglycerate → 2 pyruvate → 2 acetyl CoA → acetoacetyl CoA → 3-hydroxy butyryl CoA → polyhydroxybutyrate.

PHB may be produced either by growing the bacterium in a suitable medium which results in the accumulation of PHB. The other process is that the organism is grown on a suitable carbon source to get a large amount of biomass which after 48-72 hrs of incubation is transferred to a fresh medium without any nitrogen source and supplemented with an aliphatic organic acid. The yield of PHB may be more with the latter process. Wingender *et al.* (1999) have excellently discussed various aspects of extracellular polymeric substances of various microorganisms.

PHB formed by APB can be extracted from lyophilized dry cells by solvent extraction and determined gravimetrically and calculated as the percentage of cellular dry weight. Characterization of purified PHB can be quantitatively determined either by GC, HPLC and NMR after methanolysis in the presence of H_2SO_4.

The PHB has a number of interesting characteristics and can be used in various ways similar to many conventional synthetic plastics such as polyhydroxyvalerate (PHV) and polypropylene (PP) (**Table 11.10**) (Sasikala and

Ramana 1995b). It can be molded reinforced with inorganic fillers, spun into a fiber, or formed into a film with excellent gas barrier properties. However, PHB is relatively hard and brittle apart from its thermal instability at a temperature above the melting point (above 180ºC). At present number of bacterial polymers containing β-hydroxybutyrate and β-hydroxypentanoate with desirable properties are available for this purpose.

Table 11.10. Properties of PHB, PHV and PP

Property	PHB	PHV	PP
Crystalline melting point (˚C)	175	100	176
Crystalinity (%)	80	80	70
Molecular weight (10^5)	1-8	2	2-7
Glass transition temperature (˚C)	5-15	-16	-10
Density (g cm^{-3})	1.25	1.2	0.91
Flexurial modules (CPa)	4.0	-	1.7
Tensile strength (MPa)	40	-	38
Extension to break (%)	0	-	400
Ultraviolet resistance	Good	Good	Poor
Solvent resistance	Poor	Poor	Good
Biodegradability	+	+	-
Oxygen permeability (cm^2m atm-day)	40	-	1700

Data was taken from Howells (1982), King (1982), Winton (1985), Brandl *et al.* (1990), Hocking and Marchessault (1994) and Pearce and Marchessault (1994).

Some of the applications of PHA in medicine and industry are: 1. In surgical pins, sutures, staples and swabs, 2. Wound dressing, 3. Blood vessel replacements, 4. Stimulation of bone growth by piezoelectric properties, 5. Bone (orthopedic) replacement and plates, 6. Biodegradable carrier for the long-term dosage of drug and medicine, 7. Packing containers, bottles, wrapping bags and films, 8. Disposable items such as diapers or feminine hygiene products and 9. Biodegradable carrier for the long-term dosage of herbicides, fungicide, insecticide and fertilizers.

Biosynthesis of PHB and photoproduction of hydrogen are the two processes that compete for reducing equivalents coming from the assimilation of organic acids in the cell. The carbon source, nitrogen amount and pH are some factors that affect PHB accumulation inside the cells. The production of H_2 by APB takes place when there is a high amount of carbon and a low amount of nitrogen in the medium. The acetate-containing medium with a high concentration of ammonia is advantageous for PHB accumulation (Khatipov *et al.,* 1998). The influence of different cultural conditions for the production of PHB and polyhydroxybutyrate- co-valerate by *Rhodopsedomonas palustris* SP5212 is discussed by Mahuya *et al.* (2005). Sangkharak and Prasertsan (2008) have tried to optimize the composition of the medium for the production of PHB from halotolerant photosynthetic bacteria cultivated under aerobic-dark conditions. Waligorska *et al.* (2009) discussed the kinetics of PHB production by *Rba. sphaeroides* and observed that the accumulation of PHB is related to C/N ratio of the substratum. When a C/N ratio of 120 was used instead of C/N ratio of 6, PHB accumulation increased 30-fold. Sheng *et al.* (2010) and Frolund *et al.* (1996) recorded the production of extracellular polymeric substances by *Rhodopseudomonas acidophila* and activated sludge respectively.

Rhodobacter sphaeroides 14F (Lorrungruang *et al.,* 2006) was found to be suitable production of polyhydroxyalkanoate in two-stage dark aerobic dark cultivation. Takeuchi *et al.* (2016) have reported the increased synthesis of PHA in seawater supplemented with acetate. Further, these marine purple sulfur bacteria proved to be ideal microorganisms for polyhydroxyalkanoates production. Chowdhury *et al.* (1996) and Lu *et al.* (2009) have discussed various factors influencing polyhydroxybutyrate synthesis by marine photosynthetic bacterium *Rhodopseudomonas* sp. strain W-1S. Lenz and Marchersault (2005) have excellently discussed various aspects of biodegradable polyester biosynthesis by different bacteria.

11.15. BIOPESTICIDES

Photodynamic compounds are a group of chemicals that have been proposed as potential herbicides in recent years. 5-aminolevulinic acid (ALA) is the only naturally occurring photodynamic compound that can be used as an effective herbicide/pesticide. 5-aminolevulinic acid is known as a plant growth regulator. Several studies have determined the effect of ALA produced from APB on plant growth (Akram and Ashraf, 2013; Nunkaew *et al.,* 2014a, b), however, the mechanisms of ALA-induced growth and yield promotion have not been clarified yet. ALA was also provided at low concentrations to plants, as plant growth-promoting properties (Zhen *et al.,* 2012; Ali *et al.,* 2013). On the other hand, high ALA levels might cause oxidative stress, thus limiting plant growth (Akram and Ashraf, 2013), but as a plant growth promoter, it also increases the uptake of minerals and the synthesis of soluble sugars and proteins. ALA is also used as a precursor of the synthesis of chlorophyll, vitamin B_{12}, anti-oxidative enzymes and other metabolites that reduce the adverse effects of various abiotic stress conditions (Wongkantrakorn *et al.,* 2009; Nunkaew *et al.,* 2014b; Sakpirom *et al.,* 2017). Specifically, ALA can increase photosynthesis and thus stimulate plant growth at low concentrations of about 1–5 mg_{ALA} l^{-1} (Sakpirom *et al.,* 2017). ALA can also be used for the agricultural purpose to increase tolerance towards salinity (Wongkantrakorn *et al.,* 2009; Naeem *et al.,* 2011; Nunkaew *et al.,* 2014b), drought, temperature and low-light stress in plants (Akram and Ashraf, 2013), as well as a biodegradable herbicide (Sasikala *et al.,* 1994; Sasaki *et al.,* 2002). Anoxygenic phototrophic bacteria (APB) can accumulate and excrete a high amount of ALA into the medium. Hence, it is suitable for commercial exploitation and it is commercially produced by *Rba. sphaeroides*

(Sasikala *et al.,* 1994) . ALA also enhances the production of ATP and NADPH, which are important cofactors for CO_2-fixation (Sun *et al.,* 2009). The exogenous provision of ALA helps in the accumulation of chlorophyll, leads to an increase of photosynthetic activity (Nunkaew *et al.,* 2014a). As the commercially available ALA is very expensive for most agricultural applications, the use of ALA-producing microorganisms is regarded as a promising, economical and alternative for plant cultivation (Nunkaew *et al.,* 2014a). The ALA is also applied in the area of medicine and pharmacy products (Levy, 1995). Tangprasittipap and Prasertsan (2002) have given an overview on ALA production by photosynthetic bacteria concerning its biosynthesis and regulation as well as its application as herbicide, insecticide and growth stimulator. Recent medical applications in the field of photodynamic therapy, cancer treatment, tumor diagnosis and other clinical uses are recognized.

This aliphatic precursor for tetrapyrrole synthesis in animals, plants, algae and bacteria is harmless to humans and animals in addition to being biodegradable. Two ALA biosynthetic pathways are known in bacteria, namely the five-carbon and the ALA synthase pathway (Avissar *et al.,* 1989). In purple anoxygenic phototrophic bacteria, the synthesis of ALA is performed by the latter pathway, in a reaction catalyzed by ALA synthase that condensates glycine with succinyl-CoA (Beale, 1990; Sasaki *et al.,* 1990). PNSB has been demonstrated to produce ALA under saline or heavy metal stress conditions. For example, Kantha *et al.,* (2010), Nunkaew *et al.* (2014a), Sakpirom *et al.* (2017) reported the PNSB isolates from paddy fields and Cd/Zn contaminated soils produced 0.23-5.0 $mg_{ALA}\ l^{-1}$ under micro-aerobic light conditions. The production of ALA can also be promoted by adding the precursor molecules such as glycine and succinate to the cultivation medium. Su *et al.* (2017) reported the ALA concentrations reached 7.9 mg l^{-1} when the precursors were added, but it is low in the absence of precursors is about 4.5 $mg_{ALA}\ l^{-1}$. Anoxygenic phototrophic bacteria such as *Rcy. gelatinosus* are a potential group of organisms that can produce ALA in large amounts (Lim *et al.,* 1993). Similarly, *Rba. sphaeroides* are being exploited for large-scale production of ALA (Sasaki *et al.,* 1991). Kantha *et al.* (2010) reported the photosynthetic bacteria *Rhodopseudomonas palustris* producing ALA from the soil of organic saline paddy fields from the Northeast region of Thailand.

Among various groups of microorganisms, anoxygenic phototrophic bacteria produce a high amount of ALA. These organisms synthesize through two distinct metabolic pathways.

1. **C4 pathway**: Mainly occurs among the purple non-sulfur bacteria. In this pathway, succinyl CoA and glycine are condensed by enzyme 5-aminolevulinic acid synthase. At least two types of 5-aminolevulinic acid synthases have been isolated from *Rba. sphaeroides* with a molecular weight of 10,000 and 64,000 respectively.
2. **C5 carbon pathway**: It is recorded among purple- and green sulfur bacteria in which α-ketoglutamide oxidizing enzymes are lacking in

the citric acid cycle and ALA is formed from glutamate or α-ketoglutaric acid via a path that does not involve ALA synthase reaction.

i. **Mode of action:** ALA herbicide/ insecticide produced by APB gets converted to tetrapyrroles leading to excess formation of these compounds which act on triplet oxygen to the potentially oxidizing singlet form (Reberz *et al.,* 1988). In plants singlet O_2 oxidizes phospholipids of membranes in free radical chain reactions and damaged membranes become leaky and the plant loses sap and death occurs in 1-4hrs. The mechanism causing the death of insects is yet to be understood. APB not only produces the economical amount of ALA from Wastewater but also sufficient for practical use as an herbicide, spraying of culture broth of APB directly was found to be effective against a number of common dicotyledonous weeds in the field (Haruhiko, 1992). In addition to the herbicidal property of ALA, it is a good growth stimulator when used in low concentration and is being used in photodynamic therapy for the cure of superficial epithelial skin tumors (Peter *et al.,* 1993).

11.16. ENHANCEMENT OF CALORIFIC VALUE OF BIOGAS

Biogas technology provides an alternate source of energy in rural India. Using cattle waste and other organic wastes, energy and manure are derived. Realization of this potential and utilization of the largest cattle wealth of India resulted in the promotion of the National Biogas Programme in major areas of India in the late 1970s as an answer to the growing fuel crisis. Biogas is produced from organic wastes by the concerted action of various groups of anaerobic bacteria.

Microbial conversion of organic matter to methane has become attractive as a method of waste treatment and energy recovery. This process is anaerobic involving a variety of specialized anaerobic bacteria, such as methane bacteria which have specialized metabolism.

The various microbial groups involved in the flow of carbon from complex polymers to methane are described by McInerney *et al.* (1981). Attempts are being made all over the world to increase the calorific value of biogas evolved during anaerobic digestion by increasing the ratio of methane to CO_2 (4:1) so that it can be used for industrial purposes. Interaction of *Rsp. rubrum* and the natural flora present in cattle dung could increase the methane content in biogas generated will result in its sustainability. H_2 produced by phototrophic bacteria was utilized by hydrogen utilizing methanogens along with CO_2 and enhanced the methane content.

11.17. ENHANCEMENT OF INDIGO EXTRACTION

India has got rich traditional technologies of using natural dyes. The present need for natural dyes arises with the awareness of hazards posed by synthetic dyes. Increased demand for the aesthetically appealing natural dyes to the environment for their non-toxic and non-allergic properties. The research on indigo, a major requirement of the dyeing industry, is a milestone in the field of biological indigo extraction. The improved technologies for cultivation of *Indigofera tintori a* plant by organic farming, and the use of PPB to enhance the extraction process and to increase the quality and quantity of the dye and reducing the extraction time has become reality.

11.18. MICROBIAL ENHANCED PULPING OF SECONDARY FIBERS

In view of increased demand for paper and there is an acute need to look for alternate feedstock's and also have to introduce environmentally sound methods in the papermaking process. In addition to paper recycling, the use of secondary fibers for paper production has increased significantly. Silk-cotton (Vatsala *et al.,* 2008) and banana stem fibers (Manimaran and Vatsala 2007) are some of the alternative sources for paper production. Bio pulping and bio bleaching are cost-effective alternatives involving the use of microbes to reduce and/or replace harmful chemicals. Extraction of hemicellulose and lignin without affecting the cellulose fiber strength of paper products (Bajpai 2004, Kenealy and Jeffries 2003) is the need of the day. A new environment-friendly process was developed for paper production from silk-cotton, banana stem fibers and shredded currency notes, etc., (Vatsala, 2008) wherein *Rhodospirillum* sp isolated from silk-cotton pod was used for pulping and bleaching. This phototrophic bacterium loosened the bond between cellulose and lignin by consuming hemicellulose thereby making cellulose available for paper production.

11.19. MISCELLANEOUS APPLICATIONS

Anoxygenic phototrophic bacteria have the potential for the production of bacteriochlorophylls (Urakami and Yoshida, 1986), α-adenosyl-L-cysteine (Yamada *et al.,* 1986), nucleic acids, corrinoid, haploid (Nagumo *et al.,* 1991) and Brassino steroids (Yamaji *et al.,* 1994). Decarte *et al.* (1988) feel that APB can be used in the augmentation of methane and increased production of biogas. Advances in molecular biology and genetic engineering have made it possible to tailor anoxygenic phototrophic bacteria to produce desired molecules. Carotenoids used as a food dye were produced by a culture of *Rhodobacter sphaeroides* S1 (Qian *et al.,* 1991). α- adenosyl-L-cysteine a precursor useful in the manufacture of the therapeutic agent is produced by growing *Rhodopseudomonas* in medium containing adenosine and cysteine (Yamada *et al.,* 1986). Corrinoid compounds useful in the inhibition of cell division vitamin B_{12} deficient bacteria are being produced by *Chromatium* sp., *Rhodobacter rubrum* and *Rps. palustris* (John, 1968). These bacteria are also used in the production of silk, cockroach control, sorbic acid, optically active carboxylic acid, nitriles, amides and tetra payrolls. Bacteriochlorophyll *a* of *Rhodobacter sphaeroides* showed antiaging activity (Kim *et al.,* 2015) However, more detailed studies should be carried out to explain the anti-skin aging mechanisms of these pigments (Park *et al.,* 2008)

Watanbe *et al.* (1998) reported the production of extra polymeric substances (EPS) on the cell surface. The factors influencing the production of EPS by *Rps. acidophila*h has been discussed (Sheng *et al.,* 2006). Salma *et al.* (2007) reported a significant influence on the dietary *Rba. capsulatus* on egg-yolk cholesterol and laying hen performance. Hypocholesterolemic effect of the anoxygenic phototrophic bacteria, *Rhodopseudomonas palustris* MGU001 on hen egg-laying was also observed (Ramchander *et al.,* 2011). Cook and Tabita (1988) reported the expression of Ribulose- 1,5 phosphate carboxylase gene from *Rsp. rubrum* in *E. coli* to study the effect of oxygen on its regulation. Gomelsky and Kalpan, (1996) reported Gene encoding transcription termination rho from *Rba. sphaeroides* was cloned into *E. coli* and the deduced protein were sequenced. DNA repair genes *rec* A from *Rps. viridis* was transformed to rec- phenotype of *Rhizobium melitoli* (Chen and Michel, 1998). *E. coli* was transformed using a hem C gene of *Rba. capsulatus* coding for a porphobilinogen deaminase (Beil *et al.,* 2002). Min *et al.* (2005) reported the biological rhythms of gene expression in a purple bacterium, *Rba. sphaeroides,* by using a luciferase reporter gene system. Bacteriochlorophyll *a* of *Rba. sphaeroides* showed anti-aging activity (Kim *et al.*, 2015).

■■■■

REFERENCES

Abella, C.A. and L.J. Garcia-Gil (1992). Microbial ecology of planktonic filamentous phototrophic bacteria in holomictic freshwater lakes. *Hydrobiologia.* **243–244:** 79–86.

Adedayo, M.R., E.A. Ajiboye, J.K. Akintunde and A. Odaibo (2011). Single-cell proteins: As nutritional enhancer. *Advan. Appli. Scie. Resea.* **2(5):** 396–409.

Adesemoye, A.O., H.A. Torbert, H.A and J.W. Kloepper (2008). Enhanced plant nutrient use efficiency with PGPR and AMF in an integrated nutrient management system. *Can. J. Microbiol.***54:** 876–886.

Ahmet Lokurlu, T. Grube, B. Höhlein and D. Stolten (2003). Fuel cells for mobile and stationary applications - cost analysis for combined heat and power stations on the basis of fuel cells. *Int. J. Hydro. Ene.* **28:** 703–711.

Akiba, R., R. Usami and K. Horikoshi (1983). *Rhodopseudomonas rutila*, a new species of non-sulfur purple photosynthetic bacteria. *Int. J. Syst. Bacteriol.* **33:** 551–556.

Akkerman, I., M. Janssen, J. Rocha and R.H. Wijffels (2002). Photobiological hydrogen production: photochemical efficiency and bioreactor design. *Int. J. Hydrogen Energy.* **27:** 1195–1208.

Akkose, S., U. Gunduz, M. Yucel and I. Eroglu (2009). Effects of ammonium ion, acetate and aerobic conditions on hydrogen production and expression levels of nitrogenase genes in *Rhodobacter sphaeroides* O.U.001. *Int. J. Hydr. Ener.* **34(21):** 8818–8827.

Akram, N.A. and M. Ashraf (2013). Regulation in plant stress tolerance by a potential plant growth regulator, 5-aminolevulinic acid. *J. Plant Growth Regul.* **32:** 663–679.

Akroum-Amrouche, D., Hamza Akroum and Hakim Lounici (2019): Green hydrogen production by *Rhodobacter sphaeroides*, Energy Sources, Part A : Recovery, Utilization and Environmental Effects, DOI: 10.1080/15567036.2019.1666190.

Alber, B.E., R. Spanheimer, C. Ebenau-Jehle and G. Fuchs (2006). Study of an alternate glyoxylate cycle for acetate assimilation by *Rhodobacter sphaeroides*. *Mol. Microbiol.* **61:** 297–309.

Alexander, B. and J.F. Imhoff (2006). Communities of green sulfur bacteria in different marine and saline habitats analyzed by gene sequences of 16S rRNA and of the Fenna-Matthews-Olson protein. *Int. Microbiol.* **9:** 259–266.

Alexander, B., J.H. Andersen, R.P. Cox and J.F. Imhoff (2002). Phylogeny of green sulfur bacteria on the basis of gene sequences of 16S rRNA and of the Fenna-Matthews-Olson protein. *Arch. Microbiol.* **178:** 131–140.

Alexandre, J., Poulain and K. Dianne Newman (2009). *Rhodobacter capsulatus* catalyzes Light-Dependent Fe (II) Oxidation under anaerobic conditions as a potential detoxification mechanism. *Applied and Environmental Microbiology*. **2009:** 6639–6646.

Ali, B., C.R. Huang, Z.Y. Qi, S. Ali, M.K. Daud, X.X. Geng *et al (*2013). 5-Aminolevulinic acid ameliorates cadmium-induced morphological, biochemical, and ultrastructural changes in seedlings of oilseed rape. *Environ. Sci. Pollut. Res* .**20:** 7256–7267.

Ambler, R., P.M. Daniel, J. Hermoso, T.E. Meyer, T.G. Bartsch and. M.D. Kamen (1979). Cytochrome c_2 sequence variation among the recognized species of purple non-sulfur photosynthetic bacteria. *Nature*. **278:** 659–660.

Amezaga, J.M., A, Amtmann and C.A. Biggs (2014). Biodesalination: a case study for applications of photosynthetic bacteria in water treatment. *Plant physiology*. **164(4):** 1661–1676.

Androga, D.D., E. Ozgur, I. Eroglu, U, Gunduz and M. Yucel (2012). Photofermentative Hydrogen Production in Outdoor Conditions, Hydrogen Energy - Challenges and Perspectives, *Dragica Minic, Intech. Open*. doi:10.5772/50390.

Androga, D.D., E. Ozgur, U. Gunduz, M. Yucel and I. Eroglu (2011). Factors Affecting the Long-Term Stability of Biomass and Hydrogen Productivity in Outdoor Photo fermentation . *Int. J. Hydr. Ener.* **36(17):** 11369–11378.

Androga, D.D., P. Sevinc¸ H. Koku, M. Yucel, U. Gunduz and I. Eroglu (2014). Optimization of temperature and light intensity for improved photofermentative hydrogen production using *Rhodobacter capsulatus* DSM 1710. *Int. J. Hydr. Ener.* **39:** 2472-80.

Anil Kumar, P., P. Aparna, T.N.R. Srinivas, Ch. Sasikala and Ch.V. Ramana (2008b). *Rhodovulum kholense* sp. nov. *Int. J. Syst. Evol. Microbiol.***58:** 1723–1726.

Anil Kumar, P., P. Aparna, T.N.R. Srinivas, Ch. Sasikalaand and Ch.V. Ramana (2008d). *Rhodospirillum sulfurexigens* sp. nov., a phototrophic *alphaproteobacterium* requiring a reduced sulfur source for growth. *Int. J. Syst. Evol. Microbiol.* **58:** 2917–2920.

Anil Kumar, P., T. N. R. Srinivas, Ch. Sasikala and Ch.V. Ramana (2008a). *Allochromatium renukae* sp. nov. *Int. J. Syst. Evol. Microbiol.* **58:** 404–407.

Anil Kumar, P., T.N.R. Srinivas, Ch. Sasikala and Ch.V. Ramana (2007d). *Rhodobacter changlensis* sp. nov., a psychrotolerant, phototrophic *alphaproteobacterium* from the Himalayas of India. *Int. J. Syst. Evol. Microbiol.* **57**: 2568–2571.

Anil Kumar, P., T.N.R. Srinivas, Ch. Sasikala and Ch.V. Ramana (2007b). *Halochromatium roseum* sp. nov., a non-motile phototrophic *gammaproteobacterium* with gas vesicles and emended description of the genus *Halochromatium*. *Int. J. Syst. Evol. Microbiol.***57:** 2110–2113.

Anil Kumar, P., T.N.R. Srinivas, Ch. Sasikala Ch.V. Ramana and J.F. Imhoff (2008c). *Thiophaeococcus mangrove* gen. nov., sp. nov., a photosynthetic, marine *gammaproteobacterium* isolated from the Bhitarkanika mangrove forest of India. *Int. J. Syst. Evol. Microbiol.* **58:** 2660–2664

Anil Kumar, P., T.N.R. Srinivas, S. Takaichi, T. Maoka, Ch. Sasikala and Ch.V. Ramana (2009b). *Phaeospirillum chandramohanii* sp. nov., a phototrophic alphaproteobacterium with carotenoid glycosides. *Int. J. Syst. Evol. Microbiol.* **59:** 2089–2093

Anil Kumar, P., T.N.R., Srinivas, V. Thiel, V. Tank, C. Sasikala, C.V. Ramana and J.F. Imhoff (2009c). *Thiohalocapsa marina* sp. nov., from an Indian marine aquaculture pond. *Int. J. Syst. Evol. Microbiol.* **59:** 2333–2338.

Anil Kumar, P., T.S. Sasi Jyothsna, T.N.R. Srinivas, Ch. Sasikala, Ch.V. Ramana and J.F. Imhoff (2007c). Two novel species of marine phototrophic *gammaproteobacteria*: *Thiorhodococcus bheemlicus* sp. nov. and *Thiorhodococcus kakinadensis* sp. nov. *Int. J. Syst. Evol. Microbiol.***57:** 2458–2461.

Anil Kumar, P., T.S. Sasi Jyothsna, T.N.R. Srinivas, Ch. Sasikala, Ch.V. Ramana and J.F. Imhoff (2007a). *Marichromatium bheemlicum* sp. nov., a nondiazotrophic photosynthetic gammaproteobacterium from a marine aquaculture pond. *Int. J. Syst. Evol. Microbiol.* **57:** 1261–1265.

Anil Kumar. P., T. Srinivas, Ch, Sasikala, Ch.V. Ramana, J. Suling and J.F. Imhoff (2009a). *Prosthecochloris indica* sp. nov., a novel green sulfur bacterium from a marine aquaculture pond, Kakinada, India. *The Journal of general and Applied Microbiology.* **55:** 163–9.

Anupama, X. and P. Ravindra (2000). Value-added food: single-cell protein. *Biotechnol Advances.* **18:** 459–479.

Aoshima, M., M. Ishii and Y. Igarashi (2004b). A novel enzyme, citryl-CoA lyase, catalysing the second step of the citrate cleavage reaction in *Hydrogenobacter thermophilus* TK-6. *Mol. Microbiol.* **52:** 763–770.

Aoshima, M., M. Ishii, M. and Y. Igarashi (2004a). A novel enzyme, citryl-CoA synthetase, catalyzing the first step of the citrate cleavage reaction in *Hydrogenobacter thermophilus* TK-6. *Mol. Microbiol.* **52:** 751–761.

APHA (1985). Standard methods for the examination of Water and Wastewater, 16th edn. APHA, 1985. "Standard methods for the examination of water and Wastewater". 16th edn (USA: American Public Health Association).

Arai, T., T. Wakayama, S. Okana and H. Kitamura (1998). Open Air hydrogen Production by Photosynthetic Bacteria Used Solar Energy During Winter Seasons in Central Japan 1998: conference proceedings, June 23-26, Kona, Hawaii, USA. Tumer JA; ICBHP.

Arashida,H., T. Kugenuma, M. Watanbe and I.Maeda (2018). Nitrogen fixation in *Rhodopseudomonas palustris* co-cultured with *Bacillus subtilis* in the presence of air. *J. Biosci. Bioeng.* **127(5)**:589-593.

Aretz, W., H. Kaspari and J.H. Klemme (1978). Utilization of purines as nitrogen source by facultative phototropic bacteria. *FEMS Microbiol. Lett.* **4:** 249–253.

Arunasri, K., Ch. Sasikala, Ch.V. Ramana, J. Suling and J.F. Imhoff (2005). *Marichromatium indicum* sp. nov., a novel purple sulfur *gammaproteobacterium* from mangrove soil of Goa, India. *Int. J. Syst. Evol. Microbiol.* **55:** 673–679.

Arunasri, K., V. Venkata Ramana, C. Sproer, Ch. Sasikala and Ch.V. Ramana (2008). *Rhodobacter megalophilus* sp. nov., a phototroph from the Indian Himalayas possessing a wide temperature range for growth. *Int. J. Syst. Evol. Microbiol.* **58:** 1792-1796.

Asad M. J., M. Asghan, M. Yaqub and K. Shahzad (2000). Production of single-cell protein delignified corncob by *Arachniotus* species. *Pak. J. of Agric. Sci.* **37:** 3–4.

Asada, Y., M. Tokumoto, Y. Aihara, M. Oku, K. Ishimi, T. Wakayama, J. Miyake, M. Tomiyama and H. Kohno (2006). Hydrogen production by co-cultures of *Lactobacillus* and a photosynthetic bacterium, *Rhodobacter sphaeroides* RV. *Int. J. Hydr. Ener.* **31:** 1509.

Asao, M. and M.T. Madigan (2009). Family IV *Heliobacteriaceae* Madigan 2001, 625. In: *Bergey's manual of systematic Bacteriology*, (De Vos, P., G.M. Garrity, D. Jones, N.R. Krieg, W. Ludwig, F.A. Rainey, K.H. Schleifer and W.B. Whitman. eds.), (the Firmicutes), Second Edition, Vol. 3. *Springer, New York.* pp. 923–931.

Asao, M. and M.T. Madigan (2010). Taxonomy, phylogeny and ecology of the heliobacteria. *Photosynth Res.* **104:** 103–111.

Asao, M. and M.T. Madigan (2015). *Heliobacteriaceae*. Bergey's Manual of Systematics of Archaea and Bacteria. pp. 1–6. doi:10.1002/9781118960608.fbm00132.

Asao, M., D.O. Jung, L.A. Achenbach and M.T. Madigan (2006). *Heliorestis convoluta* sp. nov., a coiled, alkaliphilic heliobacterium from the Wadi El Natrun, Egypt. *Extremophiles.* **10:** 403–410.

Asao, M., S. Takaichi and M.T. Madigan (2007). *Thiocapsa imhoffii*, sp. nov., an alkaliphilic purple sulfur bacterium of the family *Chromatiaceae* from Soap Lake, Washington (USA). *Arch. Microbiol.* **188:** 665–675.

Asao, M., S. Takaichi and M.T. Madigan (2012). Amino acid-assimilating phototrophic heliobacteria from soda lake environments: *Heliorestis acidaminivorans* sp. nov. and "*Candidatus Heliomonas lunata*." *Extremophiles.* **16(4):** 585–595.

Ashok Pandey, C. R.S., P. Nigam, T.S. Vanete, P.S Luciana and V. Radjiskumar Mohan (2000). *Bioresource Technology.* **74:** 81-87.

Ashraf, M (2009). The biotechnological approach of improving plant salt tolerance using antioxidants as markers. *Biotechnol. Adv.***27:** 84–93.

Assawamongkholsiri, T. and A.Reungsang (2015). Photo-fermentation hydrogen production of Rhodobacter sp. KKU-PS1 isolated from an UASB reactor. *Electron J. Biotechnol.***18:** 221-30.

Atalo, K. and B.A. Gashe (1993). Protease production by a thermophilic *Bacillus* species (P-001A) degrades various kinds of fibrous proteins *Biotechnology Letters.* **15:** 1151–1156.

Atanasova, N.S., E. Roine, A. Oren, D.H. Bamford and H.M. Oksanen (2012). Global network of specific virus-host interactions in hypersaline environments. *Environ. Microbiol.* **14:** 426–440.

Atzpodien, W. and H. Bode (1970). Purification and Regulatory Properties of ATP-Sensitive Phosphofructokinase from Yeast. *Enr. J. Biochem.* **12:** 126-132.

Avtages, P., R.G. Kranz and R. Haselkorn (1995). *Mol. Gen. Genet.* **201:** 363-369.

Azad, S.A., S. Vikineswary, K.B. Ramachandran and V.C. Chong (2001). Growth and production of biomass *Rhodovulum sulfidophilum* in sardine processing wastewater. *Lett. Appl. Microbiol.***33:** 264-268.

Azad, S.A., S. Vikineswary, V.C. Chong and K.B. Ramachandran (2003). *Rhodovulum sulfidophilum* in the treatment and utilization of sardine processing wastewater. *Lett. Appl. Microbiol.* **38:** 13–18.

Azad, S.A., S. Vikineswary, V.C. Chong and K.B. Ramachandran (2004). *Rhodovulum sulfidophilum* in the treatment and utilization of sardine processing wastewater. *Letters in applied microbiology.* **38(1):** 13-18.

Azad, S.A., V.C. Chong and S. Vikineswary (2002). Phototrophic bacteria as a feed supplement for rearing *Penaeus monodon* larvae. *J.W. Aquacul. Soci.* **33:** 158-168.

Babudieri, B (1950). Natura delle cosidette "S-formen" delle leptospire. Loro identificazione con *Hyphomicrobiumvulgare* Stutzer e Hartleb. Studio di quest. Ultimo germ. Estratto dai Rendiconti dell'Istituto Superiore di Sanita. **13:** 580–591.

Bahatyrova, S., R.N. Frese, C.A. Siebert, J.D. Olsen, K.O. Van der Werf, R. van Grondelle, R.A. Niederman, P.A. Bullough, C. Otto and C.N. Hunter (2004). The native architecture of a photosynthetic membrane. *Nature (London).* **430:** 1058–62.

Bajpai, P (2004). Biological bleaching of chemical pulps. *Critical Review in Biotechnology.* **24(1):** 1–58.

Balat, M., M. Balat, E. Kırtay and H. Balat (2009). Main routes for the thermo-conversion of biomass into fuels and chemicals. Part 2: gasification systems. *Energy Convers Manage.* **50:** 3158–68.

Baldock, M.I., K. Denger and T.H.M. Smits and A.M. Cook (2007). *Roseovarius* sp. strain 217: Aerobic taurine dissimilation via acetate kinase and acetate–CoA ligase. *FEMS Microbiol. Lett.* **271:** 202–206

Balloni, W., C. Filpi and G. Florenzano (1980). In: *Algae Biomass* (G. Shelef and C.J. Soeder, eds.). Elsevier/ North–Holland Biomedical Press, Amsterdam. pp. 217–227

Balloni, W., P. Carlozzi, S. Ventura and A. Sacchi (1986). Microbial biomass for fertilizer use from the photo-anaerobic treatment of pig-wastes. *Int. Symposium on Compost.* 17–19 April, Udine.

Balloni, W., P. Carlozzi, S. Ventura, R. De Philippis and M.A. Bosco (1987). A three years experiment on the production of *Rhodopseudomonas* and *Rhodospirillum* biomass by outdoor culture on different wastes. 4th *Int. Conference on Biomass for Energy and Industry*, Orleans.

Banat, I.M., R.S. Makkar and S.S. Cameotra (2000). Potential commercial applications of microbial surfactants. Appl. Microbiol. Biotechnol. **53:** 495–508.

Bartsch, R.G (1978) Cytochromes. In: *The photosynthetic bacteria*, (Clayton, R.K. and W.R. Sistrom, eds.) Plenum Press, New York, pp. 249–279.

Barz, W. P., A. Verme´glio, F. Francia, G. Verturoli, B. A. Melandri and D. Oesterhelt (1995). Role of PufX protein in photosynthetic growth of *Rhodobacter sphaeroides*. 2. PufX is required of efficient ubiquinone/ubiquinol exchange between the reaction center Q(B) site and the cytochrome b/c1complex. *Biochemistry.***34:**15248–15258.

Barz, W.P., F. Francia, G. Venturoli, B. A. Melandri, A. Vermeglio and D. Oesterhelt (1995). Role of PufX protein in photosynthetic growth of *Rhodobacter sphaeroides*. 1. PufX is required for efficient light-driven electron transfer and photophosphorylation under anaerobic conditions. *Biochemistry.***34:**15235–15247.

Basak, N. and D. Das (2007). Microbial Biohydrogen Production by *Rhodobacter sphaeroides* O.U.001 in Photobioreactor. Proceedings of the World Congress on Engineering and Computer Science. San Francisco, USA. pp. 24–26.

Basak, N. and D. Das (2009). Photofermentative hydrogen production using purple non-sulfur bacteria *Rhodobacter sphaeroides*O.U.001 in an annular photobioreactor: a case study. *Biomass Bioenergy*. **33:** 911-9.

Basak, N., A.K. Jana, D. Das and D. Saikia (2014). Photofermentative molecular biohydrogen production by purple-non-sulfur (PNS) bacteria in various modes: the present progress and future perspective. *Int. J. Hydr. Ener*. **39:** 6853-71.

Batool, K., F. Zahra and Y. Rehman (2017) Arsenic redox transformation and plant growth promotion by purple nonsulfur bacteria *Rhodopseudomonas palustris* CS2 and *Rhodopseudomonas faecalis* SS5. *BioMed. Research Int.***2017:** 1–8.

Bauld, J. and T. D. Brock (1973). Ecological studies of *Chloroflexis*, a gliding photosynthetic bacterium. *Arch. Mikrobiol.* **92:** 267–284.

Bavendamm, W (1924). Die farblosen und roten Schwefelbakterien des Süß- und Salzwassers. G. Fischer. Jena,Germany.

Beate, V., S. Heinz, D. Miethe, R. Hermann, L. Helul, E. Horst, S. Ingrid and S. Burkhard (1991). Ger (East) Pat. DD. 293,048.

Beatty, J.T., J. Overmann, M.T. Lince, A.K. Manske, A.S. Lang, R.E. Blankenship, C.L. Van Dover, T.A. Martinson and F.G. Plumley (2005). An obligately photosynthetic bacterial anaerobe from a deep-sea hydrothermal vent. *Proc. Natl. Acad. Sci. U.S.A.* **102:** 9306–9310.

Becker, E.W (1994). *Microalgae: Biotechnology and Microbiology*, Cambridge Univ. Press, New York.

Beer-Romero, P. and H. Gest (1987). *Heliobacillus mobilis*, a peritrichously flagellated anoxy phototroph containing bacteriochlorophyll *g*. *FEMS Microbiol.Lett* .**41:** 109–114.

Beer-Romero, P. and H. Gest (1998a). *Heliobacillus* gen. nov. In: Validation of Publication of New Names and New Combinations Previously Effectively Published Outside the IJSB, List no. 65. *Int. J. Syst. Bacteriol.* **48:** 627.

Beil, F., M. Ester and X. Xu (2002). Frequent term-based text clustering. In proceedings of the Eighth ACM SIGKDD international conference on knowledge discovery and data mining, KDD'02, pp. 436-442, New York, NY, USA. ACM .

Bender, J. and P. Phillips (2004). Microbial mats for multiple applications in aquaculture and bioremediation. *Bioresour.Technol.* **94:** 229–238.

Benemann, J.R (1992). Microalgae Aquaculture Feeds. *J. Applied Phycol.***4:** 233–245.

Benemann, J.R (1996). Hydrogen biotechnology: Progress and prospects. *Nature Biotech.***14:** 1101–1103.

Benemann, J.R (1997). Feasibility analysis of photobiological hydrogen production. Int. *J. Hydro. Ene.***22:** 979–987.

Benemann, J.R (1999). The technology of biohydrogen. *BioHydrogen*, pp. 19–30.

Benjamin, P., P. Elion, T. Giuseppe and R. Riccardo (1993). *Bioresour. Technol.* **43:** 59–62.

Berg, I. A., O.I. Keppen, E.N. Krasilnikova, N.V. Ugolkova and R.N. Ivanovskii (2005). Carbon metabolism of filamentous anoxygenic phototrophic bacteria of the family *Oscillochloridaceae*. *Microbiology.* **74:** 258–264.

Berg, J.M., J.L. Tymoczo and L. Stryer (2002). *Biochemistry*, Fifth Edition, W.H. Freeman, New York.

Bergene T (1996). The efficiency and physical principles of photolysis of water by microalgae. *Int. J. Hydr. Ener.* **21:**189–194.

Berne, C., B. Allainmat and D. Garcia (2005) Tributyl phosphate degradation by Rhodopseudomonas palustris and other photosynthetic bacteria. *Biotechnol Lett.***27:** 561–566.

Berry, J.W., Uichol Kim, Thomas Minde and Doris Mok Source (1987). Comparative Studies of Acculturative Stress. International Migration Review, Migration and Health (Autumn, 1987). **21 (3):** 491–511.

Beynon, J.D., I.J. MacRae, S.L. Huston, D.C. Nelson, I.H. Segel and A.J. Fisher. (2001). Crystal structure of ATP sulfurylase from the bacterial symbiont of the hydrothermal vent tubeworm *Riftia pachyptila*. *Biochemistry.* **40:** 14509–14517.

Bick, J.A., J.J. Dennis, G.J. Zylstra, J. Nowack and T. Leustek (2000). Identification of a new class of 5'-adenylylsulfate (APS) reductases from sulfate assimilating bacteria. *J. Bacteriol.* **182:** 135–142.

Biebl, H. and G. Drews (1969). Das in-vivo Spektrum als taxonomisches Merkmal bei Untersuchungen zur Verbreitung von *Athiorhodaceae*. Zentralbl. Bakteriol. Parasitenkde. Infektionskr. *Hyg. Abt. II, Orig.* **123:** 425–452.

Biebl, H. and N. Pfennig (1978). Growth yields of green sulfur bacteria in mixed cultures with sulfur and sulfate-reducing bacteria. *Arch. Microbiol.* **117:** 9–16.

Biebl, H. and N. Pfennig (1979). CO_2-fixation by anaerobic phototrophic bacteria in lakes: A review. *Arch. Hydrobiol.* **12:**18–58.

Biebl, H. and N. Pfennig (1981). Isolation of members of the family *Rhodospirillaceae*. In: *The Prokaryotes,* (Starr, M.P., H. Stolp, H. G. Truper, A. Balows and H.G. Schlegel, eds.). New York: Springer. pp. 167–273.

Biebl, H., B.J. Tindall, R. Pukall, H. Luns-dorf, M. Allgaier and I. Wagner-Dobler (2006). *Hoeflea phototrophica* sp. nov., a novel marine aerobic alphaproteobacterium that forms bacteriochlorophyll *a*. *Int. J. Syst. Evol. Microbiol.* **56:** 821–826.

Biebl, H. (1973). Die Verbreitung der schwefelfreien Purpurbakterien im Plussee und anderen Seen Ostholsteins (Ph.D. thesis), University of Freiburg. FRG.

Bishop, D. H. L., K. P. Pandya and H. K. King (1962). Ubiquinone and vitamin K in bacteria. *Biochem. J.* **83:** 606-614.

Bittl, R., E. Schlodder, I. Geisenheimer, W. Lubitz and R.J. Cogdell (2001). Transient EPR and Absorption Studies of Carotenoid Triplet Formation in Purple Bacterial Antenna Complexes. *J. Phys. Chem. B.* **105:** 5525–5535.

Blankenship, R.E (1992). Origin and early evolution of photosynthesis. *Photosynth. Res.* **33:** 91–111.

Blankenship, R.E (2002). Molecular mechanisms of photosynthesis. *Blackwell Science,* Oxford.

Blankenship, R.E (2010). Early evolution of photosynthesis. *Plant Physiol.* **154:** 434–438.

Blasco, R. and F. Castillo (1992). Light-dependent degradation of nitrophenols by the phototrophic bacterium *Rhodobacter capsulatus* E1F1. *Appl. Environ. Microbiol.* **58:** 690–695.

Blasco, R. and F. Castillo (1993). Characterization of a nitrophenol reductase from the phototrophic bacterium *Rhodobacter capsulatus* E1F1. *Appl. Environ. Microbiol.* **59(6):** 1774–8.

Blazejak, A., J. Kuever, C. Erséus, R. Amann and N. Dubilier (2006). Phylogeny of 16S rRNA, Ribulose 1,5-Bisphosphate Carboxylase/Oxygenase and Adenosine 5′-Phosphosulfate Reductase Genes from *Gamma-* and *Alphaproteobacterial* Symbionts in Gutless Marine Worms (Oligochaeta) from Bermuda and the Bahamas. *Applied and Env. Microbiol.* **72:** 5527–36.

Blumenberg , M., B. I. Oppermann, R. Guyoneaud and W. Michaelis (2009). Hopanoid production by *Desulfovibrio bastinii* isolated from oilfield formation water. *FEMS Microbiology Letters.* **293(1):** 73–78.

Blumenberg, M., M.Kruger, K. Nauhaus, H.M. Talbot, B.I. Oppermann, R. Seifert, T. Pape, W. Michaelis (2006). Biosynthesis of hopanoids by sulfate-reducing bacteria (genus *Desulfovibria*) *Environmental Microbiology.* **8:** 1220–1227.

Boaree, N.R., J. Dove, J.J. Copper, J. Knowles and G.W. Hastings (1993). Development of a degradable composite for orthopedic use: mechanical evaluation of a hydroxyapatite-polyhydroxybutyrate composite material. *Biomaterials.* **14:** 793– 796.

Boden, R., L.P. Hutt and A.W. Rae (2017). Reclassification of Thiobacillus aquaesulis (Wood & Kelly, 1995) as *Annwoodia aquaesulis* gen. nov., comb. nov., transfer of *Thiobacillus* (Beijerinck, 1904) from the *Hydrogenophilales* to the *Nitrosomonadales*, the proposal of *Hydrogenophilalia* class. nov. within the '*Proteobacteria*' and four new families within the orders *Nitrosomonadales* and *Rhodocyclales. Int. J. Syst. Evol.Microbiol.***67:** 1191–1205.

Bodour, A., K.P. Drees and R.M. Maier (2003). Distribution of biosurfactant producing bacteria in undisturbed and contaminated arid south western soils *Appl. Environ. Microbiol.* **69:** 280–287.

Boldareva, E.N., A.A. Moskalenko, Z.K. Makhneva, T.P. Tourova, T. V. Kolganova and V.M. and Gorlenko (2009). *Rubribacterium polymorphum* gen. nov., sp. nov., a novel alkaliphilic nonsulfur purple bacterium from an Eastern Siberian soda lake. *Microbiol.* **78:** 732–740.

Boldareva, E.N., V.N. Akimov, V.A. Boychenko, I.N. Stadnichuk, A.A. Moskalenko, Z.K. Makhneva and V.M. Gorlenko (2008). *Rhodobaca barguzinensis* sp. nov., a new alkaliphilic purple non-sulfur bacterium isolated from a soda lake of the Barguzin Valley (Buryat Republic, Eastern Siberia). *Mikrobiologiia.***77:** 206–218.

Boran, E., E. Özgür, J. van der Burg, M. Yucel, U. Gunduz and I. Eroglu (2010). Biological Hydrogen Production by *Rhodobacter capsulatus* in Solar Tubular Photo Bioreactor. *J. Cleaner Product.***18(1):** S29–S35.

Borrego, C.M. and L.J. Garcia-Gil (1994). Separation of bacteriochlorophyll homologs from green photosynthetic sulfur bacteria by reversed-phase HPLC. *Photosynth. Res.* **41:** 157–163.

Bose, S. K (1963). Media for anaerobic growth of photosynthetic bacteria. In: *Bacterial photosynthesis*, (Gest, H., San Pietro, A., Vernon, 1. P. eds.), Springs, Ohio: Antioch Press. pp. 501–519.

Bosshard, H.R, M.W. Davidson and D.B. Knaff (1986). Complex formation and electron transfer between mitochondrial cytochrome c and flavocytochrome c552 from *Chromatium vinosum*. *J. Biol. Chem.***261:** 190–193.

Bosshard, P.P., R. Stettler and R. Bachofen (2000). Seasonal and spatial community dynamics in the meromictic Lake Cadagno. *Arch. Microbiol.* **174:** 168–174.

Bosshard, P.P., Y. Santini, D. Grüter, R. Stettler and R. Bachofen (2000a). Bacterial diversity and community composition in the chemocline of the meromictic alpine Lake Cadagno as revealed by 16S rDNA analysis. *FEMS Microbiol. Ecol.* **31:** 173–182

Boughanemi, S., J. Lyonnet, P. Infossi, M. Bauzan, A. Kosta, S. Lignon, M.T. Giudici-Orticoni and M. Guiral (2016). Microbial oxidative sulfur metabolism: biochemical evidence of the membrane-bound heterodisulfide reductase-like complex of the bacterium *Aquifex aeolicus*. *FEMS Microbiol. Lett.* 363.

Boutaiba, S., H. Hacène, K.A. Bidle and J.A. Maupin-Furlow (2011). Microbial diversity of the hypersaline Sidi Ameur and Himalatt salt lakes of the Algerian Sahara. *J. Arid. Environ*. **75:** 909-91.

Bradley, J.M., S.J. Marritt, M.A. Kihlken, K. Haynes, A.M. Hemmings, B.C. Berks, M.R. Cheesman and J.N. Butt (2012). Redox and chemical activities of the hemes in the sulfur oxidation pathway enzyme SoxAX. *J. Biol .Chem.* **287:** 40350–40359.

Brandl, H., R.A. Gross, R. W. Lenz and R.C. Fuller (1990). Plastics from bacteria and for bacteria: Poly(β-hydroxyalkanoates) as natural, biocompatible, and biodegradable polyesters. In: *Microbial Bioproducts. Adv. Biochem. Eng. / Biotechnol.Springer, Berlin, Heidelberg*.**41:** 77-93.

Brandl, H., R.A. Gross, R.W. Lenz and R.C. Fuller (1989). *Int. J. Biol. Macromol.* **11:** 49–55.

Brandl, M.T., B. Quinones and S.E. Lindow (2001). Heterogeneous transcription of an indole acetic acid biosynthetic gene in *Erwinia herbicola* on plant surfaces. *Proc. Natl. Acad. Sci. USA.* **98:** 3454–3459.

Brennan, L. and P. Owende (2010). Biofuels from a microalgae-a review of technologies for production, processing and extractions of biofuels and co-products. *Renew. Sustain. Energ. Rev.* **14:** 557–577.

Brinkhoff,T.,H.A. Giebel and M. Simon (2008). Diversity, ecology, and genomics of the *Roseobacter* clade: a short overview. *Arch.Microbiol.* **189:** 531–539.

Brito, J.A., K. Denkmann, I.A.C. Pereira, M. Archer and C. Dahl (2015). Thiosulfate dehydrogenase (TsdA) from *Allochromatium vinosum*: Structural and functional insights into thiosulfate oxidation. *J. Biol. Chem.* **290:** 9222–9238.

Broch-Due, M. and J.G. Ormerod (1978). Isolation of Bchl *c* mutant from *Chlorobium* with Bchl *d* by cultivation at low light intensities. *FEMS Microbiol. Lett.* **3:** 305–308.

Brock T.D (1978). The Genus *Thermus*. In: *Thermophilic Microorganisms and Life at High Temperatures*. Springer Series in Microbiology. Springer, New York, NY. pp. 72–91.

Brune, D.C (1995b). Sulfur compounds as photosynthetic electron donors. In: *Anoxygenic photosynthetic bacteria,* (Blankenship, R.E., M.T. Madigan and C.E. Bauer, eds.) Kluwer Academic Publishers, Dordrecht. pp. 847–870.

Bryant, D.A. and N.U. Frigaard (2006). Prokaryotic photosynthesis and phototrophy illuminated. *Trends Microbiol.* **14:** 488–496.

Bryant, D.A., Z. Liu, T. Li, F. Zhao, C.G. Klatt, D.M. Ward, N.U. Frigaard and J. Overmann (2012). Comparative and functional genomics of anoxygenic green bacteria from the taxa *Chlorobi, Chloroflexi* and *Acidobacteria*. In: *Functional Genomics and Evolution of Photosynthetitic Systems,* (Burnap, R.L. and W.F.J. Vermaas, eds.) *Advances in Photosynthesis and Respiration, Springer*, NewYork. **33:** 47–102.

Bryant. D.A., A.M. Costas, J.A. Maresca, A.G. Chew, C.G. Klatt, M.M. Bateson, L.J. Tallon, J. Hostetler, W.C. Nelson, J.F. Heidelberg and D.M. Ward (2007). *Candidatus Chloracidobacterium thermophilum*: anaerobic phototrophic *Acidobacterium. Science.* **317:** 523–526.

Bryantseva, I., V.M. Gorlenko, E. Kompantseva, L.A. Acenbach and M.T. Madigan (1999). *Heliorestis daurensis*, gen. nov. sp. nov., an alkaliphilic rod-to-coiled-shaped phototrophic heliobacterium from a Siberian soda lake.*Arch Microbiol.* **172(3):** 167–174.

Bryantseva, I., V.M. Gorlenko, E.I. Kompantseva, J.F. Imhoff, J. Suling and L. Mityushina (1999b). *Thiorhodospira sibrica* gen. nov. sp. nov., a new alkaliphilic purple sulfur bacterium from a Siberian soda lake. *Int. J. Syst. Bacteriol.* **49:** 697–703.

Bryantseva, I.A., T.P. Tourova, O.L. Kovaleva, N.A. Kostrikina and V.M. Gorlenko (2010). *Ectothiorhodospira magna* sp. nov., a new large alkaliphilic purple sulfur bacterium. *Mikrobiologiia.* **79:** 780–790.

Bryantseva, I.A., V.A. Gaisin and V.M. Gorlenko (2015). *Rhodobaculum claviforme* gen. nov., sp. nov., a new alkaliphilic nonsulfur purple bacterium. *Mikrobiologiia.* **84:** 247–255.

Bryantseva, I.A., V.M. Gorlenko, E.I. Kompantseva and J.F. Imhoff (2000). *Thioalkalicoccus limnaeus* gen. nov., sp. nov., a new alkaliphilic purple sulfur bacterium with bacteriochlorophyll *b. Int. J. Syst. Evol. Microbiol.***50:** 2157–2163.

Bryantseva, I.A., V.M. Gorlenko, E.I. Kompantseva, T.P. Tourova, B.B. Kuznetsov and G.A. Osipov (2000a). Alkaliphilic heliobacterium *Heliorestis baculata* sp. nov. and emended description of the genus *Heliorestis. Archives of Microbiology.* **174(4):** 283–291.

Bryantseva, I.A., V.M. Gorlenko, T.P. Tourova, B.B. Kuznetsov, A.M. Lysenko, S.A. Bykova and G.A. Osipov (2000b). *Heliobacterium sulfidophilum* sp. nov. and *Heliobacterium undosum* sp. nov., Sulfide-oxidizing heliobacteria from thermal sulfidic springs. *Microbiology.* **69(3):** 325–334.

Buccolieri, A., F. Italino, A. Dell'Atti, G. Buccolieri, L. Giotta, A. Agostiano, F. Milano and M. Trotta (2006). Testing the photosynthetic bacterium *Rhodobacter sphaeroides* as a heavy metal removal tool. *Ann. Chim.***96:** 195–203.

Budiman, P.M., T.Y. Wu, Ramanan, N. Ramakrishnan, X.W.H. Jacqueline Xiao (2014). Treatment and reuse of effluents from palm oil, pulp and paper mills as a combined substrate by using purple non-sulfur bacteria. *Ind. Eng. Chem.Res.* **53 (39):** 14921–14931.

Bulen, W.A., R.C. Burns and R.J. Lecomte (1965). Nitrogen fixation: hydrosulfite as an electron donor with cell-free preparations of *Azotobacter vinelandii* and *Rhodospirillum rubrum. Proceedings of the National Academy of Sciences of the United States of America.* **53:** 532-9.

Buranakarl, L., Fan Cheng-Ying, Kazutoshi Ito, Kazuo Izaki and Hajime Takahashi (1985). Production of Molecular Hydrogen by Photosynthetic Bacteria with Raw Starch, *Agricultural and Biological Chemistry.* **49(11):** 3339–3341.

Buranakarl, L., K. Ito, K. Izaki and H. Takahashi (1988). Purification and characterization of a raw starch digestion amylase from non-sulfur purple photosynthetic bacteria. *Enzy. Microbiol. Technol.* **10:** 173–179.

Burgess, J.G (1993). *Truns. Muter. Res.* SOC. Jpn. **10:** 16–24.

Burgess, J.G., H. Miyashita, H. Suda and T. Matsunaga (1991). *Fems. Microbiol. Lett.* **84:** 301–306.

Burgess, J.G., K. Tsubaki and T. Matsunaga (1993). *Biotechnol. Lett.* **15:** 111–114.

Burgess, J.G., R. Kawaguchi, A. Yamada and T. Matsunaga (1994). *Microbiology.* **140:** 965–970.

Cabello, P., M.D. Roldan and C. Moreno-Vivian (2004). Nitrate reduction and the nitrogen cycle in archaea. *Microbiology.* **150:** 3527–3546.

Cammack, R., M. Frey and R. Robson (2001)." Hydrogen as a fuel: Learning from Nature", *Taylor & Francis, London*, UK. 15-180.

Carlozzi, P., A. Buccioni, S. Minieri, B. Pushparaj, R. Piccardi, A. Ena and P. Cristina (2010). Production of bio-fuels (hydrogen and lipids) through photo fermentation . *Bioresource technology.* **101 (9):** 3115–3120.

Carlozzi, P., G. Padovani, P. Cinelli and A. Lazzeri (2015). An Innovative Device to Convert Olive Mill Wastewater into a Suitable Effluent for Feeding Purple Non-Sulfur Photosynthetic Bacteria. *Resources.* **4:** 621–636.

Carmona, M., M.A. Prieto, B. Galan, J.L. Garcia and E. Diaz (2008). Signaling networks and Design of pollutants biosensors, p. 97–143. In: *Microbial Biodegradation: Genomics and Molecular Biology*, (Díaz, E, ed.), Caister Academic Press, Norfolk, United Kingdom.

Carmona. R., G.P. Kraemer and C. Yarish (2006). Exploring Northeast American and Asian species of *Porphyra* for use in an integrated finfish-algal aquaculture system. *Aquaculture.* **252:** 54–65.

Carr, N.G.and G. Exell (1965). Ubiquinone concentrations in *Athiorhodaceae* grown under various environmental conditions. *Biochem. J.* **96:** 688–692.

Carrasco, C.D., J.A. Buettner and J.W. Golden (1995). Programmed DNA rearrangement of a cyanobacterial hupL gene in heterocysts. *Proc. Natl. Acad. Sci.* USA. **92:** 791–795.

Castenholz R.W. and B.K. Pierson (1981). Isolation of Members of the Family *Chloroflexaceae*. In: The Prokaryotes, (Starr, M.P., H. Stolp, H.G. Trüper, A. Balows and H.G. Schlegel, eds.), Springer, Berlin, Heidelberg.

Castenholz, R.W (1973). The possible photosynthetic use of sulfide by the filamentous phototrophic bacteria of hot springs. *Limnol. Oceanogr.* **18**: 863–876.

Castenholz, R.W (2001a). Class, ''*Chloroflexi*''. In: *Bergey's Manual of Systematic Bacteriology*, The Archaea and the deeply branching and phototrophic bacteria, (Boone, D.R., R.W. Castenholz and G.M. Garrity, eds.) Second edn, Vol 1. Springer Verlag, New York, pp 427.

Castenholz, R.W (2001b). Order II. ''*Herpetosiphonales*''. In: *Bergey's Manual of Systematic Bacteriology*, The Archaea and the deeply branching and phototrophic bacteria, (Boone D.R., R.W. Castenholz and G.M. Garrity, eds.) Second edition, Vol 1. Springer Verlag, New York, pp. 444–446.

Castenholz, R.W. and B.K. Pierson (1995). Ecology of thermophilic anoxygenic phototrophs. In: *Anoxygenic Phototrophic Bacteria*, (Blankenship, R.E., M.T. Madigan and C.E. Bauer, eds.) Kluwer Academic Publishers, Dordrecht. pp. 87–103.

Castenholz, R.W., J. Bauld and B.B. Jorgenson (1990). Anoxygenic microbial mats of hot springs: thermophilic *Chlorobium* sp. *FEMS Microbiology Letters.* **74(4):** 325–336.

Caumette, P (1984). Distribution and characterization of phototrophic bacteria isolated from the water of Bietri Bay (Ebrie Lagoon, Ivory Coast). *Can. J. Microbiol.* **30:** 273-284.

Caumette, P (1986). Phototrophic sulfur bacteria and sulfate-reducing bacteria causing red waters in a shallow brackish coastal lagoon. Prévost Lagoon, France. *FEMS Microbiol. Ecol.* **38:** 113–124.

Caumette, P (1993). Ecology and physiology of phototrophic bacteria and sulfate-reducing bacteria in marine salterns. *Experientia*. **49:** 473–481.

Caumette, P., J.F. Imhoff, J. Suling and R. Matheron (1997). *Chromatium glycolicum* sp. nov., a moderately halophilic purple sulfur bacterium that uses glycolate as substrate. *Arch. Microbiol.* **167:**11–18.

Caumette, P., R. Baulaigue and R. Matheron (1988). Characterization of *Chromatium salexigens* sp. nov., a halophilic *Chromatiaceae* isolated from Mediterranean salinas. *Syst.Appl. Microbiol.* **10:** 284–292.

Caumette, P., R. Baulaigue and R. Matheron (1988). In Validation list No. 28. *Int. J. Syst. Bacteriol.* **39:** 93-94.

Caumette, P., R. Baulaigue and R. Matheron (1991). *Thiocapsa halophila* sp. nov., a new halophilic phototrophic purple sulfur bacterium. *Arch Microbiol.* **155:** 170–176.

Caumette, P., R. Guyoneaud and J.F. Imhoff (2015). *Thiolamprovum. Bergey's Manual of Systematics of Archaea and Bacteria,* 1–6. doi:10.1002/9781118960608.gbm01122.

Caumette, P., R. Guyoneaud, J.F. Imhoff , J. Suling and V. Gorlenko (2004). *Thiocapsa marina* sp. nov., a novel, okenone-containing, purple sulfur bacterium isolated from brackish coastal and marine environments. *Int. J. Syst. Evol. Microbiol.***54:** 1031–1036.

Caumette, P., R. Guyoneaud, R. Duran, C. Cravo-Laureau and R. Matheron (2007). *Rhodobium pfennigii* sp. nov., a phototrophic purple non-sulfur bacterium with unusual bacteriochlorophyll an antenna, isolated from a brackish microbial mat on Rangiroa atoll, French Polynesia. *Int. J. Syst. Evol. Microbiol.* **57:** 1250–1255.

Caumette, P., R. Matheron, N. Raymond and J. C. Relexans (1994). Microbial mats in the hypersaline ponds of Mediterranean salterns (Salins-de-Giraud, France). *FEMS Microbiol. Ecol.* **13:** 273–286.

Celik, L., A. Ozturk and M.I. Abdullah (2012). Biodegradation of reactive red 195 azo dye by the bacterium Rhodopseudomonas palustris 51ATA. *Afr. J. Microbiol. Res.* **6(1):** 120–126.

Cetinkaya Donmez and Ayten Ozturk (1999). Biodegradation of Homocyclic and Heterocyclic Aromatic Compounds by *Rhodopseudomonas palustris* strains. *Tr. J. Biology.* **23:** 507–511.

Cetinkaya Donmez, G., A. Ozturk and L. Cakmakci (1999). Properties of the *Rhodopseudomonas palustris* strains, isolated from an alkaline lake in Turkey, Dog. *Biyoloji Dergisi.* **23:** 457–463.

Cetinkaya Donmez, G., B. Aslım and Y. Beyatlı (1999). Formation of Poly-B-hydroxybutyrate (PHB) by Photosynthetic Bacteria. G. U. Fen Bil. *Ens. Dergisi.* **12:** 605–611.

Chen, Wen-Ming, Chang, Ting-Hsuan, Sheu, Der-Shyan, Jheng, Li-Cheng, Sheu, Shih-Yi (2021). *Rhodobacter ruber* sp. nov., isolated from a freshwater pond. 10.21203/rs.3.rs-197366/v1.

Chader, S., H. Hacene and S.N. Agathos (2009). Study of hydrogen production by three strains of *Chlorella* isolated from the soil in the Algerian Sahara. *Int. J. Hydr. Ener.***34:** 4941–4946.

Chae, S.R., E.J. Hwang and H.S. Shin (2006). The single-cell protein production of *Euglena gracilis* and carbon dioxide fixation in an innovative photo-bioreactor. **97:** 322–329.

Chakravarthy, S.K., E.V.V. Ramprasad, E. Shoba, Ch. Sasikala and Ch.V. Ramana (2012). *Rhodoplanes piscinae* sp. nov. isolated from pond water. *Int. J. Syst. Evol. Microbiol.* **62:** 2828–2834.

Chakravarthy, S.K., K. Sucharitha, Ch. Sasikala and Ch. V. Ramana (2009). *Rhodovulum lacipunicei* sp. nov., an obligate sulfide-demanding phototrophic *alphaproteobacterium* isolated from a purple pond in India. *Int. J. Syst. Evol. Microbiol.* **59:** 1615–1619.

Chakravarthy, S.K.,T.N.R. Srinivas, P. Anil Kumar, Ch. Sasikala and Ch.V. Ramana (2007). *Roseospira visakhapatnamensis* sp. nov. and *Roseospira goensis* sp. nov. *Int. J. Syst. Evol. Microbiol.* **57:** 2453–2457.

Chalam, A., Ch. Sasikala, Ch. Ramana, N. Uma and P. Rao (1997). Effect of Pesticides on the Diazotrophic Growth and Nitrogenase Activity of Purple Non-sulfur Bacteria. Bulletin of environmental contamination and toxicology. **58:**463-8.

Champine, J.E. and R.L. Uffen. (1987). Regulation of anaerobic carbon monoxide oxidation activity in *Rhodocyclus gelatinosus*. *FEMS Microbiol. Lett.***44:** 307–311.

Champine, J.E. and R.L. Uffen (1987). Membrane topography of anaerobic carbon monoxide oxidation in *Rhodocyclus gelatinosus*. *J. Bacteriol.* **169:** 4784–4789.

Chang, F.Y. and C.Y. Lin (2004). Bio-hydrogen production using an up-flow anaerobic sludge blanket reactor. *Int. J. Hydr. Ener.* **29:** 33–39.

Chen, C.Y. and J.S. Chang (2006). Enhancing phototropic hydrogen production by solid-carrier assisted fermentation and internal optical-fiber illumination. *Process Biochemistry*, **41(9):** 2041–2049.

Chen, I.P. and H. Michel (1998). Cloning, sequencing and characterization of the recA gene from *Rhodopseudomonas viridis* and construction of a recA strain. *J. Bacteriol.* **180(12):** 3227–32.

Cheng, J., H. Su,J. Zhou, W. Song and K. Cen (2011). Microwave-assisted alkali pretreatment of rice straw to promote enzymatic hydrolysis and hydrogen production in dark- and photo fermentation . *Int. J. Hydr. Ener.* **36:** 2093-101.

Cherchi, C., O.H. Annalisa, E.S. Ibrahim and A.Z. Gu (2009). The implication of using different carbon Sources for denitrification in wastewater treatments. Water environment research: *A research publication of the Water Environment Federation*. **81:**788–99.

Cherni, N.E., J.V. Soloveva and V.D. Fedorova and E.N. Kondrateva (1969). The ultrastructure of cells of the species of purple sulfur bacteria. *Mikrobiologiya*. **38:** 479–484.

Chiemchaisri, C., L. Jaitrong, R. Honda, K. Fukushiand K.Yamamoto (2007). Photosynthetic bacteria pond system with infra-red transmitting filter for the treatment and recovery of organic carbon from industrial wastewater. *Water Sci. Technol.* **56(7):** 109–116.

Chiemchaisri, C., R. Honda, L. Chaitrong, K. Fukushi and K.Yamamoto (2008). Application of cleaner technology and photosynthetic wastewater treatment system in noodle processing factory. *Southeast Asian Water Environment.***2:** 327-334.

Chitapornpan, S., C. Chiemchaisri, W. Chiemchaisri, R. Honda and K. Yamamoto (2012). Photosynthetic bacteria production from food processing wastewater in sequencing batch and membrane photo-bioreactors. *Water Sci Technol.* **65 (3):** 504–512.

Chitapornpan, S., C. Chiemchaisri, W. Chiemchaisri, R. Honda and K. Yamamoto (2013). Organic carbon recovery and photosynthetic bacteria population in an anaerobic membrane photo-bioreactor treating food processing wastewater. *Bioresource Technology*. **141:** 65-74.

Chong, M.N., B. Jin, W.K.C. Christopher and C. Saint (2010). Recent developments in photocatalytic water treatment technology: A review. *Water research*. **44:** 2997–3027.

Choorit, W., P. Thanakoset, J. Thongpradistha, K. Sasaki and N. Noparatnaraporn (2002). Identification and cultivation of photosynthetic bacteria in wastewater from a concentrated latex processing factory. *Biotechnology Letters.***24(13):**1055-1058.

Chowdhury W.Q., K. Idehara, I. Maeda, F. Umeda, K. Yagi, Y. Miura and M. Tadashi (1996). Factors affecting polyhydroxybutyrate biosynthesis in the marine photosynthetic bacterium *Rhodopseudomonas* sp strain W-1S. *Appl. Biochem. Biotechnol.* **5:** 361–366.

Choudhary, D.K., A. Prakash and B.N. Johri (2007). Induced systemic resistance (ISR) in plants: mechanism of action. *Indian J. Microbiol.* **47:** 289–297.

Christensen, Q.H., N. Martin and M.C. Mansilla, D. de Mendoza and J. E. Cronan.(2011). A novel amidotransferase is required for lipoic acid cofactor assembly in *Bacillus subtilis*. *Mol. Microbiol.* **80:** 350–363.

Chumpol, S., D. Kantachote, T. Nitoda and H. Kanzaki (2018). Administration of purple nonsulfur bacteria as single-cell protein by mixing with shrimp feed to enhance growth, immune response and survival in white shrimp (*Litopenaeus vannamei*) cultivation. *Aquaculture.* **489:**85–95.

Ciambelli, P., V. Palma, A. Ruggiero and G. Iaquaniello (2009). Platinum catalysts for the low-temperature catalytic steam reforming of ethanol. *Chemical Engineering Transactions.* **17:** 19-24.

Claassen, P.A.M., T. de Vrije, K. Urbaniec and R. Grabarczyk (2010). Development of a fermentation-based process for biomass conversion to hydrogen gas. *Zuckerindustrie.* **135(4):** 218-21.

Clemente T., D. Shah , M. Tran M, D. Stark, S. Padgette, D. Dennis, K. Bruckener, A. Steinbüche and T. Mitsky.(2000). The sequence of PHA synthase gene from two strains of *Rhodospirillum rubrum* and in vivo substrate specificity of four PHA synthases across two heterologous expression systems. *Appl. Microbiol. Biotechnol.* **53:** 420–429.

Coenye, T., J. Goris, T. Spilker, P. Vandamme and J.J. LiPuma (2002). Characterization of unusual bacteria isolated from respiratory secretions of cystic fibrosis patients and description of *Inquilinus limosus* gen. nov., sp nov. *J. Clin. Microbiol.* **40:** 2062–2069.

Cogdell, R. J., A. Gall and J. Kohler (2006). The architecture and function of the light-harvesting apparatus of purple bacteria: from single molecules to in vivo membranes. *Q Rev. Biophys.* **39:** 227–324.

Cogdell, R. J., A.T. Gardiner, M. Gabrielsen, J. Southall, A.W. Roszak, N. W. Isaacs, R. Fujii and H. Hashimoto (2008). The structure of purple bacterial antenna complexes. In Photosynthetic Protein Complexes: *A Structural Approach, Fromme* P ed, pp 325–340. Wiley-Blackwell, Weinheim, Germany.

Cogdell, R. J., I. Durant, J. Valentine, J. G. Lindsay and K. Schmidt (1983). The isolation and partial characterization of the light-harvesting pigment-protein complexes of *Rhodopseudomonas acidophila. Biochem. Biophys. Acta.* **722:** 427–435.

Cogdell, R. J., N. W. Isaacs, A. A. Freer, J. Arrelano, T. D. Howard, M. Z. Papiz, A. M. Hawthornthwaite-Lawless and S. M. Prince (1997). The structure and function of the LH2 (B800-850) complex from the purple photosynthetic bacterium *Rhodopseudomonas acidophila* strain 10050. *Prog. Biophys. Mol. Biol.* **68:** 1–27.

Cogdell, R. J., N.W. Isaacs,A.A., Freer,T.D., Howard,A.T., Gardiner, S.M. Prince and M.Z. Papiz (2003). The structural basis of light-harvesting in purple bacteria. *FEBS Letters.* **555(1):**35–39.

Cogdell, R. J., P. K. Fyfe, S. J. Barrett, S. M. Prince, A. A. Freer, N. W. Isaacs, P. McGlynn and C. N. Hunter (1996). The purple bacterial photosynthetic unit. *Photosynth. Res.* **48:** 55–63.

Cogdell, R.J., J.G. Lindsay, MacDonald and G.P. Wand Reid (1980). A comparison of the constituent polypeptides of the B800-850 light-harvesting pigment-protein complex from *Rhodopseudomonassphaeroides. Biochim. Biophys. Acta.* **591:** 312–320.

Cogdell, R.J., N.W. Isaacs, T.D. Howard, K. McLuskey, N.J. Fraser and S.M. Prince (1999). How photosynthetic bacteria harvest solar energy. *J. Bacteriol.* **18:** 3869–3879.

Cohen, Y., W. E. Krumbein and M. Shilo (1977). Solar Lake (Sinai). 2: Distribution of photosynthetic microorganisms and primary production. *Limnol. Oceanogr.* **22:** 609–620.

Cohen-Bazire, G., N. Pfennig and R. Kunisawa (1964). The fine structure of green bacteria. *J. of Cell Biology.***22:** 207–225.

Cohen-Bazire, G., W.R. Sistrom and R.Y. Stanier (1957). Kinetic studies of pigment synthesis of non-sulfur bacteria. *J. Cellul. Comparat. Physiol.* **49:**25-68.

Conway, T (1992). The EntnerDoudoroff pathway: history, physiology and molecular biology. *FEMS Microbiol. Rev.* **9:** 1–27.

Cook, L.S. and F.R. Tabita (1988) Oxygen regulation of ribulose 1,5- bisphosphate carboxylase activity in *Rhodospirillum rubrum. J. Bacteriol.***170:** 5468–5472

Copeland, H.F (1956). *The Classification of Lower Organisms*, Pacific Book, Palo Alto.

Cort, J.R., U.M. Selan, A. Schulte, F. Grimm, M.A. Kennedy and C. Dahl (2008). *Allochromatium vinosum* DsrC: solution-state NMR structure, redox properties and interaction with DsrEFH, a protein essential for purple sulfur bacterial sulfur oxidation. *J. Mol. Biol.* **382:** 692–707.

Costa, S., S. Ganzerli, I. Rugiero, S. Pellizzari, P. Pedrini and E. Tamburini (2017). Potential of *Rhodobacter capsulatus* grown in Anaerobic –light or aerobic-dark conditions as a bioremediation agent for biological wastewater treatment. *Water.* **9(2):** 108. doi:10.3390/w9020108.

Cowey, C. B (1994). *Aquaculture.***124:** 1–11.

Cusanovich, M.A. and R.G. Bartsch (1969). A high potential cytochrome c from *Chromatium vinosum* chromatophores. *Biochem. Biophys. Acta.* **189:** 245–255.

Cviic, V (1960). Apparition d'eau rouge dans le Veliko Jezero (Ile de Mljet). Rapports et Procès-Verbeaux des Réunions de la Commission Internationale de l'Exploration Scientifique de la Mer Mediterranée. **15:** 79–81.

Czeczuga, B (1968). Primary production of the purple sulphuric bacteria *Thiopedia rosea* Winogr (Thiorhodaceae). *Photosynthetica.* **2:** 161–166.

Czurda, V. and E. Maresch (1937). Beitrag zur Kenntnis der Atiorhodobakterien-Gesellschaften. *Archiv Mikrobiol.* **8**: 99–124.

Dahl, C (2015). Cytoplasmic sulfur trafficking in sulfur-oxidizing prokaryotes. *Iubmb Life.* **67:** 268–274.

Dahl, C., A. Schulte, Y. Stockdreher, C. Hong, F. Grimm, J. Sander, R. Kim, S.H. Kim and D.H. Shin (2008c). Structural and molecular genetic insight into a widespread bacterial sulfur oxidation pathway. *J. Mol. Biol.* **384:** 1287–1300.

Dahl, C., B. Franz, D. Hensen A. Kesselheim and R. Zigann (2013). Sulfite oxidation in the purple sulfur bacterium *Allochromatium vinosum*: identification of SoeABC as a major player and relevance of SoxYZ in the process. *Microbiology.* **159:** 2626–2638.

Dahl, C., C.G. Friedrich and A. Kletzin (2008a). Sulfur oxidation in prokaryotes. In: *Encyclopedia of Life Sciences* (ELS), John Wiley and Sons, Ltd., Chichester, p http://www.els.net/ (doi: 10.1002/9780470015902.a0021155).

Dahl, C., R. Hell, T. Leustek and D.B. Knaff (2008b). Introduction to sulfur metabolism in phototrophic organisms. In: Hell, R., C. Dahl, D.B. Knaff, T. Leustek (eds) *Sulfur metabolism in phototrophic organisms*. Springer, Dordrecht. pp 1–14.

Dahl, C., S. Engels, A.S. Pott-Sperling, Schulte, J. Sander, Y. Lubbe, O. Deuster and D.C. Brune (2005). Novel genes of the dsr gene cluster and evidence for close interaction of Dsr proteins during sulfur oxidation in the phototrophic sulfur bacterium *Allochromatium vinosum. J. Bacteriol.***187:**1392–1404.

Das, D (2009). Advances in biohydrogen production processes: An approach towards commercialization. *Int. J. Hydr. Ener.* **34:** 7349–7357.

Das, D. and T.N. Veziroglu (2001). Hydrogen production by biological processes: A survey of the literature. *Int. J. Hydrogen. Energy.* **26:**13-28.

Das, D. and T.N. Veziroglu (2008). Advances in biological hydrogen production processes. *Int. J. Hydr. Ener.***33:** 6046-6057.

Das, D., N. Khanna and T. N. Veziroglu (2008). Recent developments in biological hydrogen production processes. *Chemical Industry and Chemical Engineering Quarterly.***14:** 57–67.

Das, D., T. Dutta, K. Nath, S. Kotay, A.M. Das and T. Veziroglu (2006). Role of Fe-hydrogenase in biological hydrogen production. *Current Science.* **90:** 1627-1637.

Dasgupta. C.N., J. Jose Gilbert, P. Lindblad, T. Heidorn, S.A. Borgvang, K. Skjanes and D. Das (2010). Recent trends on the development of photobiological processes and photobioreactors for the improvement of hydrogen production. *Int. J. Hydr. Ene.* **35(19):** 10218-38.

Daven port, C., D. W. Ussery and B. Tummler (2010). Comparative genomics of green sulfur bacteria. *Photosyn. Res.* **104:** 137–152.

David A. W. and A. Scott Ensign (2005). Photoheterotrophic Metabolism of Acrylamide by a Newly Isolated Strain of *Rhodopseudomonas palustris. Appl. Environ. Microbiol.* **71(10):** 5850-5857.

Davies, P.J (1995). Plant Hormones: Physiology, Biochemistry and Molecular Biology. Dordrecht: Springer, Netherlands.

De Bont, J.A., A. Scholten and T.A. Hansen (1981). DNA-DNA hybridization of *Rhodopseudomonas capsulata, Rhodopseudomonas sphaeroides* and *Rhodopseudomonas sulfidophila* strains. *Arch. Microbiol.* **128:** 271-274.

De Jong, G.A.H., J.A. Tang, P. Bos, S. de Vries and G.J. Kuenen (2000). Purification and characterization of a sulfite: cytochrome *c* oxidoreductase from *Thiobacillus acidophilus. J. Mol. Catal. B.***8:** 61–67.

De Jong, G.A.H., W. Hazeu, P. Bos and J.G. Kuenen (1997). Isolation of the tetrathionate hydrolase from *Thiobacillus acidophilus. Eur. J. Biochem.* **243:** 678–683.

De Lima, L., E. Ponsano and M. Pinto (2011). Cultivation of *Rubrivivax gelatinosus* in fish industry effluent for depollution and biomass production. *World Journal of Microbiology and Biotechnology.* **27(11):** 2553-2558.

De Vrije, T., A. E. Mars, M.A. Budde, M.H. Lai, C. Dijkema, P. de Waard and P.A. Claassen (2007). Glycolytic pathway and hydrogen yield studies of the extreme thermophile *Caldicellulosiruptor saccharolyticus. Appl. Microbiol. Biotechnol.* **74:** 1358–1367.

De Weerd, K.A., L. Mandelco, R.S. Tanner, C.R. Woese and I.M. Suflita (1990). *Desulfomonile tiedjei* gen. nov. and sp. nov., a novel anaerobic, dehalogenating, sulfate-reducing bacterium. *Arch. Microbiol.***54:** 23-30.

De-Bashan, L.E. and Y. Bashan (2004). Recent advances in removing phosphorus from wastewater and its future use as fertilizer (1997-2003). *Water Research.***38:**4222-46.

DeCorte, B., K. Verhaegen, P. Bossier and W. Verstraete (1988). *Appl. Microbiol. Biotechnol.* **247:** 121-133.

Dehghan-Noudeh, G., M.H. Moshafi, K. Payam and A. Farideh (2010). Studies on bioemulsifier production by *Bacillus licheniformis* PTCC 1595. *Afri. J. Biotechnol.***9:** 352-356.

Deisenhofer, J (2009). Nobel Laurates: Johan Deisenhofer (1988). Nobel Prizein Chemistry. Howard Hudges Medical Institute. http://www.hhmi.org / research / investigators / figs / deisenhofer_nobel1.html.

Deisenhofer, J. and H. Michel (1988). The Crystal Structure of the Photosynthetic Reaction Center from*Rhodopseudomonas viridis*. In: The Photosynthetic Bacterial Reaction Center (Breton J. and A. Vermēglio, eds) NATO ASI Series (Series A: Life Sciences), Vol 149. Springer, Boston, MA

Deisenhofer, J. and H. Michel (1989). The Photosynthetic Reaction Centre from the Purple Bacterium *Rhodopseudomonas viridis*. *EMBO J.***8:** 2149-2170.

Delwiche C. F. and J.D. Palmer (1996). Rampant horizontal transfer and duplication of rubisco genes in eubacteria and plastids. *Mol. Biol. Evol.* **13:**873–882.

Demirbas, A (2007) Progress and recent trends in biofuels. *Prog. Energy Conbust.Sci.* **33:** 1-18.

Demmer, J.K., H. Huang, S. Wang, U. Demmer, R.K. Thauer and U. Ermler 2015). Insights into flavin-based electron bifurcation via the NADH-dependent reduced ferredoxin: NADP oxidoreductase structure. *J. Biol. Chem.* **290:** 21985–21995.

Deng, R (2009). Food and food supplements with hypocholesterolemic effects. *Nutr. Agric.***1:** 15–24.

Denger, K., T.H.M. Smits and A.M. Cook (2006). Genome-enabled analysis of the utilization of taurine as the sole source of carbon or of nitrogen by *Rhodobacter sphaeroides* 2.4.1. *Microbiology*. **152:** 3197–3206.

Denkmann, K., F. Grein, R. Zigann, A. Siemen, J. Bergmann, S. van Helmont, A. Nicolai I.A. Pereira and C. Dahl (2012). Thiosulfate dehydrogenase: a widespread unusual acidophilic c-type cytochrome. *Environ. Microbiol.* **14:** 2673–2688.

Desai, J.D. and I.M. Banat (1997). Microbial production of biosurfactants and their commercial potential. *Microbial Mol. Biol. Rev.* **61:** 47-64.

Dhansekaran, D. S.S. Lawanya, N.T. Saha and A. Paneerselvam (2011). Production of single-cell protein from pineapple waste using yeast. *Innovat. Roman. Food Biotechnol.* **8:** 26-32.

Dhillon, A., S. Goswami, M. Riley, A. Teske and M. Sogin (2005). Domain evolution and functional diversification of sulfite reductases. *Astrobiology*. **5:** 18–29.

Dickerson, R.E (1980b). Evolution and gene transfer in purple photosynthetic bacteria. *Nature* **283**: 210–212.

Dilling, W., W. Liesack and N. Pfennig (1995). *Rhabdochromatium marinum* gen. nom. rev., sp. nov., a purple sulfur bacterium from a salt marsh microbial mat. *Arch. Microbiol.***164:** 125–131.

Dilling, W., W. Liesack and N. Pfennig (1996). *In* Validation of the publication of new names and new combinations previously effectively published outside the IJSB. List No. 56. *Int. J. Syst. Bacteriol.* **46**: 362–363.

Divyasree, B., K.V.N.S. Lakshmi, Ch. Sasikala and Ch.V. Ramana (2014). *Thiophaeococcus fuscus* sp. nov., isolated from a lagoon. *Int. J. Syst. Evol. Microbiol.***64**: 2528-2533.

Divyasree, B., K.V.N.S. Lakshmi, D. Bharti, Ch. Sasikala and Ch.V. Ramana (2016). *Rhodovulum aestuarii* sp. nov., isolated from a brackish water body. *Int. J. Syst. Evol. Microbiol.* **66:** 165–171.

Dixit, R., D. Wasiullah, R. Malaviya, K. Pandiyan, U.B. Singh, A. Sahu R. Shukla, B.P. Singh, J.P. Rai, P.K. Sharma, H. Lade, D. Paul (2015). Bioremediation of Heavy Metals from Soil and Aquatic Environment: An Overview of Principles and Criteria of Fundamental Processes. Sustainability. **7:** 2189-2212.

Do, Y.S., J. Smeenk, K.M. Broer, C.J. Kisting, R. Brown, T.J. Heindel, T.A. Bobik and A.A. DiSpirito (2007). Growth of *Rhodospirillum rubrum* on synthesis gas: conversion of CO to H_2 and poly-β-hydroxyalkanoate. *Biotechnol. Bioeng*. **97:** 279–286.

Donald, H (1970). U.S. Pat. **3:** 525-671.

Douglas G.B., M. Lurling and B.M. Spears (2016). Assessment of changes in potential nutrient limitation in an impounded river after application of lanthanum-modified bentonite.Water Res. **97:** 47–54.

Drews, G (1965). Die Isolierung Schwefelfreier Purpurbakterien. Zentralblatt fur bakteriologie, Parasitenkunde, Infektions-krankheiten and Hygiene. *Abt. l. Suppl.* **1:** 170-178.

Drews, G (1981). *Rhodospirillum salexigens*, spec. nov., an obligatory halophilic phototrophic bacterium. *Arch. Microbiol.* **130:** 325–327.

Drews, G. and J.F. Imhoff (1991). Phototrophic purple bacteria. In: Variations in Autotrophic Life, (Shively and Barton, eds.), Academic Press, London. pp. 51-97.

Drews, G. and J.R. Golecki (1995). Structure, molecular organization and biosynthesis of membranes of purple bacteria. In Anoxygenic photosynthetic bacteria, (Blankenship, R.E. and M.T. Madigan, eds.) Dordrecht: Kluwer Academic Publisher. pp. 231-257

Drews, G. and P. Giesbrecht (1966). *Rhodopseudomonas viridis*, nov. spec., ein neu isoliertes, obligat phototrophes Bakterium. *Archiv. Mikrobiol.***53(3):** 255-262.

Drider, D.1., G. Fimland, Y. Hechard, L.M. McMullen and H. Prevost (2006). The continuing story of class IIa bacteriocins. *Microbiol Mol Biol Rev.* **70(2):** 564-82.

Driessens, K., J. Liessens, S. Masduki, W. Verstraete, H. Nelis and A. DeLeenheer (1987). *Process Biochem.* **22:** 160-163.

Dubinina, G.A. and V.M. Gorlenko (1975). New filamentous photosynthetic green bacteria containing gas vacuoles. *Mikrobiologiya.* **44:** 511–517.

Duchow. E. and H.C. Douglas (1949). *Rhodomicrobium vannielii*, a new photoheterotrophic bacterium. *J. Bacteriol.* **58:** 407–416.

Duggeli, M (1924). Hydrobiologische Untersuchungen im Pioragebiet. Bakteriologische Untersuchungen am Ritomsee. *Schweiz. Zeitschr. Hydrobiol.* **2:** 65–205.

Dutton, P.L. and W.C. Evans (1978). In: *the photosynthetic bacteria* (R.K. Clayton and W.R. Sistrom, eds.), Plenum, New York. pp. 719-726.

Eckersley, K. and C.S. Dow (1980). *Rhodopseudomonas blastica* sp. nov.: A member of the *Rhodospirillaceae*. *J. Gen. Microbiol.* **119:** 465–473.

Eddie, B.J. and T.E. Hanson (2013). *Chlorobaculum tepidum* TLS displays a complex transcriptional response to sulfide addition. *J. Bacteriol.* **195:** 399–408.

Edith Ifeoma Madukasi and Guangming Zhang (2010). Microaerobic biodegradation of high organic load wastewater by phototrophic bacteria. *Afr. J. Biotechnol.* **9(25):** 3852-3860.

Edwards, G.E. and C.R. Borell (1969). Characteristics of light-dependent proton transport in cells of *Rhodospirillum rubrum*. *Biochim.Biophs. Acta.* **172:**126.

Egamberdieva, D (2009). Alleviation of salt stress by plant growth regulators and IAA-producing bacteria in wheat. *Acta. Physiol. Plant.***31:** 861–864.

Ehrenberg, C.G (1838). Die Infusionstierchen als vollkommene Organismen. Voss L, Leipzig, **I–XVIII**:1–547.

Ehrenfeld, N., G. Levicán and P. Parada (2013). Heterodisulfide reductase from *Acdidithiobacilli* is a key component involved in the metabolism of reduced inorganic sulfur compounds. *Adv. Mater. Res.* **825:** 194–197.

Ehrenreich, A. and F. Widdel (1994). Anaerobic oxidation of ferrous iron by purple bacteria, a new type of phototrophic metabolism. *Applied Environ Microbiol.* **60:** 4517-4526.

Eichler, B. and N. Pfennig (1986). Characterization of a new platelet-forming purple sulfur bacterium, *Amoebobacter pedioformis* sp. nov. *Arch. Microbiol.* **146:** 295–300.

Eichler, B. and N. Pfennig (1988). A new purple sulfur bacterium from stratified freshwater lakes, *Amoebobacter purpureus* sp. nov. *Arch Microbiol.* **149:** 395–400.

Eichler, B. and N. Pfennig (1990). Seasonal development of anoxygenic phototrophic bacteria in a holomictic drumlin lake (Schleinsee, Germany). *Arch. Hydrobiol.* **119:**369–392.

Eichler, B. and N. Pfennig (1991). Isolation and characteristics of *Thiopedia rosea* (neotype). *Arch. Microbiol.* **155:** 210-216.

Eimhjellen, K.E (1970). *Thiocapsa pfennigii* sp. nov., a new species of the phototrophic sulfur bacteria. *Arch Mikrobiol.* **73:**193–194.

Eimhjellen, K.E., H. Steensland and H. Traetteberg (1967). A *Thiococcus* sp. nov. gen., its pigments and internal membrane system. *Arch. Microbiol.* **59:** 82–92.

Eisen, J.A., K.E. Nelson, I.T. Paulsen, J.F. Heidelberg, M. Wu, R.J. Dodson, R. Deboy, M.L. Gwinn,W.C. Nelson, D.H. Haft, E.K. Hickey, J.D. Peterson, A.S. Durkin, J.L. Kolonay, F. Yang, I. Holt, L.A. Umayam, T. Mason, M. Brenner, T.P. Shea, D. Parksey, W.C. Nierman, T.V. Feldblyum, C.L. Hansen, M.B. Craven, D. Radune, J. Vamathevan, H. Khouri, O. White, T.M. Gruber, K.A. Ketchum, J.C. Venter, H. Tettelin, D.A. Bryant and C.M. Fraser (2002). The complete genome sequence of *Chlorobium tepidum* TLS, a photosynthetic, anaerobic, green sulfur bacterium. *Proceedings of the National Academy of Sciences of the USA*. **99:** 9509–9514.

Elbadry, M. and K. Elbanna (1999). The response of four rice varieties to *Rhodobacter capsulatus* at the seedling stage.*World J. Microbiol. Biotechnol.* **15:** 363-363.

Elbadry, M., A. El-Bassel and K. Elbanna (1999a). Occurrence and dynamics of phototrophic purple non-sulfur bacteria compared with other asymbiotic nitrogen fixers in ricefields of Egypt. *World J. Microbiol. Biotechnol.* **15:** 359–362.

Elbadry, M., H. Gamal-Eldin and K. Elbanna (1999b). Effects of *Rhodobacter capsulatus* inoculation in combination with graded levels of nitrogen fertilizer on growth and yield of rice in pots and lysimer experiments. *World J. Microbiol. Biotechnol.***15:** 393–395.

Elder, D.J.E. and D.J. Kelly (1994). The bacterial degradation of benzoic acid and benzenoid compounds under anaerobic conditions: Unifying trends and new perspectives. *FEMS Microbiolo. Reviews.***13:** 44 1-468.

Englund, E., J. Andersen-Ranberg, R. Miao, B.R. Hamberger, P. Lindberg (2015). Metabolic engineering of Synechocystis sp. PCC 6803 for production of the plant diterpenoid manoyl oxide. *ACS Synth. Biol.***4:** 1270-1278.

Erb, T. J., I.A. Berg, V. Brecht, M. Muller, G. Fuchs and B.E. Alber (2007). Synthesis of C5-dicarboxylic acids from C2-units involving crotonyl-CoA carboxylase/reductase: the ethyl malonyl-CoA pathway. *Proc. Natl. Acad. Sci.U.S.A.* **104:** 10631–10636.

Erb,T.J., G. Fuchs and B.E. Alber (2009) (2S)-Methylsuccinyl- CoAdehydrogenaseclosesthe ethyl malonyl-CoApathwayfor acetyl-CoAassimilation. *Mol. Microbiol.* **73:** 992–1008.

Erbakan, M., B.S. Curtis, B.T. Nixon, M. Kumar and W.R. Curtis (2015). Advancing *Rhodobacter sphaeroides* as a platform for expression of functional membrane proteins. *Protein Expression and Purification*. **115:** 109–117.

Eroglu, E., I. Eroglu, U. Gunduz and M. Yucel (2008). Effect of clay pretreatment on photofermentative hydrogen production from olive mill wastewater. *Bioresource Technology.* **99:** 6799–6808.

Eroglu, E., U. Gunduz, M. Yucel and I. Eroglu (2010). Photosynthetic bacterial growth and productivity under continuous illumination or diurnal cycles with olive mill wastewater as feedstock. *Int. J. Hyd. Ener*. **35(11):** 5293-5300.

Euzeby, J.P (1997). Revised nomenclature of specific or subspecific epithets that do not agree in gender with generic names that end in -bacteria. *Int. J. Syst. Bacteriol.* **47**: 585.

Euzeby, J.P (2011). List of prokaryotic names with standing in nomenclature. http: // www.bacterio.cict.fr/classifphyla. html. Accessed 20 July 2012

Evans, M.C.W., B.B. Buchana and D.I. Arnon (1966). A new ferredoxin-dependent carbon reduction cycle in a photosynthetic bacterium. *Proc. Natl. Acad. Sci. USA*. **55:** 928–34.

Fahad, S., S. Hussain, A.Bano, S. Saud, S. Hassan, D. Shan, et al (2015) Potential role of phytohormones and plant growth-promoting rhizobacteria in abiotic stresses: consequences for changing environment. *Environ. Sci. Pollut. Res*.**22:** 4907–4921.

Falcone, D.L. and F.R. Tabita (1991). Expression of endogenous and foreign ribulose1,5-bisphosphate carboxylase-oxygenase (RubisCO) genes in a RubisCO deletion mutant of *Rhodobacter sphaeroides*. *J. Bacteriol.* **173:** 2099–2108.

Fan, W., Y. Jia, X. Li, W. Jiang and L. Lu (2012). Phytoavailability and geospeciation of cadmium in contaminated soil remediated by *Rhodobacter sphaeroides*. *Chemosphere*.**88:** 751–756.

Fang, H. H.P., H. Liu and T. Zhang (2005). Phototrophic hydrogen production from acetate and butyrate in wastewater. *Int. J. Hydr. Ener*. **30:** 785-793.

Fang, H.H.P. and H. Liu (2002). Effect of pH on hydrogen production from glucose by a mixed culture. *Biores. Technol*.**82:** 87-93.

Fang, H.H.P., H. Zhu and T. Zhang (2006). Phototrophic hydrogen production from glucose by pure and co-cultures of *Clostridium butyricum* and *Rhodobacter sphaeroids*. *Int. J. Hydr. Ener*.**31:** 2223–2230.

Fascetti, E., D.E. Addario, O. Todini, A. Robertiello (1998). Photosynthetic hydrogen evolution with volatile organic acids derived from the fermentation of source selected municipal solid wastes. *Int. J. Hydr. Ener*. **23(9):** 753–60.

Faust, U. and P. Prave (1979). Process alternatives in SCP-fermentation. *Process Biochem.* **14(11):** 28-31.

Favinger, J., R. Stadtwald and H. Gest (1989). *Rhodospirillum centenum* sp. nov., a thermotolerant cyst-forming photosynthetic bacterium. *Antonie Van Leeuwenhoek*. **55:** 291–296.

Favinger, J., R. Stadtwald and H. Gest (1994). In: Validation of the publication of new names and new combinations previously effectively published outside the IJSB. List No. 48. *Int. J. Syst. Bacteriol.* **44**: 182–183.

Fedorov, A.S., A.A. Tsygankov, K.K. Rao and D.O. Hall (1998). Hydrogen photoproduction by *Rhodobacter sphaeroides* immobilized on polyurethane foam. *Biotechnology Letters*. **20(11):** 1007–1009

Feng, X (2016). Applied Research of Microbial Degradation of High Concentration Organic Wastewater. *Chemical Engineering Transactions*. **55:** 133-138.

Feng, Y., X. Lin, Y. Wang, J. Zhang, T. Mao, R. Yin and J. Zhu (2009). Free-air CO_2 enrichment (FACE) enhances the biodiversity of purple phototrophic bacteria in flooded paddy soil. *Plant and Soil*. **324:** 317-328.

Feng, Y., Y. Yu, Y. Wang and X. Lin (2007). Biosorption and bioreduction of trivalent aurum by photosynthetic bacteria Rhodobacter capsulatus. *Curr Microbiol*. **55(5):** 402–408.

Finneran, K. T., C.V. Johnsen and D.R. Lovley (2003.) *Rhodoferax ferrireducens* sp. nov., a psychrotolerant, facultatively anaerobic bacterium that oxidizes acetate with the reduction of Fe(III). *Int. J. Syst. Evol. Microbiol*. **53:** 669-673.

Fissler, J., C. Schirra, G.W. Kohring and F. Giffhorn (1994). Hydrogen production from aromatic acids by *Rhodopseudomonas palustris*. *Appl.Microbiol. Biol*. **41:** 395-399.

Florent, J. and L. Ninet (1979). In: *Microbial Technology* (Peppler, H.G. and D. Perlman, eds.), Second Edition. Academic Press, New York. **1:** 497-519.

Florin, L., A. Tsokoglou and T. Happe (2001). A novel type of iron hydrogenase in the green algae *Scendesmus obliquus* is linked to the photosynthetic electron transport chain. *J. Biol. Chem*,. **276:** 6125–6132.

Folt, C. L., M. J. Wevers, M. P. Yoder-Williams and R. P. Howmiller (1989). Field studies comparing growth and viability of a population of phototrophic bacteria. *Appl. Environ. Microbiol.* **55:** 78–85.

Fotiadis, D., P. Qian, A. Philippsen, P.A. Bullough, A. Engel and C.N. Hunter (2004). Structural analysis of the reaction center light-harvesting complex I photosynthetic core complex of *Rhodospirillum rubrum* using atomic force microscopy. *J. Biol. Chem.* **279:** 2063–2068.

Fowler, V.J., N. Pfennig, W. Schubert and E. Stackebrandt (1984). Towards a phylogeny of phototrophic purple sulfur bacteria: 16S ribosomal RNA oligonucleotide cataloging of 11 species of *Chromatiaceae*. *Arch. Microbiol.***139:** 382–387.

Franche, C., K. Lindstrom and C. Elmerich (2009). Nitrogen-fixing bacteria associated with leguminous and nonleguminous plants. *Plant Soil.***321:** 35–59.

Frank, H.A. and R.J. Cogdell (1993). Photochemistry of carotenoids, In: Carotenoids in photosynthesis, Young, A. and G. Britton, ed.), Chapman and Hall, London, United Kingdom, pp. 252–326.

Frank, H.A. and R.J. Cogdell (1996). Carotenoid in Photosynthesis. *Photochem. & Photobiol.* **3:** 257-264.

Frank, H.A., A. Young, G. Britton and R.J. Cogdell (1999). The photochemistry of carotenoids. Dordrecht, The Netherlands: *Kluwer Academic Publishers* .

Frank, H.A., J. Young, G. Britton and R.J. Cogdell (2000). The photochemistry of Carotenoids: Application in Biology. *Kluwer Academic Publisher* .

Franz, B., H. Lichtenberg and J. Hormes, H. Modrow, C. Dahl and A. Prange (2007). Utilization of solid elemental sulfur by the phototrophic purple sulfur bacterium *Allochromatiumvinosum*: a sulfur K-edge XANES spectroscopy study. *Microbiology.* **153:** 1268–1274.

Franz, B., T. Gehrke, H. Lichtenberg J. Hormes, C. Dahl, A. Prange (2009). Unexpected extracellular and intracellular sulfur species during the growth of *Allochromatium vinosum* with reduced sulfur compounds. *Microbiology.* **155:** 2766–2774.

Freer, A.A., S.M. Prince, K. Sauer, M.Z. Papiz, A.M. Hawthornthwaite-Lawless, G. McDermott, R.J. Cogdell and N.W. Isaacs (1996). Pigment-protein interactions and energy transfer in the antenna complex of the photosynthetic bacterium *Rhodopseudomonas acidophila*. *Structure.* **4:** 449–462.

Friedrich, C.G., Bardischewsky, F., D. Rother, A. Quentmeier and J. Fischer (2005). Prokaryotic sulfur oxidation. *Curr. Opin. Microbiol.* **8:** 253–259.

Friedrich, C.G., D. Rother, F. Bardischewsky, A. Quentmeier and J. Fischer (2001). Oxidation of reduced inorganic sulfur compounds by bacteria: the emergence of a common mechanism. *Appl. Environ. Microbiol.* **67:** 2873–2882.

Frigaard N.U., H. Li, K.J. Milks and D.A. Bryant (2004). Nine mutants of *Chlorobium tepidum* each unable to synthesize a different chlorosome protein still assemble functional chlorosomes. *J. Bacteriol.* **186:** 646–653.

Frigaard N.U., J.A. Maresca, C.E. Yunker, A.D. Jones and D.A. Bryant (2004). Genetic manipulation of carotenoid biosynthesis in the green sulfur bacterium *Chlorobium tepidum*. *J. Bacteriol.* **186:** 5210–5220.

Frigaard, N.U. and C. Dahl (2008). Sulfur metabolism in phototrophic sulfur bacteria. Adv. *Microb. Physiol.* **54:** 103–200.

Frigaard, N.U. and D.A. Bryant (2008a). Genomic and evolutionary perspectives on sulfur metabolism in green sulfur bacteria. In: *Microbial sulfur metabolism*, (Dahl, C. and C.G. Friedrich, eds.) Springer, Heidelberg, Heidelberg, pp 60–76.

Frigaard, N.U. and D.A. Bryant (2008b). Genomic insights into the sulfur metabolism of phototrophic green sulfur bacteria. In: Sulfur metabolism in phototrophic organisms, (Hell, R., C. Dahl, D.B. Knaff and T. Leustek, eds.). Springer, Dordrecht, pp. 337–355.

Frigaard, N.U., A.G.M. Chew, H. Li, J.A. Maresca and D.A. Bryant (2003). *Chlorobium tepidum*: Insights into the structure, physiology and metabolism of a green sulfur bacterium derived from the complete genome sequence. *Photosyn. Res.* **78:** 93–117.

Frolund, B., R. Palmgren, K. Keiding and P.H. Nielsen (1996). Extraction of extracellular polymers from activated sludge using a cation exchange resin. *Water Res.* **30:** 1749–1758.

Frolund, B., T. Griebe and Nielsen, Per (1995). Enzymatic-Activity in the Activated-Sludge Floc Matrix. *Appl. Microbial. Biotechnol.*. 43. 755-61. 10.1007/BF00164784.

Frydman, B. and H. Rapoport (1963). Non-Chlorophyllous Pigments of *Chlorobium Thiosulfatophilum* Chlorobiumquinone. *J. Am. Chem. SOC.* **85:**823-825.

Fuchs, G., E. Stupperich and G. Eden (1980). Autotrophic CO_2fixation in *Chlorobium limicola*. Evidence for the operation of a reductive tricarboxylic acid cycle in growing cells. *Archives ofMicrobiology*. **128:** 64–71.

Fuchs, G., M. El Said Mohamed, U. Altenschmidt, J. Koch, A. Lack, R. Brackmann, C. Lochmeyer and B. Oswald (1994). In: *Biochemistry of Microbial Degradation*, (Ratledge, C. ed.). Kluwer Academic Publishers, Dordrecht, The Netherlands, pp. 513-553.

Fuhrer, T., E. Fischer and U. Sauer (2005). Experimental identification and quantification of glucose metabolism in seven bacterial species. *J. Bacteriol.* **187:** 1581–1590.

Fuhrmann, S., J. Overmann, N. Pfennig and U. Fischer (1993). Influence of vitamin B_{12} and light on the formation of chlorosomes in green- and brown-colored *Chlorobium* species. *Archives of Microbiology*. **160:** 192–198.

Fullerton, S.W., J.S. Griffiths, A.B Merkel, M. Cheriyan, N.J. Wymer, M.J. Hutchins, C.A. Fierke, E. Toone and J.H. Naismith (2006). Mechanism of the class I KDPG aldolase. *Bioorg. Med.Chem.* **14:** 3002–3010.

Furch, T., M. Preusse, J. Tomasch, H. Zech, I. Wagner-Dobler, R. Rabus and C. Wittmann (2009). Metabolic fluxes in the central carbon metabolism of *Dinoroseobacter shibae* and *Phaeobacter gallaeciensis*, two members of the marine *Roseobacter clade*. *BMC Microbiol.* **9:** 209.

Gad El-Rab, S.M.F., A.A.F. Shoreit and Y. Fukumori (2006). Effects of cadmium stress on growth, morphology and protein expression in *Rhodobacter capsulatus* B10. *Biosci. Biotechnol. Biochem.* **70:** 2394-2402.

Gaffron, H. and J. Rubin (1942). Fermentative and photochemical production of hydrogen in algae. *J. Gen. Physiol.*, **26:** 219-240.

Gaisin, V.A., A.M. Kalashnikov, D.S. Grouzdev, M.V. Sukhacheva, B.B Kuznetsov and V.M. Gorlenko (2017). *Chloroflexus islandicus* sp. nov., a thermophilic filamentous anoxygenic phototrophic bacterium from a geyser. *Int. J. Syst. Evol. Bacteriol.***67(5):** 1381–1386.

Gall, A (1995). Purification, characterization and crystallization of a range of *Rhodospirillineae* pigment-protein complexes. Ph.D. thesis. The University of Glasgow, Glasgow, United Kingdom.

Gamal-Eldin, H. and K. Elbanna (2011). Field evidence for the potential of *Rhodobacter capsulatus* as Biofertilizer for flooded rice. *Curr Microbiol.* **62(2):** 391-395.

Garcia Costas, A.M., Y. Tsukatani, W.I. Rijpstra, S. Schouten, P.V. Welander, R.E. Summons and D.A. Bryant (2012b). Identification of the bacteriochlorophylls, carotenoids, quinones, lipids and hopanoids of *Candidatus Chloracidobacterium thermophilum*. *J. Bacteriol.***194:** 1158–1168.

Garcia, D., P. Parot, A. Vermeglio and M.T. Madigan (1986). The light-harvesting complexes of a thermophilic purple sulfur photosynthetic bacterium *Chromatium tepidum*. *Biochim. Biophys. Acta.* **850:** 390–395.

Gardiner, A.T., R.J. Cogdell and S. Takaichi (1992). The effect of growth conditions on the light-harvesting apparatus in *Rhodopseudomonas acidophila*. *Photosynth. Res.* **38:** 159-167.

Garrison, S.A (2005). Synthesis and photophysical characterization of phorphyrin containing supramolecular systems: Structural issues for porphyrin photophysics and electron transfer. Doctoral Thesis. The Graduate Faculty of The University of Akron.

Garrity, G. M., J.A. Bell and T. Lilburn (2015). *Rhodospirillales*. *Bergey's Manual of Systematics of Archaea and Bacteria*. 1–1. doi:10.1002/9781118960608.obm00073

Garrity, G. M., J.A. Bell and T. Lilburn (2015). *Betaproteobacteria class. nov. Bergey's Manual of Systematics of Archaea and Bacteria,* 1–1. doi:10.1002 / 9781118960608.cbm00042.

Garrity, G. M., J.A. Bell and T. Lilburn (2015). *Rhodobacterales ord. nov. Bergey's Manual of Systematics of Archaea and Bacteria,* 1–1. doi:10.1002/9781118960608.obm00072.

Garrity, G.M. and J.G. Holt (2001a). *Chlorobi* phy. nov. In: *Bergey's Manual of Systematic Bacteriology*, (Boone, D.R. R.W. Castenholz and G.M. Garrity, eds.), Vol. 1, Springer, New York. , pp. 601–623.

Garrity, G.M. and J.G. Holt (2001a). Phylum BVI, *Chloroflexi* phy. nov. In: *Bergey's manual of systematic bacteriology*, The Archaea and the deeply branching and phototrophic bacteria, (Boone, D.R., R.W. Castenholz, eds.), Second edition, Vol 1, Springer, New York, pp. 427–446.

Garrity, G.M. and J.G. Holt (2001b). Phylum BVII. *Thermomicrobia* phy. nov. In: *Bergey's Manual of Systematic Bacteriology*, The Archaea and the deeply branching and phototrophic Bacteria, Second edition, Vol 1, Springer, New York, pp 447–450.

Garrity, G.M., J.A. Bell and T. Liburn (2015). *Rhodospirillaceae*. *Bergey's Manual of Systematics of Archaea and Bacteria*. 1–2. doi:10.1002/9781118960608.fbm00175.

Garrity, G.M., J.A. Bell and T. Lilburn. (2005). Family X. *Rhodobiaceae* fam. nov. In: *Bergey's Manual of Systematic Bacteriology*, (Brenner, D,J., N.R. Krieg, J.T. Staley and G.M. Garrity, ed.), Second edition, Vol. 2 (The *Proteobacteria*), part C (The *Alpha,- Beta,- Delta-*, and *Epsilonproteobacteria*), Springer, New York. p. 571.

Garrity, G.M., J.A. Bell and T. Lilburn (2005). Class II. *Betaproteobacteria* class. nov. In: *Bergey's Manual of Systematic Bacteriology,* (Brenner, D.J., N.R. Krieg, J.T. Staley and G.M. Garrity, eds.), Second edition, Vol. 2 (The *Proteobacteria*), Part C (The *Alpha,- Beta,- Delta-*, and *Epsilonproteobacteria*), Springer, New York, p. 575.

Garrity, G.M., J.A. Bell and T. Lilburn (2005). Class III. *Gammaproteobacteria* class. nov. In: *Bergey's Manual of Systematic Bacteriology*, (The *Proteobacteria*), Part B (The *Gammaproteobacteria*), (Brenner, D.J., N.R. Krieg, J.T. Staley and G.M. Garrity, eds.), Second edition, Vol. 2. New York, Springer. p.1.

Garrity, G.M., J.A. Bell and T. Lilburn (2005). Family I. *Rhodobacteraceae* fam. nov. In: *Bergey's Manual of Systematic Bacteriology,* (Brenner, D.J., N.R. Krieg, J.T. Staley and G. M. Garrity, eds.), Second edition,Vol. 2 (The *Proteobacteria*), part C (The *Alpha,- Beta,- Delta-*, and *Epsilonproteobacteria*), Springer, New York, p. 161.

Garrity, G.M., J.A. Bell and T. Lilburn (2005). Family I. *Rhodocyclaceae* fam. nov. In: *Bergey's Manual of Systematic Bacteriology*, (Brenner, D.J., N.R. Krieg, J.T. Staley and

G.M. Garrity, eds.) Second edition, Vol. 2 (The *Proteobacteria*), part C (The *Alpha,- Beta,- Delta-*, and *Epsilonproteobacteria*), Springer, New York, p. 887.

Garrity, G.M., J.A. Bell and T. Lilburn (2005). Family VII. *Bradyrhizobiaceae* fam. nov. In: *Bergey's Manual of Systematic Bacteriology*, (The *Proteobacteria*), part C (The *Alpha,- Beta,- Delta-*, and *Epsilonproteobacteria*), (Brenner, D.J., N.R. Krieg, J.T. Staley and G. M. Garrity, eds.), Second edition, Vol. 2, Springer, New York. pp. 438.

Garrity, G.M., J.A. Bell and T. Lilburn (2005). Order I. *Burkholderiales* ord. nov. In: *Bergey's Manual of Systematic Bacteriology,* (Brenner, D.J., N.R. Krieg, J.T. Staley and G.M. Garrity, eds.), Second edition, Vol. 2 (The *Proteobacteria*), Part C (The *Alpha,- Beta,- Delta-*, and *Epsilonproteobacteria*), Springer, New York, p. 575.

Garrity, G.M., J.A. Bell and T. Lilburn (2005). Order III. *Rhodobacterales* ord. nov. *In*: Bergey's Manual of Systematic Bacteriology, (Brenner, D.J., N.R. Krieg, J.T. Staley and G. M. Garrity, eds.), Second edition,Vol. 2 (The *Proteobacteria*), Part C (The *Alpha,- Beta,- Delta-*, and *Epsilonproteobacteria*), Springer, New York, p. 161.

Garrity, G.M., J.A. Bell and T. Lilburn (2005). Order VI. *Rhodocyclales* ord. nov. In: *Bergey's Manual of Systematic Bacteriology*, Second edition, Vol. 2 (The *Proteobacteria*), Part C (The *Alpha,- Beta,- Delta-*, and *Epsilonproteobacteria*), Springer, New York, p. 887.

Garrity, G.M., J.A. Bell and T. Lilburn (2005). Phylum XIV. *Proteobacteria* phyl. nov. In: *Bergey's Manual of Systematic Bacteriology*, (The Proteobacteria), part B (The *Gammaproteobacteria*), (Brenner, D.J., N. R. Krieg, J. T. Staley and G. M. Garrity, eds.), Second edition, Vol. 2. *New York, Springer*. p. 1.

Garrity, G.M., J.A. Bell and T. Lilburn (2006). Validation list N° 107. *Int. J. Syst. Evol. Microbiol.***56:** 1–6.

Garrity, G.M., J.A. Bell and T. Lilburn (2015). *Bradyrhizobiaceae fam. nov. Bergey's Manual of Systematics of Archaea and Bacteria,* 1–1. doi:10.1002 / 9781118960608.fbm00165.

Garrity, G.M., J.A. Bell and T. Lilburn (2015). *Burkholderiales ord. nov. Bergey's Manual of Systematics of Archaea and Bacteria,* 1–1. doi:10.1002/9781118960608.obm00076.

Garrity, G.M., J.A. Bell and T. Lilburn (2015). *Gammaproteobacteria class. nov. Bergey's Manual of Systematics of Archaea and Bacteria,* 1–1. doi:10.1002 / 9781118960608.cbm00045.

Garrity, G.M., J.A. Bell and T. Lilburn (2015). *Hyphomicrobiaceae. Bergey's Manual of Systematics of Archaea and Bacteria,* 1–1. doi:10.1002/9781118960608.fbm00167.

Garrity, G.M., J.A. Bell and T. Lilburn (2015). *Rhodobacteraceae fam. nov. Bergey's Manual of Systematics of Archaea and Bacteria,* 1–2. doi:10.1002 / 9781118960608.fbm00173.

Garrity, G.M., J.A. Bell and T. Lilburn (2015). *Rhodobiaceae fam. nov. Bergey's Manual of Systematics of Archaea and Bacteria,* 1–1. doi:10.1002/9781118960608.fbm00172.

Garrity, G.M., J.A. Bell and T. Lilburn (2015). *Rhodocyclaceae fam. nov. Bergey's Manual of Systematics of Archaea and Bacteria,* 1–1. doi:10.1002/9781118960608.fbm00190.

Garrity, G.M., J.A. Bell and T. Lilburn (2015). *Rhodocyclales ord. nov. Bergey's Manual of Systematics of Archaea and Bacteria,* 1–1. doi:10.1002/9781118960608.obm00081.

Gautam, K. and V. Tyagi (2006). Microbial surfactants: a review. *J. Oleo. Sci.* **55:** 155-166.

Gavala, H.N., I.V. Skiadas and B.K. Ahring (2006). Biological hydrogen production in suspended and attached growth anaerobic reactor systems. *Int. J. Hydr. Ener.* **31:** 1164-1175.

Gazso, G.L (2001). The key microbial processes in the removal of toxic metals and radionuclides from the environment. *Cent. Eur. J. Occup. Environ. Med. Hung* **.7:** 178–185.

Gerhardt, M., V. Svetlichny, S. Tatyana, G. Zavarzin and M. Ringpfeil (1991). Bacterial CO utilization with H_2 production by the strictly anaerobic lithoautotrophic thermophilic bacterium *Carboxydothermus hydrogenus* DSM 6008 isolated from a hot swamp. *Fems Microbiol. Lette.* **83:** 267-271.

Gerken, U., D. Lupo, C. Tietz, J. Wrachtrup and R. Ghosh (2003). Circular symmetry of the light-harvesting1complex from *Rhodospirillum rubrum* is not perturbed by interaction with the reaction center. *Biochem.* **42:** 10354–10360.

Gest, H. and J.L. Favinger (1983). *Heliobacterium chlorum*, an anoxygenic brownish-green photosynthetic bacterium containing a 'new form of bacteriochlorophyll. *Arch Microbiol.***136:** 11–16.

Gest, H. and M.D. Kamen (1949). Photoproduction of molecular hydrogen by *Rhodospirillum rubrum. Science.* **109:** 558–559.

Gest, H., J.L. Flavinger and M.T. Madigan (1985). The exploitation of N_2-fixation capacity for the enrichment of anoxygenic photosynthetic bacteria in ecological studies. *FEMS Microbiol. Ecol.* **31:** 317-322.

Ghasem Najafpour, Ku Syahidah, Ku Ismail, Habibollah Younesi, Abdul Rahman Mohamed and Azlina Harun Kamarduiin (2004). Hydrogen as clean fuel via continuous fermentation by anaerobic photosynthetic bacteria, *Rhodospirillum rubrum. Afri. Journ. Biotechnol.* **3(10):** 503-507.

Ghirardi, L,M., A. Dubini, J. Yu and Pin-Ching Maness (2009). Photobiological Hydrogen-Producing Systems. *Chemical Society Reviews.* **38(1):** 52-61.

Ghirardi, M. L., R.K.,Togasaki and M. Seibert (1997). Oxygen sensitivity of algal H_2 production. *Appl Biochem. Biotechnol.* **63–65:** 141–151.

Ghirardi, M.L., P.W. King, M.C. Posewitz, P. Ching Maness, A. Federov, K. Kim, J. Cohen, K. Schulten and M. Seibert (2005). Approaches to developing biological H(2)-photoproducing organisms and processes. *Biochem. Soc. Trans.* **33:** 70-72.

Gibbons, N.E. and R.G.E. Murray (1978). Proposals Concerning the Higher Taxa of Bacteria. *International Journal of Systematic Bacteriology.* **28:** 1-6.

Gibson, J (1980). Phylogenetic analysis of photosynthetic bacteria based on comparison of 16S ribosomal RNA catalogs. In: *The Origins ofLife and Evolution*, (Halvorson and van Holde, eds.), Alan R. Liss, New York. pp. 97–102.

Gibson, J (2015). Chloroherpeton. Bergey's Manual of Systematics of Archaea and Bacteria, 1–4. doi:10.1002/9781118960608.gbm00375

Gibson, J., N. Pfennig and J.B. Waterbury (1984). *Chloroherpeton thalassium* gen. nov. et spec. nov., a non-filamentous, flexing and gliding green sulfur bacterium. *Archives of Microbiology.* **138:** 96–101.

Gibson, J., N. Pfennig and J.B. Waterbury (1985b). In Validationof the publication of new names and new combinations previously effectively published outside the IJSB. List No. 17. *Int. J. Syst. Bacteriol.* **35**: 223–225.

Gibson, J., W. Ludwig, E. Stackebrandt and C.R. Woese (1985). The phylogeny of the green photosynthetic bacteria: the absence of a close relationship between *Chlorobium* and *Chloroflexus. Syst. Appl. Microbiol.* **6:** 152–156.

Gich, F., J. Garcia-Gil and J. Overmann (2001). Previously unknown and phylogenetically diverse members of the green nonsulfur bacteria are indigenous to freshwater lakes *Arch. Mikrobiol.* **177:** 1–10.

Gich, F., R.L. Airs, M. Danielsen, B.J. Keely, C.A. Abella, J. Garcia-Gil, M. Miller and C.M. Borrego (2003). Characterization of the chlorosome antenna of the filamentous anoxygenic phototrophic bacterium *Chloronema* sp. Strain UdG9001. *Arch. Microbiol.***180:** 417-426.

Giesberger, G (1947). Some observations on the culture, physiology and morphology of some brown-red *Rhodospirillum* species. *Antonie van Leeuwenhoek J. Microbiol. Serol.* **13:** 135–148.

Gillis, M. and J. Deley (1980). Intra- and intergeneric similarities of the ribosomal ribonucleic acid cistrons of *Acetobacter* and *Gluconobacter. Int. J. Syst. Bacteriol.* **30:** 7-27.

Gillis, M., J. Dejonghe, A. Smet, G. Onghenae and J. De Ley (1982). Intra- and intergeneric similarities of the ribosomal ribonucleic acid cistrons in the *Rhodospirillaceae*. Abstract A-16. IV. *Intern. Symp. Photosynthetic Procaryotes.* Bombannes-Bordeaux:

Giovannoni, S.J., N.P. Revsbech, D. Ward and R.W. Castenholz (1987). Obligately phototrophic *Chloroflexus*: primary production in anaerobic hot spring microbial mats. *Arch. Microbiol.* **147:** 80–87.

Girija, K.R., Ch. Sasikala, Ch.V. Ramana, C. Spröer, S. Takaichi, V. Thiel and J.F. Imhoff (2010). *Rhodobacter johrii* sp. nov., an endospore-producing cryptic species isolated from semi-arid tropical soils. *Int. J. Syst. Evol. Microbiol.* **60:** 2099–2107.

Glaeser, J. and J. Overmann (1999). Selective enrichment and characterization of *Roseospirillum parvum*, gen. nov. and sp. nov., a new purple non-sulfur bacterium with unusual light absorption properties. *Arch. Microbiol.* **171:** 405–416.

Glaeser, J. and J. Overmann (1999). Validation List no. 80. *Int. J. Syst. Evol. Microbiol.***51:** 793-794.

Gloyna, E. F (1971). Waste stabilization ponds. World Health Organization Monograph Series No. 60. World Health Organization. Geneva, Switzerland.

Gogotov, I.N (1985). *Proc. 16th FEBS Congr. A*, pp. 385-390.

Gogtova, G.I. and M.B. Vainstein (1981) Thiosulfate oxidation in the dark by the purple sulfur bacterium *Ectothiorhodospira shaposhnikovii. Mikrobiologiya* **55:** 960–963.

Goldberg, I., V. Nadler and A. Hochman (1987). "Mechanism of nitrogenase switch-off by oxygen". *Journal of Bacteriology.* **169:** 874-879.

Goldrick, B. (2003). Food-borne diseases. *American. J. Nursing.***103:** 105-106.

Gomashe, A.V., M.A. Pounikar and P.A. Gulhane (2014). Liquid whey: a potential substrate for single-cell protein production from *Bacillus subtilis* NCIM2010. *Int. J. Life Sci.* **2(2):** 119-123.

Gomelsky, M. and S. Kaplan (1996). The *Rhodobacter sphaeroides* 2.4.1 *rho* gene: expression and genetic analysis of structure and function. *J. Bacteriol.* **178:**1946-1954.

Gonzalez, J.M., J.S. Covert, W.B. Whitman, J.R. Henriksen, F. Mayer, B. Scharf, R. Schmitt, A. Buchan, J.A. Fuhrman, R.P. Kiene and M.A. Moran (2003). *Silicibacter pomeroyi* sp. nov. and *Roseovarius nubinhibens* sp. nov., dimethyl sulfoniopropionate-demethylating bacteria from marine environments. *Int. J. Syst. Evol. Microbiol.* **53:** 1261–1269.

Gordon, J (2002). Tailoring optical systems to optimized photobioreactors. *Int. J. Hydr. Ener.* **27:** 1175-1184.

Gorlenko, V. M (2015). Ancalochloris. Bergey's Manual of Systematics of Archaea and Bacteria, 1–3. doi:10.1002/9781118960608.gbm00373.

Gorlenko, V. M., M. B. Vainstein, and V. I. Kachalkin (1978). Microbiological characteristic of Lake Mogilnoye. *Arch. Hydrobiol.***81:** 475.

Gorlenko, V.M (1970). A new phototrophic green sulfur bacterium. *Prosthecochloris aestuarii* nov. gen. nov. spec. *Z Allg Mikrobiol.* **10:** 147–149.

Gorlenko, V.M (1972). A new species of phototropic brown sulfur bacteria *Pelodictyon phaeum* nov. spec. *Mikrobiologiya.* **41:** 370–371.

Gorlenko, V.M (1988). Ecological niches of green sulfur and gliding bacteria. In: *Green photosynthetic bacteria*, (Olson, J.M., J.G. Ormerod, J. Amesz, E. Stackebrandt and H.G. Truper, eds.), Plenum, New York, pp. 257–267.

Gorlenko, V.M (2015). Prosthecochloris. Bergey's Manual of Systematics of Archaea and Bacteria, 1–6. doi:10.1002/9781118960608.gbm00377.

Gorlenko, V.M. and B.K. Pierson (2015). *Chloronema. Bergey's Manual of Systematics of Archaea and Bacteria,* 1–4. doi:10.1002/9781118960608.gbm00381.

Gorlenko, V.M. and E.V. Lebedeva (1971). New green bacteria with out-growths. *Mikrobiologiya.* **40:** 1035–1039.

Gorlenko, V.M. and J.F. Imhoff (2015). *Lamprobacter. Bergey's Manual of Systematics of Archaea and Bacteria,* 1–4. doi:10.1002/9781118960608.gbm01108.

Gorlenko, V.M. and S.A. Korotkov (1979). Morphological and physiological features of the new filamentous gliding green bacteria *Oscillochloris trichoides* nov. comb. *Izv. Akad. Nauk. SSSR. Ser. Biol.***6:** 848-857 (in Russian).

Gorlenko, V.M. and S.A. Korotkov (1989). *Oscillochloris trichoides* comb. nov. In: Validation of the Publication of New Names and new Combinations Previously Effectively Published Outside theIJSB, List no. 31. *Int. J. Syst. Bacteriol.* **39:** 495-497.

Gorlenko, V.M. and T.A. Pivovarova (1977). On the belonging of blue-green alga *Oscillaloria coerulescens* Gicklhorn, 1921 to a new genus of chlorobacteria *Oscillochloris* nov. gen. *Izv. Akad. Nauk. SSSR. Ser. Biol.* **3:** 396-409 (in Russian).

Gorlenko, V.M. and T.A. Pivovarova (1989). *Oscillochloris* gen. nov. and *Oscillochloris chrysea* comb. nov. In Validation of the Publication of New Names and New Combinations Previously Effectively Published Outside the IJSB, List no. 31. *Int. J. Syst. Bacteriol.***39:** 495-497.

Gorlenko, V.M., E.N. Chebotarev and V.I. Kachalkin (1974). Participation of microorganisms in sulfur turnover in Pomiaretzkoe Lake. *Mikrobiologiya.* **43:** 908–914.

Gorlenko, V.M., E.N. Krasilnikova, O.G. Kikina and N.Y. Tatarinova (1988). Validation of the publication of new names and new combinations previously effectively published outside the IJSB, *Int. J. Syst. Bacteriol.***38:** 220–222.

Gorlenko, V.M., E.N. Krasilnikova, O.G. Kikina and N.Yu Tatarinova (1979). The new motile purple sulfur bacteria *Lamprobacter modestohalophilus* nov. gen., nov. sp. with gas vacuoles. *Izv. Akad. Nauk SSSR. Ser. Biol.***5:** 755–767.

Gorlenko, V.M., G.A. Dubinina and S.I. Kusnetsov (1983). The ecology of aquatic microorganisms [monograph]. In: W. Ohle (Ed.) Binnengewässer. Schweizerbartsche Verlagsbuchhandlung. Stuttgart, Germany. 254.

Gorlenko, V.M., I.A. Bryantseva, E.E. Panteleeva, T.P. Tourova, T.V. Kolganova, Z.K. Makhneva and A.A. Moskalenko (2004). *Ectothiorhodosinus mongolicum* gen. nov., sp. nov., a new purple bacterium from a soda lake in Mongolia. *Microbiology.* **73:** 66–73.

Gorlenko, V.M., I.A. Bryantseva, E.E. Panteleeva, T.P. Tourova, T.V. Kolganova, Z.K. Makhneva and A.A. Moskalenko (2007). Validation List no. 116. *Int. J. Syst. Evol. Microbiol.***57:** 1371-1373.

Gorlenko, V.M., I.A. Bryantseva, O.N. Lunina and T.P. Tourova (2014). The phylogenetic position of the purple sulfur bacterium *Lamprobacter modestohalophilus* determined based on the data on new strains of the species. *Microbiology.* **83(6):** 829-837.

Gorlenko, V.M., I.A. Bryantseva, S. Rabold, T.P. Tourova, D. Rubtsova, E. Smirnova, V. Thiel and J.F. Imhoff (2009). *Ectothiorhodospira variabilis* sp. nov., an alkaliphilic and halophilic purple sulfur bacterium from soda lakes. *Int. J. Syst. Evol. Microbiol.***59:** 658-664.

Gothberg, A., M. M. Greger and B.E. Bengtsson (2002). Accumulation of heavy metals in water spinach (*Ipomoea aquatica*) cultivated in the Bangkok region, Thailand. *Environ Toxicol Chem.* **21(9):** 1934-9.

Gottschalk, G (1986). Bacterial Metabolism, 2nd edn., Springer, New York.

Gregersen, L.H., D.A. Bryant and N.U. Frigaard (2011). Mechanisms and evolution of oxidative sulfur metabolism in green sulfur bacteria. *Front. Microbiol.* **2:** 116. doi: 10.3389/fmicb.2011.00116.

Grein, F., A.R. Ramos, S.S. Venceslau and I.A.C. Pereira 2013). Unifying concepts in anaerobic respiration: Insights from dissimilatory sulfur metabolism. *Biochem. Biophys. Acta. Bioenerget.* **1827:** 145–160.

Greub, G. and D. Raoult (2003). *Rhodobacter massiliensis* sp. nov., a new amoebae-resistant species isolated from the nose of a patient. *Res. Microbiol.***154:** 631-635.

Griesbeck, C., M. Schutz, T. Schodl, S. Bathe, L. Nausch, N. Mederer, M. Vielreicher and G. Hauska (2002). Mechanism of sulfide-quinone oxidoreductase investigated using site-directed mutagenesis and sulfur analysis. *Biochemistry.* **41:** 11552–11565.

Gross, J (1991). Pigments in vegetables: Chlorophyll and Carotenoids. New York: van Nostrand Reinhold.

Grousset, F.E., J.M. Jouanneau, P. Castaing, G. Lavaux and C. Latouche (1999). A 70-YearRecord of Contamination from Industrial Activity along the Garonne River and Its Tributaries. *Estuarine Coast. Shelf Sci.* **48:** 401-414.

Gruenwald, A. and P. Grau (1980*). Czech. Pat.* 184,561.

Guan, Y.F., W. Zhang, X.J. Yu and M.C. Deng (2003). Two-stage photo hydrogen production using immobilized marine green alga *Platymonas Subcordiformis*. Marine Biotechnology: Basics and applications, Matalascaas, Spain, February 25–March 1.

Guest, J. R (1974). *J. Gen. Microbiol.* **81:** 513-515.

Gulati, A., P. Vyas, P. Rahi and R.C. Kasana (2009). Plant growth-promoting and rhizosphere competent Acinetobacter rhizosphere strain BIHB 723 from the cold desert of Himalayas. *Curr. Microbiol.***58:** 371-377.

Gupta, K.K., Kamal Rai Aneja and Deepanshu Rana.(2016). Current Status of Cow Dung as a Bioresource For Sustainable Development.*Bioresour. Bioprocess.* **3:**28.

Gupta, R.S (2012). Origin and spread of photosynthesis based upon conserved sequence features in key bacteriochlorophyll biosynthesis proteins. *Mol. Biol. Evol.* doi:10.1093/ molbev/ss145.

Gupta, R.S., P. Chander and S. George (2013). Phylogenetic framework and molecular signatures for the class *Chloroflexi* and its different clades; proposal for division of the class *Chloroflexi* class. nov. into the suborder *Chloroflexineae* subord. nov., consisting of the emended family *Oscillochloridaceae* and the family *Chloroflexaceae* fam. nov., and the suborder *Roseiflexineae*subord. nov., containing the family *Roseiflexaceae* fam. nov. *Antonie van Leeuwenhoek.* **103:** 99–119.

Gupta, R.S., P. Chander and S. George (2013). Validation List no. 151. *Int. J. Syst. Evol. Microbiol.***63:** 1577-1580.

Gupta, R.S., T. Mukhtar and B. Singh (1999). Evolutionary relationships among photosynthetic prokaryotes (*Heliobacterium chlorum*, *Chloroflexus aurantiacus*, cyanobacteria, *Chlorobium tepidum* and *Proteobacteria*): implications regarding the origin of photosynthesis. *Mol. Microbiol.* **32:** 893–906.

Guyoneaud, R., J. Suling, R. Petri, R.C. Matheron, P. Caumette, N. Pfennig and J.F. Imhoff (1998). Taxonomic rearrangements of the genera *Thiocapsa* and *Amoebobacter* on the basis of 16S rDNA sequence analyses and description of *Thiolamprovum* gen. nov. *Int. J. Syst. Bacteriol.* **48:** 957–964.

Guyoneaud, R., P. Caumette and J.F. Imhoff (2015). *Thiorhodococcus. Bergey's Manual of Systematics of Archaea and Bacteria,* 1–7. doi:10.1002/9781118960608.gbm01124.

Guyoneaud, R., R. Matheron, R. Baulaigue, K. Podeur, A. Hirschler and P. Caumette (1996). Anoxygenic phototrophic bacteria in eutrophic coastal lagoons of the French

Mediterranean and Atlantic Coast (Prévost Lagoon, Arcachon Bay, Certes Fishponds). *Hydrobiologia*. **329:** 33–43.

Guyoneaud, R., R. Matheron, W. Liesack, J.F. Imhoff, P. Caumette (1997). *Thiorhodococcus minus*, gen. nov., sp. nov., A new purple sulfur bacterium isolated from coastal lagoon sediments. *Arch Microbiol*. **168:** 16–23.

Guyoneaud, R., S. Mouné, C. Eatock, V. Bothorel, A. Hirschler-Réa, J. Willison, R. Duran W. Liesack, R. Herbert, R. Matheron and P. Caumette (2002). Characterization of three spiral-shaped purple non-sulfur bacteria isolated from coastal lagoon sediments, saline sulfur springs and microbial mats: emended description of the genus *Roseospira* and description of *Roseospira marina* sp. nov., *Roseospira navarrensis* sp. nov., and *Roseospira thiosulfatophila* sp. nov. *Arch. Microbiol*. **178:** 315–324.

Habte, M. and M. Alexander (1980). Nitrogen fixation by photosynthetic bacteria in lowland rice culture. *Appl. Environ. Mcrobiol.***39:** 342-347.

Haffen, L.N., B.K. Pierson, G.W. Francis and L.N. Halfen (1972). Carotenoids of a gliding organism containing bacteriochlorophylls. *Arch. Microbiol*. **82:** 240–246.

Haider, M. and M. EL-Hassy (2000). Biodegradation of orange and pineapple peels for production of single-cell protein. *Garyounis J. Sci.***1:** 7-23.

Hakobyan, L., L. Gabrielyan and A. Trchounian (2012). Bio-hydrogen production and the F0F1-ATPase activity of *Rhodobacter sphaeroides*: effects of various heavy metal ions. *Int. J. Hydr. Ener.***37:** 17794-800.

Hakobyan, L., L. Gabrielyan and A.Trchounian (2012). Ni (II) and Mg (II) ions as factors enhancing biohydrogen production by *Rhodobacter sphaeroides* from mineral springs. *Int J. Hydr. Ener*. **37:** 7482-6.

Hale, M. B., R.E. Blankenship and R.C. Fuller (1983).Menaquinone is the sole quinone in the facultatively aerobic green photosynthetic bacterium *Chloroflexus aurantiacus.Biochim. Biophys. Acta*. **723(3):** 376-382.

Hallenbeck, P.C (2005). Fundamentals of the fermentative production of hydrogen. *WaterSci. Technol*. **52(1–2):** 21–29.

Hallenbeck, P.C. and J.R. Benemann (2002). Biological hydrogen production; fundamentals and limiting processes. *Int. J. Hydr. Ener*. **27:** 1185 -1193.

Hallenbeck, P.C., D. Ghosh, M.T. Skonieczny, V. Yargeau (2009). Microbiological and engineering aspects of biohydrogen production. *Indian J. Microbiol.***49(1):** 48-59.

Hamzah, A., L. Tan and S. Vikineswary (2006). Biological Characterization of *Rhodomicrobium vannielii* Isolated from a Hot Spring at Gadek, Malacca, Malaysia. *Malaysian Journal of Microbiology*. 2. 10.21161/mjm.210603.

Hanada S (2014). The Phylum *Chloroflexi*, the Family *Chloroflexaceae* and the Related Phototrophic Families *Oscillochloridaceae* and *Roseiflexaceae*. In: *The Prokaryotes*, (Rosenberg, E., E.F. DeLong, S. Lory, E. Stackebrandt and F. Thompson, eds.). Springer, Berlin, Heidelberg, pp. 515-532.

Hanada, S. and B.K. Pierson (2006). The Family *Chloroflexaceae*. In: *The Prokaryotes: A Handbook on the Biology of Bacteria*, (Dworkin, M., S. Falkow, E. Rosenberg, K.H. Schleifer and E. Stackebrandt, eds.), Third Edition, Vol. 7. Springer, New York, pp. 815–842.

Hanada, S., A. Hiraishi, K. Shimada and K. Matsuura (1995a). *Chloroflexus aggregans* sp. nov., a filamentous phototrophic bacterium that forms dense cell aggregates by active gliding movement. *Int. J. Syst. Evol. Microbiol.***45:** 676–681.

Hanada, S., A. Hiraishi, K. Shimada and K. Matsuura (1995b). Isolation of *Chloroflexus* sp. and related thermophilic photosynthetic bacteria from hot springs using an improved isolation procedure. *J. Gen. Appl. Microbiol.***41:** 119–130.

Hanada, S., S. Takaichi, K. Matsuura and K. Nakamura (2002). *Roseiflexus castenholzii* gen. nov., sp. nov., a thermophilic, filamentous, photosynthetic bacterium that lackschlorosomes. *Int. J. Syst. Evol. Microbiol.***52**: 187-193.

Hankamer, B., L. Florian, R. Jens, M. Jan, P. Clemens and K. Olaf (2007). Photosynthetic biomass and H_2 production by green algae: From bioengineering to bioreactor scale-up. *Physiologia plantarum*. **131:** 10-21.

Hansen, T.A. and H. Veldkamp (1973). *Rhodopseudomonas sulfidophila*, nov. spec., a new species of the purple non-sulfur bacteria. *Arch. Mikrobiol*. **92:** 45–58.

Hansen, T.A. and J.F. Imhoff (1985). *Rhodobacter veldkampii*, a new species of phototrophic purple non-sulfur bacteria. *Int. J. Syst. Bacteriol*. **35:**115–116.

Harada, N., M. Nishiyama , S. Otsuka and S. Matsumoto (2005). Effects of Inoculation of Phototrophic Purple Bacteria on Grain Yield of Rice and Nitrogenase Activity of Paddy Soil in a Pot Experiment. *soil science & Plant Nutrition.***51(3):** 361-367.

Hartner, T., K.L. Straub and E. Kannenberg (2005). The occurrence of hopanoid lipids in anaerobic *Geobacter* species. *FEMS Microbiol. Lett.* **243:** 59–64.

Herridge, D.F., M.B. Peoples and R.M. Boddey (2008). Global inputs of biological nitrogen fixation in agricultural systems. *Plant Soil.***311:** 1–18.

Hoffmann, M.C., A. Muller, M. Fehringer, Y. Pfander, F. Narberhaus and B. Masepohl (2014).Coordinated expression of fdxD and molybdenum nitrogenase genes promotes nitrogen fixation by *Rhodobacter capsulatus* in the presence of oxygen. *J. Bacteriol.***196:** 633–640.

Haruhiko, T (Cosmo Oil Co., Ltd., Japan) (1992). Jpn. Kokai Tokkyo Koho JP 92/356,193.

Harwood, C. and J. Gibson (1988). Anaerobic and aerobic metabolism of diverse aromatic compounds by the photosynthetic bacterium *Rhodopseudomonas palustris*. *Appl. Environ. Microbiol.***54:** 712–7.

Hasebe, K. and M. Mitsumasu (1986). Jpn. Kokai Tokkyo Koho JP 611148,288.

Haskins, E.F. and T. Kihara (1967). The use of spectrophotometry in the ecological investigation of the facultatively anaerobic purple photosynthetic bacteria. *Canadian J. Microbiol*. **13:** 1283–1293.

Hatzikakidis, A.D (1953). Epochiakai ydrologikai ereynai eistas limnothalassas Mesologgiou kai Aitolikou. Anatypon ek ton praktikon tou Ellenikou Ydrobiologikou Institoutou Akademias Athenon. **6:** 85–143.

Haverkamp, T. and J.D. Schwenn (1999). Structure and function of a cysBJIH gene cluster in the purple sulfur bacterium *Thiocapsa roseopersicina*. *Microbiology*. **145:**115–125.

He, J., G. Zhang and H.F. Lu (2010). Treatment of soybean wastewater by a wild strain *Rhodobacter sphaeroides* and to produce protein under natural conditions. *Environ. Sci. Eng. China*. **4(3):** 334-339.

He, S., L. Zhong, J. Duan, Y. Feng, B. Yang and L.Yang (2017). Bioremediation of Wastewater by Iron Oxide-Biochar Nanocomposites Loaded with Photosynthetic Bacteria. *Frontiers In Microbiology*. **8:** 823. doi:10.3389/fmicb.2017.00823.

Head, I.M., W.D. Hiorns, T.M. Embley, A.J. McCarthy and J.R. Saunders (1993). The phylogeny of autotrophic ammonia-oxidizing bacteria as determined by analysis of 16S ribosomal RNA gene sequences. *J. Gen. Microbiol.***139:** 1147–1153.

Heising, S., L. Richter, W. Ludwig and B. Schink (1999). *Chlorobium ferrooxidans* sp. nov., a phototrophic green sulfur bacterium that oxidizes ferrous iron in coculture with a ''*Geospirillum*'' sp. strain. *Arch. Microbiol*. **172:** 116–124.

Hensel, G. and H.G. Truper (1976). Cysteine and S-sulfocysteine biosynthesis in phototrophic bacteria. *Arch. Microbiol*. **109:** 101–103.

Hensen, D., D. Sperling, H.G. Truper, D.C. Brune and C. Dahl (2006). Thiosulphate oxidation in the phototrophic sulphur bacterium *Allochromatium vinosum. Mol. Microbiol.* **62:** 794–810.

Henshaw, P.F. and W. Zhu (2001). Biological conversion of hydrogen sulphide to elemental sulphur in a fixed-film continuous flow photo-reactor. *Water Res.***35(15):** 3605–3610.

Hest, H., H.D. Kamen and H.M. Bregoff (1950). Studies on the metabolism of photosynthetic bacteria: V. photoproduction of hydrogen and nitrogen fixation by *Rhodospirillumrubrum. J. Biol. Chem.* **182:** 153-170.

Hiraishi, A (1994). Phylogenetic affiliations of *Rhodoferax fermentans* and related species of phototrophic bacteria as determined by automated 16S DNA sequencing. *Curr.Microbiol.* **28:** 25–28.

Hiraishi, A (1997). Transfer of bacteriochlorophyll *b*-containing phototrophic bacteria *Rhodopseudomonas viridis* and *Rhodopseudomonas sulfoviridis* to the genus *Blastochloris* gen. nov. *Int. J. Syst. Bacteriol.* **47:** 217–219.

Hiraishi, A (2017). Characterization of thermotolerant phototrophic bacteria, *Rhodoplanes tepidicaeni* sp. nov. and *Rhodoplanes azumiensis* sp. nov., isolated from a geothermal spring. *Int. J. Syst. Evol. Microbiol.***67:** 5038-5045.

Hiraishi, A. and J.F. Imhoff (2015). *Rhodoferax. Bergey's Manual of Systematics of Archaea and Bacteria,* 1–11. doi:10.1002/9781118960608.gbm00951.

Hiraishi, A. and J.F. Imhoff (2015). *Rhodoplanes. Bergey's Manual of Systematics of Archaea and Bacteria,* 1–14. doi:10.1002/9781118960608.gbm00826.

Hiraishi, A. and K. Okamura (2017). Proposal of *Rhodoplanes tepidamans* sp. nov. toaccommodate the thermotolerant phototrophic bacterium previously referred to as *Rhodoplanes*(*Rhodopseudomonas*) *cryptolactis*. *Int. J. Syst. Evol. Microbiol.***67:** 1540-1545.

Hiraishi, A. and K. Okamura (2017). *Rhodopseudomonas telluris* sp. nov., a phototrophic alphaproteobacterium isolated from paddy soil. *Int. J. Syst. Evol. Microbiol.***67:**3369-3374 .

Hiraishi, A. and Y. Hoshino (1984). Distribution of rhodoquinone in *Rhodospirillaceae* 360 and its taxonomic implications. *J. Gen Appl Microbiol.* **30:** 435-448.

Hiraishi, A. and Y. Ueda (1994a). Intrageneric structure of the genus *Rhodobacter* transfer of *Rhodobacter sulfidophilus* and related marine species to the genus *Rhodovulum* gen. nov. *Int. J. Syst. Bacteriol.* **44:**15–23.

Hiraishi, A. and Y. Ueda (1994b). *Rhodoplanes* gen. nov., a new genus of phototrophic bacteria including *Rhodopseudomonas rosea* as *Rhodoplanes roseus* comb. nov. and *Rhodoplanes elegans* sp. nov. *Int. J. Syst. Bacteriol.* **44:** 665–673.

Hiraishi, A. and Y. Ueda (1995a). Isolation and characterization of *Rhodovulum strictum* sp. nov. and some other purple non-sulfur bacteria from colored blooms in tidal and seawater pools. *Int. J. Syst. Bacteriol.* **45:** 319–326.

Hiraishi, A., K. Matsuzawa, T. Kanbe and N. Wakao (2000). *Acidisphaera rubrifaciens* gen. nov., sp. nov., an aerobic bacteriochlorophyll-containing bacterium isolated from acidic environments. *Int. J. Syst. Evol. Microbiol.* **50:** 1539-1546.

Hiraishi, A., K. Muramatsu and Y. Ueda (1996). Molecular genetic analyses of *Rhodobacter azotoformans* sp. nov. and related species of phototrophic bacteria. *Syst Appl Microbiol* **19:**168–177.

Hiraishi, A., K. Urata and T. Satoh (1995b). A new genus of marine budding phototrophic bacteria, *Rhodobium* gen. nov., which includes *Rhodobium orientis* sp. nov. and *Rhodobium marinum* comb. nov. *Int. J. Syst. Bacteriol.* **45:** 226–234.

Hiraishi, A., T.S. Santos, J. Sugiyama and K. Komagata (1992b). *Rhodopseudomonas rutila is* a later subjective synonym of *Rhodopseudomonas palustris*. *Int. J. Syst. Bacteriol.***42:**186–188.

Hiraishi, A., Y. Hoshino and H. Kitamura (1984). Isoprenoid quinone composition in the classification of *Rhodospirillaceae*. *J. Gen. Appl. Microbiol.***30:** 197-210.

Hiraishi, A., Y. Hoshino and T. Satoh (1991). *Rhodoferax fermentans* gen. nov., sp. nov., a phototrophic purple non-sulfur bacterium previously referred to as the "*Rhodocyclus* gelatinosus-like" group. *Arch.Microbiol.***155:** 330–336.

Hiraishi, A., Y. Hoshino and T. Satoh (1992). Validation List no. 40. *Int. J. Syst. Bacteriol.***42:** 191-192.

Hirayama, O. and K. Ishigaki (1988). Effects of photosynthetic bacteria feeding on the components of hen's eggs. *Bull. Fac. Agric.,* Shimane Univ. **8:** 85-88.

Hirayama, O. and Y. Katsuta (1988). Stimulation of vitaminB_{12} formation in *Rhodospirillim rubrum* G-9 BM. *Agric. Biol. Chem.* **52(11):** 2949-2951.

Hiraishi, A (1989). The occurrence of menaquinone as the sole isoprenoid quinone in the photosynthetic bacterium *Heliobacterium chlorum*. *Arch. Microbiol.* **151:** 378-379.

Hiroo Takabatake, Kiyohiko Suzuki, In-Beom Ko and Tatsuya Noike (2004). Characteristics of anaerobic ammonia removal by a mixed culture of hydrogen-producing photosynthetic bacteria. *Bioresour. Technol.* **95:** 151-158.

Hirotani, H., H. Ohigashi, M. Kobayashi, K. Koshimizu and E. Takahashi (1991). *FEMSMicrobiol. Lett.* **77:** 13-18.

Hirotani, H., M. Kobayashi, E. Takahashi, H. Yamada and T. Kosaka, T (1987). *Soil Microorg.***30:** 49-53.

Hirotani, H., Y. Agui, M. Kobayashi and E. Takahashi (1989). *Int. Symp. Waste Manage.*Probl. Agro Ind., Istanbul-Turkey, pp. 69-73.

Hiroto, N (1981). *Jpn. Kokai Tokkyo Koho* 81/08,682.

Hirschler-Rea, A., R. Matheron, C. Riffaud, S. Moune, C. Eatock, R.A. Herbert, J.C. Willison and P. Caumette (2003). Isolation and characterization of spirilloid purple phototrophic bacteria forming red layers in microbial mats of Mediterranean salterns: Description of *Halorhodospira neutriphila* sp. nov. and emendation of the genus *Halorhodospira*. *Int. J. Syst. Evol. Microbiol.* **53:** 153–163.

Hocking, P.J. and R.H. Marchessault (1994) Biopolyesters. In: *Chemistry and technology of Biodegradable Polymers* (Griffin, G.J.L. eds), Blackie Academic. NewYork. pp. 48-96.

Hoffmann, C (1949). Uber die Durchlässigkeit dünner Sandschichten für Licht. *Planta.* **37:** 48–56.

Holkenbrink, C., S. OconBarbas, A. Mellerup, H. Otaki and N.U. Frigaard (2011). Sulfur globule oxidation in green sulfur bacteria is dependent on the dissimilatory sulfite reductase system. *Microbiology.***157:** 1229–1239.

Holo, H. and R. Sirevag (1986). Autotrophic growth and CO2 fixation of *Chloroflexus aurantiacus*. *Arch. Microbiol* .**159:** 257–264.

Holt, J.G. and R.A. Lewin (1968). *Herpetosiphon aurantiacus*. *J. Bacteriol.***95:** 2407–2408.

Holt, J.G., N.R. Krieg, P.H.A. Sneath, J.T. Staley and S.T. Williams (1994). *Bergey's manual of determinative bacteriology*. 9th. Baltimore; The Williams and Wilkins Co. ISBN 0-683-00603-7.787.

Honda, R., K. Fukushi and K. Yamamoto (2006). Optimization of wastewater feeding for single-cell protein production in an anaerobic wastewater treatment process utilizing purple non-sulfur bacteria in mixed culture conditions. *J. Biotechnology*. **125:** 565-573.

Hongyu Ma, Honghui Yang, Xiaojing Zheng, Thomas Lie, Wei Yan (2020). Promoting photo-fermentative hydrogen production performance by substituting the rnf promoter in

Rhodobacter capsulatus. *Int. J. Hydr. Ener.* ISSN 0360-3199, https://doi.org/10.1016/j.ijhydene.2020.10.270.

Hoogewerf G.J, D.O. Jung and T. Madigan (2003). Evidence for limited species diversity of bacteriochlorophyll *b*-containing purple non-sulfur anoxygenic phototrophs in freshwater habitats. *FEMS Microbiology Letters*. **218 (2):** 359-364.

Hougardy, A., B.J. Tindall and J.H. Klemme (2000). *Rhodopseudomonas rhenobacensis* sp. nov., a new nitrate-reducing purple non-sulfur bacterium. *Int. J. Syst. Evol. Microbiol.* **50:** 985–992.

Howard, G (1987). The world of microbes, The Benjamin Cummings Publishing, Inc.

Howells, E.R (1982). *Chem. Ind (London),* p. 508.

Hu, X., Thorsten Ritz, Ana Damjanovic, Felix Autenrieth and Klaus Schulten (2002). Photosynthetic apparatus of purple bacteria.*Quarterly Reviews of Biophysics.***35(1):** 1–62.

Hua, J., Feng, Y., Bai, J., Yin, Y., and Lin, X (2014) Co-inoculation with AM fungus *Glomus caledonium* and the photoheterotrophic purple non-sulfur bacterium*Rhodopseudomonas palustris* results in mutual inhibition and lower arsenic accumulation of *Nicotiana tabacum* L. in an arsenic content. *Fresenius Environ. Bull.***23:** 867–874.

Huang, F., L. Ge, B. Zhang, Y. Wang, H. Tian, L.P. Zhao (2014). A fullerene colloidal suspension stimulates the growth and denitrification ability of wastewater treatment sludge-derived bacteria. Chemosphere. **108:** 411–417.

Hudson, B. S., B. E. Kohler and K. Schulten (1982). Linear polyene electronic structure and potential surfaces, p. 22–95. In E. C. Lim (ed.), Excited states, vol. 6. Academic Press, New York, N.Y.

Hugenholtz, P. and E. Stackebrandt (2004). Reclassification of *Sphaerobacter thermophilus* from the subclass *Sphaerobacteridae* in the phylum *Actinobacteria* to the class *Thermomicrobia* (emended description) in the phylum *Chloroflexi* (emended description). *Int. J. Syst. Evol. Microbiol.* **54:** 2049–2051.

Hugler, M., H. Huber, S.J. Molyneaux, C. Vetriani and S.M. Sievert (2007) Autotrophic CO_2 fixation via the reductive tricarboxylic acid cycle in different lineages within the phylum *Aquificae*: evidence for two ways of citrate cleavage. *Environ. Microbiol.* **9:** 81–92.

Hulsen, T., D.J. Batstone and J. Keller (2014). Phototrophic bacteria for nutrient recovery from domestic wastewater. *Water Research*. **50:** 18-26.

Hurse, T., U. Kappler and J. Keller (2008). Using Anoxygenic Photosynthetic Bacteria for the Removal of Sulfide from Wastewater. *In: Sulfur metabolism in phototrophic organisms. Springer*, Dordrecht. pp. 437-460.

Hyun, M. S., S. Okuda and K. Izaki (1989). Purification and characterization of a neutral serine protease from non-sulfur purple photosynthetic bacterium. *Curr. Microbiol.* **18:** 379.

Ibrahim, S., S. Vikineswary, Sujjat Al-Azad and L.L. Chong (2006). The effects of light Intensity, inoculum Size and cell immobilization on the treatment of sago effluent with *Rhodopseudomonas palustris* strain B1. *Biotechnol. Bioprocess Eng.***11:** 377-381.

Idi, A., M.H. Muhamadd Nor and M.F. Abdul Wahab (2015). Photosynthetic bacteria: an eco-friendly and cheap tool for bioremediation.*Rev. Environ. Sci. Biotechnol.* **14(2):** 271-285.

Imhoff J.F (2006d). The Family *Ectothiorhodospiraceae.* In: *The Prokaryotes*, (Dworkin, M., S. Falkow, E. Rosenberg, K.H. Schleifer, E. Stackebrandt, eds.) Springer, New York, NY. pp. 874-876.

Imhoff J.F (2014) The Family *Chlorobiaceae*. In: *The Prokaryotes,* (Rosenberg, E., E.F. DeLong, S. Lory, E. Stackebrandt and F. Thompson, eds.) *Springer,* Berlin, Heidelberg. pp. 501-514.

Imhoff, J. F (2015). Lamprocystis. Bergey's Manual of Systematics of Archaea and Bacteria, 1–7. doi:10.1002/9781118960608.gbm01109.

Imhoff, J. F (2015). Thiopedia. Bergey's Manual of Systematics of Archaea and Bacteria, 1–4. doi:10.1002/9781118960608.gbm01123.

Imhoff, J. F. and P. Caumette, P (2015). *Thiohalocapsa. Bergey's Manual of Systematics of Archaea and Bacteria,* 1–6. doi:10.1002/9781118960608.gbm01121.

Imhoff, J.F. and A. Hiraishi (2015). *Rhodobium. Bergey's Manual of Systematics of Archaea and Bacteria,* 1–6. doi:10.1002/9781118960608.gbm00849.

Imhoff, J.F (1982). Occurrence and evolutionary significance of two sulfate assimilation pathways in the *Rhodospirillaceae. Arch. Microbiol.* **132:** 197–203.

Imhoff, J.F (1983). *Rhodopseudomonas marina* sp. nov., a new marine phototropic purple bacterium. *Syst. Appl. Microbiol.* **4:** 512–521.

Imhoff, J.F (1984a). Quinones of phototropic purple bacteria. *FEMS Microbiol. Letts.* **25:** 85-89.

Imhoff, J.F (1984b). Reassignment of the genus *Ectothiorhodospira*. Pelsh 1936 to a new family, *Ectothiorhodospiraceae* fem. nov. and emended description of the *Chromatiaceae*. Bavendamm 1924. *Int. J. Syst. Bacteriol.* **134:** 338–339.

Imhoff, J.F (1988). Halophilic phototrophic bacteria. In: *Halophilic Bacteria*, (Rodriguez-Valera, F, ed.), CRC Press. Boca Raton, FL. pp. 85-108.

Imhoff, J.F (1988). Lipids, fatty acids and quinones in taxonomy and phylogeny of anoxygenic phototrophic bacteria. In: *Green Photosynthetic Bacteria*, (Olson, J.M., J. G. Ormerod, J. Amesz, E. Stackebrandt and H.G. Truper), New York: Plenum. pp. 223–232.

Imhoff, J.F (1989a). Family *Ectothiorhodospiraceae*. In: *Bergeys Manual of Systematic Bacteriology*, (Staley, Bryant, Pfennig and Holt, eds.), 1st Ed., Vol. 3, The Williams &Wilkins Co., Baltimore. pp. 1654–1658.

Imhoff, J.F (1989b). In: *Bergey's Manual of Systematic Bacteriology*, (J.T. Staley, M.P. Bryant, N. Pfennig and J.G. Holt, eds.), Williams and Wilkins, Baltimore, MD. **3:** 1668-1672.

Imhoff, J.F (1992). Taxonomy, phylogeny and general ecology of anoxygenic phototrophic bacteria. In: *Biotechnology Handbook Photosynthetic Prokaryotes,* (Can, N.G. and N. H. Mann, eds.). *Plenum Press. London,* UK. pp. 53–92.

Imhoff, J.F (1995). Taxonomy and physiology of phototrophic purple bacteria and green sulfur bacteria. In: *Anoxygenic photosynthetic bacteria* (Blankenship, R.E., M.T. Madigan and C.E. Bauer, eds.). Advances in photosynthesis and respiration, Vol 2. Springer, Dordrecht. pp. 1-15.

Imhoff, J.F (1999). A phylogenetically oriented taxonomy of anoxygenic phototrophic bacteria. In: *The Phototrophic Prokaryotes*, (Pescheck, G.A., T.W. Loffelhard and G. Schmetterer, eds.), New York: Plenum. pp. 763–774.

Imhoff, J.F (2001). Transfer of Rhodopseudomonas acidophila to the new genus Rhodoblastus as Rhodoblastus acidophilus comb. nov. Int. J. Syst. Evol. Microbiol.**51:**1863–1866.

Imhoff, J.F (2001a). True marine and halophilic anoxygenic phototrophic bacteria. *Arch Microbiol.* **176:** 243–254.

Imhoff, J.F (2001b). Transfer of *Pfennigia purpurea* Tindall 1999 (*Amoebobacter purpureus* Eichler and Pfennig 1988) to the genus *Lamprocystis* as *Lamprocystis purpurea. Int. J. Syst. Evol. Microbiol.* **51:**1699–1701.

Imhoff, J.F (2003). Phylogenetic taxonomy of the family *Chlorobiaceae* on the basis of 16S rRNA and *fmo* (Fenna-Matthews-Olson protein) gene sequences. *Int. J. Syst. Evol. Microbiol.***53:** 941–951.

Imhoff, J.F (2005). Genus 1. *Rhodospirillum* Molisch 1907, 24[AL] emend. Imhoff, Petri and Suiling 1998, 796. In: *Bergey's Manual of Systematic Bacteriology*, (Brenner, D.J., N.R. Krieg, J.T. Staley and G.M. Garrity, eds.), Second edition, Vol. 2, Part C, Springer, New York, pp. 1-6.

Imhoff, J.F (2005). Genus *Rhodopseudomonas.* In: *Bergey's Manual of Systematic Bacteriology*, (Brenner, D.J., N.R. Krieg, J.T. Staley and G.M. Garrity, eds.) Second edition, Vol. 2, New York: Springer. pp.473–476.

Imhoff, J.F (2005). Order I. *Chromatiales* ord. nov. In: *Bergey's Manual of Systematic Bacteriology*, (Brenner, D.J., N.R. Krieg, J.T. Staley and G.M. Garrity, eds.), Second edition, Vol. 2 (The *Proteobacteria*), Part B (The *Gammaproteobacteria*), Springer, New York, pp. 1-3.

Imhoff, J.F (2005a) Family I. *Chromatiaceae* Bavendamm 1924, 125AL emend. Imhoff 1984b, 339. In: *Bergey's Manual of Systematic Bacteriology*, (Brenner, D.J., N.R. Krieg, J.T. Staley and G.M. Garrity, eds), Vol. 2, Part B, Springer, New York. pp. 3–40.

Imhoff, J.F (2005b). Family II. *Ectothiorhodospiraceae* Imhoff 1984b, 339VP. In: Bergey's Manual of Systematic Bacteriology (Brenner, D.J., N.R. Krieg, J.T. Staley and G.M. Garrity, eds.), Vol. 2, Part B, Springer, New York. pp. 41–57.

Imhoff, J.F (2005c). Genus *Rhodobacter.* Imhoff, Truper and Pfennig 1984, 342VP. In: *Bergey's Manual of Systematic Bacteriology*, (Brenner, D.J., N.R. Krieg, J.T. Staley and G.M. Garrity, eds.). Second Edition, Vol. 2, Part C, Springer, New York. pp. 161–167.

Imhoff, J.F (2006a). The phototrophic *Alphaproteobacteria*. In: *The Prokaryotes.* A handbook on the biology of bacteria, (Dworkin, M., S. Falkow, E. Rosenberg, K.H. Schleifer, E. Stackebrandt, eds.). Third Edition, Vol 5, Springer, New York. pp. 41–64.

Imhoff, J.F (2006b). The phototrophic *Betaproteobacteria*. In: *The Prokaryotes*. A handbook on the biology of bacteria, (Dworkin, M., S. Falkow, E. Rosenberg, K.H. Schleifer and E. Stackebrandt, eds.) Third Edition, Vol 5, Springer, New York, pp. 593–601.

Imhoff, J.F (2006c). The *Chromatiaceae*. In: *The Prokaryotes*. A handbook on the biology of bacteria, (Dworkin, M., S. Falkow, E. Rosenberg, K.H. Schleifer, E. Stackebrandt, eds.) Third Edition, Vol 6, Springer, New York. pp. 846–873.

Imhoff, J.F (2006d). The family *Ectothiorhodospiraceae*. In: *The Prokaryotes*. A handbook on the biology of bacteria, (Dworkin, M., S. Falkow, E. Rosenberg, K.H. Schleifer, E. Stackebrandt, eds.) Third Edition, Vol 6, Springer, New York. pp. 874–886.

Imhoff, J.F (2008) Systematics of anoxygenic phototrophic bacteria. In: *Sulfur Metabolism in Phototrophic Organisms* (R. Hell, C. Dahl, D.B. Knaffand T. Leustek, eds), Advances in Photosynthesis and Respiration, Springer, New York.**27:**269–287.

Imhoff, J.F (2014). Biology of Green Sulfur Bacteria. *eLS*. doi:10.1002/9780470015902.a0000458.

Imhoff, J.F (2014). The family *Chromatiaceae.* The Prokaryotes, *pp.* 151–178. doi:10.1007/978-3-642-38922-1_295

Imhoff, J.F (2015). Allochromatium. Bergey's Manual of Systematics of Archaea and Bacteria, 1–7.doi:10.1002/9781118960608.gbm01104.

Imhoff, J.F (2015). Blastochloris. Bergey's Manual of Systematics of Archaea and Bacteria, 1–11. doi:10.1002/9781118960608.gbm00815.

Imhoff, J.F (2015). Chromatiaceae. Bergey's Manual of Systematics of Archaea and Bacteria, 1–12. doi:10.1002/9781118960608.fbm00219.

Imhoff, J.F (2015). Chromatiales ord. nov. Bergey's Manual of Systematics of Archaea and Bacteria, 1-7. doi:10.1002 / 9781118960608.obm00096 .

Imhoff, J.F (2015). Chromatium. Bergey's Manual of Systematics of Archaea and Bacteria, 1–8. doi:10.1002/9781118960608.gbm01105.

Imhoff, J.F (2015). Ectothiorhodospira. Bergey's Manual of Systematics of Archaea and Bacteria, 1–12. doi:10.1002/9781118960608.gbm01128.

Imhoff, J.F (2015). Ectothiorhodospiraceae. Bergey's Manual of Systematics of Archaea and Bacteria, 1–5. doi:10.1002/9781118960608.fbm00220.

Imhoff, J.F (2015). Halorhodospira. Bergey's Manual of Systematics of Archaea and Bacteria, 1–9. doi:10.1002/9781118960608.gbm01129.

Imhoff, J.F (2015). *Isochromatium.* Bergey's Manual of Systematics of Archaea and Bacteria, 1–6. doi:10.1002/9781118960608.gbm01107.

Imhoff, J.F (2015). Marichromatium. Bergey's Manual of Systematics of Archaea and Bacteria, 1–6. doi:10.1002/9781118960608.gbm01110.

Imhoff, J.F (2015). Phaeospirillum. Bergey's Manual of Systematics of Archaea and Bacteria, 1–7. doi:10.1002/9781118960608.gbm00894.

Imhoff, J.F (2015). Rhabdochromatium. Bergey's Manual of Systematics of Archaea and Bacteria, 1–6. doi:10.1002/9781118960608.gbm01113.

Imhoff, J.F (2015). Rhodobaca. Bergey's Manual of Systematics of Archaea and Bacteria, 1–4.doi:10.1002/9781118960608.gbm00861.

Imhoff, J.F (2015). Rhodobacter. Bergey's Manual of Systematics of Archaea and Bacteria, 1–12. doi:10.1002/9781118960608.gbm00862.

Imhoff, J.F (2015). Rhodocista. Bergey's Manual of Systematics of Archaea and Bacteria, 1–5. doi:10.1002/9781118960608.gbm00895.

Imhoff, J.F (2015). Rhodocyclus. Bergey's Manual of Systematics of Archaea and Bacteria, 1–8. doi:10.1002/9781118960608.gbm01003.

Imhoff, J.F (2015). Rhodomicrobium. Bergey's Manual of Systematics of Archaea and Bacteria, 1–12. doi:10.1002/9781118960608.gbm00825.

Imhoff, J.F (2015). Rhodopseudomonas. Bergey's Manual of Systematics of Archaea and Bacteria, 1–13. doi:10.1002/9781118960608.gbm00806.

Imhoff, J.F (2015). Rhodospira. Bergey's Manual of Systematics of Archaea and Bacteria, 1–5. doi:10.1002/9781118960608.gbm00896.

Imhoff, J.F (2015). Rhodospirillum. Bergey's Manual of Systematics of Archaea and Bactera, 1–10. doi:10.1002/9781118960608.gbm00897.

Imhoff, J.F (2015). Rhodovibrio. Bergey's Manual of Systematics of Archaea and Bacteria, 1–5. doi:10.1002/9781118960608.gbm00898.

Imhoff, J.F (2015). Rhodovulum. Bergey's Manual of Systematics of Archaea and Bacteria, 1–8.doi:10.1002/9781118960608.gbm00863.

Imhoff, J.F (2015). Thioalkalicoccus. Bergey's Manual of Systematics of Archaea and Bacteria, 1–6. doi:10.1002/9781118960608.gbm01115.

Imhoff, J.F (2015). Thiococcus. Bergey's Manual of Systematics of Archaea and Bacteria, 1–6. doi:10.1002/9781118960608.gbm01117.

Imhoff, J.F (2015). Thiocystis. Bergey's Manual of Systematics of Archaea and Bacteria, 1–1. doi:10.1002/9781118960608.gbm01118.

Imhoff, J.F (2015). Thiodictyon. Bergey's Manual of Systematics of Archaea and Bacteria, 1–4. doi:10.1002/9781118960608.gbm01119.

Imhoff, J.F (2015). Thioflavicoccus. Bergey's Manual of Systematics of Archaea and Bacteria, 1–6. doi:10.1002/9781118960608.gbm01120.

Imhoff, J.F (2015). Thiospirillum. Bergey's Manual of Systematics of Archaea and Bacteria, 1–3. doi:10.1002/9781118960608.gbm01126.

Imhoff, J.F. and H. G. Truper (1977). *Ectothiorhodospira halochloris* sp. nov., a new extremely halophilic phototrophic bacterium containing bacteriochlorophyll b. *Arch. Microbiol.***114:** 115–121.

Imhoff, J.F. and H.G. Truper (1976). Marine sponges as habitats of anaerobic phototrophic bacteria. *Microb. Ecol.* **3:** 1–9.

Imhoff, J.F. and H.G. Truper (1980). *Chromatium purpuratum* sp. nov., a new species of the *Chromatiaceae*. *Zbl. Bakt., I. Abt. Orig.* **C1:** 61–69.

Imhoff, J.F. and H.G. Truper (1981). *Ectothiorhodospira abdelmalekii* sp. nov., a new halophilic and alkaliphilic phototrophic bacterium. *Zbl. Bakt., I. Abt. Orig.* **C2:** 228–234.

Imhoff, J.F. and H.G. Truper (1989). In: *The purple non-sulfur bacteria* (Staley, J.T., M.P. Bryant, N. Pfennig and J.G. Holt eds.) *Bergey's Manual of Systematic Bacteriology*. Williams and Wilkins. Baltimore, MD. **3:** 1658–1661.

Imhoff, J.F. and H.G. Truper (1992). The genus *Rhodospirillum* and related genera. In: *The Prokaryotes*, (Balows, A., H.G. Truper, M. Dworkin, W. Harder and K.H. Schleifer eds.), Second Edition. *Springer,New York*. pp. 2141-2155.

Imhoff, J.F. and J. Overmann (2015). *Thiorhodovibrio. Bergey's Manual of Systematics of Archaea and Bacteria,* 1–6. doi:10.1002/9781118960608.gbm01125.

Imhoff, J.F. and J. Suling (1996). The phylogenetic relationship among *Ectothiorhodospiraceae*. A reevaluation of their taxonomy on the basis of 16S rDNA analysis. *Arch Microbiol.***165:** 106–113.

Imhoff, J.F. and J. Suling (1997). In Validation of the publication of new names and new combinations previously effectively published outside the IJSB. List No. 62. *Int. J. Syst. Bacteriol.* **47:** 915–916.

Imhoff, J.F. and M.T. Madigan (2015). *Thermochromatium. Bergey's Manual of Systematics of Archaea and Bacteria,* 1–6. doi:10.1002/9781118960608.gbm01114.

Imhoff, J.F. and N. Pfennig (2001). *Thioflavicoccus mobilis* gen. nov., sp. nov., a novel purple sulfur bacterium with bacteriochlorophyll *b*. *Int. J. Syst. Evol. Microbiol.* **51:** 105–110.

Imhoff, J.F. and P. Caumette (2005). Genus XIII. *Thiocapsa* Winogradsky 1888, 84AL emend Guyoneaud, Suling, Petri, Matheron, Caumette, Pfennig and Imhoff 1998b, 962. In: Bergey's manual of systematic bacteriology, (Brenner, D.J., N.R. Krieg, J.T. Staley and G.M. Garrity, eds.), Second edition, Vol 2, Springer, New York, pp 26–28.

Imhoff, J.F. and P. Caumette (2015). *Halochromatium. Bergey's Manual of Systematics of Archaea and Bacteria,* 1–6. doi:10.1002/9781118960608.gbm01106.

Imhoff, J.F. and P. Caumette (2015). *Thiocapsa. Bergey's Manual of Systematics of Archaea and Bacteria,* 1–8. doi:10.1002/9781118960608.gbm01116.

Imhoff, J.F. and U. Bias-Imhoff (1995). Lipids, quinines and fatty acids of anoxygenic phototrophic bacteria. In: *Anoxygenic Photosynthetic Bacteria* (Blankenship, R.E., M.T. Madigan and C.E. Bauer, eds.), *Kluwer Academic Publ*, Netherlands. pp.179–205.

Imhoff, J.F. and V. Thiel (2010). Taxonomy and phylogeny of *Chlorobiaceae*. *Photosynthesis Research*. **104:** 123–136.

Imhoff, J.F. and V.M. Gorlenko (2015). *Roseospira. Bergey's Manual of Systematics of Archaea and Bacteria,* 1–6. doi:10.1002/9781118960608.gbm00899.

Imhoff, J.F. and V.M. Gorlenko (2015). *Thiorhodospira. Bergey's Manual of Systematics of Archaea and Bacteria,* 1–3. doi:10.1002/9781118960608.gbm01132.

Imhoff, J.F., A. Hiraishi and J. Suiling (2005). Anoxygenic phototrophic purple bacteria. In: *Bergey's Manual of Systematic Bacteriology* (Brenner, D.J., N.R. Krieg and J.T. Staley eds.), Second edition, Volume 2, Part A, Springer, New York. pp. 119–132.

Imhoff, J.F., B.J. Tindall, W.D. Grant and H.G. Truper (1981). *Ectothiorhodospira vacuolata* sp. nov., a new phototrophic bacterium from soda lakes. *Arch. Microbiol.* **130:** 238–242.

Imhoff, J.F., D.J. Kushner, S.C. Kushwaha and M. Kates (1982). Polar lipids in phototrophic bacteria of the *Rhodospirillaceae* and *Chromatiaceae* families. *J. Bacteriol.***150:** 1192–1201.

Imhoff, J.F., F. Hashwa and H.G. Truper (1978). Isolation of extremely halophilic phototrophic bacteria from the alkaline Wadi Natrun, Egypt. *Arch. Hydrobiol.* **84:** 381–388.

Imhoff, J.F., H.G. Sahl, G.S.H. Soliman and H. Truper (1979). The Wadi Natrun: chemical composition and microbial mass developments in alkaline brine of eutrophic desert lakes. *Geomicrobiol. J.* **1:** 219-234.

Imhoff, J.F., H.G. Truper and N. Pfennig (1984). Rearrangements of the species and genera of the phototrophic purple non-sulfur bacteria. *Int. J. Syst. Bacteriol.* **34:** 340.343.

Imhoff, J.F., J. Suling and R. Petri (1998b). Phylogenetic relationships among the *Chromatiaceae*, their taxonomic reclassification and description of the new genera *Allochromatium, Halochromatium, Isochromatium, Marichromatium, Thiococcus, Thiohalocapsa* and *Thermochromatium*. *Int. J. Syst. Bacteriol.* **48:** 1129–1143.

Imhoff, J.F., R. Petri and J. Suling (1998). Reclassification of species of the spiral-shaped phototrophic purple non-sulfur bacteria of the *α-Proteobacteria*: description of the new genera *Phaeospirillum* gen. nov., *Rhodovibrio* gen. nov., *Rhodothalassium* gen. nov. and *Roseospira* gen. nov. as well as transfer of *Rhodospirillum fulvum* to *Phaeospirillum fulvum* comb. nov., of *Rhodospirillummolischianum* to *Phaeospirillum molischianum* comb. nov., of *Rhodospirillum salinarum* to *Rhodovibrio salinarum* comb. nov., of *Rhodospirillum sodomense* to *Rhodovibrio sodomensis* comb. nov., of *Rhodospirillum salexigens* to *Rhodothalassiumsalexigens* comb. nov. and of *Rhodospirillum mediosalinum* to *Roseospira mediosalina* comb. nov. *Int. J. Syst. Bacteriol.***48:** 793–798.

Isaki, K. and Y. Kamio (1992) (Snoue Brand Milki Product Co. Ltd). *Japan Kokai Tokkyo Koho JP.* **04 :** 267-277.

Ismail, K.S.K., G. Najafpour, H. Younesi, A.R. Mohamed and A.H. Kamaruddin, (2008). Biological hydrogen production from CO: Bioreactor performance. *Biochem. Eng. J.***39:** 468-477.

Ivanova, T. L., T.P. Turova and A.S. Antonov (1985). DNA-DNA and rRNA-DNA hybridization studies in the genus *Ectothiorhodospira* and other purple sulfur bacteria. *Arch. Microbiol.* **143:** 154–156.

Ivanovsky, R.N., Y.I. Fal, I.A. Berg, N.V. Ugolkova, E.N. Krasilnikova, O.L. Keppen, L.M. Zakharchuc and A.M. Zyakun (1999). Evidence for the presence of the reductive pentose phosphate cycle in a filamentous anoxygenic photosynthetic bacterium, *Oscillochloris trichoides* strain DG-6. *Microbiology.* **145:** 1743-1748.

Iyer, A., K. Mody and B. Jha (2005). Biosorption of heavy metals by a marine bacterium. *Marine Pollution Bulletin.* **50(3):** 340-343.

Jahnavi, A., M. P. Pratap Rudra and M. Ramchander (2014). Impact of various cultural conditions on photoproduction of hydrogen from Kanyakumari photosynthetic bacteria*IJMRD.* **1(7):** 361-365.

Jahnavi, A., M.P. Pratap Rudra and M. Ramchander (2014).Photoproduction of Hydrogen by Vishakapatnam Photosynthetic Bacteria and their effect on different cultural conditions.*Int. J. Research (IJR).* **1(8):** 805-812.

James, C. Ogbanna, Hitoshi Yoshi Zawa and Hidepo Tanaka (2000). Treatment of high strength organic wastewater by a mixed culture of photosynthetic microorganisms. *J. Appl. Phycol.***12:** 277-284.

Janssen, P. H. and C.G. Harfoot (1991). *Rhodopseudomonas rosea* sp. nov., a new purple nonsulfur bacterium. *Int. J. Syst. Bacteriol.* **41:** 26-30.

Jesudason, J, J. and R.H. Marchessault (1994). Synthetic Poly [(R,S)-beta.-hydroxyalkanoates] with Butyl and Hexyl Side Chains. *Macromolecules.* **27 (9):** 2595-2602.

Jhojaosadati, S.A., K. Rasoul, J. Abbas and R.S. Hamid (1999). Bioconversion of molasses stillage to protein as an economic treatment of this effluent. *Resources Conservation and Recycling.* **27 (1-2):** 125-138.

Jiang, D., X. Ge, T. Zhang, H. Liu andQ. Zhang (2016). Photo-fermentative hydrogen production from enzymatic hydrolysate of corn stalk pith with a photosynthetic consortium. *Int. J. Hyd. Ener.* **41:** 16778-85.

John, T. I (1968). *Fr. Pat.* 1,521,676.

Jonathan Willett, J.L. Smart and C. E. Bauer (2007). RegA Control of Bacteriochlorophyll and Carotenoid Synthesis in *Rhodobacter capsulatus*. *J. Bacteriol.* **189:** 7765-7773.

Jormakka, M., K. Yokoyama, T. Yano, M. Tamakoshi, S. Akimoto, T. Shimamura, P. Curmi and S. Iwata (2008). Molecular mechanism of energy conservation in polysulfide respiration. *Nat. Struct. Mol. Biol.* **15:** 730–737.

Joshi, H.M. and F.R. Tabita (1996). A global two-component signal transduction system that integrates the control of photosynthesis, carbon dioxide assimilation and nitrogen fixation. *Proc. Natl .Acad. Sci. U.S.A.* **93:** 14515–14520.

Judd, S. and C. Judd (2011). The MBR book. Principles and applications of membrane bioreactors for water and wastewater treatment. 2nd ed. Elsevier. ISBN 9780080966823 pp. 536.

Julie L. Zilles, Jordan Peccia, Myeong-Woon Kim, Chun-Hsiung Hung and R. Daniel R. Noguera (2002). Involvement of *Rhodocyclus* related Organisms in Phosphorus Removal in Full-Scale Wastewater Treatment Plants. *Appl. Environ. Microbiol.* **68:** 2763-2769.

Jung, G.Y., H.O. Jung, J.R. Kim, Y. Ahn and S. Park (1999). Isolation and characterization of *Rhodopseudomonas palustris* P4 which utilizes CO with the production of H_2. *Biotechnol. Lett.* **21:** 525–529.

Kaden, R.., C. Sproer, D. Beyer and P. Krolla-Sidenstein (2014). *Rhodoferaxsaidenbachensis* sp. nov., a psychrotolerant, very slowly growing bacterium within the family *Comamonadaceae*, proposal of the appropriate taxonomic position of *Albidiferax ferrireducens* strain T118T in the genus *Rhodoferax* and emended description of the genus *Rhodoferax.Int. J. Syst. Evol. Microbiol.* **64(4):** 1186–1193.

Kadnikov, V.V., A.V. Mardanov and O.A. Podosokorskaya (2013). Genomic analysis of *Melioribacter roseus*, facultatively anaerobic organotrophic bacterium representing a novel deep lineage within Bacteroidetes/*Chlorobi* group. *PLoS One.***8:** e53047.

Kaewsuk, J., W. Thorasampan, M. Thanuttamavong and G.T. Seo (2010). Kinetic development and evaluation of membrane sequencing batch reactor (MSBR) with mixed cultures photosynthetic bacteria for dairy wastewater treatment. *J. Environ. Manage.* **91(5):** 1161–68.

Kaiser, P (1966). Contribution a l'étude de l'écologie des bacteries photosynthetiques. *Ann. Inst. Pasteur.* **111:** 733–749.

Kamal, V.S. and R.C. Wyndham (1990). Anaerobic phototrophic metabolism of 3-chlorobenzoate by *Rhodopseudomonas palustris* WS17. *Appl. Environ. Microbiol.***56(12):** 3871–3873.

Kamen, M. D. and H. Gest (1949). Evidence for a nitrogenase system in the photosynthetic bacterium *Rhodospirillum rubrum*. *Science*. **109:** 560.

Kampf, C. and N. Pfennig (1980). Capacity of *Chromatiaceae* for chemotrophic growth. Specific respiration rates of *Thiocystis violacea* and *Chromatium vinosum*. *Arch. Microbiol.* **127:** 125–135.

Kampfer, P., R. Rossello´-Mora, H.C. Scholz, C. Welinder-Olsson, E. Falsen and H.J. Busse (2006). Description of *Pseudochrobactrum* gen. nov., with the two species *Pseudochrobactrum asaccharolyticum* sp. nov. and *Pseudochrobactrum saccharolyticum* sp. nov. *Int. J. Syst. Evol. Microbiol.***56:** 1823–1829.

Kanamori, M. and M. Kobayashi (1982). Kankyo Kagaku Sogo Kenkyusho Nenpo. **2:** 181-188.

Kantachote, D., N. Kornochalert and S. Chaiprapat (2010). The use of purple nonsulfur bacterium isolates P1 and fermented pineapple extract to treat latex rubber sheet wastewater for possible use as irrigation water. *Afr. J. Microbiol. Res.* **4:**2604–2616.

Kantachote, D., S. Torpee and K. Umsakul (2005). The potential use of anoxygenic phototrophic bacteria for treating latex rubber sheet wastewater. *Electron. J. Biotechnol.* **8:** 314–323.

Kantha, T., C. Chaiyasut, D. Kantachote, S. Sukrong and A. Muangprom (2010) Selection of photosynthetic bacteria producing 5 aminolevulinic acid from soil of organic saline paddy fields from the Northeast region of Thailand. *African J. Microbiol. Res.***4:** 1848–1855.

Kantha, T.,C. Chaiyasut, D. Kantachote, S. Sukrong and A. Muangprom (2012). Synergistic growth of lactic acid bacteria and photosynthetic bacteria for possible use as a biofertilizer. *African Journal of MicrobiologyResearch.* **6(3):**504-511.

Kantha, T., D. Kantachote, N. and Klongdee (2015). Potential of biofertilizers from selected *Rhodopseudomonas palustris* strains to assist rice (*Oryza sativa* L. subsp. indica) growth under salt stress and to reduce greenhouse gas emissions. *Annu. Microbiol.***65:** 2109–2118.

Kantachote,D., Nunkaew, T. Kantha and S. Chaiprapat (2016). Biofertilizers from *Rhodopseudomonas palustris* strains to enhance rice yields and reduce methane emissions. *Appl. Soil. Ecol.***100:** 154–161.

Kapdan, I.K. and F. Kargi (2006). Bio-hydrogen production from waste materials. *Enzyme and Microbial Technology*.**38:** 569–582.

Kappler, U. and H. Schafer (2014). Transformations of dimethylsulfide. *Met. Ions. Life Sci.* **14:** 279–313.

Kappler, U. and S. Bailey (2005). The molecular basis of intramolecular electron transfer in sulfite-oxidizing enzymes is revealed by high-resolution structure of a heterodimeric complex of the catalytic molybdopterin subunit and a c-type cytochrome subunit. *J. Biol. Chem.* **280:** 24999–25007.

Kappler, U., B. Bennett and J. Rethmeier (2000). Sulfite: cytochrome c oxidoreductase from *Thiobacillus novellus*—Purification, characterization and molecular biology of a heterodimeric member of the sulfite oxidase family. *J. Biol. Chem.***275:** 13202–13212.

Karr, E.A., W.M. Sattley, D.O. Jung, M.T. Madigan and L.A. Achenbach (2003). The remarkable diversity of phototrophic purple bacteria in a permanently frozen Antarctic Lake. *App. Env. Microbiol.* **69:** 4910-4914.

Karrasch, S., P.A. Bullough and R. Ghosh (1995). The 8.5-angstrom projection map of the light-harvesting complex-I from *Rhodospirillum rubrum* reveals a ring composed of 16 subunits. *EMBO Journal*. **14:** 631-638.

Kaspari, H (1979). *Microbiologica* (Bologna). **2:** 231-241.

Kaspari, H. and B. Werner (1986). *Arch. Microbiol.* **144:** 67-70.

Katayama, T., M. Kobayashi and A. Okuda, A (1965b). *Soil Sci. Plant Nutr* **13:** 101-106.

Katayama, T., M. Kobayashi and A. Okuda (1965). *Soil Sci. Plant Nutr.***11:** 176-179.

Kavarnos, G. J. and N.J. Turro (1986). Photosensitization by Reversible Electron Transfer: Theories, Experiment, Evidence and Example. *Chem. Rev*. **86:** 401-449.

Kawasaki, H., Y. Hoshino and K. Yamasato (1993b). Phylogenetic diversity of phototrophic purple non-sulfur bacteria in the *Proteobacteria* alpha group. *FEMS Microbiol. Lett.* **112:** 61–66.

Kawasaki, H., Y. Hoshino, A. Hirata and K. Yamasato (1993a). Is intracytoplasmic membrane structure a generic criterion? It does not coincide with phylogenetic interrelationships among phototrophic purple nonsulfur bacteria. *Arch. Microbiol.***160:** 358-362.

Kawasaki, H., Y. Hoshino, Y. Kuraishi and K. Yamasato (1992). *Rhodocista centenaria* gen. nov., sp. nov., a cyst-forming anoxygenic photosynthetic bacterium and its phylogenetic position in the *Proteobacteria* alpha group. *J. Gen. Appl. Microbiol.* **38:** 541-551.

Kawasaki, H., Y. Hoshino, Y. Kuraishi and K. Yamasato (1994). Invalidation of the publication of new names and new combinations previously effectively published outside the IJSB. List No. 48. *Int. J. Syst. Bacteriol.* **44:**182–183.

Kawasaki, K., H. Tanimoto, R. Nagasaki and A. Matsuda (2010). The filtration characteristics of hollow fiber microfiltration-Effect of various kinds of solids in the excess activated sludge. *Desalination and water treatment*. **17:** 155-159.

Kawasaki, N., F. Ogata and H. Tominaga (2010). Selective adsorption behavior of phosphate onto aluminum hydroxide gel. *Journal of Hazardous Materials*. **181(1–3):** 574-579.

Kazan, K (2013). Auxin and the integration of environmental signals into plant root development. *Ann. Bot.***112:** 1655–1665.

Kazuhisa Miyamoto, Souichi Ohta, Yoshihito Nawa, Yasuko Mori and Yoshiharu Miura (1987). Hydrogen Production by a Mixed Culture of a Green Alga, *Chlamydomonas reinhardtii* and a Photosynthetic Bacterium, *Rho do spirillum rubrum. Agri. Biologic. Chemist*. **51(5):** 1319-1324.

Kenealy, W.R. and T.W. Jeffries (2003). Enzyme processes for pulp and paper: a review of recent developments. Wood deterioration and preservation: advances in our changing world (Goodell, B., D.D. Nicholas, T.B. Schultz, eds.), In: chapter 12, Washington, DC, American Chemical Society, Distributed by Oxford University Press, 2003. ACS symposium series. **845:** 210-239.

Keppen, O. I. and V.M. Gorleko (1975). Characteristics of a new species of purple budding bacteria containing bacteriochlorophyll *b*. *Mikrobiologiya* **44:** 258-264.

Keppen, O. I., V.M. Gorlenko and B.K. Pierson (2015.) *Oscillochloris. Bergey's Manual of Systematics of Archaea and Bacteria,* 1–7. doi: 10.1002 / 9781118960608.gbm00383.

Keppen, O.I., I.A. Berg, N.V. Lebedeva (2008) *Chlorobaculum macestae* sp nov., a new green sulfur bacterium. *Mikrobiologiia*. **77:** 69–77.

Keppen, O.I., O.I. Baulina and E.N. Kondratieva (1994). *Oscillochloris trichoides* neotype strain DG-6. *Photosynth. Res*. **41:** 29–33.

Keppen, O.I., O.I. Baulina, A.M. Lysenko and E.N. Kondrateva (1993). New green bacterium belonging to the family *Chloroflexaceae*. *Mikrobiologiya*. **62:** 179–185.

Keppen, O.I., T.P. Tourova, B.B. Kuznetsov, R.N. Ivanovsky and V.M. Gorlenko (2000). Proposal of *Oscillochloridaceae* fam. nov., on the basis of a phylogenetic analysis of the filamentous anoxygenic phototrophic bacteria and emended description of *Oscillochloris* and *Oscillochloris trichoides* in comparison with further new isolates. *Int. J. Syst. Evol. Microbiol.* **50:** 1529–1537.

Kerber, N. L. and J. Cardenas (1982). Nitrate reductase from *Rhodospeudomonas sphaeroides*. *J. Bacteriol.* **150:** 1091-1097.

Kerby, R.L., S.S. Hong, S.A. Ensign, L.J. Coppoc, P.W. Ludden and G.P. Robersts (1992). Genetic and physiological characterization of the *Rhodospirillum rubrum* carbon monoxide dehydrogenase system. *J. Bacteriol.***174(16):** 5284-5294.

Kessi, J., M. Ramuz, E. Wehrli, M. Spycher and R. Bachofen (1999). Reduction of selenite and detoxification of elemental selenium by the phototrophic bacterium *Rhodospirillum rubrum*. *Appl. Environ. Microbiol.* **65:** 4734–4740.

Ketchum, P.A. and C.L. Sevilla (1973). *In vitro* formation of nitrate reductase using extracts of the nitrate reductase mutant of *Neurospora crassa*, nit-1 and *Rhodospirillumrubrum*. *J. Bacteriol.* **116:** 600–609.

Khan, I.U., N. Habib, M. Xiao, M.M. Li , Wen-Dong Xian, Mohammad Saeid Hejazi, Vahideh Tarhriz , Xiao-Yang Zhi , Wen-Jun Li (2019). *Rhodobacter thermarum* sp. nov., a novel phototrophic bacterium isolated from sediment of a hot spring. *Antonie van Leeuwenhoek*. 112: 867–875.

Khanal, S.K., W.H. Chen, L. Li and S. Sung (2006). Biohydrogen production in continuous flow reactor using mixed microbial culture. *Water Environ. Res.* **78(2):** 110-117.

Khanna S. and D.J.D. Nicholas (1982). Utilization of tetrathionate and 35S-labelled thiosulphate by washed cells of *Chlorobium vibrioforme* f. sp. *thiosulfatophilum*. *J. Gen. Microbiol.* **128:** 1027–1034.

Khanna, P., B. Rajkumar and N. Jothikumar (1992). Anoxygenic degradation of aromatic substances by *Rhodopseudomonas palustris*. *Curr. Microbiol.* **25:** 63–67.

Khatipov, E., Masato Miyake, Jun Miyake and Yasuo Asada (1998). Accumulation of poly-β-hydroxybutyrate by *Rhodobacter sphaeroides* on various carbon and nitrogen substrates. *FEMS. Microbiol. Lett.* **162:** 39-45.

Khuong, N.Q., D. Kantachote, J. Onthong and A. Sukhoom (2017) The potential of acid-resistant purple nonsulfur bacteria isolated from acid sulfate soils for reducing toxicity of Al^{3+} and Fe^{2+} using biosorption for agricultural application. *Biocatal. Agric. Biotechnol.***12:** 329–340.

Kim M. S., K.W. Moon and S.K. Lee (2001). Hydrogen production from food processing wastewater and sewage sludge by anaerobic dark fermentation combined with photo fermentation . In: Miyake J., *et al* (eds*) Biohydrogen II*. Elsevier, Amsterdam, pp. 263–272.

Kim, J. K., L. Nhat, Y.N. Chun and S.W. Kim (2008). Hydrogen production conditions from food waste by dark fermentation with *Clostridium beijerinckii* KCTC 1785. *Biotechnology and Bioprocess Engineering,***13(4):** 499–504.

Kim, J.G., J.H. Kim, H.S. Moon, C.M. Chon and J.S. Ahn (2003). Removal capacity of water plant alum sludge for phosphorus in aqueous solutions. *Chemical Speciation and Bioavailability*. **14:** 67-73.

Kim, J.S., K. Ito, K. Izaki and H. Takahashi (1987). Production of Molecular Hydrogen by a SemiContinuous Outdoor Culture of *Rhodopseudomonas sphaeroides*. *Agricultural and Biological Chemistry*. **51(4):** 1173-1174.

Kim, K.J. and B.K. Lee (2000). Mass production of *Rhodopseudomonas palustris* as diet for aquaculture. *Aquacultural Engineering.***23:** 281-293.

Kim, M.K., K.M. Choi, C.R. Yin, K.Y. Lee, W.T. Im, J.H. Lim and S. Lee (2004). Odorous swine wastewater treatment by purple non-sulfur bacteria, *Rhodopseudomonas palustris*, isolated from eutrophicated ponds. *Biotechnol. lett.* **26:** 819-822.

Kim, N.Y., T.B. Yim and H.Y. Lee (2015). Skin Anti-Aging Activities of Bacteriochlorophyll a from Photosynthetic Bacteria, *Rhodobacter sphaeroides. J. Microbiol. Biotechnol.***25(10):** 1589–1598.

Kimble, L.K. and M.T. Madigan (1992a). N_2 fixation and nitrogen metabolism in heliobacteria. *Arch. Microbiol.* **158:** 155–161.

Kimble, L.K. and M.T. Madigan (1992b). Evidence for an alternative nitrogenase system in *Heliobacterium gestii. FEMS Microbiol Lett.***100:** 255–260.

Kimble, L.K., A.K. Stevenson and M.T. Madigan (1994). Chemotrophic growth of heliobacteria in darkness. *FEMS Microbiol. Lett.* **115:** 51–55.

Kimble, L.K., L. Mandelco, C.R. Woese and M.T. Madigan (1995). *Heliobacterium modesticladum*, sp. nov., a thermophilic heliobacterium of hot springs and volcanic soils. *Arch. Microbiol.* **163:** 259–267.

King, P.P (1982). Biotechnology. An industrial view. *1. Chem. Technol. Eiotechnol (Oxford, Oxfordshire).***32 (1):** 2-8.

Kirti, K., S. Amita, S. Priti, A. Mukesh Kumar and S. Jyoti (2014). Colorful world of microbes: carotenoids and their applications. *Advances in Biology*. **2014:**13 pages. https://doi.org/10.1155/2014/837891.

Klasson, K.T., M.D. Ackerson, E.C. Clusen and J.L. Gaddy (1992b). Bioconversion of synthetic gas into liquid or gaseous fuels. *Enzyme Microb. Technol.***14:** 602-608.

Kleinig, H. and H. Reichenbach (1977). Carotenoid glucosides and menaquinones from the gliding bacterium *Herpetosiphon giganteus* Hpa2. *Arch. Microbiol.* **112**: 307–310.

Klemme, J. H., I. Chyla and M. Reuss (1980). Dissimilatory nitrate reduction by strains of the facultative phototrophic bacterium *Rhodopseudomonas palustris. FEMS Microbiol. Lett.* **9:** 137-140.

Klemme, J.H (1968). Untersuchungen zur Photoautotrophie mit molekularem Wasserstoff bei neuisolierten schwefelfreien Purpurbakterien. *Arch. Microbiol.* **64:** 29–42.

Klemme, J.H (1979). The occurrence of assimilatory nitrate reduction in phototropic bacteria of the genera *Rhodospirillum* and *Rhodopseudomonas*. *Microbiologica.* **2:** 415–420.

Klipp, W.B., B. Masepohl and A. Puhler (1988). Identification and mapping of nitrogen fixation genes of *Rhodobacter capsulatus*. Duplication of a *nifA-nifB* region. *J. Eacteriol.***170:** 693-699.

Kluyver, A.J. and C.B. van Niel (1936). Prospects for a natural system of classification of bacteria. *Zentbl. Bakteriol.Parasitenkd. Infektkrankh. Hyg. Abt. II.* **94:** 369–403.

Ko, I.B. and T. Noike (2002). Use of blue optical filters for suppression of growth of algae in hydrogen-producing non-axenic cultures of *Rhodobacter sphaeroides* RV. *Int. J. Hydr. Ener.***27:** 1297 – 1302.

Kobayashi, M (1970). *Jpn. Kokai Tokkyo Koho JP*. 70/14,091.

Kobayashi, M (1975a). *Prog. Water Technol.* **7:** 309-315.

Kobayashi, M (1975b). *Jpn. Kokai Tokkyo Koho Jp.* 75/69, 279.

Kobayashi, M (1976). In: "*Microbial Energy Conversion,*" (Schlegel, H.G. and J. Barnea,eds.). Erich Goltz KG, Gbttingen. pp. 443-453.

Kobayashi, M (1977a). *Jpn. Kokai Tokkyo Koho Jp.* 77/103, 845.

Kobayashi, M (1977b). *Jpn. Kokai Tokkyo Koho Jp.* 77/103, 846.

Kobayashi, M (1982). In: *Advances in Agricultural Microbiology*, (N.S. Subba Rao, ed.), Oxford and IBH Publishing Co., New Delhi. pp. 643-661.

Kobayashi, M. and H. Hirotani (1987). Production of antiviral substances by phototrophic bacteria. *Soil Microorg.* **30:** 43-48.

Kobayashi, M. and M. Kobayashi (1995). Waste remediation and treatment using anoxygenic phototrophic bacteria. In: *Anoxygenic phototrophic bacteria*, (Blankenship, R.E., M.T. Madigan and C.E. Bauer, eds.). Kluwer Academic Publishers. Dordrecht, The Netherlands. pp. 1269-1282.

Kobayashi, M. and M. Kondo (1984). *Int. Symp. Environ.*, 3rd, Singapore, pp. 1-6.

Kobayashi, M. and M.Z. Haque (1971). Contribution to nitrogen fixation and soil fertility by photosynthetic bacteria. *Plant Soil.* **35:** 443-456.

Kobayashi, M. and S. Kurata (1978). The mass culture and cell utilization of photosynthetic bacteria. *Process Biochem*istry. **13(9):** 27-30.

Kobayashi, M. and T. Miyazawa (1990). Jpn. Kokai Tokkyo Koho JP 90/184,399.

Kobayashi, M. and T. Yoshida (1983). *Int. Symp. Sol. Energy Bioconvers.,* Pushchino,pp. 76-87.

Kobayashi, M. and W. Ye (1986). *Fourth International Symposium on Microbial Ecology,* Ljubljana, Yugoslavia. pp.1-9.

Kobayashi, M., E. Takahashi and K. Kawaguchi (1967). Distribution of nitrogen-fixing microorganisms in paddy soils of Southeast Asia. *Soil Sci.***104:** 113-118.

Kobayashi, M., H. Oh-oka, S. Akutsu, M. Akiyama, K. Tominaga, H. Kise, F. Nishida, T. Watanabe, J. Amesz, M. Koizumi, N. Ishida and H. Kano (2000). The primary electron acceptor of green sulfur bacteria, bacteriochlorophyll 663, is chlorophyll esterified with (delta)2,6-phytadienol. *Photosyth Res.***63:** 269–280.

Kobayashi, M., K. Mochida and T. Oosaki (1975). *Jpn. Kokai Tokkyo KohoJp.*75/155,697.

Kobayashi, M., M. Kobayashi, R.E. Blankenship, M.T. Madigan and C.E. Bauer (1995). Anoxygenic photosynthetic bacteria. 1269-1282.

Kobayashi, M., N. Taketomo and Y.T. Tchan (1977). *J. Ferment. Technol.* **56:** 615-620.

Kobayashi, M., S. Okano and T. Nagi (1981). Non-Symbiotic Nitrogen Fixation. **8:** 1-8.

Kobayashi, T., S. Nobuyuki and O. Azuma (1970). *Jpn. Kokai Tokkyo KohoJP.* 70/13,434.

Koblizek, M. (2015). Ecology of aerobic anoxygenic phototrophs in aquatic environments. *FEMS Microbiol. Rev.* **39:** 854–870.

Koh, R.H. and H.G. Song (2007). Effects of application of *Rhodopseudomonas* sp. on seed germination and growth of tomato under axenic conditions. *J. Microbiol. Biotechnol.***17:** 1805-10.

Koku, H., I. Eroglu, U. Gunduz, M. Yuce and L. Turker (2003). Kinetics of biological hydrogen production by the photosynthetic bacterium *Rhodobacter sphaeroides* O.U. 001. *Int. J. Hydr. Ener.* **28:** 381 – 388.

Koku, H., I. Eroglu, U. Gunduz, M. Yucel and L. Turker (2002). Aspects of the metabolism of hydrogen production by *Rhodobacter sphaeroides. Int. J.Hydr. Ener.* **28:** 381-8.

Kompantseva, E.I (1985). *Rhodobacter euryhalinus* new species a new halophilic purple bacterial species. *Mikrobiologiya.* **54:** 974–981.

Kompantseva, E.I., A.V. Komova and N.A. Kostrikina (2010). *Rhodovulum steppense* sp. nov., an obligatory haloalkaliphilic purple non-sulfur bacterium widespread in saline soda lakes of Central Asia. *Int. J. Syst. Evol. Microbiol.***60:**1210–1214.

Kompantseva, E.I., A.V. Komova, A.A. Novikov and N.A. Kostrikina (2012). *Rhodovulum tesquicola* sp. nov., a haloalkaliphilic purple non-sulfur bacterium from brackish steppe soda lakes. *Int. J. Syst. Evol. Microbiol.* **62:** 2962–2966.

Kompantseva, E.J (1989a). A new species of budding purple bacterium: *Rhodopseudomonas julia* sp. nov. *Mikrobiologiia.* **58:** 254–259.

Kompantseva, E.J (1989b). In Validation of the publication of new names and new combinations previously effectively published outside the IJSB. List No. 29. *Int. J. Syst. Bacteriol.* **39:** 205–206.

Kompantseva, E.J. and V.M. Gorlenko (1984). A new species of moderately halophilic purple bacterium *Rhodospirillum mediosalinum* sp. nov. *Mikrobiologiia.***53:** 775–781.

Kondo, K., E. Nishihara and N. Nakata (2007). Effect of the purple non-sulfur bacterium (*Rhodobacter sphaeroides*) on the fruit quality of tomato. *Acta Hortic.* **761:** 581-587.

Kondo, T., M. Arakawa, T. Wakayama and J. Miyake (2002). Hydrogen production by combining two types of photosynthetic bacteria with different characteristics. *Int. J. Hydr. Ener.***27:** 1303-1308.

Konietzny, U. and R. Greiner (2004). Bacterial Phytase: Potential application, in vivo function and regulation of its synthesis. *Brazil. J. Microbiol.* **35:** 11-18.

Kornberg, H.L. and H.A Krebs (1957). Synthesisofcellconstituents fromC2-units by amodified tri- carboxylic acid cycle. *Nature.***179:** 988–991.

Kornochalert, N., D. Kantachote, S. Chaiprapat andS. Techkarnjanaruk (2014). Use of *Rhodopseudomonas palustris* P1 stimulated growth by fermented pineapple extract to treat latex rubber sheet wastewater to obtain single-cell protein. *Ann. Microbiol.* **64:**1021–1032.

Kosamu, I.B.M. and M. Obst (2009). The influence of picocyanobacterial photosynthesis on calcite precipitation. *Int. J. Environ. Sci. Tech.* **6:** 557-562.

Kostanjevecki, V., A. Brige, T.E. Meyer, M.A. Cusanovich, Y. Guisez and J. van Beeumen (2000). A membrane-bound flavocytochrome c-sulfide dehydrogenase from the purple phototrophic sulfur bacterium *Ectothiorhodospira vacuolata. J. Bacteriol.* **182:** 3097–3103.

Kotay, S. and D. Das (2008). Biohydrogen as a renewable energy resource—Prospects and potentials. *Int. J. Hydr. Ener.* **33:** 258-263.

Kotsopoulos, T.A., R.J. Zeng and I. Angelidaki (2006). Biohydrogen production in granular up-flow anaerobic sludge blanket (UASB) reactors with mixed cultures under hyper-thermophilic temperature (70 ºC). *Biotechnol. Bioengine.***94:** 296-302.

Koyama, Y., M. Kuki, P.O. Andersson and T. Gillbro (1996). Singlet excited states and the light-harvesting function of carotenoids in bacterial photosynthesis. *Photochem. Photobiol.* **63:** 243–256.

Kranz, R.G. and R. Haselkorn (1986). Anaerobic regulation of nitrogen-fixation genes in *Rhodopseudomonas capsulata. Proc. Natl. Acad. Sci. U.S.A* . **83(13):** 6805-6809.

Kranz, R.G., K.K. Gabbert, T.A. Locke and M.T. Madigan (1997). Polyhydroxyalkanoate production in *Rhodobacter capsulatus*: genes, mutants, expression and physiology. *Appl. Environ.Microbiol.* **63:**3003-3009 .

Krasilnikova, E.N., O.I. Keppen, V.M. Gorlenko and E.N. Kondrateva (1986). Growth of *Chloroflexus aurantiacus* on media with different organic compounds and pathways of their metabolism. *Microbiology.* **55:** 325–329.

Krasilnikova, E.N., R.N. Ivanovskii and E.N. Kondratieva (1983). Growth of purple bacteria utilizing acetate under anaerobic conditions in darkness [English trans.]. *Mikobiologiya.* **52:** 189–194.

Kredich, N.M (1992). The molecular basis for positive regulation of cys promoters in *Salmonella typhimurium* and *Escherichia coli. Mol. Microbiol.* **6:** 2747–2753.

Kredich, N.M (1996). Biosynthesis of cysteine. In: *Escherichia coli* and *Salmonella typhimurium, Cellular and molecular biology*, (Neidhardt, F.C. ed.), American Society for Microbiology, Washington, pp. 514–527.

Kriss, A.E. and E.A. Rukina (1953). Purple sulphur bacteria in deep sulfurous water of the Black Sea [in Russian]. *Doklady Akademii Nauk SSSR*. **93:** 1107–1110.

Krooneman, J., S. van den Akker, T.M.P. Gomes, L.J. Forney and J.C. Gottschal (1999). Degradation of 3-chlorobenzoate under low-oxygen conditions in pure and mixed cultures of the anoxygenic photoheterotroph *Rhodopseudomonas palustris* DCP3 and an aerobic *Alcaligenes* species. *Appl. Environ. Microbiol.* **65(1):**131–137.

Kulichevskaya, I.S., V.S. Guzev, V.M.Gorlenko, W. Liesack and S.N. Dedysh (2006). *Rhodoblastus sphagnicola* sp. nov., a novel acidophilic purple non-sulfur bacterium from Sphagnum peat bog. *Int. J. Syst. Evol. Microbiol.***56:** 1397-402.

Kumar, B.V., E.V.V. Ramprasad Ch. Sasikala and Ch.V. Ramana (2013). *Rhodopseudomonas pentothenatexigens* sp. nov. and *Rhodopseudomonas thermotolerans* sp. nov., isolated from paddy soils. *Int. J. Syst. Evol. Microbiol.* **63:** 200–207.

Kumar, N. and D. Das (2000). Enhancement of hydrogen production by *Enterobacter cloacae*IIT-BT 08. *Proc. Biochem.* **35:** 589-593.

Kumar, N., A.K. Ghosh and D. Das (2001). Redirection of biochemical pathways for the enhancement of H_2 production by *Enterobacter cloacae* IIT-BT 08. *Biotechnol. Lett.***23:** 537–541.

Kumar, V., Prashant Singh, Milko, A. Jorquera, Punesh Sangwan, Piyush Kumar, A.K. Verma, Sanjeev Agrawal (2013). Isolation of phytase-producing bacteria from Himalayan soils and their effect on growth and phosphorus uptake of Indian mustard (*Brassica juncea.*) *W. J. Microbiol. Biotechnol.* **29:** 1361-1369.

Kundu, D.K. and J.K. Ladha (1997). Effect of growing rice on nitrogen mineralization in flooded soil. *Soil. Sci. Soc. Am. J.***61:** 839–845.

Kuo, F.S, Y.H. Chien and C.J. Chen (2012). Effects of light sources on growth and carotenoid content of photosynthetic bacteria *Rhodopseudomonas palustris*. *Bioresource Technol.* **113:** 315-318.

Kurbanoglu Esabi Basaran (2001). Production of Single-Cell Protein from Ram Horn Hydrolysate. *Turk. J. Biol.***25:** 371-377, Copyright TUBITAK.

Kurbanoglu, E.B. and O.F. Algur (2002). Single-cell protein production from ram horn hydrolysate by bacteria. *Bioresour. Technol.* **85(2):** 125-129.

Kuroda, Y (1990). *Jpn. Kokai Tokkyo Koho JP*. 90/138,987.

Kurth, J., C. Dahl and J.N. Butt (2015). Catalytic protein film electrochemistry provides a direct measure of the tetrathionate/thiosulfate reduction potential. *J. Am. Chem. Soc.* **137:** 13232–13235.

Kusche, W.H (1985). Untersuchungen an Elektronen transport proteinen und zum Schwefelstoffwechsel in *Ectothiorhodospiraceae*. University of Bonn, Germany: PhD Dissertation.

Kushalatha, M., G. Vidya and K. Chandrakant (2010). Photobiodegradation of halogenated aromatic pollutants. *Adv. Bioscien. Biotechnol.* **1:** 238-240.

Kusnetzov, S. I (1970). The Microflora of Lakes and its Geochemical Activity. University of Texas Press. Austin, TX.

Kutzing, F.T (1849). *Species Algarum*, I-IV, Leipzig.

Kuykendall, L.D (2005). Order VI. *Rhizobiales* ord. nov. In: *Bergey's Manual of Systematic Bacteriology*, (Brenner, D.J., N.R. Krieg, J.T. Staley and G.M. Garrity, ed.), Second edition, Vol. 2 (The *Proteobacteria*), part C (The *Alpha,- Beta,- Delta-* and *Epsilonproteobacteria*), Springer, New York, pp. 324.

Kuykendall, L.D (2015). *Rhizobiales* ord. nov. Bergey's Manual of Systematics of Archaea and Bacteria, 1–1. doi:10.1002/9781118960608.obm00071.

Kuzmanova, S., E. Vandeska, A. Dimitrovski and D. Doneva (1989). Microbial procedures from the utilization of food industry wastes. In: DECHEMA biotechnology conference, (Behrens, D. and A.J. Griesel, eds.) Weinheim, Germany, VCH Verlagsesellschaft. *Microbiology Skopj.* **26:** 159-164.

Laguna, R., F.R. Tabita and B.E. Alber (2011). Acetate-dependent photoheterotrophic growth and the differential requirement for the Calvin-Benson-Bassham reductive pentose phosphate cycle in *Rhodobacter sphaeroides* and *Rhodopseudomonas palustris*. *Arch. Microbiol.* **193:** 151–154.

Laible, P.D., D.L. Mielke and D.K. Hanson (2009). Foreign gene expression in photosynthetic bacteria. In: The Purple Phototrophic Bacteria, (Hunter, C.N. F. Daldal, M. C. Thurnauer and J.T. Beatty, eds.), Springer Science, Dordrecht, The Netherlands. pp. 839–860.

Lakshmi, K,V., Ch. Sasikala, G.V. Ashok Kumar, R. Chandrasekaran and Ch.V. Ramana (2011a). *Phaeovibrio sulfidiphilus* gen. nov., sp. nov., phototrophic *alphaproteobacteria* isolated from brackish water. *Int. J. Syst. Evol. Microbiol.* **61:** 828–833.

Lakshmi, K.V. N.S., B. Divyasree, K. Sucharita, Ch. Sasikala and Ch.V. Ramana (2015). *Thiorhodococcus fuscus* sp. nov., isolated from a lagoon.. *Thiorhodococcus fuscus* sp. nov., isolated from a lagoon. *Int. J. Syst. Evol. Microbiol.***65:** 3938-3943.

Lakshmi, K.V., B. Divyasree, E.V.V. Ramprasad, Ch. Sasikala, Ch.V. Ramana (2014). Reclassification of *Rhodospirillum photometricum* Molisch 1907, *Rhodospirillum sulfurexigens* Anil Kumar *et al.* 2008 and *Rhodospirillum oryzae* Lakshmi *et al.* 2013 in a new genus, *Pararhodospirillum photometricum* comb. nov., *Pararhodospirillum sulfurexigens* comb. nov. and *Pararhodospirillum oryzae* comb. nov., respectively and emended description of the genus *Rhodospirillum*. *Int. J. Syst. Evol. Microbiol.* **64:** 1154–1159.

Lakshmi, K.V., Ch. Sasikala and Ch.V. Ramana (2009). *Rhodoplanes pokkaliisoli* sp. nov., a phototrophic *alphaproteobacterium* isolated from waterlogged brackish paddy soil. *Int. J. Syst. Evol. Microbiol.* **59:** 2153–2157.

Lakshmi, K.V., Ch. Sasikala, S. Takaichi and Ch.V. Ramana (2011b). *Phaeospirillum oryzae* sp. nov., a spheroplast-forming, phototrophic *alphaproteobacterium* from paddy soil. *Int. J. Syst. Evol. Microbiol.* **61:** 1656–1661.

Lakshmi, K.V., Ch. Sasikala, V. V. Ramana, E.V. Ramaprasad and Ch. V. Ramana (2011). *Rhodovulum phaeolacus* sp. nov. a phototrophic *alphaproteobacterium* isolated from a brown pond. *J. Gen. Appl. Microbiol.* **57:** 145–151.

Lakshmi, K.V.N.S., Ch. Sasikala, E.V. Ramaprasad and Ch.V. Ramana (2013). *Rhodospirillum oryzae* sp. nov., a phototrophic bacterium isolated from rhizosphere soil of paddy. *Int. J. Syst. Evol. Microbiol.***63:** 3050-3055.

Lamzin, V.S., Z. Dauter and K.S. Wilson (1995). How nature deals with stereoisomers. *Curr. Opin. Struct. Biol.* **5:** 830–836.

Lan, G.Q., N. Abdullh, S. Jalaludi and Y.W. Ho (2002). Culture conditions influencing phytase production of *Mitsuokella jalaludinii*, a new bacterial species from the rumen of cattle. *J. Appl. Microbiol.* **93:** 668-674.

Lancaster, D.R.C (2008). Structure of Reaction Centres in Anoxygenic Bacteria. In: Primary Processes of Photosynthesis-Part 2 (Renger, G., ed.) Cambridge: RSC Publishing. pp. 5-56.

Lang, A.S. and J.T. Beatty (2000). Genetic analysis of a bacterial genetic exchange element: The gene transfer agent of *Rhodobacter capsulatus*. *Proc Natl Acad SciUSA*. **97:** 859–864.

Lankester, E. Ray (1873). On a peach-coloured bacterium. *Quart. J. MICr. Sc.* **13:** 408-424.

Lankester, E. Ray (1976). Further observation on a peach- or red-colored bacterium-*Bacterium ruescens*. *Quart. J. Micro. Sci. N.S.***16:** 27-40.

Lapage, S.P., P.H.A. Sneath, E.F. Lessel Jr., V.B.D. Skerman, H.P.R. Seeliger and W.A. Clark (1992). international code of Nomenclature of Bacteria (1990) Revision. *Bacteriological Code,* American Society for Microbiology, Washington, DC.

Larimer, F.W., P. Chain, L. Hauser, J. Lamerdin, S. Malfatti, L. Do and D.A. Pelletier, J.T. Beatty, A.S. Lang, F.R. Tabita, J.L. Gibson, T.E. Hanson, C. Bobst, J.L. Torres y Torres, C. Peres, F.H. Harrison, J. Gibson and C.S. Harwood (2004). Complete genome sequence of the metabolically versatile photosynthetic bacterium *Rhodopseudomoans palustris*. *Nat. Biotechnol.* **22:** 55–61.

Larsen, H (1952). On the culture and general physiology of the green sulfur bacteria. *Journal of Bacteriol.* **64:** 187-196.

Lascelles, J. and D. Wertlieb (1971). Mutant strains of *Rhodopseudomonas spheroides* which form photosynthetic pigments aerobically in dark – growth characteristics and enzymic activities. *Biochim Biophys Acta.***226:** 328–340.

Laska, S., F. Lottspeich and A. Kletzin (2003). Membrane-bound hydrogenase and sulfur reductase of the hyperthermophilic and acidophilic archaeon *Acidianus ambivalens*. *Microbiology.***149:** 2357–2371.

Laurinavichene, T., D. Tekucheva, K. Laurinavichius and A. Tsygankov (2018). Utilization of distillery wastewater for hydrogen production in one-stage and two-stage processes involving photo fermentation. *Enzym. Microb. Technol.* **110:** 1-7.

Lauterborn, R (1913). Zur Kenntnis einiger sapropelischer Schizomyceten. *Allg. Bot. Z.* **19**: 97–100.

Law, C. J (1999). Structure and function studies on the LH1-RC core complex from a range of photosynthetic purple bacteria. Ph.D. thesis. The University of Glasgow, Glasgow, United Kingdom.

Law, C. J. and R.J. Cogdell (2008). Structure and Function of Antenna System. In *Primary Process of Photosynthesis* - Part 1 (Renger, G. ed.) Cambridge: RSC Publishing. pp. 205-260.

Lay, J.J. and T. Noike (1999). Feasibility of biological hydrogen production from the organic fraction of municipal solid waste. *Water Res.* **33:** 2579-2586.

Lazaro, C.Z., M.B. AmancioVaresche, E. LuizSilva, E (2015). Effect of inoculum concentration, pH, light intensity and lighting regime on hydrogen production by a phototrophic microbial consortium. *Renewable Energy*. **75:** 1-7.

Lee, C.M., C. Tsai, C. Yang, C.M Lee, C. Tsai and C. Yang (2010). Photohydrogen production from thermophilic aerobic digestion effluent and distillery wastewater by photosynthetic bacterium *Rhodopseudomonas palustris* WP 3-5, In: *Conference Proceedings* (Thomas Grube, eds.). 18th World Hydrogen Energy Conference 2010 - WHEC 2010. 33-35.

Lee, C.M., P.C. Chen, C.C. Wang and Y.C. Tung (2002). Photohydrogen production using purple non-sulfur bacteria with hydrogen fermentation reactor effluent. *Int. J. Hydr. Ener.* **27:** 1309-13.

Lee, K.H., R.H. Koh and H.G. Song (2008). Enhancement of growth and yield of tomato by *Rhodopseudomonas* sp. under greenhouse conditions. *J. Microbiol.* **46:** 641–646.

Lee, K.S., P.J. Lin and J.S. Chang (2006). Temperature effects on biohydrogen production in a granular sludge bed induced by activated carbon carriers. *Int. J. Hydr. Ener.* **31 :** 465-472.

Lee, S.S., T. Oh, J. Kim, J.B. Kim and H.S. Lee (2009). Bacteriocin from purple non-sulfur phototrophic bacteria, *Rhodobacter capsulatus*. *J. Bacteriol. Virolo.* **39(4):** 269-276.

Lee, Z.K., S.L. Li, P.C. Kuo, I.C. Chen, Y.M. Tien, Y.J. Huang and S.S. Cheng (2010). Thermophilic bioenergy process study on hydrogen fermentation with vegetable kitchen waste. *International Journal of Hydrogen Energy*. **35(24):** 13458– 13466.

Lee, S.K., Huu-Sheng Lur, Kai-Jiun Lo, Kuan-Chen Cheng, Chun-Chao Chuang, Shiueh-Jung Tang, Zhi-Wei Yang and Chi-Te Liu (2016). Evaluation of the effects of different liquid inoculant formulations on the survival and plant-growth-promoting efficiency of *Rhodopseudomonas palustris* strain PS3. *Appl. Microbiol. Biotechnol.* **100(18):** 7977–87.

Lenz, W. and R.H. Marchesault (2005). Bacterial polyesters: Biosynthesis, biodegradable plastics and biotechnology. *Biomacromolecules*.**6:** 1–8.

Leustek, T. and K. Saito (1999). Sulfate transport and assimilation in plants. *Plant Physiol.* **120:** 637–643.

Leustek, T., M.N. Martin, J.A. Bick and J.P. Davies (2000). Pathways and regulation of sulfur metabolism revealed through molecular and genetic studies. *Annu. Rev. Plant Physiol. Plant Mol. Biol.***51:** 141–165.

Levin, D.B., L. Pitt and M. Love (2004). Biohydrogen production: prospects and limitations to practical application. *Int. J. Hydr. Ener*. **29:** 173–185.

Levy, J.G (1995). Photodynamic therapy. *Trends Biotechnol.* **13:** 14-17.

Leyh, T.S (1993). The physical biochemistry and molecular genetics of sulfate activation. *Crit. Rev. Biochem. Mol. Biol.* **28:** 515–542.

Li, F., C.H. Hagemeier, H. Seedorf, G. Gottschalk and R.K. Thauer (2007). Re-citrate synthase from *Clostridium kluyveri* is phylogenetically related to homocitrate synthase and isopropylmalate synthase rather than to Si-citrate synthase. *J. Bacteriol.* **189:** 4299–4304.

Li, R., L. Morrison, G. Collins, A. Li and X. Zhan (2016). Simultaneous nitrate and phosphate removal from wastewater lacking organic matter through microbial oxidation of pyrrhotite coupled to nitrate reduction. Water Res. **96:** 32–41.

Li, R.Y and H.P. Herbert Fang (2008). Hydrogen production characteristics of photoheterotrophic *Rubrivivax gelatinosus* L31. *Int. J. Hydr. Energ*. **33:** 974-980.

Liao, W.B., H.L. Xiao and M.L. Zhang (2010). Effect of nitric oxide and hydrogen peroxide on adventitious root development from cuttings of ground-cover *Chrysanthemum* and associated biochemical changes. *J. Plant Growth Regul.* **29:** 338–348.

Libing Chu, Xingwen Zhang, Fenglin Yang, Xiaohui Li (2006). Treatment of domestic wastewater by using a microaerobic membrane bioreactor. *Desalination*.**189:** 181-192.

Lijinsky, W. and S.S. Epstein (1970). Nature (London). **225:** 21-23.

Lim, S. K., S.J. Kim, S. H Cha, Y.K. Oh, H.J. Rhee, M.S. Kim and J.K. Lee (2009). Complete genome sequence of *Rhodobacter sphaeroides* KD131. *J. Bacteriol.* **191:** 1118–1119.

Lim, W. J., K.M. Choi and S.Y. Hwang, S. Y (1993). *J. Microbiol. Biotechnol.* **3:** 244-251.

Liss, P (1999). Biogeochemistry: Take the shuttle – from marine algae to atmospheric chemistry. *Science*. **285:** 1217–1218.

Litchfield, J.H. (1980). Microbial protein production. *Bioscience*. **30:** 387-396.

Liu B. F., Y.R. Jin, Q.F. Cui, G.J. Xie, Y.N. Wu and N.Q. Ren (2015). Photo-fermentation hydrogen production by *Rhodopseudomonas* sp. nov strain A7 isolated from the sludge in a bioreactor. *Int. J. Hydrogen Energ*. **40:** 8661–8668.

Liu, B., N.Q. Ren, J. Ding,G.J, Xie and W.Q. Guo (2009). The effect of Ni_2 , Fe_2 and Mg_2 concentration on photo-hydrogen production by *Rhodopseudomonas faecalis* RLD-53. *Int. J. Hydr. Ener*. **34:** 721-6.

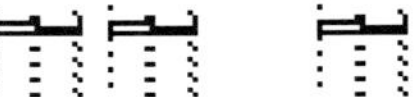

Liu, G.F, J.T. Zhou, J. Wang, Z.Y. Song and Y.Y. Qv (2006). Bacterial decolorization of azo dyes by *Rhodopseudomonas palustris*. *World J. Microbiol. Biotechnol.* **22(10):** 1069–1074.

Liu, G.P.W., T.C. Chang, L.M. Whang, C.H. Kao, P.T. Pan and S.S. Cheng (2011). Bioremediation of petroleum hydrocarbon contaminated soil: Effects of strategies and microbial community shift. *Int. Biodeter. Biodegrad.***65(8):** 1119-1127.

Liu, Y.W., K. Denkmann, K. Kosciow, C. Dahl and D.J. Kelly.(2013). Tetrathionate stimulated the growth of *Campylobacter jejuni* identifies TsdA as a new type of bi-functional tetrathionate reductase that is widely distributed in bacteria. *Mol. Microbiol.* **88:** 188.

Liu, Z. and J. Liu (2013). Evaluating bacterial community structures in oil collected from the sea surface and sediment in the northern Gulf of Mexico after the Deepwater Horizon oil spill. *Microbiology Open.***2:** 492–504.

Liu, Z., C. Yang and C. Qiao (2007). Biodegradation of p-nitrophenol and 4-chlorophenol by *Stenotrophomonas* sp. *FEMS Microbiol. Lett.***277:** 150–156.

Liu, Z., C.G. Klatt, M. Ludwig, D.B. Rusch, S.I. Jensen, M. Kuhl, D.M. Ward and D.A. Bryant (2012). *Candidatus Thermochlorobacter aerophilum,* an aerobic chlorophotoheterotrophic member of the phylum *Chlorobi* defined by metagenomics and metatranscriptomics. *ISME Journal*. **6(10):** 869-82.

Livia giottia, A. Agastiano, F. Italiano, F. Milano and M. Trotta (2006). Heavy metal ion influence on the photosynthetic growth of *Rhodobacter sphaeroides*. *Chemosphere*. **62:** 1490-9.

Lo S. C., S.H. Shih, J.J. Chang, C.Y. Wang and C.C. Huang (2012). Enhancement of photoheterotrophic biohydrogen production at elevated temperatures by the expression of a thermophilic clostridial hydrogenase. Appl. Microbiol. Biotechnol. **95**: 969–977.

Lo, Y.C., Ming-Der Bai., W.M. Chen and J.S.Chang (2008). Cellulosic hydrogen production with a sequencing bacterial hydrolysis and dark fermentation strategy. *Bioresour Technol.* **99(17):** 8299–8303.

Lo, Y.C., W.M. Chen, C.H. Hung, S.D. Chen and J.S. Chang (2008). Dark H_2 fermentation from sucrose and xylose using H_2-producing indigenous bacteria: Feasibility and kinetic studies. *Water Resea.* **42:** 827-842.

Loach, P.A. and P.S. Parkes-Loach (1995). Structure-function relationship in core light-harvesting complexes (LH1) as determined by characterization of the structural subunit and by reconstitution experiments, In: *Anoxygenic photosynthetic bacteria,* (Blankenship, R.E., M.T. Madigan and C.E. Bauer, ed.), Kluwer Academic Publishers, Dordrecht, The Netherlands, pp. 437–471.

Loeschcke, A., D. Dienst, V. Wewer, J. Hage-Hülsmann, M. Dietsch, S. Kranz-Finger, V. Huren, S. Metzger, V.B. Urlacher, T. Gigolashvili, S. Kopriva, I.M. Axmann, T. Drepper and K.E. Jaeger (2017). The photosynthetic bacteria *Rhodobacter capsulatus* and *Synechocystis* sp. PCC 6803 as new hosts for cyclic plant triterpene biosynthesis. *PLoS One*. **12(12):** e0189816.

Loo P.L (2012). Phototrophic Bacteria Grown in Palm Oil Mill Effluent As Feed For Culturing Rotifers and Marble Goby Larvae. PhD Thesis. The University of Malaya, Kuala Lumpur, Malaysia.

Lorrungruang. C., J. Martthong, K. Sasaki and N. Noparatnaraporn (2006). Selection of photosynthetic bacterium *Rhodobacter sphaeroides* 14f for polyhydroxyalkanoate production with two-stage aerobic dark cultivation. *J. Bioscie. Bioengine*. **101:** 28-131.

Lou, J., P. Sun, M. Guo, G. Wu and Y. Song (2011). Simultaneous sludge reduction and nutrient removal (SSRNR) with the interaction between Tubificidae and microorganisms: a fullscale study. *Bioresour. Technol.* **102:** 11132–11136.

Lozinsky, V.I., I.Y. Galaev, F.M. Plieva, I.N. Savina, H. Jungvid and B. Mattiasson (2003). Polymeric cryogels as promising materials of biotechnological interest. *Trends Biotechnol.* **21:** 445-451.

Lu H.F, G.M. Zhang and C.H. He (2011). The identification of a strain of photosynthetic bacteria and its effect in treating soybean wastewater. *J Harbin Inst Tech.* **43(12):** 72–76.

Lu, H., G. Zhang, T. Wan and Y. Lu (2011). Influences of light and oxygen conditions on photosynthetic bacteria macromolecule degradation: different metabolic pathways. *Bioresour Technol.* **102:** 9503–9508.

Lu, H., G. Zhang, X. Dai and C. He (2010). Photosynthetic bacteria treatment of synthetic soybean wastewater: direct degradation of macromolecules. *Bioresour Technol.* **101(19):** 7672–7674.

Lu, L., R. Ren, D. Xing and B.E. Logan (2009). Hydrogen production with effluent from ethanol–H_2-coproducing fermentation reactor using a single-chamber microbial electrolysis cell. *Biosens. Bioelectron.* **24:** 3055–3060.

Lubbe, Y.J., H.S. Youn, R. Timkovich and C. Dahl (2006). Siro(haem)amide in *Allochromatium vinosum* and relevance of DsrL and DsrN, a homolog of cobyrinic acid a,c diamide synthase for sulphur oxidation. *FEMS. Microbiol. Lett.* **261:** 194–202.

Ludden, P.W. and G.P. Roberts (1995). The Biochemistry and Genetics of Nitrogen Fixation by Photosynthetic Bacteria. In: *Anoxygenic Photosynthetic Bacteria*, (Blankenship, R.E., M.T. Madigan and C.E. Bauer, eds.), Kluwer Academic Publishers, Dordrecht, The Netherlands. pp. 929-947.

Luedin, S. M., J.F. Pothier, F. Danza, N. Storelli, N.U. Frigaard, M. Wittwer and M. Tonolla (2018). Complete genome sequence of "*Thiodictyon syntrophicum*" sp. nov. strain Cad16T , a photolithoautotrophic purple sulfur bacterium isolated from the alpine meromictic Lake Cadagno. *Standards in Genomic Sciences.* 13, (14). https://doi.org /10.1186/s40793-018-0317-z

Luo, W., X. Deng, W. Zeng and D. Zheng (2012). Treatment of wastewater from shrimp farms using a combination of fish, photosynthetic bacteria and vegetation. *Desalination and Water Treatment.* **47(1-3):** 221-227.

Mack, E.E., L. Mandelco, C.R. Woese and M.T. Madigan (1993). *Rhodospirillum sodomense*, sp. nov. A Dead Sea *Rhodospirillum* species. *Arch. Microbiol.* **160:** 363- 371.

Macpherson, A.N., J. Arellano, N.J. Fraser, R.J. Cogdell and T. Gillbro (1998). Ultra-fast energy transfer for rhodopin glucoside in the LH 2 complex from *Rhodopseudomonas acidophila* strain 10050. In G. Garab (ed.), Photosynthesis: mechanisms and effects, Vol. 1. Kluwer Academic Publishers, Dordrecht, The Netherlands. pp. 9–14.

MacRae, I.J., I.H. Segel and A.J. Fisher (2001). Crystal structure of ATP sulfurylase from *Penicillium chrysogenum*: insights into the allosteric regulation of sulfate assimilation. *Biochemistry.* **40:** 6795–6804.

Madigan, M.T (1986). *Chromatium tepidum* sp. nov., a thermophilic photosynthetic bacterium of the family *Chromatiaceae. Int. J. Syst. Bacteriol.* **36:** 222-227.

Madigan, M.T (1988). Microbiology, physiology and ecology of phototrophic bacteria.In: A. J. B. Zehnder (Ed.) *Biology of Anaerobic Microorganisms*. John Wiley. Chichester, UK. 39–111.

Madigan, M.T (2001). Family VI. ''*Heliobacteriaceae*'' Beer-Romero and Gest 1987, 113. In: *Bergey's Manual of Systematic Bacteriology*, (Boone, D.R., R.W. Castenholz and G. M. Garrity, eds.), Second Edition, Vol. 1, *New York: Springer*, pp. 625–626.

Madigan, M.T (2003). Anoxygenic phototrophic bacteria from extreme environments. *Photosynth. Res.* **76:** 157-171.

Madigan, M.T (2006). The family *Heliobacteriaceae*, In The Prokaryotes, 3rd Edn, Vol. 4, eds. M. Dworkin, E. Rosenberg, K.-H. Schleifer and E. Stackebrandt (New York: Springer), 951–964.

Madigan, M.T (2015). *Heliobacillus*. Bergey's Manual of Systematics of Archaea and Bacteria. pp. 1–5. doi:10.1002/9781118960608.gbm00633.

Madigan, M.T (2015). *Heliobacterium*. Bergey's Manual of Systematics of Archaea and Bacteria. pp. 1–4. doi:10.1002/9781118960608.gbm00634.

Madigan, M.T (2015). *Heliophilum*. Bergey's Manual of Systematics of Archaea and Bacteria. pp. 1–5. doi:10.1002/9781118960608.gbm00635.

Madigan, M.T (2015). *Heliorestis*. Bergey's Manual of Systematics of Archaea and Bacteria. pp. 1–4. doi:10.1002/9781118960608.gbm00636.

Madigan, M. andS.S. Cox (1982). Nitrogen metabolism in *Rhodopseudomonas globiformis*. *Arch. Microbiol.***133:** 6–10.

Madigan, M.T. and D.O. Jung (2009). An overview of purple bacteria: systematics, physiology and habitats. In: The Purple Phototrophic Bacteria, (Hunter, C.N. F. Daldal, M. C. Thurnauer and J. T. Beatty, eds.) (Dordrecht: Springer Academic Publishers). pp. 1–15.

Madigan, M.T. and D.O. Jung (2009). In: *The Purple Phototrophic Bacteria,* (Neil Hunter, C., Fevzi Daldal, Marion C. Thurnauer and J. Thomas Beatty, eds.), pp. 1-15.

Madigan, M.T. and J.F. Imhoff (2015). *Rhodopila. Bergey's Manual of Systematics of Archaea and Bacteria,* 1–10. doi:10.1002/9781118960608.gbm00886.

Madigan, M.T. and T.D. Brock (1975). Photosynthetic sulfide oxidation by *Chloroflexus aurantiacus*, a filamentous, photosynthetic, gliding bacterium. *J. Bacteriol.* **122:** 782–784.

Madigan, M.T., D.O. Jung, C.R. Woese and L.A. Achenbach (2000). *Rhodoferax antarcticus*, sp. nov., a moderately psychrophilic purple non-sulfur bacterium from an *Antarctic microbial mat. Arch. Microbiol.* **173:** 69-277.

Madigan, M.T., S.S. Cox and R.A. Stegeman (1984). Nitrogen fixation and nitrogenase activities in members of the family *Rhodospirillaceae.J. Bacteriol.* **157:** 73–78.

Madukasi, E., X. Dai, C. He and J. Zhou (2010). Potentials of phototrophic bacteria in treating pharmaceutical wastewater. *Int. J. Environ. Sci. Technol.* **7:** 165–174.

Madukasi, E.I. and G. Zhang (2010). Microaerobic biodegradation of high organic load wastewater by phototrophic bacteria. *Afr. J. Biotechnol.***9:** 3852-3860.

Madukasi, E.I., H. Chunhua and G. Zhang (2011). Isolation and application of a wild strain photosynthetic bacterium to environmental waste management. *Int. J. Environ. Sci. Technol.***8(3):** 513–522.

Maeda, I., H. Miyasaka, F. Umeda, M. Kawase and K. Yagi (2003). Maximization of hydrogen production ability in the high-density suspension of *Rhodovulum sulfidophilum* cells using intracellular poly (3-hydroxybutyrate) as the sole substrate. *Biotechnol. Bioeng.* **81(4):** 474–81.

Maeda, N., K. Kitano, T. Fukui, S. Ezaki, H. Atomi, K. Miki and T. Imanaka (1999). Ribulose bisphosphate carboxylase/oxygenase from the hyperthermophilic archaeon *Pyrococcus kodakaraensis* KOD1 is composed solely of large subunits and forms a pentagonal structure. *J. Mol. Biol.* **293:**57–66.

Maeda, T., V. Sanchez-Torres and T.K. Wood (2007). Enhanced hydrogen production from glucose by metabolically engineered *Escherichia coli*. *Appl Microbiol Biotechnol.* **77:** 879–90.

Maeda, T., V. Sanchez-Torres and T.K. Wood (2008a). Protein engineering of hydrogenase 3 to enhance hydrogen production. *Appl. Microbiol. Biotechnol.* **79:** 77–86.

Maggini, L., Raquez, Jean-Marie, Marega Riccardo, Jensen Jensen Ahrens. Florent Pineux, Meyer Franck, Dubois Philippe and Bonifazi Davide (2012). Magnetic Poly (vinylpyridine)-Coated Carbon Nanotubes: An Efficient Supramolecular Tool for Wastewater Purification. *Chem. Sus. Chem.* **6(2).** 10.1002/cssc.201200413.

Mahajan, A. and S. Dua (1995). A perspective on biotechnological potential. *J. Food. Sci. Technol.* **32:** 162-165.

Mahakhan, P., C. Chobvijuk, M. Ngmjarearnwong, S. Trakulnalermsai, C.S. Bucke, J. Svasti, W. Kanlayakrit and L. chitradon (2005). Molecular hydrogen production by a thermotolerant *Rubrivivax gelatinosus* using raw cassava starch as an electron donor. *Science Asia* .**31:** 415–24.

Mahuya, M., A. Patra and A.K. Paul (2005). Production of polyhydroxy butyrate and polyhydroxybutyrateco-valerate by *Rhodopseudomonas palustris* SP5212. *W. J. Microbiol.Biotechnol.* **21:** 765-769.

Mahyudin, A.R., Y. Furutani, Y. Nakashimada, T. Kakizono and N. Nishio (1997). Enhanced hydrogen production in altered mixed acid fermentation of glucose by *Enterobacter aerogenes. J. Ferment. Bioeng.* **83:** 358-363.

Mandel, M., E.R. Leadbetter, N. Pfennig and H.G. Truper (1971). The deoxyribonucleic acid-base composition of phototrophic bacteria. *Int. J. Syst. Bacteriol.***21:** 222–230.

Maness, P.C. and P.F. Weaver (1994). *Appl. Biochem. Biotechnol.* **45-46:** 395-406.

Maness, P.C. and P.F. Weaver (2002). Hydrogen production from carbon-monoxide oxidation pathway in *Rubrivivax gelatinosus.Int. J. Hydr. Ener.***27:** 1407-11.

Maness, P.C., J. Huang, S. Smolinski, V. Tek and G. Vanzin (2005). Energy generation from the CO oxidation-hydrogen production pathway in *Rubrivivax gelatinosus. Applied and Environmental Microbiology* .**71(6):** 2870-4.

Maness, P.C., S. Smolinski, A.C. Dillon, M.J. Heben and P.F. Weaver (2002). Characterization of the oxygen tolerance of a hydrogenase linked to a carbon monoxide oxidation pathway in *Rubrivivax gelatinosus. Appl Environ. Microbiol.***68:** 2633-2636.

Manimaran, A. and T.M. Vatsala (2007). Biobleaching of banana fiber pulp using *Bacillus subtilis* C O1 xylanase produced from wheat bran under solid-state cultivation. *J. Indust. Microbiol.Biotechnol.* **34(11):** 745-749.

Manisha D.R, M.P. Pratap Rudra and M. Ramchander (2015). Hydrogen production by photosynthetic bacterial consortium isolated from sewage water. *Int. J. Resear. Biochem. Biophys.* **5(1):** 10-14.

Manske, A.K., J. Glaeser, M.A.M. Kuypers and J. Overmann (2005). Physiology and phylogeny of green sulfur bacteria forming a monospecific phototrophic assemblage at a depth of 100 meters in the Black Sea. *Appl. Environ. Microbiol.* **71:** 8049–8060.

Marchessault, R. H., T.L. Bluhm, Y. Deslandes, G.K. Hamer, W.J. Orts, P.J. Sundarajan, M.G.Taylor, S. Bloembergen and D.A. Holden (1988). *Makromol. Chem., Macromol. Symp.* **1:** 235-254.

Marcia, M., J.D. Langer, D. Parcej, V. Vogel, G. Peng and H. Michel (2010b). Characterizing a monotopic membrane enzyme. Biochemical, enzymatic and crystallization studies on *Aquifex aeolicus* sulfide:quinone oxido-reductase. *Biochem. Biophys. Acta. Biomembr.* **1798:** 2114–2123.

Marcia, M., U. Ermler, G.H. Peng and H. Michael (2009). The structure of *Aquifex aeolicus* sulfide:quinone oxidoreductase, a basis to understand sulfide detoxification and respiration. *Proc. Natl. Acad. Sci. U.S.A.* **106:** 9625–9630.

Marcia, M., U. Ermler, G.H. Peng and H. Michel (2010a) A new structure-based classification of sulfide:quinoneoxidoreductases. Proteins: *Struct. Funct. Bioinf.* **78:**1073–1083.

Markov, S.A. and P.F. Weaver (2008). Bioreactors for H_2 production by Purple Non-sulfur Bacteria. *Appl. Biochem. Biotechnol.* **145:** 79-86.

Maroc, J., De Klerk, H., and Kamen, M.D. (1968). Quinones of *Athiorhodaceae. Biochim. Biophys. Acta.* **162:** 621-623.

Marnocha, C.L, A.P. Levy, D.H. Powell, T.E. Hanson and C.S. Chan (2016). Mechanisms of extracellular S(0) globule production and degradation in *Chlorobaculum tepidum* via dynamic cell-globule interactions. *Microbiology.* **162(7):** 1125–1134.

Masepohl, B. and R.G. Kranz (1978). Regulation of nitrogen fixationIn . Current Topics in Cellular Regulation. New York; London: Academic Press, pp. 179–232.

Martinez-Luque, M., M.M. Dobao and F. Castillo (1991). Characterization of the assimilatory and dissimilatory nitrate-reducing systems in *Rhodobacter*: a comparative study. *FEMS Microbiol. Lett.* **83:**329-334.

Matheron, R. and R. Baulaigue (1972). Marine phototrophic sulfur bacteria, assimilation of organic and mineral substances and influence of the NaCl content of the medium upon growth. *Arch. Mikrobiol.* **86:** 291–304.

Matsumura, M., T. Kobayashi and S. Aiba (1983). *Bur. J. Appl. Mcrobiol. Biotechnol.* **17:** 85-89.

Matsunaga T., T. Hatano, A. Yamada and M. Matsumoto (2000). Microaerobic hydrogen production by photosynthetic bacteria in a double-phase photobioreactor. *Biotechnol Bioeng.* **68(6):** 647–51.

Maturrano, L., F. Santos, R. Rossello-Mora and J. Anton (2006). Microbial diversity in Maras salterns, a hypersaline environment in the Peruvian Andes. *Appl Environ Microbiol.* **72:** 3887–3895.

Maudinas, B., M. Chemardin, E. Yovanovitch and P. Gadal (1981) Gnotobiotic cultures of rice plants up to ear stage in the absence of combined nitrogen source but in the presence of free-living nitrogen-fixing bacteria *Azotobacter vinelandii*and *Rhodopseudomonas capsulata. Plant Soil.* **60:** 85–97.

McDermott, G., S.M. Prince, A.A. Freer, A.M. Hawthornthwaite-Lawless, M.Z. Papiz, R.J. Cogdell and N.W. Isaacs (1995). Crystal-structure of an integral membrane light-harvesting complex from photosynthetic bacteria. *Nature.* **374:** 517-521.

McEwan, A.G., H.G. Wetzstein, O. Meyer, J.B. Jackson and S.J. Ferguson (1987). The periplasmic nitrate reductase of *Rhodobacter capsulatus*; purification, characterization and distinction from a single reductase for trimethylamine-N-oxide, dimethylsulphoxide and chlorate. *Arch. Microbiol.* **147:** 340–345.

McGoldrick, H.M., C.A. Roessner, E. Raux, A.D. Lawrence, K.J. Mc Lean, A.W. Munro, S. Santabarbara, S.E. Rigby, P. Heathcote, A.I. Scott and M.J. Warren (2005). Identification and characterization of a novel vitamin B_{12} (cobalamin) biosynthetic enzyme (CobZ) from *Rhodobacter capsulatus*, containing flavin, heme and Fe-S cofactors. *The Journal of biological chemistry.* **280(2):** 1086-94.

McGrath, J.E. and C.G. Harfoot (1997). Reductive dehalogenation of halocarboxylic acids by the phototrophic genera *Rhodospirillum* and *Rhodopseudomonas. Appl. Environ. Microbiol.***63(8):** 3333–3335.

McInerney, M.J., M.P. Bryant, R.B. Hespell and J.W. Costerton (1981). *Syntrophomonas wolfei* gen. nov. sp. nov., an anaerobic, syntrophic, fatty acid oxidizing bacterium. *Applied and Environmental Microbiology.* **41:** 1029-1039.

McKinlay, J.B. and C.S. Harwood (2010). Carbondioxide fixation as a central redox cofactor recycling mechanism in bacteria. *Proc. Natl. Acad. Sci. U.S.A.* **107:** 11669–11675.

McKinlay, J.B. and C.S. Harwood (2011). Calvin cycle flux, pathway constraints and substrate oxidation state together determine the H_2 biofuel yield in photoheterotrophic bacteria. *MBio.* **2:** e00323–e00310.

Mclnerney, M.J. and M.P. Bryant (1981). *Fuel Gas Production from Biomass* (Wise, D.L. ed.), Chemical Rubber Co. Press Inc., West Palm Beach, Florida. pp. 26-40.

McTavish, H., L.A. Sayavedra-Soto and D.J. Arp (1995). Substitution of *Azotobacter uinelandii* hydrogenase small-sub unit cysteines to serines can create insensitivity to inhibition by O_2 and preferentially damages H_2 oxidation over H_2 evolution. *J. Bacteriol.* **177:** 3960-3964.

Meckenstuck, R., K. Krusche, L. A. Staehelin, M. Cryklaff, R. A. Brunisholz and H. Zuber (1994). The six fold symmetry of the B880 light-harvesting membranes of *Rhodopseudomonas marina. Biol. Chem. Hoppe-Seyler.* **375:** 429-438.

Meckenstuck, R., K. Krusche, R. A. Brunisholz and H. Zuber (1992). The light-harvesting core-complex and the B820-subunit from *Rhodopseudomonas marina*. Part II. Electron microscopic characterization. *FEBS Lett.***311:** 135–138.

Melis, A. and M.R. Melnicki (2006). Integrated biological hydrogen production. *Int. J. Hydr. Ener.***31(11):** 1563-1573.

Melis, T (2005). Integrated biological hydrogen production. Proceedings, *International Hydrogen Energy Congress and ExhibitionIHEC*. Istanbul, Turkey.

Melnicki, M.R., E. Eroglu and A. Melis (2009). Changes in hydrogen production and polymer accumulation upon sulfur deprivation in purple photosynthetic bacteria. *Int. J. Hydr. Ener.***34:** 6157–6170.

Mendez-Alvarez, S., I. Esteve, R. Guerrero and N. Gaju (1996). Genomic analysis of different *Chlorobium* strains by pulsed-field gel electrophoresis and ribotyping. *Int. J. Syst. Bacteriol.* **46:** 1177–1179.

Menendez, C., Z. Bauer, H. Huber, N. Gadon, K.O. Stetter and G. Fuchs (1999). Presence of acetyl coenzyme A (CoA) carboxylase and propionyl-CoA carboxylase in autotrophic *Crenarchaeota* and indication for the operation of a 3-hydroxypropionate cycle in autotrophic carbon fixation. *Journal of Bacteriology*. **181(4):** 1088–1098.

Meulenberg, R., J.T. Pronk, J. Frank, W. Hazeu, P. Bos and J.G. Kuenen (1992a). Purification and partial characterization of a thermostable trithionate hydrolase from the acidophilic sulfur oxidizer *Thiobacillus acidophilus*. *Eur. J. Biochem.* **209:** 367–374.

Meulenberg, R., J.T. Pronk, W. Hazen, J.P. van Dijken, J. Frank, P. Bos and J.G. Kuenen (1993). Purification and partial characterization of thiosulphate dehydrogenase from *Thiobacillus acidophilus*. *J. Gen. Microbiol.* **139:** 2033–2039.

Meulenberg, R., J.T. Pronk, W. Hazeu, P. Bos and J.G. Kuenen (1992b). Oxidation of reduced sulphur compounds by intact cells of *Thiobacillus acidophilus*. *Arch. Microbiol.***157:** 161–168.

Meyer, B. and J. Kuever (2007). Molecular analysis of the distribution and phylogeny of dissimilatory adenosine-5′-phosphosulfate reductase-encoding genes (aprBA) among sulfur-oxidizing prokaryotes. *Microbiology*. **153:** 3478–3498.

Meyer, T.E., R.G. Bartsch, M.A. Cusanovich and J.H.Mathewson (1968). The cytochromes of *Chlorobium thiosulfatophilum*. *Biochim. Biophys. Acta.* **153**: 854–861.

Michalski, W.P., D.H. Hein and D.J.D. Nicholas (1986). Purification and characterization of nitrous oxide reductase from *Rhodopseudomonas sphaeroides* f.sp. *denitrificans*. *Biochim. Biophys. Acta.***872:** 50-60.

Migula, W (1900). Das System der Bakterien, Vol. 2. Gustav Fischer, Jena.

Milford, A.D., L.A. Achenbach, D.O. Jung and M.T. Madigan (2000). *Rhodobaca bogoriensis* gen. nov. and sp. nov., an alkaliphilic purple nonsulfur bacterium from African Rift Valley soda lakes. *Arch. Microbiol.* **174:**18–27.

Milford, A.D., L.A. Achenbach, D.O. Jung and M.T. Madigan (2001). In:Validation of the publication of new names and new combinations previously effectively published outside the IJSB. List No. 80. *Int. J. Syst. Bacteriol.* **51:** 793–794.

Miller, K.R (1982). Three-dimensional structure of a photosynthetic membrane. *Nature*. **300:** 53–55.

Miller, K.R., J.S. Jacob, U. Smith, S. Kolaczkowski and M.K. Bowman.(1986). *Heliobacterium chlorum*: cell organization and structure. *Arch. Microbiol.***146:** 111–114.

Min, H., H. Guo and J. Xiong (2005). Rhythmic gene expression in a purple photosynthetic bacterium, *Rhodobacter sphaeroides. FEBS Letters*. **579(3):** 808–812.

Mitsubishi Heavy Industries, Ltd., Japan (1982). Japan Kokai Tokkyo Koho JP 82121,400.

Mitsubishi, J. E. K. K (1985). Japan Kokai Tokkyo Koho JP 85118,479.

Mitsui, A., S. Kumazawa, E.J. Philips, K. J. Reddy, K. Gill, T. Matsunaga, B.R. Renuka, T. Kusumi, G. Reyes-Vasquez, K. Miyazawa, L. Haynes, H. Ikemoto, E. Duerr, C. B. Leon, D. Rosner, R. Sesco and E. Moffat (1985a). In: "*Biotechnology and Bioprocess Engineering*", (Ghose, T.K. ed.), International Union of Pure and Applied Chemistry and Indian National Science Academy United India Press, New Delhi. pp. 119-155.

Mitsui, A., T. Matsunaga, H. Ikemoto and B.R. Renuka (1985). *Developm. Indust. Microbiol.* **26:** 209-222.

Mittal, V., O. Singh, H. Nayyar, J. Kaur and R. Tewari (2008). The stimulatory effect of phosphate-solubilizing fungal strains (*Aspergillus awamori* and *Penicillium citrinum*) on the yield of chickpea (*Cicer arietinum* L. cv. GPF2). *Soil. Biol. Biochem.***40:** 718-727.

Miyake, J., N. Tomizuka and A. Kamibayashi (1982). Prolonged photo-hydrogen production by *Rhodospirillum rubrum, J. Ferment. Technol.* **60:** 199-203.

Miyamoto, K (1997). Hydrogen production. In: *Enewable Biological System for alternative sustainable energy production* (Miyamoto, K. ed.), Food and Agriculture Organization. **128:** 108.

Mizuno, O., R. Dinsdale, F.R. Hawkes, D.L. Hawkes and T. Noike (2000). Enhancement of hydrogen production from glucose by nitrogen gas sparging. *Bioresource Technol.* **73(1):** 59-65.

Mochida, K., T. Ohsaki, T. Kobayashi and R. Kimura (1975). *Jpn. Kokai Tokkyo Koho. JP* 75/27, 668.

Mochida, K., T. Ohsaki, T. Kobayashi and R. Kimura (1975). *Jpn. Kokai Tokkyo Koho. JP* 75/27, 666.

Moe, W.M., J. Yan, M.F. Nobre, M.S. da Costa and F.A. Rainey (2009). *Dehalogenimonas lykanthroporepellens* gen. nov., sp. nov., a reductively dehalogenation bacterium isolated from chlorinated solvent-contaminated groundwater. *Int. J. Syst. Evol. Microbiol.* **59:** 2692–2697.

Mohd. Ali Hassan, Yoshihito Shirai, Noriko Kusubayashi, Mohd. Ismail Abdul Karim, Kazuhiro Nakanishi, Kenji Hasimoto (1997). The production of polyhydroxyalkanoate from anaerobically treated palm oil mill effluent by *Rhodobacter sphaeroides. J. Fermenta. Bioengine.* **83(5):** 485-488.

Mokhtar M., Z.A. Aris and V. Munusamy (2009) Assessment Level of Heavy Metals in *Penaeus Monodon* and *Oreochromis* spp. in Selected Aquaculture Ponds of High Densities Development Area. *Europ. J. Scientif. Resear.* **30(3):** 348-360.

Molisch, H (1907). Die Purpurbakterien nach neuen Untersuchungen. G. Fischer, Jena.

Mondal, K., A. Kaviraj and P.K. Mukhopadhyay (2012). Effects of partial replacement of fishmeal in the diet by mulberry leaf meal on growth performance and digestive enzyme activities of Indian minor carp Labeo bata. *Int. J. Aqu. Sci.* **3(1):** 72–83.

Montano, G.L., J.S. Chan, R.E. Jarabelo, A.B. Pastor and T.E. Dela Cruz (2009). Isolation and characterization of purple non-sulfur bacteria (PNSB) from a rice paddy soil in Bulacan, Philippines. *Philippine Journal of Systematic Biology*. **3(1).** doi:10.3860/pjsb.v3i1.1013.

Moore, H.W. and K. Folkers (1966). *J. Am. Chem. SOC.* **88:** 567-570.

Moore, M.D. and S. Kaplan (1992). Identification of intrinsic high-level resistance to rare earth oxides and oxyanions in members of the class *Proteobacteria*: characterization of tellurite, selenite and rhodium sesquioxide reduction in *Rhodobacter sphaeroides*. *J. Bacteriol.* **74:** 1505-1514.

Moran, M.A., Gonzalez, J.M. and R.P. Kiene (2003) Linking a bacterial taxon to sulfur cycling in the sea: studies of the marine *Roseobacter* group. *Geomicrobiol. J.***20:** 375–388.

Mortensen, S. A., S. Vadhanavikit, K. Muratsu and K. Folkers (1990). *Int. I. TissueReact.* **12:** 155-162.

Mujahid, Md., Ch. Sasikala and Ch.V. Ramana (2010). Aniline-induced tryptophan production and identification of indole derivatives from three purple bacteria. *Curr. Microbiol.***61:** 285-290.

Mulkidjanian, A.Y., E.V. Koonin, K.S. Makarova, S.L. Mekhedov, A. Sorokin, Y.L. Wolf, A. Dufresne, F. Partensky, H. Burd, D. Kaznadzey, E. Haselkorn and M.Y. Galperin (2006). The cyanobacterial genome core and the origin of photosynthesis. *Proc. Natl. Acad. Sci. USA.***103:** 13126–13131.

Muller, F.H., T.M. Bandeiras, T. Urich, M. Teixeira, C.M. Gomes and A. Kletzin.(2004). Coupling of the pathway of sulphur oxidation to dioxygen reduction: characterization of a novel membrane-bound thiosulphate:quinone oxido-reductase. *Mol. Microbiol.* **53:** 1147–1160.

Murray, R.G.E (1984). The higher taxa, or, a place for everything…? In: Bergey's Manual of Systematic Bacteriology, (Krieg and Holt, eds.), Vol. 1. Williams and Wilkins, Baltimore. pp. 31–34.

Nadson, G.A (1906). The morphology of inferior algae. III.*Chlorobium limicola* Nads., the green chlorophyll bearing microbes. *Bull. Jard. Bot. St. Petersb.* **6:** 190.

Nagadomi, H., K. Takahashi, K. Sasaki and H.C. Yang (2000). Simultaneous removal of chemical oxygen demand and anerobi treatment of segwage water using an immobilized photosyntheic bacterium of porous ceramic plates. *W. J. Micrbiol. Biotehnol.* **16:** 57-62.

Nagadomi, H., T. Hiromitsu, K. Takeno, M. Watanabe and K. Sasaki (1999). Treatment of aquarium water by denitrifying photosynthetic bacteria using immobilized polyvinyl alcohol beads. *J. Biosci. Bioeng.* **87:** 189–193.

Nagadomi, H., T. Kitamura, M. Watanabe, K. Sasaki (2000). Simultaneous removal of chemical oxygen demand (COD), phosphate, nitrate and H_2S in the synthetic sewage wastewater using porous ceramic immobilized photosynthetic bacteria. *Biotechnology Letters*. **22(17):** 1369–1374.

Nagumo, A., F. Takanashi, H. Hojo and Y. Suzuki, 1991. *Toxicol. lett.* **58:** 309-313.

Najafpour, B.D (2007). Single-cell Protein, Biochemical Engineering and Biotechnology. **2:** 332-341.

Najafpour, G., H. Younesi and A.R. Mohamed (2004). Effect of organic substrate on hydrogen production from synthesis gas using *Rhodospirillum rubrum* in batch culture. *Biochem Eng J.***21:** 123–30.

Najafpour, G.D. and H. Younesi (2007). Bioconversion of synthesis gas to hydrogen using a light-dependent photosynthetic bacterium, *Rhodospirillum rubrum*. *World J Microbiol Biotechnol.***23(2):** 275-284.

Najafpour, G.D., H. Younesi and A.R. Mohamed (2003) Continuous hydrogen production via fermentation of synthesis gas. *Int. J. Petrol Process, Petrochem Coal Process.* **45:** 154–158

Nakamura, H (1939). Metabolism of purple bacteria. V. Hydrogen transformations in the purple bacteria, including the mutual relationships between - Thio and *Athiorhodaceae*. *Acta Phytochim* (Japan). **11:** 109-125.

Nakashimada, Y., M.A. Rachman, T. Kakizono and N. Nishio (2002). Hydrogen production of *Enterobacter aerogenes* altered by extracellular and intracellular redox states. *Int. J. Hydr. Ener.* **27:** 1399-1405.

Nakatani, T., I. Ohtsu, G. Nonaka, N.Wiriyathanawudhiwong, S. Morigasaki and H. Takagi (2012). Enhancement of thioredoxin/glutaredoxin-mediated L-cysteine synthesis from S-sulfocysteine increases L-cysteine production in *Escherichia coli*. *Microb. Cell. Fact.* **11:** 62.

Narula, N., V. Kumar, R.K. Behl, A. Deubel, A. Gransee and W. Merbach (2000). Effect of P-solubilizing *Azotobacter chroococcum* on N, P, K uptake in P-responsive wheat genotypes grown under greenhouse conditions. *J. Plant Nutr. Soil Sci.* **163:** 393-398.

Nasseri, A.T., S. Rasoul-Amini, M.H. Moromvat and Y. Ghasemi (2011). Single-cell protein, production and process. *American Journal of Food Technology*. **6(2):** 103-116.

Nath, K. and D. Das (2003). Hydrogen from biomass. *Cur Sci*. **85:** 265-271

Nath, K. and D. Das (2004). Improvement of fermentative hydrogen production: various approaches. *Appl. Microbiol. Biotechnol.* **65:** 520–529.

Nath, K. and D. Das (2009). Effect of light intensity and initial pH during hydrogen production by an integrated dark and photo fermentation process. *Int. J. Hydr. Ener.* **34:** 7497-7501.

Nath, K., A. Kumar and D. Das (2005). Hydrogen production by *Rhodobacter sphaeroides* strain O.U.001 using spent media of *Enterobacter cloacae* strain DM11. *Appl Microbiol Biotechnol*. **66:** 533–541.

Nath, K., M. Muthukumar, A. Kumar and D. Das (2008). Kinetics of Two-Stage Fermentation Process for the Production of Hydrogen. *Int. J. Hydr. Ener*. **33(4):** 1195-1203.

Negishi, M., L.G. Pedersen, E. Petrotchenko, S. Shevtsov, A. Gorokhov, Y. Kakuta and L.C. Pedersen (2001). Structure and function of sulfotransferases. *Arch. Biochem. Biophys.* **390:** 149–157.

Neik, T.S (2006). The enhancement of growth and survival of ornamental koi (*Cyprinus carpio*) larvae using phototrophic bacteria, Master thesis, Faculty of Science, University of Malaya, Kuala Lumpur, Malaysia.

Naeem, M.S., M. Rasheed, D.Liu, Z.L. Jin, D.F. Ming, K. Yoneyama, *et al* (2011). 5-Aminolevulinic acid ameliorates salinity-induced metabolic, water-related and biochemical changes in *brassica napus* L. *Acta Physiol. Plant* .**33:** 517–528.

Nepple, B.B., J. Kessi and R. Bachofen (2000). Chromate reduction by *Rhodobacter sphaeroides*. *J. Ind. Microbiol. Biotechnol*. **25:** 198-203.

Neumann, S., A. Wynen, H.G. Truper and C. Dahl (2000). Characterization of the cys gene locus from *Allochromatium vinosum* indicates an unusual sulfate assimilation pathway. *Mol. Biol. Rep*. **27:** 27–33.

Neunlist, S. and M.A. Rohmer (1985). Novel hopanoid, 30-(5'-adenosyl) hopane, from the purple non-sulphur bacterium *Rhodopseudomonas acidophila*, with possible DNA interactions. *Biochem. J.* **228(3):** 769–771.

Neunlist, S., P. Bisseret and M. Rohmer (1988). The hopanoids of the purple non sulfur bacteria *Rhodopseudomonas palustris* and *Rhodopseudomonas acidophila* and the absolute configuration of bacteriohopanetetrol. *Eur.J. Biochem.* **171:** 245–252.

Neunlist, S., O. Holst and M. Rohmer (1985). Prokaryotic triterpenoids. The hopanoids of the purple non-sulphur bacterium *Rhodomicrobium vannielii*: an aminotriol and its aminoacyl derivatives, N-tryptophanyl and N-ornithinyl aminotriol. *Eur. J. Biochem.* **147(3):** 561–568.

Nicholson, J.A.M., J.F. Stolz and B.K. Pierson (1987). Structure of a microbial mat at Great Sippewissett Marsh, Cape Cod, Massachusetts. *FEMS Microbiol Ecol.* **45:** 343–364.

Niels-Ulrik Frigaard and Donald A. Bryant (2004). Seeing green bacteria in a new light: genomics-enabled studies of the photosynthetic apparatus in green sulfur bacteria and filamentous anoxygenic phototrophic bacteria. *Arch. Microbiol.* **182:** 265–276.

Nissen, H. and I.D. Dundas (1984). *Rhodospirillum salinarum* sp. nov., a halophilic photosynthetic bacterium from a Portuguese saltern. *Arch. Microbiol.* **138:** 251– 256.

Nookongbut, P., D. Kantachote, M. Megharaj and R. Naidu (2018) Reduction in arsenic toxicity and uptake in rice (Oryza sativa L.) by As-resistant purple nonsulfur bacteria. *Environ. Sci. Pollut. Res* .**25:** 36530–36544.

Noparatnaraporn, N. and S. Nagai (1986). Selection of *Rhodobacter sphaeroides* P47 as a useful source of single-cell protein. *J. Gen. Appl. Microbiol.* **32**: 351-359.

Noparatnaraporn, N., K. Sasaki, Y. Nishiiawa and S. Nagai (1986). *Biotechnol. Lett.* **8:** 491-496.

Noparatnaraporn, N., K. Takeno and K. Sasaki (2001). Hydrogen and polyhydroxybutyrate production from organic acids by photosynthetic bacteria. In: *Biohydrogen II Pergamon*, (Miyake, J., T. Matsunaga and A. Pan Pietro, eds.), *Elsevier Science*, Amsterdam, London, New York. pp. 33-40.

Noparatnaraporn, N., S. Trakulnaleumsai, R.G. Silveira, Y. Nishizawa and S. Nagai (1987). SCP production by mixed culture of *Rhodocyclus gelatinosus* and *Rhodobacter sphaeroides* from Cassava Waste. *Journal of Fermentation Technology*. **65(1):** 11-16.

Noparatnaraporn, N., S. Trakulnaleumsai, R.G. Silveria, Y. Nishizawa and S. Nagai, N. Noparatnaraporn, S. Trakulnaleumsai and S. Duangsawat (1987b). *J. Sci. SOC.* Thai land **13:**15-27.

Noparatnaraporn, N., Y. Nishizawa, M. Hayashi and S. Nagai (1983). Single-cell protein production from cassava starch by *Rhodopseudomonas gelatinosa. J. Ferment. Technol.* **61:** 515-519.

Ntougias, S (2014). Phylogeny and ecophysiological features of prokaryotes isolated from temporary saline tidal pools. *Ann. Microbiol.* **64:** 599–609.

Nubel U., M.M. Bateson, M.T. Madigan, M. Kuhl and D.M. Ward (2001). The diversity and distribution of bacteria phylogenetically related to *Chloroflexus* in hypersaline microbial mats. *Appl. Environ. Microbiol.* **67:** 4365-71.

Numata, K., H. Abe and T. Iwata (2009). Biodegradability of Poly-(hydroxyalkanoate) Materials. *Materials (Basel).* **2(3):** 1104–1126.

Nupur, N., M.K. Saini, P K. Singh, S. Korpole, T.N.R. Srinivas, S. Takaichi and A. K. Pinnaka (2017). *Imhoffiella* gen. nov., a marine phototrophic member of the family *Chromatiaceae* including the description of *Imhoffiella purpurea* sp. nov. and the reclassification of *Thiorhodococcus bheemlicus* Anil Kumar *et al.* 2007 as *Imhoffiella bheemlica* comb. nov. *Int. J. Syst. Evol. Microbiol.* **67:** 1949-1956.

Nupur, P., T.N.R. Srinivas, S. Takaichi and P. Anil Kumar (2014). *Rhodovulum mangrovi* sp. nov., a phototrophic *alphaproteobacterium* isolated from a mangrove forest sediment sample.*Int. J. Syst. Evol. Microbiol.***64**: 3168-3173.

Nupur, P., T.N.R. Srinivas, S. Takaichi and P. Anil Kumar (2015). *Phaeobacterium nitratireducens* gen. nov., sp. nov., a phototrophic *gammaproteobacterium* isolated from a mangrove forest sediment sample. *Int. J. Syst. Evol. Microbiol.* **65:** 2357-2364.

Nunkaew, T., D. Kantachote, H. Kanzaki, T. Nitoda and R.J. Ritchie (2014a). Effects of 5-aminolevulinic acid (ALA)-containing supernatants from selected *Rhodopseudomonas palustris* strains on rice growth under NaCl stress, with mediating effects on chlorophyll, photosynthetic electron trans. *Electron J. Biotechnol.***17:** 19–26.

Nunkaew, T., D. Kantachote, T. Nitoda and H. Kanzaki (2014b). Selection of salt-tolerant purple non-sulfur bacteria producing 5-aminolevulinic acid (ALA) and reducingmethane emissions from microbial rice straw degradation. *Appl. Soil Ecol.***86:** 113–120.

Oda,Y., F.W. Larimer, P.S. Chain, S. Mal- fatti, M.V. Shin, L. Vergez, L. Hauser, M.L. Land, S. Braatsch, J.T. Beatty, D.A. Pelletier, A.L Schaefer and C.S. Harwood (2008). Multiple genome sequences reveal adaptations of a phototrophic bacterium to sediment microenvironments. *Proc. Natl. Acad. Sci. U.S.A.* **105:** 18543–18548.

Oelze, J., W. Pahlke and S. Bohm (1975). *Arch. Microbiol.***102:** 65-69.

Ogawa, T., T. Furusawa and R. Nomura, D. Seo, N. Hosoya Matsuda, H. Sakurai and K. Inoue (2008). SoxAX binding protein, a novel component of the thiosulfate-oxidizing multienzyme system in the green sulfur bacterium *Chlorobium tepidum*. *J. Bacteriol.* **190:** 6097–6110.

Ogbonna, J., Y. Hitoshi and H. Tanaka (2000). Treatment of high-strength organic wastewater by a mixed culture of photosynthetic microorganisms. *J. Appl. Phycol.* **12:** 277-284.

Oh S. Y., Y.D. Seo, B. Kim, I.Y. Kim and D.K. Cha (2016). Microbial reduction of nitrate in the presence of zero-valent iron and biochar. *Bioresour. Technol.* **200:** 891–896.

Oh, Y.K, E.H. Seol, M.S. Kim and S. Park (2004). Photoproduction of hydrogen from acetate by a chemoheterotrophic bacterium *Rhodopseudomonas palustris* P4. *Int. J. Hydrogen Energy*. **29:** 1115–21.

Oh, Y.K., E.H. Seol, J. R. Kim and S. Park (2003). Fermentative biohydrogen production by a new chemoheterotrophic bacterium *Citrobacter* sp. Y19. *Int. J. Hydr. Ener.* **28:** 1353–1359.

Oh, Y.K., H.J. Kim, S. Park, M.S. Kim and D.D.H. Ryu (2008). Metabolic flux analysis of hydrogen production pathway in *Citrobacter amalonaticus* Y19. *Int. J. Hydr. Ener.* **35:** 1471-82.

Ohta, Y. and A. Mitsui (1981). Enhancement of hydrogen production by marine *Chromatium* sp. Miami PBS 1071 grown in molecular nitrogen. In: *Advances in Biotechnology,* Vol.II (Moo-Young, M. and C.W. Robinson, eds.), pp. 303–307. Pergamon Press. Toronto.

Okamura, K., T. Hisada and A. Hiraishi (2007). Characterization of thermotolerant purple non-sulfur bacteria isolated from hot-spring *Chloroflexus* mats and the reclassification of "*Rhodopseudomonas cryptolactis*" Stadtwald-Demchick *et al.* 1990 as *Rhodoplanes cryptolactis* nom.rev., comb. nov. *J. Gen. Appl. Microbiol.* **53:** 357–361.

Okamura, K., T. Hisada, T. Kanbe and A. Hiraishi (2009a). *Rhodovastum atsumiense* gen. nov., sp. nov., a phototrophic alphaproteobacterium isolated from paddy soil. *J. Gen. Appl. Microbiol.* **55:** 43–50.

Okamura, K., T. Kanbe and A. Hiraishi (2009). *Rhodoplanes serenus* sp. nov., a purple non-sulfur bacterium isolated from pond water. *Int. J. Syst. Evol. Microbiol.* **59:** 531–535.

Okubo, Y., F. Hiroyuki and H. Akira (2006). Characterization of phototrophic purple non-sulfur Bacteria forming colored microbial mats in a swine wastewater Ditch. *Appl. Environ. Microbiol.***72 (9):** 6225-6233.

Olivares, J., E.J. Bedmar and J. Sanjuan (2013) Biological nitrogen fixation in the context of global change. *Mol. Plant-Microbe Interact.* **26:** 486–494.

Olson, J.M. and B.K. Pierson (1987). Evolution of reaction centers in photosynthetic prokaryotes. *Int. Rev. Cytol.***108:** 209–248.

Oren, A., M. Kessel and E. Stackebrandt (1989). *Ectothiorhodospira marismortui* sp. nov., an obligately anaerobic, moderately halophilic purple sulfur bacterium from a hypersaline sulfur spring on the shore of the Dead Sea. *Arch. Microbiol.* **151:** 524-529.

Ormerod, J.G., L.K. Kimble, T. Nesbakken, Y.A. Torgersen, C.R. Woese and M.T. Madigan (1996). *Heliophilum fasciatum* gen. nov., sp. nov. and *Heliobacteriumgestii* sp. nov.: endospore-forming heliobacteria from rice field soils. *Arch.Microbiol.* **165:** 226–234.

Osho, A (1995). Production of metabolites, industrial enzymes, amino acids, vitamins, single-cell protein. *J. Res.* **6:** 521-529.

O'Sullivan, L., R.P. Ross and C. Hill (2002). Potential of bacteriocin-producing lactic acid bacteria for improvements in food safety and quality. *Biochimie.* **84(5-6):** 593-604.

O-Thong, S., P. Prasertsan, N. Intrasungkha, S. Dhamwichukorn and N.K. Birkeland (2007). Improvement of biohydrogen production and treatment efficiency on palm oil mill effluent with nutrient supplementation at thermophilic conditions using an anaerobic sequencing batch reactor. *Enz. Microb.Technol.* **41:** 583-590.

Ourisson, G. and P. Albrecht (1992). Hopanoids. 1. Geohopanoids: the most abundant natural products on Earth? *Accounts of Chemical Research.* **25(9):** 398-402.

Ourisson, G., M. Rohmer and K. Poralla (1987). Prokaryotic hopanoids and other polyterpenoid sterol surrogates. *Annual Review of Microbiol.* **41:** 301–333.

Ouyang, J., Q. Liu, B. Li, J. Ao and X. Chen (2013). Proteomic analysis of differential protein expression in *Acidithiobacillus ferrooxidans* grown on ferrous iron or elemental sulfur. *Indian J. Microbiol.* **53:** 56–62.

Overmann, J (2001). Green sulfur bacteria. In: *Bergey's Manual of Systematic Bacteriology,* (Boone, D.R. and R.W. Castenholz, eds.) *Springer-Verlag,* New York, Berlin-Heidelberg. **1:** 601-605.

Overmann, J (2008). Ecology of phototrophic sulfur bacteria. In: Sulfur Metabolism in Phototrophic Organisms (R. Hell, C. Dahl, D.B. Knaff and T. Leustek, eds), *Advances in Photosynthesis and Respiration,Springer*, Dordrecht. pp. 375–396.

Overmann, J (2015). Green Sulfur Bacteria. Bergey's Manual of Systematics of Archaea and Bacteria. pp. 1–8. doi:10.1002/9781118960608.gbm00378.

Overmann, J. and C. Tuschak (1997). Phylogeny and molecular fingerprinting of green sulfur bacteria. *Arch.Microbiol.***167**: 302–309.

Overmann, J. and N. Pfennig (1989). *Pelodictyon phaeoclathratiforme* sp. nov., a new brown-colored member of the *Chlorobiaceae* forming net-like colonies. *Arch. Microbiol.* **152**: 401–406.

Overmann, J. and N. Pfennig (1990). In Validation of the publication of new names and new combinations previously effectively published outside the IJSB. List No. 33. *Int. J.Syst. Bacteriol.* **40**: 212.

Overmann, J., C. Tuschak, J. Frostl and H. Sass (1998). The ecological niche of the consortium *Pelochromatium roseum. Arch. Microbiol.***169:** 120–128.

Overmann, J., H. Cypionka and N. Pfennig (1992a) An extremely low-light-adapted phototrophic sulfur bacterium from the Black Sea. Limnol. *Oceanogr.* **37:** 150–155.

Overmann, J., J.T. Beatty and K.J. Hall (1994). Photosynthetic activity and population dynamics of *Amoebobacter purpureus* in a meromictic saline lake. *FEMS Microbiol Ecol.***15:** 309–320.

Overmann, J., J.T. Beatty and K.J. Hall (1996). Purple sulfur bacteria control the growth of aerobic heterotrophic bacterioplankton in a meromictic salt lake. *Appl. Environ. Microbiol.***62:** 3251–3258.

Overmann, J., K.J. Hall, T.G. Northcote and J.T. Beatty (1999). Grazing of the copepod *Diaptomous connexus* on purple sulphur bacteria in a meromictic salt lake. *Environ. Microbiol.* **1:** 213-221.

Overmann, J., U. Fischer and N. Pfennig (1992b). A new purple sulfur bacterium from saline littoral sediments, *Thiorhodovibrio winogradskyi* gen. nov. and sp. nov. *Arch. Microbiol.* **157:** 329–335.

Overmann, J.,U. Fischer and N. Pfennig (1993). In validation of the publication of new names and new combinations previously effectively published outside the IJSB. List No. 44. *Int. J. Syst. Bacteriol.* **43**: 188–189.

Ozgur, E., A.E. Mars, B. Peksel, A. Louwerse, M. Yucel and U. Gundu and I. Eroglu (2010). Biohydrogen production from beet molasses by sequential dark and photo fermentation . *Int. J. Hydr. Ener.* **35:** 511-7.

Ozgur, E., B. Uyar, Y. Ozturk, M. Yucel, U. Gunduz and I. Eroglu (2010). Biohydrogen production by *Rhodobacter capsulatus* on acetate at fluctuating temperatures. *Resources, Conservation and Recycling*. **54:** 310-314.

Ozkan, E., B. Uyar, E. Ozgur, M. Yucel, I. Eroglu and U. Gunduz (2012). Photofermentative hydrogen production using dark fermentationeffluent of sugar beet thick juice in outdoor conditions. *Int. J. Hydr. Ener.* **37:** 2044-2049.

Pal, S.S (1998). Interactions of an acid-tolerant strain of phosphate solubilizing bacteria with a few acid-tolerant crops. *Plant Soil.***198:** 169-177.

Pandey, A., P. Singh and L. Iyengar (2007). Bacterial decolorization and degradation of azo dyes. *Int. Biodeterior. Biodegr.* **59:** 73–84.

Panwichian, S., D. Kantachote, B. Wittayaweerasak and M. Mallavarapu (2010). Isolation of purple non-sulfur bacteria for the removal of heavy metals and sodium from contaminated shrimp ponds. *Electr. J. Biotechnol.* **13(4.)**. doi:10.2225/vol13-issue4-fulltext-8.

Panwichian, S., D. Kantachote, B. Wittayaweerasak and M. Mallavarapu (2011). Removal of heavy metals by exopolymeric substances produced by resistant purple non-sulfur bacteria isolated from contaminated shrimp ponds. *Electron. J. Biotechnol.* **14(4):** 1–13.

Panwichian, S., D. Kantachote, B. Wittayaweerasak and M. Mallavarapu (2012). The use of selected purple non-sulfur bacteria to remove heavy metals and salts from sediment and water collected from contaminated areas to decrease their phytotoxicity. *African J. Biotechnol.* **11(29):** 7434-7444.

Papiz, M. Z., S. M. Prince, A. M. Hawthornthwaite-Lawless, G. McDermott, A. A. Freer, W. W. Isaacs and R. J. Cogdell (1996). A model for the photosynthetic apparatus of purple bacteria. *Trends Plant Sci.***1:** 198–206.

Papiz, M.Z., S.M., Prince, T., Howard, R.J., Cogdell and N.W., Isaacs (2003). *J. Mol. Biol.* **326:** 1523-1538.

Parey, K., U. Demmer, E. Warkentin, A. Wynen, U. Ermler and C. Dahl (2013). Structural, biochemical and genetic characterization of ATP sulfurylase from *Allochromatium vinosum*. *PLoS One*. **8:** e74707.

Park, S.K., G.P. Page, K. Kim, D.B. Allison, M. Meydani, R. Weindruch and A.T. Prolla (2008). Differential effect of α- and γ-tocopherol supplementation in age-related transcriptional alterations in heart and brain of B6/C3H F1 mice. *J. Nutr.* **138(6):** 1010–1018.

Parkin, T.B. and T.D. Brock (1980). The effects of light quality on the growth of phototropic bacteria in lakes. *Arch. Microbiol.***125:** 19–27.

Paronyan, A. K. and A.V. Gasparyan (2009). Production of biomass of photosynthetic bacteria on base of stock-breeding and poultry wastes. *Biological Journal of Armenia.* **1:** 5–10.

Patthawaro, S. and C. Saejung (2019). Production of single-cell protein from manure as animal feed by using photosynthetic bacteria. *Microbiology Open.* **8:** e913. https://doi.org/10.1002/mbo3.913.

Pearce, R. P. and R.H. Marchessault (1994). Melting and Crystallization in Bacterial Poly(.beta.-hydroxyvalerate), PHV, and Blends with Poly (.beta.-hydroxybutyrate-co-hydroxyvalerate). *Macromolecules.* **27 (14):** 3869-3874.

Peduzzi, S., A. Welsh, A. Demarta, P. Decristophoris, R. Peduzzi, D. Hahn and M. Tonolla (2011). *Thiocystis chemoclinalis* sp. nov. and *Thiocystis cadagnonensis* sp. nov., motile purple sulfur bacteria isolated from the chemocline of a meromictic lake. *Int. J. Syst. Evol. Microbiol.***61:** 1682-1687.

Pellerin, N.B. and H. Gest (1983). Diagnostic features of the photosynthetic bacterium *Rhodopseudomonas sphaeroides*. *Curr. Microbiol.* **9:** 339–344.

Pelsh, A.D (1936). Hydrobiology of Karabugaz Bay of the Caspian Sea. Trudy Vseoyuznogo Issledovotelskie Nauchni Issledovotelskie Trudy. Institut Galurgii Leningrad. **5:** 49–126.

Pelsh, A.D (1937). Photosynthetic sulfur bacteria of the eastern reservoir of Lake Sakskoe. *Mikrobiologiya*. **6:** 1090–1100.

Perrotta, J.A. and C.S. Harwood (1994). Anaerobic metabolism of cyclohex-1-ene-1-carboxylate, a prposed intermediate of benzoate degradation by *Rhodopseudomonas palustris*. *Appl. Environ. Microbiol.* **60:** 1775-1782.

Perty, M (1852). *Zur Kenntnis kleinster lebensformen*, Vol. I–VIII, Jent and Reinert, Bern.

Peter, W., E. Riege and H. Kerl (1993). *J. Am. Acad. Dennatol.* **28:** 17-21.

Pieterse, C.M.J., C. Zamioudis, R.L.Berendsen, D.M. Weller, S.C.M. Van Wees and P.A.H.M. Bakker (2014). Induced systemic resistance by beneficial microbes. *Annu. Rev. Phytopathol.***52:** 347–375.

Petri, R. and J.F. Imhoff (2000). The relationship of nitrate-reducing bacteria on the basis of *narH* gene sequences and the congruent phylogeny of *narH*and 16S rDNA based on pure culture studies and analyses of environmental DNA. *Syst. Appl. Microbiol.* **23**: 47–57.

Petri, R. and J.F. Imhoff (2001). Genetic analysis of sea-ice bacterial communities of the Western Baltic Sea using an improved double gradient method. *Polar. Biol.***24:** 252-257.

Pfennig, N (1965). Anreicherungskulturen für rote und grüne Schwefelbakterien. Zentralbl. Bakteriol. Parasitenkd. Infektionskrankh. Hyg. Abt. 1, Orig. Suppl. 1. 179–189, 503–505.

Pfennig, N (1968). *Chlorobium phaeobacteroides* nov. spec. und *Chlorobium phaeovibrioides* nov. spec., zwei neue Arten der grünen Schwefelbakterien. *Arch. Mikrobiol.* **63:** 224–226.

Pfennig, N (1969). *Rhodopseudomonas acidophila*, sp. nov., a new species of the budding purple non-sulfur bacteria. *J. Bacteriol.***99:** 597–602.

Pfennig, N (1969b). *Rhodospirillum tenue* sp. n., a new species of the purple non-sulfur bacteria. *J. Bacteriol.* **99:** 619–620.

Pfennig, N (1974). *Rhodopseudomonas globiformis*, sp.n., a new species of the *Rhodospirillaceae*. *Arch. Microbiol.* **100:** 197–206.

Pfennig, N (1977). Phototrophic green and purple bacteria: a comparative systematic survey. *Ann. Rev. Microbiol.***31:** 275–290.

Pfennig, N (1978a). General physiology and ecology of photosynthetic bacteria. In: The Photosynthetic Bacteria, (Clayton, R.K. and W.R. Sistrom, eds.) Plenum Press, New York, pp. 3–18.

Pfennig, N (1978b). *Rhodocyclus purpureus* gen. nov. and sp. nov., a ring-shaped, vitamin B_{12}-requiring member of the family *Rhodospirillaceae*. *Int. J. Syst. Bacteriol.* **28:** 283–288.

Pfennig, N (1989). Green sulfur bacteria. In: *Bergey's Manual of Systematic Bacteriology*, (J.T. Staley, M.P. Bryant, N. Pfennig and J.G. Holt, eds.), Vol. 3, pp. 1682–1697. Baltimore: Williams and Wilkins.

Pfennig, N (1989a). Ecology of phototrophic purple and green sulfur bacteria. In: *Autotrophic Bacteria*, (Schlegel H.G. and B. Bowien, eds.) SpringerVerlag, Heidelberg. pp. 97–116.

Pfennig, N (1989b). Genus *Chromatium*. In: *Bergey's Manual of Systematic Bacteriology*, (Staley, J.T., M. P. Bryant, N. Pfennig and J. C. Holt, eds.), First edition. Williams and Wilkins. Baltimore, MD. **3:** 1639–1643.

Pfennig, N (1989d). Genus *Thiospirillum*. In: *Bergey's Manual of SystematicBacteriology*, (Staley, Bryant, Pfennig and Holt, eds.), First Edition, Vol. 3, The Williams and Wilkins Co., Baltimore. pp. 1644–1645.

Pfennig, N. and G. Cohen Bazire (1967). Some properties of the green bacterium *Pelodictyon clathratiforme*. *Archiv. Fur. Mikrobiologie*. **59:** 226-236.

Pfennig, N. and H.G. Truper (1969). Phototrophic bacteria. GSF-Bericht M32, Gesellschaft für Strahlenforschung Hmb (MünchenPuchkova, N.N., J.F. Imhoff, V.M. Gorlenko, 2000), *Thiocapsa litoralis* sp. nov., a new purple sulfur bacterium from microbial mats from the White Sea. *Int. J. Syst. Evol. Microbiol.***50:** 441–1447.

Pfennig, N. and H.G. Truper (1971a). New nomenclatural combinations in the phototrophic sulfur bacteria. *Int. J.Syst. Bacteriol.* **21**: 11–14.

Pfennig, N. and H.G. Truper (1971b). Conservation of the family name *Chromatiaceae* Bavendamm 1924 with the type genus *Chromatium* Perty 1852, request for an opinion. *Int. J. Syst. Bacteriol.* **21:** 15–16.

Pfennig, N. and H.G. Truper (1971c). Higher taxa of the phototrophic bacteria. *Int. J. Syst. Bacteriol.* **21**: 17–18.

Pfennig, N. and H.G. Truper (1971d). Type and neotype strains of the species of phototrophic bacteria maintained in pure culture. *Int . J. Syst. Bacteriol.* **21:**19–24.

Pfennig, N. and H.G. Truper (1974). The phototrophic bacteria. In: *Bergey's Manual of Determinative Bacteriology*, (Buchanan, R.E. and N.E. Gibbons, eds.) Eighth edition. Williams and Wilkins. Baltimore, MD. pp. 24–75.

Pfennig, N. and H.G. Truper (1981). Isolation of members of the families *Chromatiaceae* and *Chlorobiaceae*. In: *The prokaryotes*: A handbook on ha Université Bordeaux bitats, isolation and identification of bacteria (Starr, M.P., H. Stolp, H.G Truper, A. Balows and H.G. Schlegel, eds.) Berlin: Springer. pp. 279–289.

Pfennig, N. and H. Truper (1989). Anoxygenic Phototrophic Bacteria. In: Bergey's manual of systematic bacteriology. Staley, J., M. Bryant, N. Pfennig and J. Holt, eds.). Baltimore,MD: Williams & Wilkins, pp. 1635–1709.

Pfennig, N. and H.G. Truper (1989). Family I. *Chromatiaceae*. In: *Bergey's Manual of Systematic Bacteriology*, (Stale, Bryant, Pfennig and Holt eds.), First edition., Vol. 3. The Williams and Wilkins Co., Baltimore. pp. 1637-1653.

Pfennig, N. and H.G. Truper (1992). The family *Chromatiaceae*. In: The Prokaryotes (Balows, A., H.G. Truper, M. Dworkin, W. Harder and K.H. Schleifer, eds.), Second edition, Springer-Verlag. New York, NY. pp. 3200–3221.

Pfennig, N. and J. Overmann (2015). *Pelodictyon. Bergey's Manual of Systematics of Archaea* and *Bacteria,* 1–5. doi:10.1002/9781118960608.gbm00376.

Pfennig, N. and J. Overmann (2015). *Chlorobium. Bergey's Manual of Systematics of Archaea* and *Bacteria,* 1–9. doi:10.1002/9781118960608.gbm00374.

Pfennig, N. and K.D. Lippert (1966). Uber das Vitamin B_{12}-Bedurfnis phototropher Schwefelbakterien. *Arch Mikrobio.* **155:** 245-256.

Pfennig, N., H. Lunsdorf, J. Suling and J.F. Imhoff (1997). *Rhodospira trueperi*, gen. nov. and spec. nov., a new phototrophic *Proteobacterium* of the alpha-group. *Arch. Microbiol.* **168:** 39–45.

Pfennig, N., H. Lunsdorf, J. Suling and J.F. Imhoff (1998). Validation List no. 64. *Int. J. Syst. Bacteriol.***48:** 327-328.

Pfennig, N., N.E. Elmhjellen and S.L. Jensen (1965). *Arch. Mcrobiol.***51:** 258-266.

Phang, S.M (1990). Algal production from agro-industrial and agricultural waste in Malaysia. *Ambiosis.***19:** 415-418.

Pichinoty, F (1965). L'effet oxygkne et la biosynthkse des enzymes d'oxydorkduction bactkriens. In *Mecanismes de Regulation des ActivitPs Cellularies chez les Micro-organismes*, *Paris*: C.N.R.S. p. 507.

Pierson, B. K. and R.W. Castenholz (2015). *Heliothrix*. Bergey's Manual of Systematics of Archaea and Bacteria, 1–3. doi:10.1002/9781118960608.gbm00382.

Pierson, B.K (1994). The emergence, diversification and role of photosynthetic eubacteria. In: *Early Life on Earth: Nobel Symposium* No. 84, (Benston, S, ed.), Columbia University Press, New York, pp. 161–180

Pierson, B.K (2015). "*Chloroflexaceae*." Bergey's Manual of Systematics of Archaea and Bacteria, 1–3. doi:10.1002/9781118960608.fbm00082.

Pierson, B.K. and R.W. Castenholz (1974a). A phototrophic gliding filamentous bacterium of hot springs, *Chloroflexus aurantiacus*, gen. and sp. nov. *Archives of Microbiol.***100:** 5-24.

Pierson, B.K. and R.W. Castenholz (1974b). Studies of pigments and growth in *Chloroflexus aurantiacus*, a phototrophic filamentous bacterium. *Archives of Microbiol.***100:** 283-305.

Pierson, B.K. and R.W. Castenholz (1992). The Family *Chloroflexaceae*. In: *The Prokaryotes*, (Balows, A., H.G. Truper, M. Dworkin, W. Harder and K.H. Schleifer, eds.), Springer, New York, pp. 3754–3775.

Pierson, B.K. and R.W. Castenholz (1995). Taxonomy and physiology of filamentous anoxygenic phototrophs. In: *Anoxygenic photosynthetic bacteria*, (Blankenship, M.T., M.T. Madigan and C.E. Bauer, eds.), Kluwer Academic Publisers. Dordrecht, The Netherlands. **31:** 181-196.

Pierson, B.K. and R.W. Castenholz (2015). *Chloroflexus. Bergey's Manual of Systematics of Archaea and Bacteria,* 1–16. doi:10.1002/9781118960608.gbm00380.

Pierson, B.K. and R.W. Castenholz (2015). *Heliothrix. Bergey's Manual of Systematics of Archaea and Bacteria,* 1–3. doi:10.1002/9781118960608.gbm00382

Pierson, B.K., L.M. Keith and J.G. Leovy (1984b). Isolation of pigmentation mutants of the green filamentous photosynthetic bacterium *Chloroflexus aurantiacus*. *J. Bacteriol.***159:** 222–227.

Pierson, B.K., S.J. Giovannoni and R.W. Castenholz (1984a). Physiological ecology of a gliding bacterium containing bacteriochlorophyll *a*. *Appl. Environ. Microbiol.***47:** 576–584.

Pierson, B.K., S.J. Giovannoni, D.A. Stahl and R.W. Castenholz (1985). *Heliothrix oregonensis*, gen. nov., sp. nov., a phototrophic filamentous gliding bacterium containing bacteriochlorophyll *a*. *Arch. Microbiol.* **142:** 164–167.

Pierson, B.K., S.J. Giovannoni, D.A. Stahl and R.W. Castenholz (1986). In Validation of the publication of new names and new combinations previously effectively published outside the IJSB. List No. 20. *Int. J. Syst. Bacteriol.* **36:** 354–356.

Pietro Carlozzi., Giulia Padovani, Patrizia Cinelli and Andrea Lazzeri (2015). An Innovative Device to Convert Olive Mill Wastewater into a Suitable Effluent for Feeding Purple Non-Sulfur Photosynthetic Bacteria. *Resources.* **4:** 621-636.

Pinto, F.A.L., O. Troshina and P. Lindblad (2002). A brief look at three decades of research on cyanobacterial hydrogen evolution. *Int. Hydr. Ener.* **27:** 1209-1215.

Piper, S (2004). Continuous cultures of *Methylococcus capsulatus.* Center of Microbial Biotechnology (Biocentrum). The Technical University of Denmark. Master's thesis. *Saccharomyces cerevisiae*, PhD. Thesis. Chalmers Univ. Techno. **90:** 123-167.

Pires, R.H., A.L. Lourenco and F. Morais (2003). A novel membrane-bound respiratory complex from *Desulfovibrio desulfuricans* ATCC 27774. *Biochim Biophys Acta.* **1605:** 67–82.

Planchard, A., L. Mignot, T. Jouenne and G.A. Junter (1989). Photoproduction of molecular hydrogen by *Rhodospirillum rubrum* immobilized in composite agar layer/microporous membrane structures. *Appl. Environ. Microbiol.* **31:** 49–54.

Polle, J.E.W., S. Kanakagiri, E.S. Jin, T. Masuda and A. Melis (2002). The truncated chlorophyll antenna size of the photosystems-a practical method to improve microalgal productivity and hydrogen production in mass culture. *Int. J. Hydr. Ener.***27:** 1257-1264.

Ponsano, E.H., C.Z. Paulino and M.F. Pinto (2008). Phototrophic growth of *Rubrivivax gelatinosus* in poultry slaughterhouse waste water. *Bioresour Technol.* **99(9):** 3836–3842.

Ponsano, E.H.G., M.F. Pinto, M. Garcia Neto and P.M. Lacava (2002). Evaluation of *Rhodocyclus gelatinosus* biomass for broiler pigmentation. *J. Appli. Poult. Res.* **11:** 77-82.

Ponsano, E.H.G., P.M. Lacava and M.F. Pinto (2002). Isolation of *Rhodocyclus gelatinosus* from poultry slaughterhouse wastewater. *Brazilian Archives of Biology and Technology.* **45 (4):** 1-8.

Ponsano, E.H.G., P.M. Lacava and M.F. Pinto (2003). Chemical the composition of *Rhodocyclus gelatinosus* biomass produced in poultry slaughterhouse wastewater. *Braz. Archives Biol. Technol.* **46:** 143-147.

Prange, A., H. Engelhardt, H.G. Truper and C. Dahl (2004). The role of the sulfur globule proteins of *Allochromatium vinosum*: mutagenesis of the sulfur globule protein genes and expression studies by real-time RT PCR. *Arch. Microbiol.* **182:** 165–174.

Prasertsan, P., M. Jaturapornpipat and C. Siripatana.(1997). Utilization and treatment of tuna condensate by photosynthetic bacteria. *Pure and Applied Chemistry.* **69 (11):**2439-2445.

Prasertsan, P., Suteera Prasertsan and Aran H-Kittikun (1997). Recycling of agro-industrial wastes through cleaner technology, *Biotechnology- Encyclopedia of life support system* (EoLss), 10.

Prasertsan, P., W. Choorit and S. Suwanno (1993a). Isolation, identification and growth condition of photosynthetic bacteria found in seafood processing Wastewater. *W. J. Microbiol. Biotechnol.***9:** 590-593.

Prasertsan, P., W. Choorit and S. Suwanno (1993b). Optimization for growth of *Rhodocyclus gelatinosus* in seafood processing effluents. *World J. Microbiol. Biotechnol.* **9:** 593-596.

Prathyash Ushus, M.J., B. Divyasree, K. Lakshmi, B.P. Dave, Ch. Sasikala and Ch.V. Ramana (2017). Description of a phototrophic bacterium, *Thiorhodococcus alkaliphilus* sp. nov. *Int. J. Syst. Evol. Microbiol.* **67:** 2323-2327.

Pratt, D.C. and E. Gorham (1970). The occurrence of *Athiorhodaceae* in woodland, swamp and pond soils. *Ecology*. **51(2):** 346-349.

Prazmowski, A (1880). Untersuchung über die Entwickelungsgeschichte und Fermentwirking einiger Bacterien-Arten. *Hugo Voigt, Leipzig* .

Prevot, A.R (1953). Dictionnaire des Bactéries Pathogènes, Second Edition. Masson, Paris.

Proctor, L.M (1997). Nitrogen-fixing, photosynthetic, anaerobic bacteria associated with pelagic copepods. *Aquat. Microb. Ecol.* **12:** 105–113.

Providenti, M.A., C.A. Flemming, H. Lee and J.T. Trevors (1995). Effect of addition of rhamnolipid biosurfactants or rhamnolipid producing *Pseudomonas aeruginosa* on phenanthrene mineralization in soil slurries. *FEMS Microbiol. Ecol.***17:** 15-26.

Puchkova, N.N (1984). Green sulfur bacteria as a component of the "sulfureta" of shallow saline waters of the Crimea and nothern Caucasus. *Mikrobiologiya.* **53**: 324–328.

Puchkova, N.N. and V.M. Gorlenko (1976). New brown chlorobacteria *Prosthechloris phaeoasteroidea* nov. sp. *Mikrobiologiya.* **45:** 656–660.

Puchkova, N.N., J.F. Imhoff and V.M. Gorlenko (2000.) *Thiocapsa litoralis* sp. nov., a new purple sulfur bacterium from microbial mats from the White Sea. *Int. J. Systema. Evolution. Microbiol.* **50:** 1441–1447.

Pullerits, T. and V. Sundstrom (1996). Photosynthetic light-harvesting pigment-protein complexes. Toward understanding how and why. *Acc. Chem. Res.* **29:** 381–389.

Qian, J., J. Khandogin, A.H. West and P.F. Cook (2008). Evidence for a catalytic dyad in the active site of homocitrate synthase from *Saccharomyces cerevisiae. Biochemistry.* **47:** 6851–6858.

Qian, P., T. Yagura, Y. Koyama and R. Cogdell (2001). Isolation and Purification of the Reaction Center (RC) and the Core (RC-LH1) Complex from *Rhodobium marinum*: the LH1 Ring of the Detergent-Solubilized Core Complex Contains 32 Bacteriochlorophylls. *Plant & Cell Physiology.* **41:** 347-53. 10.1093.

Qian, X., Q. Ren and W. Yu (1991). *Weishengwuxue Zashi.* **11:** 11-15.

Quatrini, R., C. Appia-Ayme, Y. Denis, E. Jedlicki, D.S. Holmes and V. Bonnefoy (2009). Extending the models for iron and sulfur oxidation in the extreme acidophile *Acidithiobacillus ferrooxidans*. *BMC Genomics.***10:** 394.

Rabold, S., V.M. Gorlenko and J.F. Imhoff (2006). *Thiorhodococcus mannitoliphagus* sp. nov., a purple sulfur bacterium from the White Sea. *Int. J. Syst. Evol. Microbiol.***56**: 1945-1951.

Rahalkar, S. B., S.R. Joshi and N. Shivaraman (1993.) *Cum. Microbiol.* **26:** 1-9.

Rahalkar. S. B., S.R. Joshi and N. Shivaraman (1991). Biodegradation of aromatic compounds by *Rhodopseudomonas gelatinosa*. *Cum. Microbiol.* **22(3):** 155-158.

Rainey, F.A (2015). *Clostridiales. Bergey's Manual of Systematics of Archaea and Bacteria.* pp. 1–5. doi:10.1002/9781118960608.obm00059.

Raj, P.S., S.K. Chakravarthy, E.V. Ramaprasad, Ch. Sasikala and Ch. V. Ramana (2012). *Phaeospirillum tilakii* sp. nov., a phototrophic *alphaproteobacterium* isolated from aquatic sediments. *Int. J. Syst. Evol. Microbiol.* **62:** 1069–1074.

Rajani, B., R. Sunil Kumar, M. Uma Devi and J.B. Nayak (2016). Role of purple non-sulfur bacteria *Rhodopseudomonas palustris* RSOU000 and *Rhodopseudomonas thermotolerance* RSOU555 in wastewater treatment. *World Journal of Pharmacy and Pharmaceutical Sciences*. **5(8):** 1379-1387.

Rajasekhar, N., Ch. Sasikala and Ch. V. Ramana (2000). Toxicity of N-containing heterocyclic aromatic compounds and their utilization for growth by a few purple non-sulfur bacteria. *Bull. Environ. Contam. Toxicol.***65:** 375–382.

Rajkumar, B. and D. Lalithakumari (1992). *Biomed. Lett.* **47:** 337-346.

Raju, N., M. Ramchander, P. Rudra and Moinack (2014). Photoproduction of hydrogen by photosytnhetic bacteria isolated from oil-contaminated soil of Mallapur, Hyderabad, India. *Klinicka Farmakologie a Farmacie*. **28:** 4625-4628.

Raju, N., M. Ramchander, P. Rudra and Moinack (2015). Biohydrogen production by photosynthetic bacteria isolated from oil-contaminated soils of Delhi, India. *Int. J. Chem.Tech. Resear.***8:** 1365-1368.

Rajyalaxmi, K. and S. Girisham (2015). Production of Indole acetic acid by three anoxygenic phototrophic purple bacteria. *Ijpbs*. **6(4):** B274-278.

Rajyalaxmi, K., M. Ramchander, S. Girisham and S.M. Reddy (2015). Phosphate solubilization by *Allochromatium* sp. GSKRLMBKU-01 isolated from marine water of Visakhapatnam. *Int. J. Appl. Biol. Phamceu. Technol.***6(2):** 1-6.

Rajyalaxmi, K., M. Ramchander, S. Girisham and S.M. Reddy (2015). Bioremediation of slaughterhouse wastewaterby *Rhodobacter* sp. GSKRLMBKU 02. *Int. J. Appl. Biol. Phamceu. Technol.***6(3):** 176-179.

Rajyalaxmi, K., M. Ramchander, S. Girisham and S.M. Reddy (2017). Chromate reduction by purple non sulphur phototrophic bacterium *Rhodobacter* sp. GSKRLMBKU–03 isolated from pond water. *Proc. Natl. Acad. Sci., India, Sect. B Biol. Sci.* pp. 1-7. https://doi.org/10.1007/s40011-017-0939-8.

Rajyalaxmi, K., M. Ramchander, S. Girisham and S.M. Reddy (2018). Impact of cultural conditions on photoproduction of hydrogen by *Allochromatium* sp. GSKRLMBKU-01 isolated from marine water of Visakhapatnam. *Int. J. Hyd. Ener.* **43:** 6060 -6065.

Rajyalaxmi, K., S. Girisham and S.M. Reddy (2015). Production of free amino acid by three anoxygenic phototrophic bacteria. *Int. J. Pharm. Bio. Sci.* **5(2):** 130-133.

Rana, G., S. Meikap, M. Mondol and P.P. Bose (2016) Green-fertilizer, *Rhodospirillum rubrum*, for agricultural development on fly-ash without any toxic metal ion release. *Basic Res. J. Agric. Sci. Rev.***5:** 109–117.

Ramana, Ch.V. and Ch. Sasikala (2002). Light-dependent reductive degradation of nitrobenzene by *Rhodopseudomonas palustris*. *Ind. J. Microbiol.* **42:** 229-232.

Ramana, Ch.V. and Ch. Sasikala (2009). *Albidoferax*, a new genus of *Comamonadaceae* and reclassification of *Rhodoferax ferrireducens* (Finneran *et al.,* 2003) as *Albidoferax ferrireducens* comb. nov. *The J. Gen. Appl. Microbiol.* **55(4):** 301-304.

Ramana, Ch.V., Ch. Sasikala, K. Arunrasi, P.A. Kumar, T.N.R. Srinivas, S. Shivaji, J. Suling and J.F. Imhoff (2006). *Rubrivivax benzoatilyticus* sp. nov., an aromatic hydrocarbon-degrading purple beta proteobacterium. *Int. J. Syst. Evol. Microbiol.* **56:** 2157–2164.

Ramaprasad, E.V., L. Tushar, B. Dave, Ch. Sasikala and Ch.V. Ramana (2016). *Rhodovulum algae* sp. nov., isolated from an algal mat. *Int. J. Syst. Evol. Microbiol.***66**: 3367-3371.

Ramasamy, Y., C.V. Seshadri, T.M. Vatsala and S.K. Rakshit (1990). *Australian J. of Biotech.***4:** 287-289.

Ramchander Merugu, A. Sridhar Rao, D. Ramesh, S. Girisham and S.M. Reddy (2012). Production of PHB by under aerobic dark conditions by two anoxygenic phototrophic purple non sulfur bacteria isolated from tannery effluent. *Inter. J. ChemTech Resea.* **4:** 1103-1107.

Ramchander, M (2009). Studies on biotechnological potentials of some anoxygenic phototrophic bacteria. A thesis submitted to Department of Biochemistry, Kakatiya University, Warangal.

Ramchander, M., K. Rajya laxmi, S. Girisham and S.M. Reddy (2013). Chromate reduction by purple non- sulphur phototrophic bacterium *Rhodopseudomonas palustris* KU003 isolated from tannery effluents. *Int. J. Environ. Bioene.* **6:** 187-192.

Ramchander, M., M. Vasantha, M.P. Pratap Rudra, S. Girisham and S.M. Reddy (2012b). Photoproduction of hydrogen by alginate immobilized *Rhodobacter capsulatus* KU002 under sulphate and phosphate limitations. *Int. J. Environ. Bioene.* **4:** 141-146.

Ramchander, M., M. Vasantha, M.P. Pratap Rudra, S. Girisham and S.M. Reddy (2013b). Photoproduction of hydrogen by alginate immobilized cultures of *Rhodobacter capsulatus* KU002 isolated from tannery effluents. *J. Biofuels.* **4:** 56-60.

Ramchander, M., M.P. Pratap Rudra, A . Thirupathaiah, S. Girisham and S.M. Reddy (2011). Chromate reduction by a purple non sulphur phototrophic bacterium *Rhodobacter capsulatus* KU002 isolated from tannery effluents. *J. Pure. Appl. Microbiol.* **5:** 66-69.

Ramchander, M., M.P. Pratap Rudra, A. Sridhar Rao, D. Ramesh, D., B. Nageshwari, K. Rajyalaxmi, S. Girisham and S.M. Reddy (2011). Influence of different cultural conditions on photoproduction of hydrogen by *Rhodopseudomonas palustris* KU003. *ISRN Renewable Energy*. 328984-90.

Ramchander, M., M.P. Pratap Rudra, A. Thirupathaiah and N. Veerababu (2011). Hypocholesterolemic effect of the anoxygenic phototrophic bacterium *Rhopseudomonas palustris* MGU001 in hen laying eggs. *Int. J. Appl. Biol. Pharmaceu. Technol.* **2:** 463-466.

Ramchander, M., M.P. Pratap Rudra, A. Thirupathaiah, S.Girisham and S.M. Reddy (2011). Optimization of Indole Acetic Acid production by two anoxygenic phototrophic bacteria isolated from tannery effluents. *J.pure. Appl. Microbiol.* **5:** 34-37.

Ramchander, M., M.P. Pratap Rudra, B. Nageshwari, A. Sridhar Rao and D. Ramesh (2012c). Photoproduction of hydrogen under different cultural conditions by alginate immobilized *Rhodopsedomonas palustris* KU003. *ISRN Renewable Ener.* Article ID 757503, 5 pages doi: 10.5402/2012/ 757503.

Ramchander, M., M.P. Pratap Rudra, S. Girisham and S.M. Reddy (2013a). Bioproduction of hydrogen by alginate immobilized *Rhodopseudomonas acidophila* KU001. *Int. J. Chem. Engine. Appl. Sci.* **3:** 7-9.

Ramchander, M., M.P. Pratap Rudra, S. Girisham and S.M. Reddy (2012). Effect of Bioinoculation of *Rhodobacter capsulatus* KU002 on two rice varieties of India. *Int. J. Appl Biol. Pharm Techn.* **3:** 373-375.

Ramchander, M., M.P. Pratap Rudra, S. Girisham and S.M. Reddy (2012). Biotechnological applications of purple non sulphur phototrophic bacteria: a minireview. *Int. J. Applied Biology and Pharmaceu. Technol.* **3:** 376-384.

Ramchander, M., M.P. Pratap Rudra, S. Girisham and S.M. Reddy (2012).Hydrogen uptake hydrogenase activities of two anoxygenic phototrophic bacteria isolated from Leather industry effluents.*Int. J. ChemTech. Resea.* **4:** 1108-1110.

Ramchander, M., M.P. Pratap, B. Nageshwari, S. Girisham and S.M. Reddy (2012a). Factors influencing the production of hydrogen by the hydrogen by the purple non-sulfur phototrophic bacterium *Rhodopseudomonas acidophila* KU001. *Microb. Biotechnol.* **5:** 674-678.

Ramchander, M., M.S.K. Prasad, D. Vasavi, S. Girisham and S.M. Reddy (2007). Bioremediation of wastewater by two anoxygenic phototrophic bacteria. *Nat. Acad. Sci. Lett.* **30:** 223-227.

Ramchander, M., M.S.K. Prasad, D. Vasavi, S. Girisham and S.M. Reddy (2008). Production of asparaginase by four anoxigenic phototrophic bacteria isolated from leather industry effluents. *Ecol. Envi. Con.* **14:** 485-487.

Ramchander, M., M.S.K. Prasad, S. Girisham and S.M. Reddy (2008). Production of Indole acetic acid and free amino acids by two anoxygenic phototrophic bacteria. *Bioinfolet* **5:** 82-84.

Ramchander, M., M.S.K. Prasad, S. Girisham and S.M. Reddy (2008). Influence of some metals on the growth of two anoxygenic phototrophic bacteria. *Nat. Env. Pol. Tech.* **7:** 225-228.

Ramchander, M., M.S.K. Prasad, S. Girisham and S.M. Reddy (2008). Tolerance of certain pesticides by two nitrogen-fixing anoxygenic phototrophic bacteria. *Nat. Env. Pol. Tech.* **7:** 467-469.

Ramchander, M., M.S.K. Prasad, S. Girisham and S.M. Reddy (2008). Phosphate solubilization by four anoxygenic phototrophic bacteria isolated from the leather industry. *Nat. Env. Pol. Tech.* **7:** 597-599.

Ramchander, M., M.S.K. Prasad, S. Girisham and S.M. Reddy (2008). Effect of trace elements on the growth of two anoxygenic phototrophic bacteria. *Ecol. Envi. Con.* **14:** 367-369.

Ramchander, M., S. Girisham and S.M. Reddy (2010). Bioproduction of hydrogen by *Rhodobacter capsulatus* KU002 isolated from leather industry effluents. *Inter. J. Hydr. Ener.* **35:** 9591-9697.

Ramchander, M., S. Girisham and S.M. Reddy (2010). Extracellular enzymes of two anoxygenic phototrophic bacteria isolated from leather industry effluents. *Biotechnology an Indian Journal.* **4:**86-87.

Ramchander, M., S. Girisham and S.M. Reddy (2010a). Production of PHB (Polyhydroxy butyrate) by *Rhodopseudomonas palustris* KU003 under nitrogen limitation. *Int. J. Appl. Bio. and Pharm. Tech.* **2:** 686-688.

Ramchander, M., S. Girisham and S.M. Reddy (2010b). Production of PHB (Polyhydroxybutyrate) by *Rhodopseudomonas palustris* KU003 and *Rhodobacter capsulatus* KU002 under phosphate limitation. *Int. J. Appl. Biol. Pharm. Tech.* **3:** 746-748.

Ramchander, M., V. Namratha, D. Nagaraju (2015). Bioremediation of sewage wastewater by the phototrophic bacterial consortium isolated from sewage water. *Int. J. Appl. Biol. Pharma. Technol.***6:** 191-195.

Ramchander, M., Y. Prashanthi, T. Sarojini and B. Nageshwari (2014). Bioremediation of Wastewater by the anoxygenic photosynthetic bacterium *Rhodobacter sphaeroides* SMR 009. *Int. J. Resea. Environ. Sci. Technol.* **4:** 16-19.

Ramos, A.R., K.L. Keller, J.D. Wall and I.A. Cardoso Pereira (2012). The membrane QmoABC complex interacts directly with the dissimilatory adenosine 5′-phosphosulfate reductase in sulfate reducing bacteria. *Front Microbiol.* **3:** 137.

Ranjith N.K., Ch.V. Ramana and Ch. Sasikala (2011) Rubrivivaxin, a new cytotoxic and cyclooxygenase-I inhibitory metabolite from *Rubrivivax benzoatilyticus JA2. World J. Microbiol. Biotechnol.* **27:** 11–16.

Ranjith, N.K., Ch. Sasikala and Ch.V. Ramana (2007a). Rhodethrin: a novel indole terpenoid ether produced by *Rhodobacter sphaeroides* has cytotoxic and phytohormonal activities. *Biotechnol Lett.* **29:**1399–1402.

Ranjith, N.K., Ch. Sasikala and Ch.V. Ramana (2007b). Catabolism of L-phenylalanine and L-tyrosine by *Rhodobacter sphaeroides* OU5 occurs through 3, 4-dihydroxyphenylalanine. *Res. Microbiol.* **158:** 506-511.

Ranjith, N.K., Ch. Sasikala and Ch.V. Ramana (2010a). Production of phenols and alkyl gallate esters by *Rhodobacter sphaeroides* OU5. *Curr Microbiol.* **60(2):** 107–111.

Ranjith, N.K., Ch. Sasikala and Ch.V. Ramana (2010b) L-Tryptophan catabolism by *Rubrivivax benzoatiliticus JA2* occurs through indole 3-pyruvic acid pathway. *Biodegradation.* **21:** 825–832.

Ratanakhanokchai, K., J. Kaneko, Y. Kamio and K. Izaki (1992). Purification and properties of maltotetraose and maltotriose producing amylase from *Chloroflexus aurantiacus*. *Appl. Environ. Microbiol.* **58:** 2490-2494.

Raymond, J. and R.E. Blankenship (2004). Biosynthetic pathways, gene replacement and the antiquity of life. *Geobiology*. **2:** 199–203.

Raymond, J., O. Zhaxybayeva, J.P. Gogarten , S.Y. Gerdes and R.E. Blankenship (2002). Whole-genome analysis of photosynthetic prokaryotes. *Science*. **298:**1616–1620.

Raymond, J.C. and W.R. Sistrom (1969). *Ectothiorhodospira halophila*: a new species of the genus *Ectothiorhodospira. Arch Mikrobiol.***69:**121–126.

Reddy, C.S.K., G.R. Rashmi and V.C. Kalia (2003). Polyhydroxyalkanoates: an overview. *Bioreso. Technol*. **87:** 137-146.

Reed, D.W. and R.K. Clayton (1968). Isolation of a reaction fraction from *Rhodopseudomonas palustris. Biochem. Biophys. Res. Communications*. **30(5):** 471-475.

Redfearn, E. R. and R. Powls (1968). The quinones of green photosynthetic bacteria.*Biochem. J*. **106:** 50 P.

Rees, G.N., C.G. Harfoot, P.H. Janssen, L. Schoenborn, J. Kuever and H. Lunsdorf (2002). *Thiobaca trueperi* gen. nov., sp. nov., a phototrophic purple sulfur bacterium isolated from freshwater lake sediment. *Int. J. Syst. Evol. Microbiol.* **52:** 671–678.

Reis, R.S., G.J. Pacheco, A.G. Pereira and D.M.G. Freire (2013). Biosurfactants: Production and applications. In: (ed.) Chamy, R. Biodegradation- Life of Science, *InTech*. pp. 31-61.

Reith, J.H., R.H. Wijffels and H. Barten (2003). Biomethane and biohydrogen staus and perspective of biological methane and hydrogen production. Dutch Biol Hydr Foundat, The Netherlands. pp. 118-125.

Remsen, C.C (1978). The comparative subcellular architecture of photosynthetic bacteria. In: *The Photosynthetic Bacteria*, (Clayton and Sistrom, eds.), Plenum Press, New York. pp. 31–60.

Ren, N., A. Wang, G. Cao, J. Xu and L. Gaφ 2009). Bioconversion of lignocellulosic biomass to hydrogen: Potential and challenges. *Biotechnol. Adv.* **27:** 1051–1060.

Ren, N., J. Li, B. Li, Y. Wang and S. Liu (1). Biohydrogen production from molasses by anaerobic fermentation with a pilot-scale bioreactor system. *Inter. J. Hydr. Ener.***31(15):** 2147–2157.

Ren, N.Q., B.F. Liu, J. Ding, W.Q. Guo, G.L. Cao and G.J. Xie (2008). The effect of butyrate concentration on photo-hydrogen production from acetate by *Rhodopseudomonas faecalis* RLD-53. *Int. J. Hydr. Ener.* **33 :** 5981-5985.

Rene Kaden, Cathrin Sproer, Daniel Beyer and Peter Krolla-Sidenstein (2014). *Rhodoferax saidenbachensis* sp. nov., a psychrotolerant, very slowly growing bacterium within the family *Comamonadaceae*, proposal of the appropriate taxonomic position of *Albidiferax ferrireducens* strain T118T in the genus *Rhodoferax* and emended description of the genus *Rhodoferax. Int. J. System. Evolution. Microbiol.* **64:** 1186–1193.

Resnick S. and M.T. Madigan (1989). Isolation and characterization of a mildly thermophilic non-sulfur purple bacterium containing bacteriochlorophyll *b*. *EMS Microbiol. Lett*. **65 :** 165-1 31.

Rhishipinal, R. and P. Rosamma (1998). Selection of marine yeasts for the generation of single-cell protein from prawn-shell waste. *Bioresource Technol.***65 (3):** 255-256.

Robert, E. and E. Lars (1990). In: "*Highlights on Ubiquinone Research*" (Lena, G. and F. Taylor, eds.), pp. 185-190. London.

Robertson, J.B. and S.C. Edberg (1997). *Critical Reviews in Microbiol.* **23:** 143-178.

Rodrigues, L., I.M. Banat, J. Teixeira and R. Oliveira (2006). Biosurfactants: potential applications in medicine. *J. Antimicrob. Chemother.***57:** 609-618.

Rodriguez, H. and R. Gallardo (1993). Single-cell protein from bagasse pith by a mixed bacterial culture. *Acta Biotechnologica*. **13:**141-149.

Rodriguez, J., J. Hiras and T.E. Hanson (2011). Sulfite oxidation in *Chlorobaculum tepidum. Front. Microbiol.***2:** 112.

Rodriguez-Amaya, D.B (2001). A guide to carotenoid analysis in foods. Washington: ILSI Press.

Rodríguez-Valera, F., A. Ventosa, G. Juez and J.F. Imhoff (1985). Variation of environmental features and microbial populations with salt concentrations in a multi-pond solar saltern. *Microb. Ecol.***11:** 105-107.

Roeder, A,, H. Seidel and F. Giffhorn (1983). *Ger. Offen.* DE. **3:** 210-583.

Roelofsen, P. A (1934). Metabolism of pure sulfur bacteria. *Proc. Aca. Sci.* Amsterdam. **37:** 660-669.

Roeselers, G., L.J. Stal, M.C. Van Loosdrecht and G. Muyzer (2007). Development of a PCR for the detection and identification of Cyanobacterial nifD genes. *J. Microbiol. Methods.* **70:** 550-556.

Rohmer, M., P. Bouvier-Nave and G. Ourisson (1984). Distribution of hopanoid triterpenes in prokaryotes. *J. Gen. Microbiol.* **130:** 1137-1150.

Roth, F.X (1982). Microorganisms as a source of protein for animal nutrition. *In: Advances in Agricultural Microbiology*, (Subbarao, N.S. ed.), Oxford and IBH Publishing Co, New Delhi. pp. 636-672.

Ru Ying Li and Herbert H.P. Fang (2008). Hydrogen production characteristics of photoheterotrophic *Rubrivivax gelatinosus* L3. *Int. J. Hydr. Ener.* **33:** 974-980.

Sabaty, M., P. Gans and A. Vermeglio (1993). Inhibition of nitrate reduction by light and oxygen in *Rhodobacter sphaeroides* f. sp. *denitrificans*. *Arch. Microbiol.* **159:** 153-159.

Saejung, C. and P. Apaiwong (2015). Enhancement of carotenoid production in the new carotenoid producing photosynthetic bacterium *Rhodopseudomonas faecalis* PA2. *Biotechnology and Bioprocess Engineering*. **20(4):** 701–707.

Saejung, C. and Pongsathorn Salasook (2018). Recycling of sugar industry wastewater for single-cell protein production with supplemental carotenoids. *Environmental Technology*. DOI: 10.1080/09593330.2018.1491633.

Saejung, C. and T. Thammaratana (2016). Biomass recovery during municipal wastewater treatment using photosynthetic bacteria and the prospect of production of single-cell protein for feedstuff. *Environmental Technology*. DOI:10.1080/09593330.2016.1175512.

Sagir, E., S. Alipour, K. Elkahlout, H. Koku, U. Gunduz, I. Eroglu and M. Yucel (2018). Biological hydrogen production from sugar beet molasses by agar immobilized *R. capsulatus* in a panel photobioreactor. *Int. J. Hyd. Ener.* **43:** 14987–95.

Sakpirom,J., D. Kantachote, S. Siripattanakul-Ratpukdi, John McEvoy and E. Khan (2019). Simultaneous bioprecipitation of cadmium to cadmium sulfide nanoparticles and nitrogen fixation by *Rhodopseudomonas palustris* TN110. *Chemosphere*. **223:** 455–464.

Sakpirom, J., D. Kantachote, T. Nunkaew and E. Khan (2017). Characterizations of purple non sulfur bacteria isolated from paddy fields, and identification of strains with potential for plant growth promotion, greenhouse gas mitigation and heavy metal bioremediation. *Res. Microbiol.***168:** 1–10.

Salma, U., A.G. Miah, K.M. Tareq, T. Maki and H. Tsuji (2007). Effect of dietary *Rhodobacter capsulatus* on egg-yolk cholesterol and laying hen performance *J. Poultry Science.***86:** 714–719.

Sanaa Mohamed Fahmy, Gad El-Rab, Ahmed Abdel-Fattah Shoreit and Yoshihiro Fukumori (2006). Effects of cadmium stress on growth, morphology and protein expression in *Rhodobacter capsulatus* B10. *Bioscie. Biotechnol. Biochem.***70:** 2394–2402.

Sanchez, O., I. Ferrera, C. Dahl and J. Mas (2001). In vivo role of APS reductase in the purple sulfur bacterium *Allochromatium vinosum. Arch .Microbiol.* **176:** 301–305.

Sander, J. and C. Dahl (2008) Metabolism of inorganic sulfur compounds in purplebacteria. In: *Purple Bacteria* (Hunter, C.N., F. Daldal, M.C. Thurnauer and J.T.Beatty, eds), Advances in Photosynthesis and Respiration. Springer, Dordrecht.**28:** 595–622.

Sander, J. and C. Dahl (2009). Metabolism of inorganic sulfur compounds in purple bacteria. In: Purple bacteria, (Hunter, C.N., F. Daldal, M.C. Thurnauer, J.T. Beatty, eds.), Springer, Dordrecht, pp. 595–622.

Sangkharak, K. and P. Prasertsan (2008). Nutrient optimization for production of polyhydroxybutyrate from halotolerant photosynthetic bacteria cultivated under aerobic-dark condition. *Electr. J. Biotechnology.***11:** 3-8.

Sangyoka, S., A. Reungsang and S. Moonamart (2007). Repeated-batch fermentative for bio-hydrogen production from cassava starch manufacturing wastewater. *Pak. J. Biol. Sci.* **10:** 1782-1789.

Santos, A.A., S.S. Venceslau, F. Grein, W.D. Leavitt, C. Dahl, D.T. Johnson and I.A.C. Pereira (2015). A protein trisulfide couples dissimilatory sulfate reduction to energy conservation. *Science.* **350:** 1541–1545.

Sasaki, K. and S. Nagai (1979). The optimum pH and temperature for the aerobic growth of Rhodopsedomonas gelatinosa and vitamin B_{12} and ubiquinone formation on a starch medium. *J. Ferment. Technol.* **57:** 383–386.

Sasaki, K., H. Morikawa, T. Kisibe, K. Takeno, A. Mikami, T. Harada and M. Ohta (2013) Simultaneous removal of cesium and strontium using a photosynthetic bacterium, *Rhodobacter sphaeroides* SSI immobilized on porous ceramic made from waste glass. *Adv. Biosci. Biotechnol.* **04(01):** 6–13.

Sasaki, K., K. Takeno and Y. Emoto (1996). Utilization of organic acids and volatile fatty acids and hydrogen production by photosynthetic bacteria. *J. Soc. Water Environ.***19:** 63–70.

Sasaki, K., M. Watanabe and N. Noparatnaraporn (2002). Applications of photosynthetic bacteria for medical and environmental fields. *Seibutsu-kogaku.* **80:** 234–236.

Sasaki, K., M. Watanabe, T. Tanaka and T. Tanaka (2002). Biosynthesis, biotechnological production and applications of 5-aminolevulinic acid. *Appl Microbiol Biotechnol.***58:** 23-29.

Sasaki, K., M. Watanabe, Y. Suda, A. Ishizuka and N. Noparatnaraporn (2005). Applications of photosynthetic bacteria for medical fields. *J. Biosci. Bioeng.***100:** 481–488.

Sasaki, K., N. Noparatnaraporn and S. Nagai (1991). Use of photosynthetic bacteria for the production of SCP and chemicals from agroindustrial wastes. In: Martin, A.M. ed. *Bioconversion of waste materials to industrial product,* 4. New York, New York Universities Press. pp. 225-226.

Sasaki, K., N. Noparatnaraporn, M. Hayashi, Y. Nishizawa and S. Nagai (1981). Single-cell protein production by treatment of soybean wastes with *Rhodopseudomonas gelatinosa, J. Ferment. Technol.* **59:** 471-477.

Sasikala, C., C.V. Ramana and P. Raghuveer Rao (1994a). Photometabolism of heterocyclic aromatic compounds by *Rhodopseudomonas palustris* OU 11. *Appl. Environ. Microbiol.* **60(6):** 2187-2190.

Sasikala, C., C.V. Ramana and P. Raghuveer Rao (1994b). Nitrogen fixation by *Rhodopseudomonas palustris* OU 11 with aromatic compounds as carbon source/electron donors. *FEMS Mcrobiol. Lett.* **122 (1-2):** 75-78.

Sasikala, C., C.V. Ramana and P.R. Rao (1994). 5-Aminolevulinic acid: A potential herbicide/insecticide from microorganisms. *Biotechnol. Prog* .**10:** 451-459.

Sasikala, C., C.V. Ramana and P.R. Rao (1995). Regulation of simultaneous hydrogen production during growth by pH and glutamate in *Rhodobacter sphaeroides* O.U.001, *Int. J. Hydrogen Ener.* **20:** 123-126.

Sasikala, Ch. and Ch.V. Ramana (1990). Ammonia leaching by resting cells of a photosynthetic purple non-sulfur bacterium *Rhodobacter sphaeroides* OU 001. *J. Indian Inst. Sci.* **70:** 447-450.

Sasikala, Ch. and Ch.V. Ramana (1995a). Biotechnological potentials of anoxygenic phototrophic bacteria. 1. Production of single-cell protein, vitamins, ubiquinones, hormones and enzymes and use in wastewater treatment. *Adv. Appl. Microbiol.* **41:** 173-226.

Sasikala, Ch. and Ch.V. Ramana (1995b). Biotechnological potentials of anoxygenic phototrophic bacteria. 2. Biopolyesters, biopesticides, biofuel and biofertilizer. *Adv. Appl. Microbiol.* **41:** 227-278.

Sasikala, Ch. and Ch.V. Ramana (1998). Biodegradation and metabolism of unusual carbon compounds by anoxygenic phototrophic bacteria. *Adv. Microbial. Physiol.* **39:** 339–377.

Sasikala, Ch. and Ch.V. Ramana (1999). Anoxygenic phototrophic bacteria and their role in soil nutrient cycles. In Agromicrobes, *Curr. Trends Life Sci* (Jha, M. N., ed.), Today and Tomorrow‘s Printers and Publishers, New Delhi. **23:** 219-234.

Sasikala, Ch., C. V. Ramana and G.S. Prasad (1994). *World J. Micbiol. Biotech.***10:** 221-223.

Sasikala, Ch., Ch. V. Ramana and P. Raghuveer Rao (1991). Environmental regulation for optimal biomass yield and photoproduction of hydrogen by *Rhodobacter sphaerodies*.O.U.001. *Int. J. Hyd. Ene.***16:** 597-601.

Sasikala, Ch., Ch.V. Ramana, P.R. Rao and K.L. Kovacs (1993). Anoxygenic phototropic bacteria: Physiology and advances in hydrogen production technology. *Advan. Appl. Microbiol.***38:** 211-295.

Sasikala, K and C.V. Ramana (1991). Photoproduction of hydrogen from wastewater of a lactic acid fermentation plant by a purple non-sulfur photosynthetic bacterium, *Rhodobactersphaeroides* O.U.001. *Indian J. Exp. Biol.* **29:** 74-75.

Sasikala, K., C.V. Ramana and P. Raghuveer Rao (1992). *Int. J. Hydr. Ener.* **17:** 23-27.

Sasirekha, B., T. Bedashree and K.L. Champa (2012). Optimization and partial purification of extracellular phytase from *Pseudomonas aeruginosa* p6. *Euro. J. Experim. Biol.* **2 (1):** 95–104.

Sastri, M.V.C (1989). India's hydrogen energy program-a status report. *Int. J. Hydr. Ener.* **14:** 507–513.

Sato, K., T. Shiba and Y. Shioi (1989). Regulation of the biosynthesis of bacteriochlorophyll. In: *Aerobic Photosynthetic Bacteria*, (Harashima, K., T. Shiba and N. Murata, eds.). Springer Verlag. Berlin. pp. 95–124.

Sattley, W.M. and M. Madigan (2014). The Family *Heliobacteriaceae*. In: *The Prokaryotes*, (Rosenberg, E., E.F. DeLong, S. Lory, E. Stackebrandt and F. Thompson, eds.). *Springer, Berlin, Heidelberg*. pp. 185–196.

Sattley, W.M. and W.D. Swingley (2013). Properties and evolutionary implications of the heliobacterial genome. In: Genome evolution of photosynthetic bacteria, (Beatty, J.T. ed.). Advances in botanical research. Academic Press, Vol. 66, *Elsevier, San Diego*, pp. 67–97.

Sattley, W.M., M.T. Madigan, W.D. Swingley, P.C. Cheung, K.M. Clocksin, A.L. Conrad, L.C. Dejesa, B.M. Honchak, D.O. Jung, L.E. Karbach, A. Kurdoglu, S. Lahiri, S.D. Mastrian, L.E. Page, H.L. Taylor, Z.T. Wang, J. Raymond, M. Chen, R.E. Blankenship and J.W. Touchman (2008). The genome of *Heliobacterium modesticaldum*, phototrophic representative of the *Firmicutes* containing the simplest photosynthetic apparatus. *J. Bacteriol.* **190:** 4687–4696.

Sauve, V., P. Roversi, K.J. Leath, E.F. Garman, R. Antrobus, S.M. Lea and B.C. Berks (2009). The mechanism for the hydrolysis of a sulfur-sulfur bond based on the crystal structure of the thiosulfohydrolase SoxB. *J. Biol. Chem.* **284:** 21707–21718.

Sauve, V., S. Bruno, B.C. Berks and A.M. Hemmings (2007). The SoxYZ complex carries sulfur cycle intermediates on a peptide swinging arm. *J. Biol. Chem.* **282:** 23194–23204.

Sawada, H. and P.L. Rogers (1977a). Photosynthetic bacteria in waste treatment. *J. Ferment. Technol.* **55:** 297-310.

Sawada, H. and P.L. Rogers (1977b). Photosynthetic bacteria in waste treatment. J. *Ferment Technol.***55:** 311-325.

Sawada (1977). Photosynthetic bacteria in waste treatment. Role of *Rhodopseudomonas capsulata* with agricultural/industrial effluents. *J. Ferment. Technol.* **55:** 4.

Sawayama, S., K. Rao and D. Hall (1998). Nitrate and phosphate ion removal from water by *Phormidium laminosum* immobilized on hollow fibers in a photobioreactor. *Appl.Microbiol. Biotechnol.* **49(4):** 463 –468.

Sawayama, S., S. Hanada and Y. Kamagata (2000). Isolation and characterization of phototrophic bacteria growing in lighted upflow anaerobic sludge blanket reactor. *J. Bioscien. Bioengine.***89(4):** 396-399.

Say, R. F. and G. Fuchs (2010). Fructose 1,6-bisphosphate aldolase/phosphatase may be an ancestral gluconeogenic enzyme. *Nature*. **464:** 1077–1081.

Scavetta (2000). Structure of Rsrl methyltransferase, a member of the N6-adenine B class of DNA methyltransferases. *Nucleic Acids Research*. **28:** 3950-3961.

Schaub, B.E.M. and H. van Gemerden (1994). Simultaneous phototrophic and chemotrophic growth in the purple sulfur bacterium *Thiocapsa roseopersicina* M1. *FEMS. Microb. Ecol.* **13:** 185–196.

Scheuring, S., J. Busselez and D. Levy (2005). Structure of the dimeric PufX-containing core complex of *Rhodobacter blasticus* by in situ atomic force microscopy. *Journal of Biological Chemistry*,**280:** 1426–1431.

Scheuring, S., J. Seguin, S. Marco, D. Levy, B. Robert and J.L. Rigaud (2003). Nanodissection and highresolution imaging of the *Rhodopseudomonas viridis* photosynthetic core complex in native membranes by AFM. *Proc. Natl. Acad. Sci.* USA. **100:** 1690–1693.

Schmidle, W (1901). Neue Algen aus dem Gebiete des Oberrheins. *Beih Bot Zentbl.* 10: 179–180 (in German)

Schmidt, K (1978). Biosynthesis of carotenoids. In: *The Photosynthetic Bacteria*, (Plenum Clayton and Sistrom, eds.), Press, New York. pp. 729–750.

Schneider, D., G. Arp, A. Reimer, J. Reitner and R. Daniel (2013). Phylogenetic analysis of a microbialite-forming microbial mat from a hypersaline lake of the Kiritimati atoll, central Pacific. *PLoS One,* **8:** e66662.

Schroeter, J (1886). In: *Kryptogamenflora von Schlesien*, (F.J. Cohn, ed.), Band 3, Heft 3, Pilze J.U. Kern's Verlag, Breslau, pp. 1-814.

Schutz, M., I. Maldener and C. Griesbeck (1999). Sulfide-quinone reductase from *Rhodobacter capsulatus*: requirement for growth, periplasmic localization and extension of gene sequence analysis. *J. Bacteriol.* **181:** 6516–6523.

Segers, L. and W. Verstraete (1985). *Experientia*. **41:** 99-101.

Seki, H., A. Suzuki and S. Mitsueda (1998). Biosorption of Heavy Metal Ions on *Rhodobacter sphaeroides* and *Alcaligenes eutrophus* H16. *J. Colloid. Interface Sci.* **197:** 185-190.

Serdyuk, O.P., L.D. Smolygia, E.F. Kobzar and I.N. Gogotov (1993). Occurrence of palnt horomones in cells of the phototrophic bacterium *Rhodospirillum rubrum* 1R. *FEMS Microbiol Letter.* **109:** 113-116.

Serrano, W., J. Schrubbers, R. Amann and U. Fischer (2015). *Allochromatium humboldtianum* sp. nov., isolated from soft coastal sediments. *Int. J. Syst. Evol. Microbiol.***65:** 2980-2985.

Serrano, W., R. Amann and U. Fischer (2009). A new moderately thermophilic and high sulfide tolerant biotype of *Marichromatium gracile*, isolated from tidal sediment of the German Wadden Sea: *Marichromatium gracile* biotype *thermosulfidiphilum. Systematics and Appl. Microbiol.* **32:** 1-7.

Sessions, A.L., L. Zhang, P.V. Welander, D. Doughty, R.E. Summons and D.K. Newman (2013). Identification and quantification of polyfunctionalized hopanoids by high temperature gas chromatography-mass spectrometry. *Organic geochemistry*. **56:** 120–130.

Shalem Raj, P., E.V. Ramaprasad, S. Vaseef, Ch. Sasikala and Ch.V. Ramana (2013). *Rhodobacter viridis* sp. nov., a phototrophic bacterium isolated from the mud of a stream. *Int. J. Syst. Evol. Microbiol.***63:** 181–186.

Shaliza Ibrahim, S. Vikineswary, Sujjat Al-Azad and L.L. Chong (2006). The Effects of Light Intensity, Inoculum Size and Cell Immobilization on the Treatment of sago effluents with *Rhodopseudomonas palustris* Strain B1. Biotechnol. Bioprocess Eng. **11:** 377-381.

Shan Wen, H. Liu, H. He, L. Luo, X. Li, G. Zeng, Z. Zhou, W. Lou and C. Yang (2016). Treatment of anaerobically digested swine wastewater by *Rhodobacter blasticus* and *Rhodobacter capsulatus. Bioresour. Technol.* **222:** 33-38.

Shen, J. and O. Hirayama (1993). Denitrification of PVA-immobilized denitrifying photosynthetic bacterium, *Rhodobacter sphaeroides .J. Ferment. Bioeng*. **75:** 43-47.

Sheng, G.P., H.Q. Yu and Z. Yue (2006). Factors influencing the production of extra cellular polymeric substances by *Rhodopseudomonas acidophila. Int. Biodeter. Biodegrad.* **58:** 89-93

Sheng, G.P.P., H.Q.Q. Yu and X.Y.Y. Li (2010). Extracellular polymeric substances (EPS) of microbial aggregates in biological wastewater treatment systems: A review. Biotechnol. *Advan.* **28 (6):** 882–894.

Sheu, C., Zhi-Hao Li, Shih-Yi Sheu, Che-Chia Yang and Wen-Ming Chen (2020). *Tabrizicola oligotrophica* sp. nov. and *Rhodobacter tardus* sp. nov., two new species of bacteria belonging to the family *Rhodobacteraceae. Int. J. Syst. Evol. Microbiol.* **70:** 6266–6283.

Shi, X.Y. and H.Q. Yu (2005). Optimization of glutamate concentration and pH from volatile fatty acids in *Rhodopseudomonas capsulatus. Lett. Appl. Microbiol.* **40:** 401-406.

Shibata, K (1931) Carbon and Nitrogen Assimilation [In Japanese]; English translation: Gest H and Togasaki RK (eds and translators 1975), Japan Science Press, Tokyo.

Shimbo, S., Z.W. Zhang, T. Watanabe, H. Nakatsuka, N. Matsuda-Inoguchi, K. Higashikawa and M. Ikeda (2001). Cadmium and lead contents in rice and other cereal products in Japan in 1998-2000. *Sci. Total. Environ.* **5:** 165–175.

Shin, H.S., J.H. Youn and S.H. Kim (2004). Hydrogen production from food waste in anaerobic mesophilic and thermophilic acido genesis. *Inter. J. Hydr. Ener.* **29(13):** 1355–1363.

Shivali, K., Ch. Sasikala and Ch.V. Ramana (2012). MLSA barcoding of *Marichromatium* spp. and reclassification of *Marichromatium fluminis* (Sucharita et al., 2010) as

Phaeochromatium fluminis gen. nov. comb. nov. Systematic and Applied Microbiology. **35(4):**221–225.

Shoreit, A.A.M. and M.S.A. Shabeb (1994). Utilization of aromatic compounds by phototrophic purple non-sulfur bacteria. *Biodegradation*. **5:** 71-76.

Show, K., Z. Zhang, J. Tay, D. Teeliang, D. Lee and W. Jiang (2007). Production of hydrogen in a granular sludge-based anaerobic continuous stirred tank reactor. *Inter. J. Hydr. Ener.* **32(18):** 4744–4753.

Shreve, A.P., J.K. Trautman, H.A. Frank, T.G. Owens and A.C. Albrecht (1991). Femtosecond energy transfer processes in the B800-850 light-harvesting complex of *Rhodobacter sphaeroides* 2.4.1. *Biochim. Biophys. Acta.* **1058:** 280–288.

Shuman, K.E. and T.E. Hanson (2016). A sulfide: quinone oxidoreductase from *Chlorobaculum tepidum* displays unusual kinetic properties. *FEMS. Microbiol. Lett.* 363.

Siebert, C.A., P. Qian, D. Fotiadis, A. Engel, C.N. Hunter and P.A. Bullough (2004). Molecular architecture of photosynthetic membranes in *Rhodobacter sphaeroides*: the role of PufX. *EMBO*. **23:** 690–700.

Siefert, E. and N. Pfennig (1984). A convenient method to prepare neutral sulfide solution for the cultivation of phototrophic sulfur bacteria. *Arch. Microbiol.* **139:** 100-101.

Siefert, E. and V.B. Koppenhagen (1982). Studies on the vitamin B12 auxotrophy of *Rhodocyclus purpureus* and two other vitamin B_{12}-requiring purple non-sulfur bacteria. *Arch. Microbiol.* **132:** 173–178.

Siefert, E., R.L. Irgens and R.L. Pfennig (1978). Phototrophic purple and green bacteria in a sewage treatment plant. *Appl. Environ. Microbiol.***35:** 38–44.

Sievers, M. and J. Swings (2015). *Acetobacteraceae. Bergey's Manual of Systematics of Archaea and Bacteria,* 1–20. doi:10.1002/9781118960608.fbm00174.

Simon, J. and M. Kern (2008). Quinone-reactive proteins are devoid of haem b form widespread membrane-bound electron transport modules in bacterial respiration. Biochem. Soc. Trans. **36:**1011–1016.

Simon, J. and P.M. Kroneck (2013). Microbial sulfite respiration. *Adv Microb Physiol.* **62:** 45–117.

Sirevag, R (1995). Carbon metabolism in green bacteria, in Anoxygenic Photosynthetic Bacteria, eds R.E. Blankenship, M. T. Madigan and C. E. Bauer (Amsterdam: Kluwer). pp. 871–883.

Sirevag, R. and J.G. Ormerod (1977). Synthesis, storage and degradation of polyglucose in *Chlorobium thiosulfatophilum*. *Arch. Microbiol.***111:** 239–244.

Sistrom, W.R (1960). A requirement for sodium in the growth of *Rhodopseudomonas sphaeroides*. *J. Gen. Microbiol.* **22:** 778–782.

Skerman, V.B.D., V. Mc Gowan and P.H.A. Sneath (1980). Approved Lists of Bacterial Names. *Int. J. Syst. Bacteriol.* **30**: 225-420.

Smiejan, A., K.J. Wilkinson and C. Rossier (2003). Cadmium bioaccumulation by a freshwater bacterium, *Rhodospirillum rubrum*. *Environ. Sci. Tech.***37:** 701-706.

Smith, B.E (2002). Structure: Nitrogenase reveals its inner secrets, Science. **297(5587):** 1654-1655.

Smith, R. L. and M.J. Klug (1981). *Appl. Environ. Microbiol.* **42:** 116-121.

Sokolova, T. G., J.M. Gonzalez, N.A. Kostrikina, N.A. Chernyh, T.V. Slepova, E.A. Bonch-Osmolovskaya and F.T. Rob (2004a). *Thermosinus carboxydivorans* gen. nov., sp. nov., a new anaerobic, thermophilic, carbon-monoxide-oxidizing, hydrogenogenic bacterium from a hot pool of Yellowstone National Park. *Int. J. Syst. Evol. Microbiol.* **54:** 2353–2359.

Sokolova, T.G., N.A. Kostrikina, N.A. Chernyh, T.V. Kolganova, T.P. Tourova and E.A. Bonch-Osmolovskaya (2005). *Thermincola carboxydiphila* gen. nov., sp. nov., a novel anaerobic, carboxydotrophic, hydrogenogenic bacterium from a hot spring of the Lake Baikal area. *Int. J. Syst. Evol. Microbiol.* **55:** 2069–2073.

Sokolova, T., J. Hanel, R. Onyenwoke, A.L. Reysenbach, A. Banta, R. Geyer, J.M. Gonzalez, W.B. Whitman and J. Wiegel 2007). Novel chemolithotrophic, thermophilic, anaerobic bacteria *Thermolithobacter ferrireducens* gen. nov., sp. nov. and *Thermolithobacter carboxydivorans* sp. nov. *Extremophiles.* **11:** 145–157.

Sokolova, T., N. Kostrikina, N. Chernyh, T. Tourova, T. Kolganova and E. Bonch-Osmolovskaya (2002). *Carboxydocella thermautotrophica* gen. nov., sp. nov., a novel anaerobic, CO-utilizing thermophile from a Kamchatkan hot spring. *Int. J. Syst. Evol. Microbiol.* **52:** 1961–1967.

Sokolova, T.G., C. Jeanthon, N.A. Kostrikina, N.A. Chernyh, A.V. Lebedinsky, E. Stackebrandt and E.A. Bonch-Osmolovskaya (2004b). The first evidence of anaerobic CO oxidation coupled with H_2 production by a hyperthermophilic archaeon isolated from deep-sea hydrothermal vents. *Extremophiles.***8:** 317-323.

Sokolova, T.G., J.M. Gonzalez, N.A. Kostrikina, N.A. Chemyh, T.P. Tourova, C. Kato, F.A. Bonch-Osmolovskaya and F.T. Robb (2001). *Carboxydobrachium pacificum* gen. nov., sp nov., a new anaerobic thermophilic, CO-utilizing marine bacterium from Okinawa Trough. *Int. J. Syste. Evol. Microbiol.***51:** 141-149.

Soriano, S (1973). Flexibacteria. Annu. Rev. Microbiol. **27:**155–170.

Sorokin, D.Y (2003). Oxidation of inorganic sulfur compounds by obligately organotrophic bacteria. *Microbiology.* **72:** 641–653.

Sorokin, D.Y., T.P. Tourova, B.B. Kuznetsov, I.A. Briantseva and V.M. Gorlenko (2000). *Roseinatronobacter thiooxidans* gen. nov., sp. nov., a new alkaliphilic aerobic bacteriochlorophyll *a*—containing bacterium isolated from a soda lake. *Microbiology.* **69:**75–82.

Sorokin, Y.I (1970). Interrelations between sulfur and carbon turnover in a meromictic lake. *Arch. Hydrobiol.* **66:** 391–446.

Spaepen, S., J. Vanderleyden and R. Remans (2007). Indole-3-acetic acid in microbial and microorganism-plant signaling. *FEMS Microbiol. Rev.***31:** 425–448.

Spalding, M.D. and S.T. Prigge (2010). Lipoic acid metabolism in microbial pathogens. *Microbiol. Mol. Bio.l Rev.* **74:** 200–228.

Sprague, S.G. and A.R. Varga (1986). Membrane architecture of anoxygenic photosynthetic bacteria. In Staehelin, L.A. and Arntzen, *C.I.,* eds., Photosynthesis III. Photosynthetic membranes and light harvesting systems. *Encycl. of Plant Physiol.* new series, Springer-Verlag, Berlin, Vol. 19, pp. 603-619.

Spring, S (2014). Function and evolution of the Sox multienzyme complex in the marine Gamma proteobacterium *Congregibacter litoralis. ISRN Microbiol.* **2014:** 597418.

Sprules, T., K.E. Kawulka and J.C. Vederas (2004). NMR solution structure of ImB2, a protein conferring immunity to the antimicrobial activity of the type IIa bacteriocin, carnobacteriocin B2. *Biochemistry.* **43(37):** 11740-9.

Sprules, T., K.E. Kawulka, A.C. Gibbs, D.S. Wishart and J.C. Vederas (2004). NMR solution structure of the precursor for carnobacteriocin B2, an antimicrobial peptide from *Carnobacterium piscicola. Eur. J. Biochem.* **271:** 1748–1756.

Srere, P. A (1963). Citryl-CoA and the citrate condensing enzyme. *Biochim. Biophys. Acta.* **77:** 693–696.

Srinivas, A., B.V. Kumar, B. Divyasree, L. Tushar, Ch. Sasikala and Ch.V. Ramana (2014). *Rhodovulum salis* sp. nov. and *Rhodovulum viride* sp. nov., phototrophic

Alphaproteobacteria isolated from marine habitats. *Int. J. Syst. Evol. Microbiol.***64:** 957–962.

Srinivas, A., Ch. Sasikala and Ch.V. Ramana (2014). *Rhodoplanes oryzae* sp. nov., a phototrophic *alphaproteobacterium* isolated from the rhizosphere soil of paddy. *Int. J. Syst. Evol. Microbiol.***64:** 2198–2203.

Srinivas, A., K. Rahul E.V.V. Ramprasad, Ch. Sasikala and Ch.V. Ramana (2012). *Rhodovulum bhavnagarense* sp. nov., a phototrophic *alphaproteobacterium* isolated from a pink pond. *Int. J. Syst. Evol. Microbiol.* **6:** 2528-2532.

Srinivas, L.N., A. Kopuri, P. Sreevennela, R.D. Venkata (2013). Isolation and Biochemical characterization of protease isolated from *Bacillus* sp SVN12. *Int. J. Res. Pure. Appl. Microbiol.* **3:** 94-101.

Srinivas, M (2001). Studies on the microorganisms of effluents and Wastewater with reference to phototrophic bacteria and hydrogen production. Thesis submitted to the Department of Microbiology, Kakatiya University, Warangal.

Srinivas, M. and K. Jyothi Rani (2014a). Optimization of cultural conditions for biohydrogen production by a photosynthetic bacterium, *Rhodocyclus tenuis* KU 017. *A. J. Microbiol. Biotechnol. Environ.***16:** 381-388.

Srinivas, M. and K. Jyothi Rani (2014b). Enhanced bioproduction of hydrogen by alginate immobilized cells of a purple non-sulphur bacterium *Rhodocyclus tenuis* KU 017. *Int. J. Pharma. Bioscie.* **5: (B)** 887-897.

Srinivas, M., D. Vasavi, S. Girisham and S.M. Reddy (2000). Isolation and characterization of anoxygenic phototrophic bacteria form wastewater. *Scien. Cult.***66:** 339-341.

Srinivas, M., D. Vasavi, S. Girisham and S.M. Reddy (2001). Anoxygenic phototrophic bacteria of Wastewater. In: *Perspectives in Biotechnology*, (Reddy, S.M. Digambar Rao and Vidyavati, eds.), Scientific Publishers (India) Jodhpur (Raj.). pp 163-170.

Srinivas, M., D. Vasavi, S. Girisham and S.M. Reddy (2002). Production of asparaginase by four anoxygenic phototrophic bacteria. *Hind. Antibiotics Bulletin.* **44:** 34-36.

Srinivas, M., D. Vasavi, S. Girisham and S.M. Reddy (2002). Production of indole-acetic-acid by anoxygenic phototrophic bacteria. *Ind. J. Microbiol.* **42:** 215-218.

Srinivas, M., D. Vasavi, S. Girisham and S.M. Reddy (2003). Production of amylases (α and β) by anoxygenic phototrophic bacteria. *Food. Sci. Technol.* **40:** 505-508.

Srinivas, M., D. Vasavi, V. Pranitha and S. Girisham (2001). Remediation of wastewater by two anoxygenic phototrophic bacteria. In: *Frontiers in Microbial Biotechnolgy and Plant Pathology*, (Manoharachary, C., D.K. Purohit, S. Ram Reddy, M.A. Singara Charya and S. Girisham, eds.). *Scientific Publishers(India).* Jodhpur (Raj.). pp. 315-322.

Srinivas, T.N.R (2008). Cultured diversity of purple non-sulfur bacteria of diverse habitats of India. Ph.D. Thesis, Jawaharlal Nehru Technological University, Hyderabad.

Srinivas, T.N.R., P. Anil Kumar, Ch. Sasikala, Ch.V. Ramana and J.F. Imhoff (2007d). *Rhodobacter vinaykumarii* sp. nov., a marine phototrophic *alphaproteobacterium* from tidal waters and emended description of the genus *Rhodobacter*. *Int. J. Syst. Evol. Microbiol.* **57:** 1984-1987.

Srinivas, T.N.R., P. Anil Kumar, Ch. Sasikala, Ch.V. Ramana and J.F. Imhoff (2007c). *Rhodovulum visakhapatnamense* sp. nov. *Int. J. Syst. Evol. Microbiol.***57:** 1762-1764.

Srinivas, T.N.R., P.A. Kumar, C. Sasikala, Ch.V. Ramana, J. Suling and J.F. Imhoff (2006). *Rhodovulum marinum* sp. nov., a novel phototrophic purple non-sulfur *alpha-proteobacterium* from marine tides of Visakhapatnam, India. *Int. J. Syst. Evol. Microbiol.* **56:** 1651–1656.

Srinivas, T.N.R., P.A. Kumar, Ch. Sasikala and Ch. V. Ramana (2007a). *Rhodovulum imhoffii* sp. nov. *Int. J. Syst. Evol. Microbiol.* **57:** 228–232.

Srinivas, T.N.R., P.A. Kumar, Ch. Sasikala, C. Spröer and Ch. V. Ramana. (2008). *Rhodobacter ovatus* sp. nov., a phototrophic *alphaproteobacterium* isolated from a polluted pond. *Int. J. Syst. Evol. Microbiol.* **58:** 1379–1383.

Srinivas, T.N.R., P.A. Kumar, Ch. Sasikala, Ch.V. Ramana and J.F. Imhoff (2007b). *Rhodobium gokarnense* sp. nov., a novel phototrophic alphaproteobacterium from a saltern. *Int. J. Syst. Evol. Microbiol.* **57:** 923–935.

Srinivas, T.N.R., P.A. Kumar, K. Sucharita, Ch. Sasikala and Ch.V. Ramana (2009). *Allochromatium phaeobacterium* sp. nov. *Int. J. Syst. Evol. Microbiol.* **59:** 750–753.

Srinvas Munjam, D. Vasavi, S. Girisham and S.M. Reddy (2005). Production of Lipases by four anoxygenic phototrophic bacteria. *Hind. Antibiot bull.* **47-48:** 32-35.

Srinvas Munjam, S. Girisham and S.M. Reddy (2007). Production of Proteases by four anoxygenic phototrophic bacteria. *Bioinfolet.* **4:** 75-77.

Srinvas Munjam, S. Girisham and S.M. Reddy (2008). Factors influencing the biomass production of some anoxygenic purple non-sulfur phototrophic bacteria. *Bioinfolet.* **5:** 85-92.

Srinvas Munjam, S. Girisham and S.M. Reddy (2009). Production of FAA and Proteins by anoxygenic phototrophic bacteria. *Bioinfolet.* **6:** 110-114.

Srisuwun, A., N. Tantiwa, A. Kuntiya, A. Kawee-Ai, A. Manassa, C. Techapun, P. Seesuriyachan (2018). Decolorization of Reactive Red 159 by a consortium of photosynthetic bacteria using an anaerobic sequencing batch reactor (AnSBR). *Prep. Biochem. Biotechnol.* **48(4):** 303-311.

Stackebrandt, E., F.A. Rainey and N. Ward-Rainey (1996). Anoxygenic phototrophy across the phylogenetic spectrum: current understanding and future perspectives. *Arch. Microbiol.* **166:** 211–223.

Stackebrandt, E., R.G.E. Murray and H.G. Truper (1988). *Proteobacteria* classis nov., a name for the phylogenetic taxon that includes the purple bacteria and their relatives. *Int. J. Syst. Bactreriol.* **38:** 321-325.

Stackebrandt, E., V.J. Fowler, W. Schubert and J.F. Imhoff (1984). Towards phylogeny of phototrophic purple sulfur bacteria—the genus *Ectothiorhodospira*. *Arch. Microbiol.* **137:** 366–370.

Stackebrandt, E., W. Frederiksen, G. Garrity, P.A.D. Grimont, P. Kampfer, M.C.J. Maiden, X. Nesme, R. Rosselló-Mora, J. Swings, H.G. Truper, L. Vauterin, A.C. Ward and W.B. Whitman (2002) Report of the ad hoc committee for the re-evaluation of the species definition in bacteriology. *Int. J. Syst. Evol. Microbiol.* **52:**1043–1047.

Stadtwald-Demchick, R., F.R. Turner and H. Gest (1990). *Rhodopseudomonas cryptolactis* sp. nov., a new thermotolerant species of budding phototrophic purple bacteria. *FEMS Microbiol Lett.* **71:** 117–122.

Staehelin, L.A., J.R. Golecki, R.C. Fuller and G. Drews (1978). Visualization of the supramolecular architecture of chlorosomes (*Chlorobium*-type vesicles) in freeze-fractured cells of *Chloroflexus aurantiacus*. *Arch. Microbiol.* **119**: 269–277.

Stahlberg, H., J. Dubochet, H. Vogel and R. Ghosh (1998). Are the light-harvesting I complexes from *Rhodospirillum rubrum* arranged around the reaction center in a square geometry? *J. Mol. Biol.* **282:** 819–831.

Stark, W., F. Jay and K. Muehlethaler (1986). Localization of reaction center and light harvesting complexes in the photosynthetic unit of *Rhodopseudomonas viridis*. *Arc. Microbiol.* **146:** 130– 133.

Steenbergen, C.L.M. and H.J. Korthals (1982). Distribution of phototrophic microorganisms in the anaerobic and microaerophilic strata of Lake Vechten (The Netherlands): pigment analysis and role in primary production. *Limnol. Oceanogr.* **27:** 883–895.

Steenbergen, C.L.M., H.J. Korthals, A.L. Baker and C.J. Watras (1989). Microscale vertical distribution of algal and bacterial plankton in Lake Vechten (The Netherlands). *FEMS Microbiol. Ecol.* **62:** 209–220.

Steinmetz, M.A. and U. Fischer (1982). Cytochromes of green sulfur bacterium *Chlorobium vibrioforme* f. sp. *thiosulfatophilum*. Purification, characterization and sulfur metabolism. *Arch. Microbiol.* **131**: 19–26.

Stevens, P., N. Plovie, P. De Vos and J. De Ley (1986). Photoproduction of molecular hydrogen by *Rhodobacter sulfidophilus*. *Systematic and Applied Microbiology*. **8:** 19-23.

Stevenson, A.K., L.K. Kimble, C.R.Woese and M.T. Madigan (1997). Characterization of new phototrophic heliobacteria and their habitats. *Photosynth. Res.* **53:** 1–11.

Stirn, J (1971). Ecological consequences of marine pollution. *Rev. Int. Oceanogr. Med.***24:** 13–46.

Stockdreher, Y., M. Sturm, M. Josten, H.G. Sahl, N. Dobler, R. Zigann and C. Dahl (2014) New proteins involved in sulfur trafficking in the cytoplasm of *Allochromatium vinosum*. *J. Biol. Chem.* **289:** 12390–12403.

Stockdreher, Y., S.S. Venceslau, M. Josten, H.G. Sahl, I.A.C. Pereira and C. Dahl (2012). Cytoplasmic sulfurtransferases in the purple sulfur bacterium *Allochromatium vinosum*: evidence for sulfur transfer from DsrEFH to DsrC. *PLoS One*.**7:** e40785.

Straub, K.L., F.A. Rainey and F. Widdel (1999). *Rhodovulum iodosum* sp. nov. and *Rhodovulum robiginosum* sp. nov., two new marine phototrophic ferrous-iron-oxidising purple bacteria. *Int. J. Syst. Bacteriol.* **49:** 729–735.

Strauss, G.and G. Fuchs (1993). Enzymes of a novel autotrophic CO_2 fixation pathway in the phototrophic bacterium *Chloroflexus aurantiacus*, the 3-hydroxypropionate cycle. *Eur. J. Biochem.* **215(3):** 633-43.

Strnad, H., A. Lapidus, J. Paces, P, Ulbrich, C. Vlcek, V. Paces and R. Haselkorn (2010). Complete genome sequence of the photosynthetic purple non-sulfur bacterium *Rhodobacter capsulatus* SB 1003. *J. Bacteriol.* **192:** 3545–3546.

Subhash, Y. and S.S. Lee (2016). *Rhodobacter sediminis* sp. nov., isolated from lagoon sediments. *Int. J. Syst. Evol. Microbiol.* **66:** 2965-2970.7

Sublette, K. L (1991). PCT. *Int. Appl.* W091/18,661.

Subudhi, S., S.K. Mogal, N. Ram Kumar, T. Nayak, B. Lal and H. Velankar, T.A.Rajiv Kumar, Peddy V.C.Rao, Nettem V. Choudary, GiteshShah, Srigarnesh Gandham (2016). Photo fermentative hydrogen production by a new strain; *Rhodobacter sphaeroides* CNT 2A, isolated from pond sediment. *Int. J. Hydr. Ene*. **41:** 13979-85.

Sucharita, K., Ch. Sasikala and Ch.V. Ramana (2010). *Thiorhodococcus modestalkaliphilus* sp. nov. a phototrophic *gammaproteobacterium* from Chilika saltwater lagoon, India. *J. Gen. Appl. Microbiol.***56**: 93-99.

Sucharita, K., E. Shiva Kumar, Ch. Sasikala, B.B. Panda, S. Takaichi and Ch.V. Ramana (2010). *Marichromatium fluminis* sp. nov., a slightly alkaliphilic, phototrophic *gammaproteobacterium* isolated from river sediment. *Int. J. Syst. Evol. Microbiol.***60:** 1103-1107.

Suckow, R (1966). Schwefelmikrobengesellschaften der Seeund Boddengewasser von Hiddensee. Zeitschr. Allgem. *Mikrobiol.* **6:** 309–315.

Suhaimi, M., L. Vandevoorde, B. Bruyneel and W. Verstraete (1988). NH^{+4} -N recovery from ammonia-treated straw-whey ensilage and cell biomass formation by *Rhodobacter capsulatus* ATCC 23782 on silage filtrates. *Appl. Mcrobiol. Biotechnol.* **27:** 606-611.

Suhas Kumar, D. and R. Arun Kumar (2014). Characterization and identification of phytase-producing bacteria isolated from the gastrointestinal tract of four freshwater teleosts. *Annals Microbiol.* **64:** 297-306.

Summons, R.E., L.L. Jahnke, J.M. Hope and G.A. Logan (1999). 2-Methylhopanoids as biomarkers for cyanobacterial oxygenic photosynthesis. *Nature*. **400:** 554-557.

Su, P., X. Tan, C. Li, Q. Yan, Q., Peng, J., Zhang, Z., and Y. Liu, Y (2017). Photosynthetic bacterium *Rhodopseudomonas palustris* GJ-22 induces systemic resistance against viruses. *Microb. Biotechnol.***10:** 612–624.

Sun, Y.P., Z.P. Zhang and L.J. Wang (2009). Promotion of 5-aminolevulinic acid treatment on leaf photosynthesis is related toan increase of antioxidant enzyme activity in watermelon seedlings grown under shade conditions. *Photosynthetica*. **47:** 347–354.

Sun, Y., L. Sun, F. Shang and G. Yan (2016). Enhanced production of β-carotene in recombinant *Saccharomyces cerevisiae* by inverse metabolic engineering with supplementation of unsaturated fatty acids. *Process Biochemistry*. **51(5):** 568–577.

Sunayana, M. R., Ch. Sasikala and Ch.V. Ramana (2005a). Production of novel indole ester from 2-amonobenzoate by *Rhodobacter sphaeroides* OU5. *J. Ind. Microbiol. Biotech.* **32:** 41-45.

Sunayana, M.R., Ch. Sasikala and Ch.V. Ramana (2005b). Rhodestrin: a novel indole terpenoid phytohormones from *Rhodobacter sphaeroides* OU5. *Biotech Lett.***27:** 1897-1900.

Sundstrom, V. and R. Van Grondelle (1995). Kinetics of excitation transfer and trapping in purple bacteria. In: *Anoxygenic Photosynthetic Bacteria, vol. 2. Advances in Photosynthesis* (Blankenship, R. E., M.T. Madigan and C.E. Bauer, eds.). pp. 350–372. Dordrecht: Kluwer Academic Publishers.

Sundstrom, V., T. Pullerits and R. Van Grondelle (1999). Photosynthetic light-harvesting: reconciling dynamics and structure of purple bacterial LH2 reveals function of photosynthetic unit. *J. Phys. Chem. B.* **103:** 2327–2346.

Suresh, G., T. Lodha, S. Chintalapati and V.R. Chintalapati (2018). *Rhodobacter alkalitolerans* sp. nov., isolated from an alkaline brown pond. *Arch Microbiol.* **200:** 1487-1492

Suresh, G., B. Sailaja, A. Ashif, B.P. Dave, Ch. Sasikala and Ch.V. Ramana (2017). Description of *Rhodobacter azollae* sp. nov. and *Rhodobacter lacus* sp. nov. *Int. J. Syst. Evol. Microbiol.***67**: 3289-3295.

Suresh, G., T.D. Lodha, B. Indu, C. Sasikala and C.V. Ramana (2019). Taxogenomics Resolves Conflict in the Genus *Rhodobacter*: A Two and Half Decades PendingThought to Reclassify the Genus *Rhodobacter*. Front. Microbiol. 10:2480. doi: 10.3389/fmicb.2019.02480.

Suresh, G., T.D. Lodha, B. Indu, C. Sasikala and C.V. Ramana (2020). Corrigendum: Taxogenomics Resolves Conflict in the Genus *Rhodobacter*: *Front. Microbiol.* doi: 10.3389/fmicb.2020.01111.

Suresh, G., D. Kumar, A. Krishnaiah, Ch. Sasikala and Ch.V. Ramana (2020). *Rhodobacter sediminicola* sp. nov., isolated from a freshwater pond. Int. J. Syst. Evol. Microbiol. **70:**1294-1299.

Suwansaard, M (2010). Production of Hydrogen and 5-Aminolevulinic Acid by Photosynthetic Bacteria from Palm Oil Mill Effluent Authors, Maneewan Suwansaard Description, Thesis (Ph.D, Biotechnology)--Prince of Songkla University.

Suwansaard, M., W. Choorit, J.H. Zeilstra-Ryalls and P. Prasertsan (2009). Isolation of anoxygenic photosynthetic bacteria from Songkhla Lake for use in a two-staged biohydrogen production process from palm oil mill effluent. *Inter. J. Hydr. Ener.* **34(17):** 7523–7529.

Suzuki, T., A.A, Tsygankov, J. Miyake, T. Yutaka and Y. Asada (1995). Accumulation of poly-(hydroxybutyrate) by a non-sulfur photosynthetic bacterium, *Rhocobacter sphaeroides* RV at different pH. *Biotech. Lett.* **17(4):** 395–400.

Swetha Garimella, Archana Vimal, Ramchander Merugu and Awanish Kumar (2019). Experimental optimization of green hydrogen production from phototrophic bacteria *Rhodobacter sphaeroides*. *Recent Innovations in Chemical Engineering*. **12(2):** 98-109.

Swingley, W.D., R.E. Blankenship and J. Raymond (2009). In: *The Purple Phototrophic Bacteria,* (Neil Hunter, C., Fevzi Daldal, Marion C. Thurnauer and J. Thomas Beatty, eds.), pp. 17–29.

Swoager, W.C. and E.S. Lindstrom (1971). Isolation and counting of *Athiorhodaceae* with membrane filters. *Appl. Microbiol.* **22:** 683–687.

Syed, M., G. Soreanu, P. Falletta and M. Beland (2006). Removal of hydrogen sulfide from gas streaming using biological processes. A. review. *Canadian Biosystems Engineering.* **48 (2)** .

Szafer, W (1911). Zur Kenntnis der Schwefelflora im der Umgebung von Lemberg. *Bull. Int. Acad. Sci. Ser. V. Cracovie.* **160–167**

Tadesse, I., S.A. Isoaho, F.B. Green and J.A. Puhakka (2003). Removal of organics and nutrients from tannery effluent by advanced integrated wastewater pond systems technology. *Water Scien. Technol.* **48(2):** 307-314.

Takahashi, M. and S. Ichimura (1968). Vertical distribution and organic matter production of photosynthetic sulfur bacteria in Japanese lakes. *Limnol Oceanog*. **13:** 644–655.

Takaichi, S (1999). Carotenoids and carotenogenesis in anoxygenic photosynthetic bacteria. In: *The Photochemistry of Carotenoids*: Applications in Biology (Frank, H.A., A.J. Young, G. Britton and R.J. Cogdell, eds.), Kluwer Academic Publishers, Dordrecht. pp. 39-69.

Takaichi, S., H. Oh-Oka, T. Maoka, D.O. Jung and M.T. Madigan (2003). Novelcarotenoid glucoside esters from alkaliphilic heliobacteria. *Arch. Microbiol.* **179:** 95–100.

Takaichi, S., T. Maoka, M. Yamada, K. Matsuura, Y. Haikawa and S. Hanada (2001). Absence of carotenes and presence of a tertiary methoxy group in a carotenoid from a thermophilic filamentous photosynthetic bacterium *Roseiflexus castenholzii*. *Plant Cell Physiol.* **42:** 1355–1362.

Takeno, K., K. Sasaki, M. Watanabe, T. Kaneyasu and N. Nishio (1999). Removal of phosphorus from oyster farm mud sediment using a photosynthetic bacterium, *Rhodobacter sphaeroides* IL106. *J. Biosci. Bioeng*. **88(4):** 410–415.

Takeno, K., Y. Yamaoka and K. Sasaki (2005). Treatment of oil-containing sewage wastewater using immobilized photosynthetic bacteria. *W. J. of Microb. Biotech*. **21:** 1385-1391.

Takeshi, S. and H. Gotaro (1991). *Jpn. Kokai Tokkyo Koho* JP 91/147,726.

Takeuchi, D. and M. Hamada (1985). *Eur. Pat. Appl.* EP 131,456.

Takeuchi, M.H. and K. Numata (2019). Marine purple photosynthetic bacteria as Sustainable microbial production hosts. *Frontiers in Bioengineering and Biotechnology*. **7:** 258. doi: 10.3389/fbioe.2019.00258

Takeuchi, M.H., K. Morisaki and K. Numata (2016). A Screening Method for the Isolation of Polyhydroxyalkanoate-Producing Purple Non-sulfur Photosynthetic Bacteria from Natural Seawater. *Front. Microbiol.* **7:** 1509.

Talaiekhozani, A. and S. Rezania (2017). Application of photosynthetic bacteria for removal of heavy metals, macro-pollutants and dye from wastewater: A review. *J. Water Process Engineering*. **19:** 312-321.

Talbot, H.M., M. Rohmer and P. Farrimond (2007). Rapid structural elucidation of composite bacterial hopanoids by atmospheric pressure chemical ionisation liquid chromatography/ion trap mass spectrometry. *Rapid Communications in Mass Spectrometry*. **21:** 880–892.

Tanaka, H., H. Maeda, H. Suzuki, A. Kamibayashi and K. Tonomura (1982). The metabolism of thiophene-2-carboxylate by a photosynthetic bacterium. *Agric Biol Chem*. **46:** 1429–1438.

Tang, K.H. and R.E. Blankenship (2010). Both forward and reverse TCA cycles operate in green sulfur bacteria. *J. Biol. Chem*. **285:** 35848–35854.

Tang, K.H., H. Yue and R.E. Blankenship (2010b). Energy metabolism of *Heliobacterium modesticaldum* during phototrophic and chemotrophic growth. *BMC Microbiol*. **10:** 150. doi: 10.1186/1471-2180-10-150.

Tang, K.H., K. Barry, O. Chertkov, E. Dalin, C.S. Han, L.J. Hauser, B.M. Honchak, L.E. Karbach, M.L. Land, A. Lapidus, F.W. Larimer, N. Mikhailova, S. Pitluck, B.K. Pierson and R.E. Blankenship (2011). Complete genome sequence of the filamentous anoxygenic phototrophic bacterium *Chloroflexus aurantiacus*. *BMC Genomics*. **12:** 334. doi: 10.1186/1471 2164-12-334.

Tang, K.H., Y.J.Tang and R.E. Blankenship (2011). Carbon metabolic pathways in phototrophic bacteria and their broader evolutionary implications. *Fmicb*. **2:** 165.

Tang, K.H., X. Feng, W.Q. Zhuang, L. Alvarez-Cohen, R.E. Blankenship and Y.J. Tang (2010a). Carbon flow of heliobacteria is related more to clostridia than to the green sulfur bacteria. *J. Biol. Chem*. **285:** 35104–35112.

Tang, K.H., X. Feng, Y.J. Tang and R.E. Blankenship (2009a). Carbohydrate metabolism and carbon fixation in *Roseobacter denitrificans* OCh114. *PLoS ONE*. 4: e7233. doi: 10.1371/journal.pone.0007233.

Tang, Y. J., S. Yi, W.Q. Zhuang, S.H. Zinder, J.D. Keasling and L. Alvarez-Cohen (2009b). Investigation of carbon metabolism in "*Dehalococcoides ethenogenes*" strain 195 by use of isotopomer and transcriptomic analyses. *J. Bacteriol*. **191:** 5224–5231.

Tangprasittipap, A. and P. Prasertsan (2002). 5-aminolevulinic acid from photosynthetic bacteria and its applications. *Songklanakarin J. Sci. Technol*. **24(4):** 717-725.

Tanisho, S. and Y. Ishiwata (1994). Continuous hydrogen production from molasses by bacterium *Enterobacter aerogenes*. *Int. J. Hydr. Ener*. **19:** 807-812.

Tanisho, S., M. Kuromoto and N. Kadokura (1998). Effect of CO_2 removal on hydrogen production by fermentation. *Int. J. Hydr. Ener*. **23:** 559 – 563.

Tank, M. and D.A. Bryant (2015). *Chloracidobacterium thermophilum* gen. nov., sp. nov.: an anoxygenic microaerophilic chlorophotoheterotrophic acido bacterium. *Int. J. Syst. Evol. Microbiol*. **65:** 1426-1430.

Tank, M. and D.A. Bryant (2015). Nutrient requirements and growth physiology of the photoheterotrophic *Acidobacterium,Chloracidobacterium thermophilum*. *Frontiers In Microbiology*. 6: 226.

Tatyana, V.L., N.T. Darya, S.L.Kestutis, L.G. Maria,S.Michael and A.A.Tsygankov (2008). Towards the integration of dark and photo fermentative waste treatment. 1. Hydrogen photoproduction by purple bacterium *Rhodobacter capsulatus* using potential products of starch fermentation. *Int. J. Hydrogen Energy*.**33:** 7020-7026.

Taylor, W.R (1964). Light and photosynthesis in intertidal benthic diatoms. *Helgol. Wiss. Meeresunters*. **10:** 29–37.

Tayne, T.A., J.E. Cutler and D.M. Ward (1987). Use of *Chloroflexus*-specific antiserum to evaluate filamentous bacteria of a hot spring microbial mat. *Appl. Environ. Microbiol*. **53:** 1982–1984.

Tegtmeyer, B., J. Weckesser, H. Mayer and J.F. Imhoff (1985). Chemical composition of the lipopolysaccharides of *Rhodobacter sulfidophilus*, *Rhodopseudomonas acidophila* and *Rhodopseudomonas blastica*. *Arch. Microbiol*. **143:** 32–36.

Teplyakov, V.V., L.G. Gassanova, E.G. Sostina, E.V. Slepova, M. Modigell, A.I. Netrusov (2002). Lab-scale bioreactor integrated with active membrane system for hydrogen production: experience and prospects. *Int. J. Hydr. Ener.***27:** 1149–1155.

Teske, A., E. Alm, J. Regan, S. Toze, B. Rittmann and D. Stahl (1994). Evolutionary relationships among ammonia- and nitrite-oxidizing bacteria. *J. Bacteriol.* **176:** 6623–6630.

Tewari, S.K., B. Das and S. Mehrotra (2004). Cultivation of medicinal plants-tool for rural development. *J. Rural. Tech.* **3:** 147-150.

Thangaraj A. and G. Kulandaivelu (1994). Biological hydrogen photoproduction using dairy and sugarcane wastewater. *Bioresou. Technol.* **48:** 9-12.

Thauer, R. K., D. Moller-Zinkhan and A.M. Spormann (1989). Biochemistry of acetate catabolism in anaerobic chemotrophic bacteria. *Annu. Rev. Microbiol.* **43:** 43–67.

Thauer, R.K., A.K. Kaster, H. Seedorf, W. Buckel and R. Hedderich (2008). Methanogenic archaea: ecologically relevant differences in energy conservation. *Nat Rev Microbiol.***6:** 579–591.

Thauer, R.K., K. Jungermann and K. Decker (1977). Energy conservation in chemotropic anaerobic bacteria. *Bacteriol.Rev.* **41:** 100–180.

Thavasi. R., S. Jayalakshmi and I.M. Banat (2011).Application of biosurfactant produced from peanut oil cake by *Lactobacillus delbrueckii* in biodegradation of crude oil. *Biosecure Technology.* **102:** 3366-3372.

Then, J. and H.G. Truper (1981). The role of thiosulfate in sulfur metabolism of *Rhodopseudomonas globiformis. Arch Microbiol.* **130:** 143–146.

Then, J. and H.G. Truper (1984). Utilization of sulfide and elemental sulfur by *Ectothiorhodospira halochloris. Arch. Microbiol.* **139:** 295–298.

Thiel, V., M. Tank, S.C. Neulinger, L. Gehrmann, C. Dorador and J.F. Imhoff (2010). Unique communities of anoxygenic phototrophic bacteria in saline lakes of Salar de Atacama (Chile): evidence for a new phylogenetic lineage of phototrophic *Gammaproteobacteria* from pufLM gene analyses. *FEMS Microbiol. Ecol.* **74:** 510-522.

Thiemann, B. and J.F. Imhoff (1996). Differentiation of *Ectothiorhodospiraceae* based on their fatty acid composition. *Syst. Appl. Microbiol.***19:** 223–230.

Thomas Beatty, J (2002). On the natural selection and evolution of the aerobic phototrophic bacteria. *Photosynthesis Research.* **73:** 109-114.

Tindall, B. J (1999). Taxonomic note: transfer of *Amoebobacter purpureus* to the genus *Pfennigia* gen. nov., as *Pfennigia purpurea* comb. nov., on the basis of the illegitimate proposal, to make *Amoebobacter purpureus* the type species of the genus *Amoebobacter. Int. J. Syst. Bacteriol.* **49:** 1307–1308.

Toerien, D. F (1776). *S. Afi. J. Sci.* **72:** 148-149.

Tolvanen, K.E.S., P.E.P. Koskinen, A. Ylikoski, P. Ollikka, I. Hemmila, J.A. Puhakka and M. Karp (2008). Quantification of *Clostridium butyricum* with real-time PCR using lanthanide labeled probes and time-resolved fluorometry. *Int. J. Hydr. Ener.* **33:** 542-549.

Tonkovich, A.Y., J.L. Zilka, M.J. LaMont, Y. Wang and R.S. Wegeng (1999). Microchaneel reactors for fuel processing applications. I. Water gas shift reactor. *Chem. Eng. Sci.* **54:** 2947-2951.

Tran, T.H.K., T.T.A. Le, L.T. Tran, T. Ooi and S. Kinoshita (2012). Decolorization of azo dyes by purple non-sulfur bacteria. Annual report of FY 2002, The Core University Program between Japan Society for the Promotion of Science (JSPS) and National Centre for Natural Science and Technology (NCST), pp. 112–118.

Tredici, M.R (2004). Mass Production of Microalgae: Photobioreactors. *Handbook of Microalgal Culture and Applied Phycology*. pp. 178–214.

Trepanier, C., S. Parent, Y. Comeau and J. Bouvrette (2002). Phosphorus budget as a water quality management tool for closed aquatic mesocosms. *Water Res.* **36:** 1007-1017.

Truper, H.G (1968). *Ectothiorhodospira mobilis* Pelsh, a photosynthetic sulfur bacterium depositing sulfur outside the cells. *J. Bacteriol.* **95:** 910–1920.

Truper, H.G (1970). Culture and isolation of phototrophic sulfur bacteria from the marine environment. *Helgol. Wiss. Meeresunters.* **20:** 6–16.

Truper, H.G (1976). Higher taxa of the phototrophic bacteria: *Chloroflexaceae* fam. nov., a family for the gliding filamentous, phototrophic "green" bacteria. *Int. J. Syst.Bacteriol.* **26:** 74–75.

Truper, H.G (1980). Distribution and activity of phiototrophic bacteria at the marine water-sediment interface: Coloques Internationeaux du C.N.R.S. Biogéochemie de la matière organique à l'interface eau-sédiment marine. **293:** 275–285.

Truper, H.G (1987). Phototrophic bacteria (an incoherent group of prokaryotes). *A taxonomic versus phylogenetic survey.Microbiol. SEM.* **3:** 71–89.

Truper, H.G. and H. Jannasch (1968). *Chromatium buderi* nov. spec., eine neue Art der groben *Thiorhodaceae. Arch.Mikrobiol.***61:** 363–372.

Truper, H.G. and N. Pfennig (1971). Family of phototrophic Green Sulfur Bacteria: *Chlorobiaceae* Copeland, the correct family name; rejection of *Chlorobacterium* Lauterborn; and the taxonomic situation of the consortium-forming species. *Int. J. Syst. Bacteriol.* **21:** 8–10.

Truper, H.G. and S. Genovese (1968). Characterization of photosynthetic sulfur bacteria causing red water in Lake Faro (Messina, Sicily). *Limnol. Oceanogr.* **13:** 225–232.

Tsygankov, A.A., Y. Hirata, M. Miyake, Y. Asada and J. Miyake (1994). Photobioreactor with photosynthetic bacteria immobilized on porous-glass for hydrogen photoproduction, *J. Ferm. Bioeng.* **5:** 575-578.

Tsavkelova, E.A., S.Y. Klimova, T.A. Cherdyntseva and A.I. Netrusov (2006) Microbial producers of plant growth stimulators and their practical use: a review. *Appl. Biochem. Microbiol.***42:** 133–143.

Turner, J., G. Sverdrup, M.K. Mann, P.C. Maness, B. Kroposki, M, Ghirardi, R.J. Evans and D. Blake (2008). Renewable hydrogen production. *Int. J. Energy Res.* **32:** 379–407.

Tushar, L., Ch. Sasikala and Ch.V. Ramana (2014). Draft genome sequence of *Rhodomicrobium udaipurense* JA643^T with special reference to hopanoid biosynthesis. *DNA Res.* **21:** 639–647.

Uffen, R.L (1976). Anaerobic growth of a *Rhodopseudomonas* species in the dark with carbon monoxide as sole carbon and energy substrate, in *Proceedings of the National Academy of Sciences (USA).* **73:** 3298–3302.

Urakami, T. and T. Yoshida (1993). Production of ubiquinone and bacteriochlorophyll a by *Rhodobacter sphaeroides* and *Rhodobacter sulfidophilus. J. Ferment. Bioeng.***76:** 191-194.

Urdiain, M., A. Lopez-Lopez, C. Gonzalo, H.J. Busse, S. Langer, P. Kampfer and R. Rossello-Mora. (2008). Reclassification of *Rhodobium marinum* and *Rhodobium pfennigii* as *Afifella marina* gen. nov. comb. nov. and *Afifella pfennigii* comb. nov., a new genus of photoheterotrophic *Alphaproteobacteria* and emended descriptions of *Rhodobium, Rhodobium orientis* and *Rhodobium gokarnense. Syst. Appl. Microbiol.* **31:** 339–351.

Urdiain, M., A. Lopez-Lopez, C. Gonzalo, H.J. Busse, S. Langer, P. Kampfer and R. Rossello-Mora. (2009). Validation List no. 127. *Int. J. Syst. Evol. Microbiol.***59**: 923-925.

Uragami, S. and Yoshida, T (1986a). Jpn. Kokai Tokkyo Koho JP 86/192,293.

Uragami, S. and Yoshida, T (1986b). Jpn. Kokai Tokkyo Koho JP 861192,294.

Valdez-Vazquez, I. and H.M. Poggi-Varaldo (2009). Alkalinity and high total solids affecting H_2 production from the organic solid waste by anaerobic consortia. *Inter. J. Hydr. Ener.* **34(9):** 3639–3646.

van der Meer, J, R., W.M. De vos, S. Harayama and A.J.B. Zehnder (1992). *Microbiol. Rev.* **58:** 677-694.

van der Woude, B.J., M. de Boer, N.M.J. vander, F.M. van der Gelc, R.A. Prins and J.C. Gottschal (1994). Anaerobic degradation of halogenated benzoic acids by photoheterotrophic bacteria. *FEMS. Microbiol. Lett.***119(1–2):** 199–207.

van Gemerden, H. and J. Mas (1995) Ecology of purple sulfur bacteria. In: Anoxygenic Photosynthetic Bacteria (R.E. Blankenship, M.T., Madigan, C.E. Bauer, eds), Advances in Photosynthesis. *Kluwer Academic Publishers, Dordrecht.* **2:** 49–85.

van Niel, C. B (1944). The culture, general physiology, morphology and classification of the non-sulfur purple and brown bacteria. *Bacteriol. Rev.* **8:** 1–118.

van Niel, C.B (1931). On the morphology and physiology of the purple and green sulfur bacteria. *Arch Mikrobiol.* **3:** 1–112.

van Niel, C.B (1941). The bacterial photosynthesis and their importance for the general problem of photosynthesis. *Adv Enzymol.* **1:** 263-328.

van Niel, C.B (1971). Techniques for the enrichment, isolation and maintenance of photosynthetic bacteria. In: Methods in Enzmology. *A san Pietro, New York, Academic Press.* **23:** 3-28.

van Niel, E.W.J., P.A.M. Claassen and A.J.M. Stams (2003). Substrate and product inhibition of hydrogen production by the extreme thermophile, *Caldicelllosiruptor saccharolyticus. Biotechnol. Bioeng.* **81(3):** 255-262.

Vareeket, R. and K. Soytong (2015). Evaluation of photosynthesizing bacteria for the growth of rice var. RD41. *J. Agricultural Technol.* **11(8):** 2257-2261.

Vasavi, D (2001). Hydrogen-producing bacteria of Wastewater. Thesis submitted to the Department of Microbiology for the Degree of Doctor of Philosophy to Kakatiya University, Warangal.

Vasavi, D., M. Ramchander, S. Girisham and S.M. Reddy (2014). Hydrogen Production by free and immobilized cells of *Rhodopseudomonas rutila* isolated from paper mill effluents. *Int. J. Resea. Enviro. Sci. Technol.* **4:** 71-75.

Vasavi, D., M. Ramchander, S.Girisham and S.M. Reddy (2008). Remediation of wastewater using two anoxygenic phototrophic bacteria. *Ecol. Envi. Con.***14:** 363-366.

Vatsala, T (1992). Hydrogen production from (cane-molasses) stillage by *Citrobacter freundii* and its use in improving methanogenesis. *Int. J. Hydr. Ener.***17:** 923-927.

Vatsala, T.M (1990). Microbial process for photohydrogen production from cellulose in high saline water medium. U. S. Patent. 4,921,800.

Vatsala, T.M. and V. Balaji (1987). Hydrogen production from cellulose by phototrophic organisms. Alternative energy source VIII (2) Research and development (Nejat, T. Veziroglu, ed.), *Hemisphere publishing corps.* 519-527.

Vatsala, T.M. and V. Ramasamy (1989). In "*Alternative Energy Sources*" (T. N. Vezirogiu, ed.). Hemisphere Publishers, New York. **2:** 519-529.

Vatsala, T.M., S. Mohan Raj and A. Manimaran (2008). A pilot-scale study of biohydrogen production from distillery effluent using defined bacterial co-culture. *Int J. Hydr. Ener.* **33:** 5404-5415.

Venceslau, S.S., Y. Stockdreher, C. Dahl and I.A.C. Pereira (2014). The bacterial heterodisulfide DsrC is a key protein in dissimilatory sulfur metabolism. *Biochem Biophys Acta.* **1837:** 1148–1164.

Venkata Mohan, S., Y. Vijaya Bhaskar and P.N. Sarm (2007). Biohydrogen production from chemical wastewater treatment by selectively enriched anaerobic mixed consortia in biofilm configured reactor operated in periodic discontinuous batch mode. *Water Res.***41:** 2652-64.

Venkata Mohan, S., Y. Vijaya Bhaskar, P. Murali Krishna, N. Chandrasekhara Rao, V. Lalit Babu, P.N. Sarma.(2007). Biohydrogen production from chemical wastewater as substrate by selectively enriched anaerobic mixed consortia: Influence of fermentation pH and substrate composition.*Int. J. Hydr. Ener.* **32:** 2286 – 2295.

Venkata Ramana, V., Ch. Sasikala and Ch.V. Ramana (2008). *Rhodobacter maris* sp. nov., a phototrophic *alphaproteobacterium* isolated from a marine habitat of India. *Int. J. Syst. Evol. Microbiol.* **58:** 1719-1722 .

Venkata Ramana, V., Ch. Sasikala, E.V.V. Ramaprasad and Ch.V. Ramana (2010). Description of *Ectothiorhodospira salini* sp. nov. *J. Gen. Appl. Microbiol.* **56:** 313-319.

Venkata Ramana, V., P.S. Raj, L. Tushar, Ch. Sasikala and Ch.V. Ramana (2013). *Rhodomicrobium udaipurense* sp. nov., a psychrotolerant, phototrophic *alphaproteobacterium* isolated from a freshwater stream. *Int. J. Syst. Evol. Microbiol.***63:** 2684–2689.

Venkata Ramana, V., S. Kalyana Chakravarthy, E.V.V. Ramaprasad, V. Thiel, J.F. Imhoff, Ch. Sasikala and Ch.V. Ramana (2013). Emended description of the genus *Rhodothalassium* Imhoff *et al.,* 1998 and proposal of *Rhodothalassiaceae* fam. nov. and *Rhodothalassiales* ord. nov.*Systematic and Applied Microbiology.***36:** 28-32.

Venkata Ramana, V., S. Kapoor, E. Shobha, E.V. Ramprasad, Ch. Sasikala and Ch.V. Ramana (2011). *Blastochloris gulmargensis* sp. nov., isolated from an epilithic phototrophic biofilm. *Int. J. Syst. Evol. Microbiol.* **61:** 1811–1816.

Venkata Ramana, V., S.K. Chakravarthy, P.S. Raj, B.V. Kumar, E. Shobha, E.V.V. Ramaprasad, Ch. Sasikala and Ch.V. Ramana (2012). Descriptions of *Rhodopseudomonas parapalustris*sp. nov., *Rhodopseudomonas harwoodiae* sp. nov. and *Rhodopseudomonas pseudopalustris* sp. nov., and emended description of *Rhodopseudomonas palustris*. *Int. J. Syst. Evol. Microbiol.***62:** 1790–1798

Venkata Ramana.V., P. Anil Kumar and T.N.R. Srinivas (2009). *Rhodobacter aestuarii* sp. nov. a phototrophic *alphaproteobacterium* isolated from a estuarine habitat of India. *Int. J. Syst. Evol. Microbiol.* **59:** 1133-1136.

Venkataraman, C. and T.M. Vatsala (1990). Hydrogen production from whey by phototrophic bacteria. W.H.E.C. *Hydrogen Energy Progress* VIII. **2:** 781.

Venkataramani, N (2002). Industrial Plasma Torches and Applications. *Current. Science.* **83:** 3.

Ventura, S., C. Viti, R. Pastorelli and L. Giovannetti (2000). Revision of species delineation in the genus *Ectothiorhodospira*. *Int. J. Syst. Evol. Microbiol.* **50:** 583-591.

Ventura, S., L. Giovanetti, A. Gori, C. Viti and R. Materassi (1993). Total DNA restriction pattern and quinone composition of members of the family *Ectothiorhodospiraceae*. *System. Appl. Microbiol.***16:** 405–410.

Vernon, L. P (1953). Cytochrome c content of *Rhodosperillum rubrum*. *Arch. Biochem.Biophys.* **43:** 492.

Veziroglu, T.N (1987). Hydrogen technology for energy needs of human settlements. *Int. J. Hydr. Ener.* **12(2):** 99-129.

Vignais, M.V., B. Billoud and J. Meyer (2001). Classification and phylogeny of hydrogenases. *FEMS Microbiol. Rev.* **25:** 455-501.

Vignais, P. M. and B. Billoud (2007). Occurrence, classification and biological function of hydrogenases: an overview. *Chemical Reviews*. **107(10):** 4206-4272.

Vijay, S., M. R., Sunayana, N. K. Ranjith, Ch. Sasikala and Ch. V. Ramana (2006). Light-dependent transformation of aniline to indole esters by the purple bacterium *Rhodobacter sphaeroides* OU5. *Curr. Microbiol.***52:** 413-417.

Vijayaraghavan, K. and Mohd Amin Mohd Soom (2006). Trends in bio-hydrogen generation – A review. *Environmental Sciences*. **3:4:** 255-271.

Vijayaraghavan, K., D. Ahmad and C. Soning (2007). Bio-hydrogen generation from mixed fruit peel waste using anaerobic contact filter. *Int. J. Hydr. Ener.* **32(18):** 4754-4760.

Vila, X., R. Guyoneaud, X.P. Cristina, J.B. Figueras and C.A. Abella (2002). Green sulfur bacteria from hypersaline Chiprana Lake (Monegros, Spain): habitat description and phylogenetic relationship of isolated strains. *Photosyn. Res.* **71:** 165–172.

Vincenzini, M., R. Materassi, M.R. Tredici and G. Florenzano (1982) Hydrogen production by immobilized cell - I. light dependent dissimilation of organic substances by *Rhodospeudomonas palustris*, *Int. J. Hydr. Ener.* **7:** 231-236.

Vishnuvardhan, S.R., S. Aspana, L. Tushar, Ch. Sasikala and Ch.V. Ramana (2013) *Spirochaeta sphaeroplastigenens* sp. nov., a novel halo-alkaliphilic, obligately anaerobic spirochaete isolated from soda lake Lonar, India. *Int. J. Syst. Evol. Microbiol.* **63:** 2223–2228.

Visscher, P. and B. Taylor (1994). Aerobic and anaerobic degradation of a range of alkyl sulfides by a denitrifying marine bacterium. *Appl. Environ. Microbiol.* **59:** 4083-4089.

Visscher, P.T. and B.F. Taylor (1993). Organic thiols as organolithotrophic substrates for the growth of phototrophic bacteria. *Appl. Environ. Microbiol.***59:** 93–96.

Visscher, P.T. and H. van Gemerden (1991) Photoautotrophic growth of *Thiocapsa roseopersicina* on dimethyl sulfide. *FEMS Microbiol. Lett.* **81:** 247–250.

Visscher, P.T., J.W. Nijburg and H. van Gemerden (1990). Polysulfide utilization by *Thiocapsa roseopersicina*. *Arch. Microbiol.***155:** 75–81.

Vogl, K., J. Glaeser, K.R. Pfannes, G. Wanner and R. Overmann (2006). *Chlorobium chlorochromatii* sp. nov., a symbiotic green sulfur bacterium isolated from the phototrophic consortium *Chlorochromatium aggregatum*. *Arch. Microbiol.* **185:** 363–372.

Voordeckers, J. W., M.H. Do, M. Hugler, V. Ko, S.M. Sievert and C. Vetriani (2008). Culture dependent and independent analyses of 16S rRNA and ATP citrate lyase genes: a comparison of microbial communities from different black smoker chimneys on the Mid-Atlantic Ridge. *Extremophiles***12:** 627–640.

Vrati, S (1984). Single-cell protein production by photosynthetic bacteria grown on the clarified effluents of biogas plant. *Appl. Microbiol. Biotechnol.***19:** 199–202.

Wani, S.H., V. Kumar, V. Shriram and S.K. Sah (2016). Phytohormones and their metabolic engineering for abiotic stress tolerance in crop plants. *Crop. J.* **4:** 162–176.

Wahlund, T.M., C.R. Woese, R.W. Castenholz and M.T. Madigan (1991). A thermophilic green sulfur bacterium from New Zealand hot springs, *Chlorobium tepidum* sp. nov. *Arch Microbiol.* **56:** 81-90.

Wahlund, T.M., C.R. Woese, R.W. Castenholz and M.T. Madigan (1996). Validation List no. 59. *Int. J. Syst. Bacteriol.***46:** 1189-1190.

Wakao, N., T. Shiba, A. Hiraishi, M. Ito and Y. Sakurai (1993). Distribution of bacteriochlorophyll *a* in species of the genus *Acidiphilium*. *Curr. Microbiol.* **27:** 277–279.

Wakayama, T. and J. Miyake (2002). Light shade bands for the improvement of solar hydrogen production efficiency by *Rhodobacter sphaeroides* RV. *Int. J. Hydr. Ener.* **27:** 1495-500.

Waligorska, M., K. Seifert, K. Gorecki, M. Moritz and M. Laniecki (2009). Kinetic model of hydrogen generation by *Rhodobacter sphaeroides* in the presence of $NH4^+$ ions. *J. Appl. Microbiol.* **107:** 1308-18.

Wall, J.D., P.F. Weaver and H. Gest (1975). Gene transfer agents, bacteriophages and bacteriocins of *Rhodopseudomonas capsulata. Arch. Microbiol.* **105(1):** 217-224.

Wang, A., D. Sun, G. Cao, H. Wang, N. Ren, W.M. Wu and B.E. Logan (2011). Integrated hydrogen production process from cellulose by combining dark fermentation, microbial fuel cells and a microbial electrolysis cell. *Bioresou. Technol.* **102(5):** 4137–43.

Wang, A., N. Ren, Y. Shi and D.J. Lee (2008). Bioaugmented hydrogen production from microcrystalline cellulose using co-culture: *Clostridium acetobutylicum* X9 and *Ethanoligenes harbinens* B49. *Int. J. Hydr. Ener.* **33:** 912-917.

Wang, J.and W. Wan (2009). Factors influencing fermentative hydrogen production: A review. *Int. J. Hydr. Ener.***34:** 799-811.

Wang, J., G. Meng, K. Tao, M. Feng and X. Zhao (2012). Immobilization of lipases on alkyl silane-modified magnetic nanoparticles: Effect of alkyl chain length on enzyme activity. *PLoS ONE.***7:** e43478. doi: 10.1371/journal. pone.0043478

Wang, J.L. and W. Wan (2009). Experimental design methods for fermentative hydrogen production: a review. *Int .J. Hydr. Ener.***34(1):** 235–244.

Wang, J.L. and W. Wan (2009). Factors influencing fermentative hydrogen production: A review. *Int. J. Hyd. Ener.* **34(2):** 799-811.

Wang, Q., J. Lian, J. Li, R. Wang, H. Huang, B. Su and Z. Lei (2015). Highly Efficient Photocatalytic Hydrogen Production of Flower-like Cadmium Sulfide Decorated by Histidine. *Scientific reports*. **5:** 13593.

Wang, X., B. Jin and D. Mulcahy (2008). Impact of carbon and nitrogen sources on hydrogen production by a newly isolated *Clostridium butyricum* W5. *Int. J. Hydr. Ener.* **33(19):** 4998–5005.

Wang, X., H.V. Modak and F.R. Tabita (1993). Photolithoautotrophic growth and control of CO_2 fixation in *Rhodobacter sphaeroides* and *Rhodospirillum rubrum* in the absence of ribulose bisphosphate carboxylase-oxygenase. *J. Bacteriol.* **175:** 7109–7114.

Wang, X., X. Cheng, D. Sun and H. Qi (2008). Biodecolorization and partial mineralization of reactive black 5 by a strain of *Rhodopseudomonas palustris*. *J. Environ. Sci.* **20:** 1218-1225.

Wang, X., Y. Wang and X. Cheng (2015). Formation characteristics of an anoxygenic photosynthetic bacterial biofilm in a photorotating biological contactor for azo dye wastewater treatment. *J. Chem. Technol. Biotechnol.***90(1):** 176-184.

Ward D.M., J. Shiea, Y.B. Zeng, G. Dobson, S. Brassell and G. Eglinton (1989). Lipid biochemical markers and the composition of microbial mats. In: *Microbial Mats: Physiological Ecology of Benthic Microbial Communities,* (Cohen Y. and E. Rosenburgs, eds.), American Society for Microbiology, Washington, DC, pp 439–454.

Ward, D.M., M.J. Ferris, S.C. Nold and M.M. Bateson (1998). A natural view of microbial biodiversity within hot spring cyanobacterial mat communities. *Microbiol. Mol. Biol.Rev.***62:** 1353–1370.

Warthmann, Rolf, Cypionka, Heribert and Pfennig, Norbert (1992). Photoproduction of H_2 from acetate by syntrophic cocultures of green sulfur bacteria and sulfur-reducing bacteria. *Archives of Microbiology*. **157:** 343-348.

Watanabe, A., Y. Kamio, W. Kimura and K. Izaki (1993). *Biosci. Biotechnol. Biochem.* **57:** 2160-2165.

Watanabe, M., K. Kawahara, K. Sasaki and N. Noparatnaraporn (2003). Biosorption of cadmium ions using a photosynthetic bacterium, *Rhodobacter sphaeroides* S and a marine photosynthetic bacterium, *Rhodovulum* sp. and their biosorption kinetics. *J. Biosci. Bioeng.***95:** 374–378.

Watanbe M., H. Shiba, K. Sasaki, Y. Nakashimada and N. Nishio (1998). Promotion of growth and flocculation of a marine phototysntheic bacterium *Rhoodvulvum* sp. *Appl. Microbiol.Biotechnol.* **50:** 682-691.

Watson, G.M., J.P. Yu and F.R. Tabita (1999). Unusual Ribulose 1,5-Bisphosphate Carboxylase/Oxygenase of Anoxic Archaea. *J. Bacteriol.***181(5):** 1569–1575.

Weaver, P.F., L. Stephen and M. Seibert (1980). Photobiological production of hydrogen. *Solar Energy*. **24 :** 3-45.

Weckesser, J., H. Mayer and G. Shulz (1995). Anoxygenic phototrophic bacteria: model organisms for studies on cell wall macromolecules. In: Anoxygenic photosynthetic Bacteria, (Blanckenship, Madigan and Bauer, eds.), Kluwer Academic Publishing, The Netherlands. pp. 207-230.

Weckesser, J., H. Mayer, G. Drews and I. Fromme (1975). Lipophilic O-antigens containing D-Glycero-D mannoheptose as the sole neutral sugar in *Rhodopseudo-monas gelatinosa. J. Bacteriol.* **123:** 449–455.

Weckesser, J., J.G. Drews and H.D. Tauschel (1969). Zur Feinstruktur von *Rhodopseudomonas gelatinosa. Arch. Mikrobiol.***65**: 346–358.

Weissgerber, T., M. Sylvester, L. Kroninger and C. Dahl (2014a). A comparative quantitative proteome study identifies new proteins relevant for sulfur oxidation in the purple sulfur bacterium *Allochromatium vinosum. Appl. Environ. Microbiol.* **80:** 2279–2292.

Weissgerber, T., M. Watanabe, R. Hoefgen and C. Dahl (2014b). Metabolomic profiling of the purple sulfur bacterium *Allochromatium vinosum* during growth on different reduced sulfur compounds and malate. *Metabolomics* **10(6):** 1094–1112.

Weissgerber, T., N. Dobler, T. Polen, J. Latus, Y. Stockdreher and C. Dahl (2013). Genome-wide transcriptional profiling of the purple sulfur bacterium *Allochromatium vinosum* DSM 180^T during growth on different reduced sulfur compounds. *J. Bacteriol.* **195(18):** 4231–4245.

Weissgerber, T., R. Zigann, D. Bruce Y.J. Chang, J.C. Detter, C. Han, L. Hauser , C.D. Jeffries, M. Land, A.C. Munk, R. Tapia and C. Dahl (2011). Complete genome sequence of *Allochromatium vinosum* DSM 180T. *Stand Genomic Sci.* **5:** 311–330.

Weller, R., M.M. Bateson, B.K. Heimbuch, E.D. Kopczynski and D.M. Ward (1992). Uncultivated cyanobacteria, *Chloroflexus*-like inhabitants and spirochete-like inhabitants of a hot spring microbial mat. *Appl. Environ.Microbiol.* **58:** 3964–3969.

Welte, C., S. Hafner, C. Kratzer, A. Quentmeier, C.G. Friedrich and C. Dahl.(2009). Interaction between Sox proteins of two physiologically distinct bacteria and a new protein involved in thiosulfate oxidation. *FEBS Lett*. **583:** 1281–1286.

Whitelaw, M.A (2001). Growth promotion of plants inoculated with phosphate solubilizing fungi. In: *Advances in Agronomy*, (Sparks, D.L. eds.), *Academic press.* **69:** 99-151.

Widdel, F. and N. Pfennig (1981). Sporulation and further nutritional characteristics of Desulfotomaculum acetoxidans. *Arch. Microbiol.* **129:** 401-402.

Widdel, F., S. Schnell, S. Heising, A. Ehrenreich, B. Assmus and B. Schink (1993). Ferrous oxidation by anoxygenic phototrophic bacteria. *Nature*. **362:** 834-836.

Wiesmann, U (1994). *Adv. Biochem. Eng./Biotechnol.* **51:** 113-154.

Willems, A. and M. Gillis (2015). *Comamonadaceae. Bergey's Manual of Systematics of Archaea and Bacteria,* pp. 1–6. doi:10.1002/9781118960608.fbm00182 .

Willems, A., J. De Ley, M. Gillis and K. Kersters (1991).*Comamonadaceae*, a new family encompassing the acidovoransribosomal RNA complex, including *Variovoraxparadoxus* gen.nov., comb. nov., for *Alcaligenes paradoxus*(Davis 1969). *Int. J. Syst. Bacteriol.* **41**: 445–450.

Willems, A., M. Gillis and J. de Ley (1991). Transfer of *Rhodocyclus gelatinosus* to *Rubrivivax gelatinosus* gen. nov., comb nov. and phylogenetic relationships with *Leptothrix*, *Sphaerotilus natans*, *Pseudomonas saccharophila* and *Alcaligenes latus*. *Int. J. Syst. Bacteriol.***41:** 65–73.

Williams, K.P. and D.P. Kelly (2013). Proposal for a new class within the phylum *Proteobacteria*, *Acidithiobacillia* class is nov., with the type order *Acidithiobacillales* and emended description of the class *Gammaproteobacteria*. *Int. J. Syst. Evol. Microbiol.* **63:** 2901-2906.

Willison, J.C., D. Madern and P.M. Vignais (1984). Increased photoproduction of hydrogen by non-autotrophic mutants of *Rhodopseudomonas capsulata*. *Biochem. J.***219:** 593-600.

Willison, J.C., J.P. Magnin and P.M. Vignais (1987). Isolation and characterization of *Rhodobacter capsulatus* strain to lack endogenous plasmids. *Arch. Microbiol.***147:** 134–142.

Wilson, J.J. and U. Kappler (2009). Sulfite oxidation in *Sinorhizobium meliloti*. *Biochem Biophys. Acta.***1787:** 1516–1525.

Wilson, R. P (1994). *Aquaculture.***124:** 67-80.

Wingender, J., T.R. Neu and H.C. Flemming (1999). Microbial Extracellular Polymeric Substances: Characterization, Structure and Function. Springer, Berlin.

Winogradsky, S (1888). Beiträge zur Morphologie und Physiologie der Bakterien. Zur Morphologie und Physiologieder Schwefelbakterien. Arthur Felix. Leipzig, Germany. **1:** 1–120.

Winton, J.M (1985).Delayed intestinal perforation after nonpenetrating abdominal trauma. *Chem. Week Aug.* **28:**55-57.

Woese, C.R (1987). Bacterial evolution. *Microbiol. Rev.* **51:** 221-271.

Woese, C.R., E. Stackebrandt, T. Macke and G.E. Fox (1985b). A phylogenetic definition of the major eubacterial taxa. *System. Appl. Microbiol.* **6:** 143-151.

Woese, C.R., W.G. Weisburg, B.J. Paster, C.M. Hahn, R.S. Tanner, N.R. Krieg, H.P. Koops, H. Harms and E. Stackebrandt (1984b). The phylogeny of purple bacteria: the beta subdivision. *Syst. Appl. Microbiol.***5:** 327–336.

Woese, C.R., W.G. Weisburg, C.M. Hahn, B.J. Paster, L.B. Zablen, B.J. Lewis, T.J. Macke, W. Ludwig and E. Stackebrandt (1985). The phylogeny of purple bacteria: the gamma subdivision. *Syst. Appl. Microbiol.* **6:** 25–33.

Wong, W.T., Ching-Han Tseng, Shu-Hua Hsu, Huu-Sheng Lur, Chia-Wei Mo, Chu-Ning Huang, Shu-Chiung Hsu, Kung-Ta Lee and Chi-Te Liu (2014). Promoting effects of a single *Rhodopseudomonas palustris* inoculant on plant growth by Brassica rapa chinensis under low fertilizer input. *Microbes Environ.***29(3):**303-13.

Wongkantrakorn, N., Y. Sunohara and H. Matsumoto (2009). Mechanism of growth amelioration of NaCl stressed rice (*Oryza sativa* L.) by δ-aminolevulinic acid. *J. Pestic. Sci.* **34:** 89–95.

Wright, G.E. and M.T. Madigan (1991). Photocatabolism of aromatic compounds by the phototrophic purple bacterium *Rhodomicrobium vannielii*. *Appl. Environ. Microb.* **57:** 2069-2073.

Wu P, Han Z, Mo W, Wu X, Chen Z, Zhang Y, Wang Y, Cui Y, Dong Y, Sun H, Zou X (2019a). Soybean processing wastewater supported the removal of propyzamide and biochemical accumulation from wastewater by *Rhodopseudomonas capsulata*. Bioprocess Biosyst Eng. **42(8):**1375-1384.

Wu, K.J. and J.S. Chang (2007). Batch and continuous fermentative production of hydrogen with anaerobic sludge entrapped in a composite polymeric matrix. *Process Biochem.* **42:** 279-284.

Wu, P., L. Xie, W. Mo, B. Wang, H. Ge, X. Sun, Y. Tian, R. Zhao, F. Zhu, Y. Zhang and Y. Wang (2019b). The biodegradation of carbaryl in soil with *Rhodopseudomonas capsulata* in wastewater treatment effluent.*J. Environ Manage.* **249:** 109226.

Wu, P., Y. Liu, X. Song, Y. Wang, L. Sheng, H. Wang and Y. Zhang (2020). *Rhodopseudomonas sphaeroides* treating mesosulfuron-methyl waste-water.*Environ Pollut.* **262:**114166.

Wu, P., Z. Chen, Y. Zhang, Y. Wang, F. Zhu, B. Cao, L. Jin, Y. Hou, Y. Wu and N. Li (2019). Retraction notice to "Carbaryl waste-water treatment by *Rhodopseudomonas sphaeroides*". Chemosphere. **233:**597-602.

Wu, S. C., S.Z. Liou and C.M. Lee (2012). Correlation between bio-hydrogen production and polyhydroxybutyrate (PHB) synthesis by *Rhodopseudomonas palustris* WP3-5. *Bioresource Technol.* **113:** 44–50.

Wu, S.Y., C.H. Hung, C.Y. Lin, P.J. Lin, K.S. Lee, C.N. Lin, F.Y. Chang and J.S. Chang (2008). HRT dependent hydrogen production and bacterial community structure of mixed anaerobic microflora in suspended, granular and immobilized sludge systems using glucose as the carbon substrate. *Int. J. Hydr. Ener.***33:** 1542-1549.

Wulandari, A.P., J. Miyazaki, N. Kobashi, M. Nishiyama, T. Hoshino and H. Yamane (2002). Characterization of bacterial homocitrate synthase involved in lysine biosynthesis. *FEBS Lett.***522:** 35–40.

Xian, W.D., Ze-Tao Liu, Meng-Meng Li, Lan Liu, Yu-Zhen Ming, Min Xiao, Nimaichand Salam and Wen-Jun Li. *Rhodobacter flagellatus* sp. nov., a thermophilic bacterium isolated from a hot spring. *Int. J. Syst. Evol. Microbiol.* **70:** 1541–1546.

Xiao, N. and N. Jiao (2011). Formation of polyhydroxyalkanoate in aerobic anoxygenic phototrophic bacteria and its relationship to carbon source and light availability. *Appl. Environ. Microbiol.* **77(21):** 7445-50.

Xiong, J., W.M. Fischer, K. Inoue, M. Nakahara and C.E. Bauer (2000). Molecular evidence for the early evolution of photosynthesis. *Science.* **289:** 1724–1730.

Yabe, S., Y. Aiba, Y. Sakai, M. Hazaka and A. Yokota (2010). *Thermosporothrix hazakensis* gen. nov., sp. nov., isolated from compost, description of *Thermosporotrichaceae* fam. nov. within the class *Ktedonobacteria* Cavaletti *et al.* 2007 and emended description of the class *Ktedonobacteria*. *Int. J. Syst. Evol. Microbiol.* **60:** 1794–1801.

Yamada, A., T. Hatano and T. Matsunaga (1998). Conversion efficiencies of light energy to hydrogen by a novel *Rhodovulum* sp. and its uptake-hydrogenase mutant in biohydrogen. Edited by Oskar R. Zaborsky, Plenum Press, London. pp. 167-171.

Yamada, H., A. Shimizy and S. Shiozaki (1986). *Jpn.Kokai.Tokkyo Koho* 86/205,496-498.

Yamada, T., Y. Sekiguchi, S. Hanada, H. Imachi, A. Ohashi, H. Harada and Y. Kamagata (2006). *Anaerolinea thermolimosa* sp. nov., *Levilinea saccharolytica* gen. nov., sp. nov. and *Leptolinea tardivitalis* gen. nov., sp. nov., novel filamentous anaerobes and description of the new classes *Anaerolineae* class is nov. and *Caldilineae* class is nov. in the bacterial phylum *Chloroflexi*. *Int. J. Syst. Evol. Microbiol.* **56:** 1331–1340.

Yamaji, M., H. Simoto and Y. Yamamoto (1994). *Jpn. Kokai Tokkyo Koho*. JP 94/86,691.

Yamanaka, K., M. Moriyama, R. Minoshima and Y. Tsuyuki (1983). *Agric. Biol. Chem.* **47:** 1257-1267.

Yamane, T. (1992). Cultivation engineering of microbial bioplastics production. *FEMS Microbiol. Rev.* **103:** 257-264.

Yamaoka, Y., K. Takeno and N. Noparatnaraporn (2008). Oil degradation using a photosynthetic bacterium *Rhodobacter sphaeroides* S. *Japanese Journal of Water Treatment Biology.***44(1)**:29-39.

Yamaoka, Y., K. Takeno, H. Shinkawa, N. Noparatnaraporn and K. Sasaki (2008). Isolation of a thermotolerant photosynthetic bacterium, *Rhodobacter sphaeroides* strain, NAT and its capacity for oil and chemical oxygen demand removal at high temperatures. *Biosci. Biotechnol. Biochem.* **72:** 1601-3.

Yan Bin, Zhou Jiti, Wang Jing, Du Cuihong, Hou Hongman, Song Zhiyong, Bao Yongming (2004). Expression and characteristics of the gene encoding azoreductase from *Rhodobacter sphaeroides* AS1.1737. *FEMS Microbiology Letters*. **236(1):** 129–136.

Yang, P., R. Zhang, J.A. Mc Garvey and J.R. Benemann (2007). Biohydrogen production from cheese processing wastewater by anaerobic fermentation using mixed microbial communities. *Int. J. Hydr. Ener.* **32 :** 4761-4771.

Yap, C.K., A. Imail and S.G. Tan (2004). Heavy metal (Cd, Cu, Pb and Zn) concentration in the green-lipped mussel *Perna viridis* (Linnaeus) collected from contaminated and uncontaminated sites of Malaysia. *Food Chem.* **84(4):** 569-575.

Yarza, P., W. Ludwig, J. Euzeby, R. Amann, K.H. Schleifer, F.O. Glockner and R. Rosselló-Móra (2010). Update of the all-species living tree project based on 16S and 23S rRNA sequence analyses. *Syst. Appl. Microbiol.* **33:** 291–299.

Yigit, D.O., U. Gunduz, I.Turker, M. YuceL and I. Eroglu (1999). Identification of by-products in hydrogen producing bacteria: *Rhodobacter sphaeroides* O.U. 001 grown in the wastewater of sugar refinery. *J. Biotechnol.* **70 :** 125-131.

Yokoi H., S. Mori, J. Hirose, S. Hayashi and Y. Takasaki (1998). H_2 production from starch by a mixed culture of *Clostridium butyricum* and *Rhodobacter* sp. M-19, *Biotechnology Letter*. **20:** 890– 895.

Yokoi, H., T. Tokushige, J. Hirose, S. Hayashi and Y. Takasaki (1997). Hydrogen production by immobilized cells of aciduric *Enterobacter aerogenes* strain HO-39. *J. Ferment. Bioeng*. **83:** 484-491.

Yokoyama, Y (1990). *Jpn. Kokai Tokkyo Koho JP*. 90197, 395.

Yoshida A., Nishimura T., Kawaguchi H., Inui M., Yukawa H (2005). Enhanced hydrogen production from formic acid by formate hydrogen lyase overexpressing *Escherichia coli* strains. *Appl. Environ. Microbiol.* **71:** 6762–6768.

Yoshida, T. and T. Kobayashi (1992). Jpn. Kokai Tokkyo Koho JP 92/71, 696.

Younesi, H., G. Najafpour, K.S.K. Ismail, A.R. Mohamed and A.H. Kamaruddin (2008). Biohydrogen production in a continuous stirred tank bioreactor from synthesis gas by anaerobic photosynthetic bacterium: *Rhodospirillum rubrum. Bioresour. Technol.* **99:** 2612-2619.

Young, S.D., Jerod Smeenk, K.M. Broer, C.J. Kisting, Robert Brown, T.J. Heindel, T.A. Bobik and A.A. DiSpirito (2007). Growth of *Rhodospirillum rubru*m on synthesis gas: conversion of CO to H_2 and poly-beta-hydroxyalkanoate. *Biotechnol Bioeng*. **97(2):** 279–286.

Youvan, D.C., J.B. Edward, M. Alberti, H. Begusch and J.E. Hearst (1984). Nucleotide and Deduced Polypeptide Sequences of the Photosynthetic Reaction-Center, B870 Antenna and Flanking Polypeptides from *Rhodobacter capsulate.Cell,***37:** 949-957.

Yu , J., S. Ge, Y. Ding and L.Wu (1991). Shanghai Jiaotong Daxue Xuebao. **25:** 106-109.

Yurkov, V.V. and J.T. Beatty (1998). Aerobic anoxygenic phototrophic bacteria. *Microbiol. Mol. Biol. Rev.* **62:** 695-724.

Yurkov, V.V., E.N. Krasilnikova and V.M. Gorlenko (1994). Thiosulfate metabolism in the aerobic bacteriochlorophyll-*a*-containing bacteria *Erythromicrobium hydrolyticum* and *Roseococcus thiosulfatophilus*. *Microbiology*. **63:** 91–94.

Zaar, A., G. Fuchs, J.R. Golecki and J. Overmann (2003). A new purple sulfur bacterium isolated from a littoral microbial mat, *Thiorhodococcus drewsii* sp. nov. *Arch Microbiol.* **179:** 174–183.

Zabut, B., El-Kahlout, K. Yucel, M. Gunduz, U.U. Turker and L. Eriglu (2005). Hydrogen gas produciton by coupled systems of *Rhodobacter sphaeroides* O.U.001 and *Halobacterium salinarium* in photobioreactors. Proceedings of International conference of Hydrogen energy congress exhibition. IHEEC.

Zadvorny, O. A., N. A. Zorin and I. N. Gogotov (2006). Transformation of metals and metal ions by hydrogenases from photrophic bacteria. *Archives of Microbiol.***184:** 279-285.

Zahr, M., B. Fobel, H. Mayer, J.F. Imhoff, V. Campos and J. Weckesser (1992). Chemical composition of the lipopolysaccharides of *Ectothiorhodospira shaposhnikovii, Ectothiorhodospira mobilis* and *Ectothiorhodospira halophila*. *Arch. Microbiol.* **157:** 499–504.

Zander, U., A. Faust, B.U. Klink. D. de Sanctis, S. Panjikar, A. Quentmeier, F. Bardischewsky, C. G. Friedrich and A.J. Scheidig (2010). Structural basis for the oxidation of protein-bound sulfur by the sulfur cycle molybdohemo-enzyme sulfane dehydrogenase SoxCD. *J. Biol. Chem.***286:** 8349–8360.

Zarezadeh, S., N.R. Moheimani, S.N. Jenkins, T. Hulsen, H. Riahi and B.S. Mickan (2019). Microalgae and phototrophic purple bacteria for nutrient recovery from agri-industrial effluents: influences on plant growth, rhizosphere bacteria, and putative carbon- and nitrogen-cycling genes. *Front. Plant Sci.* **10:**1193. doi: 10.3389/fpls.2019.01193.

Zhang, D., H. Yang, W. Zhang, Z. Huang, S.J. Liu (2003). *Rhodocista pekingensis* sp. nov., a cyst-forming phototrophic bacterium from a municipal wastewater treatment plant. *Int. J. Syst. Evol. Microbiol.* **53:** 1111–1114.

Zhang, D., H. Yang, Z. Huang, W. Zhang, S.J. Liu (2002). *Rhodopseudomonas faecalis* sp. nov., a phototrophic bacterium isolated from an anaerobic reactor that digests chicken faeces. *Int . J. Syst. Evol. Microbiol.* **52:** 2055–2060.

Zhang, D., Y. Chen, Y. Zhao and X. Zhu (2010). New sludge pretreatment method to improve methane production in waste-activated sludge digestion. *Environ. Sci. Technol.* **44 (12):** 4802–4808.

Zhang, M.L., Y.T. Fan, Y. Xing, C.M. Pan, G.S. Zhang and J.J. Lay (2007). Enhanced biohydrogen production from cornstalk wastes with acidification pretreatment by mixed anaerobic cultures. *Biomass Bioener.* **31:** 250-254.

Zhang, Y. and R.M. Miller (1992). Enhanced octadecane dispersion and biodegradation by a *Pseudomonas* rhamnolipid surfactant (biosurfactant). *Appl. Envi. Microbiol.* **58:** 3276-3282.

Zhang, Y.H.P., B.R. Evans, J.R. Mielenz, R.C. Hopkins and M.W. Adams (2007). High-Yield hydrogen production from starch and water by a synthetic enzymatic pathway. *PLoS One*. **2 :** e456.

Zhang, Z., J. Yue, X. Zhou, Y. Jing, D. Jiang and Q. Zhang (2014). Photo-fermentative Bio-hydrogen from Agricultural Residue Enzymatic Hydrolyzate and the Enzyme Reuse. *Bioresources*. **9(2):** 2299-2310.

Zhao, S., W. Xu, W. Jiang, W. Yu, Y. Lin, T. Zhang, J. Yao, L. Zhou, Y. Zeng, H. Li, Y. Li, J. Shi, S.M. An, W., Hancock, F. He, L. Qin, J. Chin, P. Yang, X. Chen, Q. Lei, Y. Xiong and K.L. Guan (2010). Regulation of cellular metabolism by protein lysine acetylation. *Science*. **327:** 1000–1004.

Zhen, A., Z.L. Bie,Y. Huang, Z.X. Liu and Fan, M.L (2012). Effects of 5-aminolevulinic acid on the H_2O_2-content and antioxidative enzyme gene expression in NaCl treated cucumber seedlings. *Biol. Plant.* **56:** 566–570.

Zheng, G.H., L. Wang and Z.H. Kang (2010). Feasibility of biohydrogen production from tofu wastewater with glutamine auxotrophic mutant of *Rhodobacter sphaeroides/ Renew. Ener.* **35:** 2910-2913.

Zhou, Q., P. Zhang and G. Zhang (2015). Biomass and pigments production in photosynthetic bacteria wastewater treatment: Effects of light sources. *Bioresou. Technol.* **179:** 505-509.

Zhou, X.P., S. Yuan, R. C. Chen and B. Song (2014). Modeling microalgae growth in nitrogen-limited continuous culture. *Energy*. **73:** 575–580.

Zhu, H., H. Fang, T. Zhang and L. Beaudette (2007). Effect of ferrous ion on photoheterotrophic hydrogen production by *Rhodobacter sphaeroides. Inter. J. Hydr. Ener*. **32(17):** 4112–4118.

Zhu, H., S. Ueda, Y. Asada and J. Miyake (2002). Hydrogen production as a novel process of wastewater treatment-studies on tofu wastewater with entrapped *Rba. sphaeroides* and mutagenesis. *Int. J. Hydr. Ener*. **27:** 1349-1357.

Zhu, X., W. Yuan, Z. Wang, M. Zhou and J. Guan (2016). Effect of worm predation on changes in waste-activated sludge properties. *Water Environ. Res*. **88 (5):** 387–393.

Zuber, H. and R. J. Cogdell (1995). Structure and organization of purple bacterial antenna complexes In: *Anoxygenic Photosynthetic Bacteria,* Kluwer Academic Publishers Dordrecht, The Netherlands. pp. 315–348.

Zubi, W (2005). Production of single-cell protein from base hydrolyzed of date extract byproduct by the fungus *Fusarium graminearum*. M.Sc.Thesis, Benghazi: Garyounis University.

Zurrer, H. and R. Bachofen (1979). Hydrogen Production by the Photosynthetic Bacterium *Rhodospirillum rubrum. Appl. Environmental Microbiol.* **37(5):** 789-793.

■■■■